Lecture Notes in Computer Science 2110

Edited by G. Goos, J. Hartmanis and J. van Leeuwen

Springer
Berlin
Heidelberg
New York
Barcelona
Hong Kong
London
Milan
Paris
Singapore
Tokyo

Bob Hertzberger Alfons Hoekstra
Roy Williams (Eds.)

High-Performance Computing and Networking

9th International Conference, HPCN Europe 2001
Amsterdam, The Netherlands, June 25-27, 2001
Proceedings

Springer

Series Editors

Gerhard Goos, Karlsruhe University, Germany
Juris Hartmanis, Cornell University, NY, USA
Jan van Leeuwen, Utrecht University, The Netherlands

Volume Editors

Bob Hertzberger
Alfons Hoekstra
University of Amsterdam
Faculty of Mathematics, Computer Science,
Physics,and Astronomy
Kruislaan 403, 1098 SJ Amsterdam, The Netherlands
E-mail: {bob,alfons}@science.uva.nl

Roy Williams
California Institute of Technology
Caltech 158-79, Pasadena, CA 91125, USA
E-mail: roy@cacr.caltech.edu

Cataloging-in-Publication Data applied for

Die Deutsche Bibliothek - CIP-Einheitsaufnahme

High performance computing and networking : 9th international conference ; proceedings / HPCN Europe 2001, Amsterdam, The Netherlands, June 25 - 27, 2001. Bob Hertzberger ... (ed.). - Berlin ; Heidelberg ; New York ; Barcelona ; Hong Kong ; London ; Milan ; Paris ; Singapore ; Tokyo : Springer, 2001
(Lecture notes in computer science ; Vol. 2110)
ISBN 3-540-42293-5

CR Subject Classification (1998):C.2.4, D.1-2, E.4, F.2, G.1-2, J.1-2, J.3, J.6

ISSN 0302-9743
ISBN 3-540-42293-5 Springer-Verlag Berlin Heidelberg New York

Springer-Verlag Berlin Heidelberg New York
a member of BertelsmannSpringer Science+Business Media GmbH

http://www.springer.de

Typesetting: Camera-ready by author
Printed on acid-free paper SPIN 10839794 06/3142 5 4 3 2 1 0

Preface

This volume contains the proceedings of the international HPCN Europe 2001 event that was held in Amsterdam, The Netherlands, June 25–27, 2001. HPCN (High Performance Computing and Networking) Europe was organized for the first time in 1993 in Amsterdam as the result of several initiatives in Europe, the United States, and Japan. HPCN Europe events were then held in Munich (1994), Milan (1995), Brussels (1996), Vienna (1997), at Amsterdam RAI in 1998/9, and then at the University of Amsterdam, Watergraafsmeer, in 2000 and 2001.

HPCN 2001 was a conference and multi-workshop event. At the conference there were 3 conference tracks presenting 56 selected papers in the following areas: Web/Grid Applications of HPCN, End User Applications of HPCN, Computational Science, Computer Science, as well as 2 poster sessions with 30 presentations.

Three renowned speakers presented the HPCN 2001 keynote lectures: Giovanni Aloisio (U. of Lecce, Italy), on Grid Applications in Earth Observing, Paul Messina (Caltech, California, USA), on Using Terascale Grids in Science, and Jose Moreira (IBM, New York, USA) on Performance and Prospects for Java.

Since research in HPCN is progressing rapidly, newly emerging domains of this field and their present and possible future applications are covered by five thematic workshops. Over 20 well-known experts from Europe and the United States participated in the organization of the workshops and agreed to present invited lectures demonstrating the current trends in their fields of interest.

The workshop "Demonstrations of the Grid" featured a few significant speakers, and each presented a live demonstration of high-performance computing and networking, together with an explanation. The associated workshop on "Scheduling Grid Resources" covered practical aspects of getting work done using these powerful resources. The workshop "Java in High Performance Computing" covered ways of using this developer-friendly language for high-performance computing. After the first full release of the human genome a few months before, the "Bioinformatics" workshop is especially apt, and we expect this segment of computing to take up more space in the next HPCN. The workshop on "Monte-Carlo Numerical Methods" was an informative exchange on developments in this burgeoning field.

The papers included in the proceedings reflect the multidisciplinary character and broad spectrum of the field. We thank all contributors for their cooperation. Due to the high quality of almost 200 submitted contributions, the selection of papers for oral and poster presentation was not simple. We are very grateful to the reviewers for their efforts in evaluating so many papers in a very short time. The best conference papers will be published in a special issue of the journal, Future Generation Computer Systems.

The drawing up of an interesting program for the conference would not have been possible without the invaluable suggestions and contributions of the members of the HPCN 2001 Program Committee. We highly appreciate the personal effort of the members of the local organizing committee and the conference secretariat. Special thanks to Berry van Halderen and Lodewijk Bos who prepared these proceedings. We would like to express our sincere thanks to Berry van Halderen for setting up the paper submission engine, and Anne Frenkel and Joost Bijlmer for creating the web pages.

The organizers acknowledge the help of the Dutch HPCN foundation, the Dutch Organization for Scientific Research, the University of Amsterdam, Hewlett Packard Corporation, Cosinus Computing bv, and the council of the city of Amsterdam for supporting the event. Finally we thank all the attendees and contributors to the conference who made this conference and multi-workshop a high quality event!

May 2001

Bob Hertzberger
Alfons Hoekstra
Roy Williams

Organization

Event Chairman:

L.O. Hertzberger, University of Amsterdam, The Netherlands

Scientific Organization:

Roy Williams, California Institute of Technology, U.S.A.
Conference Chair
Alfons Hoekstra, University of Amsterdam, The Netherlands
Conference Co-chair

Program Committee

Alfons Hoekstra (Section Computational Science, University of Amsterdam, The Netherlands)
Hamideh Afsarmanesh (University of Amsterdam, The Netherlands)
Giovanni Aloisio (University of Lecce, Italy)
Anne Trefethen (NAG, U.K.)
Alexander Bogdanov (Institute for High Performance Computing and Databases, St. Petersburg, Russia)
Marian Bubak (Institute of Computer Science and ACC CYFRONET, AGH, Cracow, Poland)
Luis Camarinha-Matos (New University of Lisbon, Portugal)
Bastien Chopard (Depertment of Computer Science, University of Geneva, Switzerland)
Marco Danelutto (Department of Computer Science, University of Pisa, Italy)
Vincenzo DeFlorio (Kath. University Leuven, Belgium)
Denis Caromel (I.N.R.I.A. Sophia Antipolis, University of Nice, France)
Jack Dongarra (University of Tennessee, U.S.A.)
Iain Duff (Rutherford Appleton Laboratory DRAL, U.K.)
Alfred Geiger (University of Stuttgart, Germany)
Wolfgang Gentzsch (Sun Microsystems Inc., Germany)
George K. Thiruvathukal (Loyola University, Chicago, and Northwestern University, Evanston, IL, U.S.A.)
Vladimir Getov (University of Westminster, London, U.K.)
Luc Giraud (CERFACS, France)
Gregor von Laszewski (Argonne National Laboratory, France)
Ralph Gruber (Polytechnic Federal School of Lausanne, Switzerland)
Tony Hey (University of Southampton, U.K.)
Ladislav Hluchy (Institute of Informatics, Slovak Academy of Sciences, SK)

Marty Humphrey (University of Virginia, U.S.A.)
Jean Louis Pazat (IRISA, France)
Jose Moreira (IBM T. J. Watson Research Center, U.S.A.)
Peter Kacsuk (KFKI-MSZKI Research Institute, Hungary)
Heather Liddell (Queen Mary, University of London, U.K.)
Bob Madahar (BAE SYSTEMS Advanced Technology Centres, U.K.)
Tomas Margalef (Universitat Autònoma de Barcelona, Spain)
Michael Philippsen (University of Karlsruhe, Germany)
Jarek Nabrzyski (PSNC, Poland)
Wolfgang Nagel (TU Dresden, Germany)
Omer Rana (University of Cardiff, U.K.)
Phil Hatcher (University of New Hampshire, U.K.)
Jeff Reeve (University of Southampton, U.K.)
Alexander Reinefeld (ZIB Berlin, Germany)
Dirk Roose (K.U.Leuven, Belgium)
Guiseppe Serazzi (Politecnico di Milano, Italy)
Sia Zadeh (Sun Microsystems, U.S.A.)
Peter Sloot (University of Amsterdam, The Netherlands)
C.J.Kenneth Tan (The Queen's University of Belfast, U.K.)
David Taniar (Monash University, Australia)
Vladimir Getov (University of Westminster, London, U.K.)
Henk van der Vorst (Universiteit Utrecht, The Netherlands)
Roy Williams (California Institute of Technology, U.S.A.)

Referees

Hamideh Afsarmanesh
Dick van Albada
Giovanni Aloisio
Shaun Arnold
Adam Belloum
Ammar Benabdelkader
Alexander Bogdanov
Holger Brunst
Marian Bubak
Massimo Cafaro
Luis Camarinha-Matos
Denis Caromel
Bastien Chopard
Paolo Cremonesi
Marco Danelutto
Vincenzo De Florio
Vijay Dialani
Miroslav Dobrucky
Jack Dongarra
Iain Duff
Uwe Fladrich
Anne Frenkel
Cesar Garita
Alfred Geiger
Wolfgang Gentzsch
Vladimir Getov
Nguyen T. Giang
Luc Giraud
Ralph Gruber
Phil Hatcher
Tony Hey
Ladislav Hluchy
Alfons Hoekstra
Fabrice Huet

Marty Humphrey
Astalos Jan
Peter Kacsuk
Ersin Kaletas
C.J. Kenneth Tan
Gregor von Laszewski
Heather Liddell
Bob Madahar
Danelutto Marco
Tomas Margalef
Jose Moreira
Lorenzo Muttoni
Jarek Nabrzyski
Wolfgang Nagel
Juri Papay
Jean Louis Pazat
Michael Philippsen
Andy Pimentel
Norbert Podhorszki
Omer Rana
Jeff Reeve
Alexander Reinefeld
Dirk Roose
Florian Schintke
Claudia Schmidt
Stephan Seidl
Miquel Angel Senar
Guiseppe Serazzi
Jens Simon
Peter Sloot
Geoffrey Stoker
David Taniar
George K. Thiruvathukal
Anne Trefethen
Julien Vayssiere
Tran D. Viet
Dialani Vijay
Henk van der Vorst
Roy Williams
Manuela Winkler
Sia Zadeh
Nemeth Zsolt

Workshop Chairs:

Demonstrations of the GRID

Peter Kacsuk; Roy Williams

Java in High Performance Computing

Vladimir Getov; George K. Thiruvathukal

Monte Carlo Numerical Methods

C.J. Kenneth Tan; Vassil Alexandrov

Scheduling GRID Resources

Marian Bubak

BioInformatics

Harmen Bussenaker

Local Organization:

Lodewijk Bos, MC-Consultancy
Rutger Hamelynck and Marinella Vermaas, Conference Office, University of Amsterdam
Anne Frenkel, Berry van Halderen, Joost Bijlmer, Jacqueline van der Velde, University of Amsterdam

Table of Contents

I Web- and Grid-Based Applications

Session 1

Session 2

Session 3

III Computer Science

Session 1

Session 2

Session 3

IV Computational Science

Session 1

Session 2

Session 3

V Posters

VI Workshops

Java in High Performance Computing

GRID Demo

Track I

Web- and Grid-Based Applications

A Virtual Data Grid for LIGO

Ewa Deelman, Carl Kesselman
Information Sciences Institute, University of Southern California
Roy Williams
Center for Advanced Computing Research, California Institute of Technology
Albert Lazzarini, Thomas A. Prince
LIGO Experiment, California Institute of Technology
Joe Romano
Physics, University of Texas, Brownsville
Bruce Allen
Physics, University of Wisconsin

Abstract. GriPhyN (Grid Physics Network) is a large US collaboration to build grid services for large physics experiments, one of which is LIGO, a gravitational-wave observatory. This paper explains the physics and computing challenges of LIGO, and the tools that GriPhyN will build to address them. A key component needed to implement the data pipeline is a virtual data service; a system to dynamically create data products requested during the various stages. The data could possibly be already processed in a certain way, it may be in a file on a storage system, it may be cached, or it may need to be created through computation. The full elaboration of this system will allow complex data pipelines to be set up as virtual data objects, with existing data being transformed in diverse ways.

This document is a data-computing view of a large physics observatory (LIGO), with plans for implementing a concept (Virtual Data) by the GriPhyN collaboration in its support. There are three sections:

- Physics: what data is collected and what kind of analysis is performed on the data.
- The Virtual Data Grid, which describes and elaborates the concept and how it is used in support of LIGO.
- The GriPhyN Layer, which describes how the virtual data will be stored and handled with the use of the Globus Replica Catalog and the Metadata Catalog.

1 Physics

LIGO (Laser Interferometer Gravitational-Wave Observatory) [1] is a joint Caltech/MIT project to directly detect the gravitational waves predicted by Einstein's theory of relativity: oscillations in the fabric of space-time. Currently, there are two observatories in the United States, (Washington state and Louisiana), one of which recently went through the "first lock" phase, in which initial calibration data was collected. Theoretically, gravitational waves are generated by accelerating masses, however

B. Hertberger et al. (Eds.): HPCN Europe 2001, LNCS 2110, pp. 3 - 12, 2001.

they are so weak that so far they have not been directly detected (although their indirect influence has been inferred [2]). Because of the extreme weakness of the signals, large data-computing resources are needed to extract them from noise and differentiate astrophysical gravitational waves from locally-generated interference.

Some phenomena expected to produce gravitational waves are:

- Coalescence of pairs of compact objects such as black-holes and neutron stars. As they orbit, they lose energy and spiral inwards, with a characteristic "chirp" over several minutes. Estimates predict of order one observation per year for the current generation of detectors.
- Continuous-wave signals over many years, from compact objects rotating asymmetrically. Their weakness implies that deep computation will be necessary.
- Supernova explosions — the search may be triggered by observation from traditional astronomical observatories.
- "Starquakes" in neutron stars.
- Primordial signals from the very birth of the Universe.

LIGO's instruments are laser interferometers, operating in a 4km high-vacuum cavity, and can measure very small changes in length of the cavity. Because of the high sensitivity, phenomena such as seismic and acoustic noise, magnetic fields, or laser fluctuations can swamp the astrophysical signal, and must themselves be instrumented. Computing is crucial to digitally remove the spurious signals and search for significant patterns in the resulting multi-channel time-series.

The raw data is a collection of time series sampled at various frequencies (e.g., 16kHz, 16Hz, 1Hz, etc.) with the amount of data expected to be generated and catalogued each year is in the order of tens of terabytes. The data collected represents a gravitational channel (less than 1% of all data collected) and other instrumental channels produced by seismographs, acoustic detectors etc. Analysis on the data is performed in both time and frequency domains. Requirements are to be able to perform single channel analysis over a long period of time as well as multi-channel analysis over a short time period.

1.1 Data in LIGO

Each scalar time series is represented by a sequence of 2-byte integers, though some are 4-byte integers. Time is represented by GPS time, the number of seconds since an epoch in 1981, and it is therefore a 9-digit number, possibly followed by 9 more digits for nanosecond accuracy (it will become 10 digits in Sept. 2011). Data is stored in Frame files, a standard format accepted throughout the gravitational wave community. Such a file can hold a set of time-series with different frequencies, together with metadata about channel names, time intervals, frequencies, file provenance, etc. In LIGO, the Frames containing the raw data encompass an interval of time of one second and result in about 3Mb of data.

In addition to the raw time series, there are many derived data products. Channels can be combined, filtered and processed in many ways, not just in the time domain, but also in the Fourier basis, or others, such as wavelets. Knowledge is finally extracted from the data through pattern matching algorithms; these create collections of candidate events, for example, inspiral events or candidate pulsars.

1.2 LIGO Data Objects

The Ligo Data Analysis System (LDAS) [3] is a set of component services for processing and archiving LIGO data. Services deliver, clean, filter, and store data. There is high-performance parallel computing for real-time inspiral search, storage in a distributed hierarchical way, a relational database, as well as a user interface based on the Tcl language.

The LIGO data model splits data from metadata explicitly. Bulk data is stored in Frame files, as explained above, and metadata is stored in a relational database, IBM DB2. There is also an XML format called LIGO-LW (an extension of XSIL [4]) for representing annotated, structured scientific data, that is used for communication between the distributed services of LDAS.

In general, a file may contain more than one Frame, so we define the word *Frame-File*, for a file that may contain many frames. Raw data files contain only one frame, and they are named by the interferometer that produced the data (H: Hanford, L: Livingston), then the 9-digit GPS time corresponding to the beginning of the data. There is a one or two letter indication of what kind of data is in the file, (F: full, R: reduced, T: trend, etc.). So an example of a raw data frame might be `H-276354635.F`.

For long-term archiving, rather larger files are wanted than the 3-megabyte, one second raw frames, so there are collection-based files, generally as multi-frame Frame-Files. In either case, an additional attribute is in the file name saying how many frames there are, for example `H-276354635.F.n200` would be expected to contain 200 frames.

One table of the metadatabase contains FrameSets, which is an abstraction of the FrameFile concept, recognizing that a FrameFile may be stored in many places: perhaps on tape at the observatory, on disk in several places, in deep archive.

Each frame file records the names of all of the approximately one thousand channels that constitute that frame. In general, however, the name set does not change for thousands or tens of thousands of one-second frames. Therefore, we keep a ChannelSet object, which is a list of channel names together with an ID number. Thus the catalog of frames need only store the ChannelSet ID rather than the whole set of names.

The metadatabase also keeps collections of Events. An event may be a candidate for an astrophysical event such as a black-hole merger or pulsar, or it may refer to a condition of the LIGO instrument, the breaking of a feedback loop or the RF signal from a nearby lightning strike. The generic Event really has only two variables: type and significance (also called Signal to Noise Ratio, or SNR). Very significant events are examined closely, and insignificant events used for generating histograms and other statistical reports.

1.3 Computational Aspects, Pulsar Search

LDAS is designed primarily to analyze the data stream in real time to find inspiral events, and secondarily to make a long-term archive of a suitable subset of the full LIGO data stream. A primary focus of the GriPhyN effort is to use this archive for a full-scale search for continuous-wave sources. This search can use unlimited computing resources, since it can be done at essentially arbitrary depth and detail. A major objective and testbed of the GriPhyN involvement is to do this search with *back-*

fill computation (the "SETI@home" paradigm), with high-performance computers in the physics community.

If a massive, rotating ellipsoid does not have coincident rotational and inertial axes (a time-dependent quadrupole moment), then it emits gravitational radiation. However, the radiation is very weak unless the object is extremely dense, rotating quickly, and has a large quadrupole moment. While the estimates of such parameters in astrophysically-significant situations are vague, it is expected that such sources will be very faint. The search is computationally intensive primarily because it must search a large parameter space. The principle dimensions of the search space are position in the sky, frequency, and rate of change of frequency.

The search is implemented as a pipeline of data transformations. First steps are cleaning and reshaping of the data, followed by a careful removal of instrumental artifacts to make a best estimate of the actual deformation of space-time geometry at the LIGO site. From this, increasingly specialized data products are made, and new ways to calibrate and filter the raw and refined data.

The pulsar search problem in particular can be thought of as finding features in a large, very noisy image. The image is in time-frequency space, and the features are curves of almost-constant frequency — the base frequency of the pulsar modulated by the doppler shifts caused by the motion of the Earth and the pulsar itself.

The pulsar search can be parallelized by splitting the possible frequencies into bins, and each processor searching a given bin. The search involves selecting sky position and frequency slowing, and searching for statistically-significant signals. Once a pulsar source has been detected, the result is catalogued as an event data structure, which describes the pulsar's position in the sky, the signal-to-noise ratio, time etc.

1.4 Event Identification Computation Pipeline

During the search for astrophysical events a long duration, one dimensional time series is processed by a variety of filters. These filters then produce a new time series which may represent the signal to noise ratio in the data. A threshold is applied to each of the new time series in order to extract possible events, which are catalogued in the LIGO database. In order to determine if the event is significant, the raw data containing instrumentation channels needs to be re-examined. It is possible that the occurrence of the event was triggered by some phenomena such as lightning strikes, acoustic noise, seismic activity, etc. These are recorded by various instruments present in the LIGO system and can be found in the raw data channels. To eliminate their influence, the instrumental and environmental monitor channels must be examined and compared to the occurrence of the event. The location of the raw data channels can be found in the LIGO database. Since the event is pinpointed in time, only small portions of the many channels (that possibly needs to be processed) need to be examined. This computational pipeline clearly demonstrates the need for efficient indexing and processing of data in various views:

- a long time interval single channel data, such as the initial data being filtered, and
- the many channel, short time interval such as the instrument data needed to add confidence to the observation of events.

2 GriPhyN/LIGO Virtual Data Grid

GriPhyN [5] (Grid Physics Network) is a collaboration funded by the US National Science Foundation to build tools that can handle the very large (petabyte) scale data requirements of cutting-edge physics experiments. In addition to the LIGO observatory, GriPhyN works with the CMS and Atlas experiments at CERN's Large Hadron Collider [6] and the Sloan Digital Sky Survey [7].

A key component needed to implement the data pipeline is a *virtual data service*; a system to dynamically create data products requested during the various stages. The data could possibly be already processed in a certain way, it may be in a file on a storage system, it may be cached, or it may need to be created through computation. The full elaboration of this system will allow complex data pipelines to be set up as virtual data objects, with existing data being transformed in diverse ways.

2.1 Virtual Data

An example of Virtual Data is this: *"The first 1000 digits of Pi"*. It defines a data object without computing it. If this request comes in, we might already have the result in deep archive: should we get it from there, or just rerun the computation? Another example is: *"Pi to 1000 places"*. How can we decide if these are the same request even though the words are different? If someone asks for *"Pi to 30 digits"*, but we already have the first two, how can we decide that the latter can be derived easily from the former? These questions lie at the heart of the GriPhyN Virtual Data concept.

In the extreme, there is only raw data. Other requests for data, such as obtaining a single channel of data ranging over a large time interval, can be derived from the original data set. At the other extreme, every single data product that has been created (even if it represents an intermediate step not referred to again) can be archived. Clearly, neither extreme is an efficient solution; however with the use of the Virtual Data Grid (VDG) technology, one can bridge the two extremes. The raw data is of course kept and some of the derived data products are archived as well. Additionally, data can be distributed among various storage systems, providing opportunities for intelligent data retrieval and replication.

VDG will provide transparent access to virtual data products. To efficiently satisfy requests for data, the VDG needs to make decisions about the instantiation of the various objects. The following are some examples of VDG support for LIGO data:

- Raw data vs. cleaned data channels. Most likely, only the virtual data representing the most interesting clean channels should be instantiated.
- Data composed from smaller pieces of data, such as long duration frames that could have been already processed from many short duration frames.
- Time-frequency image, such as the one constructed during the pulsar search. Most likely the entire frequency-time image will not be archived. However, all its components (short power spectra) might be instantiated. The VDG can then compose the desired frequency-time images on demand.
- Interesting events. Given a strong signal representing a particularly promising event, the engineering data related to the time period of the occurrence of the

event will most likely be accessed and filtered often. In this case, the VDG might instantiate preprocessed instrumental and environmental data channels, data that might otherwise exist only in its raw form.

2.2 Simple Virtual Data Request Scenario

A VDG is defined by its Virtual Data Domain, meaning the (possibly infinite) set of all possible data objects that can be produced, together with a naming scheme ("coordinate system") for the Domain. There are functions on the domain that map a name in the domain to a data object. The VDG software is responsible for caching and replicating instantiated data objects. In our development of the LIGO VDG, we begin with a simple Cartesian product domain, and then add complexity.

Let D be the Cartesian product of a time interval with a set of channels, and we think of the LIGO data as a map from D to the value of the channel at a given time. In reality, of course, it is complicated by missing data, multiple interferometers, data recorded in different formats, and so on.

In the following, we consider requests for the data from a subdomain of the full domain. Each request is for the data from a subset of the channels for a subinterval of the full time interval. Thus, a request might be written in as:

```
T0=700004893, T1=700007847; IFO_DCDM_1, IFO_Seis_*
```

where `IFO_DCDM_1` is a channel, and `IFO_Seis_*` is a regular expression on channel names. Our first task is to create a naming scheme for the virtual data object, each name being a combination of the name of a time interval and the name of a set of channels.

This could be satisfied if there is a suitable superset file in the replica database, for example this one:

```
T0=70004000,T1=700008000;IFO_*
```

We need to be able to decide if a given Virtual Data Object (VDO) contains another, or what set of VDO's can be used to create the requested VDO. If C_i is a subset of the channels, and I_i is a subinterval, then tools could be used to combine multiple files (C_1,I_1), (C_2,I_2), ... perhaps as:

- The new file could be (C, I), where C = *union* C_i and I = *intersect* I_i, a channel union tool, or
- The new file could be (C, I), where C = *intersect* C_i and I = *union* I_i, an interval union tool.

We could thus respond to requests by composing existing files from the distributed storage to form the requested file.

To be effective, we need to develop a knowledge base of the various transformations (i.e., how they are performed, which results are temporary, and which need to be persistent). We need to know the nature of the context in which these transformations occur: something simple is the Fourier transform, but more difficult would a transformation that uses a certain code, compiled in a certain way, running on a specific machine. This description of context is needed in order to be able to execute the transformations on data sets as well as to determine how to describe them.

2.3 Generalized Virtual Data Description

The goal of the GriPhyN Virtual Data Grid system is to make it easy for an application or user to access data. As explained above, as a starting point we will service requests consisting of a range of time t_0 to t_1 (specified in GPS seconds), followed by a list of channels. However, we now extend the idea of channel to "virtual channel".

Virtual Channels

A virtual channel is a time series, like a real channel, but it may be derived from actual channels, and not correspond to a channel in the raw data. Some examples of virtual channels are:

- An actual recorded channel, "raw".
- An actual recorded channel, but downsampled or resampled to a different sampling rate.
- An arithmetic combination of channels, for example $2C_1 + 3C_2$, where C_1 and C_2 are existing channels.
- The actual channel, convolved with a particular calibration response function, and scaled. For example, the X component of the acceleration.
- The virtual channel might be computed from the actual data channels in different ways depending upon what time interval is requested (e.g., the calibrations changed, the channels were hooked up differently, etc.).
- A virtual channel could be defined in terms of transformations applied to other virtual channels.

In short, the virtual channels are a set of transformations applied to the raw data.

The set of virtual channels would be extendable by the user. As the project progresses, one may want to extend the set of virtual channels to include additional, useful transformations. Thus, if a user is willing to define a virtual channel by specifying all the necessary transformations, it will be entered in the catalog and will be available to all users, programs, and services. New channels can be created from the raw data channels by parameterized filters, for example decimation, heterodyning, whitening, principle components, autocorrelation, and so forth.

Data Naming

A crucial step in the creation of the GriPhyN Virtual Data model is the naming scheme for virtual data objects. Semantically, we might think of names as a set of keyword-value pairs, extended by transformations, perhaps something like `(T0=123,T1=456,Chan=[A,B*,C?]).pca().decimate(500)`. The first part in parentheses is the keyword-value set, the rest is a sequence of (perhaps parameterized) filters. We could also think of using names that contain an SQL query, or even names that include an entire program to be executed. The syntax could also be expressed in other ways, as XML, or with metacharacters escaped to build a posix-like file name. We could use a syntax like `protocol://virtual-data-name` to express these different syntax in one extensible syntax. However, decisions on naming Virtual Data must be premised on existing schemes described in Section 1.2.

3 GriPhyN Support

The goal of GriPhyN is to satisfy user data requests transparently. When a request for a set of virtual channels spanning a given time interval is made, the application (user program) does not need to have any knowledge about the actual data location, or even if the data has been pre-computed. GriPhyN will deliver the requested virtual channels by either retrieving existing data from long term storage, data caches containing previously requested virtual channels, or by calculating the desired channels from the available data (if possible).

When satisfying requests from users, data may be in slow or fast storage, on tape, at great distance, or on nearby spinning disk. The data may be in small pieces (~1 second) or in long contiguous intervals (~1 day), and conversion from one to another requires computational and network resources. A given request for the data from a given time interval can thus be constructed by joining many local, small files, by fetching a distant file that contains the entire interval, or by a combination of these techniques. The heart of this project is the understanding and solution of this optimization problem, and an implementation of a real data server using GriPhyN tools.

3.1 User Requests

The initial implementation of the GriPhyN system will accept requests in the semantic form:

`t0,t1; A,B,C...,`

where `t0`, `t1` is a time interval, and `A,B,C,...` are virtual channels. We assume that the order in which the channels are listed does not affect the outcome of the request. The syntax of the virtual channel is yet to be determined. In the simplest form, a virtual channel resulting from a transformation Tr_X on a channel C_y would be $Tr_X(C_y)$; additional attributes (such as transformation parameters or other input channels) can be specified as additional parameters in the list: $Tr_X(C_y, C_z; \alpha, \beta,)$, depending on the transformation. The transformation specific information will be stored in the Transformation Catalog described below.

The semantic content of the request is defined above, but not the syntax. The actual formulation for the GriPhyN-LIGO VDG will be an XML document, though the precise schema is not yet known. We are intending to implement requests as a very small document (for efficiency), but with links to context, so that a machine can create the correct environment for executing the request. We expect much of the schema development to be implemented with nested XML namespaces [8].

3.2 Data Access, Cost Performance Estimation

When a user or computer issues a request to the Virtual Data Grid, it is initially received by a *Request Manager* service and sent for processing to the *Metadata Catalog*, which provides the set of logical files that satisfies the request, if such exists. The files names are retrieved from the Metadata Catalog based on a set of attributes.

The logical files found in the Metadata Catalog are sent to the *Replica Catalog*, which maps them to a unique file, perhaps the closest in some sense of many such replicas. The information about the actual file existence and location (provided by the Replica Catalog) are passed to the Request Manager, which makes a determination about how to deliver the data.

If the requested data is present, the Request Manager still needs to determine whether it is cheaper to recalculate the data or access it. When considering the cost of referencing data, the cost of accessing various replicas of the data (if present) needs to be estimated. If the data is not present, the possibility and cost of data calculation needs to be evaluated. In order to make these decisions, the Request Manager queries the *Informa-*

tion Catalog. The latter can provide information about the available computational resources, network latencies, bandwidth, etc.

The next major service is the *Request Planner*. It is in charge of creating a plan for the execution of the transformations on a given data set and/or creating a plan for the retrieval of data from a storage system. The Request Planner has access to the system information retrieved by the Request Manager. To evaluate the cost of re-computation, the cost of the transformations needs to be known. This information, as well as the input and parameters required by a given transformation, code location, etc. are stored in the *Transformation Catalog*. The Request Planner uses information about the transformations that have been requested, to estimate the relative costs of computation vs. caching, and so on. The system would possibly keep a record of how long the various transformations took, and could use this performance history to estimate costs. This record and the analytical performance estimates will be maintained in the Transformation Catalog.

The performance data needed in the evaluation of re-computation and replica access costs (such as network performance, availability of computational resources, etc.) will be provided by other information services, which are part of the Globus toolkit [9], a software environment designed for the Grid infrastructure.

Once the request planner decides on a course of action the *Request Executor* is put in charge of carrying out the plan which involves the allocation of resources, data movement, fault monitoring, etc. The Request Executor will use the existing Globus infrastructure to access the data and the computational Grid, and run large jobs reliably with systems controlled by systems such as Condor [10]. As a result of the completion of the request, the various catalogs might need to be updated.

3.3 Proactive Data Replication

Simply, just retrieving data from the replica catalog is not sufficient. The system must take a proactive approach to creating replica files and decide whether the results of transformations will be needed again. For example, if there is a request for a single channel spanning a long time interval and the replica catalog contains only files which are multi-channel spanning short time periods, then a significant amount of processing is needed to create the requested file (many files need to be opened and a small amount of data needs to be retrieved from each of them). However, once this transformation is performed, the resulting data can be placed in the replica catalog for future use, thus reducing the cost of subsequent requests. New replicas should also be created for frequently accessed data, if accessing the data from the available locations is too costly. For example, it may be useful to replicate data that initially resides on tape to a local file system. Since the Request Manager has information about data existence, location and computation costs, it will also be responsible for making decisions about replica creation.

4 Conclusions

Although the GriPhyN project is only in its initial phase, its potential to enhance the research of individual physicists and enable their wide-spread collaboration is great. This paper presents the first step in bridging the understanding between the needs of the physics community and the research focus of computer science to further the use and deployment of grid technologies [11] on a wide scale, in the form of a Virtual Data Grid.

References

[1] Barish, B. C. and Weiss, R., Physics Today, Oct 1999, pp. 44-50; also http://www.ligo.caltech.edu/
[2] Taylor, G., and Weisberg, J. M., *Astrophys. J.* **345,** 434 (1989).
[3] Anderson, S., Blackburn, K., Lazzarini, A., Majid, W., Prince, T., Williams, R., The LIGO Data Analysis System, Proc. of the XXXIVth Recontres de Moriond, January 23-30, 1999, Les Arcs, France, also http://www.ligo.caltech.edu/docs/P/P990020-00.pdf
[4] Blackburn, K., Lazzarini, A., Prince, T., Williams, R., XSIL: Extensible Scientific Interchange Language, Lect. Notes Comput. Sci. **1593** (1999) 513-524.
[5] GriPhyN, Grid Physics Net, http://www.griphyn.org/
[6] CERN Large Hadron Collider, http://lhc.web.cern.ch/lhc/
[7] Sloan Digital Sky Survey, http://www.sdss.org/
[8] For information of XML and schemas, see http://www.xml.com/schemas
[9] Globus: http://www.globus.org
[10] Condor: http://www.cs.wisc.edu/condor/.
[11] Gridforum: http://www.gridforum.org/

A Multidisciplinary Scientific Data Portal

John V. Ashby[1], Juan C. Bicarregui[1], David R. S. Boyd[1],
Kerstin Kleese - van Dam[2], Simon C. Lambert[1],
Brian M. Matthews[1], and Kevin D. O'Neill[1]

[1]CLRC - Rutherford Appleton Laboratory, Chilton, Didcot, Oxon, UK, OX11 0QX
[2] CLRC - Daresbury Laboratory, Daresbury, Warrington, Cheshire, UK, WA4 4AD

Abstract. We describe a web-based multidisciplinary data portal, currently under construction and based on a new metadata model of scientific data, for exploring and accessing the content of, initially, the data resources held within CLRC's main laboratories. This system comprises a web-based user interface incorporating access control and a metadata catalogue, which interfaces to distributed data resources.

The Problem

The Central Laboratory of the Research Councils (CLRC) in the UK is one of Europe's largest multidisciplinary research support organisations. From its three sites, CLRC operates several large scale scientific facilities for the UK research and industrial community including accelerators, lasers, telescopes, satellites and supercomputers which all create copious quantities of data. Currently CLRC is holding data in access of 50 TB, however much more is expected in the future with the advent of new instruments, facilities (DIAMOND) and projects (DataGrid), which will lead to data volumes in access of several PB within the next 5-6 years. The data held at CLRC covers most major science areas e.g. Astronomy, Biology , Chemistry, Environmental Science, Physics. These data resources are stored in many file systems and databases, physically distributed throughout the organisation with, at present, no common way of accessing or searching them. Furthermore it is often necessary to open and read the actual data files to find out what information they contain. There is little consistency in the information, which is recorded for each dataset held and sometimes this information may not even be available on-line, but only in experimenters' logbooks. As a result of this there is no convenient way of posing queries, which span more than one data resource.

This situation could potentially lead to serious under-utilisation of these data resources or to wasteful re-generation of data. It could also thwart the development of cross-discipline research, as this requires good facilities for locating and combining relevant data across traditional disciplinary boundaries. The combination of experimental Synchrotron Radiation and Neutron Spallation data with computational Materials or Molecular Science results is seen as a valuable addition to CLRC's services, as is the possibility to combine Earth Observation data and computational

B. Hertzberger et al. (Eds.) : HPCN Europe 2001, LNCS 2110, pp. 13-22, 2001.

Climate Research results. Another important aspect for us is the integration of all our services in the newly developing e-Science/Grid environments. Here it is not only of importance to make our services 'Grid aware' but also the seamless integration with other services from other sites. This project is one of several within the e-Science program at CLRC [1].

The CLRC Data Portal

The preferred solution would have been the reuse or adaptation of an existing tool, however a distinguishing feature of CLRC's requirement is the necessary *Generality* of the Metadata model. Other metadata approaches are either usually closely associated with a particular scientific domain (e.g. CERA), or else are metadata design frameworks (e.g. RDF, XMI). Some of the initiatives studied where: XSIL from Caltech, work of the OODT group and the activities surrounding the *Dublin Core*. The Extensible Scientific Interchange Language (XSIL) is a flexible, hierarchical, extensible, transport language for scientific data objects. At a lower level than CLRC's metadata model it allows the definition of the data array and the transport of those arrays: it is used on LIGO - Laser Interferometer Gravitational-Wave Observatory (http://www.cacr.caltech.edu/SDA/xsil/index.html). The Object Oriented Data Technology group (OODT) at the Jet Propulsion Lab http://oodt.jpl.nasa.gov is also producing a generic framework for connecting XML based metadata profiles, and uses a CORBA based OO-system to provide a distributed resource location service. A good deal of activity also surrounds the *Dublin Core* metadata http://dublincore.org/. This provides a basic set of elements (15 in the original definition), but is unfortunately not detailed enough for CLRC's purposes. Elements of CLRC's model could be mapped onto the Dublin Core - an important feature for interoperability, especially with Digital Libraries.

With no apparent existing solution, it was decided to develop a web-based data portal with the aim of offering a single method of reviewing and searching the contents of all the CLRC data resources through the use of a central catalogue holding metadata about all of these resources. The structure and contents of this catalogue are based on a metadata model for representing scientific data, which is being developed by the project. In particular we see it of utmost importance not to replace the existing metadata, but rather extract the required information to represent it in a more general format to allow multidisciplinary searches. Extensive use is being made of XML and related W3C standards for representing, transferring and manipulating the metadata.

The objective of the current project is to prototype these ideas by developing a pilot implementation of the proposed system, which will enable researchers to access and search metadata about data resources held at the ISIS (Neutron Spallation) and SR (Synchrotron Radiation) accelerator facilities in CLRC.

System Architecture

The system being developed has 3 main components:

- a web-based user interface including a security subsystem
- a metadata catalogue
- data resource interfaces

In the pilot system these are integrated using standard Web protocols. Eventually it is anticipated that the system will exploit the emerging Grid infrastructure to offer a distributed interface to scientific data resources both inside and outside CLRC (please see figure 1).

The integration with other services can be facilitated on two different levels, depending on whether they have their own data portal or not. If they have single data holdings, it would be best to provide a XML wrapper for the existing metadata, which transfers the information into the multidisciplinary format developed by CLRC. If the institution already has its own brokering facility, then an exchange protocol has to be specified to interchange information and queries between the two.

Fig. 1. Architecture of the Multidisciplinary Data Portal.

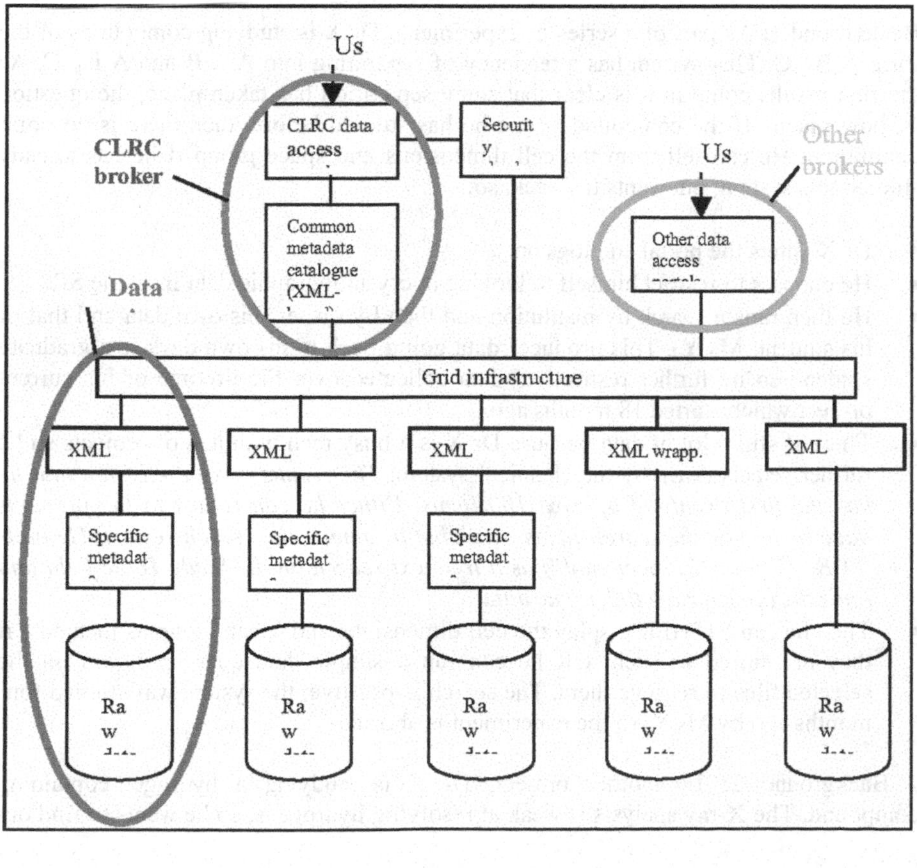

The Grid environment will help in a range of functions, e.g. authentication, connection between the data holdings and the web-based user interface/metadata catalogue, as well as the link to other brokers and portals e.g. CLRC's High Performance Computing Portal. It is also envisaged to make the data portal part of some exemplary working environments e.g. in materials science.

User Requirement Capture

To help us understand and capture the requirements of potential users of the data portal, we interviewed several experimental scientists, instrument scientists and data resource providers at each facility and also some external users. Based on their input, we constructed a series of user scenarios. These scenarios have been validated with those we interviewed and, based on them, we are developing use cases which are being used to design the user interface dialogue and to determine the required functionality of the database system underlying the metadata catalogue. Please find below a range of example scenarios.

SR Single Crystal Diffraction Scenarios

Background 1: As part of a series of experiments Dr X is studying compounds of the form $A_xB_{1-x}C$. This system has a tendency of separating into AC, B and $A_yB_{1-y}C$. As the first results come in it is clear that some separation has taken place, the question is, how much? If the compound is one he has studied before, then there is no point continuing. He can tell from the cell dimensions and space group if he has already studied this system, but wants to check, so:

- Dr X enters the portal and logs on.
- He chooses to restrict himself to looking at crystallographic data from the SR.
- He then runs a search by institution and then by owner (his own data and that of his student, Ms Y). This produces data going back to his own days as a graduate student, so he further restricts to data collected over the lifetime of his current project which started 18 months ago.
- There is still a lot of data because Dr X is a busy man with lots of projects so he further selects down by the chemical system. *The results screen tells him that he has the first twenty of approx. 100 items. Either he can return to the previous page or he has the search fields available to amend the search results. He adds "$A_xB_{1-x}C$" (or "Paracetamol") as a free text search on the Study Topic field and finds the appropriate link to the data.*
- Then he can EITHER display the cell dimensions and space group as metadata if they are stored as such, OR he can run a simple data query program on the selected files to retrieve them. The search is positive; the system was studied four months ago by Ms Y, so the experiment is aborted.

Background 2: In another project, Dr Z is studying a hydrogen-containing compound. The X-ray analysis is weak at resolving hydrogens, so he wants to find out if similar systems have been studied under similar conditions using neutrons. Then he can insert the hydrogen positions from the neutron data and the heavier atoms from his own data.

- He logs onto the portal and selects crystallographic data.
- He links off to an external database and searches there, but draws a blank.
- Returning to the CLRC Data Portal, he chooses to search the ISIS database for unpublished data.
- The search this time is by chemical system and unrestricted by owner, date, institution, etc.
- Dr Z's compound has been well studied at high temperatures, but Dr Z is interested in the low temperature properties, so he queries the metadata for any temperature conditions associated with the data. This narrows down the set of candidate datasets to two. Looking at the owners of these two he sees, that one is the highly respected Prof. Q and the other is Mr. R.. A metadata comment from the former suggests that the quality of the data is not very reliable, however, so Dr Z chooses to look at Mr. R's data.
- At this point he could start up a visualisation program on Mr R's data to pinpoint the hydrogen atoms or simply download the structure data and scan through it by hand.

ISIS Single Crystal Diffraction Scenario

Background: Dr W had student A working for her on an important experiment, unfortunately the student did not finish his placement for personal reasons and left only incomplete notes about the work he carried out. A new student is found 6 month later and Dr W now wants to establish the status of the work carried out by student A.

- Dr W goes to the portal and logs on.
- Dr W chooses ISIS crystallographic data, restricting her search to data produced by student A.
- There are three potential datasets. Selecting the first she looks to see what information is available on the RAW data. This shows that it was part of a test set of data Student A put through for calibration.
- The second set looks more promising, but when she expands the tree of files dependent on the RAW data she sees that a final structure has already been determined.
- It looks as though the third set is the right one, but to make sure she checks and finds a comment student A left for herself in the metadata. She deletes this comment and adds a fresh one that the new student B is restarting the analysis.
- She then expands the dependent tree and sees that there is already a set of .PKS and .USE files, and beneath them a set of .INT files. She looks to see which parameterizations have already been used and from that information can decide where to start the analysis.

An important observation emerges from these use cases: there are two modes of interaction required from the data portal. Firstly there is the “data browse” mode. A user such as Dr. X above, where the user wishes to search across the *studies* in a particular field, searching for work carried out on topic of interest with suitable experimental parameters. Secondly, there is the “data explore” mode, such as Dr. W. above, where the user knows already which specific data sets are of interest and wishes to explore them in detail to study their exact nature. Generally, the first mode is associated with an external user seeking information on a specific topic, and the second is associated with data generators and curators, interested in developing and preserving the data. However, a user may switch from the first mode, identifying a data set of interest, to second, exploring the nature of that data set. Accommodating these two entry points into a common metadata and user interface onto that metadata has been a crucial influence on the design.

User Interface

We applied an initial use case analysis to the collected user scenarios for three main purposes:

- To seek a high level of uniformity between them, so as to ease the development of a unified user interface;
- To detect questions of general importance about the system functionality that might need consideration before the user interface can be built;
- To provide an informal checklist for what the actual user interface provides.

We identified that there was only one complete user task: 'Search and retrieve results'. In the future there will clearly be a second one: 'Add dataset'. The sequence of steps necessary to perform the first task were identified and realized in a mock-up user interface. The results of the user requirement capture; the case analysis and the mock-up user interface were presented to the users from a range of CLRC departments and external institutes for discussion. The comments where recorded and the user interface definition accordingly amended.

Currently the user interface client for the data portal is being implemented using a standard web browser, which will accept XML. At each step in the user dialogue, users are presented with a page generated from the metadata available at that stage of their search. They are prompted to refine or broaden their enquiry. User responses are interpreted by a user interface server process and converted into queries, which are sent to a database holding the metadata. The results of each query are generated by the database as XML, which is sent to the server process, parsed, converted into a displayable page using an XSL script and presented to the user.

The user interface server will also implement security procedures to authenticate the user and to establish what authority they have to access the metadata and data. Based on this authority, they will be able to see all, part or none of the metadata in a specific area. For example, in some areas, the existence of data about certain materials obtained by a particular research group is confidential information and only members of that group would be able to see that the data existed. However this level of security is expected to be unusual and most metadata in the system will be accessible to all users.

Metadata Model and Implementation

The logical structure of the metadata in the catalogue is based on the scientific metadata model being developed in the project. This model is defined in XML using a DTD schema. The metadata catalogue is implemented using a relational database with a schema derived from the metadata model schema. This offers views of the data supporting the expected user enquiry patterns. Once the specific datasets required by the user have been identified using the available metadata, the catalogue will provide URLs for the files holding the actual data. Users can then use these links to access the data with their own applications for analysis as required.

When developing our metadata model, we aimed to provide a high-level generic model, which can be specialized to specific scientific disciplines. This was based on discussions with the ISIS and SR facilities at CLRC and initially tailored to the needs of the data holdings of those facilities. Nevertheless, the model will abstract away from the specific requirements of these facilities in order to capture scientific data from any discipline. Other influences come from the CIP metadata catalogue for Earth Observation [2], the CERA [3] metadata model for Climate Modeling Data and the DDI metadata description for Social Science data [4].

The metadata model attempts to capture scientific activities at different levels: generically, all activities are called *Studies*. Each study has an *Investigator* that describes who is undertaking the activity, and the *Study Information* that captures the details of this particular study. Studies can be of different kinds.

Programs: are studies that have a common theme, and usually a common source of funding, instigated by a principal investigator or institution. Programs can be single projects or linked sequences of projects. Each program can thus be associated (linked) with a series of sub-investigations. Programs are not expected to have direct links to data, but rather through the set of investigations within the programs.

Investigations: are studies that have links directly to data holdings. More specific types of investigations include experiments, measurements or simulations.

- *Experiments:* investigations into the physical behavior of the environment usually to test an hypothesis, typically involving an instrument operating under some instrumental settings and environmental conditions, and generating data sets in files. E.g. the subjection of a material to bombardment by x-rays of know frequency generated by the SR source at Daresbury, with the result diffraction pattern recorded.
- *Measurements:* investigations that record the state of some aspect of the environment over a sequence of point in time and space, using some passive detector. E.g. measurement of temperature at a point on the earth surface taken hourly using a thermometer of known accuracy.
- *Simulations:* investigations that test a model of part of the world, and a computer simulation of the state space of that model. This will typically involve a computer program with some initial parameters, and generate a dataset representing the result of the simulation. E.g. a computer simulation of fluid flow over a body using a specific program, with input parameters the shape of the body, and the velocity and viscosity of the fluid, generating a data set of fluid velocities

Data holdings: The metadata format given here is designed for the description of general scientific data holdings. These data holdings have three layers: the investigation, the logical data, and the physical files.

Each investigation (experiment, measurement or simulation) has a particular purpose and uses a particular experimental set up of instruments or computer systems. Experiments may be organized within larger studies or projects, which themselves may be organized into programs of linked studies.

An investigation generates raw data. This raw data can then be processed via set processing tools, forming on the way intermediate stages, which may or may not be held in the data holding. The final processing step generates the final data set. Each stage of the data process stores data in a set of physical files with a specified physical location. It is possible that there may be different versions of the data sets in the holding. Thus each *data holding* takes the form of a hierarchy: one *investigation* generates a *sequence* of logical data sets, and each data set is instantiated via a *set* of physical files. The design of the metadata model is tailored to capture such an organization of data holdings. A single metadata record in this model can provide sufficient metadata to access *all* the components of the data holding either all together or separately. Please see **Fig. 2.** Hierarchy of scientific data holdings

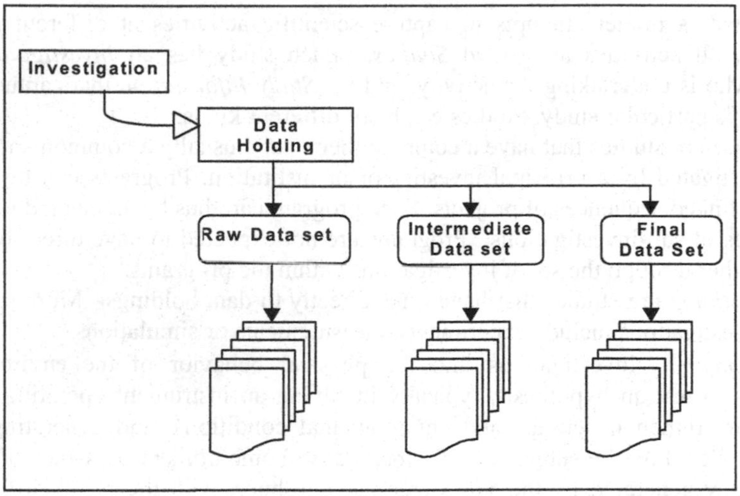

The CLRC Scientific metadata description will contain 6 major data areas, forming the top-level categorization of the metadata. These are as follows:

Additionally, each metadata record will have a unique identifier for reference. The topic gives a set of keywords relevant to the particular study, describing the subject domain with which it is concerned. It thus forms the "encyclopaedia" categorization of the study.

Metadata Category	Description
Topic	Keywords from restricted vocabulary giving the hierarchical subject categorization within each discipline; the discipline and the categorization within the discipline are taken from controlled vocabularies.
Study	The type of entry which this metadata description is capturing. Description of the study within which the dataset has been generated. Includes investigator, experimental conditions, and purpose.
Access Conditions	Access rights and conditions on the data referred to within this entry. Includes ownership and access control information.
Data Description	Description of the datasets.
Data Location	Location of data sets together with any copies, mirrors etc.
Related Material	Contextual information: domain definitional information; links to literature, related studies, user communities,

Keywords will usually come from a particular domain-specific restricted vocabulary, and correspond with some named thesaurus or glossary of terms. Thus in the Topic, we allow for not only the keyword, but also the discipline and source of the term. Thus for *each* keyword, we have three fields (the first two are optional):

Discipline	The subject domain (e.g. chemistry, astronomy, ecology etc).
Source	A pointer (such as URL) to a reference work providing the definition of the restricted vocabulary.
Keyword	The term itself.

This approach should help overcome inappropriate results being returned to the user. For example, the term *Field* has quite distinct meanings in Mathematics (an algebraic structure), Physics (the region of influence of some physical phenomenon), and Geography (a region of farmed land). Searches can be qualified by discipline to prevent results in one domain being returned in response to a query in another.

Controlled vocabularies may be arranged in hierarchies or more complicated ontologies. This structure currently does not attempt to capture such structures or reason over them in the search mechanism. This would be the subject of further development of this model.

Data Resource Interfaces

The data resources accessible through the data portal system may be located on any one of a number of data servers throughout the organisation. Interfaces between these existing data resources and the metadata catalogue are being implemented as wrappers, which will present the relevant metadata about each resource to the catalogue so it appears to the user to be part of the central catalogue. These wrappers will be implemented as XML encodings of the specific metadata relating to that resource using the metadata model schema.

For the prototype development the data collection mechanisms for SR and ISIS were analyzed based on a range of examples and the appropriate sources for the relevant metadata information were identified. The existing metadata was compared to the CLRC metadata model to determine where the relevant information needed to be extracted. Mock-up metadata sets were produced to confirm that the model would capture all relevant information.

Project Status and Future Plans

The initial development phase has finished and a working prototype is available. The metadata model has been verified with all major departments and so far proven to provide the expected flexibility. In the next phase two specific working environments will be chosen and the project will examine how to integrate the data portal into these environments. Furthermore authentication issues will be investigated and the possibility to link the DataPortal to CLRC's HPCPortal for integrated data retrieval and processing.

The longer-term goal of this work is to extend the data portal system to provide a common user interface to metadata for all the scientific data resources held in CLRC. We also envisage it being useful for locating and accessing data held in other laboratories. Where catalogues already exist in specific areas, we will not attempt to duplicate these but instead we will provide the user with a smooth connection into these domain-specific systems.

It is anticipated that the resulting system will have wide applicability across many scientific disciplines. We are keen to develop its potential in partnership with others, particularly within a European context.

References

1. CLRC e-Science Centre - http://www.escience.clrc.ac.uk/
2. CIP Metadata Model - http://158.169.50.95:10080/oii/en/gis.html#CIP
3. CERA Metadata Model - http://www.dkrz.de/forschung/reports/reports9/CERA.book.html
4. DDI Metadata Model - http://158.169.50.95:10080/oii/en/archives.html#DDI

Migratable Sockets for Dynamic Load Balancing

Marian Bubak[1,2], Dariusz Żbik[1,4], Dick van Albada[3], Kamil Iskra[3], and Peter Sloot[3]

[1] Institute of Computer Science, AGH, al. Mickiewicza 30, 30-059 Kraków, Poland
[2] Academic Computer Centre – CYFRONET, Nawojki 11, 30-950 Kraków, Poland
[3] Informatics Institute, Universiteit van Amsterdam, The Netherlands
[4] School of Banking and Management in Cracow, Kijowska 14, Kraków, Poland
`{bubak,zbik}@uci.agh.edu.pl`, `{dick,kamil,sloot}@science.uva.nl`
Phone: (+48 12) 617 39 64, *Fax:* (+48 12) 633 80 54

Abstract. This paper presents design and a prototype implementation of a network interface which may keep communication between processes during process migration and it may be used instead of the well-known socket interface. It is implemented in the user–space, and the `TCP/IP` is applied for internal communication what guarantees relatively high performance and portability. This new socket library (called *msocket*) is developed for efficient dynamic load balancing by the process migration.
Keywords: Distributed computing, sockets, process migration, load balancing, Dynamite.

1 Introduction

Process migration from one host to another is required for efficient load balancing. The processes should be moved from one host to another in a way that allows the original host to be serviced without a need to break computations. Distributed computing environments should handle the communication between processes in spite of the migration. The most popular environments like PVM and MPI do not offer this kind of functionality. One of the environments for migrating processes is Hijacking [6]. This system does not require any changes in the process code; changes are done dynamically after a process starts. The Hijacking system uses the DynInst [2], an architecture independent API for modifying the memory image of a running program. First the process to be migrated is stopped, and then the child process, named *shadow*, is created which inherits all resources used by the parent. After the migration, the processes use resources through the *shadow*. This solution is transparent but rather expensive as it uses the host where the process was initiated. Mosix [1] supports cluster computing and it is implemented on the operating system kernel level. The Mosix migration mechanism enables migration of almost any process at any time to any available host. Each running process has a *Unique Home-Node* (UHN), which is the node where the process was created. After the migration the process uses resources from the new host if possible but interaction with the environment requires communication with the UHN. Many system calls require data exchange between the new node and the UHN, and this operation is time consuming.

B. Hertzberger et al. (Eds.): HPCN Europe 2001, LNCS 2110, pp. 23–31, 2001.

One of the systems developed to support dynamic load balancing is Dynamite [4,5] which attempts to maintain an optimal task mapping in a dynamically changing environment by migrating individual tasks. Dynamite comprises monitoring, scheduling, and migration subsystems [5]. The problem is that migrating processes can not use pipes, shared memory, kernel supported threads, and socket interface.

This paper presents the concept and first implementation of a library that, besides offering the same functionality as system socket library for the TCP/IP protocols family, allows migration of the process without interrupting communication with other processes. The new library, called *msocket*, can be a substitution of the standard socket library so that no changes in the program will be required. All necessary modifications are handled at the library level so that no changes in the kernel are needed, either.

2 Requirements for Socket Migration

An essential requirement for the socket migration is that all modifications of communication libraries have to be transparent for the TCP and UDP protocols. To easily migrate a process with the socket interface in use, the modifications of the communication library have to warrant that:

- establishing new connections must be allowed regardless of the number of process migrations,
- all connections that had been established before the migration took place have to be kept, and data must not be lost.

The first requirement means that:

- migratable processes should be able to connect and keep the connection between themselves,
- migratable processes should establish and keep connections with processes which use the original socket library ,
- a client using the original socket library should be able to connect to a migratable process; the location of the migratable process should have no influence on communication establishment.

All modifications should be done in the user-space, what means that one should wrap the system calls and library calls.

3 Description of the *msocket* System

In the user code the communication through the TCP and UDP protocols uses sockets which are treated by the process as file descriptors. For this reason wrappers are used for each call which has file descriptors as arguments. In some cases this is a simple change which only translates the file descriptor number. The data sent between processes always goes through an additional layer built of wrappers of the system calls. Inside the wrappers, some control information is modified but the communication is performed by the original TCP/IP stack. This solution requires address servers which allow to find the location of a socket.

The communication with address servers is required only while establishing the connection: once the connection is established, all data goes directly betw een processes. Daemons are used to help redirect the connection. To maintain the connection while the process migrates, a mirror is kept; it receiv esdata from peer processes and redirects it to the new process location.

The mirror captures and redirects packets which were on-the-fly during the process migration. The mirror is a child of the migrating process, so it inherits all socketsused by the process before the migration. After the migration, the migrated process connects to the mirror and informs all the connection peers about its new location.

In the standard IP addressing, the address of a socket is associated with the machine where the process is running. When processes are migrating, the real address of the host changes with each migration. To become independent of the changes of real addresses, virtual addresses are used, and the form of these addresses is the same as the addresses used for the TCP and UDP communication. The address server (msmaster) tak es care of address translation and guarantees uniqueness.

After process migration, all connections have to be redirected to the new process location. The process has to leave a mirror because some data could be on-the-fly (inside netw ork stack buffers or inside active netw ork equipment like a router or switch). The mirror should be removed as soon as possible. In this case, the mirror captures and redirects the packets which were on-the-fly during the process migration.

The daemons participate in the redirection of connections. After migration, while restoring, a process has to inform its peers about its new location. The migrated process communicates with the daemon on its new host which takes care of propagating this information.

During normal work, processes use the direct connections. The daemons are not used for passing user data between hosts. Daemons should exist on each machine, and their role is limited to redirecting connections. When a process migrates, it has to inform peers about its new location. While restoring the process, daemons on all hosts which were involv ed incommunication with the migrating process should be informed about its new location. Subsequently, the daemons force processes to redirect connections.

4 Redirection of TCP Connections

For the TCP/IP sockets, there is no synchronization betw een read and write requests on both sides of the connection. After the migration, the ***msocket*** library creates a new socket (while restoring), registers it and asks the mirror to unregister the old one. Then the system call `listen()` is invok ed ona new socket. Next, the migrated process sends the *redirect request*to the local daemon. This request contains *virtual address* of the socket and the peer, *real addr ess*of the peer and the new *real address* of the socket.

The local daemon receives this request and contacts the daemon at the connection peer host. The daemon sends the redirect request to the remote daemon. A t the beginning theremote daemon opens a connection to the migrated process (the new *real address* is a part of the redirection request). Then the remote daemon tries to find the connection in its database. This entry is associated with the pid number of the process or processes which is/are the owner(s) of the socket. The remote daemon sends the request to all those processes and then passes the open file (new socket) to them. This w ay all processes share again the same soc ket. T o force the process to read the request and receiv e the file descriptor the daemon sends it a signal (using `kill()`). As a result, the process en ters the signal handler which is responsible for reading the request and modifying the internal library state. As soon as the new socket is opened by the remote daemon, it is passed to all processes which hav e the connections with the migrating process, and the remote daemon sends the response to the daemon which is running on the same host as the process after the migration. From this moment on, the new connection is established, although the connection through the mirror is still present.

At the peer (`A`) of the migrated process (see Fig. 1) one virtual file descriptor `vfd`, which is used by the process code as a socket, is associated with two real sockets `fd` and `newfd`. The first of these sockets w as created before the process migration (`fd`), while the second one (`newfd`) was created during the redirection. On the migrated process side only one real socket is used, plus the socket connected to the mirror, but the mirror connection is shared by all sockets.

When the migration takes place, the peer process (`A` in Fig. 1) could be inside of the `read()` system call and this process could still w ait for the data from the original `fd` which is connected to the mirror. This system call should be interrupted after the migration. That is why the migrated process sends a con trol request to the mirror `MS_MIRROR_CLOSE_ESTABLISHED`, instructing the mirror to close half of the connection betw een the mirror and the *peer* process. The mirror calls `shutdown()` on the `old_sock` file descriptor.

The *peer* process is an activ e part during the redirection. Each time the *msocket* library takes con trol[1] it chec ks if the connection is flushed. When the connection betw een `fd` and `old_sock` is flushed, the *peer* process closes the connection. Then `fd` can be set to the value of `newfd`. When the mirror detects the closing of the `old_sock` it sends the control message to the new process (`C`). The new process knows that this connection is completely redirected and all read and write requests will be done through the new socket (without the use of the mirror).

The process `A` may write more data than the process `C` may consume. The data coming from the mirror should be kept in a buffer. The process `C` stores the data inside the buffer associated with the appropriate socket.

Detailed analysis of the redirection of a connection is presented in Fig. 1 and in Tables 1 and 2.

[1] During all library calls and the signal handler.

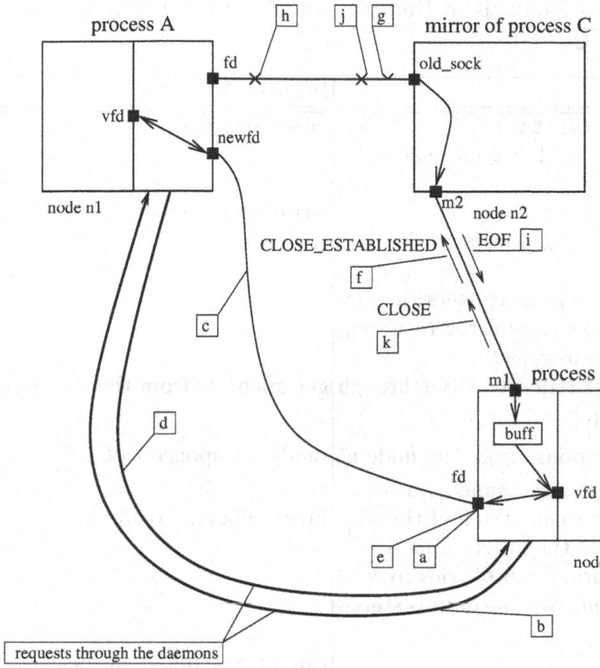

Fig. 1. Details of the Connection Redirection.

5 Implementation

The msmaster (address server) keeps t w o tables: one of them contains the *virtual address* and the time stamp of the last verification, and the second table keeps the pairs of addresses: the *virtual address* and the *real address*. The address server is decomposed into threads to simplify its structure. One of them (*cleaner*), periodically checks the database contents and removes old entries. The second one (*tcp_ listner*) takes care of communication, waits for the incoming connections from the daemons; a new thread (*tcp_ thread*) is created for each incoming connection. Each *tcp_ thread* serves a single msdaemon.

The msdaemon has a multiple thread structure. On startup, eac h daemon reads the configuration file which contains the addresses of the subnets with the address of the master server for eac h subnet. Themsdaemon also takes part in the redirection of connections. The *msocket* daemon uses tw o separate message queues. One of them is used by an application process to ask the daemon to do some activities (i.e. the address queries) while the second one is used by the daemon to force process to e.g. redirect sockets, check if the process exists and so on.

Table1. Detailed Analysis of Redirection of the Connection – Processes **A** and **C**.

T	Process **C**	Process **A**
t_0	normal `read()/write()`	`read()/write()`
t_1	receives the CKPT signal, saves its state	
t_2		writes some data
t_3	restores from file, connects to the mirror	
t_4	creates a new socket, registers the new address and asks the mirror to unregister the old address [a]	
t_5	sends a redirect request to **A** (through the daemon) [b]	get open file from the daemon
t_6	receives the response from the node n1 [d], and accepts new connection [e]	sends a response to **C** [d]
t_7	asks the mirror to close half of the connection between the mirror and **A** [f]	break blocking read
t_8	during each library call **C** tries to read data from the mirror, the data is stored in a buffer	
t_9	as above	the output buffer is flushed, **A** closes the connection [h]
t_{10}	reads rest of the data, puts it into the buffer and finally asks the mirror to exit [k]	
t_{11}	during each read request process uses the buffer first	process uses the new connection to read and write
t_{12}	when the buffer is empty process starts using the new connection	

6 Tests of Functionality and Overhead

In order to check the system development, two groups of tests have been worked out. The first group was designed to check if the planned functionality of the *msocket* is realized whereas the second group focuses on measuring the level of overhead induced by *msocket*. The main features of this system are establishing the connection and communication, so performing these operations is sufficient to test the *msocket* library.

To verify the functionality, the following tests have been performed: creation of the virtual socket, establishing connection before the migration, establishing connection after migration(s), establishing connection during migration, migration with connection in use. The tests confirmed that *msocket* library has the required functionality.

Overhead tests measure the time spent on executing the code of this library. The generic application used in all tests is a typical producer–consumer program.

Table 2. Detailed Analysis of Redirection of the Connection – Mirror and Daemon Behavior.

T	mirror	daemon n3	daemon n1
t_0	does not exist		
t_1	starts		
t_2	all incoming data are buffered in system buffers		
t_3	accepts connection from **C**, then reads data sent by **A** at t_2		
t_4	unregisters old address (this operation involves the daemon on the node n2)		
t_5		receives the request from **C**, and forwards it to the daemon on the node n1 [b]	receives the request from the daemon on the node n3, creates new socket and connects to **C** and then passes the open socket to **A** [b,c]
t_6		receives the response from the daemon on the node n1 and sends it to **C** [d]	sends a response to **C** [d]
t_7	closes half of the connection to **A** [g]		
t_8	reads data from **A** and sends it to **C** (with additional header)	periodically sends a signal to **C** to force data read from the mirror	periodically sends signal to **A** to force data flush
t_9	recognizes when the socket is closed b y**A** and sends this information to **C** [i] and closes old connection [j]	as above	as above
t_{10}	receive close request from **C** [k]	as above	as above
t_{11}	does not exist		
t_{12}			

The producer and consumer were link ed with*msocket* library or with standard system library

We have measured the transfer speed for different buffer sizes varying betw een 1 byte to 8 K bytes at the producer side. The consumer buffer was the same all the time – 10 K b ytes. The result concerning the producer was taken, so it is the test of *writing* through *msocket* link.

Results. The first test takes place between tw o hosts connected by the 10 Mbps Ethernet. In Fig. 2 the filled circle marks (•) denote measured points of the test with the *msocket* library. The empty inside circle marks (∘) are results of the test without the *msocket* library . The plot in Fig. 2 shows the transfer speed as a function of the buffer size.

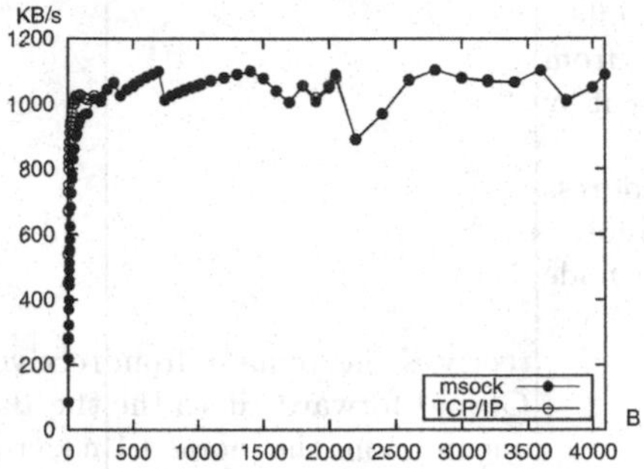

Fig. 2. T ransfer Speed vs Buffer Size for 10Mbps Ethernet.

The tests were also run on the D AS [3] distributed supercomputer built out of four clusters connected by the ATM netw ork. Clusters use Myrinet and F astEthernet as local net w orks.Our test used the 100 Mbps Ethernet. The results are similar to previously observed (see Fig. 3).

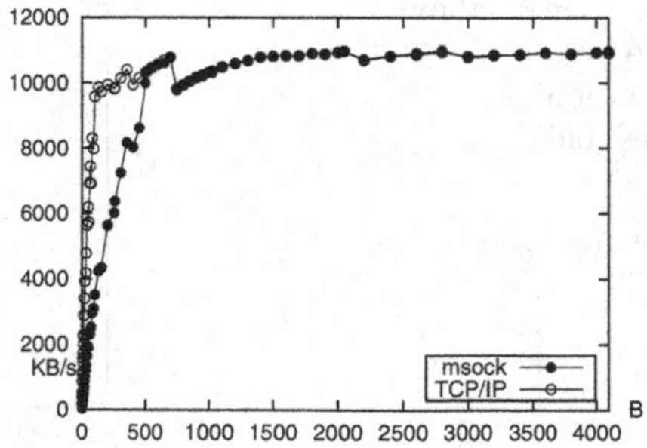

Fig. 3. T ransfer Speed vs Buffer Size – on DAS.

7 Conclusions and F utureWork

The *msocket* library does not support nonbloc king I/O operations. This kind of access to the sock et is v ery often connected with the user defined signal handler, and to support this access, it is required to create a wrapper for the user signal handler. Urgent data is also unsupported; this feature of the socket stream also requires dealing with the signal handlers. Applications cannot use the socket option calls like `setsockopt()`, `ioctl()`, `fcntl()`. In the future, this limitation may be partially removed. To do this, it is necessary to retain more information about the socket state and restore it after the migration. Unfortunately, some instances of these system calls are system dependent.

The socket and file descriptors are equivalent in the Unix systems, therefore sockets can be shared between parent and child(ren) in the same way as files. This situation is dangerous for the *msocket* library because it is impossible to migrate a process which shares a socket with other processes.

The *msocket* library is written in C and was tested on Linux systems. It is possible to port this environment to other system platforms, the number of system dependent parts is limited.

We are now working on the implementation of mpich MPI with the *msocket* library .

Acknowledgments

This research is done in the framework of the Polish-Dutch collaboration and it is partially supported by the KBN grant 8 T11C 006 15. The Authors are grateful to Dr. W. Funika for his remarks.

References

1. A. Barak, O. La'adan, and A. Shiloh. Scalable Cluster Computing with MOSIX for LINUX. In *Proceedings of Linux Expo 1999*, pages 95–100, May 1999. `http://www.mosix.org/`.
2. J.K. Hollingsworth and B. Buck. *DyninstAPI Programmer's Guide Release.* Computer Science Department University of Maryland. `http://www.cs.umd.edu/projects/dyninstAPI`.
3. The distributed ASCI supercomputer (DAS). `http://www.cs.vu.nl/das/`.
4. K.A. Iskra, F. van der Linden, Z.W. Hendrikse, B.J. Overeinder, G.D. van Albada, and P.M.A. Sloot. The implementation of Dynamite – an environment for migrating PVM tasks. *Operating Systems Review*, 34(3):40–55, July 2000.
5. G.D. van Albada, J. Clinckemaillie, A.H.L. Emmen, O. Heinz J. Gehring, F. van der Linden, B.J. Overeinder, A. Reinefeld, , and P.M.A. Sloot. Dynamite — blasting obstacles to parallel cluster computing. In Peter Sloot, Marian Bubak, Alfons Hoekstra, and Bob Hertzberger, editors, *Proceedings of High Performance Computing and Networking Europe*, volume 1593 of *Lecture Notes in Computer Sciene*, pages 300–310, Amsterdam, The Netherlands, April 1999. Springer-Verlag.
6. V.C. Zandy, B.P. Miller, and M. Livny. Process Hijacking. In *The Eighth IEEE International Symposium on High Performance Distributed Computing (HPDC'99)*, pages 177–184, Redondo Beach, California, August 1999. `http://www.cs.wisc.edu/paradyn/papers/`.

Toward Realizable Restricted Delegation in Computational Grids[1]

Geoff Stoker, Brian S. White, Ellen Stackpole, T.J. Highley, and Marty Humphrey

Department of Computer Science, University of Virginia
Charlottesville, VA 22903
{gms2w,bsw9d,els2a,tlh2b,humphrey}@cs.virginia.edu

Abstract. In a Computational Grid, or Grid, a user often requires a service to perform an action on his behalf. Currently, the user has few options but to grant the service the ability to wholly impersonate him, which opens the user to seemingly unbounded potential for security breaches if the service is malicious or errorful. To address this problem, eight approaches are explored for realizable, practical, and systematic restricted delegation, in which only a small subset of the user's rights are given to an invoked service. Challenges include determining the rights to delegate and easily implementing such delegation. Approaches are discussed in the context of Legion, an object-based infrastructure for Grids. Each approach is suited for different situations and objectives. These approaches are of practical importance to Grids because they significantly limit the degree to which users are subject to compromi*se.*

1 Introduction

A Computational Grid, or Grid is a wide-area distributed and parallel computing environment consisting of heterogeneous platforms spanning multiple administrative domains. Typical resources include supercomputers, PCs, archival storage, and specialized equipment such as electron microscopes. The goal of a Grid is to enable easy, fast, and inexpensive access to resources across an organization, state, country, or even the world. In the general model, individual users will be able to contribute to collective Grid services by creating and publicizing their particular service, for example, the specialized scheduler AppLeS [1]. Users could request action from this scheduler (perhaps at a cost) if they believed the output of the scheduler would be more efficient or predictable than some default scheduling mechanism.

In a Grid, a user may need to ask a service-providing object[2] to perform an operation on her behalf. For example, a scheduler may need to query an information service to determine on which machines a user is allowed to and has enough allocation to execute a job. The scheduler does not have the privilege to ask the

[1] This work was supported in part by the National Science Foundation grant EIA-9974968, DoD/Logicon contract 979103 (DAHC94-96-C-0008), and by the NASA Information Power Grid program.

[2] For ease of presentation, an *object* is chosen as an arbitrary abstraction of an entity in a Grid.

B. Hertzberger et al. (Eds.) : HPCN Europe 2001, LNCS 2110, pp. 32-41, 2001.

information service directly, because the information service protects its potentially proprietary information from unauthorized access. The information service would provide a user's information to a scheduler if it was acting on that user's behalf.

The conventional approach when a user must ask a service to perform some operation on her behalf is to grant *unlimited delegation*, which is to unconditionally grant the service the ability to impersonate the user. To date, restricted delegation is not used in emerging Grids because it is too difficult to design, implement, and validate except in very limited, *ad hoc* cases. While unlimited delegation is a reasonable approach in which all services can be wholly trusted by the users who wish to invoke them, it is clearly not scalable into general-purpose Grids. For delegations within a Grid, *the crucial issue is the determination of those rights that should be granted by the user to the service and the circumstances under which those rights are valid.* Delegating too many rights could lead to abuse, while delegating too few rights could prevent task completion.

Under *unlimited delegation* it is possible for a rogue object, to which unlimited rights have been delegated, to trick a trustworthy object into executing a service that it would otherwise not perform. The problem addressed in this paper is how to restrict, or eliminate, the potential security breaches possible when one object misplaces trust in another object. The problem does *not* focus on what an untrustworthy object can do directly, but rather what the untrustworthy object can get *other* trustworthy objects to do. There is a separate issue regarding how an initiator validates the correctness of a target's response, which is not directly addressed by this paper. While we acknowledge that a malicious object can "lie" in response to a request, we strive to ensure that that malicious object cannot obtain and misuse another object's rights.

There are two main contributions of this paper. First, we present a number of approaches for restricted delegation that more fully support realizable policy than existing approaches. "Realizable" in this context refers to the ability of one user to *define* and *implement* a policy by which to grant a limited set of privileges to another principal. By creating and supporting policies for restricted delegation that can be more easily determined, defined, and enforced, a number of security vulnerabilities can be reduced and even eliminated in complex computing environments such as Grids. Second, we explore the feasibility and implementation of these policies within Legion [8], which is mature software that provides a secure, fault-tolerant, high-performance infrastructure for Grids.

Section 2 gives a brief overview of Legion and its security mechanisms. In Section 3, we discuss general approaches for determining the rights to delegate when one object invokes the services of another and the applicability of each approach to Legion. Section 4 contains related work and Section 5 concludes.

2 Legion Security Overview

This work is presented in the context of Legion, an object-based Grid operating system that harnesses heterogeneous, distributed, and independently administered resources (e.g. compute nodes), presenting the user with a single, coherent environment [8]. Unlike traditional operating systems, Legion's distributed, extensible nature and user-level implementation prevent it from relying on a trusted code base or kernel. Further, there is no concept of a *superuser* in Legion. Individual objects are responsible for legislating and enforcing their own security policies; Legion does provide a default security infrastructure [4] based on access control lists (ACLs).

Legion represents resources as active objects, which are instances of classes, responsible for their creation and management. Objects communicate via asynchronous method invocations. To facilitate such communication, Legion uses Legion Object Identifiers (LOIDs) for naming. Included in the LOID is the associated object's public key. During Legion object-to-object communication, a recipient authenticates a sender through the use of its well-known LOID and encapsulated public key. Legion can also be configured to support X.509-based authentication.

Legion *credentials* are used to authenticate a user and make access control decisions. When a user authenticates to Legion the user obtains a short-lived, unforgeable credential uniquely identifying himself. A person may possess multiple credentials, signifying multiple roles. When a user invokes an operation of another object, the user's credentials are encrypted and transmitted to the target object. The recipient object unmarshalls the credentials and passes them to a layer (called "MayI") that conceptually encapsulates the object. For each credential passed, MayI determines whether key security critereia are satisfied. Authorization is determined by an ACL associated with each object; an ACL enumerates the operations on an object that are accessible to specific principals (or groups). If the signer of any of these credentials is allowed to perform the operation, the method call is permitted.

Prior to the work described in this paper, the restriction fields of a credential have been unused, because no general-purpose procedure existed by which to determine and implement policy regarding restriction. The credentials have been *bearer* credentials, implying that the possessor or bearer of a particular credential had permission to request *any* operation on behalf of the user. Bearer credentials have been valuable in the development of Legion but are becoming potential security liabilities, especially for Legion's continuing wide-scale deployment. Note that credential interception is not a concern; credentials cannot be intercepted because they are encrypted with the public key of their recipient. Rather, someone could send her credential to an object and that object could then use the credential maliciously in transactions with arbitrary objects. In the next section, we discuss approaches that can be used in general Grids and examine their practical applicability by relating these approaches to Legion.

3 Delegation Approaches

We examine three schemes of restricting delegation: restricting methods that may be called, restricting objects that may pass delegated rights or be the target of such calls, and restricting the validity of delegated rights to a given time period. In each of the following, we start with the base credential: *[The bearer of this credential may do anything on Alice's behalf forever] signed Alice.*

3.1 Method Restrictions

Instead of allowing the presenter of a bearer credential to perform any and all operations, the constructor of a bearer credential can explicitly enumerate allowable methods, either individually or categorically.

Bearer Identity Restrictions: *none*
Method Restrictions: $M_1, M_2, \ldots M_j$
Time Restrictions: *none*

Fig. 1. Bearer Credential with Method Restrictions.

Explicit Method Enumeration by Manual Inspection. A set of well-known object-to-object interactions can be manually identified to determine a remote method invocation tree rooted at a particular remote call, and a credential can list these methods. By restricting the methods that the service can call on her behalf, the user has significantly reduced the ability of an untrustworthy service to perform malicious operations, such as destroy one of her private files.

However, this approach does not require any meaningful context for the credential. This strategy is tedious to implement and inflexible. In large systems, it is extremely difficult, if not impossible, to determine remote method invocations through manual code inspection, e.g., the source code may not even be readily available. Any changes to code shared by many objects (such as Grid infrastructure support code) require that all previously computed restricted credentials be modified to reflect the changes.

Experiences applying explicit enumeration to Legion highlight the limitations of this approach. Many hours were needed to enumerate the method call tree of a simple Legion utility. This utility, like all other objects and utilities, makes use of the services provided by the Legion distributed object framework. While the number of explicit method calls made by the application was small, these calls may lead to a complex web of remote interactions. For example, if the target of a request is not actively running, Legion will need to invoke additional services to schedule and activate the object. These calls were the source of complexity in the exercise and are a likely source of inspection error. We believe that such calls are not specific to Legion, but inherent in any large-scale Grid.

When a user wishes to invoke a remote method call, the user must pass an enumeration containing *each* subsequent remote method invocation. In the cases of large call chains, this list may well be larger than the message itself. This will lead to additional network contention and message setup and transmission latencies. Further, because one monolithic signed enumeration is sent, it cannot be selectively dismantled and will be sent down each branch of the call chain, though many of the enumerated methods will have no bearing on an object or its descendants. The obvious approach is to partition this set into multiple credentials, and generate the necessary information for a particular object in the call chain to determine *when* it can discard a credential that will not be used in the future.

Compiler Automation. In order for the method enumeration approach to scale beyond a trivial number of operations, the reliance on direct human interaction must be reduced or eliminated. A bottom-up *security compiler* could annotate each remote method call with the remote method calls invoked recursively from it. When compiling an application, the compiler would examine the dependencies to other modules to union these annotations. The compiler would then automatically insert code to initialize the application's credential set in accordance with these annotations. Once the security compiler is written, it is reusable for any operation. In principle, this eliminates many of the limitations of the previous section, particularly the tedious and error-prone nature of generating credentials. If the security compiler is invoked as part of the general compilation process, then this should be viewed as an automatic mechanism with relatively little overhead.

While independence from user intervention removes the tedium of manual enumeration, it also removes the user from the policy decision altogether. That is, the security compiler blindly blesses untrusted code by enumerating the remote method invocations it performs. Under the manual approach, the user is forced to visually inspect the code and may thus discover blatant security violations. For this reason and because of the daunting complexity involved in writing compilers for all of the languages that Legion supports (C, C++, Fortran), this approach has not been pursued for Legion to date.

Method Abstractions. A higher level of abstraction than that of specific method enumeration can be used. As a first approach, methods can be coarsely described as *read* or *write* operations. Then, it is possible to construct a credential for *read* and/or *write* operations in general, in lieu of specific methods. The size and complexity of a credential can thus be significantly reduced. This approach is attractive for high-level object interactions that can be wholly classified as read-only, which seemingly need not authorize write operations (and thus reduce the scope of a compromise).

This approach is limited because it likely does not provide additional security for applications that perform 'write' operations. That is, if a particular remote method invocation entails some destructive or write operation, the credential must grant the whole 'write' capability. Because the Grid infrastructure will almost certainly require some form of 'read' capability (as is the case in Legion), this approach is reduced to an all-inclusive bearer credential. A better approach is to combine the gross categorization of the 'read' capability with specific write operations.

Method abstraction represents a good tradeoff between flexibility and ease of implementation and has been utilized in Legion. Legion was experimentally modified to directly support a 'read' credential by altering the default MayI layer. Next, the 'write' methods made during the execution of a particular tool were enumerated. The all-encompassing 'read' and enumerated 'write' methods form the credential utilized in this tool. Such credentials are currently being tested and evaluated in limited deployment. Unfortunately, such partitioning of the method space requires updating the infrastructure as remote methods are introduced and removed from the system.

3.2 Object Restrictions

In Section 3.1, the scope of attack is limited to those methods enumerated, but the methods are not associated with specific objects and thus may be invoked by *any* object acquiring the credential. By explicitly associating an object or objects with each method, greater security can be provided by restricting not only *what* may be invoked, but also *who* may invoke it or upon *whom* it may be invoked. Object restrictions are discussed as both a complementary and stand-alone measure.

Bearer Identity Restrictions: B_1, B_2, ... B_i
Method Restrictions: M_1, M_2, ... M_j
Time Restrictions: *none*

Fig. 2. Bearer Credential with Method and Object Restrictions.

Object Classes. Due to the dynamics of the system, it may not be possible for an application to ascertain the identity of all principals involved in a call chain. Furthermore, in some systems, the class or type of the object is more useful than the identity of a particular instance of that class. Therefore, one approach is to restrict the bearer identity to a particular class.

The approach of utilizing class restrictions could follow from the enumeration mechanisms explained above. Unfortunately, it would suffer many of the same setbacks. Nevertheless, when used in tandem with method enumeration, greater security is achieved than via either mechanism in isolation.

This approach is directly applicable to Legion, in which the object hierarchy is rooted at a particular well-known object. All objects are derived directly or indirectly from this class. Legion supports a *classOf* method to describe the relationships throughout the object hierarchy. This approach is currently being considered for a number of Legion operations.

Transitive Trust. A form of transitive trust can be used to address the scalability concern of passing around one all-encompassing credential, as introduced in Section 3.1. In this approach, each remote method invocation is annotated with the methods it invokes directly or those within a certain depth relative to the current position in the call chain. The annotations reflect the target of the method invocation. Unlike previous models discussed, this option would concatenate credentials signed by intermediary nodes throughout the call chain, rather than passing one monolithic list signed by the invoking principal at the root of the call chain.

Embedded metadata within the annotations might allow intermediary nodes to discard credentials that apply neither to themselves nor their descendants. Though the credential set is determined dynamically, the credentials themselves are not created dynamically. They are associated with remote method invocations as before, using either a manual or automatic approach.

While it addresses the scalability concerns of passing bloated credentials, this approach is less secure than the monolithic approaches. A target would recognize an intermediary acting on behalf of the initiator if the credentials form a chain originating at the initiator. While the initiator may specify method restrictions, there is no obvious means for a target to discern that an intermediary is acting in the intended context. For example, two objects could conspire such that the first object gives the second object the ability to execute *arbitrary* methods within the rights of any of its predecessors. To avoid such a coordinated attack, it could be required that the final target object trust each principal named in the chain in order to respect the transitivity properties of that chain. Because of this concern, the use of this approach in Legion is still under consideration.

Trusted Equivalence Classes. Another approach is to group objects into security equivalence classes (e.g. according to importance, functionality, administrative domain, etc.). Instead of specifying an object instance or class name, the name of the equivalence class would be specified as a credential restriction. A practical and important example of such an equivalence class in Legion is the "system" objects. Such objects form the core of the system and include objects representing hosts, schedulers, and storage units. Method invocations initiated from such objects need not be explicitly granted on a per-object basis in the list of rights contained in a credential. This could significantly reduce both the off-line costs and the on-line overhead of generating and sending restricted credentials, as it provides a coarse-grain abstraction. Note that it is not mandated that every user trust a collection of objects, even "system" objects — any user is free to simply not use this approach when constructing credentials.

The major problem in this approach is how to both define these groups and verify membership in these groups, because groups will most likely need to be defined on a per-user basis. Another shortcoming is that a user implicitly trusts any changes made to objects of trusted classes objects.

Trusted Application Writers and Deployers. Rather than trusting classes of objects, trust can be placed in principals that implement those classes. Implicit in this approach is the assumption that an object trusts another object if and only if it trusts the principal which 'vouches' for it. Intuitively, this is more appealing than simply trusting an object which exports a familiar interface but which may have been implemented by a malicious user.

When a user compiles an object or application and makes it available to the general Grid community, the deployment of the object or application in this approach now includes the signature of that user and the signatures attached to any dependent modules — the application is "branded" with the signature of its developer and/or compiler. When a user interacts with services in a Grid, the user sends a credential stating its trust in a specific subset of users and/or compilers. When an object *receives* a request for service, it bases access on the direct or indirect objects that made the request for service. If the request for service on behalf of a particular user is augmented with a credential stating trust in the deployer of the object from which the request arrived, then access is allowed.

In Legion, applications and objects must be *registered* before they can be executed in a distributed environment. Legion manages registered binaries, distributing them to hosts for execution on demand. Registration serves as a convenient time to bind the trust relationship. The principal initiating the registration of a particular binary would be noted as the 'owner' of that object. A user could specify her trust in particular principals or groups. This enumeration could be made explicit in a credential. This approach is appealing because of its generality and the ease with which it allows the instantiation of policy decisions. It should work well for common cases in which a large group of users trust one another (e.g. within a single administrative domain), but do not necessarily trust another group.

3.3 Time-Dependent Restrictions

The last approach is orthogonal to the previous mechanisms and may be used in conjunction with them to restrict the window of vulnerability to replay attacks. In the simplest version of this approach, a pre-determined amount of time is used as the time-out for the credential. This guarantees that the credential cannot be abused in a replay attack after a fixed time interval. Unfortunately, as the perceived slack time in the timeout value is reduced, the likelihood that the credential will need to be refreshed increases. Too much slack time opens the door for a replay attack.

Bearer Identity Restrictions: $B_1, B_2, \ldots B_i$
Method Restrictions: $M_1, M_2, \ldots M_j$
Time Restrictions: $T_1, T_2, \ldots T_k$

Fig. 3. Bearer Credential with Method, Object, and Time Restrictions.

If the initiating object will be alive during the entire operation, then it is possible to mandate that each object in the call chain perform a callback to verify that the initiator is indeed still running. This prevents a malicious object from acquiring a bearer credential and using it long after the application has exited but before a fixed timeout value has expired. Unfortunately, this doubles the number of object-to-object transactions and it does not benefit a long-lived application, whose natural execution duration provides plenty of opportunity for a man-in-the-middle to abuse privileges. Callbacks could potentially serve the alternate purpose of recording an audit trail of methods invoked under the identity of a given principal. To limit the performance degradation of indiscriminant callbacks, the principal may require notification of only 'important' method invocations.

This approach has not been used in Legion because it is impossible to predict how long an operation will require, due to the inherent unpredictability in the computational nodes and the networks. Even if this could be performed on a single case, it is not clear to what extent the results can be applied to another tool, as each tool has different functionality. It is also not clear if the increased security of short time-outs will be of any value to an average user — if more than one operation "fails" in a short time period, the user may think that Legion as a whole is experiencing errors, even though the observed behavior is the result of credential expiration.

4 Related Work

Related delegation work generally assumes that the rights to be delegated have already been decided by a separate mechanism, and it is just a matter of encoding and checking these rights. The general need for delegation is exemplified by the Digital Distributed System Security Architecture, a comprehensive collection of security services and mechanisms for general-purpose distributed systems [7]. The designers of this system note that this form of delegation is unrestricted and that, while restricted delegation "seems desirable", the useful types of restrictions are specific to each application and thus difficult to generalize. Erdos and Pato present an architecture for implementing delegation in the Distributed Computing Environment (DCE) [3]. There are two types of delegation-related restrictions that principals can use to place limitations on revealing their identity – target restrictions and delegate restrictions. Neuman shows how existing authentication systems can be used to support restricted proxies or restricted delegation [10]. In this work, there are two proxy types: a bearer proxy, which may be used by anyone, and a delegate proxy, which may only be used by an object enumerated within. The CORBA Security Specification [2] identifies five types of delegation. The types differ in terms of combining rights, such as whether the initiator's rights are combined or remain distinct from the intermediaries' rights, when a subsequent call must be performed. It is noted that no one security technology supports all options. Linn and Nystrom discuss application of attribute certificates to delegation in distributed environments [9]. Designation of an object's membership in a group is the most common use for attribute certificates. There are numerous ways in which to address delegation with attribute certificates: using a generic attribute certificate as a capability (similar to Legion's bearer credential), enumerating authorized delegates within an attribute certificate, and having an attribute certificate reference a short-term key that could be used by a delegate.

Globus [5] is a Grid toolkit being developed at the Argonne National Lab and USC/ISI. Delegation of rights is a significant problem in Globus (as in Legion), and the Globus developers do not have a solution (although other aspects of the Grid security problem are discussed in [6]). In fact, the reliance on GSS-API [13] in Globus may hinder efforts at delegation in Globus, because GSS-API does not offer a clear solution to delegation (and could impede restricted delegation). There have been some efforts in Globus to adopt and support the Generic Authorization and Access-Control API (GAA API [11]); however, GAA API focuses on mechanism and an API for authorization and does not address a systematic way in which to determine the rights that should be delegated. Independent of Globus, Akenti [12] is a Grid-based project that makes important contributions for a uniform mechanism to perform authorization in a Grid environment but does not directly address the determination of rights to be granted.

5 Conclusions

Computational Grids are a powerful new computing paradigm. Users must currently grant unlimited delegation to services that might act on their behalf. As Grids scale, this implicit trust model will allow potential compromises. Eight general approaches have been described that are valuable for restricting delegation in different situations. To show the utility of these approaches, each was described in the context of Legion, a widely-deployed infrastructure for Grids. Next steps are to determine the characteristic properties of a given situation that determine the most appropriate approach for restricted delegation. The approaches described in this paper are an important first step toward practical restricted delegation, and, as these approaches are further characterized, promise to make using Grids more secure.

References

1. Berman, F., R. Wolski, S. Figueira, J. Schopf, and G. Shao. "Application-level Scheduling on Distributed Heterogeneous Networks", in *Proceedings of Supercomputing 96*, 1996.
2. Chizmadia, David. A Quick Tour of the CORBA Security Service, http://www.omg.org /news/corbasec.htm, Reprinted from Information Security Bulletin-September 1998.
3. Erdos, M.E. and J.N. Pato. "Extending the OSF DCE Authorization System to Support Practical Delegation", *PSRG Workshop of Network and Distributed System Security*, pages 93-100, February 1993.
4. Ferrari, Adam, Frederick Knabe, Marty Humphrey, Steve Chapin, and Andrew Grimshaw. "A Flexible Security System for Metacomputing Environments." In *Seventh International Conference on High Performance Computing and Networking Europe (HPCN Europe 99)*, pages 370-380, April 1999.
5. Foster, Ian, and Carl Kesselman. "Globus: a metacomputing infrastructure toolkit". *International Journal of Supercomputer Applications*, 11(2): pages 115-128, 1997.
6. Foster, Ian, Carl Kesselman, Gene Tsudik, and Steven Tuecke. "A Security Architecture for Computational Grids." In *Proceedings of the 5th ACM Conference on Computer and Communications Security*, pages 83-92, November 1998.
7. Gasser, Morrie, Andy Goldstein , Charlie Kaufman, and Butler Lampson. "The Digital Distributed System Security Architecture." In *Proceedings of 1989 National Computer Security Conference*, 1989.
8. Grimshaw, Andrew S, Adam Ferrari, Frederick Knabe, and Marty Humphrey. "Wide-Area Computing: Resource Sharing on a Large Scale." *Computer*, 32(5): pages 29-37, May 1999.
9. Linn, J. and M. Nystrom. "Attribute Certification: An Enabling Technology for Delegation and Role-Based Controls in Distributed Environments", Proceedings of the *Fourth ACM workshop on Role-Based Access Control*, 1999, pages 121 - 130.
10. Neuman, B. Clifford. "Proxy-Based Authorization and Accounting for Distributed Systems," *Proceedings of the ICDCS'93*, May 1993.
11. Ryutov, T.V., G. Gheorghiu, and B.C. Neuman. "An Authorization Framework for Metacomputing Applications", *Cluster Computing*. Vol 2 (1999), pages 165-175.
12. Thompson, M., W. Johnston, S. Mudumbai, G. Hoo, K. Jackson, and A. Essiari. "Certificate-based Access Control for Widely Distributed Resources", Proceedings of the *Eighth Usenix Security Symposium*, August 1999.
13. Wray, J. "Generic Security Services Application Programmer Interface (GSS-API), volume 2". RFC 2078, January 1997.

Profiling Facility on a Metasystem

Carlos Figueira and Emilio Hernández*

Universidad Simón Bolívar
Departamento de Computación y Tecnología de la Información
Apartado 89000, Caracas 1080-A, Venezuela
{figueira,emilio}@ldc.usb.ve
http://suma.ldc.usb.ve

Abstract. A metasystem allows seamless access to a collection of distributed computational resources. We address the problem of obtaining useful information about the performance of applications executed on a metasystem, as well as the performance of the metasystem itself. This article describes the experiences and preliminary results on incorporating a profiling facility in a Java-based metasystem. Our case study is SUMA, a metasystem for execution of Java bytecode, with additional support for scientific computing.

1 Introduction

A metacomputer is a collection of computational resources interconnected by a network and accessible transparently. Metacomputers (or metasystems) have become popular as a natural way of harnessing distributed and heterogeneous resources available in organizations (intranets) or in the Internet.

Metacomputers, also called "computational grids" [1] might offer the following possible execution scenarios. Firstly, there may be applications that run on a single node of the metacomputer. These applications may be either sequential or multi-threaded, in which case they can be executed on multiprocessor architectures. In second place, there may be applications that run on multiple-node platforms. These platforms can be divided into tightly coupled parallel platforms (e.g. parallel platforms or workstation clusters) and distributed platforms. These can be formed by a combination of single-node and tightly coupled parallel platforms. In this case, the application uses the network for communication between parallel components, as well as parallel nodes' own interconnection networks.

In many cases, especially for high performance applications, users would like not only to run applications on a metacomputer, but also to understand the applications' behavior. The user will then try to improve the performance, modifying the application's code according to the results of the analysis of performance data captured during execution.

In [2], the general problem of performance analysis in grids is addressed. Grids are specially complex because of the high degree of heterogeneity (e.g., a

* This work was partially supported by grants from Conicit (project S1-2000000623) and from Universidad Simón Bolívar (direct support for research group GID-25)

B. Hertzberger et al. (Eds.): HPCN Europe 2001, LNCS 2110, pp. 42–51, 2001.

program consists of different components with different programming models), and because of high external perturbation. The goal is to help the user understand how and why the application behaved from performance point of view. However, the problem of relating performance data from different platforms for the same application is not considered.

In this paper, we explore the issues underlying application tuning on metacomputers, limiting ourselves to single node (sequential or multi-threaded) applications.

The rest of this document is organized as follows. Section 2 gives a general introduction to profiling, and section 3 goes through the question of profiling in metacomputers. A number of requirements to support profiling in metacomputers are proposed. Then, section 4 shows how these mechanisms are implemented in our metasystem SUMA [3]. Section 5 describes some experiments and results. Finally, in section 6, we address some conclusions and suggest directions for future research.

2 Profiling

A profile of an execution provides a user with a report containing basic information about performance, discriminated by application component, such as function, method, class, thread, module or block (i.e., sequence of instructions). The information reported in a profile can be more or less sophisticated.[1] It might include:

- Execution time, absolute or relative to total execution time. In the case of methods, measurements may include time spent in other methods invoked from this method. When included, we call it "inclusive time", otherwise we call it "exclusive" or "self" time.
- Execution number count (how many times this component was executed).
- Memory usage (e.g., maximum heap size used).
- Input/output information (time, count).

The profile can be given in terms of the application's *stack tree* [5]. In this case, resource usage is related to the sequence of active methods during execution. A profile can be used to improve the performance of an application on a particular machine. It allows the user to identify, for instance, the "most expensive components", which are those whose optimization will report the most benefit to the application's performance. Concentrating on improving this component will optimize user effort as well. Sometimes a tool can be used to help the user to analyze profiles. This performance improving activity (sometimes called *tuning* or *performance debugging*) usually goes through several iterations of:

1. Running on the target platform, while profiling the execution
2. Analyze the profile
3. Changing the source in order to improve performance

[1] See [4] for a comprehensive description of performance data.

There are basically two types of profilers, according to the data collection mechanism they implement:

- Statistical. The application is periodically interrupted, then information about the process stack is recorded
- Instrumented. The application is pre-processed to insert code snippets that collect data.

Statistical profilers are less precise but introduce low overhead. Note they can not provide component execution counts. Instrumented profilers are more accurate, but introduce more overhead. In both cases, as in every process observation activity, there is a trade off between precision and perturbation. High overhead can invalidate measurements. Trying to get the best of both worlds, some profilers mix both methods (e.g, *gprof* [6]). Regardless of the method used, a profiler collects data during execution and saves it to disk when the program terminates.[2] Another utility comes afterward to process the raw data and present it in a more convenient way to the user.

3 Profiling on a Metacomputer

From the perspective of performance analysis, the upgrade from single-node platforms to a *grid* pose challenging problems of increasing complexity. A simple approach for obtaining information about program execution is to provide a package that registers events, so the user may instrument the source code, either automatically or manually. This solution is offered by Globus [8], which provides a timing package. However, the user must deal with additional problems in order to obtain sensible information from profiles obtained from a metasystem. Among the general problems are:

- Heterogeneity. Typically, a metacomputer comprises diverse platforms, with varying architectures, operating systems, compilers, libraries, run-time support, etc. Running the same application in different nodes might result in different performance behavior.
- External perturbations. Applications usually share resources (e.g., the network) with other applications in the metacomputer.

For these reasons, profiling in a metacomputer is a significantly more complex problem than profiling in a single computer system. Important additional dimensions of this problem are explained in the following sections.

3.1 Profile Variation

Running on a metacomputer usually means letting the system scheduler choose an execution node according to user requirements and availability. Hence, in the

[2] Note some advanced tools, like Paradyn [7], collect performance data and analyze them dynamically.

tuning process described in section 2, there will likely be a different machine assigned on each iteration. The user will end up with a collection of performance data sets, one per iteration corresponding to a different execution. The question she faces is how can she know if there was an actual improvement with respect to previous iterations?

Giving an accurate answer to this question would need accurate performance models of applications and platforms belonging to the metacomputer. Developing accurate performance models is still an open problem. We must consider approximate methods in order to help the user in answering the fundamental question. At the core of the problem lies the relative power of platforms. Though in some cases there is a "clear winner" (e.g., a very powerful machine against a slim system), the best platform for an application will depend on how well they match. Reporting metrics related to the computer power as well as other parameters of execution nodes will help in comparing performance data of different executions. In other words, platforms need be assessed, for instance, by running benchmarks.

An additional level of uncertainty introduced by metacomputers is whether resources are dedicated or shared. It may be the case that some of the nodes accept more than one application at the same time. Even if nodes are dedicated, applications will probably need to access the shared network to exchange information with remote servers, clients, etc. In these cases, information about the *external* load (not due to the application) may have to be reported along with the profile.

3.2 Profiling Levels

In general, it would be desirable to identify *platform independent* and *platform dependent* performance factors. For this reason, we view the performance tuning process as a two levels, relatively differentiated, process: platform independent and platform dependent optimizations.

First level optimizations consist in identifying inefficient constructs which incurs in performance degradation almost in any platform; for example, an unnecessary and excessive number of object creation/destruction. Fixing such a problem will certainly improve performance disregard of platform. Second level optimizations look for exploiting specific platform characteristics. Examples of these are using redundant paths between nodes in a cluster of processors, rolling/unrolling loops according to local just-in-time compiler, etc. Because of the platform dependence, second level optimizations make sense when executions can be bound to a specific metacomputer's node or a family of nodes, all having similar hardware and software. In the general case, only platform independent optimizations will take place in a metacomputer.

A metacomputer must incorporate mechanisms that support reporting performance data in these two levels of detail. We will refer to them as *condensed* and *complete* profile, for first and second level optimizations, respectively. Both kinds of profile must separate as much as possible the overhead due to the metasystem (e.g., remote file access) from the application's profile.

Complete profiles allow the user to fully exploit the profiler power, as in a single machine. In this case, executions have to be associated to a particular platform, since tuning for a platform might not improve performance for a different platform. In this case, the metasystem's scheduler should permit to choose a platform for an application.

3.3 Condensed Profiles

Condensed profiles consist of a reduced set of performance information that help the user to uncover coarse level performance problems, enough to support first level optimizations. These profiles should have a number of desirable properties

- easy to obtain from common, widespread profilers. This helps to present a unified, profiler independent, profile format.
- light in resources usage, so overhead is reduced (see section 3.4)

Additionally, a metasystem is intended to offer transparent access to remote resources. Hence, it is also desirable to provide a single image of the software tools, including the profiler. There may be different profilers executing on different platforms, but ideally the user sees a single metasystem profiler. The generation of profiler traces takes place in the executing platform, while profile visualization has to be done in the client machine. The profiler traces are processed before sending the information to the client machine, where this information will be in turn processed for visualization. Hence, a **unified profile format** is needed for transport purposes across the network, that is, between the clients and the execution nodes.

3.4 Overhead on the Metasystem

Profiling on a metacomputer implies that some additional resources have to be allocated to the application. First, the execution node must provide for additional execution time (for data collection, storing and processing), memory and disk space (if data is saved locally). Additionally, the profiling data must travel back to the client machine, thus adding some overhead on the metacomputer components and network.

Probably the critical most resource is the network, which will likely be shared not only by metacomputer users, but other users as well. Hence, the profiling support mechanisms should save as much as possible network usage, at the expense of other resources like CPU, disk, etc. This will be the case for condensed profiles. On the other hand, comprehensive profiles pay the price of a higher overhead on the metasystem in order to accomplish second level optimizations.

The information collected at the metasystem level may be important for the user to understand overall behaviour. The user needs to know the proportion of time spent in metasystem operations in order to assess the effort she would have to put into program optimization.

4 Support for Profiling in SUMA

We describe how we implement mechanisms that support profiling in SUMA. First, a brief introduction of SUMA is presented.

4.1 Description of SUMA

SUMA(`http://suma.ldc.usb.ve`) is a metasystem aimed at executing high performance Java byte-code applications. It transparently redirects I/O to client machine, load classes and data from client. It provides for interactive and batch execution. SUMA is based on CORBA. It currently uses JacORB [9], a freely available CORBA implementation. The user starts the execution of programs through a client running in her machine. Once the SUMA core receives the request from the client machine, it transparently finds a platform for execution and sends a request message to that platform. An *Execution Agent* at the designated platform starts the execution of the program. This *Execution Agent* gets all classes, files and data from the client, and sends results and output to client after the execution.

4.2 A Model for Profiling in SUMA

One of SUMA's design goals is to build as much as possible on widespread, established, open technologies. Consequently, we didn't build a whole new profiler or bind to a particular profile tool. Instead, we defined mechanisms (hooks) such that it is possible, with little effort, to plug in any profiler, which we call the back-end profiler. Doing so allows us to take advantage of currently available and forthcoming profilers.

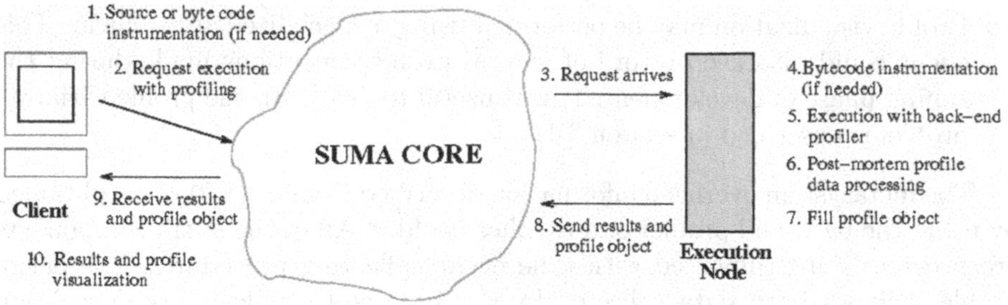

Fig. 1. Execution of an Application with Profiling Option.

Figure 1 shows the general model for the profiling process in SUMA, in terms of what happens at both ends (client/server) when an application is executed with the profiling option. The processes associated to specific back-end profilers, such as code instrumentation, are included by means of the class overloading mechanism. More detailed descriptions of the numbered steps shown in figure 1 follow:

1 Profilers usually need instrumentation at either source code or byte code level. However, instrumentation may not be necessary at all, especially if statistical profiling methods are used. If source code instrumentation is used, then it has to be done at the client side, otherwise, it could also be executed at the server side (see step 4).

2 At the client side, the profiling facility is explicitly requested by the user. In this way, the SUMA core is capable of selecting an *Execution Agent* that runs a back-end profiler. The users select the profiling level according to their performance analysis needs, as described in section 3.2.

3 After the SUMA scheduler has selected a node, both client and *Execution Agent* communicate directly to each other for class and data transference.

4 In this step only bytecode can be instrumented, because the *Execution Agent* only receives bytecode.

5 The main class is started under the control of the back-end profiler, which generates the profile data locally, in its own format. In order to meet the requirements described in 3.4, profile data should include overhead produced by the *Execution Agent* itself.

6 Some processing is necessary after program execution for preparing the data to be sent back to the client. This step is intended to cover the need for generating condensed and unified profile formats, described in section 3.3.

7 The selected data is placed in the object returned to the client in a unified profile format.

8,9 Standard output and output files are sent directly from the execution node to the client machine. The profile object is sent through SUMA, which adds metasystem overhead information. This functionality helps the user assess external overhead, as explained in section 3.4.

10 Profile visualization may be performed using a specialized SUMA client. This client could also keep record of several profile executions made during the tuning phase of development. This is useful to deal with the profile variation problems described in section 3.1.

The metasystem overhead information described in steps (8,9) is not obtained by using the back-end profiler or any other profiler. All of the SUMA components are previously instrumented with time event collecting procedures. The entire profile includes both data collected by the back-end profiler at the execution node and data collected by the SUMA core.

Condensed profiles are encapsulated in what we call the *profile object*, in the unified profile format defined for SUMA. It contains information for most relevant methods, including

- Percentage of total execution time spent in method (exclusive).
- Percentage of total execution time spent in method and in methods called from this method (inclusive).

- Number of times this method was called.
- Method's name.

Extended profile object additionally contains information collected by the back-end profiler that may be specific to some (kind of) profiler. If the user requests an extended profile object, more information is sent back to the client, for instance, the whole trace file. This information is potentially very useful, but the user must have a way to visualize such data.

At installation time, the *Execution Agent* runs a benchmark, as in Netsolve [10]. This benchmark is a basis for comparing performance data obtained from different nodes. As the user does not know which machine was selected by SUMA to execute her application, the benchmark results are always sent to the client along with the profile and relevant information available about the execution platform (e.g. architecture, operating system, JVM version, etc.). This information is also useful to deal with the performance variation problems described in section 3.1.

5 Experiments

SUMA is currently implemented on a campus-wide network. The selected server platform for the test is a dual processor Pentium III, 600 MHz, 256 MB, with Linux 2.2.14-5.0smp and Sun's JDK 1.2.2. As back-end profiler for these initial tests, we used HPROF [11], which comes with JDK version 1.2 or newer. Note HPROF does not need instrumentation, either at the client or the server side, so steps 1 and 4 described above are void.

The chosen application is *Jess* [12], a rule engine and scripting environment, which uses a considerable number of classes. As input file, we used a *Jess* script which computes factorial of 10000. Major components in global time reported by the profile are shown on table 1. The application takes 47% of the turnaround time executing at the server. As expected, a fair part of this time (48%) is devoted to remote class loading. The rest is due to the application only by a small fraction (about 2.6%). These results uncovered an unadverted problem with the application launching mechanism of SUMA.

Table 1. Measured Times (in Milliseconds) for Jess on Platform II.

Total time	Execution time (at the server)	Class loading time
16024	7542	3644

Table 2 shows an excerpt of the application profile. The most relevant methods are correctly found by the profiler, but relative importance, as well as total contribution, are not relative to the application itself but the whole universe of methods running on the *Execution Agent*, which include SUMA methods.

Table 2. Application (Jess) Profile Reported by SUMA, Platform II (excerpt).

Inclusive time (Percentage)	Method
2.61	jess/Main.main
1.12	jess/Main.execute
1.12	jess/Rete.<init>
1.12	jess/Jesp.parse
1.12	jess/Funcall.loadIntrinsics

6 Conclusions

A general model for building execution profiles in a Java-based metasystem is proposed in this paper. This profiling model allows the use of different back-end profilers, that is, profilers running at the execution servers of the metasystem. The implementation of this model on SUMA metasystem proved its suitability. Preliminary experiments show that this model can be used for both performance analysis of applications running on the metasystem and performance analysis of the metasystem itself.

There is going-on work on implementing the profiling facility for tightly coupled parallel platforms. The unified profile format needs to be extended for parallel programs.

Acknowledgement

We thank Eduardo Blanco, for his support on the experimental part.

References

1. Ian Foster and Carl Kesselman. *The Grid: Blueprint for a new Computing Infrastructure*, chapter Computational Grids, pages 367–393. Morgan Kaufmann Publishers, Inc., 1999.
2. Daniel Reed and Randy Ribler. *The Grid: Blueprint for a new Computing Infrastructure*, chapter Performance Analysis and Visualization, pages 367–393. Morgan Kaufmann Publishers, Inc., 1999.
3. Emilio Hernández, Yudith Cardinale, Carlos Figueira, and Alejandro Teruel. SUMA: A Scientific Metacomputer. In *Parallel Computing: Fundamentals and Applications. Proceedings of the International Conference ParCo99*, pages 566–573. Imperial College Press, 1999.
4. Jeffrey Hollingsworth and Bart Miller. *The Grid: Blueprint for a new Computing Infrastructure*, chapter Instrumentation and Measurement, pages 339–365. Morgan Kaufmann Publishers, Inc., 1999.
5. Robert J. Hall. Call path refinement profiling. *IEEE Transactions on Software Engineering*, June 1995.
6. S. L. Graham, P. B. Kessler, and M. K. McKusick. An execution profiler for modular programs. *Software–Practice and Experience*, 13:671–685, 1983.

7. B. P. Miller, M. D. Callaghan, J. M. Cargille, J. K. Hollingsworth, R. B. Irvin, K. L. Karavanic, K. Kunchithapadam, and Tia Newhall. The paradyn parallel performance measurement tool. *IEEE Computer*, 28:37–46, November 1995.
8. The Globus Project: Timing and Profiling. `http://www.globus.org/timing/`.
9. Gerald Brosse et al. JacORB. `http://jacorb.inf.fu-berlin.de/`.
10. Henri Casanova and Jack Dongarra. Netsolve: A network server for solving computational science problems. *The International Journal of Supercomputer Applications and High Performance Computing*, 11(3):212–223, Fall 1997.
11. JavaTM Virtual Machine Profiler Interface (JVMPI). `http://java.sun.com/j2se/1.3/docs/guide/jvmpi/jvmpi.html`.
12. Jess: the Java Expert System Shell. `http://herzberg.ca.sandia.gov/jess/`.

Utilizing Supercomputer Power from Your Desktop

Bert C. Schultheiss and Erik H. Baalbergen

[1] National Aerospace Laboratory NLR
Voorsterweg 31, 8316 PR Marknesse, The Netherlands
schulth@nlr.nl
[2] National Aerospace Laboratory NLR
Anthony Fokkerweg 2, 1059 CM Amsterdam, The Netherlands
baalber@nlr.nl

Abstract. Potentially unlimited computing power is provided today by computing facilities available via the Internet. Unfortunately, the facilities exhibit a variety of non-uniform interfaces to the users. Each facility has its own operating system, its own policy with respect to access and accounting, its own set of tools, and its own data. To utilize the computing power, a user has to select an appropriate facility, and has to know how to access and operate the facility, which tools to use, what options to provide for these tools, how to execute tools, how to submit a job for execution, etc..
The *Superbroker* infrastructure gives easy access to HPCN facilities in the Netherlands through a web browser, thereby taking care of the networking and system details. The user is provided with a single working environment, which gives easy access to the available, usually remote resources as if they were present on a single computer (a "metacomputer"). Using the desktop's native software only, the user can easily browse the working environment, start tools, and submit jobs using point-and-click and drag-and-drop operations. The working environment may be tailored for particular end users and application areas. This paper presents the design and implementation of the infrastructure.

1 Introduction

Potentially unlimited computing power is provided today by computing facilities available via the Internet. Unfortunately, the facilities exhibit a variety of non-uniform interfaces to the users. Each facility has its own operating system, its own policy with respect to access and accounting, its own set of tools, and its own data. To utilize the computing power, a user has to select an appropriate facility, and has to know how to access and operate the facility, which tools to use, what options to provide for these tools, how to execute tools, how to submit a job for execution, etc..

Ideally, the remote computing resources should be accessible without any technical barrier: without the need to install extra software and accessible

B. Hertzberger et al. (Eds.): HPCN Europe 2001, LNCS 2110, pp. 52–61, 2001.

through a simple graphical user interface (GUI) with a look and feel the user is familiar with. Put in technical terms, we want the high performance computers to operate as *applications servers* that can be accessed by an end-user through a *thin client*.

First, this paper presents the requirements, design, and implementation of a general infrastructure for web-access to remote computing facilities. The key elements of the implementation are the GUI consisting of Java applets, an HTTP daemon with Java servlets, and SPINEware ([7]) object servers with CORBA interfaces. The implementation raises issues such as the communication between the Java applets and the HTTP daemon servlets, the communication between the HTTP daemon servlets and the SPINEware object servers, and launching SPINEware servers when necessary. These topics are covered in the present paper.

Next, a case study demonstrates how this general infrastructure is applied to provide the *Superbroker* working environment, for accessing the public Dutch HPCN facilities.

2 Requirements

Figure 1 presents an overview of the required functionality for the Superbroker working environment: a user easily and uniformly accesses tools and data on the HPCN facilities as if these are available from a single computer. The decisive

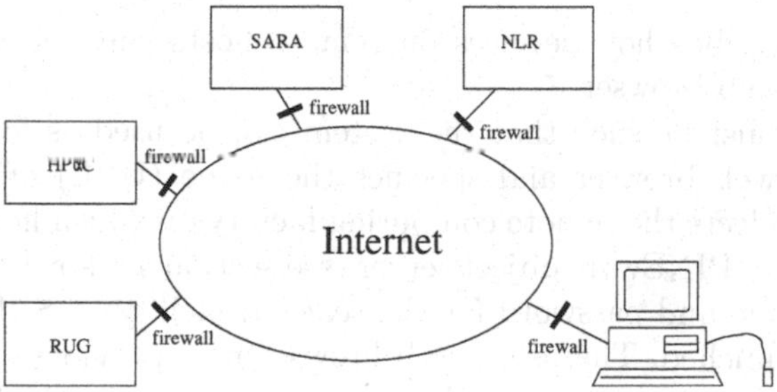

Fig. 1. Usage of SPINEware to Access Remote HPCN Facilities.

requirements for the web-based access to the HPCN facilities are:

- Provide access to the HPCN facilities through a single network address.
- Provide access to the HPCN facilities without the need for installation of extra software on the local desktop. Also, the software running on the local desktop shall be portable. We have translated these requirements into: the only required software is a *Java 2 capable* web browser.
- Provide the user with a simple, predefined working environment which allows the user to browse through and access the available tools, data, and information.

- Provide support for the integration of tools in a working environment. The tools vary from compilers to Computational Fluid Dynamic applications. Tools may require a set of input parameters and files.
- Offer support for transportation of code and data files, including the required build (i.e., *make*) instructions, between the end user and the HPCN facility, if possible in compressed or encrypted form.
- Provide support for starting up (remote) jobs on the HPCN facilities.
- At job start-up, the job parameters can be specified, allowing the user to, e.g., limit the duration of a job.
- The user interface shall give a fast response (say within 0.2 seconds). Note that the action as specified using the web interface may take some time, depending on network delay.
- An authentication mechanism using user name and password shall be used. The user must be able to change the password. We translated this requirement into: the system must use the user name password mechanism of the HPCN facilities that are accessed. Data communication shall be secure.
- In order to enable users protected by a firewall to access a remote HPCN server, the communication between the user's user interface and the remote HPCN server shall be based on a widely available protocol such as HTTP or HTTPS.

3 Design

This section describes how services on remote hosts can be accessed using a locally running web browser.

The design must be such that the system can be used as follows. The user starts a native web browser and specifies the name (URL) of the login html page. The user selects the remote computing facility on which he or she wants to launch the initial SPINEware object server (see section 4.1 for details), and specifies the user name and password for the selected facility. A SPINEware object server will be launched. The user's web browser now provides a single working environment, which gives access to the available resources as if they were stored on a single computer (a "metacomputer") [7]. The user can browse the working environment, start tools, and submit jobs using point-and-click and drag-and-drop operations. In addition to browsing through information and launching tools on the specified host, the user may open browsers for accessing resources on other hosts as well. Data can simply be moved and copied by dragging and dropping icons from one browser window to another. The working environment may be tailored for particular end users and application areas.

The architecture of the system is shown in figure 2. The figure displays a 3-tier model:

- The Java applets that implement the GUI.
- SPINEware object servers that carry out the requested actions and that manage the resources involved.

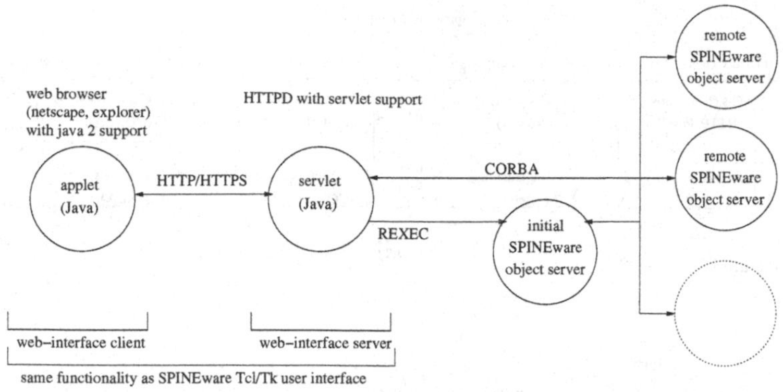

Fig. 2. SPINEware Architecture.

– A web-interface server that enables the user to obtain the applets, and that functions as a middle-tier in the communication between the applets and the SPINEware object servers.

The technical details are discussed in section 4. Figure 3 shows two possible architectures for the communication layer between the SPINEware object servers and the local GUI applets.

– Architecture I is the 3-tier architecture as presented and implemented. This architecture is required when using Java applets, because applets can only reconnect to the HTTP daemon from which they were obtained. An advantage of this architecture is that we can use our own communication layer between the Java applets and the web interface server. At this moment, we simply use the HTTP protocol or the Secure HTTP (HTTPS) protocol. HTTPS is available by default from the Java libraries. However, the HTTP daemon has to support it. Within the *Superbroker* project (see section 5), an Apache HTTP Daemon ([6]) with HTTPS support is used. Another advantage of the 3-tier architecture is that it is possible to monitor which users are logged on at any moment and what the users are doing, because all communication is done via the same web-interface server. An administration applet enables the administrator to do the monitoring. A disadvantage of a 3-tier system is the communication overhead.
– Architecture II shows a web-interface client which directly communicates with the SPINEware object servers. This is only possible when running the web-interface client as a stand-alone program instead of Java applets due to security restrictions imposed on Java applets. Note that architecture II does not fulfill the "only a Java 2 capable browser is required" requirement.

As an obvious design principle, we have clearly separated the communication layer from the implementation of the GUI.

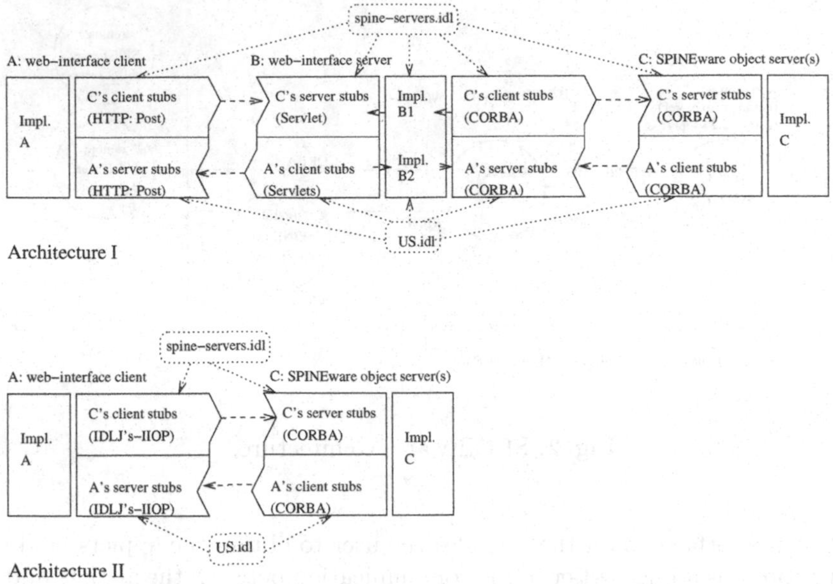

Fig. 3. Architectures I and II.

4 Implementation

This section presents the technical details of the 3-tier model: SPINEware object servers, the web-interface server, and the Java applets.

4.1 SPINEware Object Servers

SPINEware exhibits an object-based model. In a working environment, all resources available from the interconnected hosts are modeled as objects. SPINEware provides "basic" object classes such as File, Directory (for a native system's directory or folder), ObjectFolder (a folder for organizing objects), Trashcan ("recycle bin"), Printer, AtomicTool (wrapped, ready-for-execute programs), Job (for managing the execution of AtomicTools), Workflow (for chaining tools, thereby passing output files from one tool as input files for another), and the GUI object named UserShell. Starting from the basic object classes, new classes may be added - whether or not inheriting from existing classes - and existing classes may be modified. The access operations on the objects are defined in terms of methods. Examples are *Edit* for File, *View* for Directory and ObjectFolder, and *Execute* for AtomicTool.

Each object in SPINEware has a (world-wide) unique object identifier, which contains the class ("type"), the Internet name of the host it resides on, and a host-unique identifier. The URL naming scheme is used to identify a SPINEware object. An object identifier is usually written as a so-called "SPIRL", a *SPI*NE

*R*esource *L*ocator. For example, directory */home/user/src* on host *desk12.nlr.nl* can be identified using the SPIRL

```
spine://Directory@desk12.nlr.nl//home/user/src
```

SPIRLs are also used for specifying method invocations, such as

```
spine://Directory:View@desk12.nlr.nl//home/user/src
```

which instructs SPINEware to display the contents of the specified directory on the user's desktop. SPINEware provides a *SPIRL broker* utility, which enables method invocations from within tools, scripts, and command-line interpreters.

The objects available from a host via a SPINEware working environment are managed by and accessible via a *SPINEware (object) server* running on the host on behalf of the user and the working environment. When a session with a working environment is started, one SPINEware object server is started initially. This server will handle all requests from the user interface. A minimum (i.e., single-host) session with a working environment involves at least a user interface and the initial SPINEware object server for accessing the host's resources as objects. If an object on another host is accessed, the initial SPINEware object server starts ("on demand") a SPINEware object server on the other host, and relays all subsequent method invocations (and the corresponding results) involving objects on that host.

The communication among the SPINEware objects is based on CORBA, and is implemented using an off-the-shelf CORBA implementation, *ILU* [4]. This product supports CORBA-based inter-object communication over local networks as well as the Internet, and is capable of being used in combination with firewalls.

4.2 Web-Interface Server

The web-interface server is the middle tier of the system, see figure 2. This middle tier is implemented using an HTTP daemon (*httpd*) with Java servlets.

The servlets are invoked by applets (see section 4.3) when the GUI wants to invoke an object method via the SPINEware object server. For example, if the user wants to see the contents of a job queue, the *getContents* method of the Job object is invoked.

When the user logs on to the system, the *Login servlet* is invoked. The Login servlet takes care of publishing a new instance of the SPINEware UserShell object to the CORBA naming service for the user who logs on, and launches a SPINEware object server on the specified host using *rexec* and the user name and password provided by the end user. For example, if the SPINEware Tool Editor (started upon invocation of the *Edit* method of the AtomicTool object) needs to display an error message, it invokes the *ErrorMessage* method of the UserShell object. This results in invocation of the ErrorMessage implementation in the web-interface server via its CORBA interface. The web-interface server ensures that the correct applet code is run. This applet method invocation is, due to security restrictions on applets, implemented by a blocking poll for a command by the applet.

The CORBA interface of the webinterface server has been built using Java's *idlj*. So, the protocol used between the SPINEware object server and the webinterface server is CORBA's *IIOP*.

4.3 Web-Interface Java Applets

The web-interface client (i.e. the GUI) is implemented by Java 2 applets. All communication with the web-interface server is implemented using HTTP or HTTPS *Post* commands. Consequently, if a firewall is between the end user's system and the web-interface server, this poses no problem as long as the firewall allows HTTP (or HTTPS). Also, the applets always initiate the communication. As explained in the previous section, the Java applets poll the web-interface server for commands to be executed.

The webserver will send an `SC_UNAUTHORIZED` message if the user name and password are not specified in the servlet request, and thus obtains the username and password. In response, the web browser will display a login window, prompting the user to specify the user name and password of the system he or she wants to access initially.

Any customization information for the GUI (e.g., which browsers to open, colors, character sets, etc.) will be stored at the accessed remote host, and will be obtained from the SPINEware object server at login.

To access local files from the GUI, Java Web Start [8] is supported. The GUI could also be run as a Java standalone program instead of using the web browser, thus enabling access to the local file system.

5 A Case Study: Superbroker

The managing institutes of the major HPCN facilities in the Netherlands (NLR, SARA, HPC-RuG, and HPaC/TUD) have developed a uniform infrastructure named *Superbroker*, that provides users from small and medium-sized enterprises (SMEs), government, and industry, with easy access to the HPCN facilities in the Netherlands and support for using those facilities. The activities for realizing this infrastructure have been carried out in the scope of the HPCN 2000 Infrastructure project, which was funded by the Dutch HPCN Foundation.

The Superbroker web site (`http://www.superbroker.nl`) provides the user with access to information on the HPCN facilities, enables the user to obtain accounts on the HPCN facilities, and offers working environments to facilitate access to data and tools on the HPCN facilities. Figure 4 shows the Superbroker architecture: it is a specialization of the general SPINEware architecture as shown in figure 2.

To interactively access an HPCN facility with a predefined working environment, start a web browser on your local workstation or PC, and provide the *Superbroker* URL. After successful registration and login, a file browser for the top-level ObjectFolder of the working environment pops up, as shown in figure 5. From this very moment on, you may browse through the working environment

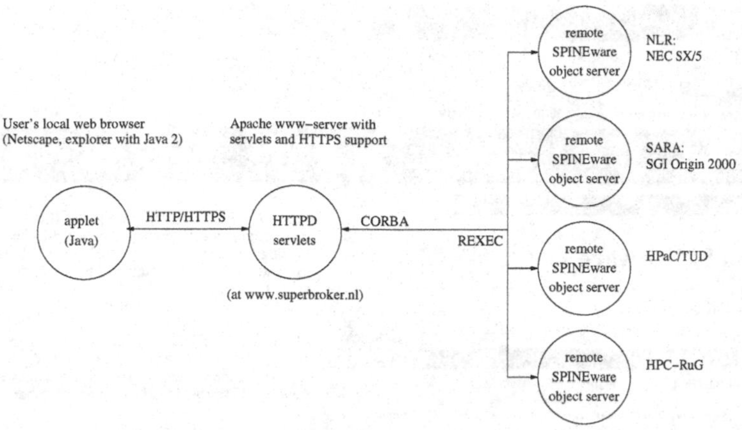

Fig. 4. Access to HPCN Facilities through a Web Interface Using SPINEware.

by opening Directories and ObjectFolders via double clicking on icons. The tree-shaped organization of the working environment together with the on-line help information will help you to find your way around in the working environment and to locate and use the resources you need.

The *Superbroker* presently provides means to upload program code and data, to build executable programs and tools from the program code on the HPCN facilities, to run existing as well as your own programs, and to download results. The SPINEware object classes Queue (for Job queues) and Workflow (for chaining tools) offer useful possible extensions for working environments based on the Superbroker concept. The Queue class supports specification of jobs to be run on supercomputers, and to be managed by supercomputer-native job management systems, such as NQS and Codine, both of which are supported by SPINEware. The Workflow class supports definition of tool chains, which allow the use of tools and data involved to be specified in the form of a graph and to be reused by other users as well.

6 Conclusions

This paper described a general infrastructure for easy and uniform access to remote computing facilities from the desktop computer, using the desktop's native web browser as GUI. This infrastructure, SPINEware, has been applied to instantiate the *HPCN 2000 Superbroker* working environment.

The GUI is simply started by entering the URL of the start page in the native web browser. The GUI has the look and feel of a graphical file-system browser, enabling the user to browse through and access the available resources, such as files, directories/folders, and tools. Tools can be started, and jobs can be submitted using intuitive point-and-click and drag-and-drop operations. SPINEware takes care of the system details.

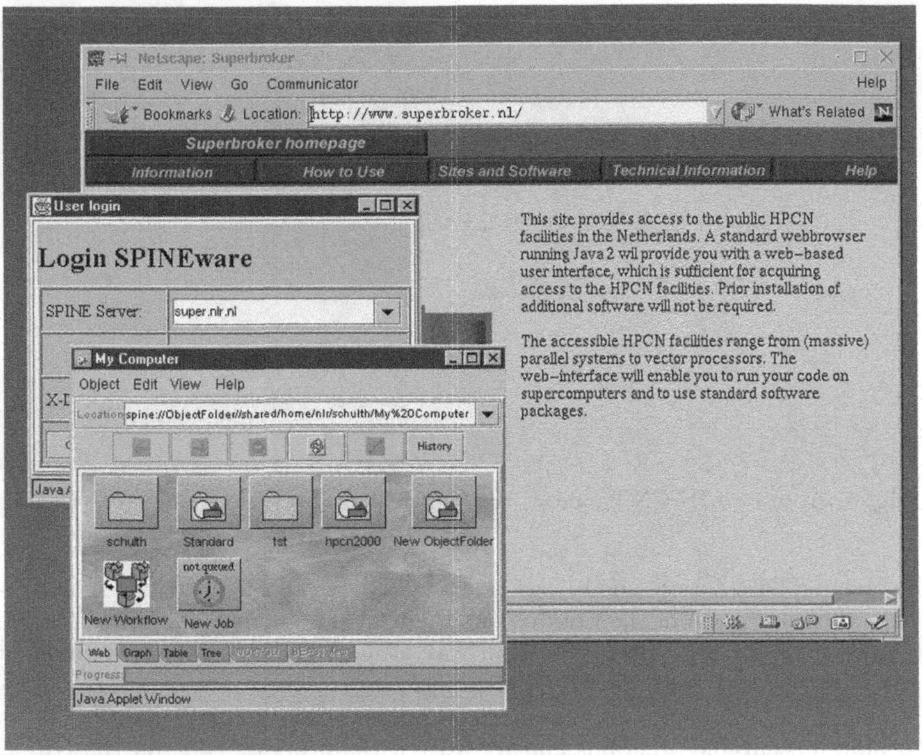

Fig. 5. Accessing HPCN Facilities through the Superbroker Web Site.

The end user only requires a Java-2 enabled web browser to access the defined working environments. The end user is able to enhance the working environments by integrating new tools, or adding new tool chains.

Behind the screen, a web-interface server (an HTTP daemon with Java servlets) functions as an intermediate between the end user's GUI (Java applets) and the SPINEware object servers on the remote hosts. Communication between the web-interface server and the Java applets is based on HTTP or HTTPS, thus minimizing firewall problems. The web-interface server communicates with the SPINEware object servers through CORBA. SPINEware object servers are automatically launched on the hosts that are accessed during a session.

References

1. Robert Orfali, Dan Harkey, Client/Server programming with Java and Corba, ISBN 0-471-16351-1
2. Bruce Eckel, Thinking in Java, ISBN 0-13-659723-8, `http://www.bruceeckel.com`.
3. Java Servlets and Serialization with RMI, `http://developer.java.sun.com/developer/technicalArticles/RMI/rmi/`.
4. ILU 2.0alpha14 Reference Manual, Xerox Corporation. Can be obtained via `ftp://beta.xerox.com/pub/ilu/ilu.html`.
5. J. Hunter, W. Crawford, Java Servlet Programming, ISBN 1-56592-391-X, O'Reilly & Associates.
6. Apache HTTP server project, `http://www.apache.org/httpd.html`.
7. E.H. Baalbergen and H. van der Ven, SPINEware - a framework for user-oriented and tailorable metacomputers, in: Future Generation Computer Systems 15 (1999) pp. 549-558.
8. Java Web Start, `http://java.sun.com/products/javawebstart/`.

Tertiary Storage System for Index-Based Retrieving of Video Sequences

Darin Nikolow[1], Renata Słota[1], Jacek Kitowski[1,2], Piotr Nyczyk[1], and Janusz Otfinowski[3,4]

[1] Institute of Computer Science, AGH, Cracow, Poland
[2] Academic Computer Centre CYFRONET AGH, Cracow, Poland
[3] Collegium Medicum Jagellonian University, Cracow, Poland
[4] Cracow Rehabilitation Centre, Thraumatology Department, Cracow, Poland
{darin,rena,kito}@uci.agh.edu.pl
Phone: (+48 12) 6173964, *Fax:* (+48 12) 6338054

Abstract. In this paper we present a research which is aimed at developing a tertiary storage system architecture for efficient index-based retrieving of video sequences. During the research we built such a system by creating dedicated tertiary storage management software called VTSS. The system is intended to be a part of a multimedia database system. We also present preliminary experimental results from a prototype implementation.

1 Introduction

There is a lot of hardware and software research on tertiary storage systems driven by the increasing amount of archival data and the need to explore, process and search for the useful information. The problem of efficient video sequence retrieval from the tertiary storage devices arises in the context of multimedia database systems (MBS). There is a broad scope of potential MBS applications, for example MPEG data could be used in advanced simulations (e.g. [1]), in virtual reality and virtual laboratories (e.g. [2]), in telemedicine and many others.

The software applications which manage the tertiary storage devices usually do not consider the type of data objects stored, although it would be useful for applications which need access to separate video sequences (video fragments). Video files are complex data objects and retrieving a video fragment is not a simple task.

The authors of this article take part in research on Distributed Medical Information System within the PARMED project [3]. In order to share large stuff of images and video clips (videos) with many medical institutions for scientific and educational purposes, we provide in this project a large virtual database of medical video and image data. Users of the medical multimedia database usually do not need to view the whole videos, frequently they access fragments of them. For example they may need to view only short sequences of implantation of the artificial hip joint resulting from different medical operations (thus from different

B. Hertzberger et al. (Eds.): HPCN Europe 2001, LNCS 2110, pp. 62–71, 2001.

videos). In PARMED project the data kept in virtual database is stored locally, near to the its owner. Due to large amount of data and the rare access to them the data are stored on tertiary storage.

The purpose of the presented research is to develop a tertiary storage system architecture for efficient index-based retrieving of video fragments. The system is intended to be a part of a multimedia database system.

Our first approach to the problem was to build a middleware layer (called Multimedia Storage and Retrieval System (MMSRS) [4]) on top of the UniTree [5] commercial software which is used for the tertiary storage management. The second approach presented in this paper is to build the tertiary storage system by creating dedicated tertiary storage management software called Video Tertiary Storage System (VTSS).

The rest of the paper is organised as follows. The second section presents related works. The third section describes VTSS in detail. In the next section we present some preliminary comparison tests between VTSS and MMSRS. Conclusions and some future plans are shown in the last section.

2 Related Works

Efficient storage and retrieval processing of data on remote tertiary storage system is one of the major challenges that needs to be addressed. In recent years research on linking distributed computational, networked and storage resources has been appearing, like Grid/Globus initiatives [6,7], with immediate applications, e.g. in particle physics [8,9]. Some aspects of the tertiary storage research are: distributed parallel storage systems, file systems, digital libraries, high performance I/O systems, replication and load balancing. There are examples of commercial softwares managing tertiary storage systems giving access to fragments of files (e.g. HPSS [10]). Nevertheless many researches study the access to fragments of data rather than to fragments of files. DataCutter [11] provides support for subsetting very large scientific datasets on archival storage systems. Memik, et al. have developed a run-time library for tape-resident data called APRIL [12] based on HPSS and MPI-IO [13]. It allows programmers to access data located on tape via a convenient interface expressed in terms of arrays and array portions rather than files and offsets. They use a sub-filing strategy to reduce the latency. Holtman, at al. [14] study potential benefits of object granularity in the mass storage system. The architecture is based on transparently re-mapping objects from large chunk files to smaller files according to the application access pattern.

3 Video Tertiary Storage System

The architecture of VTSS is shown in Fig. 1. The system consists of two main daemons: Repository Daemon (REPD) and Tertiary File Manager Daemon (TFMD). REPD controls the Automated Media Libraries (AML) and is responsible for mounting and dismounting required tapes. Information about files stored on VTSS is kept in the file *filedb*. Information about tapes usage is stored in the file *tapedb*. TFMD manages the files *filedb* and *tapedb* and transfers files from the removable media devices to Client. Each video stored on VTSS has an index which helps in fast frame locating in an MPEG file. This index is called *frame-to-byte index*.

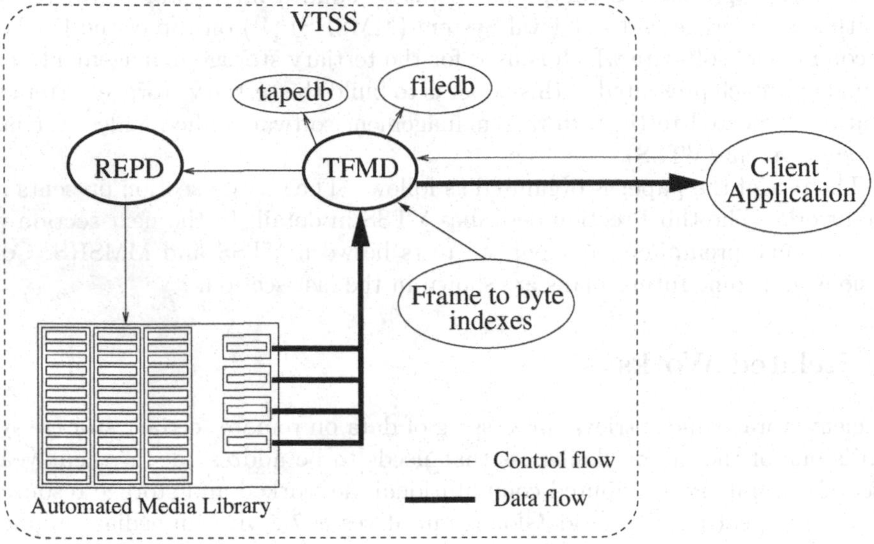

Fig. 1. Architecture of the Video Tertiary Storage System.

Client requests to VTSS can be of the following kinds:

- write newfile to VTSS,
- read a file fragment from VTSS,
- delete a file from VTSS.

The most often used request is expected to be the read fragment request. The fragment range is defined in the frame units. The following statements describe what happens inside VTSS when such a request is received:

1. TFMD calculates, using *frame-to-byte index* for the requested video, the start and end byte offsets and finds in *filedb* the tape identifier where the requested video file resides and sends mount tape request to REPD,
2. REPD tries to mount that tape and sends back to TFMD the device special file name where the tape is mounted, or an error message in the case of failure,
3. When TFMD receives back the device special file it positions the tape at the desired block. The block number to position at is calculated based on the block number found in *filedb*, the block size used to write the file and the

start byte offset of the fragment. When the tape is positioned the transfer to client begins.

The daemons are written in C and use sockets to communicate each other.

3.1 Repository Daemon

REPD interprets the following commands (requests):

- **use <tapeid>** - mount a tape with identifier label **tapeid** in a drive,
- **idle <tapeid>** - mark a tape idle, which means it could be dismounted when the drive it occupies will be needed.

Processing the **use** command may require starting of tasks which take a lot of time (e.g. few minutes). These tasks are:

- **move** - move a removable medium from a drive to a slot and vice versa,
- **unmount** - prepare a medium in a drive for eject,
- **read label** - read an internal label of a medium.

For each drive one **unmount** or **read label** task can run independently. One **move** task per AML can be processed at the same time, assuming that AML has one robot arm.

REPD handles multiple mount requests (**use** commands) by starting if necessary, new processes for the above long lasting tasks. During the execution of the mentioned processes REPD is waiting for or processing another request. When the given process is finished it sends back to REPD the result and REPD goes on with the request.

The simplified algorithm of the useproc() function of REPD is shown in Fig. 2. It processes **use** requests. The function tries to mount the desired tape in one of the drives and sends back as result:

- string containing the device name of the drive where the tape is mounted, upon success, or
- an error message, upon failure.

The *tape request queued* condition checks if a request for the same tape is being processed. The *tape mounted* condition checks if the tape is in the *tape in use* (tiu) list. *find tape* locates the tape in AML and returns the slot or the drive number where the tape resides. The *online* condition checks if the drive is allowed to be used by VTSS (*online*). Usually all of the available drives are marked online, but in some cases some drives might be used by other applications. *tiu add* adds a tape to the tiu list and *tiu delete* deletes a tape from the list.

Typical scenario of the **use <tapeid>** request is the following (assuming that there is a free drive):

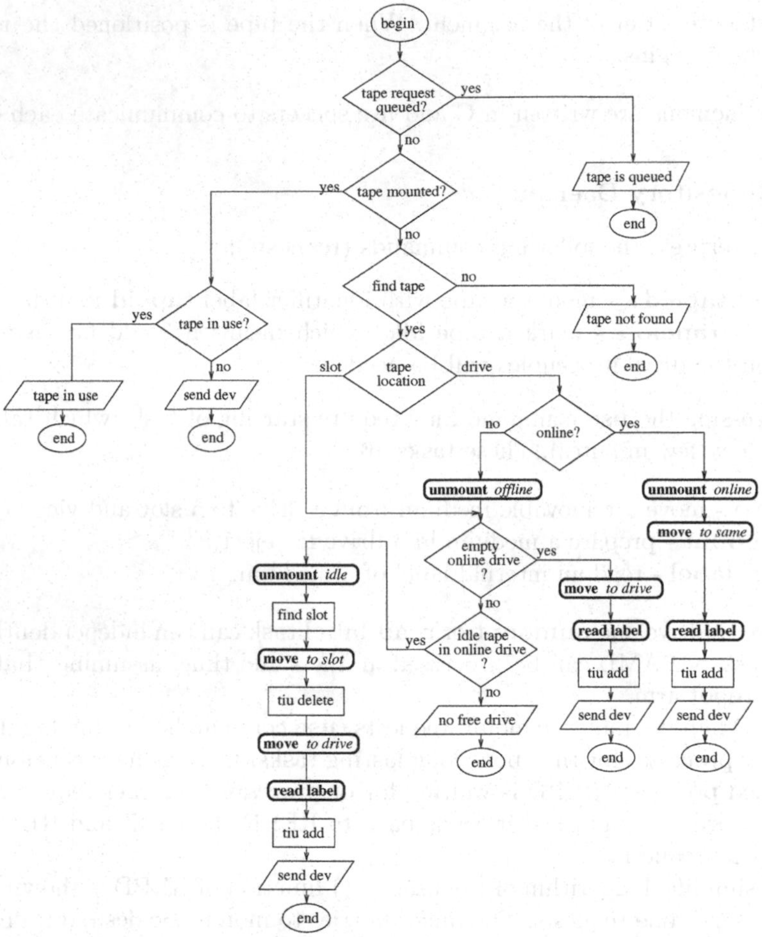

Fig. 2. Simplified useproc() Algorithm.

1. a tape labelled **tapeid** is moved by the robot arm from its slot to a free drive,
2. the tape internal label is read and checked versus **tapeid**,
3. the device string is sent back to TFMD.

If there is no empty drive then the appropriate **unmount** and **move** processes are started to free a drive or "*no free drive*" message is sent when there are no tapes to unmount (because all are busy).

3.2 Tertiary File Manager Daemon

A file in VTSS is identified by a number called **fileid**.

Records in *filedb* describe each file, i.e. on which tape the file resides, at what position on the tape the file is stored, the file attributes like read and write permitions, size, owner, etc. *tapedb* contains information about the tapes - tape label, copy number, blocks used, etc. Each file stored on VTSS can have up to four copies. Each copy is physically stored on a different medium for safety reasons.

TFMD interprets the following commands:

- **new** - store a new file on tape(s). TFMD chooses an unused **fileid** and sends it back to the client application when the file is stored,
- **get** <**fileid**> <**start**> <**end**> - retrieve and send to the client application the given file fragment from the tape,
- **put** <**fileid**> - store a file **fileid** on tape; it is used to update a file while preserving the **fileid**,
- **delete** <**fileid**> - remove a file **fileid** from the system.

A video file fragment is described by the start and end frame numbers. If the two frame numbers are set to 0 then the whole file is being transferred.

In order to serve multiple requests TFMD forks a new process for each request. Future versions will probably use POSIX threads or light-weight processes.

4 VTSS and MMSRS Performance Comparison

As mentioned before our first approach to tertiary storage management was MMSRS. Both systems were designed for the same purpose (efficient access to video fragments) but using different approaches. In order to find out which aproach is better in certain situations we present test results for the both systems. Below we present a short description of MMSRS (more details can be found in [4]) and performance tests of VTSS and MMSRS.

4.1 Multimedia Storage and Retrieval System

Contrary to VTSS the architecture of MMSRS is based upon the UniTree HSM software. The system consists of Automated Media Library (AML) and of the following software components: the AML managing system (UniTree HSM), the MPEG extension for HSM (MEH) and a public domain WWW server. MEH is a middleware which we developed and added on the top of UniTree to achieve efficient access to video fragments.

Because of the file granularity imposed by the UniTree HSM software and our requirements MEH needs to cut the videos into pieces of similar size and store them as different files into the UniTree file system. The cutting substantially reduces the start-up delay. MEH receives from the client the name of the video and the frame range (start frame and end frame). According to the range, it computes which video pieces (subfiles) will be needed to produce the output MPEG stream. The video fragment is requested by the client using HTTP.

4.2 Test Results

In these experiments we measure the tertiary storage system performance (i.e. the startup latency and the transfer time) excluding the disk cache influence. In the case of MMSRS this is done by explicitly purging out from the disk cache the files needed for the tests. VTSS itself does not have any disk cache. Anyway some caching must be implemented at the client application side for better start-up performance. This aspect is discussed in more detail in the next section.

The experiments were done using an ATL tape library with DLT2000 drives. This tape library is applied for development purposes only and since it is a little bit outdated, the positioning times and transfer rates might be disappointing (load time: 60 seconds, position time 0-120 seconds, transfer rate 1.25 MB/s). We plan to repeat the tests in real workload environment with a DLT7000 ATL library. The server running the testing software and connected to the library was HP 9000 class with two processors. In order to have the same file positions on tapes for those two different management systems we stored the same video on each system forcing them to start writing on a new tape.

The drives have been set empty before starting a retrieval of a video fragment to examine the worst case of the idle library, i.e. the tape is not mounted but still there is an empty drive to mount in.

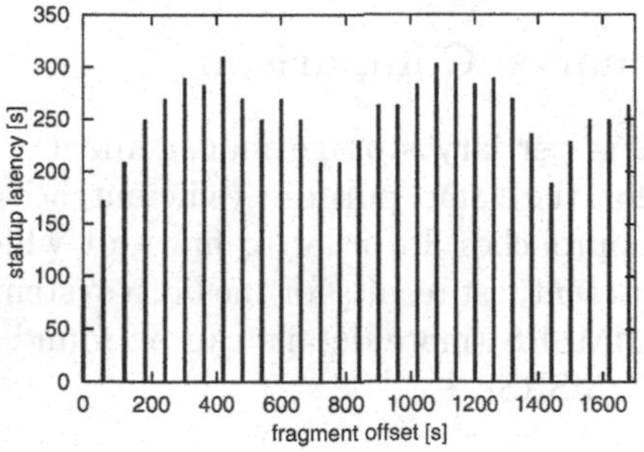

Fig. 3. MMSRS Startup Latency.

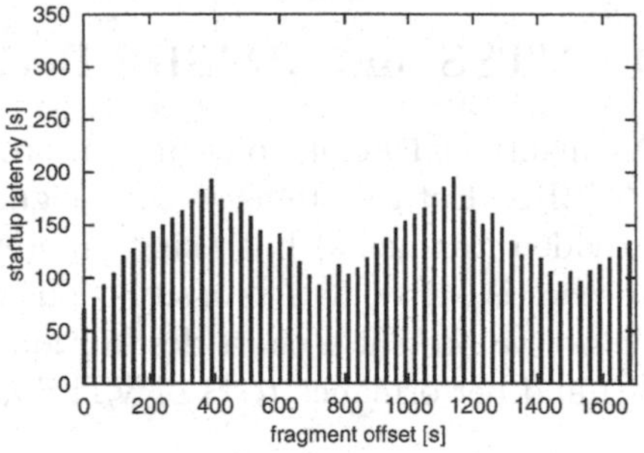

Fig. 4. VTSS Startup Latency.

In Figs. 3, 4 we present the startup latency depending on the position of the first byte of the required video sequence according to the begin of the video (video fragment offset - in seconds). The saw characteristics observed is due to the serpentine recording on the DLT tape.

In Fig. 5 the transfer time according to the length of the video sequence required is shown. The stepping characteristic is due to the UniTree feature that the file must be staged in advance, prior to the reading operation. For the VTSS case (see Fig. 6) the characteristic is linearly increasing due to the lack of any disk cache.

In Figs. 7 and 8 the minimal transfer rates for a given video fragment depending on its offset are presented. The minimal transfer rate is chosen from transfer

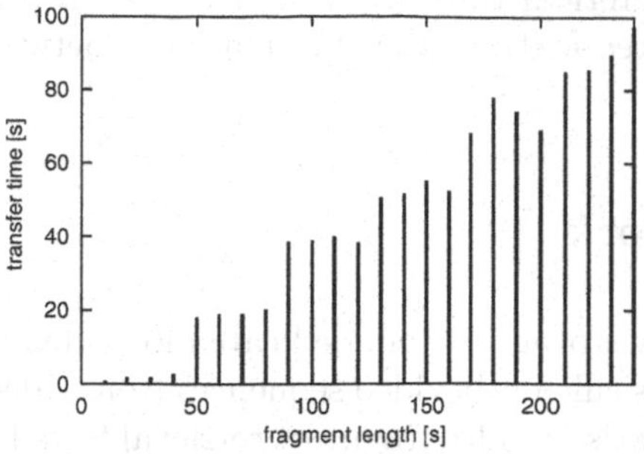

Fig. 5. MMSRS Transfer Time.

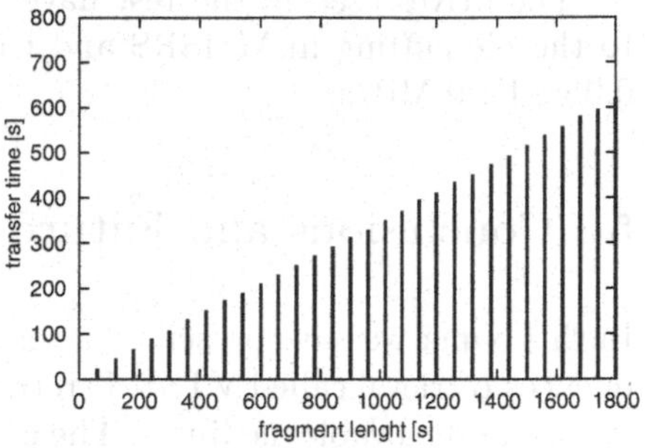

Fig. 6. VTSS Transfer Time.

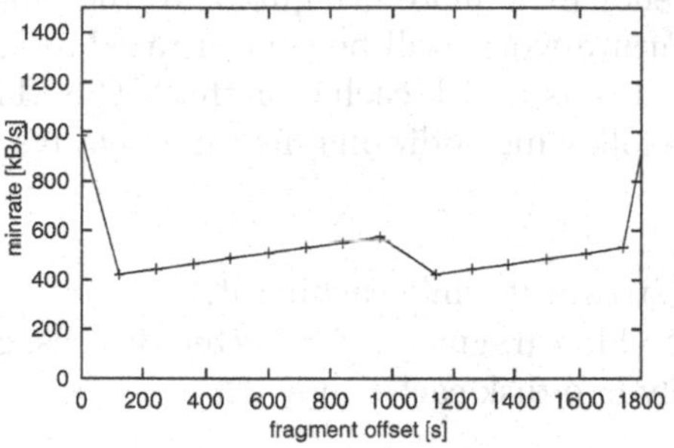

Fig. 7. MMSRS Minimal Rate.

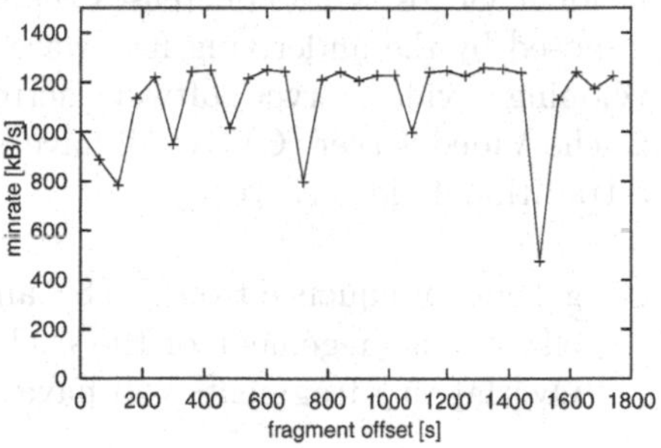

Fig. 8. VTSS Minimal Rate.

rates being calculated each time a network packet is received by a client. The transfer rate at certain moment is calculated as the ratio of the number of bytes received so far and the time it has taken to transfer them. The minimal rate must be greater than the bit rate of the video stream to display it smoothly. In the case of MMSRS it depends on where in the subfile the fragment starts. As closer to the end less time remains to prefetch next subfile, thus reducing the rate. In the case of VTSS the rate reduces when fragment starts near to the end of track, because the DLT drive needs to change the direction of tape movement, which costs about 3 sec. The down peaks indicate such situations.

The startup latency is independent of the length of the required video sequence. In the case of MMSRS it depends on the subfile size and the fragment offset. For 16 MB subfiles and for the offset equal to 0 seconds it is about 90 seconds. In the case of VTSS startup latency depends only on the fragment offset and for 0 seconds offset is about 70 seconds.

In the case of transferring a video of 30 minutes long stored in traditional way (not split) total transfer time consisting of startup latency (718 seconds) and transfer time takes 853 seconds. The same video stored on MMSRS with 16 MB subfiles is transferred entirely in 800 seconds with startup latency of 90 seconds.

The drives used in the test have theoretical throughput of 1.25 MB/s. Due to the file cutting in MMSRS and UniTree features the throughput is between 0.9 and 1.0 MB/s.

5 Conclusions and Future Work

In this paper we have presented a tertiary storage system dedicated for video sequences retrieval called VTSS. It retrieves efficiently video sequences from videos stored on removable mediums. The comparison video fragment retrieval tests between VTSS, MMSRS and the bare UniTree file system shows the advantage of VTSS in terms of startup latency and transfer time for non-cached videos. MMSRS, using sub-filing strategy, significantly decreases the startup latency of tape resident videos. VTSS decreases the latency even more bringing it to the limits imposed by the underlying hardware. Future works will be concentrated on developing a video server software acting also as a disk cache for the VTSS. The Cache Video Server (CVS) will have the following additional features relative to a traditional video server:

- get video sequence from VTSS and stream it while caching it,
- efficient management of the cached video fragments for better disk usage. Overlapping fragments will have only one disk cache copy.

We plan to use the source code of a video server software [15] developed at the University of Klagenfurt as base for out CVS. We also plan to add to the VTSS functionality a new feature which will inform the client application when the video sequence will start displaying. In the case of DLT tapes we plan to use the model for estimating access time of serpentine tape drives presented in [16].

Acknowledgements

We gratefully acknowledge contributions of Professor Lászlo Böszörményi and Dr. Harald Kosch from Institute of Information Technology, University of Klagenfurt (Austria).

The work has been supported by Scientific and Technological Cooperation Joint Project between Poland and Austria: KBN-OEADD grant No. A:8N/1999-2000 and by KBN grant No. 8 T11C 006 15.

References

1. Belleman, R.G., and Sloot, P.M.A., "The Design of Dynamics Exploration Environments for Computational Steering Simulations", in: Bubak, M., Mościński, J., and Noga, M., Proc. SGI Users' Conference, Oct. 11-14, 2000, Kraków, pp. 57-74, ACK Cyfronet-AGH.

2. Afsarmanesh, H., Benabdelkader, A., Kaletas, E.C., Garita, C., and Herzberger, L.O., "Towards a Multi-layer Architecture for Scientific Virtual Laboratories", in: Bubak, M., Afsarmanesh, H., Williams, R., Hertberger, B., (Eds.), Proc. Int. Conf. High Performance Computing and Networking, Amsterdam, May 8-10, 2000, Lecture Notes in Computer Science 1823, pp. 163-176, Springer, 2000.
3. Kosch, H., Słota, R., Böszörményi, L., Kitowski, J., Otfinowski, J., Wójcik, P., "A Distributed Medical Information System for Multimedia Data - The First Year's Experience of the PARAMED Project" in: Bubak, M., Afsarmanesh, H., Williams, R., Hertberger, B., (Eds.), Proc. Int. Conf. High Performance Computing and Networking, Amsterdam, May 8-10, 2000, Lecture Notes in Computer Science 1823, pp. 543-546, Springer, 2000.
4. Słota, R., Kosch, H., Nikolow, D., Pogoda, M., Breidler, K., Podlipnig, S., "MMSRS - Multimedia Storage and Retrieval System for a Distributed Mediacal Information System" in: Bubak, M., Afsarmanesh, H., Williams, R., Hertberger, B., (Eds.), Proc. Int. Conf. High Performance Computing and Networking, Amsterdam, May 8-10, 2000, Lecture Notes in Computer Science 1823, pp. 517-524, Springer, 2000.
5. UniTree Software, `http://www.unitree.com/`.
6. Foster, I., and Kesselman, C., (eds.), The Grid: Blueprint for a New Computing Infrastructure, Morgan Kaufmann Publ., USA, 1999.
7. The Globus Project, `http://www.globus.org/`.
8. The CERN DataGrid Project, `http://www.cern.ch/grid/`.
9. The Particle Physics Data Grid (PPDG), `http://www.cacr.caltech.edu/ppdg/`
10. Coyne, R.A., Hulen, H., Watson, R., "The High Performance Storage System", Proc. of *Supercomputing 93*, Portland, USA, November 1993.
11. Beynon, M., Ferreira, R., Kurc, T., Sussman, A., and Saltz, J., "DataCutter: Middleware for Filtering Very Large Scientific Datasets on Archival Storage Systems", Proc. of *Eighth NASA Goddard Conference on Mass Storage Systems and Technologies and the Seventeenth IEEE Symposium on Mass Storage Systems*, Maryland, USA, March 27-30, 2000, pp. 119-133.
12. Memik, G., Kandemir, M.T., Choudhary, A., Taylor, V.E., "April: A Run-Time Library for Tape-Resident Data", Proc. of *Eighth NASA Goddard Conference on Mass Storage Systems and Technologies and the Seventeenth IEEE Symposium on Mass Storage Systems*, Maryland, USA, March 27-30, 2000, pp. 61-74.
13. Corbett, P., Fietelson, D., Fineberg, S., Hsu, Y., Nitzberg, B., Prost, J., Snir, M., Traversat, B., and Wong, P., "Overview of the MPI-IO parallel I/O interface", Proc. of *Third Workshop on I/O in Paral. and Distr. Sys.*, Santa Barbara, USA, April 1995.
14. Holtman, K., Stok, P., Willers, I., "Towards Mass Storage Systems with Object Granularity", Proc. of *Eighth NASA Goddard Conference on Mass Storage Systems and Technologies and the Seventeenth IEEE Symposium on Mass Storage Systems*, Maryland, USA, March 27-30, 2000, pp. 135-149.
15. Breidler K., Kosch, H., Böszörményi, L., "The Parallel Video Server SESAME-KB" - short talk on *3rd DAPSYS'2000* Balatonfüred, Lake Balaton, Hungary, September 10th-13th, 2000.
16. Sandstå, O., Midstraum, R., "Improving the Access Time Performance of Serpentine Tape Drives", Proc. of *16th IEEE Symposium on Mass Storage Systems the 7th NASA Goddard Conference on Mass Storage Systems and Technologies*, San Diego, California, USA, March 1999, pp. 542-591.

Data and Metadata Collections for Scientific Applications

Arcot K. Rajasekar and Reagan W. Moore

San Diego Supercomputer Center
University of California, San Diego, California, USA
{sekar,moore}@sdsc.edu

Abstract. The internet has provided a means to share scientific data across groups and disciplines for integrated research extending beyond the local computing environment. But the organization and curation of data pose challenges due to their sensitive nature (where data needs to be protected from unauthorized usage) as well as their heterogeneity and large volume, both in size and number. Moreover, the importance of metadata is coming to the fore, as a means of not only discovering datasets of interest but also for organizational purposes. SDSC has developed data management systems to facilitate use of published digital objects. The associated infrastructure includes persistent archives for managing technology evolution, data handling systems for collection-based access to data, collection management systems for organizing information catalogs, digital library services for manipulating data sets, and data grids for federating multiple collections. The infrastructure components provide systems for digital object management, information management, and knowledge management. We discuss examples of the application of the technology, including distributed collections and data grids for astronomical sky surveys, high energy physics data collections, ecology, and art image digital libraries.

1 Introduction

Scientific disciplines are assembling large quantities of primary source data for use by researchers[1,2]. The data, assembled by researchers located at separate institutions are typically managed on multiple administration domains and are stored on heterogeneous storage systems. The challenge is to facilitate the organization of these primary data resources into collections without compromising local control, while maintaining the privacy of data where required. At the same time, in order to simplify usage, middleware is needed to support uniform access to the data sets, including APIs for direct application discovery and manipulation of the data, command line interfaces for accessing data sets from scripts, and web GUIs for interactive browsing and presentation of data sets. The middleware needs to work on platforms ranging from supercomputers to desk-top systems while providing a uniform interface for researchers.

The organization and sharing of data is facilitated by metadata that are assembled and/or derived as part of the scientific data collections. When a single

B. Hertzberger et al. (Eds.): HPCN Europe 2001, LNCS 2110, pp. 72–80, 2001.

researcher (or a small group) assembles and uses data, files stored in a well-defined directory structure provides a way to organize data. But with global sharing one needs a better mechanism for discovering data. Metadata (information about data) provides a means for discovering data objects as well as providing other useful information about the data objects such as experimental parameters, creation conditions, etc. Many disciplines store their metadata as part of file headers or in separate files. But these approaches do not allow for ease of querying the metadata. One needs a uniform access mechanism for storing and querying metadata across disciplines and within disciplines.

The development of infrastructure to support the publication of scientific data and metadata must recognize that information repositories and knowledge bases are both needed. One can differentiate between infrastructure components that provide:

- Data repositories for storing digital objects that are either simulation application output or remote sensing data. The digital objects are representations of reality, generated either through a hardware remote sensing device or by execution of an application.
- Information repositories for storing attributes about the digital objects. The attributes are typically stored as metadata in a catalog or database.
- Knowledge bases for characterizing relationships between sets of metadata. An example is rule-based ontology mapping[3] that provides the ability to correlate information stored in multiple metadata catalogs.

A scientific data publication system will need to support ingestion of digital objects, querying of metadata catalogs to identify objects of interest, and integration of responses across multiple information repositories. Fortunately, a rapid convergence of information management technology and data handling systems is occurring for enabling such intelligent scientific data collections. It is becoming possible to provide mechanisms for the publication of scientific data for use by an entire research community. The approach used at the San Diego Supercomputer Center is to organize distributed data sets through creation of a logical collection[4] . The ownership of the data sets is assigned to the collection, and a data handling system is used to create, move, copy, replicate, partition, and read collection data sets. Since all accesses to the collection data sets are done through the data handling system, it then becomes possible to put the data sets under strict management control, and implement features such as access control lists, usage audit trails, replica management, and persistent identifiers.

Effectively, a distributed collection can be created in which the local resources remain under the control of the local site, but the data sets are managed by the global logical collection. Researchers authenticate themselves to the collection, and the collection in turn authenticates itself to the distributed storage systems on which the data sets reside. The collection manages the access control lists for each data set independently of the local site. The local resources are effectively encapsulated into a collection service, removing the need for researchers to have user accounts at each site where the data sets are stored.

The data handling system serves as an interoperability mechanism for managing storage systems. Instead of directly storing digital objects in an archive or file system, the interposition of a data handling system allows the creation of a collection that spans multiple storage systems. It is then possible to automate the creation of a replica in an archival storage system, cache a copy of a digital object onto a local disk, and support the remote manipulation of the digital object. The creation of data handling systems for collection-based access to published scientific data sets makes it possible to automate all data management tasks. In turn, this makes it possible to support data mining against collections of data sets, including comparisons between simulation and measurement, and statistical analysis of the properties of multiple data sets. Data set handling systems can be characterized as interoperability mechanisms that integrate local data resources into global resources. The interoperability mechanisms include

- inter-domain authentication,
- transparent protocol conversion for access to all storage systems,
- global persistent identifiers that are location and protocol independent,
- replica management for cached and archived copies,
- data aggregation container technology to optimize archival storage performance for small datasets and co-locate data sets for optimal access, and
- tools for uniform collection management across file systems, databases, and archives.

2 Data Handling Infrastructure

The data management infrastructure is based upon technology from multiple communities that are independently developing archival storage systems, parallel and XML-based[5] database management systems, digital library services, distributed computing environments, and persistent archives. The combination of these systems is resulting in the ability to describe, manage, access, and build very large distributed scientific data collections. Several key factors are driving the technology convergence:

- Development of an appropriate information exchange protocol and information tagging model. The ability to tag information content makes it possible to directly manipulate information. The extensible Markup Language (XML) provides a common information model for tagging data set context and provenance. Document Type Definitions (and related organizational methods such as XML Schema) provide a way to organize the tagged attributes. Currently, each scientific discipline is developing their own markup language (set of attributes) for describing their domain-specific information. The library community has developed generic attribute sets such as the Dublin core to describe provenance information. The combination of the Dublin core metadata and discipline specific metadata can be used to describe scientific data sets.

- Differentiation between the physical organization of a collection (conceptually the table structures used to store attributes in object-relational databases) and the logical organization of a collection (the schema). If both contexts are published, it becomes possible to automate the generation of the SQL commands used to query relational databases. For XML-based collections, the emergence of XML Matching and Structuring languages[6] and the XQuery standard makes it possible to construct queries based upon specification of attributes within XML DTDs. Thus attribute-based identification of data sets no longer requires the ability to generate SQL or XQL commands from within an application.
- Differentiation of the organization and access mechanisms for a logical collection from the organization and access mechanisms required by a particular storage system. Conceptually, data handling systems store data in storage systems rather than storage devices. By keeping the collection context independent of the physical storage devices, and providing interoperability mechanisms for data movement between storage systems, logical data set collections can be created across any type of storage system. Existing data collections can be transparently incorporated into the logical collection. The only requirement is that the logical collection be given access control permissions for the local data sets. The data handling system becomes the unifying middleware for access to distributed data sets.
- Differentiation of the management of information repositories from the storage of metadata into a catalog. Information management systems provide the ability to manage databases. It is then possible to migrate metadata catalogs between database instantiations, extend the schema used to organize the catalogs, and export metadata as XML or HTML formatted files.

The metadata used to describe collections and to describe datasets in collections can be separated into multiple layers. One can identify metadata that at the lowest level provides information about the physical characteristics of the dataset (or collection) such as its location, size, ownership, etc. At the other extreme, we may have metadata that is particular to a dataset which is relevant only to one or a small group of researchers; that is metadata that is very specific to the application or group that created or used the datasets. We have identified a layered architecture for this metadata characterization:

Kernel. Kernel Metadata can be seen as the basic metadata that can be used for integrating the levels of metadata that are at the higher levels. This set of metadata also includes information about how the metadata is organized, information about catalogs, information about how two sets of metadata can be interoperated, etc.

System-Centric. System-centric metadata stores information that has systemic information. Information at this level may be about data replicas, location, size, ownership, access control information as well as information about resources being used (computation platforms used for deriving the data, methods used, etc), as well as information about users and storage resources.

Standard. Standardized metadata includes metadata that implements agreed upon standards within the digital library community, such as the Dublin core, MARC, cross-walks, etc.

Domain-Centric. Domain-centric metadata includes metadata that implements agreed upon standards inside (and possibly across) scientific disciplines. Such standards might be the ones used by an image consortium (AMICO), used by the ecological community (EML), used for astronomical instruments (AIML), used in chemistry (CML), etc.

Application-Level. Application-level metadata includes information that is particular to an application, experiment, group or individual researcher. This may also include information about how to present a particular dataset as well as creation or derivation notes by individual researchers.

The ability to manipulate data sets through collection-based access mechanisms enables the federation of data collections and the creation of persistent archives. Federation is enabled by publishing the schema used to organize a collection as an XML DTD. Information discovery can then be done through queries based upon the semi-structured representation of the collection attributes provided by the XML DTD. Distributed queries across multiple collections can be accomplished by mapping between the multiple DTDs, either through use of rule-based ontology mapping, or token-based attribute mapping. Persistent archives can be enabled by archiving the context that defines both the physical and logical collection organization along with the data sets that comprise the collection[7]. The collection context can then be used to recreate the collection on new database technology through an instantiation program. This makes it possible to migrate a collection forward in time onto new technology. The collection description is instantiated on the new technology, while the data sets remain on the physical storage resource. The collection instantiation program is updated as database technology evolves, while the archived data remain under the control of the data handling system. As the archive technology evolves, new drivers are added to the data handling system to interoperate with the new data access protocols.

The implementation of information management technology needs to build upon the information models and manipulation capabilities that are coming from the Digital Library community, and the remote data access and procedure execution support that is coming from the distributed computing community. The Data Access Working Group of the Grid Forum[8] is promoting the development of standard implementation practices for the construction of grids. Grids are inherently distributed systems that tie together data, compute, and visualization resources. Researchers rely on the grid to support all aspects of information management and data manipulation. An end-to-end system provides support for:

Knowledge discovery, the ability to identify relationships between digital objects stored in different discipline collections

Information discovery, the ability to query across multiple information repositories to identify data sets of interest

Data handling, the ability to read data from a remote site for use within an application

Remote processing, the ability to filter or subset a data set before transmission over the network

Publication, the ability to add data sets to collections for use by other researchers

Analysis, the ability to use data in scientific simulations, or for data mining, or for creation of new data collections

These services are implemented as middleware that hide the complexity of the diverse distributed heterogeneous resources that comprise data and compute grids[9]. The services provide four key functionalities or transparencies that simplify the complexity of accessing distributed heterogeneous systems.

Name transparency. Unique names for data sets are needed to guarantee a specific data set can be found and retrieved. However, it is not possible to know the unique name of every data set that can be accessed within a data grid (potentially billions of objects). Attribute based access is used so that any data set can be identified either by data handling system attributes, or Dublin core provenance attributes, or discipline specific attributes.

Location transparency. Given the identification of a desired data set, a data handling system manages interactions with the possibly remote data set. The actual location of the data set can be maintained as part of the data handling system attributes. This makes it possible to automate remote data access. When data sets are replicated across multiple sites, attribute-based access is essential to allow the data handling system to retrieve the "closest" copy.

Protocol transparency. Data grids provide access to heterogeneous data resources, including file systems, databases, and archives. The data handling system can use attributes stored in the collection catalog to determine the particular access protocol required to retrieve the desired data set. For heterogeneous systems, servers can be installed on each storage resource to automate the protocol conversion. Then an application can access objects stored in a database or in an archive through a uniform user interface.

Time transparency. At least five mechanisms can be used to minimize retrieval time for distributed objects: data caching, data replication, data aggregation, parallel I/O, and remote data filtering. Each of these mechanisms can be automated as part of the data handling system. Data caching can be automated by having the data handling system pull data from the remote archive to a local data cache. Data replication across multiple storage resources can be used to minimize wide area network traffic. Data aggregation through the use of containers can be used to minimize access latency to archives or remote storage systems. Parallel I/O can be used to minimize the time needed to transfer a large data set. Remote data filtering can be used to minimize the amount of data that must be moved. This latter capability requires the ability to support remote procedure execution at the storage resource.

3 Application

A collection-based data management system has the following software infrastructure layers:

Data grid for federation of access to multiple data collections and digital libraries

Digital library to provide services for discovering, manipulating, and presenting data from collections

Data collection to provide support for extensible, dynamically changing organizations of data sets

Data handling system to provide persistent IDs for collection-based access to data sets

Persistent archive to provide collection-based storage of data sets, with the ability to handle evolution of the software infrastructure.

The essential infrastructure component is the data handling system. It is possible to use data handling systems to assemble distributed data collections, integrate digital libraries with archival storage systems, federate multiple collections into a data grid, and create persistent archives. An example that encompasses all of these cases is the 2- Micron All Sky Survey image archive. The 2MASS survey is an astronomy project led by the University of Massachusetts and the California Institute of Technology (CalTech) to assemble a catalog of all stellar objects that are visible at the 2-micron wavelength. The goal of the project is to provide a catalog that lists attributes of each object, such as brightness and location. The final catalog can contain as many as 2 billion stars and 200 million galaxies. Of interest to astronomers is the ability to analyze the images of all the galaxies. This is a massive data analysis problem since there are a total of 5 million images comprising 10 terabytes of data.

A collaboration between IPAC at Caltech and the NPACI program at SDSC is building an image catalog of all of the 2MASS observations. A digital library is being created at Caltech that records which image contains each galaxy. The images are sorted into 147,000 containers to co-locate all images from the same area in the sky. The image collection is then replicated between two archives to provide disaster recovery. The SDSC Storage Resource Broker[10] is used as the data handling system to provide access to archives at Caltech and SDSC where the images are replicated.

The usage model supports the following access scenario:

- Astronomers access the catalog at Caltech to identify galaxy types of interest.
- Digital library procedures determine which images need to be retrieved.
- The data handling system maps the image to the appropriate container, retrieves the container from the archive, and caches the container on a disk.
- The desired image is then read from the disk and returned to the user through the digital library.

Since the images are accessed through the data handling system, the desired images can be retrieved from either archive depending upon load or availability. If the container has already been migrated to disk cache, the data handling system can immediately retrieve the image from disk avoiding the access latency inherent in reading from the archive tape system. If one archive is inaccessible, the data handling system automatically defaults to the alternate storage system. If data is migrated to alternate storage systems, the persistent identifier remains the same and the data handling system adds the location and protocol access metadata for the new storage system. The system incorporates persistent archive technology, data handling systems, collection management tools, and digital library services in order to support analysis of galactic images.

Given the ability to access a 10-terabyte collection that contains a complete sky survey at the 2-micron wavelength, it is then possible to do data intensive analyses on terascale computer platforms . The data rates that are expected for a teraflops-capable computer will be over 2-Gigabytes per second from disk. It will be possible to read the entire collection in 2 hours. During this time period, a billion operations can be done on each of the five million images. Effectively, the entire survey can be analyzed in a "typical" computation on a teraflops computer.

4 Summary

Data intensive computing is facilitated by the organization of scientific data into collections that can then be processed by the scientific community. In the long run, the utility of scientific computation will be measured by the publication of the results of the computation into collections for use by the rest of the community. Digital objects that remain as local files on a researcher's workstation will be of little use to the scientific discipline. The utility of digital objects will be directly related to the specification of their context through membership in a scientific data collection. At this point, the fundamental access paradigm will shift to reading and writing data from collections, with applications using APIs to discover, access, and manipulate collection-based data. Each discipline will use their data repositories, information catalogs, and knowledge bases to provide direct access to all of the primary data sources for their domain.

Acknowledgments

The topics presented in this report were developed by the Data Intensive Computing Environment Group at the San Diego Supercomputer Center. DICE group members include Chaitan Baru, Amarnath Gupta, Bertram Ludaescher, Richard Marciano, Michael Wan, and Ilya Zaslavsky. The data management technology has been developed through multiple federally sponsored projects, including DARPA project F19628-95- C-0194 "Massive Data Analysis Systems," the DARPA/USPTO project F19628-96-C-0020 "Distributed Object Computation Testbed," the Data Intensive Computing thrust area of the NSF project ASC 96-19020 "National Partnership for Advanced Computational Infrastructure," the NASA Information Power Grid project, and the DOE ASCI/ASAP project "Data Visualization Corridor." The development of persistent archives was supported by a NARA extension to the DARPA/USPTO Distributed Object Computation Testbed, project F19628-96-C-0020. The development of the Knowledge Network for BioComplexity is supported by NSF under a KDI grant DEB 99-80154.

References

1. Moore, R., C. Baru, P. Bourne, M. Ellisman, S. Karin, A. Rajasekar, S. Young: Information Based Computing. Proceedings of the Workshop on Research Directions for the Next Generation Internet, May, 1997.
2. Jones, M.B: Web-based Data Management, In "Data and Information Management in the Ecological Sciences: A Resource Guide", eds. W.K. Michener, J.H. Porter, S.G. Stafford, LTER Network Office, University of New Mexico, Albuquerque, New Mexico, 1998.
3. Ludaescher, B., A. Gupta, M. E. Martone: Model-Based Information Integration in a Neuroscience Mediator System. 26th Intl. Conference on Very Large Databases (demonstration track), September, 2000.
4. NPACI Data Intensive Computing Environment thrust area. `http://www.npaci.edu/DICE/`.
5. Extensible Markup Language (XML). `http://www.w3.org/XML/`.
6. Baru, C., V. Chu, A. Gupta, B. Ludascher, R. Marciano, Y. Papakonstantinou, and P. Velikhov: XML-Based Information Mediation for Digital Libraries. ACM Conf. On Digital Libraries (exhibition program), 1999.
7. Moore, R., C. Baru, A. Rajasekar, B. Ludascher, R. Marciano, M. Wan, W. Schroeder, and A. Gupta: Collection-Based Persistent Digital Archives - Part 1, D-Lib Magazine, (`http://www.dlib.org/`) March 2000.
8. Grid Forum Remote Data Access Working Group. `http://www.sdsc.edu/GridForum/RemoteData/`.
9. Moore, R., C. Baru, A. Rajasekar, R. Marciano, M. Wan: Data Intensive Computing, In "The Grid: Blueprint for a New Computing Infrastructure", eds. I. Foster and C. Kesselman. Morgan Kaufmann, San Francisco, 1999.
10. Baru, C., R, Moore, A. Rajasekar, M. Wan: The SDSC Storage Resource Broker. Proc. CASCON'98 Conference, 1998 (see also `http://www.npaci.edu/DICE/SRB`).

The VLAM-G Abstract Machine: A Data and Process Handling System on the Grid

A. Belloum[1], Z.W. Hendrikse[1], D.L. Groep[2], E.C. Kaletas[1], A.W. van Halderen[1], H. Afsarmanesh[1], and L.O. Hertzberger[1]

[1] University of Amsterdam, Computer Science Department
Kruislaan 403, 1089 SJ Amsterdam, The Netherlands
{adam,zegerh,kaletas,berry,bob}@science.uva.nl

[2] National Institute for Nuclear and High Energy Physics NIKHEF
P.O. Box 41882, 1009 DB Amsterdam, The Netherlands
{davidg}@nikhef.nl

Abstract. This paper presents the architecture of the *Virtual Laboratory Abstract-Machine* (VL-AM), a data and process handling system on the Grid. This system has been developed within the context of the Virtual Laboratory (VL) project, which aims at building an environment for experimental science. The VL-AM solves problems commonly faces when addressing process and data handling in a heterogeneous environment. Although existing *Grid technologies* already provide suitable solutions most of the time, considerable knowledge and experience outside of the application domain is often required. The VL-AM makes these grid technologies readily available to a broad community of scientists.
Moreover, it extends the current data Grid architecture by introducing an object-oriented database model handling complex queries on scientific information.

1 Introduction

Grid technology provides a persistent platform for high-performance and high-throughput applications using geographically dispersed resources. It constitutes a potential solution for effective operation in large-scale, multi-institutional, wide area environments, called *Virtual Organizations* [1,2].

Various research groups are currently investigating Grid technologies, focusing on scientific applications. In an effort to outline requirements on Grid technology, Oldfield [3] has gathered information on a large number of Grid projects. These projects cover a wide range of scientific applications, including (bio)medics, physics, astronomy and visualization. Intensive use of large data sets was a common feature of all projects involved. Therefore a scalable and powerful mechanism for data manipulation is required, known as *Data-grid*.

The Data-Grid is defined as a layer offering services related to data access and data manipulation. Within the grid-community Data-Grid systems are generally agreed to contain: Data handling, Remote processing, Publication, Information discovery, and Analysis [3].

B. Hertzberger et al. (Eds.): HPCN Europe 2001, LNCS 2110, pp. 81–93, 2001.

One of the advantages of Data-Grid systems is their ability to support location-independent data access, *i.e.* low-level mechanisms are hidden for the user. Five levels of "transparency" have been identified by the Data Access Working Group[1]: name, location, protocol, time and the single sign-on transparency [4].

A proposal for a generic Data-Grid architecture is currently being studied by the Data Access Working Group, offering protocol-, location- and time-transparency. The so-called "Data Handling System" provides an interface connecting the separate components of this data access system and offering access to it, see Figure 1.

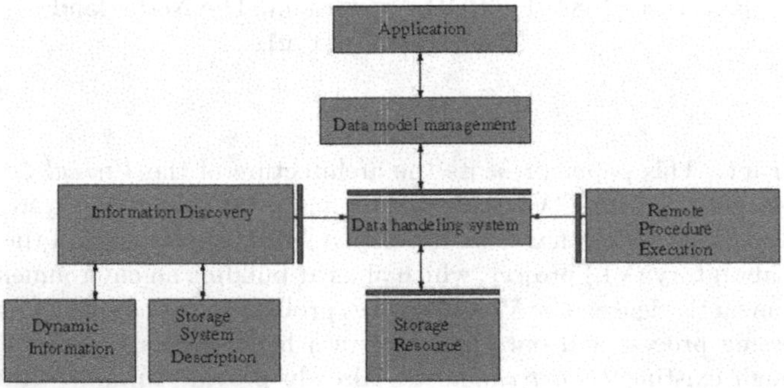

Fig. 1. The general Data-Grid architecture as proposed by the Data Access Working Group[5]. Note the central position of the Data handling system.

At this moment, the Globus Toolkit[2] provides the leading Grid-technology and is becoming a *de facto* standard in this area. It offers a variety of tools which may be used to build an efficient grid testbed quickly [5]. However, since the Globus toolkit constitutes a low-level middle-tier, it is not intended to be harnessed directly by a broad scientific community. An additional layer is needed to bring Globus technology to scientists, the most important group of Grid users thus far.

The purpose of the VL-Abstract Machine (VL-AM) is to bridge this gap. If we envisage a general Grid architecture as a four-tier model (application, application toolkit, Grid-services and Grid-fabric tier), the VL-AM is located in the third — the application toolkit — tier.

The VL-AM will initially focus on four application domains:

- chemo-physical analysis of surfaces at micrometer scales,
- bio molecular genome expression studies,
- a biomedical visualization workbench, and
- an analysis work bench for electronic fee collection.

[1] A subgroup of the Global Grid Forum, see `http://www.gridforum.org/`.
[2] `http://www.globus.org/`

In the context of the VL, experiments are defined by selecting processing elements (also referred to as modules) and connected these in such a way as to represent a data-flow; the topology of an experiment may be represented by a data-flow graph (DFG). As such, the modules comprising the data-flow have a strong analogy to those used in IRIS explorer[3]. Processing elements[4] are independent of each other and in general may have been developed by different persons. Therefore the VL-AM Run Time System (RTS) should offer an implementation of (most of) the transparency layers and a platform enabling modules to perform data processing and computations.

In addition to the VL-AM RTS, a VL-AM front end has been designed to form the main interface by which external users can access VL compute resources, data elements and software repositories. The VL-AM front end queries other VL components on behalf of the end-user to collect all information needed to facilitate his work within the VL environment. The collected information is presented to the end-users through a graphical user interface (GUI).

The remainder of this paper is organized as follows: Section 2 describes the way in which experiments are performed. The architecture for the VL-AM designed to support experimentation is addressed in 3. In Section 4, the main components of the VL-AM front end are presented. Section 5 focuses on the design of the VL-AM RTS. Thereafter Section 6 highlights some important features of related projects. Finally Section 7 presents some conclusions and future prospects.

2 Experimenting in VL

The VL [6] designs and develops an open, flexible and configurable laboratory framework. This includes the hardware and software to enable scientists and engineers to work on their problems collectively by experimentation. Developments are steered initially by the aforementioned scientific application domains.

Studying these applications domains has allowed us to identify a set of generic aspects valid for all of these scientific domains. By combining these generic features that characterize scientific experiments with the characteristics of case-studies, it has been possible to develop a database model for the Experimentation Environment (EE). The EE data model [7] focuses on storage and retrieval of information to support the generic definition of the steps and the information representing a scientific study, in order to investigate and reproduce experiments. The EE data model allows any ordering of processes and data elements, which enable one to construct an arbitrary process-data flow. The EE data model has been applied to build the information management system for the chemo-physical surface analysis lab (MACS) [8] and the EXPRESSIVE [9] (genome expression) application databases.

Within the context of the VL, experiments are thus embedded in the context of a *study*. A study is a series of steps intended to solve a particular problem

[3] `http://www.nag.co.uk/Welcome_IEC.html`

[4] Separate modules may be grouped to form a larger so-called aggregate-module.

in an application domain, and corresponds to a process flow. As such, a study generally comprises experiments performed using the VL-AM RTS, process steps performed outside the scope of the RTS – *e.g.* manually – and the description of data elements related to objects or process steps.

A *process flow* is the result of investigating the established work flow in a specific application domain, *e.g.*, space-resolved surface analysis or genome expression studies. Such a process flow is shown in Fig. 2. This particular flow describes chemo-physical analysis of samples using space-resolved techniques: material analysis of complex surfaces (MACS) [8]. A similar process flow exists for bio genetic experiments [9]. The context of the operational steps (represented by ovals) is provided by entity meta data (represented by boxes). For one-time experiments, a trivial process flow should be defined, *i.e.*, consisting of one operational step only, and limited contextual meta data associated with the result.

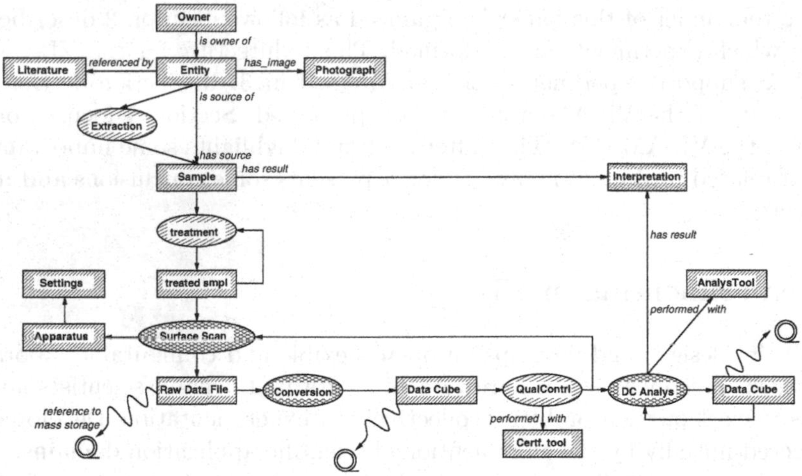

Fig. 2. Typical steps in a surface-analysis study, as represented by the MACS [8] process flow. The hatched rectangles represent meta data associated with objects (either physical objects or bulk data on mass storage). Hatched ovals represent operations: cross-hatched ovals are performed using the VL-AM Run Time System, the single-hatched ovals represent meta data associated with manually performed process steps.

A process flow template (PFT) is used to represent a process flow of the study to the scientist. The scientist is guided through the process flow using context-sensitive interaction. each PFT has a starting point; in the case of MACS this is the 'Entity', see figure 2. At first, interaction is possible with the 'Entity' box only. The object characteristics (meta data) should be supplied by the end user, *e.g.*, via a fill-in form. Once the 'Entity' data is supplied, the directly related items (here: Literature, Extraction, Sample, Photograph and Owner) become available for interaction. Boxes require object meta data to be supplied, process steps require a description of the procedure applied to particular objects. In the case of sample extraction, the process is performed outside of the VL-AM scope and details should be supplied (by means of a fill-in form).

Cross hatched ovals represent process steps performed by the VL-AM RTS. Selection of such a process step by a user will initiate a VL experiment, refer to section 5. Any resulting meta data from a process step performed by the VL-AM RTS is captured. This context consists of the VL-AM RTS experiment, constituting a process step, input and output data objects and position of a box within the PFT. This meta data is optionally added to by the scientist by means of a fill-in form. Note that meta data are stored in the application database, whereas any resulting 'base data sets' may be stored outside of this context, depending on the application.

3 VL-AM Architecture

The VL-AM is the heart of the VL. It should support the functionality provided to the VL to the users, hiding details related to the middleware layer that require non-application specific knowledge. Its proposed architecture is represented in Figure 3. It consists of four major components:

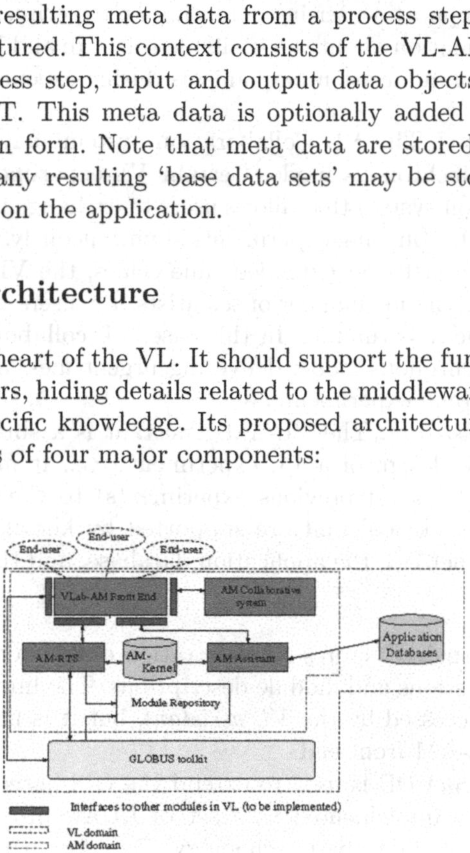

Fig. 3. Representation of the VL-AM Architecture and Its Components.

- a front end,
- a run time system (VL-AM RTS),
- a collaboration system (VL-AM Collaboration), and
- an assistant (VL-AM Assistant).

These components are supported by an information management system that includes the Kernel DB and application databases.

The VL-AM front end and the VL-AM RTS form the core of the VL-AM. Together they take care of the execution of tasks. The VL-AM front end is the only component with which the VL-users interact: it provides the user interface (UI) to the VL-AM. It delegates VL users commands to the various components and relays the results back to the user. The VL-AM RTS is responsible for scheduling, dispatching and executing the tasks comprising an experiment. These components are discussed in Sections 4 and 5.

The two remaining are not directly involved in the execution of the VL-experiments: they support a user in composing an experiment by providing the templates and information about the available software and hardware resources, and by providing communication mechanisms during execution.

- *VL Collaboration* The VL Collaboration system may operate in a synchronous or asynchronous mode. Certain VL-experiments require a real-time collaboration system that allows users to exchange ideas, share working space, and monitor ongoing experiments simultaneously. On the other hand, experiments may last over extended time scales, the VL collaboration system supports a dynamic number of scientists, *i.e.* users may join or leave an experiment while it is running. In this case, VL collaboration system operates in an asynchronous mode, delivering urgent messages to disconnected members of ongoing experiments.
- *The VL-AM Assistant:* The VL-AM Assistant is a subsystem that assists users during the design of a VL experiment, *e.g.*, it may provide module definitions or PFT's (of previous experiments) to the VL-AM front end. Decisions of the VL assistant are supported by knowledge gathered from the VL-AM Kernel DB, the application database, and the Grid Information Service (GIS).

The VL-AM Kernel DB stores administration data, such as the topology of experiments, user sessions and module descriptions. The information in the VL-AM Kernel DB is accessed by the *VL assistant*, but it is presented to the user by means of the VL-AM front end.

The VL-AM Kernel DB is used to extend the GIS-based resource discovery subsystem, currently implemented as a set of LDAP directory services. It is based on object-oriented database technology. Therefore it improves on support for complex queries (as opposed to a hierarchical data model, inherent to LDAP).

A database model for the VL-AM Kernel DB has been developed to keep track of the necessary information for the VL-AM RTS to execute properly. Amongst others, this includes user profile data, processing-module information and experiment definitions [10].

As stated above, an experiment may be represented by a DFG, composed of modules that are connected by their I/O ports. This implies the definition of two main data types for the VL-AM Kernel DB data model:

- the *Experiment Topology* defines a specific set of modules and their interconnects, which make up a VL-experiment. The topology is stored in the VL-AM Kernel database to be reused and to analyze results of a particular experiment at another time.

– the *Modules* represent the processing elements and are centrally registered by a VL administrator. Every module contains a number of *ports*. A developer should summarize its functionality, and specify the data-types of the ports and the resource requirements needed during execution.

4 VL-AM Front End

Since the VL-AM front end provides the interface between end-users and the constituent components of the VL middleware, it is required to be user friendly and to provide seamless access to the available features of the VL. Given the wide range of scientific disciplines, it is essential that the VL-AM front end is designed such that it can easily be adapted to include either a new scientific domain or a new group of users.

Its modular design therefore combines *Grid technology* components for authorization, resource discovery, and monitoring with existing *commodity components* that are more focused on the collaboration, editing and science-portal areas.

Commodity Grid Toolkits (CoGs) provide an appropriate match between similar concepts in both technologies [11]. The tools already developed for Java [12] provide an apt basis for supporting the diverse current and future end-user groups and applications.

The VL-AM front end consists of:

1. *Resource discovery component*, enabling users to make an inventory of all the available software, hardware and information resources. The information needed to provide such a service will be offered by a collection of meta-data stored either in the AM-Kernel database or in the Globus GIS or meta data replica catalogue. A connecting layer contains interfaces to the subsystems managing the meta data directories.
2. *VL execution and resource monitoring*: Two ways of monitoring are defined: on the one hand, application-level parameters and module state may be monitored [13], on the other hand process and resource status by using Grid monitoring tools. The monitoring is carried out using the VL-AM front end, although, in the case of resource monitoring, Grid monitoring tools can also be used directly.
3. *VL Experiment editing and execution*: Editing is performed using an appropriate science portal. All portals provide a drag-and-drop GUI (like the one shown in Figure 4). Processing elements, selected from a list supplied by the VL Assistant appear on the editing pane as a box with pending connection endings. Processing elements that are specific to a certain scientific discipline appear in a particular science portal only, generic processing elements may be selected using any portal. Connections are established by drawing lines between output and input ports. In addition, the experiment editing interface may automatically generate a module skeleton appropriate to the users needs, if the user want to add new functionality. Thereafter a user can add his own code to this skeleton (module skeletons are described in section 5).

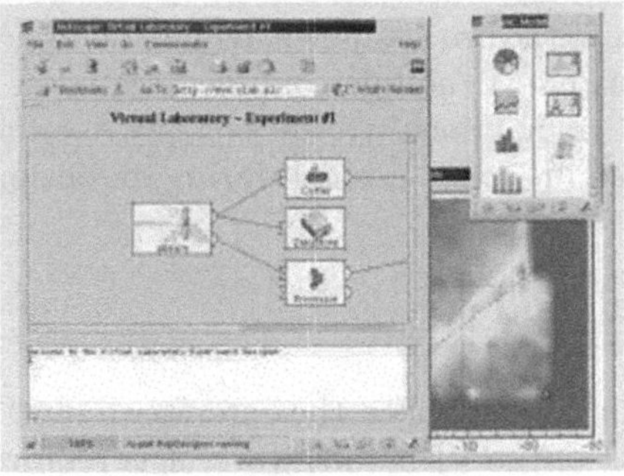

Fig. 4. VL-AM Front End Editing "ergonomics".

The VL Experiment editing component generates a graph like representation of an experiment and forwards it to the VL-AM RTS. While editing an experiment, a VL end-user is assisted by the system. Depending on the study context, a relevant topology of a previously conducted experiment may be suggested by the VL Assistant.

5 VL-AM RTS

The VL-AM RTS executes the graph it receives from the VL-AM front end. It has to provide a mechanism to transfer data between modules. These data-streams are established between the appropriate modules. The modules are instantiated and parameter and state values cached or propagated by the VL-AM RTS.

5.1 Design

From case studies (*e.g.* the chemo-physical surface laboratory [14]), it becomes clear that at least three connection types have to be supported:

- interactions among processing elements located on the same machine,
- interactions among two or more machines and
- interactions with external devices (where some of the processing elements are bound to a specific machine).

Moreover, the chemo-physical case study has not only shown the need for point-to-point connections, but also the need for connections shared by several modules, in the form of a fan-out. Since interactions with external devices are involved, QoS has to be dealt with properly.

Processing elements are developed independently. Communication is established using uni-directional ports. In order to make sure that only 'proper' connections exist, processing elements communicate via a strongly typed communication mechanism. Data types constitute the interface with which developers of modules must communicate with the 'outside world', *i.e.* with which modules must exchange their data. This will eliminate the possibility of an *architectural mismatch*, a well know problem in software composing design [15,16].

The sole use of unidirectional, typed streams intentionally precludes the possibility for 'out-of-band' communication between modules. Such communication is only allowed between the VL-AM RTS and a module. Predefined, asynchronous events can be sent from the VL-AM RTS to any module (parameters) and from any processing module to the VL-AM RTS (state).

Modules are continuously active, *i.e.* once instantiated they enter an continuous processing loop, blocked by waiting for arrival of data on the input ports. The communication mechanism allows modules to communicate in a transparent way: they do not address their peers directly, but merely output their data to the abstract port(s).

The implementation of the VL-AM RTS would potentially suffer from a tedious development path, need it be built from scratch. Fortunately, existing Grid technologies provide suitable middleware for such a development. In particular, the Globus toolkit [1] provides various secure and efficient communication mechanisms that can be used for an implementation of many of the VL-AM RTS components. The use of Globus is further motivated by the fact that it is a *de facto* standard toolkit for Grid-computing.

Using Globus toolkit components, we take advantage of all the new software currently being developed around it. For instance, the mapping of an experiment data-flow graph (DFG) onto the available computing resources can be alleviated by using the Globus resource management module GRAM and information from the Grid Information Service GIS [5]. In addition, tools developed for forecasting the network bandwidth and host load such as the *Network Weather Service* [17] can be incorporated.

5.2 Implementation

Implementation of the VL-AM RTS on a stand-alone resource could be based on several inter-process communication (IPC) mechanisms such as sockets, streams or shared memory. Within a Grid environment, the Globus toolkit offers two possibilities to implement wide-area IPC: Globus IO[6], (mimicking socket communication) and GridFTP[18].

We have chosen to use the latter for the VL-AM. Although GridFTP is a more extensive protocol for data transfer, its merits outweigh its liabilities.

GridFTP is based on the FTP protocol[7]. It focuses on high-speed transport of large data sets, or equivalently a large number of small data sets. Key features include a privacy-enhanced control connection, seamless parallelism (multiple streams implementing a single stream), striped transfers and automatic window and data-buffer size tuning. Data transfer is memory-to-memory based, and optionally authenticated and encrypted.

[5] For information on the GIS, the reader is referred to `http://www.globus.org/mds/`
[6] `http://www.globus.org/v1.1/io/globus_io.html`
[7] IETF Internet Standard STD0009

Using GridFTP as a data transport mechanism in the VL-AM RTS has significant advantages:

– efficient, high-performance data transfer across heterogeneous systems,
– third party arbitrated communication inherent in the protocol (in our case, the AM RTS acts as an arbitrator in the inter-module communication protocol; ports are assigned symbolic names corresponding to quasi-URIs for GridFTP access),
– uniformity between modules and external storage resources like HPSS's producing data (the latter running GridFTP as a standard service).
– an elaborate API exists together with an implementation as part of the Globus toolkit.

However, for small packets of data exchanged, the use of GridFTP may induce additional overhead in the communications channel.

6 Related Work

Currently several projects are aiming at a redefinition of the scientific computing environment by introducing and integrating technologies needed for remote operation, collaboration and information sharing. The Distributed Collaboratory Experiment Environments (DCEE) program of the US Department of Energy involves a large set of such projects. In the following paragraphs we summarize briefly some of the important features of the DCEE projects:

The Spectro-Microscopy Laboratory. [8] In this collaboration scientists at the University of Wisconsin-Milwaukee remotely operate the synchrotron-radiation instruments at the Berkeley Lab Advanced Light Source, Beamline 7. A remote experiment monitoring and control facility has been developed within this project. It utilizes a server that collects requests on data from experiment and interfaces to the hardware specialized in collecting data. Client programs provide a user interface and allow researchers to design an experiment, monitor its progress and analyze results. The client software may be both accessed locally or remotely to conduct experiments. The prototype includes a.o. remote teleconferencing system that allows users to control remote audio-visual transmissions and a Electronic Notebook.

LabSpace[9]**:** This project focuses on building a virtual shared space with persistence and history. LabSpace led to a number of developments which have not only allowed remote access to scientific devices over the Internet, but also allowed scientists at Argonne to meet colleagues in virtual rooms. The prototypes produced within LabSpace are composed of a set of software components for communication between multiple networked entities using a brokering system. This system has been used to support collaboration from desktop to desktop over a LAN as well as from CAVE to CAVE over a WAN (a CAVE is an immersive 3D virtual reality environment).

[8] `http://www-itg.lbl.gov/~deba/ALS.DCEE/design.doc/design.doc.html`
[9] `http://www-fp.mcs.anl.gov/division/labspace/`

Remote Control for Fusion[10]: Lawrence Livermore National Laboratory, Oak Ridge National Laboratory, the Princeton Plasma Physics Laboratory, and General Atomics are building the "Distributed Computing Testbed for a Remote Experimental Environment (REE)," to enable remote operations at the General Atomics' D-IIID tokamak. This project focuses on the development of web-enabled control and collaboration tools. It has also produced a distributed computing environment (DCE) that provides naming, security and distributed file services. The conventional inter-process-communication system (IPCS) have been converted to use authenticated DCE remote procedure calls (RPC) to provide a secure service for communication among tasks running over a wide area network.

Environmental and Molecular Sciences Collaboratory. The testbed developed at Pacific Northwest Laboratories (PNL) is based on instrumentation being developed for the Environmental Molecular Sciences Laboratory project at PNL. The goals of the Environmental Molecular Sciences Collaboratory are:

- to develop an understanding of scientific collaboration in terms of the tasks that distributed collaborators undertake, and the communications required to complete these tasks;
- to develop/integrate a suite of electronic collaboration tools to support more effective distributed scientific collaboration; and
- to use these tools for project analysis, testing, and deployment.

Security for Open, Distributed Laboratory Environments.[11] This project developed at LBNL (Lawrence Berkeley National Laboratory) aims at the development of a security model and architecture that is intended to provide general, scalable, and effective security services in open and highly distributed network environments. It specializes on on-line scientific instrument systems. The same level of, and expressiveness of, access control that is available to a local human controller of information and facilities, and the same authority, delegation, individual responsibility and accountability, and expressiveness of policy that one sees in specific environments in scientific organizations. The security model is based on public-key and signed certificate.

The InterGroup Protocol Suite: The project at U.C. Santa Barbara is developing an integrated suite of multicast communication protocols that will support multi-party collaboration over the Internet. The collaboratory environment requires communication services ranging from "unreliable, unordered" message delivery (for basic video-conferencing) to "reliable, source-ordered" message delivery (for data dissemination) to "reliable, group-ordered" message delivery (for coordinating activities and maintaining data consistency) [19].

---

[10] `http://www.es.net/hypertext/collaborations/REE/REE.html`

[10] `http://www.emsl.pnl.gov:2080/docs/collab/CollabHome.html`

[11] `http://www-itg.lbl.gov/SecurityArch/`

7 Conclusions

The VL Abstract Machine (VL-AM) is middleware to make scientific experiments on a computational Grid accessible to a wide range of users. Through its front end, it provides science portals, specifically hiding details related to computing from the end-user scientist. Moreover, it will promote collaboration by offering real time (synchronous and asynchronous) collaboration tools. Since both the VL-AM Run-Time System (RTS) and VL-AM front end are based on tools from Globus, this project can incorporate new Grid developments.

The design of the VL-AM has been presented in this paper. Six main components and their mutual interaction has been outlined. An extensive meta-data driven system (active meta-data) based on application process-flows has been proposed (templates), in order to ease the execution of an experiment and to impose uniformity.

A first prototype of the VL-AM will support filters developed for the initial application domains: chemo-physical micrometer analysis, mass- and database storage, external device controllers and visualization. The basic experimental building blocks, modules, may be aggregated to form complex experiments using visual programming techniques. Once these are instantiated, the VL-AM RTS provides a transparent execution platform. Another important feature of the VL-AM is the data-typing of the I/O-streams.

In the future, the VL-AM will be extended with an intelligent agent, assisting scientists in various ways. It will collect data from the Grid layer, the Abstract-Machine layer and the application layer to search for an optimal allocation of the computing resources. Moreover, it will assist both during design and analysis of an experiment (data mining) and perform statistics on module resource consumption. In doing so, it will learn from the successes and failures of previous experiments, and utilize that knowledge to the benefit of other users.

References

1. I. Foster and C. Kesselman. Globus: A toolkit-based grid architecture. In *The Grid: Blueprint for a Future Computing Infrastructure*, pages 259–278. MORGAN-KAUFMANN, 1998.
2. Ian Foster, Carl Kesseleman, and Steven Tuecke. The anatomy if the grid: Enabling scalable virtual organizations. *International Journal on Supercomputer Applications, 2001*, (To be published), `http://www.globus.org/research/papers/anatomy.pdf` 2001.
3. Ron Oldfield. Summary of existing data grid. Draft `http://www.cs.dartmouth.edu/~raoldfi/grid-white/paper.html`, Grid Forum Data Acces Working Group, 2000.
4. Reagan Moore. Grid forum data access working group. Draft `http://www.sdsc.edu/GridForum/RemoteData/Papers/`, Grid Forum, 2000.
5. I. Foster and C. Kesselman. The Globus project: A status report. 21100081 In *Proceedings of the Heterogeneous Computing Workshop*, pages 4–18. IEEE-P, 1998.

6. H. Afsarmanash and et al. Towards a multi-layer architecture for scientific virtual laboratories. In *Proceedings of High Performance Computing and Networking Conference*, pages 163–176, Amsterdam, The Netherlands, May 2000.
7. E.C. Kaletas and H. Afsarmanesh. Virtual laboratory experimentation environment data model. Technical Report CS-2001-01, University of Amsterdam, 2001.
8. Anne Frenkel, G. Eijkel, H. Afsarmanash, and L.O. Hertzberger. Information management for physics applications in the vl environment. Technical Report CS-2001-03, University of Amsterdam, 2000.
9. E.C. Kaletas and et al. Expressive - a database for gene expression experiments. Technical Report CS-2001-02, University of Amsterdam, 2000.
10. E.C. Kaletas, A. Belloum, D. Group, and Z. Hendrikse. Vl-am kernel db: Database model. Technical Report UvA/VL-AM/TN07, University of Amsterdam, 2000.
11. Gregor Von Laszewski, Ian Foster, Jarek Gawor, Warren Smith, and Steven Tuecke. Cog kits: A bridge between commodity distributed-computing and high-performance grids. In *Proceedings of the ACM Java Grande 2000*, San Francisco, June 2000.
12. Gregor Von Laszewski, et, and al. Cog high-level components. White paper, Argonne National Laboratory, 2000.
13. A.W. van Halderen. User guide for vlab developers. Technical Report UvA/VLab-AM/TN05, University of Amsterdam, 2000.
14. David Groep, Jo Van Den Brand, Henk Jan Bulten, Gert Eijkel, Massi Ferro-Luzzi, Ron Heeren, Victor Klos, Michael Kolsteini, Edmar Stoks, and Ronald Vis. Analysis of complex surfaces in the virtual laboratory. Technical Report -Draft Version-, national Institute for Nuclear and High Energy Physics (The Netherlands), 2000.
15. A. Abd-Allah and B. Boehm. Reasoning about the composition of heterogeneous architecture. Technical Report UCS-CSE-95-503, University of Southern California, 1995.
16. David Garlan, Robert Allen, and John Ockerbloom. Architectural mismatch: Why reuse is so hard. *IEEE software*, Vol. 12(No. 6), November 1995.
17. Rich Wolski, Neil T. Sprin, and Jim Hayes. The network weather service: A distributed resources preformance forecasting services for the metacomputing. Technical Report TR-CS98-599, UCSD, 1998.
18. GlobusProject. Gridftp: Universal data transfer for the grid. White paper `http://www.globus.org/datagrid/deliverables/default.asp`, Grid Forum Data Acces Working Group, 2000.
19. Kim Potter Kihlstrom, Louise E. Moser, and P. M. Melliar-Smith. The secure ring protocols for securing group communication. In *Proceedings of the IEEE 31st Hawaii International Conference on System Sciences*, pages 317–326, Kona Hawaii, January 1998.

Optimal Caching Policies for Web Objects

M.D. Hamilton[1], P. McKee[2], and I. Mitrani[1]

[1] University of Newcastle upon Tyne, NE1, 7RU
[2] BT Research Laboratories, Martlesham Heath, IP5 3RE

Abstract. Web objects are stored in a local cache in order to reduce the cost of future accesses. The cache size is limited, so it is necessary to have a policy for replacing old objects with new ones. This paper examines the problem of constructing optimal or near-optimal replacement policies. Different objects are accessed with different frequencies, and have different retrieval costs when not in the cache. In addition, an object which remains in the cache for a long period may become out of date as its original version is modified. The optimization problem which takes these factors into account is formulated as a Markov decision process. For small cache sizes and small numbers of objects, an optimal replacement policy can be computed by solving the corresponding dynamic programming equation. That solution is compared with a number of heuristics, some of which achieve a near-optimal performance.

1 Introduction

It is well known that the adoption of proxy caching can lead to significant performance improvements in accessing the web. Being able to fetch objects from a local cache tends to (a) speed up the average access operation, (b) lighten the load on the web server and (c) reduce the traffic among distant nodes. However, since the storage space used for caching is limited, it eventually becomes necessary to remove old objects when storing new ones. Under those circumstances, in order to reap the full benefits of caching, it is important to make the right choices about which objects to keep and which to discard, i.e., to have an efficient replacement policy.

The policy normally employed for web caching is LRU (replace the Least Recently Used object). There is quite an extensive literature examining the relative advantages and disadvantages of that, and other caching policies. Several versions of LRU, with or without considerations of object size, and algorithms based purely on size, have been evaluated empirically by means of trace-driven or stochastic simulations (e.g., see Abrams *et al.* [1] and Williams *et al.* [8]). Other proposed replacement algorithms have taken into account the download latency (Wooster and Abrams, [9]) and certain cost factors (Cao and Irani [4] and Lorenzetti *et al.* [6]). Those varying approaches have been compared to LRU and to each other, mostly using the cache hit ratio as a measure of performance.

We wish to consider the caching problem in the context of a system where different objects are accessed with different frequencies, have different retrieval

B. Hertzberger et al. (Eds.): HPCN Europe 2001, LNCS 2110, pp. 94–103, 2001.

costs, and may also become out of date while residing in the cache. Our purpose is to formulate the sequence of replacement actions as a Markov decision process, and to seek to minimize the total cost incurred over a finite or an infinite planning horizon. This allows us, at least in principle, to determine the optimal policy numerically. In practice, such a computation is feasible only when the total number of objects accessed, and the number that can be stored in the cache, are quite small. However, even if the optimal policy cannot be characterised easily, and therefore cannot be implemented in a practical system, there are good reasons for computing it numerically, whenever possible. As well as providing a lower bound on the performance that can be expected, knowledge of the optimal policy offers insights into the behaviour of various heuristics. Thus, we shall be able to make quantitative statements about propositions such as 'policy X is close to optimal', rather than 'policy X is better than policy Y'. To our knowledge, this has not been done before.

The model is described in section 2. The dynamic programming formulation is discussed in section 3, while section 4 presents the results obtained from a number of experiments.

2 The Model

Consider a system where K distinct objects, numbered $1, 2, \ldots, K$, each having a (remote) home location, may be requested via a local server. At most N of those objects ($N < K$) may be cached at any one time. When an object is requested, the contents of the cache are inspected to see whether it is present and, if so, a copy is returned to the user. If the object is not present, then it is fetched from its home location, is cached, and a copy is returned to the user. If, prior to that operation, the cache was full, then one of the stored objects is removed and is replaced by the new one. Thus, the cache goes through an initial period during which N different objects are requested and stored; after that, it always contains N objects. The system is illustrated in figure 1.

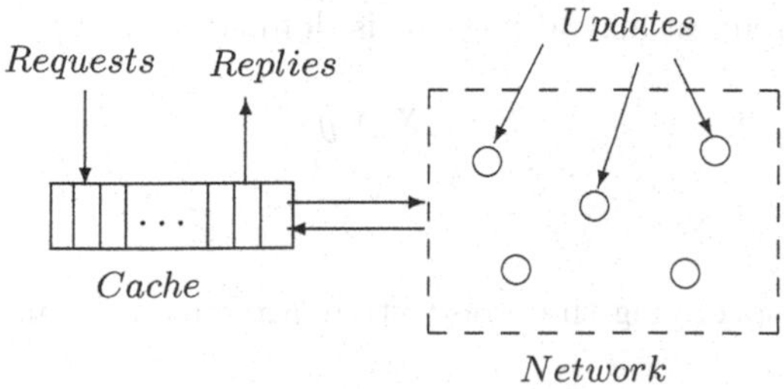

Fig. 1. Objects Requested, Fetched, Stored and Replaced in a Cache.

Requests arrive into the system according to a Poisson process with rate λ. The requested objects form a Markov chain with transition matrix $[q_{i,j}]_{i,j=1}^{K}$. That is, the probability that object j will be requested next, given that object i was requested last, is equal to $q_{i,j}$ $(i, j = 1, 2, \ldots, K)$.

Object i is modified from time to time at its home location. Those updates occur according to an independent Poisson process with rate μ_i, $i = 1, 2, \ldots, K$. Thus, if object i is present in the cache, it can be in one of two states, 'old' or 'new', depending on whether it has been modified at its home location since the time it was fetched, or not. A request for a new object is called a 'good hit'; a request for a old object is a 'bad hit'; a request for an object which is not in the cache, is a 'miss'. It is assumed that when a bad hit occurs, the user has some mechanism for detecting that the object returned is old, and requires it to be fetched again from the home location. Hence, a bad hit can be worse than a miss, because it involves an extra overhead. However, following any request – good hit, bad hit or miss – the requested object is new (until it is next updated, or removed from the cache).

The system is observed at consecutive request arrival instants. For the purposes of the model, the handling of a request (including any fetching and storing that may take place) is assumed to be instantaneous. However, the acual network delays, processing times and any other resources involved in handling a request are taken into account by means of appropriate costs. In particular,

A good hit for object i incurs a cost a_i, $i = 1, 2, \ldots, K$;
A bad hit for object i incurs a cost b_i, $i = 1, 2, \ldots, K$;
A miss for object i incurs a cost c_i, $i = 1, 2, \ldots, K$.

Typically, but not necessarily, these costs satisfy $a_i < c_i < b_i$.

Costs incurred in the future may carry less weight than current ones. If a discount rate, α, is assumed, then any cost incurred m steps into the future is discounted by a factor α^m $(m = 1, 2, \ldots; 0 \leq \alpha \leq 1)$. Setting $\alpha = 0$ implies that all future costs are disregarded; only the current operation is important. When $\alpha = 1$, a future cost, no matter how distant, carries the same weight as a current one.

The system can thus be modelled as a Markov Decision Process, whose state at (*just before*) a request arrival instant, is defined as a tripple

$$S = (X, \delta, y) , \tag{1}$$

where

X is a vector specifying the objects that are currently in the cache:

$$X = (x_1, x_2, \ldots, x_N) ;$$

x_1 is the most recently requested object, x_2 is the second most recently requested object, ..., x_N is the least recently requested object;

δ is a vector indicating whether the objects in the cache are new or old:

$$\delta = (\delta_1, \delta_2, \ldots, \delta_N) ;$$

δ_j takes value 0 if object x_j is new and 1 if x_j is old ($j = 1, 2, \ldots, N$);
y is the object being requested ($y = 1, 2, \ldots, K$).

The cost incurred in state S, $C(S)$, is given by

$$C(S) = \begin{cases} a_y \text{ if } \exists j \,:\, y = x_j \,,\; \delta_j = 0 \\ b_y \text{ if } \exists j \,:\, y = x_j \,,\; \delta_j = 1 \\ c_y \text{ if } y \notin X \end{cases} . \tag{2}$$

When the new request is a miss (i.e., $y \notin X$), one has to decide which object to remove from the cache in order to store y. If object x_k is removed, we say that 'action k' is taken ($k = 1, 2, \ldots, N$). No object is removed following a hit; this is denoted for convenience as 'action 0'. Of course, whatever action is taken, y becomes the most recently requested object in the cache and the other elements of X are renumbered accordingly.

The new vector which describes the content of the cache after action k is taken in state S, is denoted by X'_k ($k = 0, 1, \ldots, N$).

Now consider the system state, S', at the next request instant, given that at the last one it was S. If the last request was for object y, the next one will be for object z with probability $q_{y,z}$. If action k was taken last time, then the content of the cache immediately after the last request, and hence the one seen by the next arrival, will be X'_k. Any object that was old, and was left in the cache, will remain old. However, some of the objects that were new immediately after the last request instant (including the requested object, y), may become old by the time of the next arrival. In particular, if object i was new in the cache immediately after the last request, it will become old with probability $\mu_i/(\lambda + \mu_i)$, and will remain new with probability $\lambda/(\lambda + \mu_i)$, independently of all other objects.

Thus we can obtain the probability that the next indicator vector will be δ', given that the last state was S and action k was taken. Denote that probability by $p_k(S, \delta')$.

The above discussion can be summarised by saying that if action k is taken in state $S = (X, \delta, y)$, then the next state will be $S' = (X'_k, \delta', z)$ with probability $q_{y,z} p_k(S, \delta')$.

3 Dynamic Programming Equations

The objective of an optimal caching policy is to minimize the total cost incurred over some planning horizon, which may be finite or infinite. Consider the finite horizon optimization first. Let $V_n(S)$ be the *minimal* average total cost incurred in satisfying n consecutive requests (including the current one), given that the current state is S. That cost function satisfies the following dynamic programming equation ($n = 1, 2, \ldots$):

$$V_n(X, \delta, y) = C(X, \delta, y) + \alpha \min_k \left[\sum_{z=1}^{K} q_{y,z} \sum_{\delta'} p_k(S, \delta') V_{n-1}(X'_k, \delta', z) \right] , \quad (3)$$

where $C(X, \delta, y)$ is given by (2) and δ' spans all indicator vectors that can occur when new objects in X'_k become old. The minimum in the right-hand side of (3) is taken over the actions, k, that are possible in state S; note that when $y \in X$, only action 0 is possible.

The initial conditions for (3) are $V_0(S) = 0$, for any S. Applying the recurrences for $n = 1, 2, \ldots$, one can determine the minimal cost for any given state and planning horizon. The actions which achieve that minimal cost constitute the optimal replacement policy. Since the number of steps is finite, the discount factor can be equal to 1. However, the computational complexity of the solution increases very quickly with the number of objects, K, and the size of the cache, N.

In the case of an infinite horizon optimization, the objective is to compute the minimal average total cost, $V(S)$, incurred in satisfying the present, and all future requests, given that the current state is S. For that cost to be finite, we must have $\alpha < 1$. The replacement policy is assumed to be stationary, i.e. its actions depend only on the current state.

The dynamic programming equation for $V(S)$ is the stationary version of (3):

$$V(X, \delta, y) = C(X, \delta, y) + \alpha \min_k \left[\sum_{z=1}^{K} q_{y,z} \sum_{\delta'} p_k(S, \delta') V(X'_k, \delta', z) \right] . \quad (4)$$

The solution of (4), and hence the optimal policy, can be found by the so called 'policy improvement algorithm' (e.g., see [5]). Its computational complexity is governed mainly by the need to solve a set of simultaneous equations (which can be very expensive when the state space is large), and by the number of policy improvements that have to be made on the way to the optimum. The run time of the algorithm can often be reduced by choosing a good policy as the initial guess. Also, instead of solving the simultaneous equations exactly, say by Gaussian elimination, one could find an approximate solution by applying a few steps of an iterative schema. This last approach is particularly efficient when the discount factor α is small.

Another approach to computing the optimal stationary policy is to solve the finite horizon recurrence equations (3), for increasing values of n, until the total cost converges and the policy stops changing. The number of steps needed for convergence increases with the discount factor but, provided the latter is not too close close to 1, is still reasonably small.

The finite horizon method turned out to be considerably more efficient, for our system, than the policy improvement one. It was therefore employed in all experiments.

4 Numerical Results

In order to get some idea about the distribution of requested objects, a data file containing real cache statistics [10] was examined. Out of more than 9000 objects logged, the vast majority were requested fewer than 10 times each; very few objects were requested more than 50 times each. A frequency histogram of those requests is displayed in figure 2. To reduce the number of objects, the latter are grouped in 23 batches of 400 objects in each. The batches are listed in order of popularity. The figure shows that more than 50% of all requests were for batch 1; most batches received small fractions of the requests.

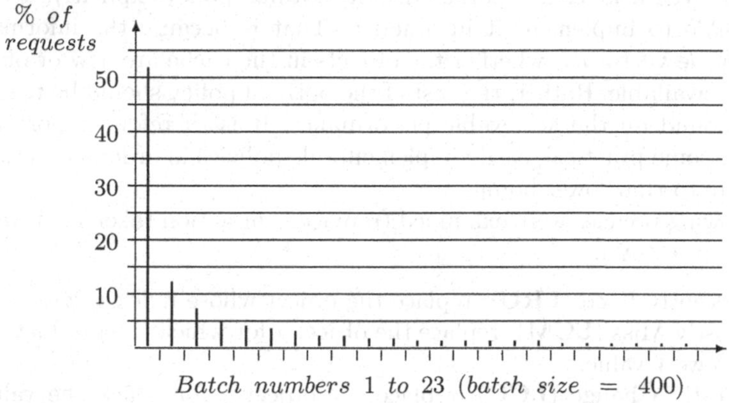

Fig. 2. Frequency Histogram of Requested Objects.

In view of those observations, and to keep the state space to a tractable size, we decided to model a system with $K = 6$ objects (they could in fact represent batches of objects), of which one is roughly six times more likely to be requested than the others. In our case, that is object 6. In other words, we assume that

$$q_{i,1} = q_{i,2} = q_{i,3} = q_{i,4} = q_{i,5} = 0.091 \ ; \ q_{i,6} = 0.545 \,,$$

for all i.

The cache has room for $N = 3$ objects. Hence, the Markov Decision Process S has 5760 possible states (120 states for the vector X, 8 states for the vector δ, 6 values for y). The arrival rate of requests was fixed at $\lambda = 0.5$. The rates at which different objects become old were also fixed, with the most popular object being most frequently updated:

$$\mu_1 = 0.01 \ ; \ \mu_2 = 0.015 \ ; \ \mu_3 = 0.02 \ ; \ \mu_4 = 0.025 \ ; \ \mu_5 = 0.03 \ ; \ \mu_6 = 0.05 \,.$$

Thus, the experiments that were carried out were distinguished by their different cost structures, i.e. by the values of a_i, b_i and c_i. For any given replacement policy, f, the performance measure of interest is the steady-state average cost, V_f, that it achieves over the infinite horizon. This is defined as

$$V_f = \sum_S V_f(S)\pi_f(S) , \tag{5}$$

where $\pi_f(S)$ is the stationary probability of state S when policy f is in operation.

The solution of the dynamic programming equations yields the optimal policy, as a mapping from states, S, to actions, k. It also provides the expected cost of the policy in each state. The mapping and costs are then fed into a computer simulation program which produces an estimate for the right-hand side of (5).

Note that even if we could characterise the optimal policy explicitly, it would not be possible to implement it in practice. That is because the information contained in the vector δ – whether the objects in the cache are new or old – is not normally available. Rather, the cost of the optimal policy should be regarded as a lower bound on the achievable performance. It is therefore important to evaluate also some practical, easily implementable policies in order to check how close they are to that lower bound.

The following policies were examined (replacement action taken in state $S = (X, \delta, y)$, when $y \notin X$):

- Least Recently Used (**LRU**): replace the object whose index is x_N.
- Least Costly Miss (**LCM**): replace the object j for which the product $q_{y,j}c_j$ has the lowest value.
- Least Costly Change (**LCC**): replace the object j for which the value of $q_{y,j}c_j/(\mu_j b_j)$ is lowest.
- Least Expected Cost (**LEC**): replace the object j for which the expression

$$LEC_j = q_{y,j}c_j + \sum_{i \in X, i \neq j} q_{y,i} \left[a_i \frac{\lambda}{\lambda + \mu_i} + b_i \frac{\mu_i}{\lambda + \mu_i} \right] ,$$

  has the lowest value.

For the LCM, LCC and LEC policies, if more than one objects have the same lowest value, the replacement is choosen among them at random.

In the first experiment, the costs of a good hit, bad hit and miss were the same for all objects: $a_i = 5$, $b_i = 300$, $c_i = 100$. The average costs of the optimal and heuristic policies, plotted against the discount factor, α, are displayed in figure 3.

In this example, there is no difference between the LCC and LEC policies, and almost none between them and LRU. Those three policies are within 20% of the optimum, while LCM is significantly worse.

For the second experiment, all parameters are kept the same, except that the cost of a bad hit is increased from 300 to 2000, for all objects. The results are illustrated in figure 4. We observe that all heuristic policies are now clearly differentiated.

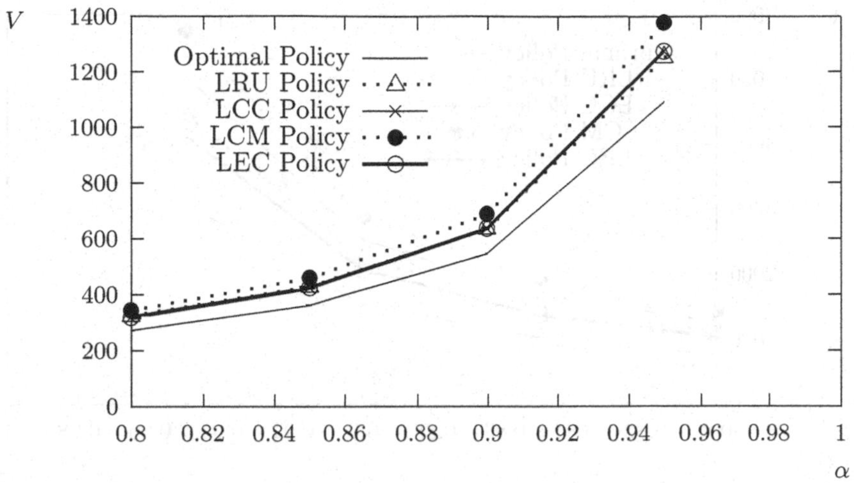

Fig. 3. Average Costs of Optimal and Heuristic Policies: Experiment 1.

The best heuristic policy is LEC. The performance it achieves is within 30% of the optimum, despite the fact that it does not know which objects in the cache are new and which are old. LRU is next, while LCC and LCM are considerably poorer.

In experiment 3, all objects have different costs of bad hits:

$$b_1 = 500 \ ; \ b_2 = 400 \ ; \ b_3 = 300 \ ; \ b_4 = 200 \ ; \ b_5 = 100 \ ; \ b_6 = 2000 \ .$$

The other parameters are the same. Note that the object most likely to be requested (object 6), is also the most frequently updated, and the most expensive if found old in the cache.

The results for experiment 3 are displayed in figure 5.

We observe that the performances of LEC and LCC are nearly identical, and within about 5% of the optimum. LRU is more than 30% costlier than the optimal policy, and LCM is even worse. These results illustrate an intuitive but nevertheless important point, namely that when the costs of bad hits are high, any policy which does not take into account the rates at which remote objects are updated is bound to be inefficient.

5 Conclusion

We believe that formulating the caching problem as a Markov decision process, and being able to compute the optimal replacement policy, provides valuable insights into the performance of practically implementable heuristics. The fact that only systems with small numbers of objects are numerically tractable, does

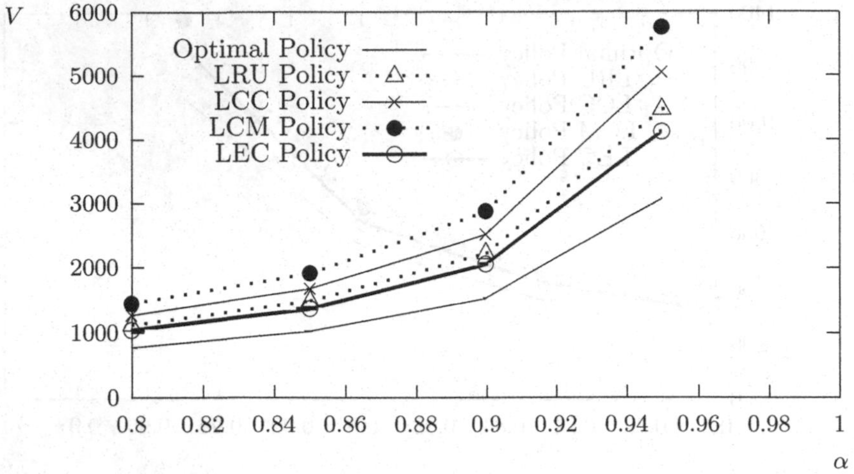

Fig. 4. Average Costs of Optimal and Heuristic Policies: Experiment 2.

not invalidate those insights. One should bear in mind that the term 'object' may in fact denote a block of objects.

The following findings are confirmed by all experiments.

- The LEC policy appears to make good use of the available parameters. It performs consistently well, both when the objects are similar, and when they are different.
- The LRU policy performs quite well when the objects are similar, not so well when they are different. The reverse behaviour is exhibited by LCC.
- The LCM policy, which is the simplest to implement (it ignores the update rates, μ_i, and the costs of bad hits), performs too poorly to be recommended.

Acknowledgement

This work was carried out as part of the project "Computer Systems Operating Over Wide Area Networks", funded by British Telecom Research Laboratories.

References

1. M. Abrams, C.R. Stanbridge, G. Abdulla, S. Williams and E.A. Fox, "Caching Proxies: Limitations and Potentials", Procs. WWW-4 Conference, Boston, 1995.
2. D. Blackwell, "Discounted dynamic programming", *Ann. Math. Stat.*, **26**, 226-235, 1965.
3. C. Derman, *Finite State Markovian Decision Processes*, Acedemic Press, New York, 1970.

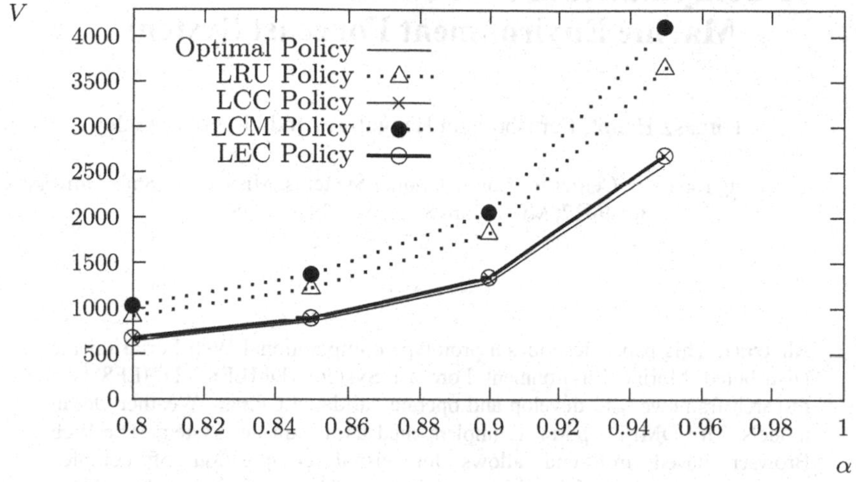

Fig. 5. Average Costs of Optimal and Heuristic Policies: Experiment 3.

4. P.Cao and S. Irani, "Cost-Aware WWW Proxy Caching Algorithms", Procs. Usenix Symposium on Internet Technology and Systems, Monterey, 1997.
5. S.E. Dreyfus and A.M. Law, *The Art and Theory of Dynamic Programming*, Acedemic Press, New York, 1977.
6. P. Lorenzetti, L. Rizzo and L. Vicisano, "Replacement Policies for a Proxy Cache", URL: `http://www.iet.unipi/luigi/research.html`, 1997.
7. S.M. Ross, *Introduction to Stochastic Dynamic Programming*, Acedemic Press, New York, 1983.
8. S. Williams, M. Abrams, C.R. Stanbridge, G. Abdulla and E.A. Fox, "Removal Policies for Network Caches for WWW Documents", Procs. ACM SIGCOM96, Stanford, 1996. Conference, Boston, 1995.
9. R. Wooster and M. Abrams, "Proxy Caching and Page Load Delays", Procs. WWW-6 Conference, Santa Clara, 1998.
10. National Laboratory for Applied Network Research (NLANR). URL: `http://www.ircache.net/Cache/Statistics/` (NSF grants NCR-9616602 and NCR-9521745).

A Computational Web Portal for the Distributed Marine Environment Forecast System

Tomasz Haupt, Purushotham Bangalore, and Gregory Henley

Engineering Research Center for Computational Systems, Mississippi State University
Box 9627, Mississippi State, MS 39762, USA
haupt@erc.msstate.edu

Abstract. This paper describes a prototype computational Web Portal for the Distributed Marine Environment Forecast System (DMEFS). DMEFS is a research framework to develop and operate validated Climate-Weather-Ocean models. The DMEFS portal is implemented as a multitier system. The Web Browser based front-end allows for visual composition of complex computational tasks and transition of these tasks for operational use, as well as for analysis of the results. The middle-tier is split into Java Server Pages based Web tier responsible for request processing and dynamic generation of responses, and Enterprise Java Beans based application server. The EJB container is populated by a hierarchy of entity beans representing the state of the system or acting as proxies of services rendered by the back end. Operations on the entity beans are implemented as session beans. This design makes it possible to separate user requests (in terms of application independent task specification) from the back end resource allocation (through platform independent resource specification). The middle-tier provides a transparent mapping between task and resource specification, hiding complexity of the heterogeneous back-end system from the user. The DMEFS portal provides a seamless and secure access to the computational resources as well as provides a simpler and easy to use interface to setup, stage, and deploy complex scientific and engineering simulations.

Introduction

The IT revolution currently unfolding is driven by three key factors: rapidly increasing power of ever more affordable computers (PCs and numerous other connected devices), rapidly increasing bandwidth of communication channels (broadband era), and intuitive, simple-to-use interfaces for non-technical individuals (web browsers). Computational power constantly is opening new opportunities for numerical simulations, in turn opening opportunities for new science. This computational power is expected to be a low cost alternative for design and validation, boosting efficiency of manufacturing, and be a reliable source of forecasts as well as all the other claimed and realized advantages for academic computing. This constant demand for faster and faster compute servers drives vendors to introduce more and more sophisticated, scalable architectures including scalable interconnects. Generally, the user is delivered what he or she expects in terms of

B. Hertzberger et al. (Eds.) : HPCN Europe 2001, LNCS 2110, pp. 104-113, 2001.

significant fraction of peak performance, but only with the loss of portability from the previous generation of systems. It is difficult and costly to port a legacy application to a new scalable platform, because a large portion of the cost comes from user re-training and re-validation of the application, since even the end user (consumer rather than code developer) is exposed to all nitty-gritty details of the system on which applications are run. In such circumstances, coupling several codes into a single complex application is in most cases prohibitively hard.

The phenomenon of the Internet with constantly increasing quality of connection opens another opportunity for numerical simulations: access to geographically remote data during runtime. The data may come from diverse sources: from remote file systems, remote databases, remote mass storage systems, web repositories, online instruments and data acquisition systems (microscopes, telescopes, medical instruments, remote sensors) and others. However, as in the case of accessing new computers, exploiting new data sources requires specific expertise in order to identify and access relevant data sets, above and beyond the issues of data formats and technologies needed to process any associated metadata. Again, the main problem is that the end user is exposed to all technical details with which he or she may or may not be familiar.

By way of contrast, industry has undertaken enormous efforts to develop easy user interfaces that hide complexity of underlying systems. Through Web Portals the user has access to variety of services such as weather forecasts, stock market quotes and on-line trading, calendars, e-mail, auctions, air travel reservations and ticket purchasing, and many others yet to be imagined. It is the simplicity of the interface that hides all implementation details from the user that has contributed to the unprecedented success of the idea of a Web Browser, including its rapid rate of adoption.

In this work we extend the notion of the Web Portal to provide web access to *remote computational resources* (that is, hardware, software and data), thereby simplifying currently difficult to comprehend and changing interfaces and emerging protocols. More specifically, we design a system that hides complexity of the inherently heterogeneous and distributed computing environment, and through the familiar Web Browser interface, allow the user to specify the computational problem in terms with which he or she is comfortable. Numerical/scientific/engineering computing will be moved from the realm of unrewarding complexity management by transparently managing staging of data, allocation of resources, monitoring application progress, and simplifying the analysis of results. This effort is fully complementary to meta-computing (whether Globus[1], Legion[2], Egrid[3], or the Grid Forum[4]), that supports the globalization of resource access, without necessarily addressing complexity of use.

Distributed Marine Environment Forecast System

Distributed Marine Environment Forecast System [5] (DMEFS) is an open framework to simulate the littoral environments across many temporal and spatial scales that will accelerate the evolution of timely and accurate forecasting. This is to be achieved by adaptation of distributed scalable computational technology into oceanic and meteorological predictions (Climate-Weather-Ocean models), and incorporating the latest advances in solution schemes, grid generation, and scientific visualizations. DMEFS is expected to provide a means for substantially reducing the time to develop, prototype, test, and validate simulation models as well as support a genuine, synergistic collaboration among the scientists, the software engineers, and the operational users. To demonstrate the forecast system, a proof-of-concept system for the Gulf of Mexico is being developed that couples atmosphere and ocean model.

The goal of the DMEFS project, however, is not building the models, doing the modeling research, or producing operational forecasts, but rather building the underlying infrastructure to facilitate it in a distributed High Performance Computing (HPC) environment. The resulting system must provide an environment for model development and validation as well as decision support. The model developers are expected to be computer savvy domain specialists. On the other hand, operational users who routinely run the simulations to produce daily forecasts have only a limited knowledge on how the simulations actually work, while the decision support is typically interested only in accessing the end results. As the domain expertise level varies from one user category to another, so does their equipment: from the high-end development environment to a portable personal computer.

Support for Complex Multistep Computational Tasks

We can satisfy these requirements by implementing the DMEFS system as a computational Web Portal. By using a standard Web browser to deploy the front end of the system, we place the user in an environment he or she is already familiar with and the software (the browser) is already available to the user. Through the portal we provide access to remote resources, hiding complexity of the underlying systems.

There are several projects that address web access to remote resources. A representative sample of these are WebSubmit[6], HotPage[7], and Teraweb[8]. They provide simple (easy to use) interfaces that allow users to select an application, select a target machine, submit the job, and monitor the job's progress. One can shortly describe such a system as a web-accessible Unix prompt augmented with simple tools such as batch script generator, file transfer, and tools to monitor status of machines. As straightforward as these are, they simplify access to high performance resources significantly, and such tools are evidently well received by users.

As demanded by DMEFS, our approach goes far beyond the simplistic functionality just mentioned. Most of the Climate-Weather-Ocean simulations comprise many steps in a pipeline (and this is also true for many other application areas). These steps, or "atomic tasks" include importing, preprocessing and validating external data such as CAD geometry or bathymetry data, and subsequent grid generation. In many cases, several jobs in sequence (with data-flow type dependencies) or concurrently must be executed to obtain final results. For example, one code may generate boundary conditions for another (typical in littoral waters simulations). Such simulations require both system and application specific knowledge to configure complex applications (understanding the sequence of steps, data formats, and tools for conversions, running applications on different platforms, modifying shell, batch or perl scripts, etc.), and it is labor intensive, requiring highly skilled user labor. Currently the user is responsible for managing the execution of each step and the flow of information through the simulation pipeline. Hundreds of files may be created as part of the process with the user being responsible for determining/remembering which files are the proper output files to be used as input to the next tool, running the tool to translate the files to be suitable for input to the next stage. Therefore, it is crucial to introduce a framework where the user can compose a complex application from atomic tasks, and save them for later reuse or share them with other users, thereby providing validity in the same sense, as does a well-designed, repeatable scientific experiment.

Sharing of the saved configurations leads to yet another important concept: separation of the process of application development, deployment, and use. Once a complex application is configured, it can be easily transitioned for "operational" use such as routine runs of a forecast system. This also provides a domain-specific working environment for researchers and engineers not directly involved in the code design or development.

Platform Independence

It is important that the computational task defined by the user is expressed in a platform independent way. The separation of the task specification from the resource allocation simplifies the user interface by hiding the details of the heterogeneous back end and automating the process of task definition. The user simply selects the target machine(s), and the middleware seamlessly stage files and codes as needed, selects or generates batch scripts, selects right version of tools such as data format converters, and so forth. In many cases, the user does not have to know, or care, where the processing actually takes place, leaving the selection of resources entirely to the portal's middle-tier discretion.

The platform independence is achieved by introducing the notion of proxies representing actual codes and data. For example, an application proxy possesses information on how to submit the corresponding code on all machines where the application is available. This information comprises the application descriptor. The format of the application descriptor resembles a Unix manual page, and is encoded in XML. It includes information on location of the executable, describes its arguments and switches, expected input files and lists output files generated, and optionally the location of a template batch script. A complex, multistep task is built of a number of atomic tasks, each represented by a proxy, with a task descriptor that defines relationships between the atomic tasks. Before submitting, the task must be resolved, that is, each atomic task must be assigned the target machine, either explicitly by the user, or implicitly by a resource broker. In addition, the location of the input files, final destination of the output files, as well as values of arguments and switches must be provided, either through the user interface or using defaults (the latter in an operational environment). The resolved task is then submitted: for each atomic task the input files are staged, the executable is submitted, and as soon as the execution completes, the output files are transferred to specified destination, if any.

The application descriptors are entered to the system through a registration process (e.g., an HTML form). This way we can easily integrate both legacy and new user applications into the portal. The task descriptor is generated as the response to the user interaction with the portal. Figure 1 shows an example screen dump of the user session. This particular screen allows the model developer to register a new application (New button), or select pre-registered application either from the shared pool of applications, or from the user private area (All button), resolve the selected application (Select button), view (View button) and modify (Edit button) already configured applications, add (Add button) them to the task descriptor and establish relationship between them (Link button). The Submit button allows submitting the task. Its progress can be monitored using the Status button.

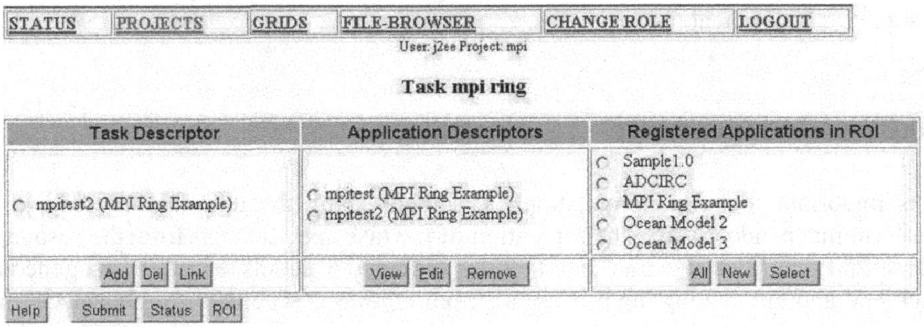

Fig. 1. A screen dump of a user session. The right column of the table lists all registered applications that can be shared between users. The resolved applications appear in the middle column. In this example, the same application (the MPI Ring Example) is configured twice, for two different target machines. One of them is added to the task descriptor.

Persistence

The task along with all its descriptors is persistent. It can be accessed again at a later time, resubmitted, modified, cloned and made available for other users, and in particular, it can be transitioned into operations. We achieve this by implementing the middle tier as a shared collection of persistent components (representing applications, data, and tools) and services using Enterprise JavaBeans (EJB) technology.

The EJB model introduces two types of components (or Beans in Java jargon): persistent entity beans and session beans (stateless or maintaining a conversational state). We use the entity beans to represent components such as users, projects, tasks, and application proxies. Our entity beans are organized into arbitrary complex hierarchies (referred to as contexts). This way we can build a multi-user system, where each user can customize his or her environment, the user can organize his or her work into independent projects (Figure 2), and most importantly, the user can compose a complex task by adding multiple applications (atomic tasks) into a task context (Figure 1).

The session beans operate on the entity beans. For example, the Context Manager bean creates and destroys contexts, as well as providing access to data stored within the selected context. The Submit Task bean, as its name suggests, submits the task for execution. As described above, each task must be resolved by selecting input files, selecting target machine, generating a batch script, and so forth. In the simplest case, the user provides necessary information by filling an html form generated by the Portal. However, much of this functionality can be automated. Currently we have a set of additional session beans that provide the user additional assistance with batch script generation - hiding platform and scheduler specific information. The user can access a remote file system, database, and other data sources to select input data, and the system can automatically stage them on the target machine prior to execution. In addition, the user can access all the data he or she used before, which can also be copied or moved between contexts and modified using a text editor. Finally, the user can use a tool to generate a new data set, for example a grid generator. If necessary, the user can use existing preprocessors for format conversion. The session beans that provide access to back end resources, the high performance computational systems, remote file systems, databases and instruments implements the grid interface of the Grid Forum, as described in our earlier papers[9,10], and we collectively call them the Enterprise Computational Services.

Middle Tier Application Server

The overall design of our middle-tier server is shown in figure 3. The EJB container is populated by a hierarchy of entity beans representing the state of the system or acting as proxies of services rendered by the back end. Operations on the entity beans are implemented as session beans. This design makes it possible to separate user requests (in terms of application independent task specification) from the back end resource allocation (through platform independent resource specification). The middle-tier provides a transparent mapping between task and resource specification, hiding complexity of the heterogeneous back-end system from the user. Persistency of the middle-tier object allows for reuse once configured complex tasks and thus transition them into operational use.

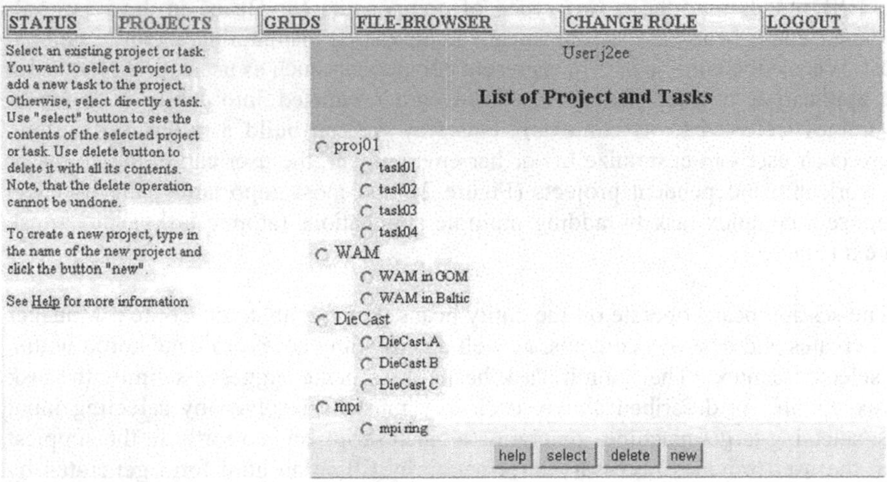

Fig. 2. A screen dump of a user session. The user organizes his or her work into projects. Each project comprise one or more tasks composed of applications (as shown in Fig. 1).

As shown in Figure 3, model developers bring new application and tools into the system by the registration process. Once registered and resolved applications (application contexts) are then used to build complex multistep task. The configured task can be transitioned for operational use just by cloning the task context. The results of routinely run tasks are available by querying the database.

Web Access

The middle-tier server can be accessed directly by EJB clients through Java RMI or CORBA. However, for reasons discussed in the beginning of this paper, we have chosen to provide the Web access through the middle-tier. This leads us to introducing additional tier placed between the front-end and the application (EJB) tier: the Web tier. It is implemented using a Web server supporting Java Server Pages (JSP). JSP processes the user requests (html forms) and through Java Beans that implement our EJB object interfaces they control contexts and invoke methods of the session beans. This way we are able to separate the application dependent part of the project (the DMEFS front end, generated dynamically in the JSP tier) from application independent Enterprise Computational Services in the EJB tier.

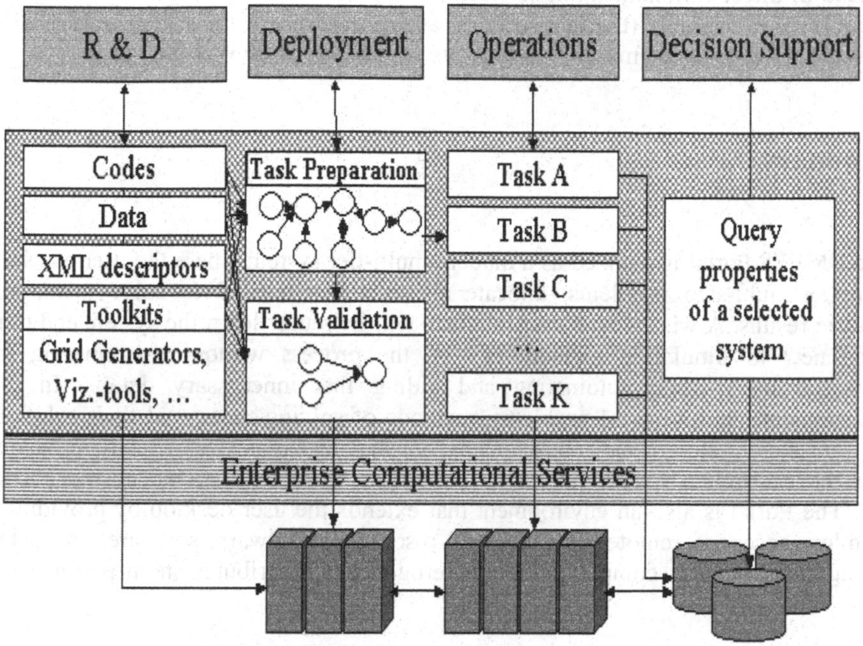

Fig. 3. DMEFS functionality: the model developers register applications, data and the tools the systems. These components are used to define complex tasks. Once the tasks are configured they are transitioned into operations for routine runs. The results of these runs become a part of the decision support system. The R&D, Deployment, Operations and Decision support front ends provide different views of the same middle-tier system that maps the user requests onto back end resource allocations through the Enterprise Computational Services.

Security

The challenge of building a Computational Web Portal is to define the architecture of such a multitier system that allows integrating existing commodity components into a single, user-friendly, web accessible system without compromising security and integrity of the computational resources. The security of the remote access has been aggressively addressed by the Grid Forum, taking the PKI-based Globus Security Services[11] as the starting point. The HotPage project[7] takes advantage of this approach in conjunction with SSL-based communication between the front end (Web Browser) and the middle-tier server.

The DMEFS system is to be delivered in an environment secured by the Kerberos Network Authentication Protocol[12]. Since we are not aware of any kerberized version of the HTTP protocol, we tunnel HTTP requests through kerberized CORBA channel (SECIOP), using a commercial implementation of secure Object Request Broker (ORB), ORBAsec SL2[13] from Adiron Secure System Design. That is, instead of direct communication between the Web Browser and the Web Server, the client invokes remotely (through secure channel) methods of the kerberized CORBA server that translates them into http requests processed by the Web Server.

Summary

The DMEFS Portal is realized as a modern multi-tier system where the user can state complex multi-step problems, allocate all resources needed to solve them, and analyze results. A wizard type user interface will help user learn the system and train users new to simulation methodology on the process without encumbering the advanced user, while automating and hiding the unnecessary details. In this environment, definitions of problems, methods of solving them, and their solutions are persistently stored and, consequently, viewed and reused at later time, shared between researchers and engineers, and transitioned for operational or educational use. The Portal is also an environment that extends the user desktop by providing a seamless access to remote computational resources (hardware, software, and data), hiding from the user complexity of a heterogeneous, distributed, high performance back end.

The middle-tier is split into Java Server Pages based Web tier responsible for request processing and dynamic generation of responses, and Enterprise Java Beans based application server. The EJB container is populated by a hierarchy of entity beans representing the state of the system or acting as proxies of services rendered by the back end. Operations on the entity beans are implemented as session beans. This design makes it possible to separate user requests (in terms of application independent task specification) from the back end resource allocation (through platform independent resource specification). The middle-tier provides a transparent mapping between task and resource specification, hiding complexity of the heterogeneous back-end system from the user. Persistency of the middle-tier object allows for reuse once configured complex tasks and thus transition them into operational use.

The project is in its initial phase. Having built the framework, we plan to extend its functionality by adding support for remote visualizations, grid generation and other tools. In particular, we are interested in a better integration of our system with the desktop resources. This can be achieved by extending of use of CORBA in the front-end, beyond current use for security purposes. Also, we plan to better integrate our system with the emerging Grid Interface of the Global Grid Forum, taking advantages of its information services. In addition, we are interested in extending our framework to support event driven simulations.

References

1. Globus Metacomputing Toolkit, home page: http://www.globus.org
2. Legion Worldwide Virtual Computer, home page: http://legion.viginia.edu
3. European Grid Forum, home page: http://www.egrid.org
4. Global Grid Forum, home page: http://www.gridforum.org
5. Distributed Marine Environment Forecast System, home page: http://www.erc.msstate.edu/~haupt/DMEFS
6. WebSubmit: AWeb-based Interface to High-Performance Computing Resources, home page: http://www.itl.nist.gov/div895/sasg/websubmit/websubmit.html
7. HotPage, home page: https://hotpage.npaci.edu
8. Teraweb, home page: http://www.arc.umn.edu/structure/
9. Tomasz Haupt, Erol Akarsu, Geoffrey Fox, "WebFlow: AFramework for Web Based Metacomputing", Future Generation Computer Systems, 16 (2000) 445-451, also in the proceedings of High-Performance Computing and Networking, Amsterdam'99
10. Tomasz Haupt, Erol Akarsu, Geoffrey Fox and Choon-Han Youn, "The Gateway system: uniform web based access to remote resources", Concurrency: Practice and Experience, 12 (2000) 629; also in proceedings of ACM 1999 Java Grande, San Francisco '99.
11. A Security Architecture for Computational Grids. I. Foster, C. Kesselman, G. Tsudik, S. Tuecke, Proc. 5th ACM Conference on Computer and Communications Security Conference, pg. 83-92, 1998.
12. Kerberos Network Authentication Protocol, home page: http://web.mit.edu/kerberos/www/
13. Adiron Secure System Design, home page: http://www.adiron.com/orbasecsl2.html

Role of Aging, Frequency, and Size in Web Cache Replacement Policies

Ludmila Cherkasova[1] and Gianfranco Ciardo[2]

[1] Hewlett-Packard Labs, 1501 Page Mill Road
Palo Alto, CA 94303, USA
cherkasova@hpl.hp.com

[2] CS Dept., College of William and Mary
Williamsburg, VA 23187-8795, USA
ciardo@cs.wm.edu

Abstract. Document caching on is used to improve Web performance. An efficient caching policy keeps popular documents in the cache and replaces rarely used ones. The latest web cache replacement policies incorporate the document size, frequency, and age in the decision process. The recently-proposed and very popular Greedy-Dual-Size (GDS) policy is based on document size and has an elegant aging mechanism. Similarly, the Greedy-Dual-Frequency (GDF) policy takes into account file frequency and exploits the aging mechanism to deal with cache pollution. The efficiency of a cache replacement policy can be evaluated along two popular metrics: file hit ratio and byte hit ratio. Using four different web server logs, we show that GDS-like replacement policies emphasizing size yield the best file hit ratio but typically show poor byte hit ratio, while GDF-like replacement policies emphasizing frequency have better byte hit ratio but result in worse file hit ratio. In this paper, we propose a generalization of Greedy-Dual-Frequency-Size policy which allows to balance the emphasis on size vs. frequency. We perform a sensitivity study to derive the impact of size and frequency on file and byte hit ratio, identifying parameters that aim at optimizing both metrics.

1 Introduction

Many replacement policies for web caches have been proposed [1,3,4,5,6,7,8,9]. Some of them are quite simple and easy to implement, while others are heavily parametrized or have aspects that do not allow for an efficient implementation (thus they can exhibit good results that serve as theoretical bounds on the best practically achievable performance). Two essential features distinguish web caching from conventional caching in the computer systems: (i) the HTTP protocol supports whole file transfers, thus a web cache can satisfy a request only if the entire file is cached, and (ii) documents stored in a web cache are of different sizes, while CPU and disk caches deal with uniform-size pages.

One key to good web cache performance is an efficient cache replacement policy to determine which files should be removed from cache to store newly requested documents. Further improvements can be achieved when such a policy is combined with a decision about whether a document is worth caching at all.

B. Hertzberger et al. (Eds.): HPCN Europe 2001, LNCS 2110, pp. 114–123, 2001.

A very good survey of currently-known replacement policies for web documents can be found in [4], which surveys ten different policies, comments on their efficiency and implementation details, and proposes a new algorithm, Greedy-Dual-Size (GDS), as a solution for the web proxy replacement strategy. The GDS policy incorporates document size, cost, and an elegant aging mechanism in the decision process, and shows superior performance compared to previous caching policies. In [1], the GDS policy was extended, taking into consideration the document frequency, resulting in the Greedy-Dual-Frequency (GDF) policy, which considers a document's frequency plus the aging mechanism, and the Greedy-Dual-Frequency-Size (GDFS), which also considers a document's size.

The typical measure of web cache efficiency is the *(file) hit ratio*: the fraction of times (over all accesses) the file was found in the cache. Since files are of different size, a complementary metric is also important, the *byte hit ratio*: the fraction of "bytes" returned from the cache among all the bytes accessed. The file hit ratio strongly affects the response time of a "typical" file, since request corresponding to a file miss require substantially more time to be satisfied. The byte miss also affects the response time as well, in particular that of "large" files, since the time to satisfy such a request has a component that is essentially linear in the size of the file; furthermore, a large byte miss ratio also indicates the need for a larger bandwidth between cache and permanent file repository.

The interesting outcome of paper [1] was that, while GDFS achieves the best file hit ratio, it yields a modest byte hit ratio. Conversely, GDF results in the best byte hit ratio at the price of a worse file hit ratio. The natural question to ask is then how the emphasis on document size or frequency (and the related aging mechanism) impact the performance of the replacement policy: do "intermediate" replacement policies exist that take into account size, frequency, and aging, and optimize both metrics? In this paper, we partially answer this question in a positive way by proposing a generalization of GDFS, which allows to emphasize (or de-emphasize) the size, frequency, or both parameters. Through a sensitivity study, we derive the impact of size and frequency on the file and byte hit ratio.

We intentionally leave unspecified the possible location of the cache: at the client, at the server, or at the network. The workload traces for our simulations come from four different popular web sites. We try to exploit the specific web workload features to derive general observations about the role of document size, frequency and related aging mechanism in web cache replacement policies. We use trace-driven simulation to evaluate these effects.

The results from our simulation study show that de-emphasizing the impact of size in GDF leads to a family of replacement policies with excellent performance in terms of both file and and byte hit ratio, while emphasizing document frequency has a similar (but weaker, and more workload-sensitive) impact.

2 Related Work and Background

The original Greedy-Dual algorithm introduced by Young [10] deals with the case when pages in a cache (memory) have the same size but have different costs to fetch them from secondary storage. The algorithm associates a "value" H_p

with each cached page p. When p is first brought into the cache, H_p is defined as the non-negative cost to fetch it. When a replacement needs to be made, the page p^* with the lowest value $H^* = \min_p\{H_p\}$ is removed from the cache, and any other page p in the cache reduces its value H_p by H^*. If a page p is accessed again, its current value H_p is restored to the original cost of fetching it. Thus, the value of a recently-accessed page retains a larger fraction of its original cost compared to pages that have not been accessed for a long time; also, the pages with the lowest values, hence most likely to be replaced, are either those "least expensive" ones to bring into the cache or those that have not been accessed for a long time. This algorithm can be efficiently implemented using a priority queue and keeping the offset value for future settings of H via a *Clock* parameter (aging mechanism), as Section 3 describes in detail.

Since web caching is concerned with storing documents of different size, Cao and Irani [4] extended the Greedy-Dual algorithm to deal with variable size documents by setting H to *cost/size* where *cost* is, as before, the cost of fetching the document while *size* is the size of the document in bytes, resulting in the Greedy-Dual-Size (GDS) algorithm. If the *cost* function for each document is set uniformly to one, larger documents have a smaller initial H value than smaller ones, and are likely to be replaced if they are not referenced again in the near future. This maximizes the file hit ratio, as, for this measure, it is always preferable to free a given amount of space by replacing one large document (and miss this one document if it is referenced again) than many small documents (and miss many of those documents when they are requested again). ¿From now on, we use a constant cost function of one and concentrate on the role of document *size* and *frequency* in optimizing the replacement policy.

GDS does have one shortcoming: it does not take into account how many times a document has been accessed in the past. For example, let us consider how GDS handles hit and miss for two different documents p and q of the same size s. When these documents are initially brought into the cache they receive the same value $H_p = H_q = 1/s$, even if p might have been accessed n times in the past, while q might have been accessed for a first time; in a worst-case scenario p could then be replaced next, instead of q. In [1], the GDS algorithm was refined to reflect file access patterns and incorporate file *frequency* count in the computation of the initial value: $H = \textit{frequency}/\textit{size}$. This policy is called the Greedy-Dual-Frequency-Size (GDFS) algorithm. Another important derivation related to introducing the frequency count in combination with GDS policy is the direct extension of the original Greedy-Dual algorithm with a frequency count: $H = \textit{frequency}$. This policy is called Greedy-Dual-Frequency (GDF) algorithm.

Often, a high file hit ratio is preferable because it allows a greater number of clients requests to be satisfied out of cache and minimizes the average request latency. However, it is also desirable to minimize the disk accesses or outside network traffic, thus it is important that the caching policy results in a high byte hit ratio as well. In fact, we will show that these two metrics are somewhat in contrast and that it is difficult for one strategy to maximize both.

3 GDFS Cache Replacement Policy: Formal Definition

We now formally describe the GDFS algorithm (and its special cases, GDS and GDF). We assume that the cache has size *Total* bytes, and that *Used* bytes (initially 0) are already in use to store files. With each file f in the cache we associate a "frequency" $Fr(f)$ counting how many times f was accessed since the last time it entered the cache. We also maintain a priority queue for the files in the cache. When a file f is inserted into this queue, it is given priority $Pr(f)$ computed in the following way:

$$Pr(f) = Clock + Fr(f)/Size(f) \tag{1}$$

where $Clock$ is a running queue "clock" that starts at 0 and is updated, for each replaced (evicted) file $f_{evicted}$, to its priority in the queue, $Pr(f_{evicted})$; $Fr(f)$ is the frequency count of file f, initialized to 1 if a request for f is a miss (i.e., f is not in the cache), and incremented by one if a request for f results in a hit (i.e., f is present in the cache); and $Size(f)$ is the file size, in bytes. Now, let us describe the caching policy as a whole, when file f is requested.

1. If the request for f is a hit, f is served out of cache and:
   - *Used* and *Clock* do not change.
   - $Fr(f)$ is increased by one.
   - $Pr(f)$ is updated using Eq. 1 and f is moved accordingly in the queue.
2. If the request for f is a miss, we need to decide whether to cache f or not:
   - $Fr(f)$ is set to one.
   - $Pr(f)$ is computed using Eq. 1 and f is enqueued accordingly.
   - *Used* is increased by $Size(f)$.

   Then, one of the following two situations takes place:
   - If $Used \leq Total$, file f is cached, and this completes the updates.
   - If $Used > Total$, not all files fit in the cache. First, we identify the smallest set $\{f_1, f_2, \ldots, f_k\}$ of files to evict, which have the lowest priority and satisfy $Used - \sum_{i=1}^{k} Size(f_i) \leq Total$. Then:
     (a) If f is not among $f_1, f_2, \ldots, f_k$:
       i. $Clock$ is set to $\max_{i=1}^{k} Pr(f_i)$.
       ii. *Used* is decreased by $\sum_{i=1}^{k} Size(f_i)$.
       iii. $f_1, f_2, \ldots, f_k$ are evicted.
       iv. f is cached.
     (b) If f is instead among $f_1, f_2, \ldots, f_k$, it is simply not cached and removed from the priority queue, while none of the files already in the cache is evicted. This happens when the value of $Pr(f)$ is so low that it would put f (if cached) among the first candidates for replacement, e.g., when the file size is very large – thus the proposed procedure will automatically limit the cases when such files are cached.

We note that the above description applies also to the GDS and GDF policies, except that, in GDS, there is no need to keep track of the frequency $Fr(f)$ while, in the GDF policy, we use the constant 1 instead of $Size(f)$ in Eq. 1.

Let us now consider some properties of GDFS. Among documents with similar size and age (in the cache), the more frequent ones have a larger key, thus a better chance to remain in a cache, compared with those rarely accessed. Among documents with similar frequency and age, the smaller ones have a larger key compared to the large ones, thus GDFS tends to replace large documents first, to minimize the number of evicted documents, and thus maximize the file hit ratio. The value of *Clock* increases monotonically (any time a document is replaced). Since the priority of files that have not been accessed for a long time was computed with an old (hence smaller) value of *Clock*, at some point, the *Clock* value gets high enough that any new document is inserted behind these "long-time-not-accessed" files, even if they have a high frequency count, thus it can cause their eviction. This "aging" mechanism avoids "web cache pollution".

4 Data Collection Sites

In our simulation study, we used four access logs from very different servers:
HP WebHosting site (WH), which provides service to internal customers. Our logs cover a four-month period, from April to July, 1999. For our analysis, we chose the month of May, which represents well the specifics of the site.
OpenView site (www.openview.hp.com, OV), which provides complete coverage on OpenView solutions from HP: product descriptions, white papers, demos illustrating products usage, software packages, business related events, etc. The log covers a duration of 2.5 months, from the end of November, 1999 to the middle of February, 2000.
External HPLabs site (www.hpl.hp.com, HPL), which provides information about the HP Laboratories, current projects, research directions, and job openings. It also provides access to an archive of published HPLabs research reports and hosts a collection of personal web pages. The access log was collected during February, 2000.
HP site (www.hp.com, HPC), which provides diverse information about HP: HP business news, major HP events, detailed coverage of the most software and hardware products, and the press related news. The access log covers a few hours[1] during February, 2000, and is a composition of multiple access logs collected on several web servers supporting the HP.com site (sorted by time).

The access log records information about all the requests processed by the server. Each line from the access log describes a single request for a document (file), specifying the name of the host machine making the request, the time the request was made, the filename of the requested document, and size in bytes of the reply. The entry also provides the information about the server's response to this request. Since the *successful* responses with code 200 are responsible for all of the documents (files) transferred by the server, we concentrate our analysis only on those responses. The following table summarizes the characteristics of the reduced access logs:

Log Characteristics	WH	OV	HPL	HPC
Duration	1 month	2.5 months	1 month	few hours
Number of Requests	952,300	3,423,225	1,877,490	14,825,457
Combined Size, or Working Set (MB)	865.8	5,970.8	1,607.1	4,396.2
Total Bytes Transferred (GB)	21.3	1,079.0	43.3	72.3
Number of Unique Files	17,489	10,253	21,651	114,388

[1] As this is business-sensitive data, we cannot be more specific.

The four access logs correspond to very different workloads. WH, OV, and HPL had somewhat comparable number of requests (if normalized per month), while HPC had three orders of magnitude heavier traffic. If we compare the characteristics of OV and HPC, there is a drastic difference in the number of accessed files and their cumulative sizes (working sets). OV's working set is the largest of the four sites considered, while its file set (number of accessed files) is the smallest one: it is more than 10 times smaller than the number of files accessed on HPC. In spite of comparable number of requests (normalized per month) for WH, OV, and HPL, the amount of bytes transferred by OV is almost 20 times greater than for WH and HPL, but still an order of magnitude less than the bytes transferred by HPC.

5 Basic Simulation Results

We now present a comparison of *Least-Recently-Used* (LRU), GDS, GDFS and GDF on a trace-driven simulation using our access logs. Fig. 1 compares GDS, GDFS, GDF and LRU according to both file and byte miss ratio (a lower line on the graph corresponds to a policy with better performance). On the X-axis, we use the cache size as a percentage of the trace's working set. This normalization helps to compare the caching policies performance over different traces.

The first interesting observation is how consistent the results are across all four traces. GDFS and GDS show the best file miss ratio, significantly outperforming GDF and LRU for this metric. However, when considering the byte miss ratio, GDS performs much worse than either GDF or LRU. The explanation is that large files are always "first victims" for eviction, and *Clock* is advanced very slowly, so that even if a large file is accessed on a regular basis, it is likely to be repeatedly evicted and reinserted in the priority queue. GDFS incorporates the frequency count in its decision making, so popular large files have a better chance of remaining in the queue without being evicted very frequently. Incorporating the frequency in the formula for the priority has also another interesting side effect: the *Clock* is now advanced faster, thus recently-accessed files are inserted further away from the beginning of the priority queue, speeding-up the eviction of "long time not accessed" files. GDFS demonstrates substantially improved byte miss ratio compared to GDS across all traces except HPL, where the improvement is minor.

LRU replaces the least recently requested file. This traditional policy is the most often used in practice and has worked well for CPU caches and virtual memory systems. However it does not work as well for web caches because web workloads exhibit different traffic pattern: web workloads have a very small temporal locality, and a large portion of web traffic is due to "one-timers" — files accessed once. GDF incorporates frequency in the decision making, trying to keep more popular files and replacing the rarely used ones, while files with similar frequency are ordered accordingly to their age. The *Clock* is advanced much faster, helping with the eviction of "long time not accessed" files. However, GDF does not take into account the file size and results in a higher file miss penalty.

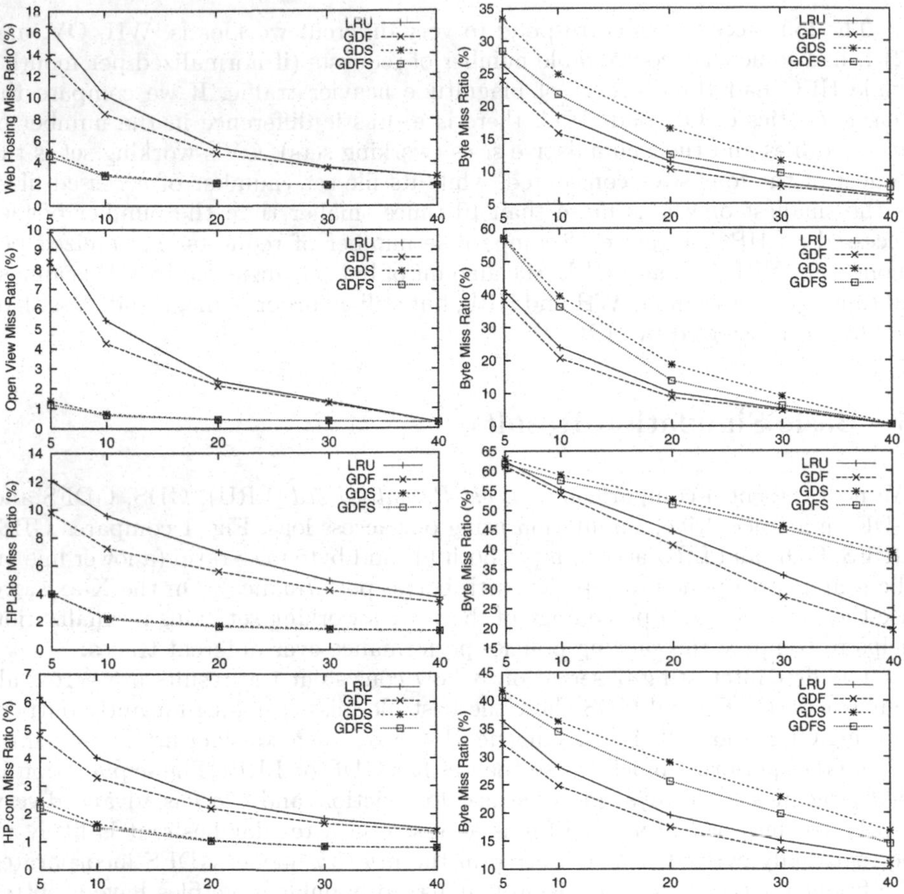

Fig. 1. File and byte miss ratio as a function of cache size (in % of the working set).

6 Generalized GDFS Policy and Its Simulation Results

The *Clock* in GDFS is updated for each evicted file $f_{evicted}$ to the priority of this file, $Pr(f_{evicted})$. In such a way, the clock increases monotonically, but at a very slow pace. Devising a faster increasing clock mechanism leads to a replacement strategy with features closer to LRU, i.e., the strategy where age has greater impact than size and frequency. An analogous reasoning applies to $Size(f)$ and $Fr(f)$: if one uses $Fr(f)^2$ instead of $Fr(f)$, the impact of frequency is stressed more than that of size; if one uses $log(Size(f))$ instead of $Size(f)$, the impact of size is stressed less than that of frequency.

With this idea in mind, we propose a generalization of GDFS (g-GDFS),

$$Pr(f) = Clock + Fr(f)^{\alpha} / Size(f)^{\beta} \tag{2}$$

where α and β are rational numbers. Setting α or β above one emphasizes the role of the correspondent parameter; setting it below one de-emphasizes it.

Impact of Emphasizing Frequency in g-GDFS. Introducing frequency in GDS had a strong positive impact on byte miss ratio and an additional slight improvement in the already excellent file miss ratio demonstrated by GDS. Led by this observation, we would like to understand whether g-GDFS can further improve performance by increasing the impact of the file frequency over file size in Eq. 2, for example setting $\alpha = 2, 5, 10$. Fig. 2 shows a comparison of GDF, GDFS, and g-GDFS with $\alpha = 2, 5, 10$ (and $\beta = 1$). The simulation shows that, indeed, the additional emphasis on the frequency parameter in g-GDFS improves the byte miss ratio (except for HPL, for which we already observed in Section 5 very small improvements due to the introduction of the frequency parameter). However, the improvements in the byte miss ratio come at the price of a worse file hit ratio. Clearly, the idea of having a different impact for frequency and size is sound, however, frequency is dynamic parameter that can change significantly over time. Special care should be taken to prevent $Fr(f)^{\alpha}$ from overflow in the priority computation. As we can see, the impact is workload dependent.

Impact of Deemphasizing Size in g-GDFS. The question is then whether we can achieve better results by de-emphasizing the role of size against that of frequency instead. If this hypothesis leads to good results, an additional benefit is ease of implementation, since the file size is a constant parameter (per file), unlike the dynamic frequency count. We then consider g-GDFS where we decrease the impact of size over frequency in Eq. 2, by using $\beta = 0.1, 0.3, 0.5$ (and $\alpha = 1$), and compare it with GDF and GDFS, in Fig. 3. The simulation results fully support our expectations: indeed, the decreased impact of file size parameter in g-GDFS improves significantly the byte miss ratio (and, at last, also for HPL). For example, g-GDFS with $\beta = 0.3$ has a byte miss ratio almost equal to that of GDF, while its file miss ratio is improved two-to-three times compared to GDF. The g-GDFS policy with decreased impact of file size parameter shows close to perfect performance under both metrics: file miss ratio and byte miss ratio.

7 Conclusion and Future Work

We introduced the generalized Greedy-Dual-Size-Frequency caching policy aimed at maximizing both file and byte hit ratios in web caches. The g-GDFS policy incorporates in a simple way the most important characteristics of each file: size, file access frequency, and age (time of the last access). Using four different web server logs, we studied the effect of size and frequency (and the related aging mechanism) on the performance of web caching policies. The simulation results show that GDS-like replacement policies emphasizing the document size yield the best file hit ratio, but typically show poor byte hit ratio, while GDF-like replacement policies, exercising frequency, have better byte hit ratio, but result in worse file hit ratio. We analyzed the performance of g-GDFS policy, which allows to emphasize (or de-emphasize) size or frequency (or both) parameters, and performed a sensitivity study to derive the impact of size and frequency on file hit ratio and byte hit ratio, showing that decreased impact of file size over file frequency leads to a replacement policy with close to perfect performance in both metrics: file hit ratio and byte hit ratio.

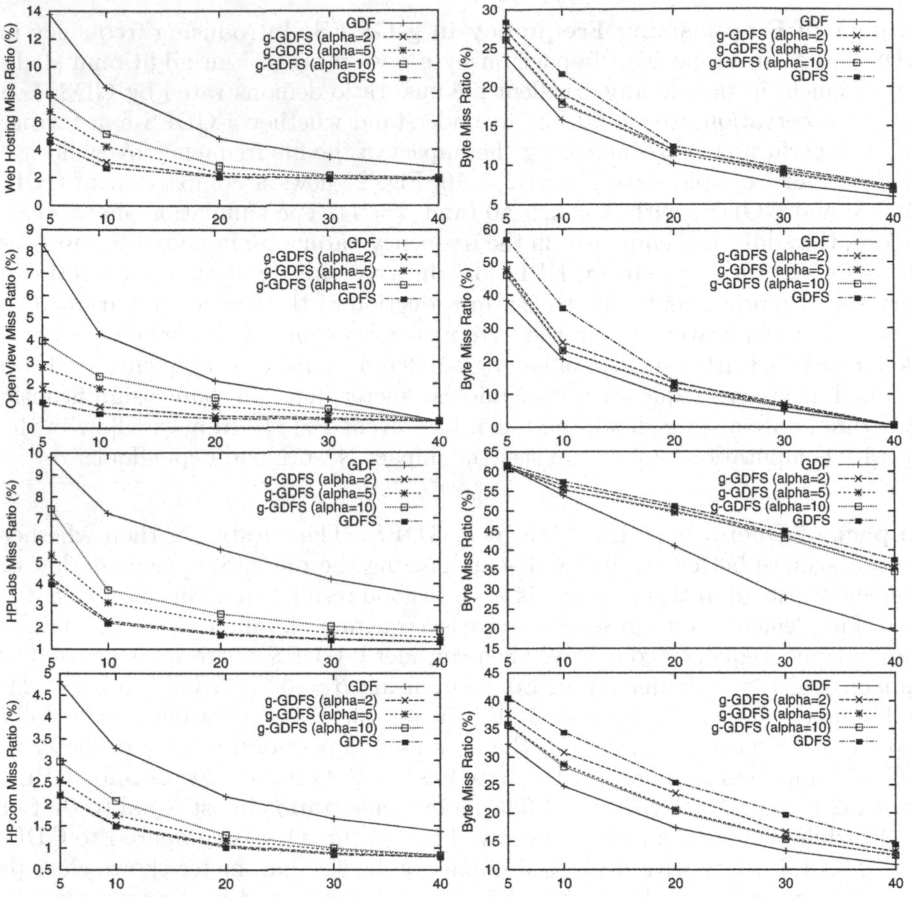

Fig. 2. File and byte miss ratio for new g-GDFS policy: $\alpha = 2, 5, 10$.

The interesting future research question is to derive heuristics that tie the g-GDFS parameters (in particular, β) to a workload characterization. Some promising work in this direction has been done in [6].

References

1. M. Arlitt, L. Cherkasova, J. Dilley, R. Friedrich, T. Jin: Evaluating Content Management Techniques for Web Proxy Caches. In Proceedings of the 2nd Workshop on Internet Server Performance WISP'99, May, 1999, Atlanta, Georgia.
2. M. Arlitt, C.Williamson: Trace-Driven Simulation of Document Caching Strategies for Internet Web Servers. The Society for Computer Simulation. Simulation Journal, vol. 68, No. 1, pp23-33, January 1997.
3. M.Abrams, C.Stanbridge, G.Abdulla, S.Williams, E.Fox: Caching Proxies: Limitation and Potentials. WWW-4, Boston Conference, December, 1995.
4. P.Cao, S.Irani: Cost Aware WWW Proxy Caching Algorithms. Proceedings of USENIX Symposium on Internet Technologies and Systems (USITS), Monterey, CA, pp.193-206, December 1997.

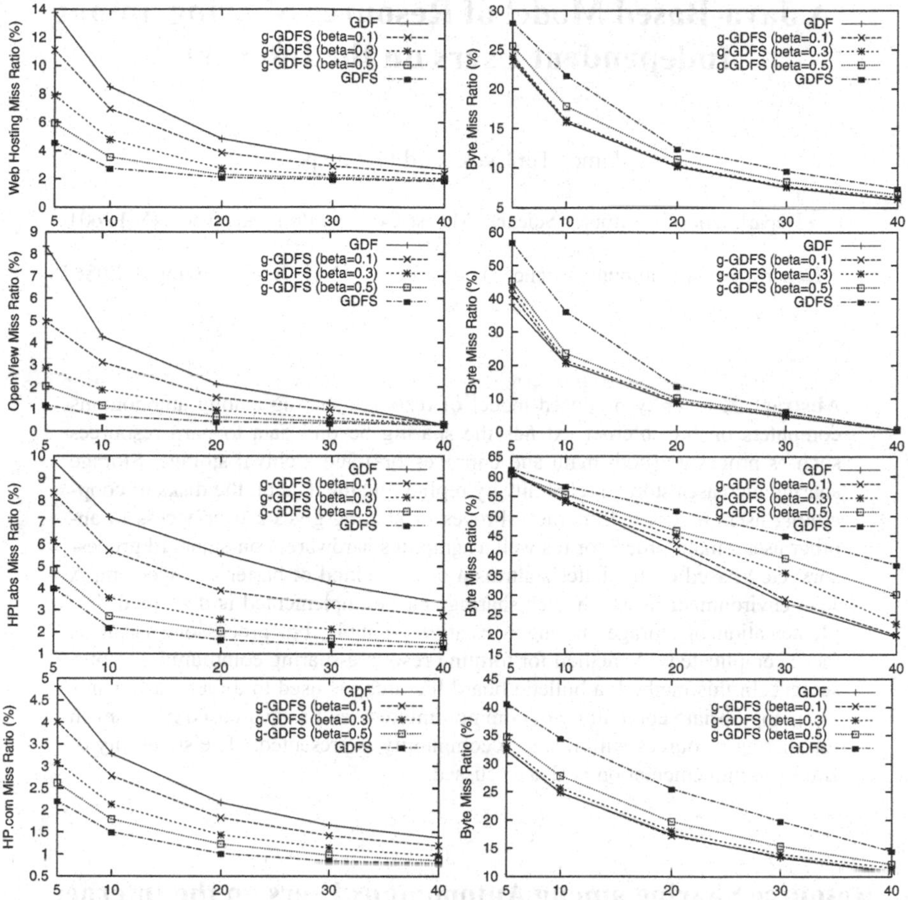

Fig. 3. File and byte miss ratio for new g-GDFS policy: $\beta = 0.1, 0.3, 0.5$.

5. S. Jin, A. Bestavros. Popularity-Aware GreedyDual-Size Web Proxy Caching Algorithms, Technical Report of Boston University, 2000-011, August 21, 1999.
6. S. Jin, A. Bestavros. GreedyDual* Web Caching Algorithm: Exploiting the Two Sources of Temporal Locality in Web Request Streams, Technical Report of Boston University, 1999-009, August 21, April 4, 2000.
7. P.Lorensetti, L.Rizzo, L.Vicisano. Replacement Policies for Proxy Cache. Manuscript, 1997.
8. S.Williams, M.Abrams, C.Stanbridge, G.Abdulla, E.Fox: Removal Policies in Network Caches for World-Wide Web Documents. In Proceedings of the ACM Sigcomm96, August, 1996, Stanford University.
9. R.Wooster, M.Abrams: Proxy Caching the estimates Page Load Delays. In proceedings of 6th International World Wide Web Conference, 1997.
10. N.Young: On-line caching as cache size varies. In the 2nd Annual ACM-SIAM Symposium on Discrete Algorithms, 241-250,1991.

A Java-Based Model of Resource Sharing among Independent Users on the Internet

James TenEyck[1] and G. Sampath[2]

[1] Department of Computer Science, Marist College, Poughkeepsie NY 12601
james.teneyck@marist.edu
[2] Department of Computer Science, The College of New Jersey, Ewing, NJ 08628
sampath@tcnj.edu

Abstract. A recently proposed model of resource sharing among autonomous computers on the internet extends the sharing beyond data to hard resources such as processor (both main and coprocessors) and archival storage. Storage sharing increases storage reliability by replicating the data on the disks of cooperating users across the internet. Processor sharing gives a user access to another user's main processor (as well as graphics hardware, non-standard processors, etc.) based on the latter's altruism or some kind of barter arrangement. A Java environment in which such sharing can be implemented is described. Implementation of storage sharing is relatively simple, but processor sharing is more complicated. A method for forming resource-sharing communities is presented. In this method, a bulletin board hierarchy is used to direct participants to an appropriate community of common interest, and a Java-based mechanism for sharing resources within such a community is presented. The suitability of Jini to an implementation is also discussed.

1 Resource Sharing among Autonomous Users on the Internet

With the rise of the internet and the world-wide web, new paradigms of computing have become necessary [1]. There are three dominant views of the internet. The first is that of a virtual source of information that can be tapped into to access large amounts of data without the need for authentication. The second is that of a metacomputer [5] that can be used to solve large problems with more computing power than would be available at a single site on the net. In this view the objective is the execution of a specific compute-intensive (for example, combinatorial optimization or simulation) or data-intensive task (for example, multimedia libraries). It is usually realized in the setting of an organization or a research group cooperating in that task. The third view is that of a group of individuals cooperating in carrying out a task such as product design or document production. In all three views, users are involved in sharing resources in the form of data, code, or processors, with the sharing implemented in a structured way through sometimes involved protocols and database concurrency mechanisms. Incidentally, shared physical objects need not be limited to computing resources. For example, a recent paper describes a form of virtual sharing that creates the illusion of a shared physical space [2] in which a telemanipulation

B. Hertzberger et al. (Eds.) : HPCN Europe 2001, LNCS 2110, pp. 124-132, 2001.

system allows a user to move objects locally and cause a similar movement of similar objects at a distant location. Suggested users include architects and planners who could use the system to do layout modeling and design.

A particularly widespread form of internet resource sharing occurs in the distributed object-oriented paradigm, where sharing based on the client-server model of distributed objects [9] results in code reuse and aids the software process. For a recent summary of work in this area, see the survey by Lewandowski [4]. Objects are usually code that is downloaded by the client for execution (for example applets), or, as has been recently suggested, uploaded to the server for execution on a remote machine [6]. Storage sharing within an organization over the internet has also been studied recently [7].

The internet, however, works on different principles than an organizational intranet or multinets, which are sustained by conventional notions of networking such as authentication and process coordination. Individual users form, usually without design, virtual groups that are embedded in the internet and interact in an almost unrestricted fashion, often with no or only marginal protocols beyond the underlying TCP/IP communication standards. Furthermore, such sharing does not have to be limited to exchange of data and messages, and it is advantageous to extend it to hardware. In this spirit, a model of cooperative consent-based use of resources residing in autonomous single computers on the Internet was proposed recently [8]. In the model, sharing goes beyond the exchange of information to unused (or underused) processor cycles, archival storage, and possibly specialized coprocessors. Thus the internet is considered to be a mass of loosely connected processors and coprocessors with (theoretically) unlimited storage space whose physical realization is the collection of hard drives on the autonomous systems. A software architecture for such sharing was presented along with a discussion of related issues, including privacy and security. Such sharing was considered to be largely in a non-profit mode although this does not preclude the application of the concept to more formal organizations such as intranets and more structured interactions such as are found in academic settings and among professional organizations. In the present work an environment that could be used to implement such a model is described in some detail, and the suitability of Jini [3] for this purpose is discussed.

In Section 2, a Java-based software architecture for the model is described. In Section 3, protocols for sharing processor cycles and excess storage capacity are described. We discuss methods for locating computing resources donated by autonomous users on the internet and consider methods for keeping this information current. Section 4 considers the use of Jini technology for implementing these protocols.

2 A Java Model of Resource Sharing

A formal view of sharing is presented in Figure 1. Virtual subnets exist by design or by one-on-one interaction across the internet in which users share information as well as hard resources. A resource-sharing layer is superimposed on top of the transport layer. The Resource Acquisition Layer involves a handshaking protocol in which the terms of the sharing arrangement are determined. There are two sharing components:

processor sharing and data storage sharing. In the standard Java environment any such relationship between users must ensure that the security of the host system is not jeopardized. The Java Runtime Layer supports the execution of a guest program on a host machine. It must provide streams over which objects may be transmitted and security for the host machine. In the Java security model, a guest process is not able to read or write files on the host machine, execute system commands that can alter the file structure of the host or cause the host to terminate its execution, or override the security manager provided in the protocol suite. This is too restrictive for supporting full processor sharing. As shown later some of these restrictions are easy to relax by a simple extension of the Java security model. The Application Layer adds management and accounting functions to the protocol suite.

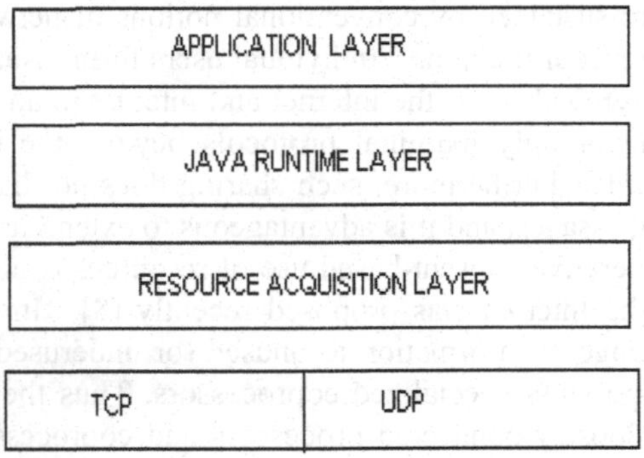

Fig. 1. A Java Model of Resource Sharing.

3 Protocols for Sharing

Potentially every computer on the internet can serve as a host or client processor. For any particular computer to assume either role, it would first have to install the necessary client/server application software, and then either locate an appropriate processor host or indicate its availability to serve in that capacity.

3.1 Location of Available Resources

A "bulletin board" facility provides a location where users can download the necessary application software and receive instructions for locating and registering with a processor broker. A bulletin board or newsgroup facility can also be the place to form user groups. User groups would need to indicate the average capacity of the various computing resources that are available within the group and the terms for becoming a group member. User groups could range from closed communities that require an application/approval process for joining to drop-in communities that allow transients to share resources on a one-time basis. User groups will be further defined by the method of compensation for acquisition of remote resources. Compensation can range from for-profit to barter arrangements to purely altruistic sharing.

Once a prospective host has located a resource broker from the bulletin board or otherwise, it would need to send a host-available packet indicating:

- which resources it was willing to share
- available capacity of these resources
- access restrictions, and
- compensation requirements.

Regular updates refresh the tables of the resource broker and alert it to any change in the status of the remote host. Failure to receive a regular update would lead to a timeout for the table entry and cause the resource broker to drop the host from its list of possible resource donors.

The resource broker regularly tabulates the information on its donor hosts and passes it along to the bulletin board services with which it is affiliated. This serves to reconfirm it as a resource broker site and maintain the currency of the information about resource capacity for user groups in the bulletin board service.

Donor hosts would be most available during the hours when they were not in use by their owner. During this time, machines that were left turned on would be running a screen saver program. The timing mechanism that triggers the start of this program could also trigger the sending of a host-available packet. Long pauses between keystrokes also trigger the start of the screen saver program, so the fraction of the resources made available for sharing could be incrementally increased by subsequent host-available updates in a manner similar to the slow-start, and additive increase/multiplicative decrease algorithms for determining available bandwidth in TCP. The frequency with which the screen saver program is initiated and the duration of its run-time indicate the extent to which the owner is using the available processing capability of the machine.

3.2 Resource Acquisition by Clients

Service-request packets are sent by a client to a service broker. A request for service would include:

- the level of service desired
- the estimated duration of the job (in fixed reference CPU time units)
- an indication if other resources are needed (such as storage or co-processor time) and, if so, the type and amount of any such need, and
- the type of compensation the client is offering

The broker forwards a copy of this request to any donor offering such service. If no such donor were available, the broker would return a service-request-reject packet to the requesting client indicating the largest set of the above parameters that could be satisfied.

Resource donors that receive a service-request packet may accept the request and return a service-request-response packet to the soliciting client. Such a packet would include the values of each of the above parameters offered by the host. These may differ from the values requested, but the broker will only forward the request to hosts generally offering the type and level of service being requested. As service-request-response packets arrive, the client will either reject the offer (by ignoring it) or accept with a connect-request packet that restates the offered parameter values. The client may also choose to connect to more than one donor. Each connection is confirmed by the donor returning a connect-confirm packet to the client.

Shared processing can take several forms including:

- a system of shared distributed objects with donor hosts willing to allow autonomous clients access to these programming resources, or
- special needs requests for enhanced processing capability in which the client supplies both the program and the data that it is to operate upon to the donor host.

In the first case, the problem is how to set up such an environment in a population of autonomous users with non-dedicated resource servers. In an intranet, corporate or institutional software resources are distributed among dedicated servers within the constrained environment. Distributed systems can be designed, publicized, and made exclusively available to clients within the organization. In our autonomous model, there is no centralized design and no organizational or geographical restriction on participation. Programmers across the internet create objects for commercial, instructional, or recreational interest. In order to make these resources generally available, it is necessary to bring them to the attention of the larger community and, in some cases, to provide the needed client classes to prospective users. A hierarchical bulletin board system, where the top level points to sites of particular broad interest and lower levels provide the specific details for locating the resource, is an attractive consideration for such a resource locator, but maintenance of such a system poses severe problems. Given the environment, CORBA or Java RMI provides the framework for managing the necessary communication.

The second form of processor sharing is more straightforward, but requires security constraints to protect the donor host from improper access or malicious intent. In asynchronous service, the client program and data might have to be stored at the donor host until processor resources become available. This would mean that the host machine would need to create a directory for storing guest files. This directory would be accessible only by the host machine and none of the executable files inside would be able to access, create, or delete files in any other directory. A program may also need to perform file I/O. It would need to read, create, and write files within its own directory. This presents two dangers:

- arbitrarily large files that consume a disproportionate amount of storage could be created, and
- a relaxation of the restrictions on the privileges granted the guest process could compromise the security of the host system.

The first concern could be alleviated by negotiating a restriction on the amount of storage allocated to a guest process during the initial handshake. Security concerns necessitate that programs within this directory run only under a security manager that prevents access or appropriation of any additional storage.

3.3 Disk Storage Sharing

Volunteer groups willing to share storage need to set up a bulletin board site where donors/tenants register and download the necessary communication software. The registration process sets up an "account" for the new user in a resource broker. The account contains the IP address of the client, the credit limit (in megabytes) that the user could borrow from the user community, the amount of storage deposited in and withdrawn from the network, and a link to the record of transactions. The credit limit for a user is proportional to the amount of storage that the user is willing to donate to the shared community minus the amount of community storage being used by the member.

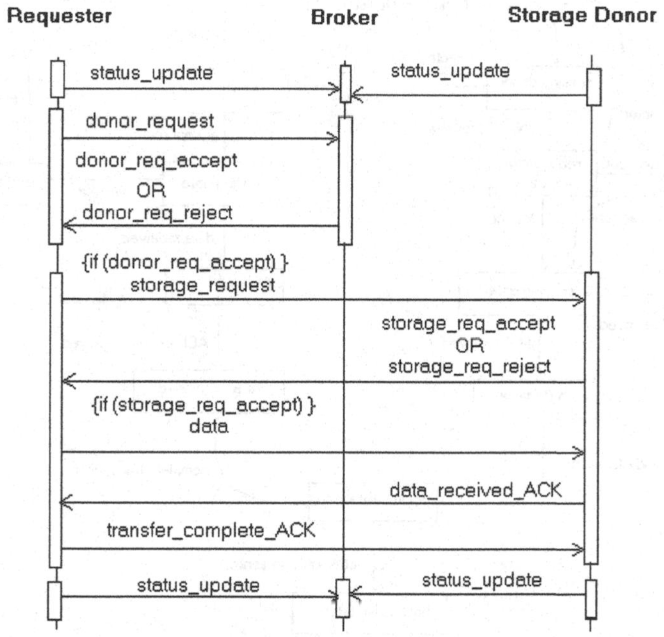

Fig. 2. Packet Exchange Sequence for Storage Acquisition.

After registration, user information is added to the tables of the resource broker. Requests for storage are directed to the broker, who verifies the authenticity and credit of the requester. For every valid request, the resource broker identifies an active donor with sufficient storage reserve and returns a packet with the IP address of the donor to the requester. An exchange of packets between the requester and donor establishes the details of the storage contract, and precedes the transmission of the data to the donor host. This exchange is shown in Figure 2 above.

The state transition diagram for a host engaging in a storage transfer transaction shown in Figure 3 below details the steps involved in completing a storage transfer to a donor host. Hosts from the storage sharing community indicate that they are active listeners by sending a status update packet to the resource broker. Status update packets will be generated whenever a host connects to the internet, and be repeated regularly to maintain state information in the broker. Arrival of a status update packet prompts the broker to update the account of the source host and create a table entry in its list of active donors. A command to disconnect from the internet will trigger sending a status update packet to the resource broker with an indication that the host has become inactive.

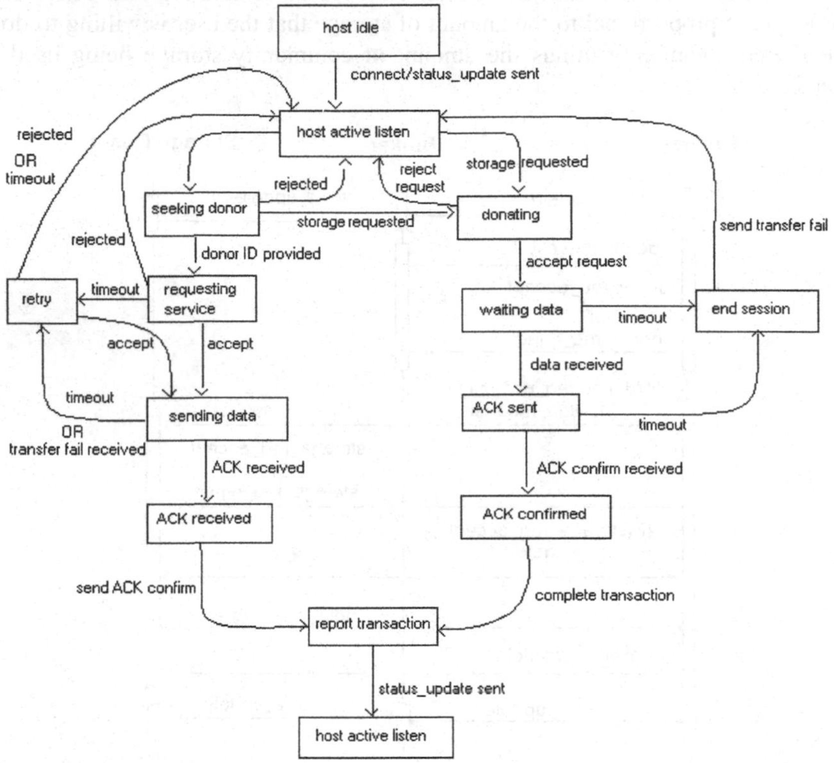

Fig. 3. State Transition Diagram for a Host in a Storage Transfer Transaction.

If a host is interrupted in its process of finding a donor for a data transfer that it is initiating by a request to become a donor, it drops its own request and assumes a state of "donating" or becoming a donor host. Transactions are successfully completed when both parties report the transaction to the resource broker. A transaction record at the resource broker confirms the existence of a duplicate copy of the data for the requester that is now a "tenant" at a remote host.

3.3.1 Resource Shedding

A donor host may need to reclaim storage from a tenant. We classify the storage arrangement as either protected or unprotected, and describe two levels of obligation for the network to maintain the data that is being shed by the donor. Data is classified unprotected when the original is considered secure. Multiple backups may exist, or the original may be reclaimed from an alternate source. Shedding unprotected data may be done by simply informing the tenant host that the storage arrangement is being unilaterally terminated. It is then up to the tenant to find a new donor willing to accept a new copy of this data.

Shedding protected data may jeopardize its existence. There may be circumstances in which all copies of that data may be lost if the copy in the donor is simply dropped. It is the community as a whole that is responsible for maintaining the existence of any protected data file. This file must be transferred to another donor host within the community before it can be dropped by the initial donor and this transfer should be transparent to the tenant. The donor host that is shedding protected data must first go through a process of transferring this data to a new donor before overwriting the reclaimed storage. The tenant is notified at the end of the transfer of the new address for its data.

Tenant hosts must be informed of shedding of either protected or unprotected data, but may not be active when this shedding occurs. The problem of notifying the former tenant is solved by having the resource broker hold a message for the tenant that is delivered when it next becomes active. When a status update message comes from a host, the source IP address is compared to a list of addresses for which a message is queued. If the address compares to one in this list, that message is directed to the source.

4 Implementing Resource Sharing in the Jini Programming Environment

The proposed model is aided considerably by the availability in Sun's Jini project of Java technology for sharing distributed services among a "federated" group of users. Jini technology enhances the communication services of RMI [3] by providing an open-ended set of interfaces called service protocols. Jini offers a lookup service that is capable of tracking available services and also uses an activation process to enable recovery after a system crash or restart. Lookup may occur serendipitously or through registration, with different protocols available in Jini. Furthermore, services may create and announce their attributes, which a user may scan to determine their usefulness and/or appropriateness. While the emphasis in the proposed model is on an altruistic sharing mode, as mentioned earlier, in practice constraints have to be placed on usage time and size of a shared resource to discourage idling or overuse (processor) and squatting (storage). In this context, the leasing services and the landlord paradigm discussed in [3] appear to be useful.

It is not clear at this time, however, what impact the licensing requirements associated with Jini would have upon an implementation involving the independent non-institutional use that is the focus of the proposed model, given that Jini is viewed in the large as a community sharing arrangement. Sharing services in a loosely defined, autonomous environment presents configuration challenges that a more controlled environment may minimize, but the open-ended protocols of the Jini architecture provide a basis for developing an appropriate framework for resource sharing across the internet.

Acknowledgement

The authors thank an anonymous referee for helpful remarks on the use of Jini technology.

References

1. T. Berners-Lee. . World-wide Computer. Communs. ACM **40** (2) (1997) 76-82.
2. S. Brave, H. Ishii, and A. Dahley. Tangible interfaces for remote collaboration and communication. In: Proceedings of the 1998 ACM Conference on Computer-Supported Cooperative Work (1998) 169-178.
3. W. K. Edwards. *Core Jini*. Prentice-Hall, Upper Saddle River (1999).
4. Scott M. Lewandowski. Frameworks for component-based client-server computing. ACM Computing Surveys **30** (1), 3-27 (1998).
5. T. L. Lewis. The next 10000_2 years. Parts I and II. IEEE Computer, April 1996, 64-70, May 1996, 78-86.
6. Mahmoud, Qusay H. The Web as a global computing platform. In Proceedings of the 7th International Conference on High Performance Computing and Networking Europe (HPCN99): Lecture Notes in Computer Science **1593**, 281-300. Springer-Verlag, Berlin, 1999.
7. C. J. Patten, K. A. Hawick, and J. F. Hercus. Towards a scalable metacomputing storage service. In: Proceedings of the 7th International Conference on High Performance Computing and Networking Europe (HPCN99): Lecture Notes in Computer Science **1593**, 350-359. Springer-Verlag, Berlin, 1999.
8. G. Sampath and J. TenEyck. Resource sharing among autonomous computers on the Internet. To appear in the Proceedings of the Global Technology Conference, Dallas (Texas), June 10-12, 2001.
9. P. Sridharan. *Java Network Programming*. Prentice-Hall, Upper Saddle River (1997).

The GRB Library: Grid Computing with Globus in C

Giovanni Aloisio, Massimo Cafaro, Euro Blasi,
Lucio De Paolis, and Italo Epicoco

ISUFI / High Performance Computing Centre
University of Lecce, Italy
{giovanni.aloisio,massimo.cafaro}@unile.it
{euro.blasi,lucio.depaolis,italo.epicoco}@unile.it

Abstract. In this paper we describe a library layered on top of basic Globus services. The library provides high level services, can be used to develop both web-based and desktop grid applications, it is relatively small and very easy to use. We show its usefulness in the context of a web-based Grid Resource Broker developed using the library as a building block, and in the context of a metacomputing experiment demonstrated at the SuperComputing 2000 conference.

1 Introduction

In the last few years, a number of interesting projects like Globus [1], Legion [2] and UNICORE [3] developed the software infrastructure needed for grid computing [4]. Several grid applications have been developed since then, using the tools and libraries available. The "first generation" grid applications were highly successful in demonstrating grid computing effectiveness, fostering subsequent research in the field, but the lack of consensus about the kind of services a grid should have been providing, and about what programming models should have been used, proved to be a large barrier that hindered grid applications development.

To reach a wider community of grid applications developers and users we need to make easier the transition from desktop computing to the grid. One way to attract the users is to provide them with a comfortable access to the grid: indeed, current trends include the development of computing portals [5], to allow trusted users a seamless access to the grid through the Web.

With regard to the applications developers, an increasing number of tools developers is now working toward high-level tools and libraries layered on existing grid services. That way application development costs decrease because enhanced services are available and can be readily incorporated into existing legacy codes or new applications.

In this paper we shall concentrate on the use of the Globus Toolkit as grid middleware, and describe a library for grid programming layered on top of Globus. The paper is organized as follows. Section 2 introduces the library design and section 3 its implementation. Section 4 presents the Grid Resource Broker, a web-based grid envi-

B. Hertzberger et al. (Eds.) : HPCN Europe 2001, LNCS 2110, pp. 133-139, 2001.

ronment built using the GRB library. We recall related work in the area and conclude the paper in section 5.

2 Library Design

In this section we present the main ideas that guided us through the process of designing our library. First of all, we selected as grid middleware the Globus Toolkit, because it is already deployed at many US academic and supercomputing sites, and is rapidly becoming the "de facto" standard for grid computing at a growing numbers of sites all over the world. Moreover, the Globus Toolkit is released under a public license that allows also commercial use and redistribution of the source code.

Using Globus we can transparently benefit from uniform access to distributed resources controlled by different schedulers and from both white and yellow pages information services whose aim is information publishing and discovery. A security infrastructure based on Public Key technologies protects network transactions using X.509 version three digital certificates.

In the design phase, we were faced with the problem of selecting a number of basic services among those provided by Globus, to be enhanced in the library. We decided to concentrate the attention on the most used services, because past experiences with several groups of Globus users revealed that only a small number of core services are effectively used by a vast majority of users. These services include:

1. Remote job submission for interactive, batch and parameter sweep jobs;
2. Job status control;
3. Resource discovery and selection through queries to GRIS and GIIS servers;
4. Remote file management using GSI FTP servers (Globus Security Infrastructure FTP).

Another requirement was to keep small the number of functions composing our library, in order to reduce the time needed to learn how to use them. Moreover, we also planned to make the library suitable for the development of web-based and desktop grid applications. Finally, the primary requirement was to provide users with enhanced Globus services. Using our library, the users can write code to:

1. Submit interactive jobs on a remote machine staging both the executable and the input file(s).
2. Submit batch or parameter sweep jobs on a remote machine. Again, we provide support for automatic staging of executable and input file(s). Moreover, the output file(s) can also be transferred automatically upon job completion to another machine. Information related to the job is stored in a request file, to be used when the users want to enquiry about the job status.
3. Check a batch or parameter sweep job status. A Globus job can be either PENDING, ACTIVE, SUSPENDED, FAILED or DONE. If PENDING, the job is waiting to be executed sitting in a queue; during execution the job is ACTIVE and can

temporarily become SUSPENDED if managed by a preemptive scheduler. If something goes wrong the job is marked FAILED, while in case of normal termination its status becomes DONE. Moreover, we signal to the user when the job was started and how, including the command line and the arguments given. It's worth noting here that Globus provides only a url which uniquely identifies a job.

4. Query a GRIS server to determine the features of a specific computational resource. Each machine running Globus can be queried about its features because a small LDAP server called GRIS (Grid Resource Information Service) is up and listens for connections on the IANA registered port 2135. The GRIS stores static and dynamic information related to hardware and system software, and can be thought of as a white pages service.
5. Query a GIIS server to find computational resources matching specified criteria. An institution or organization may decide to setup a GIIS service (Grid Index Information Service). Like the GRIS, the GIIS is an LDAP server that collects information maintained in each GRIS server available in the institution or organization. Thus, we can exploit the GIIS server like a yellow pages service to allow users to find, if available, computing resources with specific features, e.g. amount of memmemory, number of CPUs, connectivity, operating system, load etc. Although Globus provides a command line tool called *grid-info-search*, the tool itself can only deal with strings, meaning that users can not compare numerical quantities using standard relational operators, because in this case the lexicographical order will be used in place of the numerical one. Since numerical quantities are stored in GRIS and GIIS servers as strings, it is meaningless to search for machines with at least 256 Mb of memory using the Globus tool. Our functions instead do automatically proper conversions.

3 Implementation

The GRB library is made of a small number of functions; it is based on Globus version 1.1.3 or 1.1.4 and requires also the Expect library. The implementation of some of the functions is a bit tricky, due to our commitment to support web-based grid computing in the form of CGIs. Some of the functions in our library take as one of their parameters a pointer to an external, user defined function. This is an output function that the user can specialize to output data to the console, properly formatted as HTML for web based applications or even dedicated to insert text in graphical widgets.

We begin the discussion about the implementation considering first how to authenticate to the grid. Computational grids based on the Globus Toolkit require that users authenticate once to the grid (single sign on) using a X.509v3 digital certificate before doing anything else. However, the user's certificate is only used to generate a *proxy*, which acts on behalf of the user (implementing a restricted form of delegation) and lasts for a specified number of hours. The two functions *grb_generate_proxy* and *grb_destroy_proxy* are used to create a valid proxy to start a grid session and to remove it later. Since Globus does not provide an API for proxy management, we call the *grid-proxy-init* and *grid-proxy-destroy* tools. However, to allow job submission from the web, two issues need to be solved. We must create a suitable user environment, switching from the traditional web-server nobody user to the real Globus user and adding some required Globus environmental variables to the user's environment. As a matter of fact, simply switching from nobody to the Globus user ID using the unix setuid function does not replace the nobody user's environment with the Globus user's one. The function *grb_create_env* that does the job is thus required for web based grid applications.

The other issue is related to the use of some Globus tools like *grid-proxy-init* and *Globusrun* from the web. We can not simply fork/exec these Globus commands from a CGI, it will not work because of the way Globus uses ttys. Instead, we need to drive it from a pseudo tty. This is why we use the Expect library.

Some of the functions in our library exploits the information stored in a file called the user's profile. This file contains information about the computational resources on the grid that the user can access. For each machine the user supplies the hostname (required to contact the Globus resource manager and GSI-FTP server), the pathname to her shell (required to access the user's environment on the remote machine), and a float number representing the node cost per hour in her currency (can be used for job scheduling). We plan to add more information in the future to allow for more complex scheduling algorithms. A number of functions in the GRB library deals with user's profile management (e.g. for adding, removing of updating a grid resource dynamically).

We now come to job submission. The GRB library supports interactive, batch and parameter sweep jobs. We decided to use the *Globusrun* tool instead of the GRAM APIs for interactive job submission in the *grb_job_submit* function because this tool supports automatic redirection of job's output. For batch submission we used the GRAM APIs to write both a low level blocking (the function returns only when the job to be submitted is terminated) and non blocking version (the function submits the job and returns immediately) called respectively *grb_batch_wait* and *grb_batch_nowait*.
The function *grb_job_submit_batch* provides additional support for automatic staging of executable and input file(s) using internally *grb_gsiftp_copy*. We recall here the possibility to start even a graphical client by setting properly the *display_ip_addr* parameter, so that the user just need to authorize the remote host to redirect the display on her machine using the "xhost" command on Unix machines.

Finally, parameter sweep jobs can be submitted using *grb_job_submit_parameter*. A number of users need to run the same executable with different input to do parameter study. We assume that the user's input files are called *input-1*, *input-2*,...,*input-n*. Moreover, we assume that all of the input files are stored on the same machine and allow output files to be transferred on a machine possibly different from the machines hosting the executable and input files. GSI-FTP is used if needed for automatic staging of executable and input/output files.

Two functions are provided to check a batch or parameter sweep job status: *grb_batch_job_status* and *grb_parameter_job_status*. These functions read the information about submitted jobs that is stored in requests files at job submission time and contact remote Globus jobmanagers to inquiry about job status.

Information about grid resources can be obtained by querying Globus GRIS and GIIS servers. The functions *grb_search_gris* and *grb_search_giis* contact these servers using the LDAP protocol. Simple resource brokers can be built easily by searching Globus LDAP servers to acquire the features of grid resources available in the user's profile, comparing them to some user's specified criteria in order to dynamically select one or more computational resources suitable for a batch or parameter sweep job, and submitting to the resources with a best match. This is the approach we adopted for our web based Grid resource Broker.

4 The Grid Resource Broker

As an example of use of the GRB library, we developed a web-based Grid Resource Broker [6][7]. This is a computing portal that allows trusted users to create and handle computational grids on the fly exploiting a simple and friendly GUI. The Grid Resource Broker provides location-transparent secure access to Globus services. Users do not need to learn Globus commands, nor to rewrite their existing legacy codes.

The use of the library results in a dramatic decrease of the time needed to develop these functionalities; the main effort reduces to handling HTML forms, demanding to the library the issues related to grid programming.

Our portal architecture is a standard three-tier model. The first tier is a client browser that can securely communicate to a web server on the second tier over an HTTPS connection. The web server exploits Globus as grid middleware to make available to its clients a number of grid services on the third tier, the computational grid. The Globus toolkit provides the mechanisms for submitting jobs to remote resources independently of the resource Schedulers, querying for static and dynamic information about resources composing the computational grid using the LDAP API, and a secure PKI infrastructure that uses X.509v3 digital certificates.

There are no restrictions on what systems/sites could be served by GRB, because of the way user profiles are handled. As a matter of fact, GRB can be accessed regardless of system/geographical location and Grid Resources can be added/removed dynamically. In order to use GRB, a user must apply to the ISUFI/HPCC (University of Lecce) to get an account, to the Globus Certification Authority to get a certificate, and she must properly setup her Globus environment on the machine running the web server. We assume that Globus v1.1.3 or v1.1.4 and GSI-FTP is installed and listening on port 2811 on each one of the computing resources the user adds to her profile; moreover we assume that the GRIS server is up and running on the default port 2135 on each computing resource and that it can be queried starting from the distinguished name "o=Grid". The user's client browser must be configured to accept cookies.

To start using the GRB, a user authenticate herself to the system by means of her login name on the GRB web site and her PEM pass phrase (Privacy Enhanced Mail) that protects the globus X.509v3 certificate. The transaction exploits the HTTPS (SSL on top of HTTP) protocol to avoid transmitting the user's PEM pass phrase in clear over an insecure channel. Once authenticated, a user can start a GRB session that will last for at most twelve hours or until she invalidate her session using the logout tool.

A preliminary version of our library was also used in the metacomputing experiment (see Fig. 1) demonstrated at the Supercomputing 2000 conference held in Dallas. In particular, the functions related to resource discovery and selection in the context of the Egrid testbed [8] were used to allow Cactus [9] to query the testbed information service.

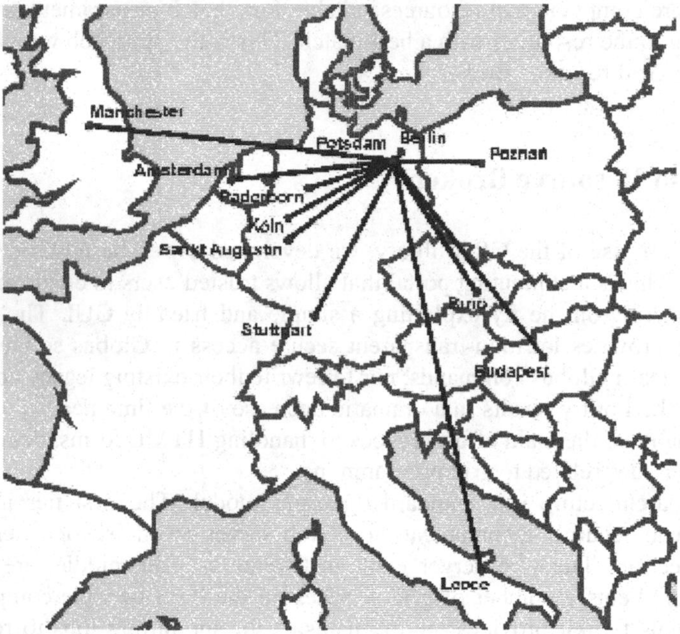

Fig. 1. The Egrid Testbed.

5 Related Work and Conclusions

The authors are not aware - as of this writing - of any projects aimed at releasing source code for enhanced Globus services. Anyway, the following two projects have provided useful software to help develop computing portals that provide access to the grid through basic Globus services:

1. SDSC GridPort Toolkit [10];
2. NLANR Grid Portal Development Kit [11].

The Grid Portal Toolkit (GridPort) provides information services that portals can access and incorporate; it leverages standard, portable technologies. GridPort makes use of advanced web, security and grid technologies such as Public Key Infrastructure and Globus to provide secure, interactive services.

Server-side Perl/CGI scripts build HTML pages, and a combination of HTML/JavaScript is used on the client side. Portals built on top of GridPort allow users job execution and data management through a comfortable web interface. Examples of portals built using GridPort are HotPage, LAPK, NBCR Heart and GAMESS.

The Grid Portal Development Kit provides access to Grid services exploiting Java Server Pages (JSP) and Java Beans. Java Server Pages invoke Bean methods to provide authentication services, management of user profiles, job submission, etc. The GPDK Java beans build on top of the Globus Java Commodity Grid (CoG) Toolkit [12].

GPDK Java beans present to web developers an high level interface to the CoG kit. Moreover, GPDK take advantage of the Myproxy package, so that users can gain access to remote resources from anywhere without requiring their certificate and private key to be located on the web browser machine. The Myproxy server is responsible for maintaing user's delegated credentials, proxies, that can be securely retrieved by a portal for later use.

A library layered on top of basic Globus services was presented in the paper. The library was designed to provide enhanced grid services and can be used to develop both web-based and desktop grid applications. Moreover, it is relatively small and very easy to use. We showed its usefulness in the context of a web-based Grid Resource Broker developed using the library as a building block, and in the context of a metacomputing experiment demonstrated at the SuperComputing 2000 conference.

References

1. I.Foster and K.Kesselman, "GLOBUS: a Metacomputing Infrastructure Toolkit", Int. J. Supercomputing Applications (1997), 115-28
2. A.S. Grimshaw, W.A. Wulf, J.C. French, A.C. Weaver and P.F. Reynolds Jr., "The Legion Vision of a Worldwide Virtual Computer", CACM 40 (1997)
3. J. Almond, D.Snelling, "UNICORE: uniform access to supercomputing as an element of electronic commerce", FGCS Volume 15 (1999), Numbers 5-6, October 1999, pp. 539-548
4. I.Foster and C.Kesselman (eds.), The Grid: Blueprint for a new Computing Infrastructure, (Morgan Kaufmann Publishers, 1998) ISBN 1-55860-475-8
5. http://www.computingportals.org
6. http//sara.unile.it/grb
7. G.Aloisio, E.Blasi, M.Cafaro, I.Epicoco, The Grid Resource Broker, a ubiquitous grid computing framework, submitted to Journal of Scientific Programming
8. G.Aloisio, M.Cafaro, E.Seidel, J.Nabrzyski, A.Reinefeld, T.Kielmann, P.Kacsuk et al., Early experiences with the Egrid testbed, to appear in Proceedings of CCGRID2001, Brisbane, Australia
9. G.Allen,W.Benger,T.Goodale,H.Hege, G.Lanfermann, A.Merzky, T.Radke, and E.Seidel, The CactusCode: A Problem Solving Environment for the Grid. In Proc.High Performance Distributed Computing (HPDC-2000), pp 253–260, IEEE Computer Society, 2000
10. SDSC GridPort Toolkit, https://gridport.npaci.edu
11. NLANR Grid Portal Development Kit, http://dast.nlanr.net/Features/GridPortal
12. G. von Laszewski, I. Foster, J. Gawor, W. Smith, and S. Tuecke, "CoG Kits: A Bridge between Commodity Distributed Computing and High-Performance Grids", accepted to the ACM 2000 Java Grande Conference, 2000

Certificate Use for Supporting Merging and Splitting of Computational Environments*

Paul A. Gray[1] and Vaidy S. Sunderam[2]

[1] University of Northern Iowa, Dept. of Mathematics
Cedar Falls, IA, 50614-0506, USA
gray@cns.uni.edu

[2] Emory University, Dept. of Mathematics and Comp. Sci.
Atlanta, GA, 30322, USA
vss@mathcs.emory.edu

Abstract. In recent years, numerous projects focusing on utilization of geographically-distributed environments have called upon certificates and certificate authorities for authentication, for validation, and for assigning access privileges to groups. This paper describes a prototype established for utilizing certificates in the IceT environment for facilitating merging of resource groups. This prototype extends this paradigm to leverage significantly upon certificate revocation lists to permit predictable, dynamic splittings of groups.
Keywords: Merging virtual machines, splitting virtual machines, certificate revocation lists, soft-installation of processes.

1 Introduction

With increasing frequency, high-performance networks are being used to couple the resources of geographically-distributed institutions. One of the most prominent paradigms being adopted is that of the "Grid," a computational grid system [5]. The Grid provides a transparent large-scaled persistent resource infrastructure. Authorized users are able to "plug into" the Grid and, once connected, utilize the resources therein.

The IceT project focuses on a smaller, more dynamic environment; where several colleagues might join together to form a short-term resource alliance. An analogy would be that users do not "plug into" resources in IceT, but rather resources and environments are "plugged together." Localized groups of resources are collectively made available for distributed computations, process migration, distributed data mining, etc. When use of the combined resource pool is completed or resources are reclaimed by a colleague, the virtual environments are re-aligned about the surviving groups. The emphasis here is on the more dynamic creation of a collective environment brought about by the merging of independent resources, and in the sharing or "soft-installation" of processes onto complement resources.

* Research supported by NSF grant ACI-9872167 and University of Northern Iowa's Graduate College.

B. Hertzberger et al. (Eds.): HPCN Europe 2001, LNCS 2110, pp. 141–150, 2001.

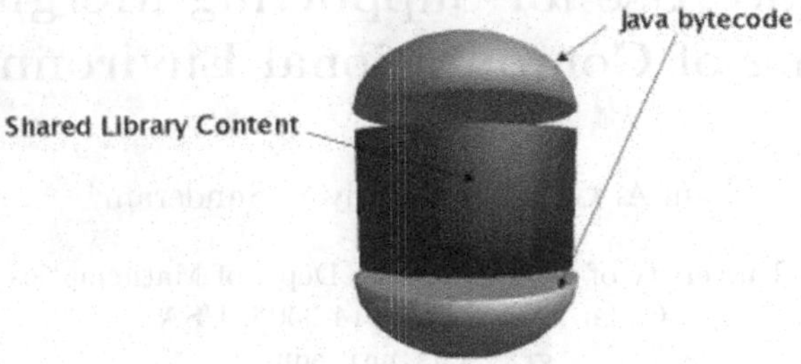

Fig. 1. IceT Processes consist of Java-Wrapped native libraries. The Java-wrapper serves as a portable packaging for detecting the appropriate native library to link with at runtime.

IceT utilizes Java and Java-wrapped native codes using the Java Native Interface (JNI, [14]) to achieve portability and performance (Figure 1). Our development to this point has shown that Java-wrapped native libraries provide a high amount of portability and performance (see, for example, [8], [7]). To facilitate process migration, IceT utilizes "IceT Repositories" for supplying Java-wrapped native codes based upon architectures and operating systems, such as depicted in Figure 2.

Example 1. Consider this simple agent-based scenario. Bob and Alice form a resource alliance. Once the alliance is formed, Bob would like to know the type and composition of resources that Alice has contributed to the alliance but has no a-priori knowledge of the resource complement before the alliance is formed. To this end, Bob uses IceT to *spawn* discovery agents upon Alice's resources in order to find platforms appropriate for subsequent use for a specific distributed computation – perhaps looking for Linux or Solaris computers with at least 256MB of memory on 100Mbit networks that are also capable of running PVM-based processes.

Alice's resources are very heterogeneous, with multiple operating systems and architectures. Since Bob's IceT resource discovery agent consists of a Java-wrapped front end that utilizes shared libraries and dynamic-link libraries to probe the resource's capabilities, the IceT environment detects which shared or dynamic link library is required, obtains the appropriate library from the IceT repository, links the Java front end with the library, and creates the agent process. The IceT agents utilize IceT's message-passing API to communicate the results of their findings back to Bob.

Several basic aspects of security arise from the above example.

1. The composition of Alice's resources might be privileged information. As such, when Bob's agents report their findings back to Bob, Alice should be able to insist that this information is encrypted such that only Bob would be able to decipher the information.

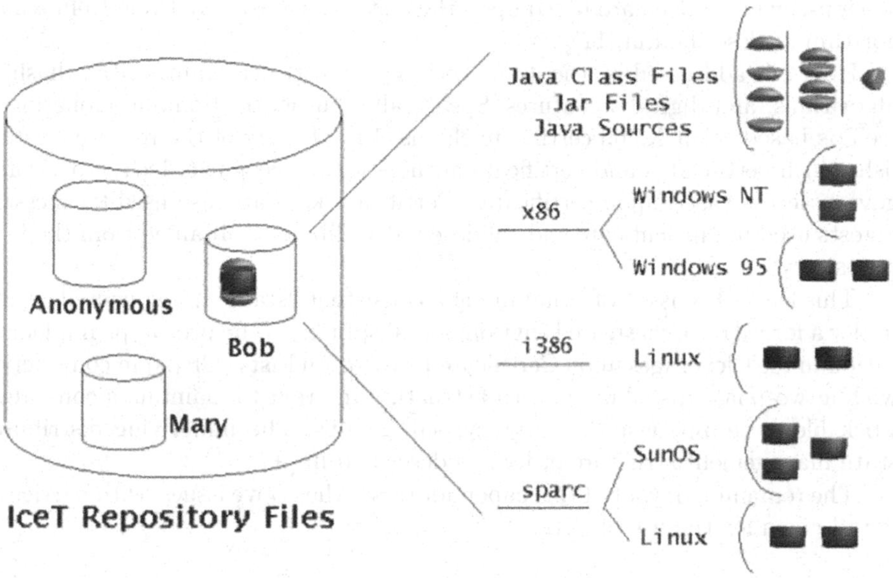

Fig. 2. IceT repositories store IceT processes. Users store Java-based front-ends and supporting native library formats for architectures that might be called upon to run the process. Access to the files in the repository are based upon certificate-based authentication.

2. In the alliance between Alice and Bob, Alice has opened her resources up to the local execution of *native* code that has been obtained from a remote IceT repository – perhaps not even originating from one of Bob's resources. Alice will need to be able to *authenticate* that it is truly Bob that has entered into the alliance, to *authorize* use of native libraries in local processes that were started by Bob, and to *validate* that the shared libraries that are being loaded and executed have not been altered by a third party.
3. In the event that the alliance is terminated (gracefully or otherwise), Alice's local processes should be unaffected. Furthermore, Alice should be able to revoke some or all of Bob's privileges and terminate any remaining rogue processes; or likewise preserve Bob's privileges and permit any of Bob's processes that remain to continue executing until Bob again re-enters into the alliance.

The issue reflected above in item 1 has been addressed in depth in projects such as Globus [4], Legion [10], Akenti [15], and Secure PVM [3]. IceT is developing a similar protocol based upon public key exchange and the subsequent generation of a private (session) key for encryption of communication over a possibly insecure transport layer. For example, in IceT, when multiple parties (more than two, that is) are enrolled in the alliance, the public key exchange and session-key generation are based upon the distributed n-party Diffie-Hellman algorithm as described in [11].

Item 2 is addressed by the use of certificates, certificate authorities, hashing algorithms, and digital signatures. Specifically, authentication and ciphering in IceT is based upon X.509 certificate chains [12]. Owners of the resources establish which certificates and certificate authorities to recognize. IceT repositories govern access based upon certificates. Certificate keys are also used for message digests used to authenticate and validate native libraries obtained from the IceT repository.

This leaves the issue of maintaining a consistent "state" in segmented groups under a forced or orchestrated environmental splitting. The prototype implementation in IceT leverages upon Certificate Revocation Lists (CRLs) in conjunction with a two-phase distributed control structure in order to maintain a consistent, workable state amongst the surviving sub groups. The underlying distributed state management structure of IceT is described in [9].

The remaining layout of this paper addresses the above issues, with particular detail given for the use of CRLs.

2 Using X.509 Certificates

In IceT, participants (users) are identified and authenticated using keys held in certificate form. Key use in IceT is closely aligned to that of standard certificate usage based upon SSL [6]. As such, *Certificate Authorities* (CA) come into play. A CA is often a trusted party that is responsible for speaking to the authenticity of a key. In IceT, it is often the case where the user holds self-certified keys. Typical IceT use allows for users to act as their own certificate authorities. This increases the flexibility of the model, but to the detriment of overall security inasmuch as the authenticity of one's identity is ultimately only as trustworthy as the trust placed in the certificate authority.

The assumption here is that colleagues that merge their resources together by entering into resource-sharing alliances are more intimately familiar with the person on the other other side of the network. Authentication of a colleagues credentials could be supported by previous communication, web-based key repositories, etc.

Example 2. Continuing with the scenario presented in Example 1... In the protocol for merging Bob and Alice's resources, Alice and Bob exchange their certificates. Alice is presented a certificate from Bob that has been self-certified. At this time, Alice – with the help of the IceT environment – is able to examine the certificate authority(ies) that have certified the authenticity of the key contained in the certificate. By default, Alice accepts several certificate authorities without question, but Bob is not one of them. Alice is then presented with information about the certificate and given the opportunity to deny or accept the certificate as valid.

Alice checks Bob's ssl-secured web site at his institution where various hashing algorithm results (MD5, SHA, etc [13]) of his certificate are posted. Among the information about the certificate that Alice was presented is the MD5 sum. This result matches the MD5 sum listed on Bob's web site. Alice then accepts this as sufficient authentication of Bob's certificate.

At the same time, Bob is presented with Alice's certificate. Alice's certificate has been certified by a CA that, in turn, has been certified by the CA run by the National Computational Science Alliance, NCSA. The CA established by the NCSA is well-known to Bob and accepted by default as a reputable CA. While Bob is unfamiliar with Alice's intermediate CA, the certification of the intermediate CA by the NCSA CA is deemed sufficient authentication.

Once certificates have been exchanged, merging of resources commences (Figure 3). Session keys may be generated for session encryption. Typical algorithms such as Diffie-Hellman [2] are called upon for session key generation.

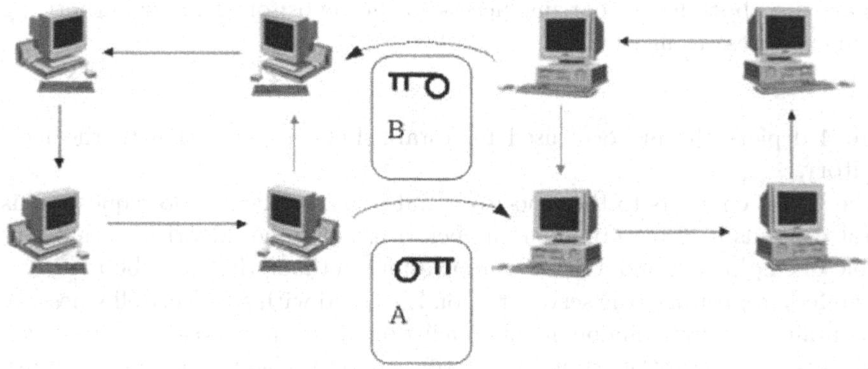

Fig. 3. Certificates are used to establish credentials when two or more IceT environments merge.

Certificates are also used for gaining access to and for acquiring files from IceT repositories. In order for a task to be spawned on a remote resource, the Java-based wrapper for the task as well as the appropriate form of the shared or dynamic link library must be located, soft-installed, and instantiated.

The protocol used to establish contact with an IceT repository is similar to the protocol used to establish a secure socket connection, with modifications to allow the use of certificate chains belonging to the merged collection of resources.

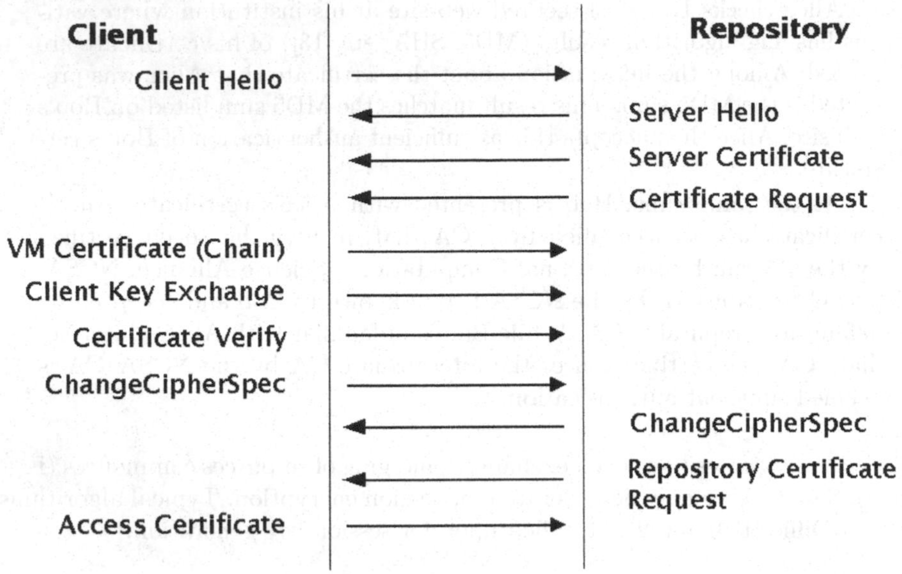

Fig. 4. The protocol established for accessing files in an IceT repository. Certificates are used both for authorizing access to the repository and for authorizing subsequent access to files.

Figure 4 depicts the protocol used for establishing a connection to the IceT repository.

The Client connects to the repository and issues a ClientHello request. This request contains a 32-byte random number, a list of cipher algorithms that it is capable of employing, and a list of compression methods that may be used.

The IceT repository (the server) responds in kind with a ServerHello message that contains a 32-byte random number, a list of ciphers and a list of compression algorithms. The server then submits a certificate to the client in order to establish the credentials of the repository. The server then issues a CertficateRequest message to the client, which contains a list of acceptable Distinguished Names (DN) that are recognized as credible. That is, the client must respond with a certificate that has been certified by one of the listed Distinguished Names.

The client responds in kind with a certificate chain associated with the combined certificates of the merged collection of resources, and initiates a session key exchange. This is followed-up with a request to verify the certificate and to subsequently utilize a specified cipher to encrypt subsequent communication.

If the IceT repository accepts the certificate as valid, a response to the change in cipher is issued, followed by a request for the certificate corresponding to the owner of the repository files is issued to the client. At this point, the client has authenticated themselves sufficiently to access the repository. However, another valid certificate is required to access files on the repository, as the owner of the IceT processes residing within the repository is allowed to impose access restrictions on the files therein.

For example, Alice may have placed her files in the repository, but only for subsequent access by Bob. If Mary also has files residing at this repository, her certificate is sufficient to gain access to the repository, but not sufficient to gain access to Bob's files. Mary *would* be able to access any file in the repository granted "anonymous" usage by the owners of the files, but would not be able to access Bob's files unless she could present a certificate that has been certified by a CA acceptable to Bob. This protocol also allows groups of resources to access files belonging to an outside party provided that one of the certificates in the certificate chain is permitted access to the files by the owner of the files. For instance, if Bob grants access to his files to Alice, and Alice and Mary have merged resources, Mary can used Alice's certificate that was exchanced in the merging of Alice and Mary's resources to gain access to Bob's files.

3 The Use of Certificate Revocation Lists

One critical aspect of the IceT implementation that distinguishes it from similarly-described projects, most notably the Globus GSI [1] and Akenti [15], is IceT's strong dependence on certificate revocation lists (CRLs). In IceT, the duration of one's privileges are not presumed to exist throughout the lifetime of the certificates that were used to gain access. While certificate lifetimes are typically on the order of months and years, access privileges in IceT are to be relinquished at the close of the collaborative resource alliance. An alliance lasting minutes, hours, or days would not be uncommon. Privileges are determined on a per-alliance (at every merging event) basis.

When two or more groups of resources are merged together, the IceT environment sets up a certificate revocation list service for each group. While access to resources is authenticated using certificates, that access is also revoked the instant the certificate that was used to gain access is revoked (Figure 6). Further, any tasks or pending operations that were granted privileges are able to have those privileges revoked when a splitting of the environment occurs.

Note also, that upon an environmental splitting, one's IceT environment may permit tasks and operations initiated by the suceeded group of resources to continue if it is desirable to do so. Such is the case where the splitting is caused by network problems or whenever re-merging is expected to occur in the future.

Certificate revocation lists are also used to govern access at the IceT repositories. Users are able to post files to the repository and to dynamically govern subsequent access based upon certificates. When resources merge together, the repository may be updated to reflect additional access privileges by certificates representing the complement set of resources. A user would post access updates to the access list to the repository that oversees the files. This access request has a "Check CRL" property set, so that the user's CRL is queried prior to servicing any file requests by users on the complement sets of resources.

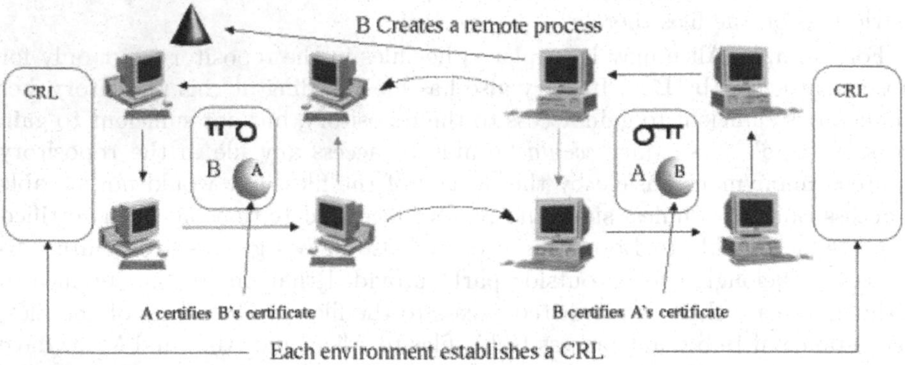

Fig. 5. The merging process involves each resource group acting as a certificate authority that scrutinizes the certificate being presented. If it passes the scrutiny, it is certified using the local certificate authority infrastructure. Privileges are then extended to the complement set of resources that permit the creation of local process on behalf of the remote resource group.

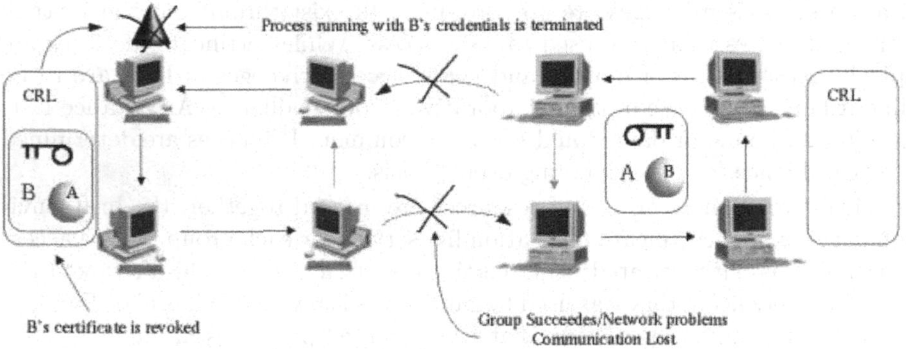

Fig. 6. If a sub-group of resources secedes, the certificate that was used to gain access is revoked. This, in turn, causes processes that were running locally under those credentials to be terminated.

Example 3. Alice and Bob are in the process of merging their resources. Part of this merging process consists of the exchanging of certificates (as in Figure 3). Once Bob receives and authenticates Alice's certificate, he certifies the certificate using his own CA utilities. Bob would then direct the IceT repository to allow Alice to access his files that are stored on that repository, and would direct the file server to query his local CRL each and every time Alice attempts to accesses his files.

Once Alice and Bob dissolve their alliance, Alice's certificate (that was certified by Bob's CA) is revoked by Bob and reflected in Bob's CRL. Since the repository is directed to query Bob's CRL list whenever Alice attempts to access Bob's files, attempts by Alice to access Bob's files at the repository would be denied.

The incorporation of certificate revocation lists into the authentication and privilege model of IceT in this manner provides a flexible mechanism for short-term sharing of resources, files, and applications. Another benefit derived from using certificate revocation lists is the added functionality that allows segmented environments and processes, that remain following a forced or orchestrated splitting of resources, to remain in tact or to follow a well-defined termination procedure. This greatly enhances the ability to maintain resilience in the presence of multiple, cascading resource or network failures, and is the subject of ongoing investigations.

4 Conclusions and Future Work

In this paper, we have presented the protocol for the use of X.509 certificates and certificate revocation lists in the IceT environment. The main use of these certificates is for facilitation of a dynamic, on-demand merging and splitting of resource groups. X.509 certificate chains provide us with a mechanism for identifying resources and authenticating users that have merged their resources together in a collective alliance. Certificate revocation lists provide the ability to have a well-defined behavioral model that restricts access and privileges when groups of resources split.

Ongoing investigations include benchmarks associated with the use of certificates. This includes (a.) the overhead introduced by the dependence upon certificate revocation lists for accessing files in the IceT repositories, (b.) the overhead associated with encrypting communication channels between groups of resources, and (c.) the overhead associated with the re-authentication of groups following a splitting of the environment. A long-term objective is to examine fault-tolerance issues associated with cascading failures within the environment and the recovery process.

References

1. Butler, R., Engert, D., Foster, I., Kesselman, C., Tuecke, S., Volmer, J., and Welch, V. Design and deployment of a national-scale auhentication infrastructure. *IEEE Computer*, 12 (2000), 60–66.
2. Diffie, W., and Hellman, M. New directions in cryptography. *IEEE Transactions on Information Theory IT*, 22 (Nov. 1976), 644–654.
3. Dunigan, T. H., and Venugopal, N. Secure PVM. Tech. Rep. TM-13203, Oak Ridge National Laboratories, Aug. 1996.

4. Foster, I., and Kesselman, C. Globus: A Metacomputing Infrastructure Toolkit. *International Journal of Supercomputing Applications* (May 1997).
5. Foster, I., and Kesselman, C. *The Grid, Blueprint for a new computing infrastructure.* Morgan Kaufmann Publishers, Inc., 1998.
6. Freier, A. O., Karlton, P., and Kocher, P. C. The SSL Protocol, version 3.0. Netscape Communications, Internet Draft. Available on-line at `http://www.netscape.com/eng/ssl3/`, Nov. 1996.
7. Getov, V., Gray, P., and Sunderam, V. MPI and Java-MPI: Contrasts and comparisons of low-level commnication performance. In *Proceedings of Supercomputing 99* (Nov. 1999).
8. Getov, V., Gray, P., and Sunderam, V. Aspects of portability and distributed execution for JNI-wrapped code. *Concurrency: Practice and Experience 20*, 11 (200), 1039–1050. ISSN 1040-3108.
9. Gray, P., Sunderam, V., Geist, A., and Scott, S. Bringing cross-cluster functionality to processes through the merging and splitting of virtual environments. Submitted to JavaGrande/ISCOPE-2001 conference, Jan. 2001.
10. Grimshaw, A., Wulf, W., and French, J. Legion: The Next Logical Step Toward a Nationwide Virtual Computer. Tech. rep., University of Virginia, 1994. T-R Number CS-94-21.
11. Ingemarsson, I., Tang, D., and Wong, C. A Conference Key Distribution System. *IEEE Transactions on Information Theory 28*, 5 (Sept. 1982), 714 – 720.
12. International Telecommunication Union. X.509: Information technology - open systems interconnection –the directory: Public-key and attribute certificate frameowrks, Mar. 2000. ITU-T Recommendation. To be published.
13. Kaufman, C., Perlman, R., and Speciner, M. *Network Security: Private communication in a public world.* Prentice Hall, Upper Saddle River, New Jersey 07458, 1995.
14. Sheng Liang. *The Java Native Method Interface: Programming Guide and Reference.* Addison–Wesley, 1998.
15. Thompson, M., Johnston, W., Mudambai, S., Hoo, G., Jackson, K., and Essiari, A. Certificate-based access control for widely distributed resources. In *Proceedings of the Eighth Usenix Security Symposium* (Aug. 99).

Data Management for Grid Environments

Heinz Stockinger[1], Omer F. Rana[2], Reagan Moore[3], and Andre Merzky[4]

[1] CERN, Switzerland, heinz.stockinger@cern.ch
[2] Cardiff University, UK, o.f.rana@cs.cf.ac.uk
[3] San Diego Supercomputer Center, USA, moore@sdsc.edu
[4] Konrad Zuse Zentrum, Berlin, Germany, merzky@zib.de

Abstract. An overview of research and development challenges for managing data in Grid environments is provided. We relate issues in data management at various levels of abstraction, from resources storing data sets, to metadata required to manage access to such data sets. A common set of services is defined as a possible way to manage the diverse set of data resources that could be part of a Grid environment.

1 Foundations

To identify data management needs in Grid environments an assessment can be based on existing components associated with data management, and one can view the Grid as integrating these. In this approach, the emphasis lies on providing interfaces between existing data storage and manipulation systems, to enable systems from different vendors and research groups to work seamlessly. The alternative approach is based on assessing application needs and requirements, and identifying missing functionality. We try to take a middle ground between these two approaches, and identify a set of common services, that may be suitable for both. At present, there are at least three communities that require access to distributed data sources: (1) Digital libraries (and distributed data collections). Digital libraries provide services for manipulating, presenting, discovering, browsing, and displaying digital objects. (2) Grid environments for processing distributed data, with applications ranging from distributed visualisation, to knowledge discovery and management. (3) Persistent archives for maintaining collections while the underlying technology changes. In this context, one must be able to deal with legacy systems that maintain such archives, and enable the migration, or wrapping, of these systems as new technology becomes available.

Hence, an architecture that is used to support Grid based environments should be consistent with the architectures needed to support digital libraries and persistent archives. An important common theme in all of these communities is the need to provide a uniform Application Programming Interface (API) for managing and accessing data in distributed sources. Consequently, specialised operations are needed to manage and manipulate digital objects with varying degrees of granularity. A digital object may be stored in a file system, as an object in an object-oriented database, as a Binary Large OBject (BLOB) in an object-relational database, or as a file in an archive, and should still utilise a

B. Hertzberger et al. (Eds.): HPCN Europe 2001, LNCS 2110, pp. 151–160, 2001.

common API. Hence, the concept of a data handling system that automates the management of digital objects stored in distributed data sources is the key concept in managing data in Grid environments.

Our objective is to identify a taxonomy for data management in scientific computing, to elicit requirements of applications that utilise Grid infrastructure, and to provide a means of classifying existing systems. The paper aims to complement work identified in [1], and identify building blocks that could be used to implement data management functions. The criteria needed for managing data in Grid applications are outlined, and based on the notion of (1) local services that must be provided by a given storage resource, (2) global services that need to be provided within a wider context to enable a better sharing of resources and data. It is intended that each data management and storage resource must subscribe to a global service, and must support some or all of the APIs required by such global services. Hence, a data storage resource is said to be 'Grid-enabled' if it can access and interact with these global services. We see two important considerations that distinguish current usage of data storage resources from Grid-enabled use – 'diversity' in operations and mechanisms, and 'performance' tolerance across mechanisms and resources.

2 Data Management - A Wider Perspective

The ability to process and manage data involves a number of common operations, the extent of which depends on the application. These operations include: *Data Pre-Processing and Formating* for translating raw data into a form that can be usefully analysed. Data processing may involve transforming a data set into a pre-defined range (for numeric data), and identifying (and sometimes filling in) missing data, for instance. The data processing stage is generally part of the 'data quality' check, to ensure that subsequent analysis of the data will lead to meaningful results. For numerical data, missing data may be handled using statistical techniques, such as an average value replacing a missing element. Metadata is often used in this context, for translating data from one form to another. Metadata can correspond to the structure of a data source, such as a database schema, which enables multiple data sources to be integrated. Alternatively, metadata may be summary data which identifies the principle features of the data being analysed, corresponding to some summary statistics. Generally, summary statistics have been generated for numeric data, however extensions of these approaches to data that is symbolic is a useful current extension.

Data Fusion for combining different types of data sources, to provide a unified data set. Data fusion generally requires a pre-processing stage as a necessity, in order for data generated by multiple experiments to be efficiently integrated. An alternative to fusion is Data Splitting, where a single data set is divided to facilitate processing of each sub-set in parallel, possibly though the use of filters which extract parts of the original data set based on pre-defined criteria.

Data Storage involves the recording of data on various media, ranging from disks to tapes, which can differ in their capacity and 'intelligence'. Data storage can involve data migration and replication between different storage media, based on a Hierarchical Storage Management (HSM) system, which vary based on access speed to storage capacity. As regards replication, large amounts of data might be transferred over the network (local or wide area) which imposes particular problems and restrictions. Specialised applications, such as scientific visualisation, require specialised data storage to enable data to be shuffled between the application program and secondary (or even tertiary) storage. Data storage hardware and software also differ quite significantly. Hardware resources can include RAID drives, where support is provided for stripping data across multiple disks, and parity support to ensure that lost data can either be reconstructed, or migrated when a disk fails. Large scale data storage units include HPSS (from IBM) and products from FileTek and AMPEX.

Data Analysis can range from analysing trends in pre-recorded data for hypothesis testing, to checking for data quality and filling in missing data. Data analysis is an important aspect of data management, and has been successfully employed in various scientific applications. Analysis approaches can range from evolutionary computing approaches such as neural networks and genetic algorithms, rule based approaches based on predicate/propositional logic to Case Based Reasoning systems, to statistical approaches such as regression analysis. The data analysis approach generally requires a prior data preparation (preprocessing) stage.

Query Estimation and Optimisation is essential if the data analysis is done on large amounts of data in a multi-user environment. Based on the input gained by a single user query, an estimate for how long it takes to transfer the required data to the computational unit may be made.

Visualisation, Navigation and Steering is the emerging area within data management that can range in complexity from output display on desktop machines, to specialised visualisation and (semi-) immersive environments such as ImmersaDesk and CAVE. Visualisation tools such as IRIS Explorer/Data Explorer have been widely used in the scientific community, and provide a useful way to both generate new applications, and for visualising the results of these applications. The next stage - providing computational steering support, will enable scientists to interact with their simulation in real time, and dynamically 'steer' the simulation towards a particular parameter space. Visualisation therefore becomes an enabler in creating and managing new types of scientific experiments, rather than as a passive means for viewing simulation output.

Data management is therefore a unified process that involves a number of stages, and it is important to view it as a whole. Each individual stage within the process has its own family of products and algorithms. To enable Grid-enabled devices to utilise these services, it is important to distinguish services between (1) management services, and (2) support and application services. Management services relate to operations and mechanisms offered within each storage resource, and global services with which the resource interacts - these are identified in Section 3. Support and application services relate to higher level operations which undertake correlations or aggregations on the stored data. These can be implemented in different ways, and are based on particular application needs and user preferences. To initiate a discussion, we identify some categories of such services in Section 4.

3 Support for Data Storage and Access

Support for data management can be divided into a number of services, each of which may be offered within a single system, or may constitute an aggregate set of operations from multiple systems, by multiple vendors. The primary objective in the context of Grid based systems is to support device and vendor heterogeneity, subject to some additional set of constraints, generally related to performance - which might not hold for data intensive high throughput applications. The criteria for categorising storage devices are: *Policy* is related to the division of services that should be directly supported within a storage device, and those that are required to be implemented by a user. The objective in many cases would be to provide a large number of services directly within the storage device. However, this may not be practical or useful from other perspectives, such as access or search time. A compromise is required between functionality and performance/throughput, and the Storage Policy should identify this. As far as possible, the Policy should be exposed to the user, and design decisions undertaken by a manufacturer should be made clear.

Operations identify the kinds of services that are required as a minimum within every Storage resource. These can be limited to 'read' and 'write' operations, or may also include additional services such as 'address lookup', 'access properties' (such as size of file, transfer rate), and 'error handling'. In each case, a minimal subset should be supported and identified. This 'Operations' set is particularly relevant when dealing with heterogeneity of the storage devices.

State Management relates to support for transactions and failure in a storage resource. Hence, a service to maintain and manage the state of a resource is important to enable the resource to be re-started in case of failure. State management can be undertaken within every resource, or a standard service for check-pointing the state of a resource may be provided.

Mechanism identifies how the operations supported within a storage resource are actually implemented. Each resource may implement read and write operations in a different way, and in some cases, this mechanism may need to be exposed to the user. In specialised tape storage devices, such as the ADT from Sony, intelligent mechanisms are provided to maintain a record of head movements during data access. These records may be made accessible to an external user or application, to enable head management by an application. To support Grid applications, it is important that storage resources should enable multiple mechanisms to co-exist, and not rely on the existence of a particular mechanism within a resource.

Errors/Exceptions can relate to a particular resource, or to data management within a group of resources. Hence, error handling may be undertaken locally, and be specific to mechanisms and operations supported within a given resource. However, each resource should also support error handling at a global level, for groups of resources being utilised within a given application.

Structure can relate to the physical organisation of a data storage resource, or the structure of the contents of the resource. Examples of the former case include number of disks, number of heads, access and transfer rates, and other physical characteristics of the storage resource. Structure of the contents may be specified using a database schema, which defines how content may be accessed or managed. This structure may therefore range from the structure of the file system, to data type based description of the contents. Structure information is important to global services, and for information services which utilise multiple data storage resources.

Accessing and transferring data from secondary and tertiary storage forms an important operation in data management. Generally, the data transfer needs to be undertaken from devices which have different access times, and support different access APIs and mechanisms. Data storage structure may also differ, requiring metadata support for describing this structure, a distinction is made between the structure of a data storage device and its content. This distinction is useful to maintain flexibility in the storage structure and in the managed data, enabling tools from multiple vendors to be used.

3.1 The Minimum Unit and Access Patterns

Access patterns for scientific computing applications are significantly different from business or commercial computing. In business and commercial computing, data access generally involves access to single units of data, often in random order. In scientific computing, access is generally more regular, such as within a loop of a numerical calculation, for instance. However, data access patterns are often very difficult to determine as regards the high throughput applications in HEP. Data structures in scientific high performance applications generally involve bulk data transfers based on arrays. Array accesses can be regular, generally as block, cyclic or block cyclic, or it may be irregular based on irregular strides across an array dimension. Access patterns can be for access to groups of data items, or to a group of files. The unit of transfer is generally determined by the storage device, ranging from single or multiple data items in database management systems such as RasDaMan [13], to file systems such as NFS or AFS, hierarchical storage systems such as the High Performance Storage System (HPSS), and network caches such as the Distributed Parallel Storage System (DPSS). In order to define standardised services, it is also useful to identify the basic unit of data transfer and access. This unit is dependent on the types of information processing services that utilise the stored data. We identify two types of data units: *Primitive types:* A primitive type is a float, integer or character that can be stored within a resource. The unit of storage is dependent on the type of programming language or application that makes use of the storage resource, and the storage medium being employed. Groups of such types may also be considered as primitive types, depending on the storage resource, and include arrays, images or binary objects. *Files:* An alternative unit of storage, not based on the type of processing or any associated semantics, may be an uninterpreted sequence of bytes – a file. The file may reside in a database or a conventional file system.

3.2 Support for Metadata Management

Metadata could relate to a hierarchical scheme for locating storage resources (such as LDAP), properties of resources that are externally visible, permissions and access rights, and information about the stored content. Developing a consensus on the desired representation for relationships is also important in the context of Grids, although this is not likely to be achieved in the short term. Relationships can have multiple types, including semantic/functional, spatial/structural, temporal/procedural. These relationships can be used to provide semantic operability between different databases, support type conversion between different object oriented systems, and manage ownership of distributed data objects. The ISO 13250 Topic Maps standard (defined below) is one candidate for managing relationships between data sources.

Metadata is also important for tagging attributes of data sets, such as digital objects (the bit streams that represent the data). The emergence of a standard tagging language, XML, has made it feasible to characterise information independently of the data bit stream. XML Document Type Definitions (DTDs) provide a way to structure the tagged attributes. The advantage of XML DTDs is that they support semi-structured organisations of data. This encompasses under one standard representation both unstructured sets, ordered lists, hierarchical graphs, and trees.

Digital objects that are created for undertaking a given analysis, may be re-used within another type of analysis, depending on their granularity. An approach that encapsulates each digital data set into a digital object, and then makes that object a member of an object class is forcing undue constraints. Conceptually, one should be able to manage the attributes that are required to define an object independently from the bits comprising the data set. This means it is possible to build constructors, in which data sets are turned into the particular object structure required by the chosen class. Data sets can be re-purposed if the class attributes are stored in a catalogue. Supporting and maintaining such catalogues then becomes crucial for the re-use of data sets. This requires templates for constructing objects from data sets. An example is the DataCutter system [9]. An XML DTD is used to define the structure of the data set. The DataCutter proxies are designed to read the associated DTD, and then process the data set based upon the structure defined within the DTD. This makes it possible to transform the structure of the data set for use by a particular object class, if the transformation is known to the XML DTD that represents the object class.

3.3 Standards

There are three standards which are of interest for the provision of data storage. One is the IEEE Reference Model for Open Storage Systems Interconnections (OSSI), previously known as the IEEE Mass Storage Reference Model, and the ISO standard for a Reference Model for Open Archival Information Systems, which is still under development. Whereas the IEEE standard is mainly concerned with architecture, interface and terminology specifications and standards, the ISO standard focuses more on necessary operational issues and interactions between different parts of an data archives. The ISO Topic Maps standard, on the other hand, deals with the contents of a data resource, and deriving an association between terms in the stored data. In this respect both the IEE and the

ISO standards can be seen as complementary. The first two of these descriptions are taken from Kleese [6].

IEEE's Open Storage Systems Interconnection (OSSI). This standard started out as the IEEE Mass Storage Reference Model in the 1980s, and is very much focused on technical details of mass storage systems, and contains specifications for storage media, drive technology and data management software. Recently, organisational functions have been added, and the description of connections and interactions with other storage systems have been stressed. Nowadays the Reference Model for OSSI provides a framework for the co-ordination of standards development for storage system interconnection, and provides a common perspective for existing standards. The descriptions used are independent of existing technologies and applications and are therefore flexible enough to accommodate advanced technologies and the expansion of user demands.

ISO's Open Archival Information System (OAIS). The OAIS standard aims to provide a framework for the operation of long term archives which serve a well specified community. Hereby issues like data submission, data storage and data dissemination are discussed. Every function is seen in its entirety, as not only describing technical details, but also human interventions and roles. For the purpose of this standard it has been decided that the information that is maintained needs long-term preservation, even if the OAIS itself will not exist through the whole time span. The OAIS standard addresses the issue of ingestion, encapsulation of data objects with attributes, and storage. OAIS does not, however, address the issue of technology obsolescence. Obsolescence can be handled by providing interoperability support between systems that support the migration onto new technology.

ISO 13250 Topic Maps standard provides a standardised notation for representing information about the structure of information resources used to define topics, and the relationships between topics, to support interoperability. A set of one or more interrelated documents that employs the notation defined by this International Standard is called a 'topic map'. In general, the structural information conveyed by topic maps includes: (1) groupings of addressable information objects around topics (occurrences), and (2) relationships between topics (associations). A topic map defines a multidimensional topic space – a space in which the locations are topics, and in which the distances between topics are measurable in terms of the number of intervening topics which must be visited in order to get from one topic to another, and the kinds of relationships that define the path from one topic to another, if any, through the intervening topics, if any.

4 Support and Application Services

Some communities like HEP are confronted with Grid-enabled data management issues more directly than computing intensive high performance applications that store only a small amount of data compared to their computational effort. Application specific concerns can subsequently be handled by developers directly, or by software libraries specific to a resource. We evaluate two domains of interest in the context of data management for Grid applications, each of which requires data access and analysis from multiple sources, maintained by different authorities, offering different levels of access privileges.

HEP: The HEP user community is distributed almost all around the globe. As for the next generation experiments starting in 2005 at CERN, large computing intensive applications run on several hundreds or even thousand CPUs in different computer centres and produce roughly 1 Petabyte of persistently stored data per year over 10 to 15 years. In particular, the collisions of particles in detectors produce large amounts of raw data that are processed and then transformed into reconstructed data. Once data is stored in disk subsystems and mass storage systems, it has to be replicated to hierarchically organised regional centres. This requires secure and fast migration and replication techniques which can be satisfied with protocol implementations like the GridFTP [12]. Once data is in place, distributed data analysis can be done by physicists at different regional and local centres. An important feature of the data is that about 90% is read-only data. This results in a simplification for replica synchronisation, as well as providing limited concurrency problems. However, the complexity for storing and replicating data is still high. Recently, at CERN the DataGrid [2] project has been initiated that deals with the management of these large amounts of data [5], and also involves job scheduling and application monitoring. The project does not only cover the High Energy Physics community, but also earth observation and bio-informatics. Thus, the research and development will serve several data intensive scientific communities.

Digital Sky Survey: The use of the Storage Resource Broker (SRB) [11] at SDSC as a data handling system, for the creation of an image collection for the 2MASS 2-micron all sky survey. The survey has 5 million images, comprising 10 Terabytes of data that need to be sorted from the temporal order as seen by the telescope, to a spatial order for images co-located in the same region of the sky. The data is read from an archive at Caltech, sorted into 140,000 containers for archival storage at SDSC, and accessed from a digital library at Caltech. When the collection is complete, the images will be replicated in the Caltech archive, and provided for use by the entire astronomy community.

5 Supporting Primitive and Global Services

We identify core services which should be supported on all storage resources to be employed for Grid-enabled applications. The definition of these services does not pre-suppose any particular implementation or mechanism. Such operations either occur very frequently, or are required to support the minimal functionality in Grid-enabled applications.

Data access operations: These operations include 'read' or 'write' to support data access and update. Such operations must also be supported by addressing information to locate data within the store. Such data access operations may also support access to collections of primitive data units through specialised operations.

Location transparency and global name space: These operations are used to discover the location of a data source, and to keep the location of a data source independent of its access method. Interoperability across heterogeneous systems is required for both federation of legacy data stores, as well as for migration of data stores onto new technology over time. A global name space may be provided through a catalogue. CERN, for instance, has defined a global Objectivity namespace, but has not yet provided the mechanisms to automate the management of the namespace. Digital objects may also be aggregated into containers – based on a different similarity criteria. Aggregation in containers is one way to also manage namespaces, providing a logical view of object location, as long as the data handling system is able to map from the desired object into the physical container. Also containers eliminate latencies when accessing multiple objects within the same container. This requires that the data handling system migrate the container to a disk cache, and support accesses against the cached container for multiple accesses.

Previledge and security operations: These operations are required to enable systems or individuals to access particular data sources, without having an account on the system hosting the data source. In Grid environments, it will not be possible to have an account on every storage system that may hold data of interest. To manage large distributed collections of data, the digital objects which act as data stores, must be owned by the collection, with access control lists managing permission for access independently of the local storage system. In this approach, the collection owns the data that is stored on each local storage system, meaning the collection must have a user ID on each storage system. Access control may be supported through a catalogue.

Persistence and Replication: Within data Grids large amounts of data need to be replicated to distributed sites over the wide area. When accessing such replicated data sources, it is difficult to utilise URLs as persistent identifiers, because the URL encapsulates the storage location and access protocol within the identifier. Similarly, it is not possible to use PURLs [7] because while a mapping to a unique identifier can be preserved, the mapping is not automatically updated when an object is moved. Persistence may be managed by moving all data objects within a data handling system. This makes it possible to update the storage location, the access protocol, and data object attributes entirely under application control. All manual intervention associated with maintaining the persistent global namespace has been eliminated.

Check-pointing and state management: A central service may be supported for recording the state of a transaction. An application user could subscribe to such a check-point service, to ensure that, on failure, it would be possible to re-build state and re-start the operation. This is not possible for all transactions however, and pre-defined points where a check-point is viable need to be identified by the user application.

6 Conclusion

In Grid computing and especially computing intensive data Grids, data management is a vital issue and has not been addressed to a large extend in high performance computing. In high throughput computing, several scientific domains have chosen the Grid as a basic part for a distributed computing model. Data management issues often need to be addressed in similar ways which requires protocols and standards.

References

1. A. Chervenak, I. Foster, C. Kesselman, C. Salisbury and S. Tuecke, The Data Grid: Towards an Architecture for the Distributed Management and Analysis of Large Scientific Datasets, See Web site at: `http://www.globus.org/`, 1999.
2. DataGrid Project. See web site at: `http://www.cern.ch/grid/`, 2001.
3. Global Grid Forum. See web site at `http://www.gridforum.org`, 2001.
4. GriPhyN Project. See web site at: `http://www.griphyn.org`, 2001.
5. W. Hoschek, J. Jaen-Martinez, A. Samar, H. Stockinger, K. Stockinger, Data Management in an International Data Grid Project, *1st IEEE, ACM International Workshop on Grid Computing (Grid'2000)*, Bangalore, India, December 2000.
6. K. Kleese, Data Management for High Performance Computing Users in the UK, *5th Cray/SGI MPP Workshop*, CINECA, Bologna, Italy, September 1999.
7. Persistent Uniform Resource Locator. See web site at: `http://purl.oclc.org/`, 2001.
8. Ron Oldfield, Summary of Existing Data Grids, white paper draft, Grid Forum, Remote Data Access Group. `http://www.gridforum.org`, 2000.
9. J. Saltz et al. The DataCutter Project: Middleware for Filtering Large Archival Scientific Datasets in a Grid Environment, University of Maryland, 2000. See web site at: `http://www.cs.umd.edu/projects/hpsl/ResearchAreas/DataCutter.htm`
10. H. Stockinger, K. Stockinger, E. Schikuta, I. Willers, Towards a Cost Model for Distributed and Replicated Data Stores, *9th Euromicro Workshop on Parallel and Distributed Processing (PDP 2001)*, IEEE Computer Society Press, Mantova, Italy, February 2001.
11. Storage Resource Broker Project. See web site at `http://www.npaci.edu/DICE/SRB/`, 2000.
12. The Globus Project - White Paper. "GridFTP: Universal Data Transfer for the Grid", September 2000. See Web site at: `http://www.globus.org/`
13. The RasDaMan Project. See web site at: `http://www.forwiss.tu-muenchen.de/~rasdaman/`, 2000.
14. "XML: Extensible Markup Language". See web site at: `http://www.xml.com/`
15. See Web site at: `http://www.ccds.org/RP9905/RP9905.html`

Mobile Agents for Distributed and Dynamically Balanced Optimization Applications

Rocco Aversa, Beniamino Di Martino,
Nicola Mazzocca, and Salvatore Venticinque

Dipartimento di Ingegneria dell' Informazione - 2nd University of Naples - Italy
Real Casa dell'Annunziata - via Roma, 29
81031 Aversa (CE) - ITALY
{beniamino.dimartino,n.mazzocca}@unina.it
{rocco.aversa,salvatore.venticinque}@unina2.it

Abstract. The Mobile Agent paradigm can increase the flexibility in the creation of distributed applications (and the restructuring of sequential applications for distributed systems), and can in particular provide with a robust framework for managing dynamical workload balancing. In this paper we show how the restructuring of a sequential code implementing an irregularly structured application, a combinatorial optimization performed with Branch & Bound (B&B) technique, with adoption of the mobile agent model, allows for yielding a dynamically load-balanced distributed version.
The application of the mobile agent model is discussed, with respect to the solutions adopted for knowledge sharing, communication, load balancing, and termination condition.

1 Introduction

The design of High Performance programs running on heterogeneous distributed architectures asks for simple and effective load balancing mechanisms, able to exploit the different performance characteristics of the computing nodes and communication networks. This issue certainly arises with irregular computations, but even the performance figures of regular computations can be hardly affected when executed on heterogeneous loosely coupled distributed architectures.

The Mobile Agents model [8], besides to represent an effective alternative to the Client-Server paradigm in several application fields such as e-commerce, brokering, distributed information retrieval, telecommunication services [6], has the potential to provide a flexible framework to face the challenges of High Performance Computing, especially when targeted towards heterogeneous distributed architectures. Several characteristics of potential benefit for scientific distributed computing can be provided by the adoption of the mobile agent technology, as shown in the literature [2,3,4]: they range from network load reduction, heterogeneity, dynamic adaptivity, fault-tolerance to portability to paradigm-oriented development. Mobile agents meets the requirements of the heterogeneous distributed computing since the agents are naturally heterogeneous, they reduce the network load and overcome network latency by means of the mechanism of

B. Hertzberger et al. (Eds.): HPCN Europe 2001, LNCS 2110, pp. 161–170, 2001.

the migration, they adapt dynamically to the computing platform through the migration and cloning, and finally the different tasks of a sequential algorithm can be embedded into different mobile agents thus simplifying the parallelization of the sequential code.

We show, through a case-study, how the restructuring of a sequential code implementing an irregular algorithm, with adoption of the mobile agent model, allows to yield a dynamically load-balanced distributed version of the algorithm without completely rethinking its structure, reusing a great deal of the sequential code and trying to exploit the heterogeneity of the target computing architecture.

The chosen application performs a combinatorial optimization, with the Branch & Bound (B&B) technique. Branch & Bound (B&B) is a technique widely used to solve combinatorial optimization problems, which occur frequently in physics and engineering science. B&B is a typical example of techniques for irregularly structured problems.

We adopted a Mobile Agent programming framework developed by the IBM Japan research center, the *Aglet mobile agents Workbench* [7]. We have augmented the Aglet Workbench by providing (1) an extension of the *multicast* communication primitive, and (2) by designing and implementing a *dynamic workload balancing* service.

The remainder of the paper proceeds as follows: a conceptual description of B&B technique, together with advantages and problems arising within a parallel framework is presented in Section 2. Section 3 describes the Aglet mobile agents Workbench and the set of extensions we added to it. In Section 4 we describe how a dynamically balanced version of an Branch & Bound optimization application is obtained by means of the extended Aglet workbench. Finally (sec. 5) we give some concluding remarks.

2 The Branch and Bound Technique

A *discrete optimization problem* consists in searching the optimal value (maximum or minimum) of a function $f : \boldsymbol{x} \in \mathcal{Z}^n \rightarrow \mathcal{R}$, and the solution $\boldsymbol{x} = \{x_1, \ldots, x_n\}$ in which the function's value is optimal. $f(\boldsymbol{x})$ is said *cost function*, and its domain is generally defined by means of a set of m constraints on the points of the definition space. Constraints are generally expressed by a set of inequalities:

$$\sum_{i=1}^{n} a_{i,j} x_i \leq b_j \qquad \forall j \in \{1, \ldots, m\} \tag{1}$$

and they define the set of feasible values for the x_i variables (the *solutions space* of the problem).

Branch & Bound is a class of methods solving such problems according to a *divide & conquer* strategy. The initial solution space is recursively divided in subspaces, until attaining to the individual solutions; such a recursive division can be represented by a (abstract) tree: the nodes of this tree represent the solution subspaces obtained by dividing the parent subspace, the leaf nodes represent the solutions of the problem, and the tree traversal represents the recursive operation of dividing and conquering the problem.

The method is enumerative, but it aims to a non-exhaustive scanning of the solutions space. This goal is achieved by estimating the best feasible solution for each subproblem, without expanding the tree node, or trying to prove that there are no feasible solutions for a subproblem, whose value is better than the *current* best value. (It is assumed that a best feasible solution *estimation function* has been devised, to be computed for each subproblem.) This latter situation corresponds to the so called *pruning* of a search subtree.

B&B algorithms can be parallelized at a *fine* or *coarse grain* level. The fine grain parallelization involves the computations related to each subproblem, such as the computation of the estimation function, or the verification of constraints defining feasible solutions. The coarse grain parallelization involves the overall tree traversal: there are several computation processes concurrently traversing a different branch of the search tree. In this case the effects of the parallelism are not limited to a speed up of the algorithmic steps. Indeed, the search tree explored is generally different from the one traversed by the sequential algorithm. As a result, the resolution time can be lower, even though the number of explored nodes can be greater than in the sequential case. As it has been demonstrated in [5], this approach exhibits *anomalies*, so that the parallelization does not guarantee an improvement in the performance. For most practical problems, however, the bigger the problem size, the larger are the benefits of the parallel traversal of the search tree.

Parallel B&B algorithms can be categorized on the basis of four features [10]:

1. how information on the global state of the computation is shared among processors (we refer to such information as the *knowledge* generated by the algorithm);
2. how the knowledge is utilized by each process;
3. how the workload is divided among the processes;
4. how the processes communicate and synchronize among them.

The knowledge is *global* if there is a single common repository for the state of the computation at any moment, accessed by all processes, otherwise it is *local*. In this latter case, processes have their own knowledge bases, which must be kept consistent to a certain degree to speed up the tree traversal and to balance the workload.

With respect to the knowledge use, the algorithms are characterized by: (1) the *reaction strategy* of processes to the knowledge update (it can range from an instantaneous reaction, to ignore it until the next decision is to be taken); (2) the *dominance rule* among nodes (a node *dominates* another if its solutions best value is better than the lower bound on the solutions of the other): it can be *partial* or *global*, if a node can be eliminated only by a dominant node belonging to the same subtree traversed by a process, or by any other dominant node; (3) the *search strategy*, that can be *breadth*, *depth* or *best* first.

With regard to the workload division, if all generated subproblems are stored in a common knowledge base, to each process that becomes idle the most promising subproblem is assigned (on the basis of an heuristic evaluation). If the state of the computation is distributed among processes (local knowledge), then a *workload balancing* strategy has to be established, consisting of a relocation of subproblems not yet traversed.

3 Extensions to the Aglet Mobile Agent System

The Aglet Workbench is a framework for programming mobile networks agents in Java developed by IBM Japan research group [7].

An *aglet* (agile applet) is a lightweight Java object that can move to any remote host that supports the Java Virtual Machine. An Aglet server program (*Tahiti*) provides the agents with an execution environment that allows for an aglet to be created and disposed, to be halted and dispatched to another host belonging to the computing environment, to be cloned, and, of course, to communicate with all the other aglets.

We augmented the Aglet Workbench by providing (1) an extension of the *multicast* communication primitive, and (2) by designing and implementing a *dynamic workload balancing* service.

3.1 Extension of the Multicast Primitive

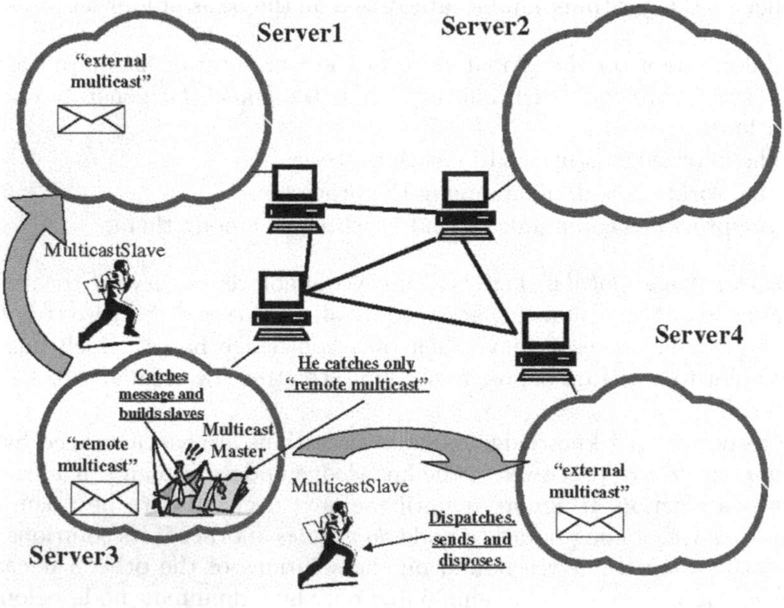

Fig. 1. The *Multicast* Extension to the Aglet System.

The Aglet Workbench supports the *multicast* primitive, but only within a local context: an *aglet* that requires to receive a multicast message with a specific label needs to subscribe to that kind of *multicast* message. We extended this mechanism by allowing a *multicast* message to reach both the local and remote *aglets* using a *multicast aglet server* running on each host. Such aglet server performs the following actions (fig. 3.1):

- subscribes to a specific *multicast* message labeled "remote multicast";
- captures in a local context a "remote multicast" message;
- dispatches a *multicast slave aglet* to each host of the computing environment with a copy of the *multicast* message.

Each *multicast slave aglet* on its arrival to destination:

- sends in a local context the *multicast* message but with a new label ("external multicast"), so allowing any *aglet* on that host to receive the message;
- disposes itself.

To summarize, an *aglet* that wishes to send a remote multicast, must label its message with the label "remote multicast", while an *aglet* interested to receive multicast messages from remote *aglets* has to subscribe to a *multicast* message labeled "external multicast".

3.2 The Dynamic Workload Balancing Service

We augmented the Aglet workbench with an extension providing a service for dynamic workload balancing that can be easily customizable to any user application developed within the Workbench.

Fig. 3.2 shows at a glance the framework we have implemented.

The framework is composed of a *coordinator agent* which controls the load balancing and an Aglet class, `Aglet_LC`; this latter must be inherited by the user class.

The coordinator agent communicates, by message passing, only with the `Aglet_LC` class support. In order to use the support, the user class must override some methods inherited by the superclass `Aglet_LC` and set some variables. The implementation of these functions depends on the specific user application.

The coordinator manages a first list of registered workers and a second list of available free hosts. When the user Agent's execution starts, the coordinator is created and executes a polling interaction with the working agents registered with it, in order to know the state of their computation, that is the percentage of computation performed, with respect to the computation amount assigned to it. The registered agent's references are stored in a vector, ordered according to their computation state; the ordering of this vector thus represents the relative speed of each worker with respect to the others. It also gives a representation of the state of the computation.

A load unbalance event occurs when:

1. a worker ends the computation assigned to it, and becomes idle;
2. the slowest worker has completed a percentage of the computation amount assigned to it which is far below the percentage completed by the fastest worker (determined by a fixed threshold on the difference of the two percentage).

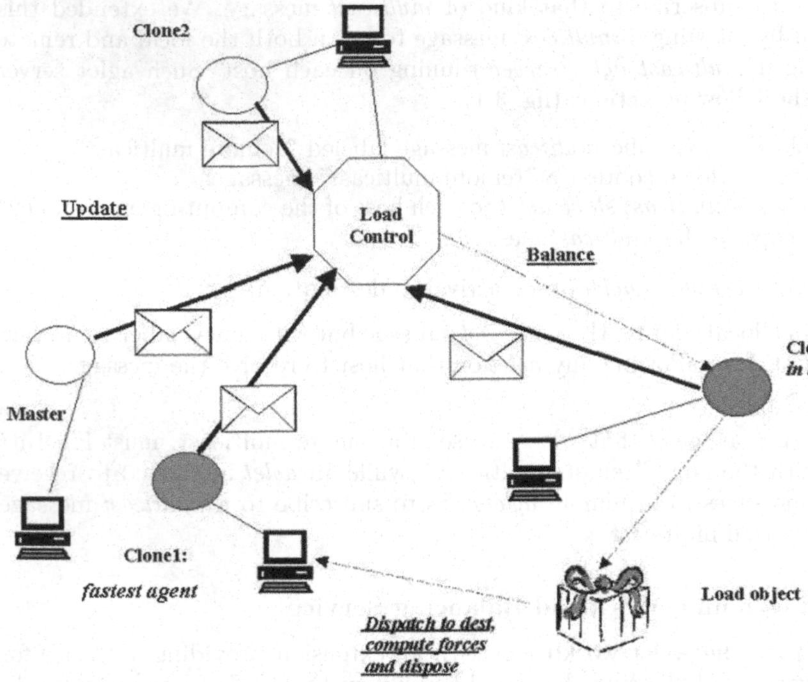

Fig. 2. Graphical Description of the Workload Balancing Strategy Using a *Coordinator* Agent.

In the first case, the idle worker notifies the coordinator of its availability to receive workload; the coordinator then asks the slowest worker (according to the vector of computation states) to send a portion of its workload to the idle worker.

In the second case, the following two situations can occur:

1. the list of available hosts is not empty: the coordinator then creates a generic slave agent and dispatches it to the target machine, then asks the slowest worker to send a part of its load to the slave.
2. the list of available hosts is empty: the coordinator then asks the slowest worker to send part of its load to the fastest worker.

The coordinator periodically updates the vector of the computation states, by executing a polling at a fixed, user-defined, frequency.

In the following some more details are provided on how the workload is represented and exchanged. Part of load is sent from an agent to another by means of a message carrying an object belonging to a class defined by the user (`UserWorkload`), implementing the `Serializable` (which is a standard Java interface) and the `Load` interface. The data declared in class `UserWorkload` represent the load and the information to elaborate it, while the method `exec` of the `Load` interface defines the way to elaborate the data. Such a method is called by the receiver to perform the computation. The user can define different kinds of `UserWorkload` classes, according to the kind of computation to send (in different points of the program we can treat different problems) or according to the destination agent (slave needs all data and code, while worker can reuse its code or its data).

Finally we describe the actions to be performed by the user of the framework, in order to specialize the balancing algorithm and to grant its correctness. Some methods have to be overrided, in order to specify how:

- to calculate and communicate the percentage of work done;
- to create and prepare the load assigned to the fast/idle worker or to the slave;
- to specify if the fast/idle worker or the slave must return the result of its computation;
- to elaborate the results eventually received from the fast/idle worker or the slave.

To personalize the behavior of the coordinator the user must provide the following information:

- the frequency rate of the polling for updating the vector of computation states;
- the threshold difference between the percentages of computations done which triggers the load transfer among the workers;
- all the URLs of the hosts of the cluster.

To personalize the behavior of the workers the user can set some variables inherited from the `Aglet_LC` class. In particular he can choose:

- if a worker wants to receive any load from the others;
- if it needs to be given back any result possibly computed by the workload passed to another worker agent.

4 The Distributed Mobile-Agent Based B&B

The algorithm, whose distributed implementation we present as case study, solves the $(0-1)$ *knapsack problem*, which can be stated as follows [9]:

$$minimize \sum_{i=1}^{n} c_i x_i \tag{2}$$

$$with \; x_i \in \{0,1\} \;\; \forall i \in \{1,\ldots,n\}$$

$$subject\ to \sum_{i=1}^{n} a_{i,j} x_i \geq b_j \;\; \forall j \in \{1,\ldots,m\} \tag{3}$$

where $a_{i,j}$ and b_j are positive integers.

The distributed implementation, carried out using the Aglet workbench augmented with the above described extensions, presents the following characteristics, with respect to the categorization of B&B algorithms presented above: (1) each agent process takes, while being dispatched from a host to an other, its own knowledge base (local knowledge); (2) agents react instantaneously to knowledge updates, because the agent's main function is a thread having a priority lower than the message handler; (3) a global dominance rule is adopted, since the current optimal value is broadcasted to all agent processes as soon as it is updated; (4) the search strategy is depth-first, as it is more suited in the case of local knowledge.

The algorithm consists of the following phases:

- the first agent decomposes the assigned problem in P different subproblems (where P is the number of available computing nodes) and clones itself P times. Each clone takes a single subproblem and dispatches itself on a target machine.
- Each worker explores its own subtree with depth-first strategy and updates its current best value, sending it to the other agents through a multicast communication, when it generates a feasible solution.
- Each agent receive the local current best value, it compares this value with the global best one and eventually update it.

About the load balancing, every clone gives to the coordinator the availability to receive any load from the slow workers. The `UserWorkload` class contains an array of subproblems while the method `exec` of the `Load` interface defines how the receiver worker can store the subproblems in its own array. Thus, when the load unbalance is detected, the slowest worker activated by the coordinator creates an object of the `UserWorkload` class (by means of a constructor that determines the number of subproblems to dispatch), and send it to the fastest agent. The first agent reads the data input and write the final results after recognizing the validity of the termination condition. This agent receives the idle message from all other workers, sends a dispose multicast message and terminates.

5 Experimental Results

The case study implementation has been experimentally evaluated, executing it on a local area Ethernet connected heterogeneous cluster of 8 nodes: 4 Intel Pentium III (800 Mhz) and 4 Intel Pentium Celeron (633 Mhz) PCs, equipped with WindowsNT operating systems.

Two versions of the distributed implementation are compared, one (`balanced`) making use the dynamical workload balancing service, and the unbalanced one (`unbalanced`).

In table. 1 the (percentaged) ratio of speed-ups of the two versions is reported, with varying the processors' number. The figures refer to a problem size of 40 variables and 10 constraints, at varying the processors number from 2 to 8.

The figures show that the load balanced version performs better than the unbalanced version for all the values of the parameters considered.

Table 1. The (percentaged) ratio of speed-ups of the two versions ((`unbalanced` / `balanced`) with varying the processors' number from 2 to 8.

Num. Proc.	4	6	8
speed-up ratio	0.96 %	1.90 %	1.2 %

The effect of load balancing on the improvement of workers' processor utilization is shown by Fig. 5, which shows the (percentaged) mean utilization time ratio of the workers, for the balanced and unbalanced versions, together with their standard deviation.

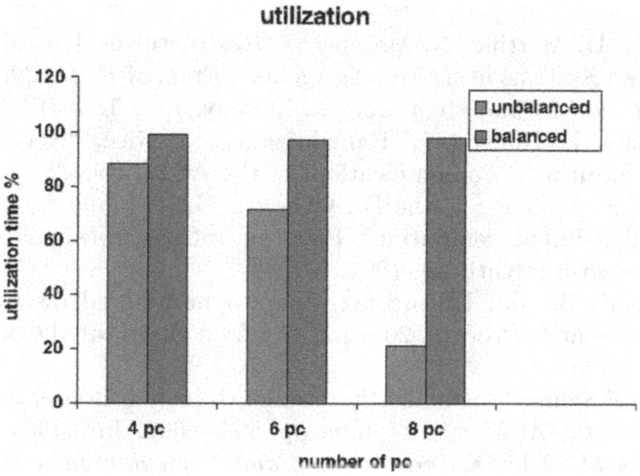

Fig. 3. Mean Utilization Time (Percentaged) of the Workers, for the Balanced and Unbalanced Versions.

The figure shows that the balancing is effective regardless of the number of processors utilized: the mean utilization time is always near to 100%; in the unbalanced version instead the mean utilization time worsens with increasing the number of processors: this means that the mean idle time of processes increases with the number of processors utilized.

6 Conclusions

In this paper we have addressed the problem of an effective utilization of the mobile agent model for High Performance Distributed Computing, particularly focusing on the dynamical workload balancing problem in irregular applications. We have shown, through a case-study, how the design of a dynamical workload strategy, based on the cloning and migration feasibility of the agent model, and coupled with coordination and reactivity properties of the balancing algorithm, provides promising results over an heterogeneously distributed architecture. In addition the restructuring of the sequential code implementing the case, with adoption of the mobile agent model, was achieved by reusing a great deal of the sequential code's structure.

Acknowledgments

We would like to thank the anonymous referees for their precise, insightful and helpful suggestions.

References

1. R. Aversa, B. Di Martino, N. Mazzocca, "Restructuring Irregular Computations for Distributed Systems using Mobile Agents", Proc. of *PARA2000, Int. Workshop on Applied Parallel Computing*, Bergen, Norway, June 18 - 21, 2000.
2. T. Drashansky, E. Houstis, N. Ramakrishnan, J. Rice, "Networked Agents for Scientific Computing", Communications of the ACM, vol. 42, n. 3, March 1999.
3. Gray R., Kotz D., Nog S., Rus D., Cybenko G., "Mobile agents: the next generation in distributed computing" Proc. of Int. Symposium on Parallel Algorithms/Architecture Synthesis, 1997.
4. H. Kuang, L.F. Bic, M. Dillencourt, "Paradigm-oriented distributed computing using mobile agents", Proc. of. 20th Int. Conf. on Distributed Computing Systems, 2000.
5. H.T. Lai and S.Sahni, "Anomalies in Parallel Branch & Bound Algorithms", *Communications of the ACM*, vol. 27, n. 6, pp. 594 – 602, Jun. 1984.
6. D. Lange and M. Oshima, *Programming and Deploying Java Mobile Agents with Aglets*, Addison-Wesley, Reading (MA), 1998.
7. D. Lange and M. Oshima, "Seven good reasons for Mobile Agents", Communications of the ACM, vol. 42, n. 3, March 1999.
8. V. A. Pham, A. Karmouch, "Mobile software agents: an overview", IEEE Communications Magazine, Vol. 36(7), July 1998, pp. 26 -37.
9. C. Ribeiro, "Parallel Computer Models and Combinatorial Algorithms", *Annals of Discrete Mathematics*, North-Holland, pp. 325 – 364, 1987.
10. H.W.J. Trienekens, "Parallel Branch&Bound Algorithms", *Ph.D. Thesis at Erasmus Universiteit-Rotterdam*, Nov. 1990.

Track II

End User Applications

A Comparison of Three Parallel Algorithms for 3D Reconstruction

Dan C. Marinescu, Yongchang Ji, and Robert E. Lynch

Department of Computer Sciences, Purdue University
West Lafayette, IN 47907
dcm,ji,rel@cs.purdue.edu

Abstract. In this paper we report on three parallel algorithms used for 3D reconstruction of asymmetric objects from their 2D projections. We discuss their computational, communication, I/O, and space requirements and present some performance data.

1 Introduction and Motivation

Very often data intensive computations demand computing resources, CPU cycles, main memory, and secondary storage, well beyond those available on sequential computers. To solve data intensive problems using parallel systems, either clusters of PCs or high performance systems, we have to design families of parallel algorithms and select the most suitable one for the specific problem size, the architecture of the target system, the performance of its input-output sub-system, and the amount of resources available, including primary memory. On a system with fast processors but with limited memory we would choose an algorithms whose implementation has a relatively small memory footprint, even though we may need to carry out redundant computations. The algorithm for a system with large memory and slow I/O will attempt to minimize the number of I/O operations. Such space-time tradeoffs are common in the design of efficient parallel algorithms.

We are primarily interested in the application of 3D reconstruction of an asymmetric object from 2D projections to atomic structure determination of biological macromolecules like viruses. The goal is to determine the 3D electron density map of a virus structure. Once an electron density map at high resolution is available, structural biologists are able to place groups of atoms on this electron density "cloud", the so-called atomic modeling.

In cryo-TEM 2D projections of a virus are gathered experimentally and their orientations are calculated through an iterative process. Given a set of N 2D projections and their orientations, we carry out the reconstruction in the Fourier domain. First, we perform a 2D FFT of each projection. Each projection corresponds to a plane in the Fourier domain and each plane intersects the 3D volume at non-integral grid points. Our goal is to determine the values of the Fourier transform of the electron density of the virus at every point of the 3D lattice. Finally, a Fourier synthesis allows us to compute the electron density.

B. Hertzberger et al. (Eds.): HPCN Europe 2001, LNCS 2110, pp. 173–182, 2001.

The basic algorithm for the reconstruction of asymmetric objects in Cartesian coordinates is presented in [Lyn97] and [Lyn99] and illustrated in Figure 1. An improvement of the algorithm that reduces the number of arithmetic operations by two orders of magnitude is outlined in [Lyn00], and a detailed presentation and analysis of the algorithms together with preliminary experimental results are given in [Lyn00a].

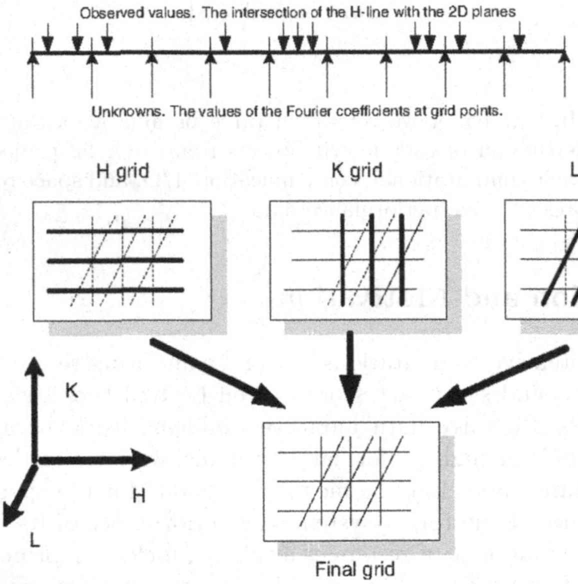

Fig. 1. One-dimensional interpolation on a line parallel with the axes of the 3D grid in the Fourier domain. The values of the function at grid points (arrows pointing upwards) are unknown, the ones corresponding to arrows pointing downwards are the known, they are the observed values. We can determine by interpolation the values at grid points for all h-lines and construct an H-grid. Similarly we obtain the K-grid and the L-grid.

The amount of experimental data for the reconstruction of an asymmetric object is considerably larger than the one necessary for a symmetric one. While a typical reconstruction of the shell of an icosahedral virus at, say, 20Å resolution may require a few hundreds projections, e.g., 300, the reconstruction of an asymmetric object of the same size and at the same resolution would require 60 times more data, e.g., 18,000 projections.

Cryo-TEM is appealing to structural biologists because crystallizing a virus is sometimes impossible and even when possible, it is technically more difficult than preparing samples for microscopy. Thus the desire to increase the resolution of Cryo-EM methods to the 5Å range. Recently, successful results in the 7–7.5Å range have been reported, [Böt97], [Con97]. But increasing the resolution of the 3D reconstruction process requires more experimental data. It is estimated that the number of views to obtain high resolution electron density maps from Cryo-EM micrographs should increase by two order of magnitude from the current hundreds to tens of thousands.

The amount of experimental data further increases when structural studies of even larger virus-antibody complexes are attempted. The use of larger zero fill ratios to improve the accuracy of 3D reconstruction [Lyn00a] increases the number of arithmetic operations and the amount of space needed for reconstruction.

Thus it is not unrealistic to expect an increase in the volume of experimental data for high resolution asymmetric objects by three to four orders of magnitude in the near future. Even though nowadays faster processors and larger amounts of primary and secondary storage are available at a relatively low cost, the 3D reconstruction of asymmetric objects at high resolution requires computing resources, CPU cycles, and primary and secondary storage, well beyond those provided by a single system. Thus the need for parallel algorithms.

The paper is organized as follows. In Section 2 we describe three parallel algorithms and present a succinct analysis of their computational and communication complexity as well as their space requirements. Then, in Section 3 we discuss performance results.

2 Basic Concepts: Algorithms and Notations

In this section, we present three algorithms for reconstruction which have different space, communication, I/O, and computational requirements. These algorithms apply to the case that the parallel system does not support parallel I/O operations. To avoid the increased time which results from I/O contention, we use a Coordinator-Worker paradigm. The coordinator (a processor), C, reads the the input images and orientations and then distributes them to the workers, $W_i, 1 \leq i \leq P-1$; C uses gather operations to collect the final result.

In addition to using (x, y, z) to denote a grid point in 'real space' and (h, k, ℓ) to denote a grid point in 'reciprocal space', we use the following notations:

$N \times N$: the number of pixels in one projection.

$K, \quad K \geq 1$: zero fill aspect ratio; the $N \times N$ pixel array is put into a $KN \times KN$ array with zero fill before the DFT is computed.

P : the number of processors.

V : the number of projections.

$v = \lceil V/P \rceil$: the number of projections processed by W_i in Algorithms I and III.

$v' = V - (P-1)v$: the number of projections processed by C in Algorithms I and III.

q' : the number of projections processed by one processor in Step 2 of Algorithm III.

$q = Pq'$: the number of projections read in one I/O operation in Algorithm III.

$q_{stages} = V/q$: the number of pipeline stages in Algorithm III.

R : the number of grid in reciprocal space to obtain the desired resolution.

n : the order of the symmetry group; one projection can be used n times with different orientations.

h-slab : complex DFT values, $\mathcal{F}(h,k,\ell)$, for $(KN)^2$ grid points (k,ℓ) for each of $KN/2P$ grid lines h.

z-slab : complex DFT values, $\mathtt{F}(h,y,z)$, for KN^2 grid points (h,y) for each of N/P grid lines z.

Algorithm I. In Algorithm I, there are no replicated computations and there is no overlapping of communication with computations. Each processor must have space to hold the entire complex 3D DFT $\mathcal{F}(h,k,\ell)$. Thus the algorithm is space-intensive.

The algorithm consists of the followings steps:

1. C reads all V projections, keeps v' projections, and scatters a different group of v projections to each processor W_i, $1 \leq i \leq P-1$.
2. C reads the orientation file and scatters a group of v orientations to the W_i, $1 \leq i \leq P-1$, so that each processor has the orientation of each projection assigned to it.
3. Each processor performs a 2D DFT for each of the v (or v') projections assigned to it.
4. Each processor uses the entire 3D volume in the reciprocal space, and uses the 2D DFT of each projection assigned to it to compute values at grid points scattered throughout the entire volume. It does so by intersecting each projection with the h, k, and ℓ lines. Then it estimates of the 3D DFT of the electron density at grid points (h,k,ℓ) near the plane of each projection and inside the hemisphere of radius R;
5. If the object has n-fold symmetry, then the interpolation of Step 4 is repeated for each of the other $n-1$ symmetrically related orientations of the 2D DFT of each projection.
6. The P processors perform a global exchange. At the end, the i-th processor has its h-slab of interpolated values, as computed by it and by each of the other the processors. It averages these values and obtains the final estimate of the 3D DFT at all grid points in its h-slab and inside the hemisphere of radius R.
7. The i-th processor performs KN/P inverse 2D DFTs in its h-slab: $\mathcal{F}(h,k,\ell) \rightarrow \mathbf{F}(h,y,z)$,
8. The P processors perform a global exchange and each ends up with its z-slab.
9. Each processor performs $(KN)^2/P$ 1D DFTs to obtain the final estimate of the electron density in its z-slab: $\mathbf{F}(h,y,z) \rightarrow f(x,y,z)$.
10. Each W_i, $1 \leq i \leq P-1$, sends its slab of electron density to C, which writes it onto the output file.

The computation complexity for each processor by Algorithm I is:
$O(\lceil V/P \rceil \, (KN)^2 \log(KN) + n \lceil V/P \rceil \, R^2 + KNR^2 + (KN)^3 \log(KN)/P\,)$
The space required for each processor is: $O(\,\lceil V/P \rceil \; (KN)^2 \; + \; (KN)^3\,)$

Algorithm I performs two *scatter* operations at the beginning, followed by two *scatter-gather* operation and one *gather* operation at the end. The communications complexity of Algorithm I is: $O(\ V N^2\ +\ (K N)^3 P\ )$

Algorithm II. This algorithm attempts to minimize the amount of space. Each processor calculates the 2D DFT of each projection. The algorithm is computation intensive, computations are duplicated to reduce space and communication requirements. There is no overlapping of communication with computations.

1. C reads the orientation file.
2. C reads one projection at a time and broadcasts it to W_i, $1 \leq i \leq P-1$.
3. C broadcasts the values of the orientation so that each W_i, $1 \leq i \leq P-1$, has the orientation of the projection assigned to it.
4. Each processor performs a 2D DFT for the projection.
5. Each processor maintains a copy of the h-slab allocated to it and uses the 2D DFT by each projections to fill out this slab. It does so by intersecting each projection with the h, k, and ℓ lines in its h-slab and inside the hemisphere of radius R
6. If the object exhibits any type of symmetry, Step 5 is repeated for all the $n-1$ other projections related by symmetry to the one obtained experimentally.
7. Steps 2 to 6 are repeated until all projections have been processed. At the end of these steps, each processor has all the values inside the h-slab allocated to it.
8. The i-th processor performs KN/P inverse 2D DFTs in its h-slab: $\mathcal{F}(h,k,\ell) \rightarrow \mathbf{F}(h,y,z)$.
9. The P processors perform a global exchange and each ends up with its z-slab.
10. Each processor performs $(K N)^2/P$ 1D DFTs to obtain the final estimate of the electron density in its z-slab: $\mathbf{F}(h,y,z) \rightarrow f(x,y,z)$.
11. Each W_i, $1 \leq i \leq P-1$, sends its slab of electron density to C, which writes it onto the output file.

The computation complexity of algorithm II is:
$O(\ V\,(K N)^2 \log(K N)\ +\ n V R K N/P\ +\ (K N)^3 \log(K N)/P\ )$

The space required for each processor is: $O(\ (K N)^2 + (K N)^3/P\ )$

Algorithm II performs two *broadcast* operations at the beginning, followed by one *scatter-gather* operation and one *gather* operation at the end. The communications complexity of Algorithm II is: $O(\ P V N^2 + (K N)^3\ )$

Algorithm III. This pipelined Algorithm reduces the amount of space and allows overlapping of I/O and computations. No computations are duplicated.

1. C reads the orientation file.
2. C reads groups of q projections at a time and scatters them in groups of size q' to W_i, $1 \leq i \leq P-1$.
3. C scatters the values of the orientations so that each W_i, $1 \leq i \leq P-1$, has the orientation of each projection assigned to it.

4. After zero fill, each processor performs a 2D DFT for each of the v projections assigned to it.
5. Each processor uses the information provided by the subset of projections to compute grid points. By intersecting each projection with the h, k, ℓ lines, it obtains all $\mathcal{F}(h,k,\ell)$ with (h,k,ℓ) within half a grid spacing from the projection. It stores the $\{h,k,\ell,\mathcal{F}(h,k,\ell)\}$ 4-tuples.
6. If the object exhibits n-fold symmetry, Step 5 is repeated, using the 2D DFT of Step 4, and each of the $n-1$ symmetrically related orientations.
7. Each processor retains those $\{h,k,\ell,\mathcal{F}(h,k,\ell)\}$ 4-tuples within the h-slab allocated to it and exchanges each of the others with the the corresponding processors, then each processor convert the 4-tuples value in to its own h-slab grid value inside the hemisphere of radius R.
8. Steps 2 to 7 are repeated until all input projections are processed. At that time, at the end of Step 7, each processor has all the values of the 3D DFT at the grid points inside its h-slab.
9. The i-th processor performs KN/P inverse 2D DFTs in its h-slab: $\mathcal{F}(h,k,\ell) \rightarrow \mathbf{F}(h,y,z)$,
10. The P processors perform a global exchange and each ends up with its z-slab.
11. Each processor performs $(K\,N)^2/P$ 1D DFTs to obtain the final estimate of the electron density in its z-slab: $\mathbf{F}(h,y,z) \rightarrow f(x,y,z)$.
12. Each W_i, $1 \leq i \leq P-1$, sends its slab of electron density to C, which writes it onto the output file.

The computation complexity of Algorithm III is:
$O(\,\lceil V/P\rceil\,(K\,N)^2\,\log(K\,N) + n\,\lceil V/P\rceil\,(R^2{+}(K\,N)^2) + (K\,N)^3\log(K\,N)/P\,)$
The space required for each processor is: $O(\,n\,q'\,(K\,N)^2 + (K\,N)^3/P\,)$

Algorithm III performs two *scatter* operations at the beginning, followed by $q_{stages}+1$ *scatter-gather* operations and one *gather* operation at the end. Recall that q_{stages} is the number of pipeline stages. The communications complexity of Algorithm III is: $O(\,V\,(K\,N)^2\;+\;(K\,N)^3\,)$

3 Performance Data

Recently we implemented a version of the three algorithms on a high performance parallel system, an IBM SP2 system with 64 nodes, each having four-processors. The processors are POWER III, running at 375 Mhz. Each node has 4 GBytes of main storage and has a 36.4 GByte disk. The 4 processors in each node share the node's main memory and communicate, using MPI, with local processors and with processors in a different node.

The eleven sets of experimental virus data are listed in Table 1. We measured the execution time and recorded the memory usage when running in 1,2,4, 8, 16, and 32 processors. Table 2 presents the execution time and space needed for Algorithm I. Table 3 presents the ratio of the execution time and space for Algorithms II and III versus Algorithm I. The execution times reported do not include the time to write the 3D electron density onto the output file. Figure 2 illustrate the speedups for the three algorithms.

Table 1. The 11 Problems Used to Test the Parallel 3D. reconstruction program on SP2.

Problem	Virus	Pixels	Views	Symmetry
A	Polyomavirus (Papovaviruses)	99 × 99	60 × 60	Icosahedral
B	Paramecium Bursaria Chlorella Virus (1)	281 × 281	107 × 60	Icosahedral
C	Auravirus (Alphaviruses)	331 × 331	1940 × 60	Icosahedral
D	Chilo Iridiscent Virus	343 × 343	667 × 60	Icosahedral
E	Herpes Virus	511 × 511	173 × 60	Icosahedral
F	Ross River Virus (Alphaviruses)	131 × 131	1777 × 10	Dihedral
G	Paramecium Bursaria Chlorella Virus (1)	359 × 359	948 × 60	Icosahedral
H	Bacteriophage Phi29	191 × 191	609 × 10	Dihedral
I	Sindbis virus (Alphaviruses)	221 × 221	389 × 60	Icosahedral
J	Sindbis virus (Alphaviruses)	221 × 221	643 × 60	Icosahedral
K	Polyomavirus (Papovaviruses)	69 × 69	158 × 60	Icosahedral

Table 2. The Execution Time (Seconds) and Memory Requirements (Mbytes) of Algorithm I.

	Processor	A	B	C	D	E	F	G	H	I	J	K
Time	1	16.4	1126.1	–	–	–	437.2	–	470.9	519.5	812.7	28.0
	2	8.4	566.4	3553.7	972.7	–	223.4	2335.0	241.0	262.8	413.6	14.1
	4	4.5	287.6	1801.2	498.7	–	115.8	1186.3	126.3	136.0	210.5	7.4
	8	2.6	149.6	968.2	265.4	376.9	64.2	617.7	72.1	71.0	110.0	3.9
	16	1.6	78.9	493.3	146.8	266.4	38.3	337.5	39.7	39.5	59.7	2.2
	32	1.2	46.6	275.9	84.2	132.5	21.8	186.7	22.3	23.7	35.1	1.5
Space	1	11.6	266.3	–	–	–	27.2	–	84.1	129.5	129.5	3.9
	2	5.8	133.6	217.8	242.4	–	27.2	277.9	84.1	65.1	65.1	2.0
	4	4.9	111.2	181.4	201.9	–	27.6	231.5	84.1	54.1	54.1	1.7
	8	4.9	110.9	181.3	201.8	667.2	28.4	231.3	84.1	54.0	54.0	1.6
	16	4.9	110.9	181.3	201.8	667.2	30.1	231.3	84.1	54.0	54.0	1.6
	32	4.9	110.9	181.3	201.8	667.2	33.4	231.3	84.1	54.0	54.0	1.6

The memory requirements of Algorithm I made it impossible to run Problems C, D, and G with less than two processors nor to run Problem E with less than 8 processors. The most computational intensive phases of Algorithm I are the 2D FFT analysis, the 3D FFT synthesis, and the 2D interpolation. The amount of memory used varied only slightly when the number of nodes increased, the largest problem, Problem E required slightly less than 700 MBytes of memory per node while problem A required less than 5 Mbytes on each of the 32 processors. The speedups for Algorithm I ranged from a low of 13.7 for Problem A to around 24 for Problems B, E, and J for 32 processors. The small size of Problems A and K limited their speedups.

Algorithm II is considerably more frugal in terms of memory than Algorithm I. Moreover, the memory required scaled down almost linearly when the

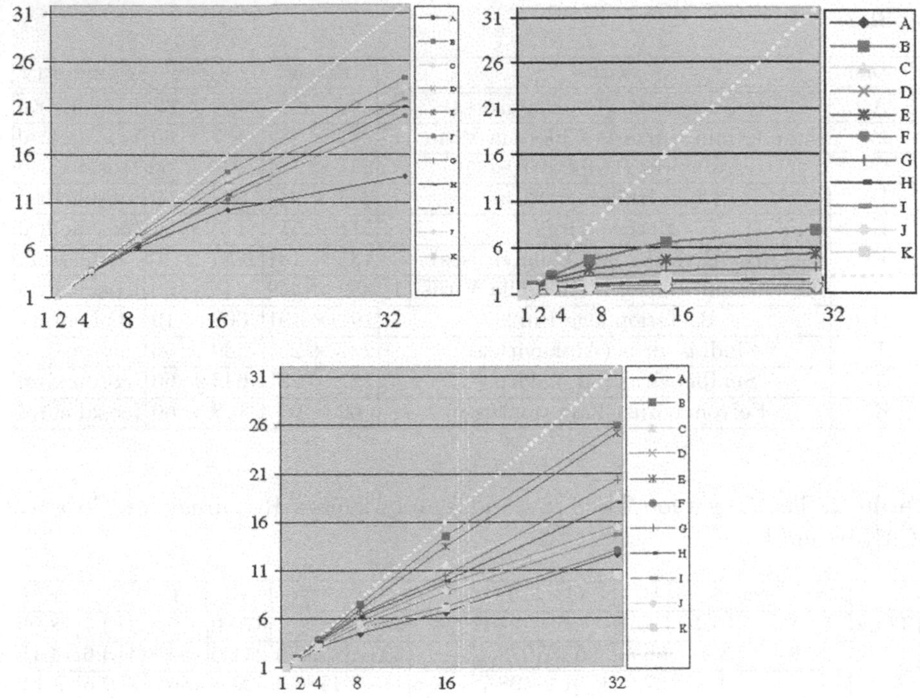

Fig. 2. The Speedup of Algorithms I (top left) II (top right) and III (bottom).

number of processors increased, The largest execution time was about 6600 seconds for Problem C with one processor but it decreased to only 3400 for 32 processors. The speedups of Algorithm II were abysmal, in the range of 2–8 for 32 processors. For eight out of the eleven structures the speedup was less than 3 on 32 nodes.

Algorithm III seems the most balanced. Indeed its memory requirements scaled down almost linearly with the number of processors and at the same time its speedups reached a maximum value.

Table 3 shows that the algorithms behave differently relative to one another, for different problems. Algorithm III is constantly better than the others in terms of execution time and speedup with the exception of Problem F where Algorithm I seems to scale slightly better. Algorithm II does not scale well in terms of execution time, it can be much as 12.4 times slower than Algorithm I for execution on 32 nodes.

Two sets of experimental data, Problems I and J, are related to the same virus, Sindbis; for the first we have a set of 389 projections and for the second we have 643, an increase of about 65%. The corresponding increase in the execution time using 32 processors is about 48% for Algorithm I and 44% for Algorithm

Table 3. Time and Memory of Algorithms II and III Compared with Algorithm I.

Problem	Processor	Algorithm II vs Algorithm I		Algorithm III vs Algorithm I	
		Time II/I	Memory II/I	Time III/I	Memory III/I
A	1	0.83	0.69	0.70	0.69
	2	1.19	0.69	0.80	0.92
	4	1.71	0.41	0.82	0.88
	8	2.46	0.22	0.96	0.90
	16	3.56	0.12	1.06	1
	32	4.25	0.06	0.75	1
B	1	0.97	0.67	0.90	0.67
	2	1.07	0.67	0.91	0.67
	4	1.24	0.41	0.91	0.41
	8	1.54	0.21	0.91	0.21
	16	2.11	0.10	0.89	0.10
	32	2.94	0.05	0.84	0.10
C	1	–	–	–	–
	2	1.41	0.67	0.69	0.67
	4	2.32	0.40	0.72	0.40
	8	3.87	0.20	0.74	0.20
	16	7.14	0.10	0.84	0.10
	32	12.38	0.06	0.93	0.09
F	1	0.98	0.83	0.85	0.83
	2	1.56	0.42	0.95	0.42
	4	2.58	0.20	1	0.28
	8	4.17	0.10	1.13	0.22
	16	6.64	0.05	1.29	0.20
	32	11.33	0.03	1.28	0.20
I	1	0.87	0.67	0.70	0.67
	2	1.35	0.67	0.79	0.67
	4	2.16	0.40	0.84	0.40
	8	3.58	0.20	0.96	0.33
	16	5.91	0.10	1.05	0.32
	32	9.48	0.05	1.05	0.35
J	1	0.86	0.67	0.69	0.67
	2	1.36	0.67	0.76	0.67
	4	2.25	0.40	0.83	0.40
	8	3.78	0.20	0.90	0.33
	16	6.42	0.10	1.05	0.32
	32	10.50	0.05	1.03	0.35

III. Thus both algorithms scale well in terms of the number of projections. This is very important because the number of projections necessary for the reconstruction of an asymmetric object is almost two orders of magnitude larger than the one for an object with icosahedral symmetry, as discussed in Sect. 1.

4 Conclusions

In this paper we present three parallel algorithms used for 3D reconstruction of asymmetric objects from their projections. The three algorithms use different strategies and data structures for distributing the data and the computations among the nodes of a parallel system. These differences affect the amount of space required and the execution time. Algorithm II is the most frugal in terms of space and Algorithm I exhibits the best speedups. The experimental results presented in Section 3 show that Algorithm III provides optimal space-time tradeoffs, its memory requirements are nearly the same as those of Algorithm II and at the same time, the speedups are very close to the ones of Algorithm I.

Acknowledgments

The authors are grateful to several colleagues for their valuable contributions. Hong Lin contributed to the optimized 3D reconstruction method. Timothy S. Baker and Michael G. Rossmann from the Structural Biology Group at Purdue provided many insightful comments. Wei Zhang and Xiaodong Yan, graduate students in the Biology Department at Purdue University, shared their data with us. The research reported in this paper was partially supported by the National Science Foundation grants MCB 9527131 and DBI 9986316, by the Scalable I/O Initiative, and by a grant from the Intel Corporation.

References

Böt97. Böttcher, B., S. A. Wynne, and R. A. Crowther, "Determination of the fold of the core protein of hepatitis B virus by electron cryomicroscopy", Nature (London) 386, 88–91, 1997.

Con97. Conway, J. F., N. Cheng, A. Zlomick, P. T. Wingfield, S. J. Stahl, and A. C. Steven, "Visualization of a 4-helix bundle in the hepatitis B virus capsid by cryo-electron microscopy", Nature (London) 386, 91–94, 1997.

Lyn97. Lynch, R. E., and D. C. Marinescu, "Parallel 3D reconstruction of spherical virus particles from digitized images of entire electron micrographs using Cartesian coordinates and Fourier analysis", CSD-TR #97-042, Department of Computer Sciences, Purdue University, 1997.

Lyn99. Lynch, R. E., D. C. Marinescu, H. Lin, and T. S. Baker, "Parallel algorithms for 3D reconstruction of asymmetric objects from electron micrographs," Proc. IPPS/SPDP, IEEE Press, 632–637, 1999.

Lyn00. Lynch, R. E., H. Lin, and D. C. Marinescu, "An efficient algorithm for parallel 3D reconstruction of asymmetric objects from electron micrographs," Proc. Euro-Par 2000, Lecture Notes in Computer Science, vol 1900, 481–490, 2000.

Lyn00a. Lynch, R. E., H. Lin, D. C. Marinescu, and Y. Ji "An Algorithm for parallel 3D reconstruction of asymmetric objects from electron micrographs", (submitted), 2000.

Parallel 3D Adaptive Compressible Navier-Stokes Solver in GeoFEM with Dynamic Load-Balancing by DRAMA Library

Kengo Nakajima[1], Jochen Fingberg[2], and Hiroshi Okuda[3]

[1] Research Organization for Information Science and Technology (RIST)
2-2-54 Naka-Meguro, Meguro-ku, Tokyo 153-0061, Japan
nakajima@tokyo.rist.or.jp
[2] C&C Research Laboratories, NEC Europe Ltd.
Rathausallee 10, D-53757, St.Augstin, Germany
fingberg@ccrl-nece.technopark.gmd.de
[3] Department of Qantum Engineering and System Sciences
The University of Tokyo
7-3-1 Hongo, Bunkyo-ku, Tokyo 113-8656, Japan
okuda@garlic.q.t.u-tokyo.ac.jp

Abstract. Grid adaptation is a very useful method for applications with unstructured meshes but requires dynamic load-balancing for efficient parallel computation. In this study, a parallel 3D compressible Navier-Stokes code with adaptive hybrid meshes (epHYBRID) and parallel adaptation procedure (pADAPT) have been developed on GeoFEM parallel platform. The DRAMA library has been integrated into the pADAPT module to solve the load-balancing problem. The entire code system has been evaluated under various types of conditions on Pentium clusters and Hitachi SR2201. Results show that DRAMA library provides accurate load-balancing for parallel mesh adaptation in pADAPT and excellent parallel efficiency in the Navier-Stokes computations in epHYBRID.

1 Introduction

Adaptive methods in applications with unstructured meshes have evolved as efficient tools for obtaining numerical solution without a priori knowledge of the details of the nature of the underlying physics. But these methods cause severe load imbalance among processors in parallel computations. Recently, various types of methods for dynamic load-balancing in parallel mesh adaptation have been developed[1][2][3].

In 1997, the Science and Technology Agency of Japan (STA) began a five-year project to develop a new supercomputer, the Earth Simulator. The goal is the development of both hardware and software for earth science simulations. The present study is conducted as a part of the research on a parallel finite element platform for solid earth simulation, named GeoFEM[4].

In this study, a parallel 3D compressible Navier-Stokes code with adaptive hybrid meshes (epHYBRID) and parallel mesh adaptation module (pADAPT) have been developed on the GeoFEM parallel platform. A repartitioning tool based on the DRAMA library[5] that provides dynamic load-balancing and complete data migration

B. Hertzberger et al. (Eds.) : HPCN Europe 2001, LNCS 2110, pp. 183-193, 2001.

has been integrated into the pADAPT module. In the following section of this paper, we outline the numerical method used in epHYBRID, and the parallel adaptation and load-balancing algorithm in pADAPT/DRAMA. Finally, the extended GeoFEM data structures for parallel mesh adaptation are described.

The entire code system (Figure 1) has been tested with the simulation of the supersonic flow around a spherical body on Pentium cluster and Hitachi SR2201. Various types of repartitioning methods in the DRAMA library have been evaluated.

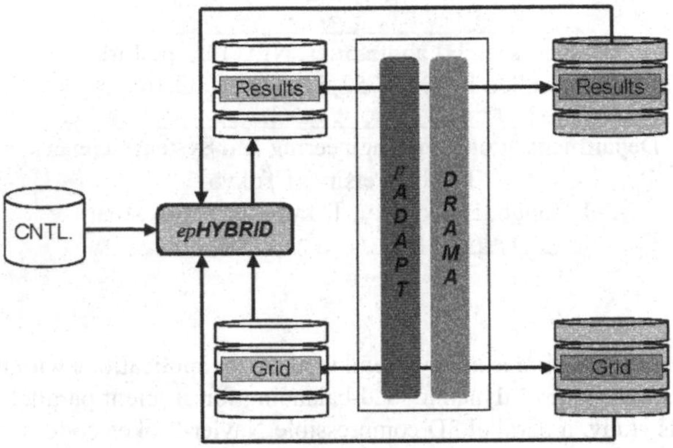

Fig. 1. epHYBRID and pADAPT/DRAMA Coupled System.

2 Parallel 3D Compressible Navier-Stokes Solver : epHYBRID

The epHYBRID code for parallel 3D compressible Navier-Stokes simulation is based on a sequential version of program which was originally developed for single CPU workstations by one of the authors[6] for the simulation of the external flow around airplanes. An edge-based finite-volume method with unstructured prismatic/tetrahedral hybrid meshes suitable for complicated geometry is applied. The solution is marched in time using a Taylor series expansion following the Lax-Wendroff approach. Although the original program was written in Fortran 77, the newly developed parallel version is written in Fortran 90 to exploit its dynamic memory management features and uses the message passing interface (MPI) for communication.

In the hybrid mesh system, the surface of the model is covered with triangles, which provide geometric flexibility, while the structure of the mesh in the direction normal to the surface provides thin prismatic elements suitable for the viscous region. The outermost layer of the prismatic mesh is then used as the inner boundary surface for a tetrahedral mesh, which covers the rest of the computational domain. Tetrahedral meshes are also suitable for connecting different prismatic regions. Figure 2 shows an example of the hybrid meshes around a sphere. Details of the numerical algorithms are described in [6].

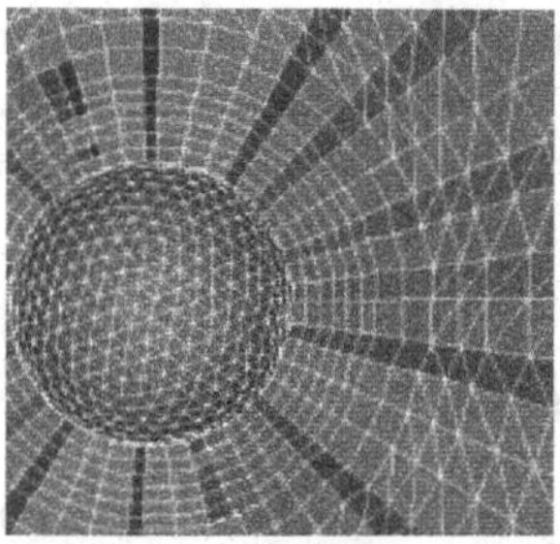

Fig. 2. Example of the Prismatic/Tetrahedral Hybrid Meshes.

3 Parallel Mesh Adaptation and Dynamic Load-Balancing Module: pADAPT/DRAMA

3.1 pADAPT

A dynamic adaptation algorithm developed for 3D unstructured meshes[6] has been parallelized on GeoFEM parallel platform. The algorithm is capable of simultaneous refinement and coarsening of the appropriate regions in the flow domain.

The adaptation algorithm is guided by a feature detector that senses regions with significant changes in flow properties, such as shock waves, separations and wakes. Velocity differences and gradients are used for feature detection and threshold parameters are set in order to identify the regions to be refined or coarsened. The details of the method for feature detection used in this study are described in [7]. In the present implementation, the feature detector marks edges.

The prisms are refined directionally in order to preserve the structure of the mesh along the normal-to-surface direction. The prismatic mesh refinement proceeds by dividing only the *lateral* edges and faces. Faces are refined either by *quadtree* or *binary* division. The resulting surface triangulation is replicated in each successive layer of the prismatic mesh as illustrated in Figure 3. As is seen from this figure, the prismatic mesh refinement preserves the structure of the initial mesh in the direction normal to the surface.

The tetrahedral elements constitute the area of mesh dominated by inviscid flow features which do not exhibit the directionality as is generally seen in the viscous region. Hence, the tetrahedral meshes are refined isotropically, employing *octree*, *quadtree* and *binary* divisions as shown in Figure 4 according to the number of marked edges.

In order to avoid excessive mesh skewness, repeated *binary* and *quadtree* divisions of tetrahedra and *binary* divisions of prisms are not allowed. Furthermore, in order to avoid sudden changes in mesh size, the mesh refinement algorithm also limits the maximum difference in embedding level between neighboring elements less than two.

Figure 5 shows the outline of the parallel mesh adaptation algorithm in pADAPT. Underlined functions use the DRAMA library and its data migration capability developed for this study.

3.2 DRAMA and Data Migration

The DRAMA library, originally developed within the European Commission funded project with the same name, supports dynamic load-balancing for parallel message passing, mesh-based applications. For a general overview see [8]. The library was evaluated with industrial FE codes and is further developed in ongoing research collaborations.

The core library functions perform a parallel computation of a mesh re-allocation that will re-balance the costs of the application code based on an adjustable, rich cost model. The DRAMA library contains geometric (RCB: Recursive Coordinate Bisection), topological (graph) and local improvement (direct mesh migration) methods and allows to use leading parallel graph partitioning packages such as METIS[9] and JOSTLE[10] through internal interfaces. DRAMA is open source, which is freely downloadable from the web-site in [5].

The DRAMA internal data structures have been designed to be suitable for adaptive applications (i.e. double numbering). The DRAMA library is a load-balancing tool that performs data migration for elements and nodes as described by the DRAMA mesh structure. It supports the application to complete the data migration by old/new and new/old element/node numbering relations. Especially for adaptive codes this is a considerable task involving the reconstruction of the entire grid hierarchy. Routines for mesh conversion and data migration have been developed to integrate the DRAMA library in the pADAPT module of the adaptive GeoFEM environment. The resulting code structure is shown in Figure 1.

4 Distributed Data Structures for Parallel Mesh Adaptation

A proper definition of the layout of the distributed data structures is very important for the efficiency of parallel computations with unstructured meshes. Although the epHYBRID code adopts an edge-based formulation, GeoFEM local data structures which are node-based with overlapping elements[4][11] has been adopted in this study. This data structure with internal/external/boundary nodes and communication tables described in Figure 6 provides excellent parallel efficiency[12].

Some additional information for mesh adaptation and grid hierarchy has been added to the original static GeoFEM data structure. In order to conform with the DRAMA library interface and the data migration procedure, *double-numbering* of nodes, elements and edges has been implemented where items are identified by 2 types of ID (original partition and local ID)[5].

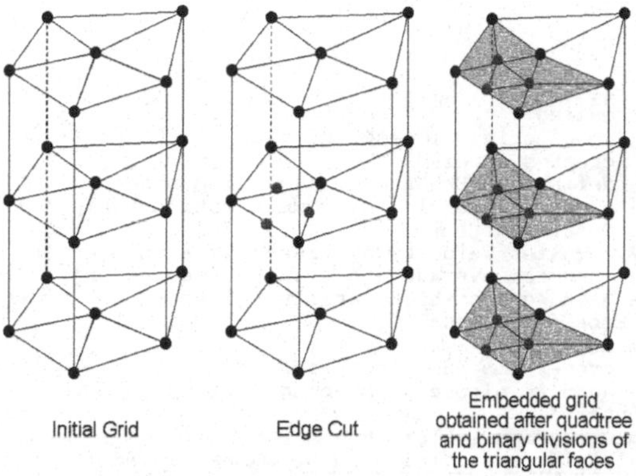

Fig. 3. Directional Refinement of Prisms Based on *quadtree* and *binary* Divisions of the Triangular Faces on the Wall.

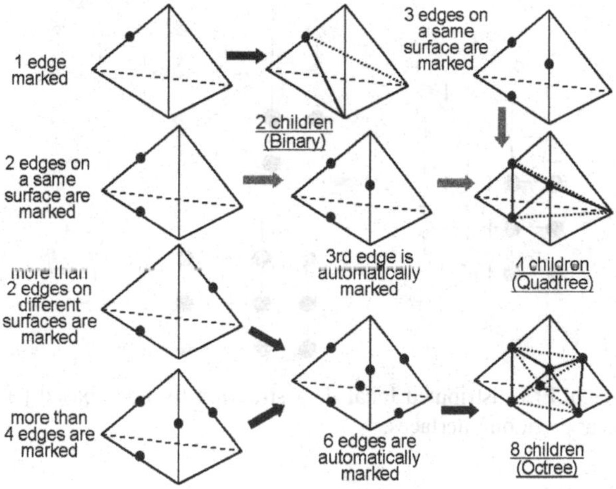

Fig. 4. Refinement Strategies for a Tetrahedron (*binary*, *quadtree* and *octree*).

```
(1) Pre-Processing
        - reads original mesh and result files
        - creates edges
        - defines INTERNAL edges and elements *
        - creates edge/element communication tables*
(2) Feature Detection
        - computes Velocity gradient/difference across the edges
        - computes average and standard deviation*
        - MARKs edges which satisfy criterion
(3) Extend Embedded Zones*
(4) Grid Smoothing*
        - proper embedding patterns
        - adjusts element embedding level around each node
(5) New Pointers*
(6) New Communication Table*
(7) Load Balancing/Repartitioning by DRAMA Library
(8) Data Migration
(9) Output                                    * : w/communication
```

Fig. 5. Parallel mesh adaptation/dynamic load-balancing/data migration procedure in pADAPT/DRAMA coupled system (underlined items are added to the pADAPT module).

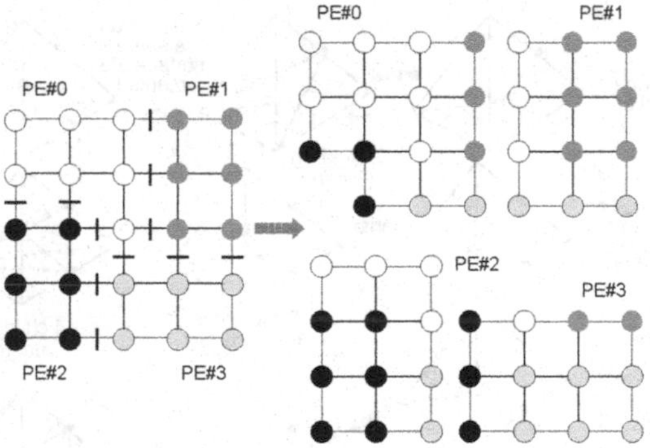

Fig. 6. Example of GeoFEM distributed local data structure by node-based partitioning with overlapping elements at partition interfaces.

5 Numerical Examples

Numerical simulations of the supersonic flow ($M=1.40$, $Re=10^{6}$) around a sphere have been conducted under various types of configurations. In Figure 7, a spherical bow shock can be observed upstream the body. It shows the Mach number distribution in very coarse initial meshes, 1-level and 2-level adapted meshes. The shock is very sharply captured by 2-level adapted meshes. Computations are executed on the 32-processor LAMP Pentium cluster[13] operated by NEC-Europe and the 1024-processor Hitachi SR2201 computer at the University of Tokyo, Japan.

In these examples, grid adaptation is required only several times during entire computations. Therefore, computation time for grid adaptation and dynamic load-balancing is almost negligible compared to time for Navier-Stokes simulation. Computational and parallel efficiency of the grid adaptation, dynamic load-balancing and data migration have not been evaluated here.

5.1 Parallel Performance of epHYBRID without Adaptation

Parallel performance of epHYBRID code was evaluated using globally fine prismatic meshes without adaptation using 2 to 256 processors on both the LAMP cluster and the SR2201 computer. In these computations, the problem size for each processor was approximately kept fixed up to the 48 PE case. Ranging from 48 to 256 PEs, the entire problem size was held constant. GeoFEM's RCB method and METIS have been applied as initial static partitioning method.

The results are summarized in Table 1. The unit elapsed user execution time (including both computation and communication time) for each iteration stays almost constant up to 256 processor case and parallel efficiency of the epHYBRID is almost perfect.

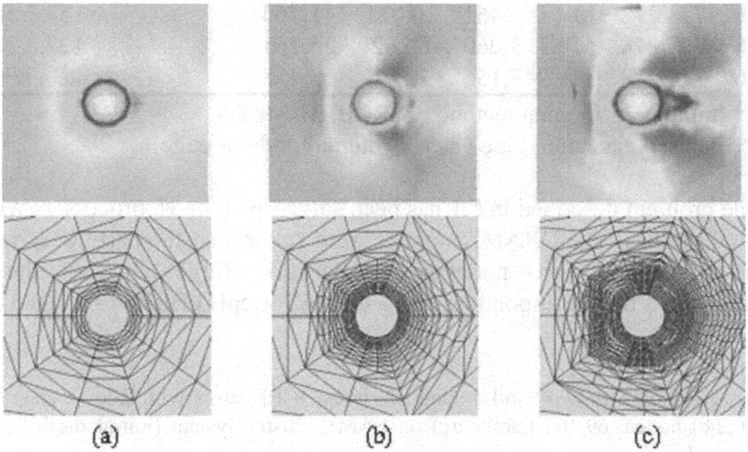

Fig. 7. Supersonic flow around a spherical body ($M=1.40$, $Re=10^{6}$). Mach number distribution and meshes (a) Initial mesh (546 nodes, 2,880 tetrahedra), (b) 1-level adapted mesh (2,614 nodes, 16,628 tetrahedra), (c) 2-level adapted mesh (10,240 nodes, 69,462 tetrahedra)

5.2 Comparison of Repartitioning Methods

As is described in 3., the DRAMA library offers various types of repartitioning methods (for instance : graph-based (PARMETIS or PJOSTLE) and geometry-based (RCB)). Here, we compare the effect of different repartitioning methods on the computational efficiency of the resulting meshes. The following repartitioning methods in the DRAMA library have been considered :

- No Repartitioning
- PJOSTLE

- PARMETIS k-way
- RCB Simple
- RCB Bucket (edgecut reduced)

Table 1. Hypersonic flow around a spherical body, 2-256 PE cases with globally fine prismatic meshes on the LAMP cluster system and Hitachi SR2201.

PE #	Total Node #	P. M. (1)*	Total Edge Cut #	Max. Internal Node #	Max. Edge #	LAMP Time(2)* (μsec.) (Node/Edge)	SR2201 Time(2)* (μsec.) (Node/Edge)
2	33,306	R	2,518	16,653	66,553	144.7/36.21	95.48/23.89
4	64,050	R	9,585	16,013	65,799	144.9/35.26	118.0/28.72
8	133,146	R	15,347	16,644	67,268	150.2/37.16	135.8/33.59
16	256,050	R	52,904	16,004	67,048	171.8/41.02	108.7/25.95
32	532,505	R	136,975	16,641	71,306	-	123.2/28.75
48	778,278	M	110,106	16,700	68,399	-	124.6/30.41
64	778,278	M	127,621	12,525	51,735	-	135.7/32.86
80	778,278	M	142,461	10,021	41,765	-	158.7/38.12
128	778,278	M	179,060	6,262	26,251	-	127.8/30.48
256	778,278	M	247,155	3,131	13,458	-	130.9/30.47

(1)*: Initial p_artitioning method : R-RCB, M-METIS.
(2)*: Elapsed execution time / step / (internal node or edge).

The same problem described in 5.1 has been solved on 8 or 16 processors with purely tetrahedral meshes. The DRAMA options were set so that the partitioner would balance the number of internal nodes in each partition. Table 2.-4. show the resulting distributions and the corresponding elapsed time for epHYBRID (averaged for 1,000 steps).

Table 2. Hypersonic flow around a spherical body, 8 PE cases with 2-level adapted meshes (total: 10,240 nodes, 69,462 tetrahedra) on LAMP cluster system (initial mesh : 546 nodes, 2,880 tetrahedra)

Repartitioning Methods	Internal Node Number (min/max)	Total Edge Cut	Edge Number (min/max)	Time(1)* (sec.)
No Repartition	561/2,335	11,224	4,918/17,639	619
PJOSTLE	1,274/1,286	7,293	9,248/9,883	354
PARMETIS k-way	1,267/1,293	7,679	9,258/10,222	363
RCB Simple	1,280/1,280	12,106	10,426/10,605	389
RCB Bucket	1,280/1,280	11,603	10,479/10,971	399

(1)*: Elapsed execution time for 1,000 time steps (averaged).

Figure 8 shows the partitioning after 2-level adaptation for the 8 processor case displayed by the parallel version of GPPView[14] tool developed within the GeoFEM project. Without repartitioning, load imbalance among the processors is severe especially after 2-level adaptation. Among the 4 repartitioning methods, PJOSTLE provided the best quality from the viewpoint of the performance of the epHYBRID code because resulting edge-cuts and edges in each partition are the fewest.

Table 3. Hypersonic flow around a spherical body, 16 PE cases with 1-level adapted meshes (total : 47,074 nodes, 306,236 tetrahedra) on LAMP cluster system (initial mesh : 16,050 nodes, 92,160 tetrahedra).

Repartitioning Methods	Internal Node Number (min/max)	Total Edge Cut	Edge Number (min/max)	Time(1)* (sec.)
No Repartition	1,343/2,335	39,888	10,576/48,495	1,683
PJOSTLE	2,929/2,961	25,085	21,089/22,233	874
PARMETIS k-way	2,905/2,984	26,274	21,201/22,630	880
RCB Simple	2,942/2,943	41,980	22,520/23,090	899
RCB Bucket	2,942/2,943	37,192	21,231/23,269	926

(1)*: Elapsed execution time for 1,000 time steps (averaged).

Table 4. Hypersonic flow around a spherical body, 16 PE cases with 2-level adapted meshes (total : 163,537 nodes, 1,116,700 tetrahedra) on LAMP cluster system (initial mesh : 16,050 nodes, 92,160 tetrahedra).

Repartitioning Methods	Internal Node Number (min/max)	Total Edge Cut	Edge Number (min/max)	Time(1)* (sec.)
No Repartition	6,621/20,842	101,178	50,386/152,059	5,384
PJOSTLE	10,195/10,260	55,663	73,262/ 75,540	2,982
RCB Bucket	10,221/10,222	100,462	82,799/ 85,819	3,227

(1)*: Elapsed execution time for 1,000 time steps (averaged).

6 Conclusion

In this study, a parallel 3D compressible Navier-Stokes code with adaptive hybrid meshes (epHYBRID) and parallel mesh adaptation module (pADAPT) have been developed on GeoFEM parallel platform. The DRAMA library has been integrated in the pADAPT module and the data migration procedure has been added. Entire code system has been tested with the simulation of the supersonic flow around a spherical body on Pentium cluster and Hitachi SR2201 computer. We found that the epHYBRID code showed excellent parallel efficiency with dynamic load-balancing. Various types of repartitioning methods in the DRAMA library have been evaluated. Among these methods, PJOSTLE provided the best mesh partitioning quality from the viewpoint of the efficiency of the epHYBRID code.

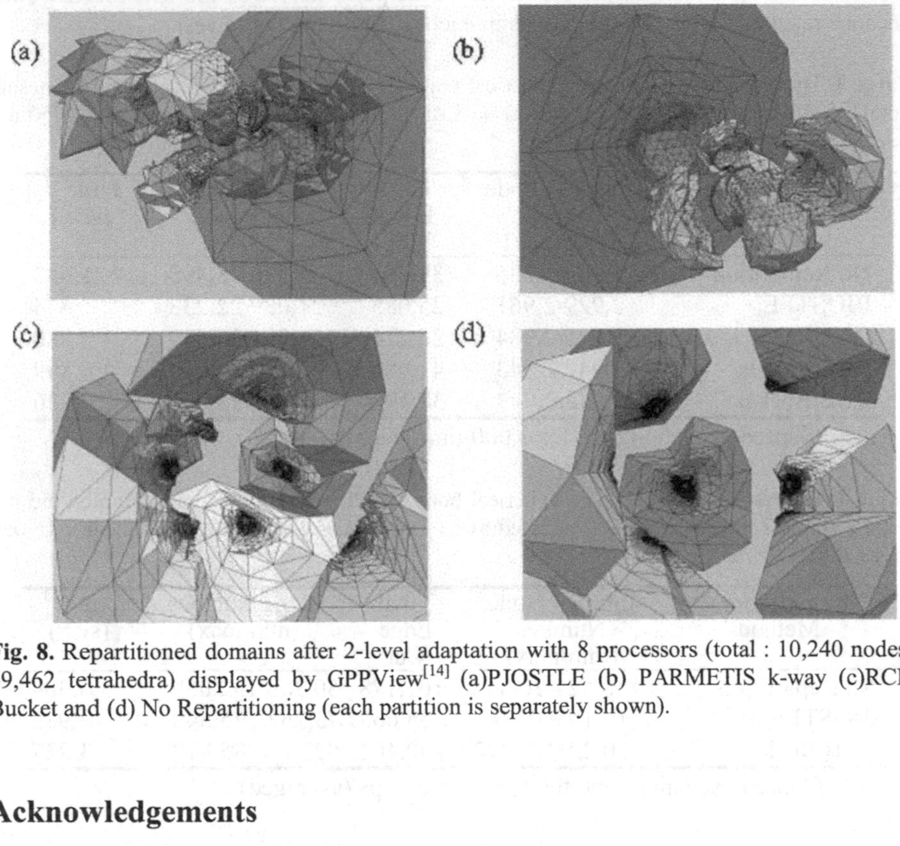

Fig. 8. Repartitioned domains after 2-level adaptation with 8 processors (total : 10,240 nodes, 69,462 tetrahedra) displayed by GPPView[14] (a)PJOSTLE (b) PARMETIS k-way (c)RCB Bucket and (d) No Repartitioning (each partition is separately shown).

Acknowledgements

This study is a part of the project "Solid Earth Platform for Large Scale Computation" funded by the Ministry of Education, Culture, Sports, Science and Technology, Japan through its "Special Promoting Funds of Science & Technology." Furthermore the authors would like to thank Professor Yasumasa Kanada (Computing Center, University of Tokyo) for fruitful discussions on high performance computing and Dr. Yoshitaka Wada (RIST) for providing a test version of GPPView for hierarchical grid systems prior to the official release.

References

1. Oliker, L., Biswas, R. : Parallelization of a Dynamic Unstructured Application using Three Leading Paradigms, SC99 Proceedings, Portland, Oregon, USA (1999)
2. Vidwans, A., Kallinderis, Y., Venkatakrishnan, V. : Parallel Dynamic Load-Balancing Algorithm for Three-Dimensional Adaptive Unstructured Grids, AIAA Journal 32 (1995), pp.497-505.
3. Shepard, M.S., Flaherty, J.E., Bottasso, C.L., de Cougny, H.L., Ozturan, C., Simone, M.L.: Parallel automatic adaptive analysis, Parallel Computing 23 (1997), pp.1327-1347.
4. GeoFEM Web Site : http://geofem.tokyo.rist.or.jp

5. DRAMA Web Site : http://www.ccrl-nece.de/DRAMA
6. Parthasarathy, V., Kallinderis, Y., Nakajima, K. : Hybrid Adaptation Method and Directional Viscous Multigrid with Prismatic-Tetrahedral Meshes, AIAA Paper 95-0670 (1995)
7. Kallinderis, Y., Baron, J.R. : A New Adaptive Algorithm for Turbulent Flows, Computers and Fluids Journal 21 (1992) pp.77-96.
8. Basermann, A., Clinckemaillie, J., Coupez, T., Fingberg, J., Digonnet, H., Ducloux, R., Gratien, J.M., Hartmann, U., Lonsdale, G., Maerten, B., Roose, D., Walshaw, C. : Dynamic load balancing of finite element applications with the DRAMA library, Appl. Math. Modelling 25 (2000) pp.83-98.
9. METIS Web Site : http://www-users.cs.umn.edu/karypis/metis/
10. JOSTLE Web Site : http://www.gre.ac.uk/Jostle
11. Nakajima, K., Okuda, H. : Parallel Iterative Solvers with Localized ILU Preconditioning for Unstructured Grids on Workstation Clusters, International Journal for Computational Fluid Dynamics 12 (1999) pp.315-322.
12. Garatani, K., Nakamura, H., Okuda, H., Yagawa, G. : GeoFEM : High Performance Parallel FEM for Solid Earth, HPCN Europe 1999, Amsterdam, The Netherlands, Lecture Notes in Computer Science 1593 (1999) pp.133-140
13. LAMP Web Site : http://www.ccrl-nece.technopark.gmd.de/ maciej/LAMP/LAMP.html
14. GPPView Web Site : http://www.tokyo.rist.or.jp/GPPView/

Parallel Models for Reactive Scattering Calculations

Valentina Piermarini, Leonardo Pacifici,
Stefano Crocchianti, and Antonio Laganà

Dipartimento di Chimica, Università di Perugia
Via Elce di Sotto, 8, 06123 Perugia, Italy

Abstract. The task of articulating some computer programs aimed at calculating reaction probabilities and reactive cross sections of elementary atom diatom reactions as concurrent computational processes is discussed. Various parallelization issues concerned with memory and CPU requirements of the different parallel models when applied to two classes of approach to the integration of the Schrödinger equation describing atom diatom elementary reactions are addressed. Particular attention is given to the importance of computational granularity for the choice of the model.

1 Introduction

Demand for chemical computational applications is shifting towards realistic simulations of systems of practical relevance for modern technologies. This is increasingly made possible not only by the advance in hardware performances due to the constant enhancement of both CPU speed and memory capacity, but also by the advance in software performances due to the increasing use of concurrent computing (vector, parallel and distributed). Most of these technology advances are not transparent to the user since they necessitate an *ad hoc* structuring of the computer codes. In fact, while vectorization is, at present, a problem largely solved at machine level, this is not the case for parallel and distributed computing. There are, indeed, fairly popular languages (such as HPF[1]) incorporating parallel structures (mainly of data type), message passing libraries[2,3] and sets of parallelized routines of general use[4]. However, there are no universally accepted standards and no fully or semi automatic tools for parallelizing complex scientific applications. To this end, it is vital in molecular modeling and other computational chemical applications to develop parallel models for simple systems and to extend them to more complex molecules. This implies a revisitation of old algorithms and the engineering of new ones (more suited for concurrent computing) to develop innovative approaches to the calculation of chemical structures and processes as well as to the realistic simulation of chemical experiments.

The key step of this process is the mapping of the physical problem into a mathematical model for which a computational procedure suitable for paral-

B. Hertzberger et al. (Eds.): HPCN Europe 2001, LNCS 2110, pp. 194–203, 2001.

lelization can be adopted. The correspondence between physical and computational parameters is, unfortunately, not unique. Therefore, the achievement of computational efficiency is strictly related to both the chosen theoretical approach and the adopted numerical implementation. The simplest way of singling out possible correspondences between the numerical formulation of a theoretical approach and the efficiency of related parallel algorithms is to analyze existing computational procedures and measure the performance of their parallel implementation on various computer architectures [5].

The present paper deals with the parallel structuring of the section of SIMBEX [6] (a computer simulation of molecular beam experiments) devoted to the calculation of scattering **S** matrices and related reaction probabilities recently embodied in a problem solving environment (PSE) [7]. The performance of the global simulation heavily relies on the performance of the individual components.

2 Quantum Approaches to Reactive Probabilities, Cross Sections, and Rate Coefficients

Physical observables of reactive processes are rate coefficients k and cross sections σ. Rate coefficients k can be worked out from quantum cross sections σ by first averaging over initial (i) and final (f) internal energy states,

$$k(T) = \sum_{if} k_{if}(T) \tag{1}$$

where T is the temperature of the system and k_{if} the state-to-state rate coefficients. State-to-state rate coefficients can be evaluated by properly integrating the state-to-state cross section $\sigma_{if}(E_{tr})$ over the translational energy E_{tr}

$$k_{if} = \left(\frac{8}{\pi\mu k_B^3 T^3}\right)^{\frac{1}{2}} \int_0^\infty E_{tr}\sigma_{i,f}(E_{tr}) e^{-\frac{E_{tr}}{k_B T}} dE_{tr} \tag{2}$$

In turn, the state-to-state cross section σ_{if} can be evaluated by summing over all the contributing values of the total angular momentum quantum number J and the detailed reactive probabilities $P_{i,f}^J(E_{tr})$ of the proper parity

$$\sigma_{i,f}(E_{tr}) = \frac{\pi}{k_i^2} \sum_J (2J+1) P_{i,f}^J(E_{tr}) \tag{3}$$

where $P_{i,f}^J(E_{tr})$ is the sum over Λ of the square modulus of the related $S_{if}^{J\Lambda}$ **S** matrix elements. For an atom diatom reaction the $S_{if}^{J\Lambda}$ matrix elements can be calculated by integrating the time dependent Schrödinger equation

$$i\hbar \frac{\partial}{\partial t} \Psi^{J\Lambda}(R, r, \Theta, t) = \hat{H} \Psi^{J\Lambda}(R, r, \Theta, t) \tag{4}$$

where $\Psi^{J\Lambda}(R, r, \Theta, t)$ is the time dependent J, Λ partial wavefunction, R is the atom-diatom center of mass distance, r is the diatom internuclear distance,

Θ is the included angle, and t is time. In general, equation 4, after separating both the motion of the center of mass and the motion of the rotating rigid body, has 3N-5 dimensions of which 3N-6 are spatial and the remaining one is time (N is the number of atoms of the system). In the followings we shall focus on the treatment of atom diatom reactions, the simplest prototype for elementary reactive systems. Atom diatom reactions are, in fact, those for which not only the formalism is fully developed but also the computational machinery has been firmly established.

A first class of approaches are the time independent ones for which equation 4 is solved by factoring out time t, defining a proper combination of internuclear distances as a suitable continuity variable (reaction coordinate), expanding the time independent partial wave in the local (at reaction coordinate grid-points) eigenfunctions of the remaining (usually orthogonal) coordinates and integrating the resulting set of coupled differential equations in the reaction coordinate [8].

An alternative class of approaches are the time dependent ones. These approaches do not factor out time, that is taken as a continuity variable, and integrate equation 4 over time by letting the three dimensional wavepacket describing the reactants (at a particular quantum vibrotational vj state and with a certain amount of translational energy) evolve to reach the product region [9].

3 The Time Independent Computational Procedure

The computational procedure associated with time independent approaches aimed at calculating a single J value **S** matrix is usually articulated into several computer programs. The first program is devoted to the construction of the local basis sets in which at each grid-point of the reaction coordinate the time independent component of the time dependent partial wavefunction is expanded. To this end the first program carries out the calculation of the eigenvalues and of the two dimensional eigenfunctions at all the grid-points of the reaction coordinate. It also evaluates the overlaps of these eigenfunctions at the common border of each pair of adjacent grid points and constructs the coupling matrix for the set of coupled differential equations.

The second program carries out the integration of the coupled differential equations over the reaction coordinate. To this end the program builds the solution by propagating the logarithmic derivative of the wavefunction through the grid-points of the reaction coordinate by making use of quantities calculated by the first program. These two programs are the most time and memory demanding components of the whole computational procedure. They are followed by a set of small and fast programs performing the necessary transformations of the numerical solution of the coupled differential equations into the **S** matrix.

The scheme of the first program is:

```
Input general data
Calculate quantities of common use
LOOP on reaction coordinate grid-points
```

```
        LOOP on Λ
            Construct the local primitive basis functions
            Calculate the local eigenvalues and surface functions
            Store on disk the eigenvalues
            IF(not first grid-point)
                THEN
                    Calculate overlaps with previous grid-point functions
                    Store on disk the overlap matrix
            END IF
            Calculate the coupling matrix
        END the Λ loop
        Store on disk the coupling matrix
    END the sector loop
```

Fig. 1. Scheme of the First Program.

The sequential version of the code consists of two nested loops: the outer loop runs over the reaction coordinate grid-points; the inner loop runs over the values of the $J+1$ projections Λ of J (the calculation is performed at a reference $\overline{J}$). The central section of the code calculates the local eigenvalues and surface functions at each point of the reaction coordinate for all the values of the projection Λ that varies from 0 to $\overline{J}$. In the same section, overlaps between surface functions calculated at neighboring grid-points and the various contributions to the coupling matrix are calculated. These quantities and sector eigenvalues are stored on disk for use by the second program.

The scheme of the second program is:

```
Input general data
Read from disk data stored by the first program
Calculate quantities of common use
LOOP on energy
    Embed the energy dependence into the coupling matrix
    LOOP over reaction coordinate grid-points
        Single step-propagate the fixed J logarithmic derivative matrix
    END the sector loop
    Calculate and store the final logarithmic derivative matrix elements on disk
END the energy loop
```

Fig. 2. Scheme of the Second Program.

As apparent from the scheme, the second program consists of two nested loops: the outer loop runs over the energy values at which the propagation has to be performed, the inner loop runs over the reaction coordinate grid-points.

The central section of this code propagates along the reaction coordinate the fixed J solution under the form of logarithmic derivatives iterating over energy at fixed parity and total angular momentum quantum number for all the allowed values of Λ until the asymptote is reached. Final values of the logarithmic derivative matrix elements are stored on disk for use by subsequent programs.

4 The Parallelization of the Time Independent Code

The two programs allow the exploitation of parallelism at various levels since the related formulation of the problem is based upon several exact decomposition steps. For both programs, in fact, there are parameters like the total angular momentum quantum number J which is exactly decoupled. However, it is quite often necessary to separate the various iterations on decoupled quantities into separate programs in order to reduce the size of the code and have easier access to the queues of Scientific Computer Centers. In this case, a finer grain parallelism needs to be implemented.

As an example, the first program that calculates eigenfunctions and assembles the coupling matrix can be parallelized at the level of individual reaction coordinate grid-points. For reduced dimensionality atom diatom treatments, eigenfunction calculations are a one dimensional eigenvalue problem. Because of this, several grid-point calculations can be grouped in a single block that is still small enough to be easily dealt by a single processor and its local memory. This makes it convenient to structure the code as a static task farm in which each processor is assigned the calculation for a block of grid-points[10]. However, to allow the construction of the overlap integral between functions belonging to different grid-points and, as a consequence, the assemblage of the coupling matrix, the calculation of the eigenfunctions of the last grid-point of a block has to be repeated when performing the same calculation for the first grid-point of the subsequent block.

On the contrary, when carrying out a full dimensional treatment, the eigenfunction calculation is a two dimensional problem (for this reason these eigenfunctions are usually called surface functions) that is solved differently according to the grid point being considered. This leads to a significant load unbalance when distributing for parallelization blocks of grid-point calculations. Accordingly, the static task farm model adopted for the reduced dimensionality treatment is not appropriate in this case. A better choice is a dynamic task farm assigning each grid-point calculation to a different processor. Obviously, this implies that at each grid-point of the surface functions calculated at the previous point is repeated [11]. Accordingly, the parallel model implemented in the full dimensional treatment is a dynamic task farm in which the master process, after reading input data, sends to every worker the value of the grid-point for which the eigenfunctions have to be calculated. Once the calculation is performed, related eigenvectors are stored on disk for use by the processor dealing with the similar calculation for the next grid-point when it has finished its own surface function calculation. Then the node retrieves from disk the eigenvectors of the previous grid-point and reconstructs the eigenfunctions at the quadrature points of the current one. To prevent attempts to read not yet stored information, nodes are synchronized before reading from disk. This is accomplished by defining a subset of the MPI communicator that groups together all the workers and puts an MPI barrier. After reading the necessary information from disk, calculations are completed by evaluating the coupling matrix.

This parallel model has been used to calculate reactive probabilities for the Li + HF reaction[12]. Typical single run calculations were performed on the Cray T3E of the EPCC Center (Edinburgh, U.K.) using 32 and 64 processors. The program parallelizes and scales very well. The maximum load imbalance among the various nodes is of the order of 10% in all the test runs we have performed marking a definite improvement over the static task farm model similar to the one used for the reduced dimensionality calculations.

The propagation of the solution carried out by the second program has a recursive structure (since it builds the local value of the logarithmic derivative of the wavefunction from its value at the previous grid-point). This makes it unsuitable for a parallelization at grid-point level. Fortunately, its memory requirements are smaller and the code can be parallelized over energy using a dynamic task farm model by distributing fixed energy propagations for a given value of J. In this implementation the master process reads from disk and broadcasts to all nodes the data required to carry out the propagation. The work is then assigned to the workers by sending the energy value for which the propagation has to be carried out. At the end of the propagation the worker stores on disk the solution matrix and gets ready to receive the next energy value, if any. When all the propagations have been carried to an end the application is closed. This parallel model leads to good speedups whose main weakness is a possible load imbalance associated with the large time consumption of the individual tasks.

5 The Time Dependent Computational Procedure

The time-dependent method used in this work collocates the wavepacket on a grid of R and r (or R' and r') coordinates and expands it on a basis set for the angular coordinate. By making simple transformations the propagation in time of the wavepacket is performed only on its real component[9,13].

At the very beginning ($t = 0$), for a given value of the total angular momentum quantum number J and its projection Λ on the z axis of a body fixed frame, the system wavefunction $\Psi^{J\Lambda}$ is defined as

$$\Psi^{J\Lambda}(R, r, \Theta; t) = \left(\frac{8\alpha}{\pi}\right)^{\frac{1}{4}} e^{-\alpha(R-R_0)^2} \cdot e^{-ik(R-R_0)} \cdot \varphi_{vj}^{BC}(r) \cdot P_j^{\Lambda}(\Theta). \tag{5}$$

where R, r and Θ are the Jacobi internal coordinates of the reactant atom diatom system. In eq. 5, $e^{-ik(R-R_0)}$ is a phase factor which gives the initial wave packet a relative kinetic energy towards the interaction region, $\varphi_{vj}^{BC}(r)$ is the initial diatomic molecule BC wavefunction (for the vibrotational state vj) expressed in the Jacobi coordinates of the reactant arrangement, $P_j^{\Lambda}(\Theta)$ is the normalized associated Legendre polynomial and k is the wavevector which determines the relative kinetic energy of the collisional partners [9]. In this way, the wavefunction is defined for a given accessible state of the reactants and a given collisional energy range. To move the wavepacket out of the reactant region, the integration in time of the time-dependent Schrödinger equation 4 is performed by the program TIDEP. This is the most time consuming task of the code in which using a discrete variable (DVR) method (or other equivalent approaches) some matrix operations are performed during the time step-propagation of the wavepacket. After each step-propagation the wavepacket is expanded at the analysis line in terms of the final diatomic molecule AB wavefunction $\varphi_{v'j'}^{AB}(r')$ where r', R' and Θ' are the Jacobi coordinates of the product arrangement. The time dependent coefficients $C_{vj\Lambda,v'j'\Lambda'}^{J}(t)$ of such an expansion are then saved on disk for further use by the subsequent program TIDAN. In fact, TIDAN works out the energy-dependent **S** matrix elements by half Fourier transforming the time-dependent coefficients $C_{vj\Lambda,v'j'\Lambda'}^{J}(t)$.

6 The Parallelization of the Time Dependent Code

As can be easily understood, parallelization efforts for the time dependent procedure were concentrated on the TIDEP code. This code is, in fact, the most time consuming component of the computational procedure (the propagation step has to be iterated for about $10^4 \div 10^5$ times). The structure of TIDEP is:

```
Read input data: v, j, k, masses, ...
Perform preliminary calculations
LOOP on J
    LOOP on t
        LOOP on Λ
            Perform time step-propagation
            Perform the asymptotic analysis
            Store C(t) coefficients
        END loop on Λ
    END loop on t
END loop on J
Calculate final quantities
Print outputs
```

As can be seen from the scheme given above, calculations are performed at fixed value of the vibrotational (vj) quantum number of the reactant diatom as well as for a given range of translational energy and a single J value. The most natural way of parallelizing the code is to distribute the calculation for a given vibrotational state, for a given interval of translational energy and a given value of J using a task farm dynamically assigning the computational workload.

To carry out the calculations the value of the physical parameters was chosen to be that of the $O(^1D)+HCl$ atom diatom reaction[14]. Accordingly, the mass values were chosen to be 15.9949 amu for O, 1.00783 amu for H atom and 34.96885 amu for Cl. The energy range covered by the calculation is approximately 1 eV, the initial vibrotational state used for the test was $v = 0$ and $j = 0$. The potential energy surface used for the calculations is described in ref. [14] where other details are also given. A gridsize of 127×119 points in R' and r' was used while the angular part was expanded over 80 basis functions. Time propagation was iterated for 40000 steps to properly diffuse the wavepacket. To evaluate a vibrational state selected rate coefficient, the calculation needs to be performed for the whole accessible translational energy values, all the reactant rotational states j populated at the temperature considered and all the contributing total angular momentum quantum numbers J (convergence occurs at values larger than 100). This not only increases enormously the computational work but also induces a high imbalance of the load since a single J calculation is $J+1$ times larger than that of $J = 0$ (for each J value there are $J+1$ projections Λ).

The next lower level of parallelization is the one distributing also fixed Λ calculations [15]. However, opposite to the J distribution (J is a good quantum number), the parallelization of fixed Λ calculations is not natural. Λ, in fact, is not a good quantum number and to decouple related calculations one has to introduce physical constraints of the centrifugal sudden type (*i.e.* the projection of J on the z axis of the body fixed frame remains constant during the collision). The parallel code performs separately the step-propagation of the wavepacket for blocks of fixed Λ values and then recombines the contributions of the various blocks at the end of the propagation step. This leads to a decomposition of the domain of the wavepacket that counterbalances the increase of the size of the matrices with J. Such a parallel model was first tested [16] on the Cray T3E of EPCC (Edinburgh, UK) for the simplest case of $J = 0$ and $J = 1$. In the present more general version of the code I/O bottlenecks are eliminated by adopting a task farm model. In this model node zero acts as a master and takes care of performing preliminary calculations as well as of distributing to the workers the fixed J and Λ propagations. Calculations are carried out in pairs of J values (together with all the related $J + 1$ component of Λ). This sets a limit to the maximum value of the total angular momentum quantum number that can be handled by the program (the maximum value of J has to be at least 3 units lower than the number of processors).

To evaluate the performances of the model, the calculations were performed on the Origin 3800 of Cineca (Bologna, I) using the already mentioned set of parameters, except for the number of angular basis functions that was reduced to 10. The average fixed J fixed Λ computing time increases gradually from 4400 s to 4680 s when J increases from 0 to 5. The rise in execution time indicates that there is, indeed, an increase in communication time when J gets larger and the number of Λ projections increases. However, its small value indicates also that the parallel model adopted is highly efficient.

7 Conclusions

In this paper an example of how progress in modeling chemical processes (such as those intervening in atmospheric phenomena, material technology, pollution and combustion) needs to be based on parallel hardware is examined. The computing power offered by these machines seems to be an ideal ground for a fast development of computational procedures able to deal with chemical applications.

However, as shown in the paper, most often, to make the implementation of these complex codes on parallel architectures efficient, a reasonable compromise between the computational granularity and the coarseness of the parallelization level has to be worked out. The case investigated here (that is the design of an efficient parallel implementation of both time independent and time dependent approaches to chemical reactivity) shows once more that the process of parallelizing a code has to be based on a deep knowledge of the physics of the problem and the mastering of its mathematical foundations. The efficiency of the models discussed in this paper, changes, in fact, depending on the value of the parameters used and on the type of approximations introduced. This emphasizes the need for working out powerful tools enabling scientists to embody parallel structures into their computational applications without caring about the physical allocation of workload and the communication pattern among the processors [17].

Acknowledgments

Thanks are due to Cineca (Bologna, Italy), EPCC (Edinburgh, U.K.) and CESCA-CEPBA (Barcelona, Spain) for computer time grants. We thank also CNR, ASI and MURST for funding the present research.

References

1. High Performance Fortran Forum, Scientific Programming 2(1), (1993) John Wiley and Sons
2. Sunderam, V.S.: PVM: A framework for parallel distributed computing. Concurrency: practice and experience. 2(4) (1990) 315-339; Geist, G.A., Sunderam, V.S.: Network based concurrent computing on the PVM system. Concurrency: practice and experience. 4(4) (1992) 293-311; Beguelin, A., Dongarra, J., Geist, G.A., Manchek, R., Sunderam, V.S.: A user's guide to PVM Parallel Virtual Machine. Oak Ridge National Laboratory, Tennessee (1992)

3. Message Passing Interface Forum, Int. J. of Supercomputer Applications 8(3/4), 1994; Smir, M., Otto, S., Huss-Ledermam, S., Walker, D., Dongarra, J.: MPI: The complete reference. MIT Press (1996)
4. Whaley, R. C., Walker, D., Petitet, A., Ostrouchov, S., Dongarra, J., Choi, J.: A Proposal for a Set of Parallel Basic Linear Algebra Subprograms. Lecture Notes in Computer Science **1041**, (1996) 107-114
5. Laganà, A. : Innovative computing and detailed properties of elementary reactions using time independent approaches. Comp. Phys. Comm., **116** (1999) 1–16; Laganà, A., Crocchianti, S., Bolloni, A., Piermarini, V., Baraglia, R., Ferrini, R., Laforenza, D.: Computational granularity and parallel models to scale up reactive scattering calculations. Comp. Phys. Comm. **128** (2000) 295-314
6. Gervasi, O., Cicoria, D., Laganà, A., Baraglia, R.: Animazione e calcolo parallelo per lo studio delle reazioni elementari. Pixel **10** (1994) 19–26
7. Gallopoulos, S., Houstis, E, Rice, J.: Computer as Thinker/Doer: Problem-Solving Environments for Computational Science. IEEE Computational Science and Engineering, Summer (1994)
8. Laganà, A., Riganelli, A.: Computational reaction and molecular dynamics: from simple systems and rigorous methods to complex systems and approximate methods. Lecture Notes in Chemistry **75** (2000) 1-12
9. Balint-Kurti, G. G.: Time dependent quantum approaches to chemical reactivity. Lecture Notes in Chemistry **75** (2000) 74–87
10. Baraglia, R., Laforenza, D., Laganà, A.: Parallelization strategies for a reduced dimensionality calculation of quantum reactive scattering cross sections on a hypercube machine. Lecture Notes in Computer Science **919** (1995) 554–561. Baraglia, R., Ferrini, R., Laforenza, D., Laganà, A.: An optimized task-farm model to integrate reduced dimensionality Schrödinger equations on distributed memory architectures. Future Generation of Computer Systems **15** (1999) 497–512
11. Bolloni, A., Riganelli, A., Crocchianti, S., Laganà, A.: Parallel quantum scattering calculations applied to the dynamics of elementary reactions. Lecture Notes on Computer Science **1497** (1998) 331-337; Laganà, A, Crocchianti, S., Ochoa de Aspuru, G., Gargano, R., Parker, G.A.: Parallel time independent quantum calculations of atom diatom reactivity. Lecture Notes in Computer Science **1041** (1995) 361-370.
12. Laganà, A, Bolloni, A., Crocchianti, S.: Quantum isotopic effects and reaction mechanism: the Li+HF reaction. Phys. Chem. Chem. Phys. **2** (2000) 535-540
13. Balint-Kurti, G.G., Dixon, R. N., Marston. C. C.: Grid methods for solving the Shrödinger equation and time dependent quantum dynamics of molecular photofragmentation and reactive scattering processes. International Reviews in Physical Chemistry **111(2)** (1992) 317-344
14. V. Piermarini, V., Balint-Kurti, G.G., Gray, S., Gogtas, F., Hernandez, M.L., Laganà, A.: Wavepacket calculation of cross sections, product state distributions and branching ratios for the $O(^1D)+$ HCl reaction. J. Phys. Chem (submitted)
15. Goldfield, E.M., Gray, S.K.: Mapping Coriolis-coupled quantum dynamics onto parallel computer architectures. Comp. Phys. Commun. **98** 1-14
16. Piermarini, V., Laganà, A., Smith, L., Balint-Kurti, G. G., Allan, R. J.: Parallelism and granularity in time dependent approaches to reactive scattering calculations. PDPTA **5** (2000) 2879-2884
17. Vanneschi, M.: Heterogeneous High Performance Computing environment. Lecture Notes in Computer Science **1470** (1998) 21-34

The Use of Intrinsic Properties of Physical System for Derivation of High-Performance Computational Algorithms

Alexander V. Bogdanov and Elena N. Stankova

Institute for High Performance Computing and Data Bases, Fontanka, 118
198005, St-Petersburgh, Russia
bogdanov@hm.csa.ru, lena@fn.csa.ru

Abstract. We discuss some new approach for derivation of computational algorithms for certain types of evolution equations. The main idea of the algorithm is to make functional transformation of variables on the base of symmetry properties of pertinent physical system to make it quasi-diagonal. The approach is illustrated with the help of two important examples – the scattering in molecular system with elementary chemical reactions and the system of chemical kinetics for the many components system. The use of proposed approach show substantial speed-up over the standard algorithms and seems to be more effective with the increase of the size of the problem.

1 Introduction

The increase of the power of computers, used for critical problem analysis nowadays, is surprising but even more surprising is the fact, that for many important problems it is not supported by dramatic increase of computational possibilities, especially when the number of nodes in computations is high. After some analysis it does not seem so strange since with the increase of number of parallel computational processes the price, which we pay for exchange of data between processes, is becoming more and more heavy load, that make the increase of number of nodes ineffective. That is why it is very important in deriving the algorithms to minimize the data exchange between computational processes. We feel, that in current situation it is useful to recover some ideas used in optimization of codes for transputers – to take into account the symmetries of the physical problem while mapping it to corresponding algorithm and computer architecture. It is easy to do it for evolution operator and so we shall illustrate our main ideas on the examples, where the physical meaning of the proposed procedure is quite transparent and at the same time the proposed approach really make it possible to get on large scale problems substantial speed-ups over the standard algorithms.

B. Hertzberger et al. (Eds.) : HPCN Europe 2001, LNCS 2110, pp. 204-210, 2001.

2 The Main Idea of Approach

It is easy to understand bases of the proposed algorithm on the example of the evolution equation of the type

$$iDu/Dt = Hu, \tag{1}$$

with H being some operator, which we will suppose to be Hermitian. Any standard approach normally will transform it to the linear system ordinary differential equations of the type

$$i dv/dt = \boldsymbol{H}\mathbf{v}, \tag{2}$$

with **v** being the large vector and ***H*** being symmetric matrix. The problem is not difficult for any size of **v**, if ***H*** is almost diagonal, but for many important situations it is not the case. From the point of view of the theory of dynamic systems [1] large nondiagonal members in ***H*** mean bad choice of representation for Eq. (1), although often such representation is forced by physical considerations.

Usually it is not difficult to find the transformation, which will make ***H*** quasidiagonal, some of the beautiful approaches in the theory of nonlinear equations [2] can have even natural physical background. At the same time some problems may cause the reverse transformation to original variables. Here we shall study two problems, for which it is not necessary to make reverse transformation since the solution of Eq. (1) is distribution function, used for computation of average values, which can be as effectively done in new variables, as in old ones.

3 The Equations of Nonequilibrium Kinetics

The system of kinetic equations, describing the relaxation in molecular systems [3], consists in realistic cases of thousands of equations of the type

$$Dc(\mathbf{j},t)/Dt = I(\mathbf{j-1},t) - I(\mathbf{j},t), \tag{3}$$

with **j** being the multi-index, describing molecular states and I being the molecular current in index space of the form I(**j**,t) = ***K***(**j**,**j+1**) c(**j**,t) – ***K***(**j+1**,**j**)c(**j+1**,t).

The problems with Eq.(3) actually come from two factors – there are large nondiagonal members, corresponding to important physical transitions, and values of I(**j**,t) are very large with difference between them in r.h.s. of Eq. (3) is relatively small. The situation becomes dramatic if you start integration with thermal equilibrium, when all I's are equal.

To overcome those difficulties it is useful to introduce new variables [4] f(**j**,t) = c(**j+1**,t)/c(**j**,t) a(**j**,t) with a(**j**,t) being the ratio of to rate constants K(**j**,**j+1**) and K(**j+1**,**j**). F's are so called slow variables, which become constants at equilibrium conditions. The equations for f's are [4]

$$df(\mathbf{j},t)/dt = R(f(\mathbf{j},t)) + H(\mathbf{j},t)f(\mathbf{j},t) + S(\mathbf{j})f(\mathbf{j},t)\,, \qquad (4)$$

with R being the quadric relaxational term, that is diagonal in **j**, H is the source term, proportional to hydrodynamic gradients, and S is the source of population change, the only term, that is nondiagonal in f(**j**,t). The main advantage of Eq.(4) is, that not only the sum of three terms in r.h.s. (4) is small, but they are small separately and it is easy to determine their relative value beforehand. More than that, it is important, that the major contribution in r.h.s. of (4) is diagonal, that opens interesting opportunities for parallel algorithm.

For illustration of proposed approach we have taken HF relaxation in mixture He-HF in nozzle flow with initial temperature of 300k [4]. To simulate this flow we must solve standard Euler equations together with kinetic equations (3). It is not very effective demonstration, since number of states, that should be taken into account is only 80 at such temperature, at the same time the relaxation of HF is rather slow, so one needs a lot of time for convergence of flow solution. Simulation was done on 8 processor linux cluster with slow 40Mb links to illuminate importance of communications between computational processes.

The complete simulation of initial problem took about 20 min. on 8 processor cluster. The use of relaxational system (4) instead of (3) reduces that time to 2.4 min.
One can reduce that time another order of magnitude (i.e. 12 times) if we shall take into account , that in nozzle flow leading term in (4) is H and we can drop in first approximation nondiagonal terms S(j). Fig. 1 shows, that difference between exact result and diagonal approximation is relatively small to support the use of approximate approach in technical applications.

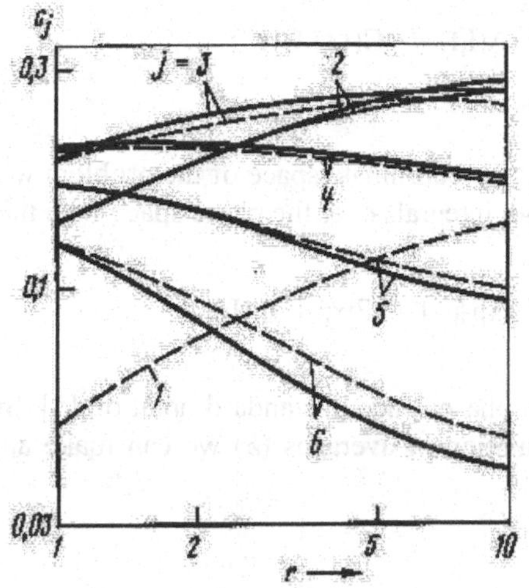

Fig. 1. The difference between exact (solid lines) and diagonal approximation (dashed lines) populations for He-HF mixture nozzle flow.

4 Inelastic Scattering Amplitude

Inelastic scattering calculations were a challenge for computational physics from the early days of computer science. For this problem in Eq.(2) with the usual problem of large dimension of $\boldsymbol{H}$ with leading offdiagonal terms we must integrate over infinite time interval with nontrivial asymptotical boundary conditions. Among many candidates for alternative approach one of the most promising is the reduction of the original formulation to Feynman's path integral. One of the firsts such effective algorithms was proposed in 1991 [5]. But due to some mathematical properties of path integrals this approach could be effectively used only for finite time interval evolution calculation. Several attempts were made to overcome those difficulties by describing the evolution in classical terms and solving large time problems by classical means. A lot of work was done in establishing rigorous relationship between classical and quantum tools for description of system evolution [6]. Although formal mathematical solution of the problem was given in 1979 [7], and computational algorithm was proposed in 1986 [8], only recently it was possible to realize it on large systems [9].

The main result of [7] was the functional transformation to so called interaction coordinates, that reduces Hamiltonian function of the problem to actually responsible for described process, reduces integration interval to finite one corresponding to interaction region, transforms asymptotical boundary conditions to standard ones and make it possible to get the expression directly to scattering amplitude as an average over Green's functions of the problem but in mixed representation and in interaction coordinates.

$$\mathbf{A(i,f) = \langle G(i,t)\ G(t,f)\rangle}\ . \tag{5}$$

The average is taken over the coordinate space of the problem with Green's functions determined in terms of path integrals over the phase space with the weights of the type

$$\mathrm{Exp}\{i \int X dP\} \exp\{-i \int H dt\}. \tag{6}$$

i.e. the classical action in phase space in standard notation [5]. Since we use Green's functions only for computation of averages (5) we can make any phase coordinates transformations of the type

$$H(P,X) \rightarrow \quad \mathit{H}(X,\partial F/\partial X) + \partial F/\partial t\ . \tag{7}$$

With F being the Generator of the transformation. It is convenient to choose F as a solution of certain equation [8] that guarantees the possibility of evaluation of path integral with the weight (6). In that case instead of computation of path integral we have to solve four times the equation for F, that is more convenient since it is partial differential equation of the first order [7].

So in general case computation of scattering amplitude in our approach is reduced to computation of average, i.e. of integral over the coordinate space and solution for every coordinate of four partial differential equations of the first order for F. It is clear, that such formalism gives an ideal parallel algorithm, since we can do the solution of equations for different points independently. More, than that, we can make the same sort of diagonal approximation as in previous example and choose one average solution for all four generators F. This is the equivalent of so called average trajectory approximation in semiclassical approach.

For illustration the rate constant for one of the most popular test processes – vibration deexcitation of first vibration state of Nitrogen molecule in collision with Nitrogen[10]. The results of exact computations and diagonal approximation are plotted with some classical calculations in Fig2. Again the difference is unimportant for practical purposes, but the difference in computer time here is even more impressive – the diagonal approximation takes 109 times less, than exact computation.

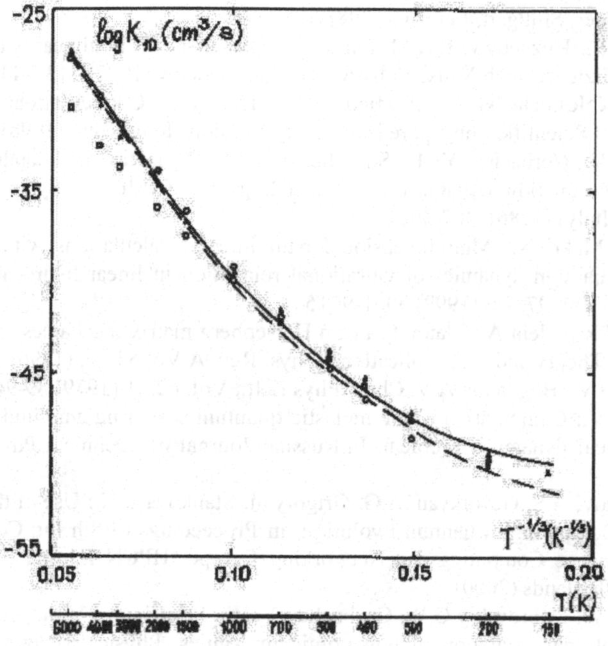

Fig. 2. De-excitation rate of first vibration state of Nitrogen in pure Nitrogen – the result of exact quantum computations (solid line), diagonal approximation (dashed line) and classical trajectory computations (dots and arrows).

5 Conclusions

We have shown that the use of some physical considerations makes it possible to derive some new algorithms for solution of the evolution equations for physical variables like distribution function. With those algorithms we can reduce the needed computer time orders of magnitude, go to substantially larger number of processor and work out approximate methods, which can be used for mass computations in technical applications.

References

1. Ebeling W., Freund J., Schweitzer F.: Komplexe Strukturen: Entropic und Information.B.G. Teubner, Stuttgart, Leipzig (1998)
2. Zakharov, V.E., Kuznetzov, E.A.: Hamiltonian formalism for nonlinear waves. In: Russian Uspekhi Fizicheskhich Nauk, Vol. 167. Nauka, Moscow (1997) 1137–1167
3. Itkin A.L., Kolesnichenko E.G.: Microscopic Theory of Condensation in Gases and Plasma. World Scientific, Singapore New Jersey London Hong Kong (1998)
4. Bogdanov A.V., Gorbachev Yu.E., Strelchenya V.M.: Relaxation and condensation processes influence on flow dynamics. in Proccedings of XV Intern. Symp. Rarefied Gas Dyn. Grado (Italy) (1986) 407-408
5. Topaler M., Makri N.: Multidimensional path integral calculations with quasidiabatic propagators: Qantum dynamics of vibrational relaxation in linear hydrocarbon chains. J.Chem.Phys. Vol. 97, 12, (1992) 9001-9015
6. Greenberg W.R., Klein A., Zlatev I.: From Heisenberg matrix mechanics to semiclassical quantization: Theory and first applications. Phys. Rev. A Vol.54, 3 , (1996) 1820-1836.
7. Dubrovskiy G.V., Bogdanov A.V. Chem.Phys.Lett., Vol. 62, 1 (1979) 89-94.
8. Bogdanov A.V.: Computation of the inelastic quantum scattering amplitude via the solution of classical dynamical problem. In:Russian Journal of Technical Physics, 7 (1986) 1409-1411.
9. A.V. Bogdanov, A.S. Gevorkyan, A.G. Grigoryan, Stankova E.N.: Use of the Internet for Distributed Computing of Qantum Evolution. in Proccedings of 8th Int. Conference on High Performance Computing and Networking Europe (HPCN Europe 2000), Amsterdam, The Netherlands (2000)
10. Bogdanov A.V., Dubrovskiy G.V., Gorbachev Yu.E., Strelchenya V.M.: Theory of vibration and rotational excitation of polyatomic molecules. Physics Reposrts, Vol.181, 3, (1989) 123-206.

Parallel DEM Simulations of Granular Materials

J.-A. Ferrez and Th.M. Liebling

Chair of Operations Research, Department of Mathematics
Swiss Federal Institute of Technology, Lausanne (EPFL)
Jean-Albert.Ferrez@epfl.ch, Thomas.Liebling@epfl.ch

Abstract. Computer simulations of granular materials are often based on the Distinct Element Method (DEM) where each grain is considered individually. Since large quantities of grains are required to perform realistic experiments, high performance computing is mandatory. This paper presents the basis of the DEM, a sequential algorithm for spherical grains in 3D and the adaptations to obtain a parallel version of that algorithm. Visualization is also discussed, as the drawing and animation of large sets of grains require special techniques and state-of-the-art graphics hardware. Finally, some applications are presented.

1 Introduction

Granular materials are commonly encountered in nature, in various industrial processes, and in everyday life: landslides and avalanches, erosion, raw minerals extraction and transport, cereal storage, powder mixing in chemistry or pharmaceutics, railroad beds, concrete or embankments in civil engineering, and many more. They have been called a fourth state of matter different from the classic solid, liquid and gas. As such, they exhibit specific phenomena that call for better understanding. Experimental studies are being conducted, but numerical simulation is increasingly seen as a promising way to understand and predict their behavior. Such simulations have become common in the design and optimization of industrial processes [1, 2].

The Distinct Element Method (DEM) is a natural tool used in computer simulations of discontinuous phenomena. By tracking the evolution of each element, particle or grain separately, it allows to take into account their local behavior while providing global measures. Classical examples include molecular dynamics and granular media simulations. However, DEM simulations require very large number of elements in order to accurately reproduce the global behavior of the material. High performance infrastructures are mandatory not only for the actual computation but also for post-processing and especially visualization of the results.

This paper presents a brief overview of the DEM, focusing on two key issues, namely the efficient detection of grain interactions and the accurate underlying physical models. A parallel version of the DEM algorithm is then presented. Visualization is discussed in section 5 and we conclude by showing some example applications.

B. Hertzberger et al. (Eds.): HPCN Europe 2001, LNCS 2110, pp. 211–220, 2001.

2 The Distinct Element Method

The DEM scheme tracks every particle in the system: position, velocity, acceleration, orientation and spin are maintained and updated whenever the particle interacts with external elements (gravitation, contact with walls...) or with other particles. It is not our intent to go into the details of the DEM here, but rather just point out the two computationally intensive issues that arise in this context: find where the interactions occur and apply a suitable numerical model. The interested reader will find a more complete coverage of the DEM in [3, 4].

2.1 Contact Detection

Localizing contact points among large sets of grains is not a trivial task. Some sort of neighborhood must be defined for each grain in order to reduce the number of potential contacts to be tested. In the much simplified case where all grains are spherical, a dynamic weighted Delaunay triangulation built on the centers of the grains has proven to be highly efficient, robust and versatile. An example of such a triangulation in two dimensions is shown in Fig. 1. This collision detection scheme works by enclosing each sphere in its own convex polyhedric cell, thus forming a spatial decomposition know as the Power Diagram [5, 6]. Its dual structure, the regular (or weighted Delaunay) triangulation [7, 8] provides an efficient way to monitor adjacency of the cells. Since spherical grains may collide only if their respective cells are adjacent, collision detection is reduced to a simple traversal of the edges of the Delaunay triangulation. A detailed discussion of this method can be found in [9]

Although this scheme reduces the algorithmic complexity of the contact detection in most cases from $O(n^2)$ to $O(n)$, the fairly large constant hidden by the $O(n)$ due to many geometric predicates evaluations means it only shows its strength for large cases involving 100'000 and more particles.

2.2 Numerical Contact Models

Soft-body contact models are based on the discretization of time into very short intervals (10^{-5}s to 10^{-7}s). During those intervals, the grains are allowed to slightly overlap, and the characteristics of the overlap are used to model the contact.

Forces acting on the grains at each contact point are decomposed into a normal and a tangential component, both of which were traditionally expressed as the sum of an elastic term (ensures repulsion between the grains) and a viscous term (provides energy dissipation):

$$F = k\xi + c\frac{d\xi}{dt}$$

where k and c are linear coefficients for elastic and viscous part, and ξ is the size of overlap caused by the contact. This is roughly equivalent to a double spring and dashpot system [10].

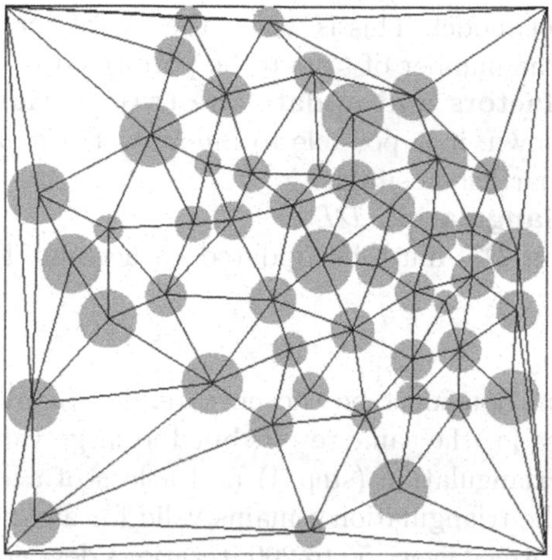

Fig. 1. The triangulation used to detect collisions, shown here in 2D. Contacts may only occur along an edge of the triangulation, thus the neighborhood of each grain is given by the adjacency defined by the triangulation.

However, those simple models are usually hard to fine tune, because they rely on parameters (k and c) that are difficult to link with the physics of the contact. If a grain is in contact with more than one other grain — which is the dominant case in packed materials — the overlaps may become much too large, up to the size of the grains themselves!

Newer numerical models are being designed that address those issues. Elastic and viscous components of the force are not linear anymore, and the history of the contact is accounted for in order to replicate more accurately the reloading mechanism of the contacts (hysteresis) [11]. One general consequence of those new models is the increased amount of computation required to obtain a more precise result.

3 The DEM Algorithm

We give here the details of the algorithm for DEM simulations of spherical grains in 3D using soft-body contact models. The use of the dynamic Delaunay triangulation as neighborhood function implies some changes in the general scheme of a distinct element simulation algorithm: steps 1 and 4 would not appear in a straight-forward method, while step 2 would require another way to find the collisions.

1. **Construct the initial Delaunay triangulation $\mathcal{DT}$**
   This is the startup overhead, incurred only once.
2. **Apply the numerical contact model at every collision**
   Along each edge (i, j) of $\mathcal{DT}$, if spheres i and j overlap, then apply the

numerical contact model. This is where the use of the triangulation shows its strength, as the number of tests to be performed is low.

3. **Add external factors and update the trajectories**
   All forces are known, it is possible to integrate the equations of motion to obtain the new particle positions.
4. **Update the triangulation $\mathcal{DT}$**
   The triangulation $\mathcal{DT}$ must be updated to account for the new particle positions.
5. **Go back to 2**

The iteration duration must be chosen carefully. Small values will require many more iterations for the same result, but too large values might introduce instabilities in the triangulation (step 4) and a loss of accuracy in the model behavior (step 2). The triangulation remains valid for a certain amount of time, so step 4 is only performed every 50 to 200 iterations depending on the volatility of the material.

4 The Parallel DEM Algorithm

Most high performance computers nowadays are parallel machines and thus require parallel codes based on parallel versions of the algorithms. The DEM algorithm presented above is well suited to a medium grain parallelism. Each iteration is divided in two main phases:

- In step 2, compute the forces along each edge of the triangulation as a function of the position (and velocity, spin,...) of the grains.
- In step 3, compute the positions (and velocity, spin,...) of the grains as a function of the forces in the incident edges and external forces.

These two phases can not be interleaved, due to data dependencies. However, each phase exhibits a very high degree of parallelism since each force, resp. the position of each grain, may be computed independently[1]. With this scheme, workload balancing follows closely data repartition on the processing elements. It is fairly easy to divide the simulated area into layers and assign the grains and edges in a layer to a given processor, as shown in Fig. 2. That processor is then responsible for updating all values in that layer, eventually reading values from a neighbor layer.

As the grains move during the course of the simulation, it may be necessary to reform the layers in order to maximize data locality. This necessitates a responsibility transfer for one given grain or edge from one processor to another, and is easily achieved as part of the triangulation update in step 4.

Most of step 4 deals with checking the validity of the triangulation through the computation of geometric predicates. That can again be performed in parallel

[1] We assume here that contentions due to concurrent memory accesses to a read-only variable have no impact. This is true for the validity of the method, but may influence the memory throughput and therefore the overall performance.

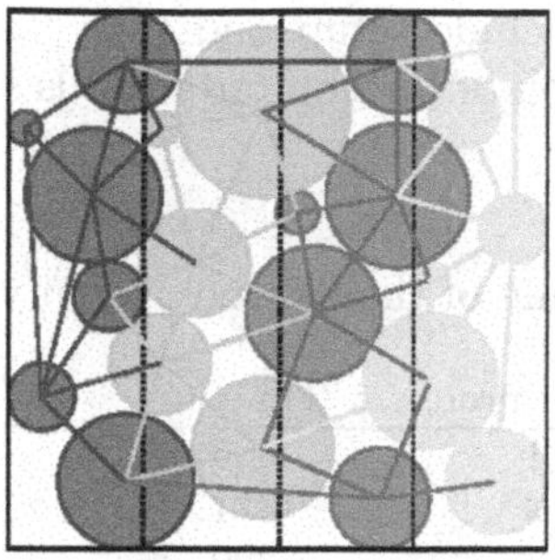

Fig. 2. The simulated area is divided in layers that are assigned to processors.

for each layer. The short phase of topological changes and layers redistribution that takes place if the triangulation is not valid anymore is performed in a sequential block, as is the initial construction of the triangulation (step 1).

At this stage comes the distinction between shared-memory and distributed-memory architectures. If the parallel machine offers uniform memory access to every processor[2], the distribution of elements (grains and edges) is just an assignment, which greatly simplifies the problem. Previous experience [12] has shown, however, that such a blind scheme is defeated on single image, Non Uniform Memory Access (NUMA-SMP) machines[3]. In this case, care must be taken that a given element reside in the local memory of the responsible processor. Access to elements of neighbor layers is still transparent but slower, and one wants to minimize the number of such accesses. Finally, if no transparent remote memory access is provided[4], either by the hardware or by an intermediate library, data locality is even more important since every access to a neighbor layer requires an explicit data exchange between the two processors under the form of a message passing.

Aside from reducing computation time, parallel machines are sometimes required to satisfy huge memory needs. This is not a problem with the current models since a standard workstation with 512Mb of memory will easily host a simulation of 100'000 spherical grains. However, more complex geometries or contact models that keep a history of the collision might change this situation.

Müller wrote several simulation codes for various 2D cases [13]. We developed a distributed memory version of one of them for the Cray T3D system [14]. The current 3D simulation code exists in single or multi threaded versions and runs on various shared memory computers, including an 80 CPU SGI Origin server. A distributed memory version is in development.

The performance of the shared memory parallel code has been tested on an SMP PC with 4 Pentium III Xeon processors. Speedup compared to the sequential code running on the same machine is given in Table 1. These results

[2] Such as SMP PCs or workstations with few processors.
[3] Such as SGI Origin 2000 with 8 or more processors
[4] This is typically the case with clusters of machines running MPI.

are wall clock time and include setup and I/O. They are in accordance with the fact that a small portion of the computation is still performed in sequential mode.

Table 1. Timing of the Shared Memory Parallel Code.

Nb of proc.	Time [s]	Speedup
1	505	-
2	311	1.62
4	220	2.29

5 Visualization

Interactive graphical representation of the results of large scale DEM simulations is a tedious task. Most visualization packages can efficiently handle large time-dependent 3D meshed data such as fluid flows or heat transfers, but in the DEM context it is usually necessary to draw each individual element. Furthermore, using spheres greatly reduces the complexity of the computation, but has a side effect since drawing a realistic 3D sphere takes much more graphical power than drawing a cube. Two trends emerge in the transformation of the computational results into visual experience:

High resolution images of instantaneous snapshots of the simulation are rendered using various techniques such as ray-tracing. Eventually series of such images are assembled into video clips. This approach is suited for static and non-interactive representation such as book or poster illustrations, web pages, etc. The main drawback — aside from a considerable post-processing effort — is the lack of interactivity. It is not possible to view a certain situation under a different angle, to perform projections, clippings, boundary removal, etc. without repeating the whole process. The scripting capabilities of packages such as AVS™ help automate these steps.

Online interactive visualization is possible on tiny cases containing 100 to 1000 spheres. Besides comfortably studying small cases during the development of the method, this allows for example to track only a few selected particles or to restrict the visualization to a very small portion such as the opening of a hopper. However, the lack of a global vision in these cases is a strong obstacle to the global comprehension of the phenomenon. We have used the VRML scene description language for this purpose, many examples are posted on the web at `http://rosowww.epfl.ch/jaf/3dwdt/`. However, VRML induces too much overhead and large cases will probably require direct OpenGL rendering.

Intensive development is underway in this area and one sees at the same time the availability of inexpensive yet powerful graphic cards on the average desktop PC, and the development of extremely powerful visualization centers,

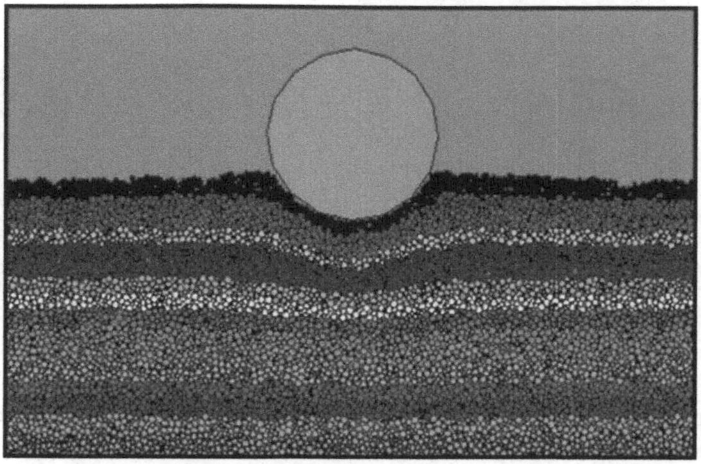

Fig. 3. Rock Fall on an Embankment.

the latest trend being *remote visualization*, where the frames are computed by a specialized graphic server and rendered to the desktop through high bandwidth networks. We are collaborating with specialists at the SGI European Headquarters in Cortaillod (CH) to improve interactive visualization of large scale DEM simulations.

6 Applications in Granular Media Simulations

We give here some example of the simulation we carry out with the algorithms and codes discussed above. More examples, including short animations, can be found on our Web page at `http://rosowww.epfl.ch/jaf/3dwdt/`.

6.1 Rockfall Impact on an Embankment

Fig. 3 shows a 2D simulation of the impact of a rock falling down on an embankment composed of about 20'000 grains. The colors show the layers and their deformation after the impact. This simulation is the replication of an experiment conducted at the Department of Civil Engineering, EPFL. This close collaboration allowed us to compare our simulations with real experiments, thus calibrating and validating the models.

We have repeated this simulation with the 3D model as shown in Fig. 4. Besides confirming the results obtained in 2D, the 3D model allows a wider range of simulations, including oblique impacts.

6.2 Powder Packings

Of crucial importance to practitioners is the control of the density of powders mixing. Theoretical results are known when the grains are all of equal size, the

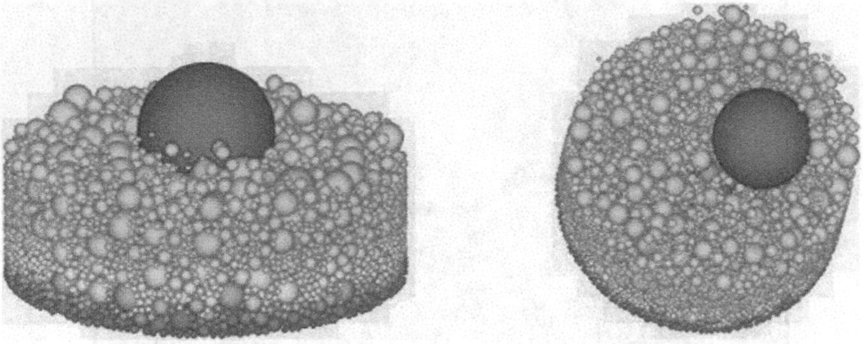

Fig. 4. Impacts on a 3D granular bed. The bed is composed of approximately 12'000 grains with a size ratio of 1:4. The falling block is 20 times larger than the small grains. On the *left* the impact is vertical, similar to the 2D case above. On the *right*, the impact is oblique and creates an asymmetric crater.

optimal density in this case is slightly larger than 74%. No such results exist for general distributions of sizes, and computer simulations help determine bounds for the optimal densities. Fig. 5 illustrates such a case with about 25'000 grains with caliber ratio 35:5:1. However, using only one large grain is not enough to draw robust conclusions in this case, and larger simulations are underway.

One way to improve the density of such a granular packing is to apply vibrations. A test was performed with approximately 10'000 spherical grains of diameter 2mm ±0.02mm in a vertical cylinder. Vertical vibrations (tapping) of amplitude of 2mm were applied at 2Hz, 5Hz and 10Hz for 10 seconds. Fig. 6 shows snapshots of the three cases. At 2Hz, the initial layers identified by the colors are still clearly visible. At 5Hz, more rearrangements occur but the packing remains compact. At 10Hz, much more energy is brought to the system, the individual grains are able to move more freely and convection effects appear.

7 Conclusion

The Distinct Element Method was a major break through in that it first allowed to perform computer simulations of mostly unknown, discontinuous phenomena. The addition of an efficient collision detection mechanism opens new opportunities, since larger simulations are usually needed to better match physical reality. Neighborhoods based on the weighted Delaunay triangulation represent such an efficient, robust and versatile collision detection mechanism. Both the DEM and the triangulation based neighborhood are well suited to parallel computing, which is increasingly required to benefit from recent HPC infrastructures. Further efforts are required, though, in the physical modeling of the contacts in granular assemblies and in the visualization of the simulation results. Yet, cur-

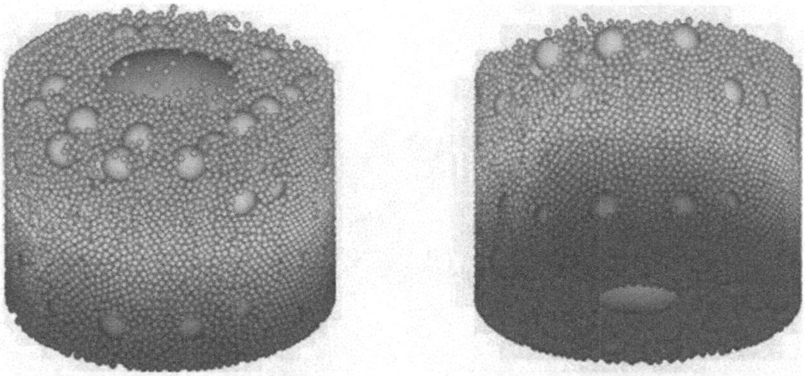

Fig. 5. Top and bottom views of a cylindric packing where 76 medium and about 25'000 small grains fill up the space around a large grain.

rent knowledge, techniques and codes allow us to study various key phenomena such as shock absorption and dense packings of irregular grains.

References

[1] P. W. Cleary and M. L. Sawley. Three-dimensional modelling of industrial granular flows. In *Second International Conference on CFD in the Minerals and Process Industries*. CSIRO, Melbourne, Australia, 1999.

[2] J.-A. Ferrez, L. Pournin, and Th. M. Liebling. Distinct element computer simulations for optimal packings of 3d spheres. Final Report, Simboules project, second year, 2000.

[3] G. C. Barker. Computer simulations of granular materials. In A. Mehta, editor, *Granular Matter: An interdisciplinary approach*. Springer-Verlag, 1994.

[4] J.-A. Ferrez, D. Müller, and Th. M. Liebling. Dynamic triangulations for granular media simulations. In K. R. Mecke and D. Stoyan, editors, *Statistical Physics and Spatial Statistics*, Lecture Notes in Physics. Springer, 2000.

[5] F. Aurenhammer. Power diagrams: properties, algorithms and applications. *SIAM J. Comput.*, 16(1):78–96, 1987.

[6] L. J. Guibas and L. Zhang. Euclidean proximity and power diagrams. In *Proc. 10th Canadian Conference on Computational Geometry*, 1998. `http://graphics.stanford.EDU/~lizhang/interests.html`.

[7] J.-D. Boissonnat and M. Yvinec. *Géométrie Algorithmique*. Ediscience, 1995. Published in english as Algorithmic Geometry, Cambridge University Press, 1998.

[8] H. Edelsbrunner and N. R. Shah. Incremental topological flipping works for regular triangulations. *Algorithmica*, 15:223–241, 1996.

[9] J.-A. Ferrez and Th. M. Liebling. Dynamic triangulations for efficient collision detection among spheres with applications in granular media simulations. *Submitted to Phil. Mag. B*, 2001.

[10] P. A. Cundall and O. D. L. Strack. A discrete numerical model for granular assemblies. *Géotechnique*, 29(1), 1979.

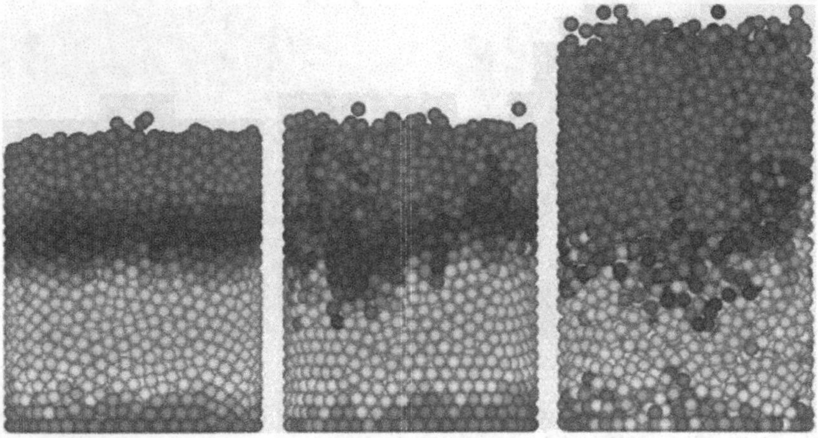

Fig. 6. Snapshots at Various Frequencies: 2Hz (left), 5Hz (center) and 10Hz (right).

[11] L. Pournin, Th. M. Liebling, and A. Mocellin. A new molecular dynamics force model for better control of energy dissipation in dem simulations of dense granular media. *in preparation*, 2001.

[12] J.-A. Ferrez, K. Fukuda, and Th. M. Liebling. Parallel computation of the diameter of a graph. In J. Schaeffer, editor, *High Performance Computing Systems and Applications*, pages 283–296. Kluwer Academic Publishers, 1998.

[13] D. Müller. *Techniques informatiques efficaces pour la simulation de mileux granulaires par des méthodes d'éléments distincts.* Thèse No 1545, EPFL, 1996.

[14] J.-A. Ferrez, D. Müller, and Th. M. Liebling. Parallel implementation of a distinct element method for granular media simulation on the Cray T3D. *EPFL Supercomputing Review*, 8, 1996. Online at `http://sawww.epfl.ch/SIC/SA/publications/SCR96/scr8-page4.html`.

[15] J.-A. Ferrez and Th. M. Liebling. Using dynamic triangulations in distinct element simulations. In M. Deville and R. Owens, editors, *Proceedings of the 16th IMACS World Congress*, August 2000.

A Generic Support for Distributed Deliberations

Jacques Lonchamp and Fabrice Muller

LORIA, BP 254
54500 Vandoeuvre-lès-Nancy, France
{jloncham,fmuller}@loria.fr

Abstract. Under the umbrella term "deliberation", we classify a wide range of process fragments, more or less spontaneous, structured, and complex. They can include both individual and collective activities, both synchronous and asynchronous phases. But in every case, at the heart of a deliberation is the negotiation of some result by exchanging arguments. This paper describes a comprehensive and flexible generic support for distributed deliberations, providing collective problem solving support. First, the paper defines the most important requirements through a real life demonstrative example: the assessment of solutions in a concurrent engineering setting. Then, the paper emphasizes the main design choices and describes the current Java prototype.

1 Introduction

Due to the possibility of distributed working, new forms of organization have recently emerged. For instance, the concept of virtual enterprise refers to a temporary alliance of enterprises, created for taking advantage of a market opportunity [2]. Large multinational research projects are other examples of this kind of temporary networks. Within such distributed and temporary alliances, participants frequently launch deliberations about shared issues. Under the umbrella term "deliberation", we classify a wide range of *process fragments*, which can be more or less spontaneous, structured, and complex. They can include both individual and collective activities, both synchronous and asynchronous phases. But in every case, *at the heart of a deliberation is the negotiation of some result by exchanging arguments*. In the simplest form, that we call "*open deliberations*", users just define a problem, propose solutions, and argue about them synchronously or asynchronously. Classical examples of *structured deliberation types* are the collective elaboration of some artifact (e.g. a list of ideas), and the collective assessment of an individually proposed artifact (e.g. a code inspection). In the second section, we will study in greater detail a real life example of a complex deliberation type in the concurrent engineering field.

We claim that no comprehensive and flexible computer support for distributed deliberations is currently available. Simple groupware tools only support a particular functionality or range of functionalities (e.g. e-mail, chat, application sharing, shared editors or whiteboards, workflow management,)..For many cooperative situations, and in particular for distributed deliberations, several forms of cooperation are required, such as synchronous and asynchronous work (and the many forms in

B. Hertzberger et al. (Eds.) : HPCN Europe 2001, LNCS 2110, pp. 221-230, 2001.

between), individual and collaborative work, explicit and implicit (social) coordination, implicit and explicit interaction. Therefore, a variety of fragmented tools would be necessary to support them, and the burden of switching between the different tools would be imposed to end users [1]. Most groupware development platforms suffer from a similar restricted applicability, because they are based on some specific technology, such as asynchronous messaging ([5, 12]) or multipoint communication and consistency management of replicated data ([8, 18]). Some more general approaches advocate for a compositional solution, where a dedicated system is built by picking and mixing small chunks of groupware functionality, and possibly by developing new ones [16]. But concrete implementations are currently missing.

We defend the idea of *systems oriented towards a specific pattern of cooperation,* such as distributed deliberations*, but providing a comprehensive and flexible generic support for all aspects of that kind of situation.* Genericity is necessary because it is not possible to design ad hoc systems from scratch for each specific case. Many similar applications share a large set of common functionalities that should not be developed over and over again. But we do not know enough about group work to design a generic environment that fits all kinds of cooperative situations. Some kind of specialization is required, not oriented towards a specific functionality, or a specific technology, but preferably towards some specific pattern of cooperation. *Our project aims at providing such a comprehensive and flexible generic support for distributed deliberations, providing collective problem solving support.* By definition, such a computerized support should satisfy the following basic requirements: (R1) it should be generic for supporting the wide range of deliberation types, (R2) it should support distributed work, (R3) it should support structured processes with both individual and collective activities, and their coordination, (R4) it should support both synchronous and asynchronous phases, (R5) it should support the core argumentation and decision activity, (R6) it should provide various channels for both informal and formal communication, (R7) it should provide process guidance, group awareness, argumentation and decision making assistance, (R8) it should be easy to install and use. Some less obvious requirements are discussed later, on the basis of a real life example.

The rest of the paper is organised as follows. Section 2 introduces the demonstrative example, taken in the concurrent engineering field. Section 3 summarizes our approach and its main design choices. Section 4 describes DOTS ('Decision Oriented Task Support'), our current prototype system, its architecture, and its usage through the asynchronous and synchronous clients. Section 5 ends the paper with the conclusions and discusses some further work.

2 A Demonstrative Example

Concurrent Engineering (CE) refers to the same organizational process described as "simultaneous engineering", "parallel engineering", or "multi-disciplinary team approach": everyone contributing to the final product, from conceptual design to production and maintenance, is required to participate in the project from its very inception [14]. In CE, the team nature of the design activity implies that many design solutions are negotiated through argumentation processes.

Cognitive ergonomics research work has been conducted, during the definition phase of an aeronautical design project, to study solution assessment processes which take place currently in informal multi-speciality meetings [13]. In the next future, large projects will be conducted within international alliances and many assessment processes could take benefit from a computer support. In a solution assessment process, an initial solution proposed by one speciality (e.g. structural design) is discussed. If this initial solution is not accepted by the other specialities (e.g. production), a negotiation follows in order to reach a consensus. One or several alternative solutions can be proposed, which are in turn assessed. Meetings analysis show three basic assessment modes: analytical assessment mode, i.e. systematic assessment according to design constraints, comparative assessment mode, i.e. systematic comparison between alternative solutions, analogical assessment, i.e. transfer of knowledge acquired on a previous solution in order to assess the current solution. Some combined modes are also possible, such as comparative/analytical, or analogical/analytical modes. Arguments are of different nature. Due to the nature of the task many arguments explicitly or implicitly make reference to design constraints. Constraints can be weighted differently according to the specialities. Some arguments can take the status of 'argument of authority', depending on specific factors: the expertise of the author and his/her role, the status of the speciality that expresses it, or the shared nature of the knowledge to which it refers. The process organization is defined opportunistically by the participants, depending on the circumstances. A usual temporal pattern starts with the analytical assessment of an initial solution. If no consensus is reached, a comparative or/and an analogical assessment follows. If this step does not lead to a consensus, arguments of authority are often used for trying to close the process.

From this cognitive analysis, we can derive several additional requirements for a computerized deliberation support: (R9) it should support dynamic process model refinement (i.e. during execution), (R10) it should manage a range of artifacts (e.g. problems, solutions, design constraints), (R11) it should provide a flexible argumentation and decision scheme, (R12) it should support role definition, (R13) it should facilitate reuse of artifacts and argumentation (for analogical reasoning).

3 Main Design Choices

3.1 A Meta Model for Describing Deliberation Models

Our deliberation support system is a generic process-centred system (R1). For building a dedicated environment, a deliberation model, describing a set of types and their relations, is written with a specific process modelling language. This model is instantiated, partly with an instantiation tool, for the instances that are necessary for starting the deliberation, and partly during the model execution. Our process modelling language is defined by a meta model, which extends classical workflow process meta models with a decision-oriented view: we call our approach *"fine grain and decision-oriented task modelling"*. Such a meta model includes (see [9] for more details):

- *a process view*: with deliberation and phase types (e.g. the 'Analytical assessment' phase type in the CE example); each deliberation instance is a network of phase instances, with precedence relationships and AND/OR split and join operators;
- *a product view* (R10): with artefact, component, and application tool types; artefact types specialize text, list, graph, or image generic types; component types specialize list element, graph node, or graph vertex generic types; application tool types mirror the specialization of artefact types, with specializations of text viewer, list viewer, or graph viewer types (e.g. 'Problem', 'Solution', 'Constraint', and 'Evaluation' artefact types, 'Solution viewer', and 'Evaluation summariser' application tool types in the CE example); each phase type grants access to some application tool types;
- *an organisational view* (R12): with role and actor types (e.g. 'Structural designer' and 'Production engineer' role types in the CE example);
- *a decision-oriented view* (R5), describing collaborative work at a fine grain level: each phase type is defined internally by a set of issue types, which must be solved either individually or collectively (R3). Each issue type is characterised by a set of option types, which describe the different possible resolutions. At run time, one option is chosen trough an argumentation process (section 3.3). Each option type can trigger, when it is chosen, some operation type, which can change some artefact, or change the process state (e.g. termination of a phase), or change the process definition itself (section 3.2) (e.g. 'Evaluate a solution for a constraint', with 'Satisfy' and 'Contravene' option types in the CE example).

3.2 Dynamic Deliberation Model Refinement

A model designer who builds the process model for a complex deliberation has often to describe several strategies for performing the task. In the CE example we can distinguish analytical, comparative, and analogical assessment strategies. Each one can be described within a dedicated phase type. The choice between them takes place at execution time, within a model refinement phase. The model refinement phase type includes *a model refinement issue type, with one option type for each strategy*. At run time, one of these option is chosen (either individually by a kind of 'deliberation manager', or collectively); the corresponding strategy is deployed, by instantiating new phase instances (R9). *By this way, artefact and process elaboration are similar, and the system can provide the same assistance and guidance for both aspects.*

3.3 Argumentation and Decision Support

Participants can argue about the different options of an issue, and give qualitative preferences between arguments. At every stage of the argumentation, the system computes a preferred option, through *an argumentative reasoning technique* similar to those described in [6, 7]. The issue, its options, the arguments 'for' and 'against' the options, the arguments 'for' and 'against' the arguments form an argumentation tree.

A score and a status (active, inactive), which derive from the score, characterise each node of the tree. The score of a father node is the sum of the weights of its active child nodes which are 'for' the father minus the sum of the weights of its active child nodes which are 'against' the father. If the score is positive, the node becomes active, otherwise, it becomes inactive. Only status propagates in the tree because scores have no global meaning. Without preferences constraints all nodes have the same weight. Leaves are always active. The preferred option(s) computed by the system has(have) the maximum score. Preference constraints are qualitative preferences between arguments. The importance of one argument (the source) is compared with the importance of the other (the destination). The aim of constraints is to define different weights for arguments in a very flexible and revisable way (R11). To each constraint is attached a "constraint issue" with three positions: 'More important than', 'Less important than', 'Equally important than'. Thus, all constraints are subject to debate and can change their value dynamically. A constraint is active if both its source and its destination arguments are active, if one of its option is chosen (with a score strictly higher than the others), and if it is consistent with the other constraints. Constraint consistency is evaluated (with a path consistency algorithm) each time a constraint is created or becomes inactive. The weight of all arguments having the same issue as grand father (global constraint), or the same argument as father (local constraint), is computed by a heuristics which gives a medium weight for arguments not participating to constraints, a higher weight for arguments which dominate other arguments and a lower weight for those which are dominated by other arguments [9].

Argumentation is useful both for individual and for collective issues. For individual issues, arguments explain some rationale (e.g. in the CE example, an individually proposed solution or a new constraint should come with some rationale). For collective issues, argumentation prepares the resolution. In most cases the decision is kept separated from the argumentation result. Each issue type has a resolution mode property, which defines how issue instances are solved. For instance, 'autocratic without justification', when an actor playing a given role can take the decision, 'autocratic with justification', when the decision maker has to complement the argumentation in order to make his/her choice equal to the preferred choice of the system, 'democratic', when the preferred choice of the system is chosen. *In most cases, the aim is to make structured communication among the participants easier, and not to automate the decision making.*

3.4 Synchronous and Asynchronous Work Support

In a typical deliberation, some result is negotiated collectively. At the beginning of the process, participants can express their opinion asynchronously. Synchronous sessions are useful when the argumentation becomes complex and *when no rapid convergence occurs*. Our deliberation system supports both working modes and smooth transitions between them. The model designer can describe how synchronous sessions are organized. Synchronous sessions can be managed as artefacts, i.e. created through an issue resolution, by a given role, and accessed by the other participants through application tools (meeting list viewer or meeting reminder). The organization process itself can be supported through issue types such as 'Call for organization',

'Answer organization', 'Call for participation': the operation types which are triggered by these issue types can use direct communication channels (R6) (e.g. dynamically built e-mails). This organization process can be described in a 'Meeting organization' phase type, whose instances run in parallel with other asynchronous phases.

The system also provides *awareness indicators (R7) for deciding which issues require a synchronous discussion*. These indicators are computed through an analysis of the valuated argumentation tree and are directed towards participation (who has already participated? when ?), divergence of opinions (who supports which option? by correlating participant interactions and option score evolutions), stability (is the preferred option computed by the system stable or not?). During asynchronous sessions, another awareness facility answers the 'what's new?' question, highlighting entities which have changed since the end of the previous connection of the same user. During synchronous sessions, several synchronous communication channels (chat, whiteboard, vote widget) are provided (R6). They can be used for instance for 'meta discussion' about the session (e.g. decision to skip to the next issue).

4 DOTS Prototype

4.1 DOTS Architecture and Tools

DOTS ('Decision Oriented Task Support') is our current prototype. It is completely developed in java for portability reasons (R8) and runs on the Internet (R2). It has an hybrid architecture with a traditional client/server architecture for asynchronous work and direct inter-client communication for synchronous work. The server is built on top of an open source object base (www.sourceforge.net/projects/storedobjects) which stores each model and all the corresponding instances (R13). The synchronous infrastructure (sessions, channels, clients, tokens) is provided by Java Shared Data Toolkit (www.java.sun.com/products/java-media/jsdt/index.html). Each sub system can be used independently. When used together, a single user (playing the pre-defined 'Moderator' role) has to access the database, for extracting already asynchronously started argumentation and for storing the results of the synchronous session.

DOTS comes with a complete development environment. All tools are developed in Java. First, a deliberation task model is written with DOTS modelling language and DOTS IDE (Integrated Development Environment). This model is compiled into Java classes by the IDE. Together with the Java classes of the generic kernel, they constitute a dedicated environment. All these classes are compiled with a Java compiler and a persistence post processor to become a persistent executable environment. Thanks to DOTS Instantiation tool, initial objects are created in the object base (actors, deliberation instances, initial documents, etc). The instantiated model can also be checked for completion by DOTS static analyser. Execution can start through the asynchronous client, which is independent of any deliberation model. The synchronous client is called from the asynchronous client (a slightly modified standalone version is also available). End users only manipulate the clients and the instantiation tool. Development of new deliberation models is performed by specialists (e.g. tool providers), who master the task modelling language. End users interact with DOTS through the very intuitive interface of the clients (R8).

4.2 DOTS Usage

A participant who wants to use DOTS, has to start the asynchronous client, to log in and to choose among the ongoing deliberation where he/she plays a role. The asynchronous client is organized with five tabbed panels (see Fig.1). The ‘action panel’ is the place where interactions occur through a tool bar and two scrollable textual panels: the log panel and the notification panel. With the tool bar participants can perform all ‘constructive’ actions (start a phase, raise an issue, give an argument, give a constraint, solve an issue, login, logout). Their availability depends on the participant role and the deliberation state. The interaction takes place through dynamically generated panels for selecting and typing information, with possibly some static guidance retrieved in the task model. Results (and error messages) are displayed in the log panel. The guidance buttons of the tool bar answer questions ‘What’s new?’, ‘What can I do?’, ‘Who is logged?’, and ‘Which issues are conflicting?’ (i.e. have currently no preferred option). The tool buttons give access to the available application tools (for instance for viewing the artefacts), to the internal e-mail tool, and to the synchronous client. Several participants can work occasionally at the same time. The notification panel shows all constructive actions from the other connected participants. The ‘task model info’ tabbed panel gives information about the deliberation model (phase and issue types), either textually, or graphically. The ‘issue instances’ tabbed panel describes issue instances, either textually, or graphically, or with a tree-like presentation (see Fig.2). Non solved issues can be selected. Issues can be depicted with and without their nested constraint issues. An analysis panel shows the participation and divergence metrics. The ‘other instances’ tabbed panel gives detailed information about the deliberation, its phases, and all the arguments and constraints. Most graphical representations use colours to highlight important features (e.g. coloured icons for active components and preferred options within argumentation trees, coloured icon for the ongoing phase in the deliberation graph, coloured font for all recently modified instances). Moreover, if a graphical representation becomes out of date, due to other participants activity, its background becomes brightly coloured.

When the synchronous client is launched from the asynchronous one, the role of the actor in the real time argumentation session (‘moderator’ or ‘participant’) is automatically retrieved from the task model, and the user interface reflects that role. The moderator can choose which issue must be extracted from the database and when they must be sent to all other participants of the synchronous session. Only one issue at a time is unlocked by the moderator for discussion. Participants can add new arguments, new constraints, and the system re-evaluates the argumentation tree. Latecomers receive automatically the current status of the argumentation and participant can leave at any moment. The moderator can use a vote tool for negotiating the termination of an issue. A chat and a whiteboard are also available for direct communication between any subset of the participants. Activity argumentation measures (participation and divergence) are graphically displayed in the 'statistics' tabbed panel.

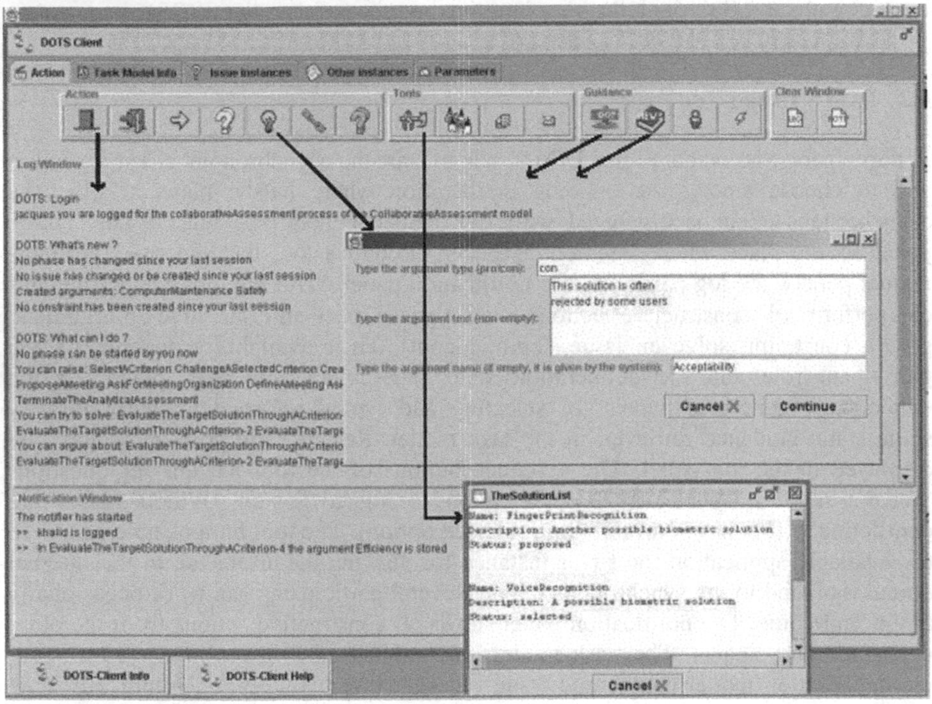

Fig. 1. The asynchronous client, with its *tool bar*, its *log and notification panels*, *'The Solution List' application tool*, and one of the *panels for creating a new argument*.

5 Conclusion

In new forms of distributed organisations, many issues are discussed through deliberation processes. These processes can vary in complexity and formality. This paper describes a comprehensive and generic support for a wide range of distributed deliberation types. The system integrates concepts and techniques coming from different domains: workflow management systems, synchronous and asynchronous groupware, argumentation and decision making systems. It provides many helpful features, such as a mechanism for discussing and choosing at run time the right strategy for performing the task, and an assistance for deciding to launch synchronous sessions, through participation, divergence, and stability indicators. However, the system remains simple to use and to deploy, as a distributed Java application on the Internet.

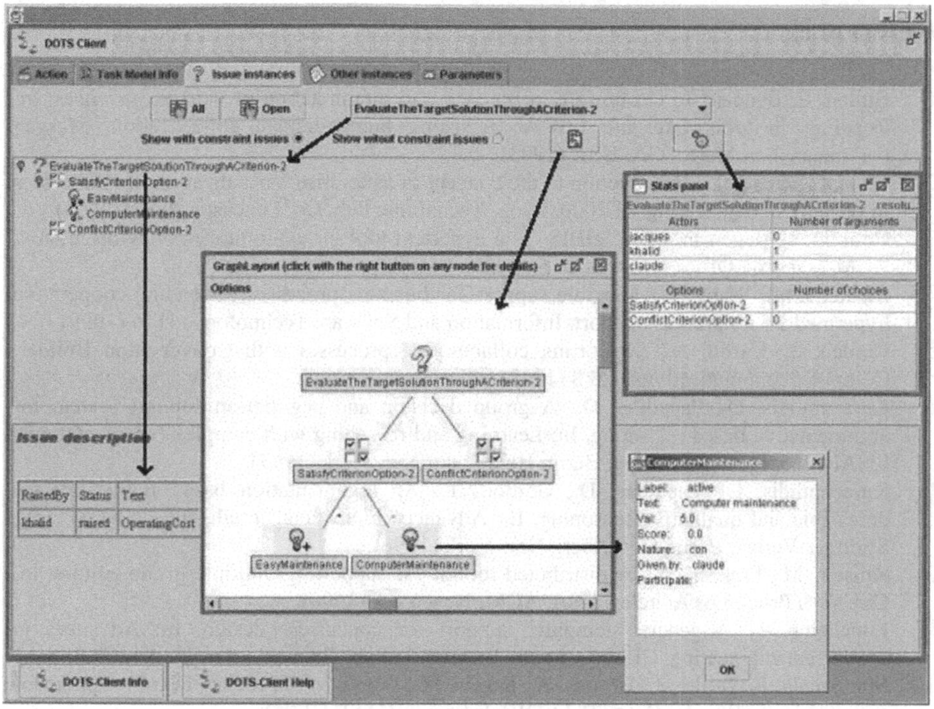

Fig. 2. An issue instance description, in a *tree-like form* and in a *graphical form*, with a *node detail*, and a part of the *statistical information* about the issue.

CHIPS [4] is the closest prototype from DOTS that we know, mixing flexible process support and synchronous/asynchronous collaboration support. CHIPS has a broader scope (flexible business processes), but provides no built-in support for argumentation and decision. Some aspects of our approach are also rooted into older collaborative projects, such as Conversation Builder [5], Cognoter [15], and gIBIS [3]. DOTS may supersede many dedicated systems, such argumentation tools [6, 7], inspection tools [10], and generic inspection environments [17, 11].

Several aspects need further work. At the conceptual level, the meaning of an issue definition can be ambiguous or not sufficiently precise for reaching a conclusion. The argumentation scheme could be extended for permitting the transformation of an issue definition. A sequence of transformations could give a kind of bargaining process, with offer and counter-offer moves. At the system level, the integration with distributed document management system on the Internet should be considered. Currently, DOTS provides no support for managing the shared artifacts. Secondly, if reuse of artefacts and deliberations (R13) is theoretically possible by keeping all terminated deliberations in the object base, we feel that a better alternative could be to feed the enterprise memory with these deliberations, in a more human readable form. A bi-directional interface with enterprise memory systems (for instance based on XML technologies), could be a very helpful extension to the current system.

References

1. Bullen, C, Bennett, J.: Groupware in practice: An interpretation of work experiences. In: Readings in groupware and CSCW: Assisting human-human collaboration, Morgan Kaufmann, San Mateo, CA, USA (1993).
2. Chan, S., Zhang, L.: Introduction to the concept of Enterprise Virtualization. In: Advances in concurrent engineering CE2000, Lyon, Technomic Pub. Co, Lancaster, USA (2000).
3. Conklin, J., Begeman, M.L.: gIBIS – a hypertext tool for exploratory policy discussion. ACM Trans. on Office Information Systems, 4 (1988).
4. Haake, J..M., Wang, W.: Flexible support for business processes: extending cooperative hypermedia with process support. Information and Software Technology, 41, 6 (1999).
5. Kaplan, S., Caroll, A.: Supporting collaborative processes with Conversation Builder. Computer communications, 15, 8 (1992).
6. Karacapilidis, D., Papadias, D.: A group decision and negotiation support system for argumentation based reasoning. In: Learning and reasoning with complex representations, LNAI 1266, Springer-Verlag, Berlin Heidelberg New York (1997).
7. Karacapilidis, D., Papadias, D., Gordon, T.: An argumentation based framework for defeasible and qualitative reasoning. In: Advances in artificial intelligence, LNCS 1159, Springer-Verlag, Berlin Heidelberg New York (1996).
8. Knister, M., Prakash, A.: A distributed toolkit for supporting multiple group editors. In: CSCW90 Proc., Los Angeles, USA, ACM, New York (1990).
9. Lonchamp, J.: A generic computer support for concurrent design. In: Advances in concurrent engineering, CE2000, Lyon, Technomic Pub. Co, Lancaster, USA (2000).
10. Macdonald, F., Miller, J., Brooks, A., Roper, M., Wood, M.: A review of tool support for software inspection. In: Proc. 7th Int. Workshop on CASE (1995).
11. Macdonald, F., Miller, J.: Automatic generic support for software inspection. In: Proc. 10th Int. Qality Week, San Francisco (1997).
12. Malone, T, Lai, K., Fry, C. : Experiments with Oval, a radically tailorable tool for cooperative work. In: Proc. of CSCW92 , Toronto, Canada, ACM, New York (1992).
13. Martin, G., Détienne, F., Lavigne, E.: Negotiation in collaborative assessment of design solutions: an empirical study on a Concurrent Engineering process. In: Advances in Concurrent Engineering, CE2000, Lyon, Technomic Pub. Co, Lancaster, USA (2000).
14. Prasad, B., Advances in Concurrent Engineering: Editor's note, CE2000, Lyon, Technomic Pub. Co, Lancaster, USA (2000).
15. Stefik, M., Foster, G., Bobrow, D. et al.: Beyond the chalkboard: computer support for collaboration and problem solving in meetings. CACM, 1 (1987).
16. ter Hofte, G.: Working Apart Together – Foundations for Component Groupware. Telematica Instituut, Enschede, The Netherlands (1998).
17. Tjahjono, D. : Building software review systems using CSRS. Tech. Report ICS-TR-95-06, University of Hawaii, Honolulu (1995).
18. Trevor, J., Rodden, T., Blair, G.: COLA: a lightweight platform for CSCW. Computer Supported Cooperative Work, 3 (1995).

Using Streaming and Parallelization Techniques for 3D Visualization in a High-Performance Computing and Networking Environment

S. Olbrich[1], H. Pralle

Institute for Computer Networks and Distributed Systems (RVS) /
Regional Scientific Computing Center for Lower Saxony (RRZN)
University of Hannover

S. Raasch

Institute for Meteorology and Climatology (IMUK)
University of Hannover

Abstract. Currently available massively parallel supercomputers provide sufficient performance to simulate multi-dimensional, multi-variable problems in high resolution. However, the visualization of the large amounts of result data cannot be handled by traditional methods, where postprocessing modules are usually coupled to the raw data source – either by files or by data flow. Due to significant bottlenecks of the storage and communication resources, efficient techniques for data extraction and preprocessing at the source have to be developed to get a balanced, scalable system and the feasibility of a „Virtual Laboratory" scenario, where the user interacts with a multi-modal, tele-immersive virtual reality environment.

In this paper we describe an efficient, distributed system approach to support three-dimensional, interactive exploration of complex results of scientific computing. Our processing chain consists of the following networked instances:

1. Creation of geometric 3D objects, such as isosurfaces, orthogonal slicers or particle sets, which illustrate the behaviour of the raw data. Our efficient visualization approach allows to handle large result data sets of simulation frameworks. It is based on processing every result data part corresponding to the domain decomposition of the parallelized simulation at the location of computation, and then collecting and exporting the generated 3D primitives. This is supported by special postprocessing routines, which provide filtering and mapping functions.
2. Storage of the generated sequence of 3D files on a separate „3D Streaming Server", which provides – controlled via „Real Time Streaming Protocol" (RTSP) – play-out capabilities for continuous 3D media streams.
3. Presentation of such 3D scene sequences as animations in a virtual reality environment. The virtual objects are embedded in a WWW page by using an advanced 3D viewer plugin, and taking advantage of high-quality rendering, stereoscopic displays and interactive navigation and tracking devices.

For requirement analysis, evaluation, and functionality demonstration purposes we have choosen an example application, the simulation of unsteady fluid flows.

---

1. *Author's address:* RRZN/RVS, Schloßwender Str. 5, D-30159 Hannover, Germany. *E-mail:* olbrich@rvs.uni-hannover.de. *WWW:* http://www.rvs.uni-hannover.de.

B. Hertzberger et al. (Eds.): HPCN Europe 2001, LNCS 2110, pp. 231-240, 2001.

1 Introduction and Motivation

Most available visualization tools are designed for postprocessing of raw data, such as multidimensional scalar or vector fields. The task of visualization generally applies data extraction, mapping to 3D objects, and 3D rendering techniques, which could be considered as modules of a visualization pipeline. It is realized either in a separate application which is coupled to the simulation process by a data flow or by stored data files (e. g. AVS [24], COVISE [19], PV3 [6], SCIRUN [9], VIS-5D [7]), or in toolkit libraries which provide appropriate functions as a part of the simulation process (e. g. VTK [20]). Combined concepts for distributed applications are available, too.

All these tools show limitations (e. g. performance of postprocessing, communication, and presentation methods, quality and functionality of presentation and interaction techniques, support of multiple platforms), which can lead to insufficient utilization of expensive resources or prohibit certain fields of applications. This is particularly observed in high-performance computing, such as high-resolution simulations of unsteady fluid flow [10]. The calculation of such multi-dimensional, multi-variable field data results with sufficient accuracy and in an acceptable time frame already represents a „Grand Challenge" problem. However, the addition or integration of visualization turns out to be a demanding job as well, especially if it is intended to build an efficient, balanced, scalable process chain.

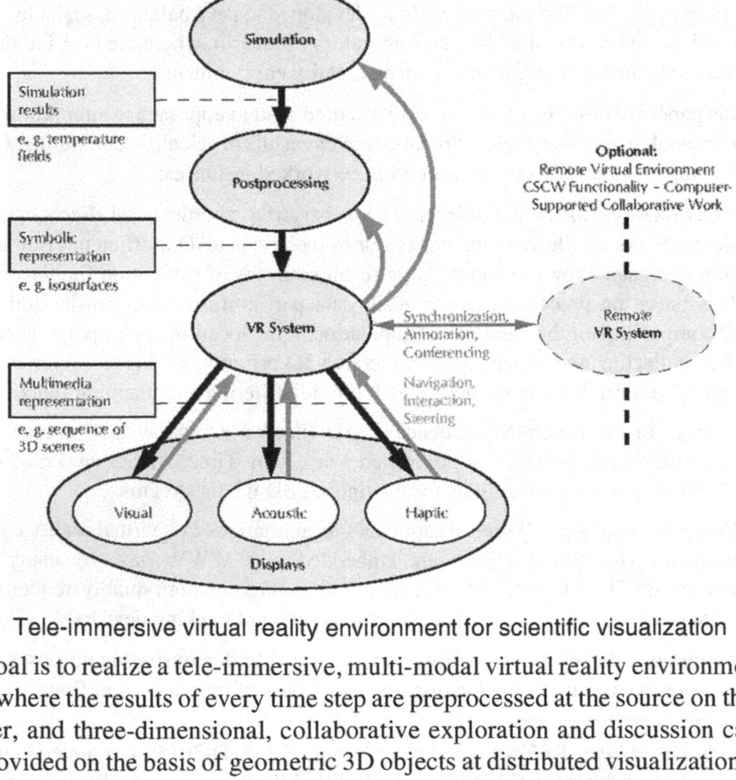

Figure 1 Tele-immersive virtual reality environment for scientific visualization

Our goal is to realize a tele-immersive, multi-modal virtual reality environment (see *figure 1*), where the results of every time step are preprocessed at the source on the compute server, and three-dimensional, collaborative exploration and discussion capabilities are provided on the basis of geometric 3D objects at distributed visualization clients [1][5][11].

Extensions for steering the supercomputer simulation will lead to a „Virtual Laboratory“ scenario, serving for several purposes, for example: better understanding of complex relationships, getting important design parameters of expensive or dangerous installations, or advanced education.

2 Requirements

Studies about certain atmospheric phenomena currently carried out by the IMUK [18], show needs for discretization of the considered three-dimensional, time-dependant field to 1000 x 1000 x 200 grid points. Two-dimensional domain decomposition leads to a significant speed-up on massively parallel supercomputers, and allows to handle the large demands of main memory: about 37 GByte (1.6 GByte per each 64-bit variable, 23 variables) of state data fields, which are updated every time step. To explore the characteristics of the results, typically about 3600 time steps have to be analysed, corresponding to a duration of 3 minutes real time of the investigated phenomenon. This scenario results in 3600 x 1.6 Gbyte = 5.8 TByte generated data per scalar variable. Its computation takes about 12 hours on an SGI/Cray T3E (Von Neumann Institute for Computing, Jülich, Germany), using 400 processor elements (PEs).
The traditional approach – which involves transfer or/and storage of the raw results, and interactively exploring on a dedicated postprocessing system (see A in *figure 2*) – is impossible to apply in this case. Currently availabe networking, storage, and postprocessing resources are not sufficient to handle this large amount of data. Since we typically expect to visualize one or two scalar fields and one three-dimensional vector field, this leads to an amount of 29 TByte. The transfer time for this amount of data would be about 64 hours for a 1 Gbit/s network infrastructure.

Considering the abstract visualization pipeline, four basic cases for subdivision can be distinguished (see *figure 2*). Generally, at each step the data volume is reduced, and the level of interactive control as well. In order to take advantage of hardware-supported 3D rendering, we choose approach (C), which supports client-side, interactive navigation of 3D scenes in a virtual reality environment. We have developed an efficient „3D Streaming System“ [14][15] with optimized networking and rendering capabilities, which could be applied and extended for this purpose (see *section 3.2*).

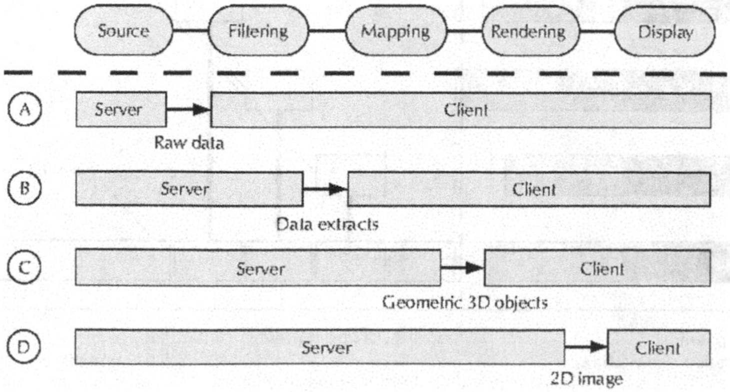

Figure 2 Different approaches of partitioning the visualization pipeline

Compared with approaches (A) and (B), we also get significant data compression. The complexity of visualized symbolic representations, such as isosurfaces or streamlines, increases just with $O(n^2)$ or $O(n)$, respectively, with increasing resolution n of the simulation volume grid, in contrast to $O(n^3)$ for the raw result data. Data flow and computation requirements at the filtering and mapping stages are high, but several algorithms can be applied on parts of a volume data set independently. Visualization tasks like computing the geometry of isosurfaces of each subdomain can be executed as part of the numerical model on the respective processor in parallel.

3 Concept and Implementation

Our concept consists of the following steps:

1. Efficient postprocessing of computation results delivers geometric 3D objects, represented in our binary 3D data format „DVR". Our approach and its implementation are described in *section 3.1*.
2. Buffering and viewing the sequence of 3D scenes (see *section 3.2*):
    - A *3D streaming server* provides buffer space resources and play-out features, which are basedo nt he *Real Time Streaming Protocol* (RTSP) [22] and an optimized transport protocol for 3D scenes („DVRP" – DVR Protocol) [15].
    - A high-performance *3D viewer*, embeddable into WWW browsers such as Netscape Communicator as a plugin („DocShow-VR") [14].

In order to get high utilization of the computing and networking resources, we have defined a binary representation for 3D objects („DVR format"). It is used for the 3D files and streams in our networked scenario. By designing this special binary format, we avoid the client-side bottleneck from decoding, parsing, traversing, and optimizing operations, which are required for standard formats such as VRML (Virtual Reality Modeling Language – ISO/IEC 14772), before 3D rendering could be started.

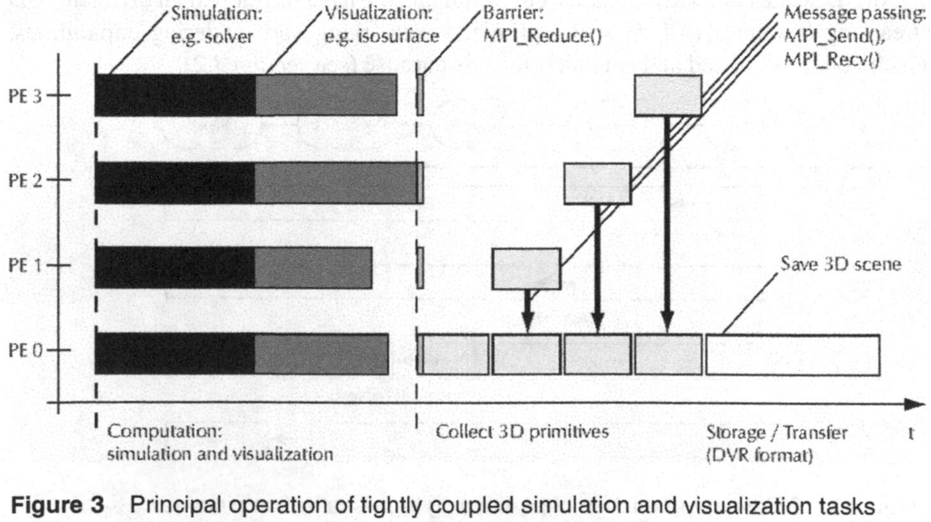

Figure 3 Principal operation of tightly coupled simulation and visualization tasks

3.1 Efficient Creation of 3D Objects

In order to take advantage of interactive 3D navigation and stereoscopic viewing techniques at the client side, we have to generate graphics primitives, such as polygons, lines or points in 3D space. These 3D objects are created to represent the characteristics of the simulation results. Typical examples for 3D visualization of three-dimensional scalar or vector data sets are isosurfaces, coloured slicers, streamlines, or particle sets. For parameter variations, such as time steps in the simulation or threshold values for isosurface generation, we get a series of 3D scenes.

As illustrated in *figure 3*, our idea was to apply the postprocessing of the raw data to 3D geometries as part of an MPI-based [3], massively parallel simulation. In this way we could apply conventional sequentially implemented filtering and mapping algorithms to the part results on every individual processor, corresponding to the domain composition of the main computation task.

We have implemented a portable C library, which is callable from Fortran. This „libDVRP" provides a set of volume visualization functions, which could be applied to an orthogonal grid, such as an isosurface generator and orthogonal, colored slicers. After experimenting with the universal visualization routines from VTK [20], we integrated a special-purpose isosurface generator, an efficient implementation of the marching cube algorithm [12], taken from *polyreduce*[1].

The in-memory generated 3D objects (polygons, attributed with normal vectors and optional colors) are stored as „DVR format" (see next section) on local files or immediately transfered to remote locations over IP network (FTP protocol, using libftp, or RTSP protocol, using our 3D streaming system in a „recording" mode).

In the current implementation, the collection and output tasks potentially cause significant bottlenecks, since they block the computation. The limitation of the scalability depends on the data volumes and available data rates, which means high-speed networking infrastructure (such as gigabit ethernet, which is used in our current projects) is required to increase efficiency. A further improvement should be achieved by parallel I/O and/or overlapping it with the simulation and visualization steps in a pipelined implementation. Compression of 3D geometry could reduce the data volume significantly, but would require additional computation time for encoding [2].

3.2 3D Streaming System

By using a „3D Streaming System" [15], sequences of high-quality 3D scenes can be efficiently stored and played out in a high-performance networking and graphics environment. The complete architecture consists of three instances, which could be distributed over IP networks and operated either in a pipelined mode or asynchronously.

1. A **3D generator** creates 3D objects and sends them to a *3D streaming server.*
2. A **3D streaming server** supports receiving and storage of 3D object sequences, which are handled as a media stream (similar to a recording mode on a video streaming server), or delivering existing 3D streams to a *3D viewer* – optional

1. Polyreduce, by Jesper James; http://hendrix.ei.dtu.dk/software/software.html (see also: http://wuarchive.wustl.edu/graphics/graphics/misc/marching-cubes/)

synchronously (similar to live video streaming) or asynchronously (similar to video on demand).

3. A **3D viewer** provides communication, presentation and interaction capabilities.

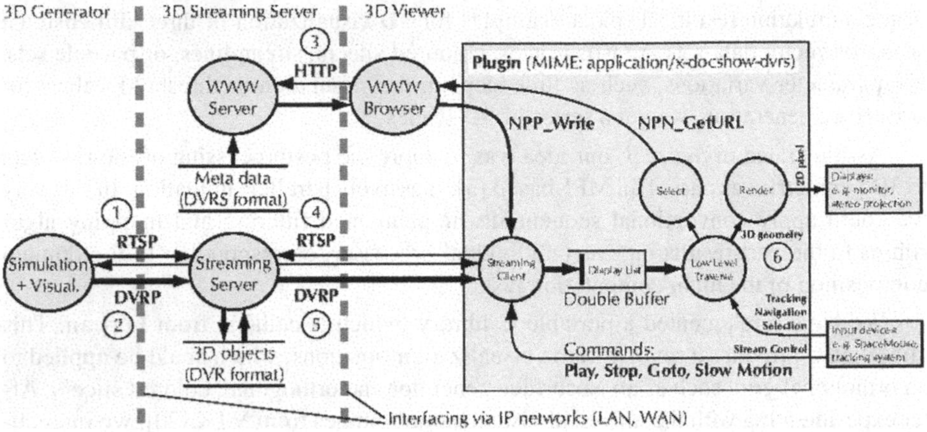

Figure 4 3D streaming system for visualization of complex results of high-performance computing in a virtual reality environment

The operating sequence is characterized by the following steps (*figure 4*):

3D Generator → 3D Streaming Server

1. The simulation and visualization software includes a client, which connects to the 3D streaming server. Control commands are transmitted over RTSP.
2. A sequence of 3D scenes (DVR format) is transmitted over TCP/IP („DVRP").

3D Streaming Server → 3D Viewer (synchronous or asynchronous to 1, 2)

3. A „DVRS" meta file (MIME type: `application/x-docshow-vrs`, file extension: `.dvrs`) is fetched from a WWW server by a WWW browser via HTTP. The request could be initiated by refering an appropriate URL immediately, or indirectly by an EMBED tag in an HTML page.
4. After the appropriate DVRS plugin is loaded by the WWW browser, it establishes a connection to the 3D streaming server, based on the reference information (IP address, port number) and attributes (frame rate, etc.) contained in the DVRS meta data file.
5. The 3D streaming server reads 3D scenes from files in DVR format and delivers them to the client, interleaved by additional delimiting DVR data elements.
6. The 3D viewer reads the first 3D scenes, and executes some further actions:
   - Reading 3D data (in a separate thread):
     Transfer each 3D scene from the streaming connection into a memory buffer.
   - 3D Rendering:
     Transformation of the 3D objects, based on the current virtual camera position, orientation, and view angle, to the display device (optional stereoscopic).

- 3D navigation:
  Modification of the virtual camera parameters, according to the control of the respective input device (keyboard, mouse, space mouse, head tracking).
- RTSP (Real Time Streaming Protocol) handling – VCR metaphor:
  Control handling, e. g. instruct the server to stop, to play-out from a new position with a modified frame rate, or to deliver an alternate data set.

The user interface provides several popup menus for configuration and a dialogue window for streaming control (see *figure 5*).

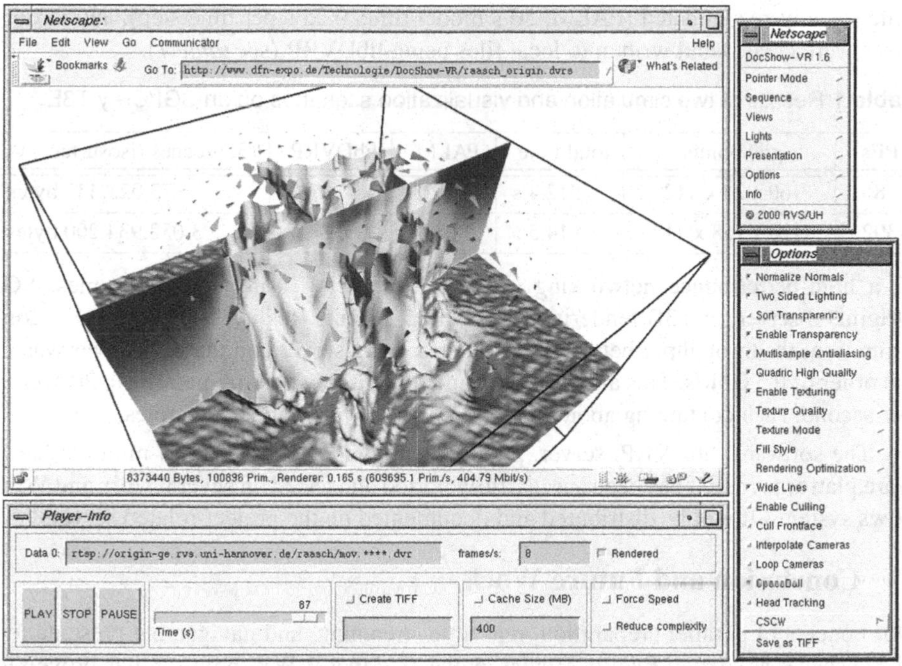

Figure 5 DocShow-VR plugin in a Netscape configuration – popup menus and streaming dialogue in an example application: Oceanic convection [17]

4 Experiences from an Example Application

From our first evaluation experiments, here we show the results of three alternatives that were tested in a simple scenario containing 40 x 40 x 40 grid points:

1. Simulation only (Large-Eddy Simulation Model – PALM-1).
2. Simulation and integrated isosurface visualization of the temperature field by VTK method *VTKContourFilter* on each processor element (PE) independently, with output on DVR files via our VTK extension *DVRExporter*, and collecting step by UNIX command *cat*, resulting in a DVR file containing a 3D scene.
3. Simulation and integrated visualization by our DVRP library, which encapsulate (a) isosurface generation (optimized marching cube algorithm for regular grid), (b) collection of generated 3D polygons from each PE to one contiguous graphics primitive list in memory via MPI, and (c) output of 3D scene on DVR file.

Using 4 PEs on an SGI/Cray T3E at the RRZN (Hannover), every second time step was visualized – 600 3D scenes were generated. The following times for a simple case study with 1200 simulation time steps demonstrate the success of our approach.

1. 181 s.
2. 872 s – simulation: 179 s, visualization (VTK): 390 s, collection (cat): 303 s.
3. 209 s – simulation: 180 s, visualization, etc. (DVRP): 29 s.

Two further scenarios with increasing size have been investigated on an SGI/Cray T3E at the ZIB (Berlin). In these cases, 20 x 24 x 112 grid points are handled per PE, 200 time steps were simulated (PALM, 50 s model time, 0.25 s per time step), and isosurfaces were created and written to local files using libDVRP (see *table 1*).

Table 1 Results of two simulation and visualization scenarios on an SGI/Cray T3E

PEs	grid points	total time	PALM	libDVRP	3D scenes (isosurfaces)
8	160 x 24 x 112	512.4 s	95 %	5 %	73.021.432 bytes
392	1120 x 168 x 112	914.5 s	63 %	37 %	3.052.934.200 bytes

In a high-performance networking (Gigabit Ethernet – using Jumbo Frames, SGI Origin200 server) and 3D rendering (SGI Onyx2 Infinite Reality, HP J6000 fx10) configuration, the throughput between the 3D streaming server and the 3D viewer was in the order of 0.5 Gbit/s. This allows to transmit and display approximately 10–20 scenes per second, each containing about 100.000 polygons in practical examples.

The software (libDVRP, server, plugin) is implemented to support multiple hardware platforms, and it has been successfully ported and tested on several Unix and Windows systems. It will be distributed and documented on the project-related servers[1,2].

5 Conclusion and Future Work

Our concept of parallel preparation, real time streaming, and navigatable presentation of 3D scenes as *Virtual Reality Movies*, embedded in a WWW browser, has proven to be powerful. We have successfully demonstrated our distributed system approach in an unsteady volume visualization environment. It provides an efficient method to explore supercomputer simulation results on a 3D graphics workstation in a high-performance network, with slight additional load of the simulation. As such, smooth on-the-fly animations are possible, supported by stereoscopic viewing, interactive 3D navigation, and repeatedly showing parts of the sequence that was already generated.

A back channel for the purpose of steering the simulation is currently developed. It is based on RTSP extensions that are transmitted through the streaming server into the running simulation. The steering parameters can be controlled by a dynamically configured user interface at the 3D viewer.

In the challenging high-resolution scenarios (e. g.: 1000 x 1000 x 200 grid points, as stated in *section 2*), we expect that it is required to reduce the complexity of the gen-

1. http://www.dfn-expo.de/Technologie/DocShow-VR/
2. http://www.rvs.uni-hannover.de/projekte/tele-immersion/

erated geometry, otherwise the scenes cannot be handled with interactive navigation rates. For performance reasons, algorithms for direct extraction of multiresolution isosurfaces from volume data [16] are probably better suited for this purpose than post-processing by mesh reduction algorithms [4][8][21][23], which are usually very time consuming.

Advanced application scenarios require the integration of computer-supported collaborative functionalities – such as synchronization of viewing parameters, annotations, and video-conferencing in a multi-user environment – and the integration of further media types – such as video and audio streams – as parts of an immersive virtual reality presentation. To achieve this, we intend to extend the presented streaming system to support a distributed multi-viewer configuration and to integrate video conferencing components.

Furthermore, we plan to integrate interactive, three-dimensional cursor and annotation features, not only to support collaborative sessions, but also to create automatically running presentations. An „authoring on-the-fly" process will record a sequence of real time events – such as: navigation in virtual space and in the time-axis, development of annotations, movement of a tele-pointer (location, orientation and scaling) – as time stamps in a journal file. This can be used later as a control data stream for synchronous play-out in conjunction with the prepared 3D scene sequence and the video and audio media streams.

Acknowledgement

This work is partly funded by the DFN-Verein (German Research Network), with funds from the BMBF (German Federal Ministry for Education and Research), and it is also sponsored by HP. The authors wish to thank A. von Berg (RVS) for the discussion about the high-performance network issues and configuration of the testbed network.

References

[1] Corrie, B., Sitsky, D., Mackerras, P.: *Integrating High Performance Computing and Virtual Environments.* Proceedings of the Seventh Parallel Computing Workshop, Canberra, Australia, September 1997.

[2] Deering, M.: *Geometric Compression.* Proceedings of ACM SIGGRAPH 1995.

[3] Gropp, W., Lusk, E., Skjellum, A.: *Using MPI – Portable Parallel Programming with the Message-Passing Interface.* MIT Press, 1999.

[4] Gumbold, S., Straßer, W.: *Real Time Compression of Triangle Mesh Connectivity.* Proceedings of ACM SIGGRAPH 1998.

[5] Haase, H., Dai, F., Strassner, J., Göbel, M.: *Immersive Investigation of Scientific Data.* In: Nielson, G. et al (eds.): Scientific Visualization – Overviews, Methodologies and Techniques, IEEE Computer Society Press, 1997.

[6] Haimes, R.: *pV3: A Distributed System for Large-Scale Unsteady CFD Visualization.* AIAA Paper 94-0321, Reno NV, Jan. 1994. (http://raphael.mit.edu/pv3/pv3.html)

[7] Hibbard, W., Santek, D.: *The VIS-5D System for Easy Interactive Visualization.* Proceedings of IEEE Visualization 1990. (http://www.ssec.wisc.edu/~billh/vis5d.html)

[8] Hoppe, H., DeRose, T., Duchamp, T., McDonald, J., Stuetzle, W.: *Mesh optimization*. Proceedings of ACM SIGGRAPH 1993.

[9] Johnson, C. R., Parker, S. G., Weinstein, D.: *Large-scale Computational Science Applications using the SCIRun Problem Solving Environment*. Proceedings of Supercomputer 2000. (http://www.sci.utah.edu/publications/super00_final.pdf)

[10] Lane, D. E.: *Scientific Visualization of Large-Scale Unsteady Fluid Flows*. In: Nielson, G. M. et al (eds.): Scientific Visualization – Overviews, Methodologies and Techniques, IEEE Computer Society Press, 1997.

[11] Leigh, J. et al: *Visualization in Teleimmersive Environments*. IEEE Computer, Vol. 32, No. 12, December 1999. (http://www.evl.uic.edu/aej/papers/computer99.pdf)

[12] Lorensen, W. E., Cline, H. E.: *Marching cubes: A high resolution 3D surface reconstruction algorithm.* Computer Graphics (Proceedings of SIGGRAPH 1987), Vol. 21, No. 3, July 1987. (http://www.cse.fsu.edu/~erlebach/courses/sciviz/papers/volumetric_papers/PAPERS/LORENSEN/MC/INDEX.HTM)

[13] Neider, J., Davis, T., Woo, M.: *OpenGL Programming Guide – The Official Guide to Learning OpenGL, Release 1*. Addison-Wesley, 1993.

[14] Olbrich, S., Pralle, H.: *High-Performance Online Presentation of Complex 3D Scenes*. In: van As, H. R. (Ed.): High Performance Networking. Kluwer Academic Publishers, 1998. (http://www.dfn-expo.de/Technologie/DocShow-VR/Vortraege/hpn98/)

[15] Olbrich, S., Pralle, H.: *Virtual Reality Movies – Real Time Streaming of 3D Objects*. Computer Networks, Vol. 31, No. 21, November 1999 – The Challenge of Gigabit Networking, Elsevier. (http://www.terena.nl/tnnc/2A/2A1/2A1.pdf)

[16] Poston, T., Wong, T., Heng, Ph.: *Multiresolution Isosurface Extraction with Adaptive Skeleton Climbing*. Computer Graphics Forum (Proceedings of EUROGRAPHICS 1998), Vol. 17, No. 3, September 1998.

[17] Raasch, S., Etling, D.: *Modeling Deep Ocean Convection: Large Eddy Simulation in Comparison with Laboratory Experiments*. J. Phys. Oceanogr., 1998, Vol. 28, 1796–1802.

[18] Raasch, S., Schröter, M., Ketelsen, K., Olbrich, S.: *A Large-Eddy Simulation Model for Massively Parallel Computers – Model Design and Scalability*. Fifth European SGI/Cray MPP Workshop, Bologna (Italy), September 1999. (http://www.cineca.it/mpp-workshop/fullpapers/raasch.ps)

[19] Rantzau, D., Frank, K., Lang, U., Rainer, D., Wössner, U.: *COVISE in the CUBE: An Environment for Analyzing Large and Complex Simulation Data*. In: Proceedings of the 2nd Workshop on Immersive Projection Technology 1998. (http://www.hlrs.de/people/rantzau/ipt98_rantzau.pdf)

[20] Schroeder, W., Martin, K., Lorensen, B.: *The Visualization Toolkit*. 2nd Edition. Prentice Hall, 1997. (VTK: see also http://www.kitware.com)

[21] Schroeder, W., Jonathan, A., Lorensen, W.: *Decimation of triangle meshes*. Computer Graphics (Proceedings of SIGGRAPH 1992), July 1992.

[22] Schulzrinne, H., Rao, A., Lanphier, R.: *Real Time Streaming Protocol (RTSP)*. RFC 2326, April 1998.

[23] Taubin, G., Rossignac, J.: *Geometric Compression Through Topologic Surgery*. ACM Transactions on Graphics, Vol. 17, No. 2, April 1998.

[24] Upson, C., Faulhaber, T., Kamins, D., Laidlaw, D., Schlegel, D., Vroom, J., Gurwitz, R., van Dam, A.: *The Application Visualization System: A Computational Environment for Scientific Visualization*. IEEE Computer Graphics and Applications, July 1989. (AVS: see also http://www.avs.com)

Performance Evaluation of Parallel *GroupBy-Before-Join* Query Processing in High Performance Database Systems

David Taniar[1], J.Wenny Rahayu[2], and Hero Ekonomosa[3]

[1)] Monash University, School of Business Systems, Vic 3800, Australia
David.Taniar@infotech.monash.edu.au
[2)] La Trobe University, Department of Computer Science and Engineering, Australia
wenny@cs.latrobe.edu.au
[3)] University Tenaga Nasional, Department of Computer Science &IT, Malaysia
hero@uniten.edu.my

Abstract. Strategic decision making process uses a lot of GroupBy clauses and join operations queries. As the source of information in this type of application to these queries is commonly very huge, then parallelization of GroupBy-Join queries is unavoidable in order to speed up query processing time. In this paper, we investigate three parallelization techniques for GroupBy-Join queries, particularly the queries where the group-by clause can be performed before the join operation. We subsequently call this query "*GroupBy-Before-Join*" queries. Performance evaluation of the three parallel processing methods is also carried out and presented here.

1 Introduction

Queries involving group-by are very common in database processing, especially in data for integrated decision making process like in On-Line Analytical Processing (OLAP), and Data Warehouse [1,3]. Queries containing aggregate functions summarize a large set of records based on the designated grouping. The input set of records may be derived from multiple tables using a join operation. As the data repository containing group-by queries normally are very huge, consequently, effective optimization of aggregate functions has the potential to result in huge performance gains. In this paper, we would like to focus on the use of parallel query processing techniques in queries, whereby group-by clauses is performed before join operations. We refer this query as "*GroupBy-Before-Join*"query.

The work presented in this paper is actually a part of final stage of a larger project on parallel aggregate query processing. Parallelization of "*GroupBy-Before-Join*" queries is the third and the final stage of the project. The first stage of this project dealt with parallelization of group-by queries on single tables. The results are reported at PART'2000 conference [8]. The second stage focused on parallelization of GroupBy-Join queries where the GroupBy attributes are different from the Join attributes with a consequence that the join operation must be carried out first and then the group-by operation. We have published the outcome of the second stage at HPCAsia'2000 conference [9]. In the third and final stage, which is the main focus of

B. Hertzberger et al. (Eds.) : HPCN Europe 2001, LNCS 2110, pp. 241-250, 2001.

this paper, concentrates on GroupBy-Join queries (like in stage two), but the join attribute is the same as the group-by attribute resulting that the group-by operation can be performed first before the join for optimization purposes.

2 GroupBy Queries: A Background

GroupBy-Join queries in SQ can be divided into two broad categories; whereby the first is group-by is carried out after join and secondly is before join. In either category, aggregate functions are normally involved in the query. To illustrate these types of GroupBy-Join queries, we use the following tables from a Suppliers-Parts-Projects database:

```
SUPPLIER (S#, Sname, Status, City)
PARTS (P#, Pname, Colour, Weight, Price, City)
PROJECT (J#, Jname, City, Budget)
SHIPMENT (S#, P#, J#, Qty)
```

For simplicity of description and without loss of generality, we consider queries that involve only one aggregation function and a single join. The queries on Figure 1 give an illustration of GroupBy-Join queries.

QUERY 1	QUERY 2:
Select PARTS.City, AVG(Qty) From PARTS, SHIPMENT Where **PARTS.P# = SHIPMENT.P#** Group By **PARTS.City** Having AVG(Qty)>500 AND AVG(Qty)<1000	Select PROJECT.J#, PROJECT.Jname, SUM(Qty) From PROJECT, SHIPMENT Where **PROJECT.J# = SHIPMENT.J#** Group By **PROJECT.J#,** **PROJECT.Jname** Having SUM(Qty)>1000

Figure 1.Types of GroupBy-Join Qeries.

In query 1, two tables are joined to produce a single table, and this table becomes an input to the group-by operation. The purpose of Qery 1 is to group the part shipment by their city locations and select the cities with average quantity of shipment between 500 and 1000." Another example is Qery 2, in which it retrieves project numbers, names, and total quantity of shipments for each project having the total shipments quantity of more than 1000."

The main difference between Qery 1 and Qery 2 above lies in the join and group-by attributes. In Qery 2, the join attribute is also one of the group-by attributes. This is not the case with Qery 1, where the join attribute is totally different from the group-by attribute. This difference is especially a critical factor in processing GroupBy-Join queries, as there are decisions to be made in which operation should be performed first: the group-by or the join operation.

When the join attribute and the group-by attribute are different as shown in Qery 1, there will be no choice but to invoke the join operation first, and then the group-by operation. However, when the join attribute and the group-by attribute is the same as shown in Qery 2 (e.g. attribute *J#* of both Project and Shipment tables), it is expected that the group-by operation is carried out first, and then the join operation. Hence, we call the latter query (e.g. Query 2) " *GroupBy-Before-Join*" query.

In Query 2, all Shipment records are grouped based on the *J#* attribute. After grouping this, the result is joined with table Project. As known widely, join is a more expensive operation than group-by, and it would be beneficial to reduce the join relation sizes by applying the group-by first. Generally, group-by operation should always precede join whenever possible. Early processing of the group-by before join reduces the overall execution time as stated in the general query optimization rule where unary operations are always executed before binary operations if possible. The semantic issues about aggregate functions and join and the conditions under which group-by would be performed before join can be found in literatures [2,4,6,10].

In this paper, we focus on cases where group-by operation is performed before the join operation. Therefore, we will use Query 2 as a running example throughout this paper.

3 Parallel Algorithms for "GroupBy-Before-Join" Query

In order to study the behavior of the algorithms, in this section we describe three parallel algorithms for *GroupBy-Before-Join* query processing, namely: *Early Distribution* scheme, *Early GroupBy with Partitioning* scheme, and *Early GroupBy with Replication* scheme.

3.1 Early Distribution Scheme

The *Early Distribution* scheme is influenced by the practice of parallel join algorithms, where raw records are first partitioned/distributed and allocated to each processor, and then each processor performs its operation [5]. This scheme is motivated by fast message passing multi processor systems.

The Early Distribution scheme is divided into two phases: *distribution* phase and *group-by-join* phase. Using Query 2, the two tables to be joined are Project and Shipment based on attribute *J#*, and the group-by will be based on table Shipment. For simplicity of notation, the table which becomes the basis for group-by is called table *R* (e.g. table Shipment), and the other table is called table *S* (e.g. table Project). For now on, we will refer them as tables *R* and *S*.

In the *distribution* phase, raw records from both tables (i.e. tables *R* and *S*) are distributed based on the join/group-by attribute according to a data partitioning function. An example of a partitioning function is to allocate each processor with project numbers ranging on certain values. For example, project numbers (i.e. attribute *J#*) *p1* to *p99* go to processor 1, project numbers *p100-p199* to processor 2, project numbers *p200-p299* to processor 3, and so on. We need to emphasize that the two tables *R* and *S* are both distributed. As a result, for example, processor 1 will have records from the Shipment table with *J#* between *p1* and *p99*, inclusive, as well as records from the Project table with *J# p1-p99*. This distribution scheme is commonly used in parallel join, where raw records are partitioned into buckets based on an adopted partitioning scheme like the above range partitioning [5].

Once the distribution is completed, each processor will have records within certain groups identified by the group-by/join attribute. Subsequently, the *second* phase (the *group-by-join* phase) groups records of table *R* based on the group-by attribute and calculates the aggregate values on each group. Aggregating in each processor can be carried out through a sort or a hash function. After table *R* is grouped in each processor, it is joined with table *S* in the same processor. After joining, each processor will have a local query result. The final query result is a union of all sub-results produced by each processor.

Figure 2 shows an illustration of the Early Distribution scheme. Notice that partitioning is done to the raw records of both tables *R* and *S*, and aggregate operation of table *R* and join with table *S* in each processor is carried out after the distribution phase.

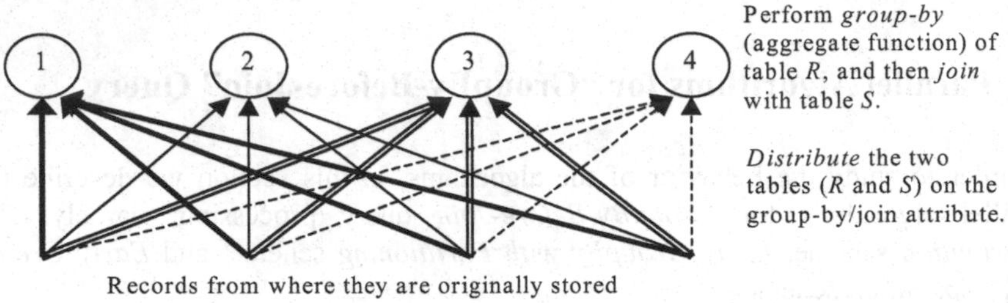
1
2
3
4
Perform group-by (aggregate function) of table R, and then join with table S.
Distribute the two tables (R and S) on the group-by/join attribute.
Records from where they are originally stored

Figure 2. Early Distribution Scheme.

There are several things need to be highlighted from this scheme. *First*, the grouping is still performed before the join (although after the distribution). This is to conform with an optimization rule for such kind of queries that group-by clause must be carried out before the join in order to achieve more efficient query processing time. *Second*, the distribution of records from both tables can be expensive, as all raw records are distributed and no prior filtering is done to either table. It becomes more desirable if grouping (and aggregation function) is carried out even before the distribution, in order to reduce the distribution cost especially of table *R*. This leads to the next schemes called *Early GroupBy* schemes for reducing the communication costs during distribution phase. There are two variations of the *Early GroupBy* schemes, each of which will be discussed in the following two sections.

3.2 Early GroupBy with Partitioning Scheme

As the name states, the *Early GroupBy* scheme performs the group by operation first before anything else (e.g. distribution). The *Early GroupBy with Partitioning* scheme is divided into three phases: (*i*) *local grouping* phase, (*ii*) *distribution* phase, and (*iii*) *final grouping and join* phase.

In the *local grouping* phase, each processor performs its group-by operation and calculates its local aggregate values on records of table *R*. In this phase each processor groups local records *R* according to the designated group-by attribute and performs the aggregate function. Using the same example as that in the previous section, one processor may produce, for example, (*p1*, 5000) and (*p140*, 8000), and another processor (*p100*, 7000) and (*p140*, 4000). The numerical figures indicate the SUM(Qty) of each project.

In the second phase (i.e. *distribution* phase), the results of local aggregates from each processor, together with records of table *S*, are distributed to all processors according to a partitioning function. The partitioning function is based on the join/group-by attribute, which in this case is attribute *J#* of tables Project and Shipment. Again using the same partitioning function in the previous section, *J#* of *p1-p99* are to go to processor 1, *J#* of *p100-p199* to processor 2, and so on.

In the third phase (i.e. *final grouping and join* phase), two operations are carried out, particularly; final aggregate or grouping of *R*, and join it with *S*. The final grouping can be carried out by merging all temporary results obtained in each processor. The way it works can be explained as follows. After local aggregates are formulated in each processor, each processor then distributes each of the groups to another processor depending on the adopted distribution function. Once the distribution of local results based on a particular distribution function is completed, global aggregation in each processor is simply done by merging all identical project number (*J#*) into one aggregate value. For example, processor 2 will merge (*p140*, 8000) from one processor and (*p140*, 4000) from another to produce (*p140*, 12000) which is the final aggregate value for this project number.

Global aggregation can be tricky depending on the complexity of the aggregate functions used in actual query. If, for example, an AVG function was used instead of SUM in Query 2, calculating an average value based on temporary averages must taken into account the actual raw records involved in each node. Therefore, for these kinds of aggregate functions, local aggregate must also produce number of raw records in each processor although they are not specified in the query. This is needed for the global aggregation to produce correct values. For example, one processor may produce (*p140*, 8000, 5) and the other (*p140*, 4000, 1). After distribution, suppose processor 2 received all *p140* records, the average for project *p140* is calculated by dividing the sum of the two quantities (e.g. 8000 and 4000) and the total shipment records for that project. (i.e. (8000+4000)/(5+1) = 2000). The total shipments in each project are needed to be determined in each processor although it is not specified in the query.

After global aggregation results are obtained, it is then joined table *S* in each processor. Figure 3 shows an illustration of this scheme.

There are several things worth noting. *First*, records *R* in each processor are aggregated/grouped before distributing them. Consequently, communication costs associated with table *R* can be expected to reduce depending on the group by selectivity factor. This scheme is expected to improve the *Early Distribution* scheme. *Second*, we observe that if the number of groups is less than the number of available processors, not all processors can be exploited; reducing the capability of parallelism. And *finally*, records from table *S* in each processor are all distributed during the second phase. In other words, there is no filtering mechanism applied to *S* prior to distribution. This can be inefficient particularly if *S* is very large. To avoid the problem of distributing *S*, we will introduce another scheme. This is introduced in the next section.

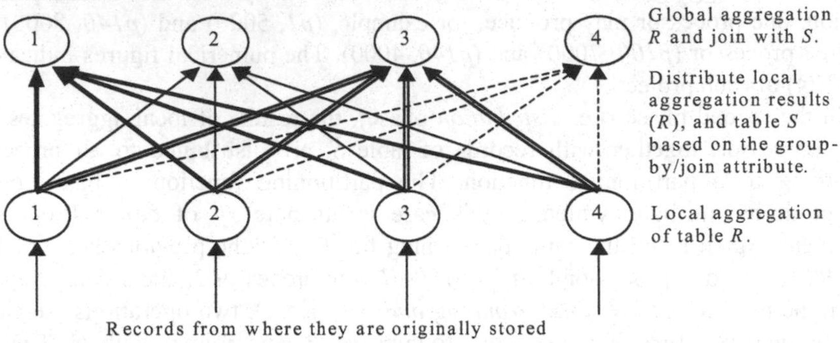
Global aggregation
R and join with *S*.
Distribute local aggregation results (*R*), and table *S* based on the group-by/join attribute.
Local aggregation of table *R*.
Records from where they are originally stored
1
2
3
4

Figure 3. Early GroupBy with *Partitioning* Scheme.

3.3 Early GroupBy with Replication Scheme

The *Early GroupBy with Replication* scheme is similar to the Early GroupBy with Partitioning scheme. The similarity is due to the group-by processing to be done before the distribution phase. However, the difference is pointed by the keyword "*with Replication*" in this scheme, as opposed to " *with Partitioning*". The Early GroupBy with Replication scheme, which is also divided into three phases, works as follows.

In the first phase, that is the *local grouping* phase is exactly the same as that of the Early GroupBy with Partitioning scheme. In each processor, local aggregate is performed to table *R*.

The main difference is in phase two. Using the "*with Replication*"scheme, the local aggregate results obtained from each processor are replicated to all processors. Table *S* is not at all moved from where they are originally stored.

In the third phase, the *final grouping and join* phase, is basically similar to that of the "*with Partitioning*" scheme. That is local aggregates from all processors are merged to obtain global aggregate, and then joined with *S*. Looking into more details we can find a difference between the two Early GroupBy schemes. In the "*with Replication*" scheme, after the replication phase, each processor will have local aggregate results from all processors. Consequently, processing global aggregates in each processor will produce the same results, and this can be inefficient as no parallelism is employed. However, joining and global aggregation processes can be done at the same time. First, hash local aggregate results from *R* to obtain global aggregate values, and then hash and probe the fragment of table *S* to produce final query result. The wasting is really that many of the global aggregate results will have no match with local table *S* in each processor.

Figure 4 gives a graphical illustration of the scheme. It looks very similar to Figure 3, except that in the replication phase, the arrows are shown thicker to emphasize the fact that local aggregate results from each processor are replicated to all processors, not distributed.

Apart from the facts that the non-group by table (table *S*) is not distributed and the local aggregate results of table *R* are replicated, assuming that table *S* is uniformly distributed to all processors initially (that is round-robin data placement is adopted in storing records *S*), there will be no skew problem in the joining phase. This is not the case with the previous two schemes, as distribution is done during the process, and this can create skewness depending on the partitioning attribute values.

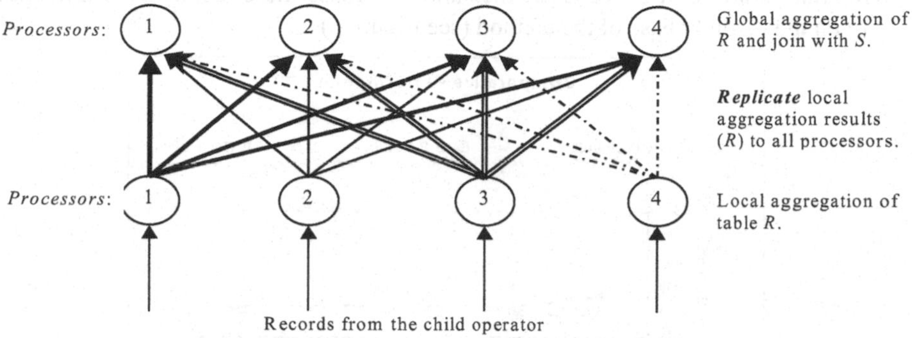

Figure 4. Early GroupBy with *Replication* Scheme.

4 Performance Evaluation

In order to study the behavior of the three methods described in this paper and to compare performance of these methods, we carried out a sensitivity analysis through a simulation. A sensitivity analysis is performed by varying performance parameters. The parameters that were varied consist of faster CPU, faster disk, faster network, faster bigger memory and number of processors. The parameters used include processor speed of 450 *Mips*, disk speed which is estimated to 3.5 *ms*, and message latency per page of 1.3 *ms*.

4.1 Result of Experiment

The graphs in Figure 5 show a comparative performance between the three parallel methods by varying the Group By selectivity ratio (i.e. number of groups produced by the query). The selectivity ratio is varied from 0.0000001 to 0.01. With 100 million records as input, the selectivity of 0.0000001 produces 10 groups, whereas the other end of selectivity ratio of 0.01 produces 1 million groups. The machine in the experiment consists of 64 processors. The graphs also show the results when variation on parameters was applied.

Using the *Early Distribution* method, more extensive data processing occurs during the first phase. In the first phase, the raw records are scanned and then distributed equally to each processor based on certain arrangement. Therefore the major cost of the method is on the scanning, loading and transferring data to every processor, and after that each processor is loaded with equal task and grouped data.

Therefore, in the second phase the total cost is minor, unless the maximum capacity of each processor exceeded. As the consequence, although the process of data transfer, aggregation and join and other processes occur after that, there are just resulting minor to the total cost. This conforms to the graph that shows the total performance is not much affected. Even when faster processors (4 times), faster communications network (4 times) and bigger memory (10 times) and less processors than the original configuration are used in the experimentation, because these factors have influences more on the second phase of the method (see Figure 6).

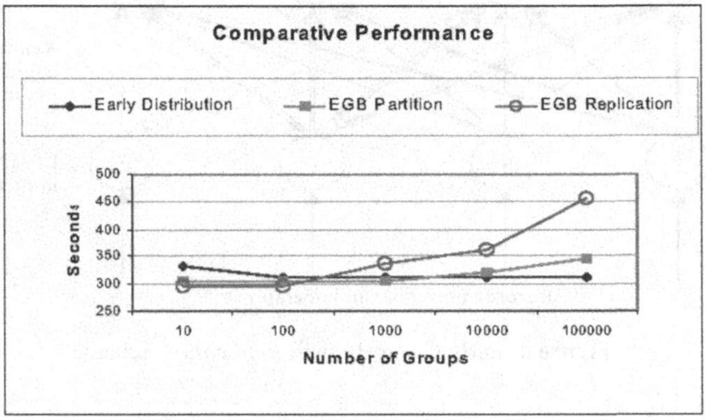

Figure 5. Comparative Performance.

In the *Early Group-By Partitioning* method, the major cost still holds by data scanning and loading. Besides that, the second major data cost is data transfer and reading/writing of overflow of bucket process. This is due to the process of early grouping occurs here. In general, almost similar to Early Distribution Method, by applying faster processors, network and bigger memory in this method does not give much impact to overall performance. However the method's performance is almost steady given number of groups increasing, except when the group number reaches 1000 then the performance starts to decline. This is conforming to the logic, that when number of group produced increasing, meaning that data volume of second and third phase to process is also increasing (also see Figure 6).

Using the *Early Group By Replication (EGB Replication)* method, the result is different compared to two previous methods. In this method, the major costs exist in all three phases of the method. The major costs are the data scanning and loading, data transfer and aggregate and join processes. During the first phase, the data scanning and loading is the major cost. This is as the result of the early grouping process. Then during the second phase, data transfer contributes in decreasing the performance, as all data is sent to all processor for replication. Finally during the third phase, also pooling all data to all processor is decreasing the performance, especially if number of groups growing.

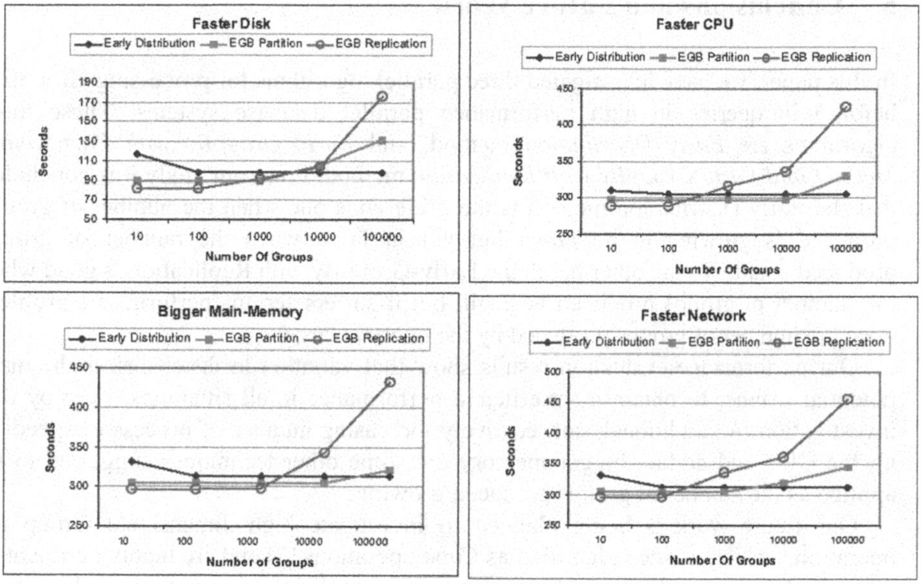

Figure 6. Varying By Selectivity Ratio.

4.2 Discussions

Comparing these three methods, there should be a combination solution for some certain situations. Early distribution shows the best performance if the number of groups produced growing to be large, especially above 1000 groups. However if the number of groups is smaller than 1000, then the choice should go for Early-Group-By Replication method. This is a logic consequence as more or huge number of groups produced, then the cost of, firstly, transferring the data to each processor and secondly the data replication process in second to third phases becomes crucial factors.

As the variation applied, when faster disk is applied, overall performance of the all methods improves significantly (almost 3 times faster). Figure 6 also shows the factors that can potentially improve performance when the number of groups produced increasing. This can be seen in particular there is a delay in the decrease of the performance when the CPU and bigger memory is applied. The delay is from number of groups 1000, as mentioned earlier, to 10000 when the performance starts to decrease. Difference case occurs when faster disk is applied, it improves up to 3 times of overall performance, which is quite substantial. There may be a good attempt to try, by combining of faster disk and either faster CPU or bigger memory to improve overall performance while also delay the decrease as the number of groups involved increasing.

5 Conclusions and Future Work

In this paper, we have investigated three parallel algorithms for processing GroupBy-before-join queries in high performance parallel database systems. These three algorithms are *Early Distribution* method, and *Early GroupBy with Partitioning Method and Early GroupBy with Replication* method. From our study it is concluded that the Early Distribution method is the preference one when the number of groups produced is growing to be large, but not in favor when the number of groups produced small. On the other hand, the Early-GroupBy with Replication is good when the number of groups produced is small, but it suffers serious performance problem once the number of groups produced by the query is large.

Our performance evaluation results show that variation in faster disk is the main potential manner to obtain most efficient performance in all situations taken by this investigation. As additional, consecutively increasing number of processor, speeding up the CPU and adding bigger memory are some other techniques suggested to be applied as the number of groups produced growing.

Our future work is being planned to investigate high dimensional Group By operations, which is often identified as *Cube* operations [7] and are highly pertinent to data warehousing applications. Since this type of applications normally involves large amount of data, parallelism is necessary in order to keep the performance level acceptable.

References

1. Bedell J.A. "Outstanding Challenges in OLAP", *Proceedings of 14th International Conference on Data Engineering*, 1998.
2. Bultzingsloewen G., "Translating and optimizing SQ queries having aggregate", *Proceedings of the 13th International Conference on Very Large Data Bases*, 1987.
3. Datta A. and Moon B., "A case for parallelism in data warehousing and OLAP", *Proc. of 9th International Workshop on Database and Expert Systems Applications*, 1998.
4. Dayal U., Of nests and trees: a unified approach to processing queries that contain nested subqueries, aggregates, and quantifiers," *Proceedings of the 13th International Conference on Very Large Data Bases*, Brighton, UK, 1987.
5. DeWitt, D.J. and Gray, J., Parallel Database Systems: The Future of High Performance Database Systems," *Communication of the ACM*, vol. 35, no. 6, pp. 85-98, 1992.
6. Kim, W., On optimizing an SQ-like nested query," *ACM TODS*, 7(3), Sept. 1982.
7. Ramakrishnan, R., *Database Management Systems*, McGraw Hill, 1998.
8. Taniar, D., and Rahayu, J.W., "Parallel Processing of Aggregate Qeries in a Cluster Architecture", *Proc. of the 7th Australasian Conf. on Parallel and Real-Time Systems PART'2000*, Springer-Verlag, 2000.
9. Taniar, D., Jiang, Y., Liu, K.H., and Leung, C.H.C., 'Aggregate-Join Qery Processing in Parallel Database Systems," *Proc. of The 4th Intl Conf on High Performance Computing in Asia-Pacific HPC-Asia2000*, vol. 2, IEEE CS Press, pp. 824-829, 2000.
10. Yan W. P. and P. Larson, Performing group-by before join," *Proceedings of the International Conference on Data Engineering*, 1994.

Design and Implementation of an RPC-Based ARC Kernel

L. Aruna, Yamini Sharma, and Rushikesh K. Joshi

Department of Computer Science and Engineering
Indian Institute of Technology, Bombay
Powai, Mumbai - 400 076, India
rkj@cse.iitb.ac.in

Abstract. Anonymous Remote Computing (ARC) is a programming paradigm for parallel and distributed computing on workstation clusters. Workstation clusters are characterized by heterogeneity, node/link failures and changing loads. Typically, a parallel program may not have any control over the changing load patterns. Stealing idle cycles on such systems require that parallel programs should adapt themselves dynamically to changing load patterns. We present a design and implementation of an RPC-based ARC kernel supporting parallel programming through ARC Function Calls in such an environment. ARC Function Calls in a C program are executed on anonymous remote machines making the distribution transparent to the parallel programmer. A *Horse Power Factor (HPF)* primitive characterizes load and speed for the use of task distribution in a parallel program. The kernel supports fault tolerance by awarding failed tasks to available nodes. Nodes can join and leave dynamically at any time during execution. The kernel was designed using object oriented techniques and implemented as a collection of collaborating RPC servers running on a a Linux cluster. The performance and overheads of implementation have also been discussed.

1 Introduction

Anonymous Remote Computing (ARC) [1] is a paradigm for parallel and distributed computing addressing engineering issues specific to workstation clusters. In contrast to explicit message passing approaches such as PVM [2], it supports coarse grain parallelism in terms of non-blocking *ARC Function Calls (AFCs)*. For the programmer, an AFC appears nearly the same as a sequential function call but with non-blocking semantics. The runtime kernel executes an AFC invocation on an anonymous remote node. A failure of a node running an AFC invocation is tolerated by restarting the AFC invocation on an available node. An AFC may be executed on the host node itself if no remote nodes are available. A design and implementation of an RPC-based ARC kernel is presented in this paper. We used Object Oriented Analysis and Design techniques to model the kernel. The implementation is in terms of RPC-based server processes.

A non-blocking AFC is a unit of parallel computation providing capability to model task-based parallelism with no intertask communication. Communication

B. Hertzberger et al. (Eds.): HPCN Europe 2001, LNCS 2110, pp. 251–259, 2001.

of data is possible at task boundaries. With respect to task distribution, the kernel addresses issues in both problem domain and execution domain. In the problem domain, the tasks may be generated during runtime, whereas, in the execution domain, the loads on the machines participating in parallel execution may change dynamically. The parallel programmer may not have control over the dynamically changing load. This scenario is typical of an academic cluster on which sequential users work at their convenient times. The ARC model targets parallel programming in this context, where dedicated clusters are unlikely to be available. Simulated annealing on the Traveling Salesman Problem (TSP) has been chosen as a demonstration problem for dynamic task distribution in presence of load. The dynamic distribution is performed at the host program by continuously partitioning the task space into smaller fragments and invoking AFC on them, while AFCs that complete execution generate new tasks.

2 Architecture of the RPC-Based ARC Kernel

The kernel has been designed using Object Oriented Analysis and Design techniques and implemented in terms of RPC servers spread over the cluster of workstations participating in anonymous remote computing. Figure 1 shows the architecture of the distributed ARC kernel. Each machine runs an instance of the RPC server. Asynchronous communication within a node is performed through message queues. The abstractions supported by the kernel fall into interface categories as described below.

Join and Leave Interface: An RPC server may be started or shut down at any time during execution. This interface projects the dynamic join and leave protocol for the use of other RPC servers. Initially, the system starts with one server and subsequently more servers can be added to the system or removed.

Host Interface: The host program is composed of AFCs which are compiled through an ARC translator to low level ARC system calls. These calls are implemented in terms of RPC invocations sent to RPC servers. An RPC server provides the host interface to allow an executing ARC program to communicate with it. The host interface consists of functions for obtaining the horse power factor values, gaining access to anonymous remote machines and posting AFCs for remote execution. Results are communicated to the host program by the corresponding RPC server through a message queue.

Failure Detection Interface: A machine may fail during execution and the failure may be detected leading to a subsequent recovery action. A ring of monitor processes is employed to detect failures. A ring formation is depicted in Figure 1. A monitoring process located on a machine monitors an RPC server located on a different machine. Upon failure detection, the ring is adjusted dynamically. The RPC servers provide an interface, which the monitoring processes invoke through a broadcast when it detects a failure. The server determines the next node to be monitored in the ring omitting the failed node. The monitor is informed about this node through a message queue. The executing host program may re-invoke the AFC depending on the fault tolerance characteristics indicated in the source program. Thus, re-executions are managed by the executing host program through the host interface discussed above.

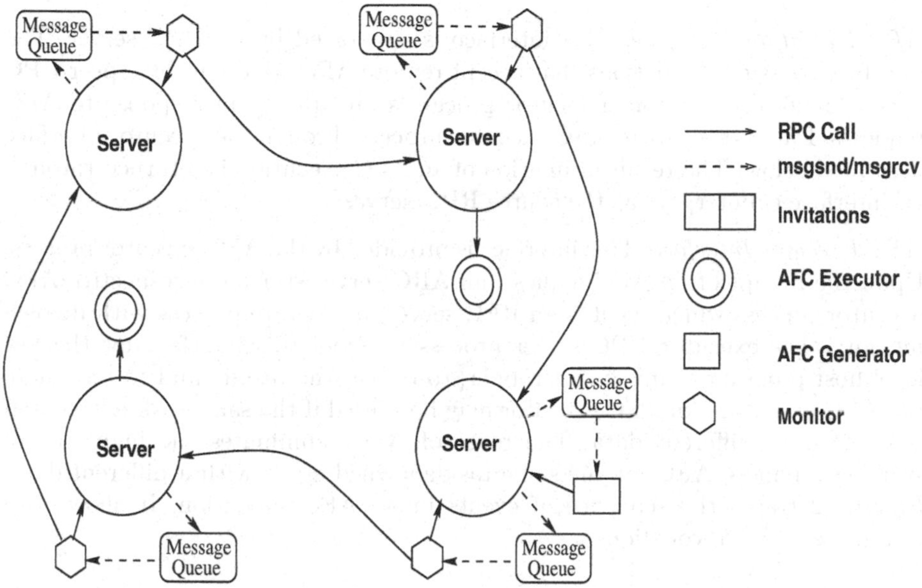

Fig. 1. The Distributed ARC Kernel.

It can be noted that the communication from an RPC server to the local monitor and the local host program is performed through message queues since the monitor and the host are both modeled as clients rather than RPC servers.

Load Sensing Interface: Before invoking an AFC, the host program may want to obtain a measure of load and speed on the anonymous machine where the task is likely to migrate and execute. An indicator called *Horse Power Factor (HPF)* is provided by the ARC paradigm. The HPF normalizes the speed and load on the machines. Thus, if a program receives the same HPF values from two different anonymous machines, the task is expected to take the same time for execution irrespective of the heterogeneity (leading to variations in speed) or the load on the machines involved. An RPC server catering to an HPF request by its client parallel program needs to obtain the HPF values from remote RPC servers. This interface provides the HPF functionalities for the use of other RPC servers.

Locking Protocol Interface: Before sending an AFC invocation for execution to a remote machine, a lock on the remote machine has to be acquired by the sending machine. A remote machine may typically be interested in inviting a few AFCs for execution. The locking mechanism guarantees that the available computational power is used in mutually exclusive manner. The locking protocol extends the Ricart and Agarawala [6] algorithm for mutual exclusion. The RPC servers provide an interface to facilitate the implementation of this locking protocol. Besides, the interface also provides for *lock release* and *lock read* functions respectively to release an already acquired lock or to obtain the availability status.

AFC Execution Interface: This interface is supported by an RPC server. The interface consists of functions that accept remote AFC code sent by a peer RPC server for local execution. Executor processes are started in response to AFC requests. Data is sent to an AFC executor process through the executor interface described below. The result intimation of an AFC execution is returned through an interface counterpart at the source RPC server.

AFC Executor Interface: This interface is provided by the AFC executor process. Upon the receipt of an AFC request, an ARC server starts the respective AFC executor process which itself is an RPC server, a separate process with its own memory. The executor RPC server process provides this interface for the low level host program to invoke a wrapper, to collect the results and to terminate the AFC executor. An AFC executor may be reused if the same task is executed possibly on a different data. The reuse of AFCs eliminates the high cost of starting a remote AFC for the same task spawned again with a different data. Figure 2 traces the sequence of events in an AFC execution. It shows two interleaved AFC invocations.

3 Resource Allocation Protocol

The Ricart and Agarawala algorithm [6] is adapted to handle resource allocation. When a remote machine desires to participate, it deposits *invitations* with the ARC kernel. A lock indicates one invitation for AFC execution. At a time, there may be multiple parallel programmers executing their tasks on the distributed ARC kernel. The algorithm guarantees that the available invitations are used in a mutually exclusive manner. The programs need to obtain locks on these invitations. The ARC system may not defer a reply for a lock request if an invitation is not available. An access or no-access reply is sent to enable an RPC server to execute a task locally if the programmer desires so and if a remote lock cannot be granted.

4 Dynamic Join/Leave Protocol

When a new RPC server joins the system, a unique sequence number is assigned for the joining server. The joining node broadcasts a subscribe request in the network. The last joined node in the system is designated as *publisher*. It replies to the subscribe request of the new server. The list of the participant machines and a sequence number higher than the existing number is provided to the new joining server. Upon a no reply to the broadcast, the initiator resumes as the

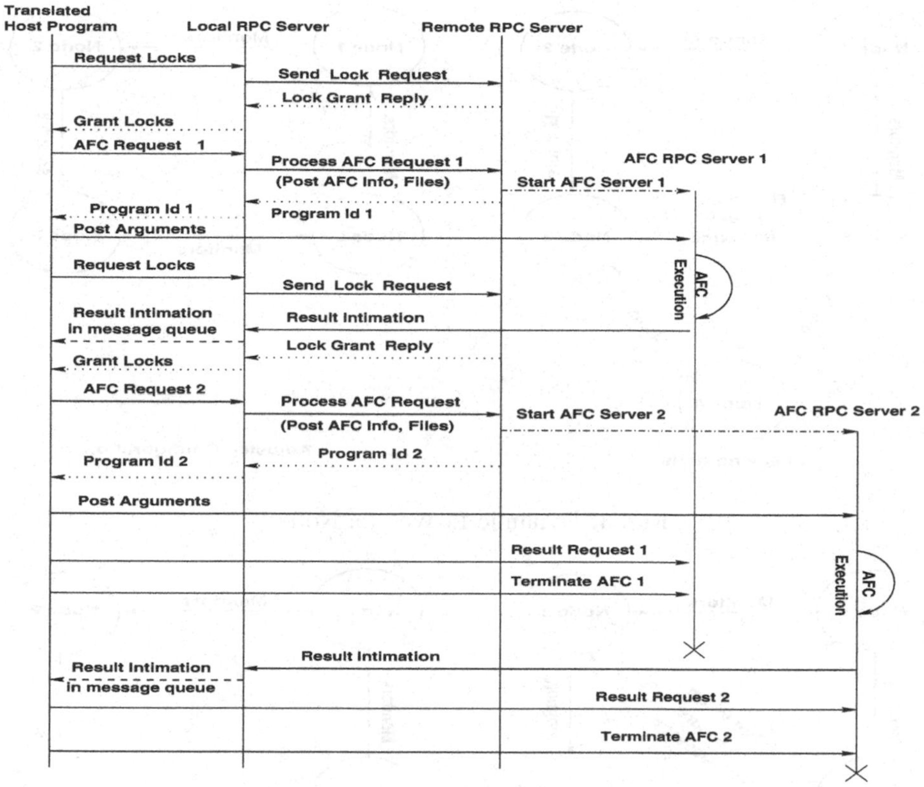

Fig. 2. An ARC Function Call Execution.

first node in the system. The new server is also put in the participant list of all the other existing servers. This is achieved with a broadcast by the newly joined machine. During the join protocol, the monitor ring is also established. The very first RPC server monitors itself and as new nodes join the execution, the ring is expanded.

A server willing to leave ARC computation sends a leave notification through a broadcast. The participant RPC servers mark the broadcasting server in their participant list as unavailable. The publisher role is assigned to a new server if the server leaving the computation happens to be the publisher. Similarly, the monitoring ring is also adjusted. Figure 3 traces the sequence of events in a dynamic leave scenario. Similarly, upon detection of a failure, the failed RPC server is granted a leave by the alive RPC servers. The leave protocol in the case of failures is captured in Figure 4.

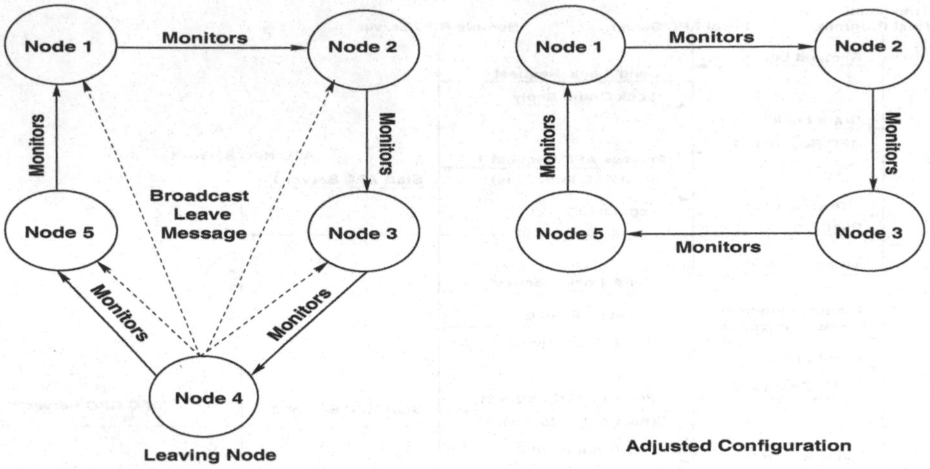

Fig. 3. Dynamic Leave of a Node.

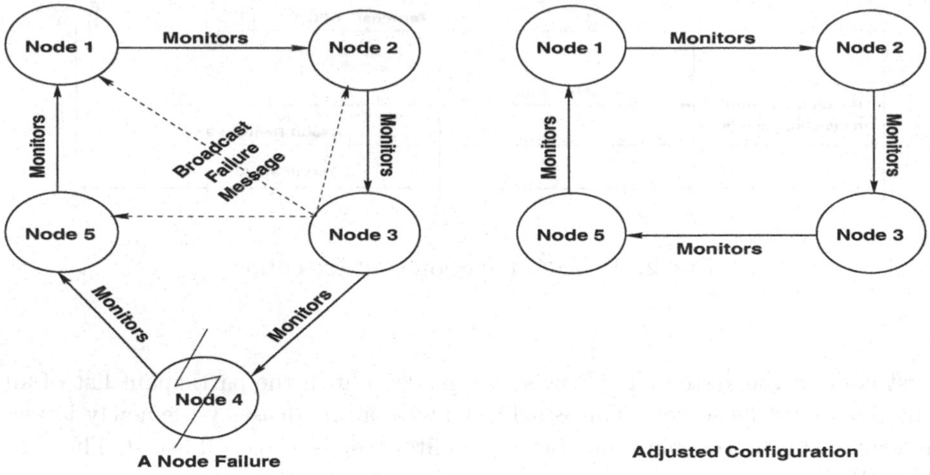

Fig. 4. Dynamic Leave of a Failed Node.

5 Horse Power Factor

An HPF represents the processing capability of the corresponding machine at the time of HPF measurement. Machines in the network are evaluated by a benchmark program to obtain a relative *speed index*. The speed index is static, but the HPF is dynamic since it also applies the *load index*. The load index is obtained from average number of tasks in run queues of the operating system, which is typically an output of the *w* command. A test was conducted using a synthetic load [5] to study the correlation between the load factors reported by the *w* and actual slowdown factor.

6 Dynamic Task Distribution

Since the parallel program is executed in presence of load and heterogeneity, it needs to adopt a dynamic task decomposition scheme. The parallel program can obtain a vector of HPFs based on which it can invoke AFCs with appropriately tuned task sizes. We have chosen a variant of simulated annealing algorithm described in [4] to solve the Traveling Salesman Problem on a network of 666 cities.

Initially, the algorithm starts at the host node with an initial task space consisting of randomly generated paths. The host obtains locks for remote execution and also obtains the associated HPF vector. A portion of the initial task space, αT, is decomposed proportionately. Decomposed tasks are sent to AFCs as function arguments. An AFC performs simulated annealing on the paths that it receives for a given temperature step and chooses best paths for the next iteration. As nodes complete their AFCs, the solutions are returned to the host node where the task space is altered. The host program has the ability to block on *any* result, which usually unblocks upon receipt of results from the first AFC that completes execution. The host continues to decompose a portion of the available task space based on HPF values that it obtains before decomposition.

The task space in the ARC approach is located at the host parallel program which awards it to anonymous remote nodes through AFC invocations. This is an example of a source driven distribution model, whereas in Linda's tuple space [3], the destination nodes may pick up their tasks leading to a destination driven task distribution scheme.

7 Performance and Overheads

The tests were conducted on a cluster of PIII based workstations running LINUX. The overheads of an AFC invocation, results for no load and load tests and communication overheads are discussed in this section.

Table 1. AFC Overheads.

Obtain Lock (ms)	Remote NULL AFC (ms)	Local NULL AFC (ms)	NULL RPC (ms)
225	4000	320	0.25

AFC Overheads: Table 1 compares the overheads of obtaining a lock and an AFC invocation w.r.t. an RPC. It can be seen that the overheads are high due to broadcasts, RPC sequences implementing the protocols and startup times for remote AFC executor process.

No Load Speedup Test: The test was conducted on no load during night hours when minimal sequential activity was expected on the network. The simulated annealing ran on a fully connected 666 cities graph. Table 2 reports the speedup and efficiency figures upto 5 processors. The initial temperature was chosen to be 10000 dropped in steps of 25, 50 and 75 each, with 2000 iterations at each temperature step. Although a static distribution would have performed better, the dynamic task distribution was used during this test to bring out the overheads involved in it.

Table 2. No Load Speedup Test.

Temp Step 25	No of Machines	Time required (secs)	Speedup	Efficiency %
Sequential Execution Time 3297 secs	2	1817	1.81	90.50
	3	1236	2.66	86.66
	4	981.94	3.35	83.75
	5	809.38	4.07	81.4
Temp Step 50	No of Machines	Time required (secs)	Speedup	Efficiency %
Sequential Execution Time 1653.85 secs	2	934.04	1.77	88.50
	3	633.78	2.60	86.66
	4	512.7	3.23	80.75
	5	406.97	4.02	80.4
Temp Step 75	No of Machines	Time required (secs)	Speedup	Efficiency %
Sequential Execution Time 1103.15 secs	2	631.72	1.74	87
	3	435.08	2.53	84
	4	363.35	3.03	75
	5	302.87	3.65	73

Horse Power Utilization (HPU): For computing speedup and efficiency figures, the sequential time must be obtained on a node of the same architecture as that of nodes running parallel program. In presence of heterogeneous nodes, the speedup figure cannot be computed. The *Horse Power Utilization* [1] is computed with real times in presence of load and heterogeneity. Tables 3 shows the HPU figures on a loaded cluster. The HPU was computed as follows:

$$HPU = \frac{Total\ remote\ startup\ time + Total\ remote\ execution\ time}{Total\ time\ for\ program\ execution * Number\ of\ processors}$$

Communication Overheads: Table 4 reports the communication overhead in terms of percentage of the total execution time of the algorithm. It can be seen that if the task is partitioned in smaller fragments, the overheads increase, but it is necessary in order to utilize the available computing power apart from considering dynamic load variations on the machines.

8 Conclusion

The design and implementation of an RPC-based ARC kernel was discussed. The kernel is fully distributed and consists of collaborating RPC server processes. The RPC based servers can be modeled as objects exporting a set of interfaces. The design was presented in terms of the interface abstractions and the dynamic

Table 3. HPU Load Test.

Temp Step	Number of Machines	No. of AFC Postings	Total Remote Computation Time (secs)	Total Time (secs)	HPU (%)
25	2	13	3585	1823	98.32
	3	17	6034	2247	89.51
	4	20	3952	1168.12	84.58
	5	23	4900	1167.9	83.97
50	2	12	2728	1400	97.42
	3	16	2251	814.2	92.15
	4	20	2618	801.1	81.7
	5	23	2253	617.44	73.12
75	2	12	1736.65	936	92.27
	3	17	1500	5598	89.44
	4	20	1400	401.09	87.2
	5	23	1897.68	494.28	76.78

Table 4. Communication Overheads.

No of Nodes	No of AFC Postings	Total Time (sec)	Communication Time(sec)	Overhead %
2	13	1817.5	21.08	1.15
3	18	1236.83	29.219	2.3
4	22	1168.12	34.37	2.94
5	21	809.38	32.62	4

behavior was analyzed through sequence diagrams. A scheme for dynamic load distribution was discussed and the results on load and no load conditions were presented. The implementation incurs overheads due to the protocols that lead to heterogeneity, fault tolerance, load adaptability and transparency to implementation.

References

1. Rushikesh K Joshi, D. Janaki Ram.:Anonymous Remote Computing, A paradigm for Parallel Programming on interconnected Workstations, IEEE Computer, 25(1):75-90, January 1994.
2. Jack Dongarra, Weicheng Jiang, Robert Manchek, Vaidy Sunderam, Al Geist, Adam Beguelin.: PVM 3 User's Guide and Reference Manual, MIT Press, 1994.
3. David Gelernter, Nicholas Carriero.: The S/Net's Linda Kernel, ACM Transactions on Computer Systems, 4(2):110–129, May 1996.
4. S.W. Otto, O.C. Martin.: Combining Simulated Annealing with Local Search Heuristics, Annals of Operations Research, 63:57–75, 1996.
5. Benjamin Wah, Pankaj Mehra.: Synthetic Workload Generation for Load Balancing Experiments, IEEE Parallel and Distributed Technology, 3(2):4–19, Summer 1995.
6. G. Ricart, A.K. Agrawala.: An Optimal Algorithm for Mutual Exclusion in Computer Networks, ACM Communications, 24(1):9–17, January 1981.

Track III

Computer Science

Application-Controlled Coherence Protocols for Scope Consistent Software DSMs

Cicero Roberto de Oliveira Galdino[2] and Alba Cristina Magalhaes Alves de Melo[1]

[1] Department of Computer Science, Campus Universitario - Asa Norte, Caixa Postal 4466
University of Brasilia, Brasilia – DF, CEP 70910-900, Brazil
albamm@cic.unb.br

[2] Medical Informatics Group, Sarah Hospital Network – SMHS 301 Conjunto A, Asa Sul
Braslia – DF, CEP 70335-901, Brazil
galdino@bsb.sarah.br

Abstract. To use the shared memory programming paradigm in distributed architectures where there is no physically shared memory, an abstraction must be created. This abstraction is known as Distributed Shared Memory (DSM). To reduce communication costs, DSM systems usually replicate data. This approach generates a coherence problem, which is generally solved by a memory coherence protocol. Unfortunately, it seems that there is no coherence protocol that achieves good performance for a large set of applications since the most appropriate coherence protocol depends on how the application accesses data. For this reason, it is interesting for a DSM system to provide multiple coherence protocols. This article presents and evaluates a low-overhead mechanism that allows a DSM application to choose among multiple coherence protocols. This mechanism was incorporated in JIAJIA, a DSM system that implements scope consistency with a write-invalidate protocol. Our results on some benchmarks show a significant reduction on the number of messages exchanged, leading to better performance results.

1 Introduction

In order to make shared memory programming possible in distributed architectures, we must create a shared memory abstraction that parallel processes can access. This abstraction is called Distributed Shared Memory (DSM). The first DSM systems tried to give parallel programmers the same guarantees they had when programming uniprocessors. It has been observed that providing such a strong memory consistency model creates a huge coherence overhead, slowing down the parallel application and bringing frequently the system into a thrashing state [14]. To alleviate this problem, researchers have proposed to relax some consistency conditions, thus creating new shared memory behaviours that are different from the traditional uniprocessor one.

In the shared memory programming paradigm, synchronisation operations must be used every time processes want to restrict the order in which memory operations should be performed. Using this fact, hybrid Memory Consistency Models guarantee that processors only have a consistent view of the shared memory at synchronisation

B. Hertzberger et al. (Eds.) : HPCN Europe 2001, LNCS 2110, pp. 263-272, 2001.

time. This allows a great overlapping of basic read and write memory accesses that can potentially lead to considerable performance gains. By now, the most popular hybrid memory consistency models for DSM systems are Lazy Release Consistency (LRC) [10] and Scope Consistency (ScC) [9].

While the memory consistency model defines when consistency should be ensured, coherence protocols define how it should be done. Indeed, there are many coherence protocols that can be used to implement the same memory consistency model. For instance, LRC can be implemented in software either with distributed diffs [10] or with home-based protocols [8]. LRC can be also implemented with hardware support (AURC) [3].

It is important for a DSM system to provide multiple coherence protocols and to use the protocol that best matches the reference pattern exhibited by an application at a particular moment. Recently, many coherence protocols were proposed that analyse the access patterns of a parallel application and dynamically choose the protocol that is best suited. Examples of such adaptive protocols are ADSM [13], ATMK [1] and Tapeworm [11]. In our opinion, adaptive protocols are often very effective but they do have some drawbacks. First, the detailed analysis of access patterns adds a non-negligible overhead to the protocol execution time [13]. Second, these protocols need an amount of time to determine that the access pattern has changed. Thus, there is a period of time when the application is already referencing shared data in a different way but the protocol is not aware of it. These two factors contribute to slowdown the application.

There is also a recent work that allows an application to annotate the code, giving some hints about the access pattern [12]. This approach is based on the fact that DSM systems are typically used by programmers developing long-run applications, which are very performance-sensitive. This kind of programmer is usually capable to modify his/her application code in order to achieve high performance [6].

The goal of this paper is to propose and evaluate a low-overhead coherence mechanism for scope-consistent software DSMs. In our mechanism, user-level code annotations are used to indicate which coherence protocol must be used. The low overhead of our mechanism comes from the fact that the programmer specifies directly which coherence protocol must be used. In our prototype, we modified JIAJIA [7], a scope consistent software DSM system, to include a new coherence protocol based on write-update while keeping the original protocol, that is invalidation-based.

Our experimental results obtained on a cluster of 8 Pentium III with some popular benchmarks (LU, TSP, Matrix Multiplication, SOR and EP) show that, as expected, LU behaves better with write-invalidate protocols and TSP behaves better on write-update. Our results also show that changing the protocol at run-time in a processor basis gives the best result for applications where processors do not execute the same code.

The rest of this paper is organised as follows. Section 2 presents some memory coherence protocols. Section 3 describes the JIAJIA software DSM system. Section 4 presents our memory coherence mechanism. Section 5 shows some experimental results. Related work in the area of memory coherence protocols for DSM systems is discussed in section 6. Finally, section 7 concludes the paper.

2 Memory Coherence Protocols for DSM Systems

There are two basic coherence protocols that are traditionally used to solve the cache coherence problem: write-invalidate and write-update protocols [5]. In general, at consistency time, a write-invalidate protocol guarantees that there is only one copy of the data in the system. If these data are further needed, they are fetched from the only node that has a valid copy. On the other hand, a write-update protocol allows many copies of the same data to be valid at consistency time and generally guarantees that updates will be done atomically and totally ordered.

It is well known that write-invalidate protocols should be used when accesses to shared data will not be issued in a near future by the other processors. On the contrary, write-update protocols are often preferred when many processors simultaneously read and write the same data. Although most of DSM systems use write-invalidate based protocols (TreadMarks [10], AURC [3], ATMK [1]), some recent DSM systems offer also write-update protocols (ADSM [13], Brazos [17]).

Another characteristic that is important in a cache coherence protocol for DSM systems is the number of simultaneous writers allowed. The most intuitive approach allows only one processor to write data at a given moment (single-writer protocol). However, as the unity of consistency is often a page, false sharing can occur when two or more processors want to access independent variables that belong to the same page. To reduce the effects of false sharing, many DSM systems use multiple-writer protocols, allowing many processors to have write access to the same page simultaneously and then merging the multiple versions of the page when a consistent view is required.

There are two basic approaches used to manage the information needed to execute coherence protocols in page-based DSM systems: home-based and homeless. In home-based systems, each page is assigned to a node (home-node) that concentrates all modifications made to the page. Every time an updated version of the page is needed, it is sufficient to contact the home node in order to fetch the page. In the homeless approach, each processor that modifies a page maintains such modifications locally. In order to obtain an updated version of the page, a node must collect the modifications that are distributed all over the system. Modifications are kept by each node and garbage collection is required.

Home-based protocols do have some advantages. First, each access fault requires only communication with the home-node. Second, since modifications are eagerly applied at the home-node, there is no need for additional control structures such as twin pages or diffs. However, as modifications are eagerly sent to the home node, such protocols generally require additional messages. Also, on an access fault, homeless protocols fetch only the modifications made to the page (diffs) while home-based protocols fetch the whole page [7].

3 The JIAJIA Software DSM System

JIAJIA is a software DSM system proposed by Hu, Shi and Tang [7]. It implements the Scope Consistency memory model with a write-invalidate multiple-writer home-based protocol. In JIAJIA, the shared memory is distributed among the nodes in a NUMA-architecture basis. Each shared page has a home node. A page is always present in its home node and it is also copied to remote nodes on an access fault. There is a fixed number of remote pages that can be placed at the memory of a remote node. When this part of memory is full, a replacement algorithm is executed.

Scope Consistency is a memory model that requires that consistency must be guaranteed when a process acquires a lock or when it reaches a synchronisation barrier. In the first case, consistency is maintained in a per-lock basis, i.e., only the shared variables that were modified on the critical section guarded by lock *l* are guaranteed to be updated when a process acquires lock *l*. On a synchronisation barrier, however, consistency is globally maintained and all processes are guaranteed to see all past modifications to the shared data [8].

In order to implement Scope Consistency, JIAJIA statically assigns each lock to a lock manager. The functions that implement lock acquire, lock release and synchronisation barrier in JIAJIA are *jia_lock*, *jia_unlock* and *jia_barrier*, respectively [15].

On a release access, the releaser sends all modifications performed inside the critical section to the home node of each modified page. The home node applies all modifications to its own copy and sends an acknowledgment back to the releaser. When all acknowledgements arrive, the releaser sends a message containing the numbers of the pages modified inside the critical section (write notices) to the lock manager [8].

On an acquire access, the acquirer sends an ACQ message to the lock manager. When the lock manager decides that the lock can be granted to the acquirer, it responds with a lock granting message that contains all write notices associated with that lock. Upon receiving this message, the acquirer invalidates all pages that have write notices associated, since their contents are no longer valid (figure 1).

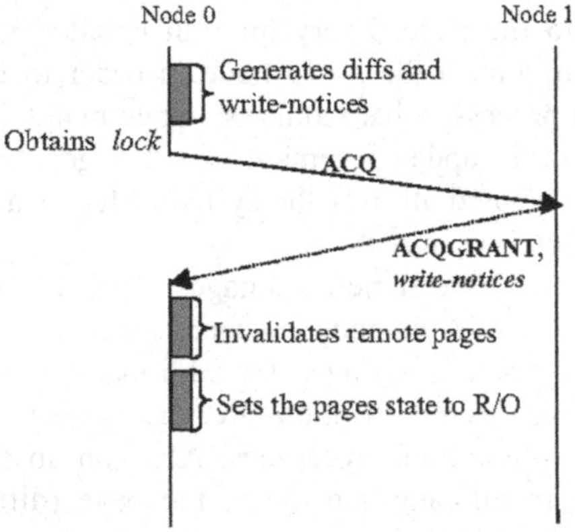

Fig. 1. Lock Acquire Operation at the JIAJIA Software DSM System.

On a barrier access, the arriving process generates the diffs of all pages that were modified since the last barrier access. Then, it sends the diffs to the respective home nodes. The home nodes receive the diffs and send an acknowledgement to the arriving process. Then, the arriving process sends a BARR message containing the write notices of all modified pages to the owner of the barrier. When all processes arrive at the barrier, the owner of the barrier sends back a message BARRGRANT containing the write notices of all pages modified since the last barrier. Upon receiving this message, the processor invalidates the pages contained in the write notices and continues the execution (figure 2).

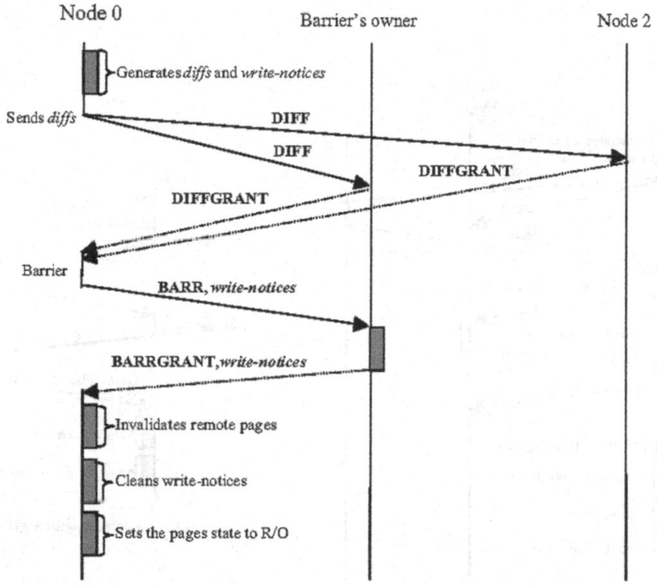

Fig. 2. Barrier Synchronization at the JIAJIA Software DSM System.

JIAJIA also offers some optional features such as home migration and load balancing, among others. These features can be activated and de-activated by the function *jia_config (option, value)*, where *option* is the feature and *value* can be either ON or OFF. At the beginning of the execution, all features are set to OFF [15].

4 The Proposed Cache Coherence Mechanism

As explained in section 3, the JIAJIA DSM system uses a write-invalidate coherence protocol to ensure consistency. It is well known that this kind of protocol is not appropriate for applications that eagerly read and write shared data, since there is a great probability for the invalidated pages to be accessed in a near future [6]. For this reason, we propose the inclusion of a write-update protocol in the JIAJIA system as an alternative to the write-invalidate protocol.

The write-update protocol was implemented as a new feature to JIAJIA, called W_UPDT. The use of this protocol is activated in a per-processor basis by the primitive *jia_config*(W_UPDT, ON) and de-activated by *jia_config*(W_UPDT, OFF).

The write-invalidate and write-update protocols differ basically on the way that modifications are propagated. As JIAJIA implements scope consistency, coherence operations occur always at the beginning of a critical section, i.e., on a lock acquire (*jia_lock*) or on a synchronisation barrier (*jia_barrier*). Thus, in order to implement the write-update protocol, the functions *jia_lock* and *jia_barrier* were modified. The new behaviour of these functions is illustrated in figure 3.

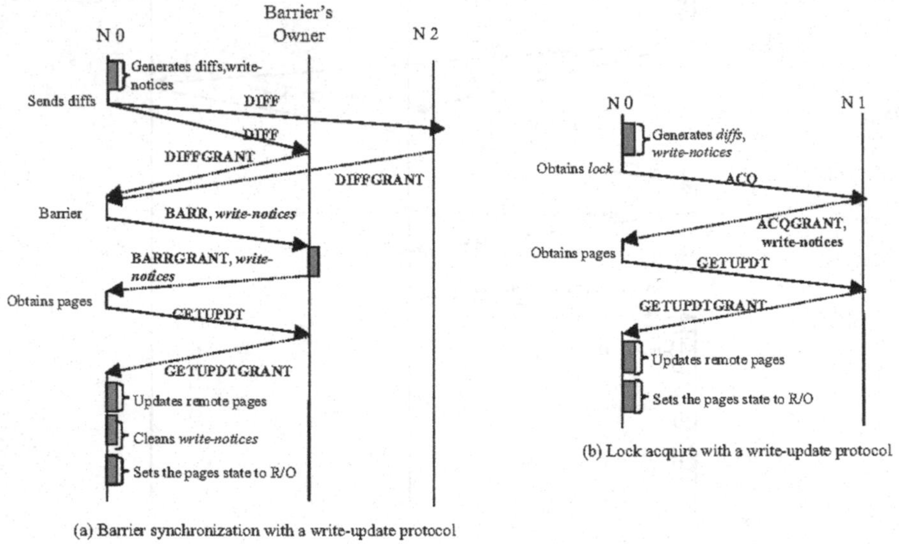

Fig. 3. Inclusion of the Write-Update Protocol at the JIAJIA Software DSM System.

When a process receives a lock grant or a barrier grant, it also receives the write-notices describing the pages associated with that lock that were modified by other processors. These pages are no longer valid and the write-invalidate protocol implemented in JIAJIA invalidates all of these pages that are locally cached. In order to implement the write-update protocol, we fetch all pages that have write-notices associated with the lock and cache them locally. To do this, a message GETUPDT containing the numbers of the obsolete pages is sent to the respective home nodes. The home nodes collect the pages, concatenate them in messages (by now, it is possible to send up to nine pages in a single message) and send them back. The process that issues a *jia_lock* or a *jia_barrier* only continues execution after all pages have arrived.

5 Experimental Results

To evaluate the gains of our multi-protocol mechanism, we ran our experiments on a dedicated cluster of 8 Pentium III 550 MHz, with 128 MB RAM connected by a 100Mbps Ethernet switch. The JIAJIA software DSM system ran on top of Conectiva Linux 5.0. Our results were obtained with 5 popular DSM benchmarks: LU from the SPLASH-2 benchmark suite [16], EP from NAS Parallel Benchmarks [2], SOR and TSP from Rice University [10] and MatMult from the JIAJIA team [7]. Table 1 shows the problem sizes and execution times with 2, 4 and 8 processors.

Table 1. Execution times in seconds for applications using write-invalidate (WI) and write-update (WU) protocols.

Application	Size	2 processors		4 processors		8 processors	
		WI	WU	WI	WU	WI	WU
EP	2^{28}	215,90	215,90	108,50	108,50	54,81	54,81
LU	1024 x 1024	62,72	64,22	34,25	36,12	-	-
MatMult	1024 x 1024	11,83	11,83	8,17	8,17	5,14	5,13
SOR	1024 x 1024	20,93	20,92	15,01	15,05	12,17	12,23
TSP	20 cities	10,87	10,55	7,40	6,81	7,70	6,76

EP is a program used to generate Gaussian randomic diverging pairs [2]. The only synchronization needed in this program is at the end of the program, when the results are collected. For this reason, the performance of both protocols is roughly the same.

LU is used to factor a dense matrix into the product of a lower triangular and upper triangular matrix. The dense matrix is divided into a matrix N x N containing B x B blocs, where n≫B, in order to explore temporal locality on the elements of each sub-matrix [16]. Synchronization is achieved exclusively by barriers, which are positioned at the end of each computation step. As the LU application does not have an access pattern that involves intensive sharing of data, the measured performance with the write-invalidate protocol was better than the one obtained with the write-update one. This occurs despite the considerable reduction in the number of messages exchanged by the write-update protocol (figure 4(b)) because the write-update protocol requests and applies a great number of pages (figure 4(a)) that are never accessed, generating an unnecessary overhead that causes a slowdown on the application execution.

Fig 4. Pages and messages exchanged by LU with write-invalidate and write-update protocols.

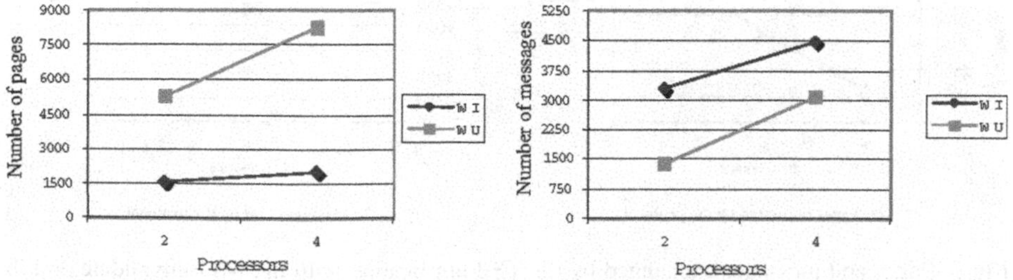

(a) Pages demanded by the protocol

(b) Messages sent by the protocol

MatMult is a program that multiplies two matrixes. The matrixes are partitioned and distributed uniformly among the processors [7]. All write accesses are made locally and only part of the read accesses are remote. Modified data remain at the home node; thus, there is no need for coherence operations. That's why the measured performance is the same with both protocols.

SOR is a program that solves partial differential equations using the Red-Block Successive Over-Relaxation method [10]. The red and black arrays are allocated in shared memory and divided into equal size bands of rows, which are distributed among the processors. Synchronisation is achieved exclusively by barriers and there are two barriers for each computation step to obtain the new values of the red and the black array, respectively. When we first ran this benchmark with a pure write-update protocol, the performance results were very bad. This happened because the initialisation of the red and black arrays is done exclusively by the first processor. Using a write-update protocol, all shared pages were received by this processor at each synchronisation barrier. We decided, thus, to use a hybrid strategy, where the first processor uses the write-invalidate protocol while the other ones use the write-update one. The measured performance of this approach was similar of the performance obtained with the write-invalidate protocol. This gain in performance over a pure write-update protocol shows the potential of choosing a protocol in a per-processor basis.

The TSP program solves the Travelling Salesman Problem using a branch-and-bound algorithm. The main shared structures are a pool of partially evaluated tours, a priority queue containing pointers to tours at the pool, a stack of pointers to tour structures that can be reused and the current shortest tour [10]. This application exhibits intensive sharing patterns since individual tours can be modified by multiple processors. Synchronisation is achieved by locks. Using the write-update protocol, we obtained a great reduction (almost 50%) on the number of messages exchanged (figure 5(b)). Although the number of pages requested is slightly smaller with the write-invalidate protocol (figure 5(a)), this was not sufficient to decrease performance results. The execution times obtained with the write-update protocol were 2.94% smaller for the 2-processor case, 7.97% smaller with 4 processors and 12.20% smaller with 8 processors.

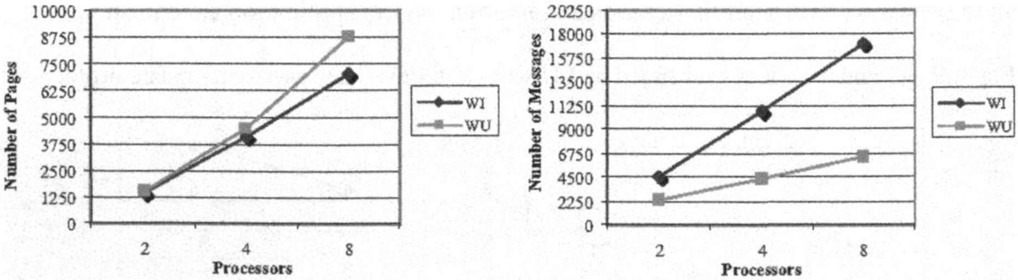

(a) Pages requested by the protocol

(b) Messages sent by the protocol

Fig 5. Pages and messages exchanged by the TSP application with the write-invalidate and the write-update protocol.

6 Related Works

The use of multiple coherence protocols to achieve better performance in DSM systems was already proposed in some earlier works. Munin [4] is one of the first DSM systems that proposed multiple coherence protocols. It implements the Release Consistency memory model and offers 5 different protocols. The association between the protocol and the shared variable is made at compile time. ATMK [1] was proposed for a LRC system (TreadMarks) and it adapts dynamically and automatically between single-writer and multiple-writer. The modifications are propagated by invalidations. ADSM [13] was also implemented in TreadMarks and it adapts between single and multiple-writer and also between write-invalidate and write-update. Brazos [17] is a Scope Consistent DSM system that uses a hybrid protocol, where a write-update protocol is used for variables that are guarded by locks and a write-invalidate protocol is used at synchronisation barriers.

Like our proposal, Munin allows the programmer to associate variables and protocols, but this is made statically at compile time. ATMK and ADSM automatically adapt the protocol to the observed access pattern. The memory model used is LRC, and that makes these systems different from our approach. Among these systems, Brazos is the only one that implements Scope Consistency. It differs from our mechanism because the association between protocol and variable is fixed and depends on where the variable is accessed (inside or outside the critical section).

As far as we know, our mechanism is the first one that provides application-controlled write-invalidate and write-update coherence protocols for Scope Consistent DSM systems.

7 Conclusions and Future Work

It is well known that access patterns to shared data determine the best protocol to be used in a DSM system. As long as a program executes, access patterns dynamically change, and a mechanism must be used to provide the most appropriate protocol for a given access pattern at a given time.

In this paper, we proposed and evaluated a mechanism that allows multiple coherence protocols for Scope Consistent DSM systems. As the programmer is the responsible to determine the instant where the protocol must be changed, our mechanism has a very low overhead. We were also able to reduce the number of messages exchanged by sending only one update request message per home-node and by concatenating pages in fewer messages. Our results show that applications that eagerly share data can benefit from the write-update protocol. That was the case for TSP.

As future work, we intend to improve our mechanism, offering the programmer the choice of a protocol in a per-variable basis. Also, the mechanism will be tested in real applications in order to evaluate if it is adequate to the programmer's needs.

Acknowledgements

The authors would like to thank Dr. Shi and his team for letting us use and modify the source code of JIAJIA v. 2.1.

References

1. Amza C., Cox A., Dwarkakas S., Zwaenenpoel W.: Software DSM Protocols that Adapt between Single Writer and Multiple Writer, Proc. of HPCA'97, p. 261-271, February, 1997.
2. Bailey D., Barton J., Lasinski T., Simon H.: The NAS Parallel Benchmarks, Technical Report 103863, NASA, July, 1993.
3. Blumrich M., Li K., Alpert R., Dubnicki C., Felten E.: A Virtual Memory Mapped Network Interface for the SHRIMP Multicomputer, Proc of ISCA'94, p. 142-153, April, 1994.
4. Carter J., Efficient Distributed Shared Memory Based on Multi-Protocol Release Consistency. PhD Thesis, Rice University, January, 1994.
5. Culler D., Singh J., Gupta A.: Parallel Computer Architecture: A Hardware/Software Approach, Morgan Kaufmann, 1998.
6. Gharachorloo K.: The Plight of Software Distributed Shared Memory, Proc. Of the 1st ACM International Workshop on Software DSM, Rhodes, Greece, June, 1999.
7. Hu W., Shi W., Tang Z.: JIAJIA: An SVM System Based on A New Cache Coherence Protocol. In Proc. of HPCN99, LNCS 1593, pp. 463-472, Springer-Verlag, April, 1999.
8. Iftode L.: Home-Based Shared Virtual Memory, PhD Thesis, June, 1998, Princeton University, Dept of Computer Science, 132 pages.
9. Iftode L., Singh J., Li K.: Scope Consistency: Bridging the Gap Between Release Consistency and Entry Consistency, Proc. Of the 8th ACM SPAA'96, June, 1996, pages 277-287.
10. Keleher P., Cox A., Dwarkakas S., Zwaenenpoel W.: TreadMarks: Distributed Shared Memory on Standard Workstations and Operating Systems", Proc. of USENIX, January, 1994, pages 115-132.
11. Keleher P: Tapeworm: High-Level Abstraction of Shared Accesses, Proc of the 3rd Symposium on Operating Systems Design and Implementation, February, 1999.
12. Lee J., Jhon C.: Reducing Coherence Overhead of Barrier Synchronization in Software DSMs, Proc. Of the 1998 ACM Int. Conf. on Supercomputing, November, 1998.
13. Monnerat L., Bianchinni R.: Efficiently Adapting to Sharing Patterns in Software DSMs, Proc. Of the 4th HPCA'98, February, 1998.
14. Mosberger D.: Memory Consistency Models, Operating Systems Review, p. 18-26, 1993.
15. Shi W.: Improving the Performance of DSM Systems, PhD Thesis, Chinese Academy of Sciences, November, 1999.
16. Singh J., Weber W., Gupta A.: SPLASH: Stanford Parallel Applications for Shared Memory, Computer Architecture News, 20(1):5-44, March, 1992.
17. Speight E., Bennett J.: Reducing Coherence-Related Communication in Software Distributed Shared Memory Systems, Technical Report ECE-TR-98-03, Rice University,1998.

Source Code and Task Graphs in Program Optimization

Welf Löwe, Wolf Zimmermann*, Sven Dickert, and Jörn Eisenbiegler

Institut für Programmstrukturen und Datenorganisation
Universität Karlsruhe, 76128 Karlsruhe, Germany
loewe@ipd.info.uni-karlsruhe.de

Abstract We present scheduling techniques considering the source code as well as the task graph of a parallel program. This approach allows to chose the proper scheduling algorithm and to estimate the execution time of the target program. We demonstrate this on stencil computations and their optimizations.

1 Introduction

For many parallel programs, the communication behavior only depends on the size of the problem and not on the actual input. Using this property for translation and optimization improves the efficiency of the generated code dramatically. Moreover, programmers may focus on the inherent parallelism of the problem and disregard properties of target machines. They can use a synchronous, shared memory programming model; neither data alignment nor mapping of processes onto processors is explicitly required in the source code. Data and processes are distributed automatically.

For optimizations, a cost model is required reflecting latency for point-to-point-communication in the network, overhead of communication on processors themselves, and the network bandwidth. The LogP-machine [2] models these communication costs with parameters *L*atency, *o*verhead, and *g*ap (which is actually the inverse of the bandwidth per processor). In addition to L, o, and g, parameter P describes the number of processors. These parameters have been determined for a number of parallel machines including the CM-5 [2], the IBM SP1 machine [3], a network of workstations and a powerXplorer [9]. Here, we extend this list by the IBM RS/6000 SP. All practical runtime measurements confirmed their LogP-based predictions. In this paper, we consider a variant of the LogP-model that takes into account message size, i.e. the parameters L, o, and g are functions on the message size. Fig. 1 shows the communication costs for the IBM RS/6000 SP depending on the message length.

To choose a proper scheduling strategy, a compiler should consider structural properties of the source code and granularity of the program on a specific target machine. The former is obtained easily by considering the program structure, the latter requires the task graph of the program.

* Now with the Institut für Informatik, Martin-Luther-Universität Halle-Wittenberg, 06099 Halle, Germany, E-mail: zimmer@informatik.uni-halle.de.

B. Hertzberger et al. (Eds.): HPCN Europe 2001, LNCS 2110, pp. 273–282, 2001.

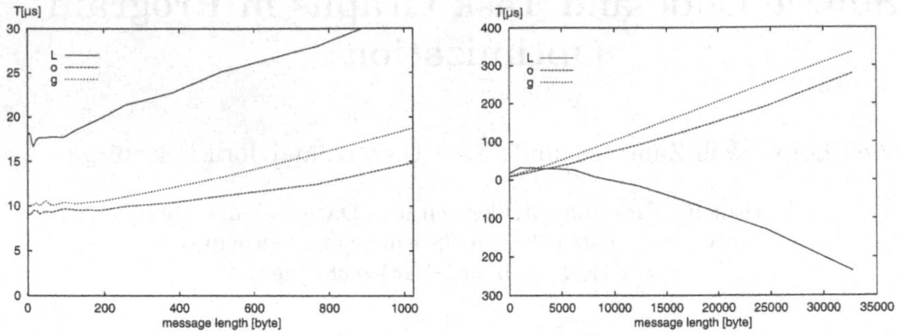

Figure 1. LogP-Parameter for the IBM RS/6000 SP.

The program class of stencil computations is defined by structural properties of source codes. We map task graphs generated from stencil computations to LogP machines optimizing their runtime in the LogP cost model. Efficient LogP programs are also efficient on real parallel machines. Hence our approach provides a general optimization for stencil computation programs. We confirm our results in practice for a linear wave simulation implementation on an IBM RS/6000 SP.

2 Basic Definitions

We can model the execution of programs on an input x by a family task-graphs $G_x = (V_x, E_x, \tau_x)$. The tasks $v \in V_x$ model local computations without access to the shared memory, $\tau(v)$ is the execution time of task v on the target machine, and there is a directed edge from v to w iff v writes a value into the shared memory that is read later by task w. Therefore, task-graphs are always acyclic. G_x does not always depend on the actual input x. In many cases of practical relevance it only depends on the problem size n. We call these programs *oblivious* and denote their task graphs by G_n. In the following, we consider oblivious programs and write G instead of G_n if n is arbitrary but fixed. The *height* of a task v, denoted by $h(v)$, is the length of the longest path from an input task (corresponding to a node with in-degree 0) to v. The set of tasks of a task graph G can be partitioned as follows: A *layer* Λ^s of a task-graph $G = (V, E)$ is the set of tasks $\Lambda^s = \{v \in V : h(v) = s\}$.

A *LogP-schedule* is a set of sequences of computations, send, and receive operations and their starting times corresponding to the tasks and edges of the task-graph. For each task, its predecessors must be computed either on the same processor or received from other predecessors. The schedules must guarantee the following constraints: (i) sending and receiving a message of size k takes time $o(k)$, (ii) a send (receive) operation must not start earlier than $g(k)$ time units before the perceeding send (receive) operationon the same processor, (iii) a receive must correspond to a send at least $L(k) + o(k)$ time units earlier in order to avoid waiting times, (iv) computing a process v takes time $\tau(v)$, and (v) a correct LogP-schedule of a task-graph G must compute all tasks at least once.

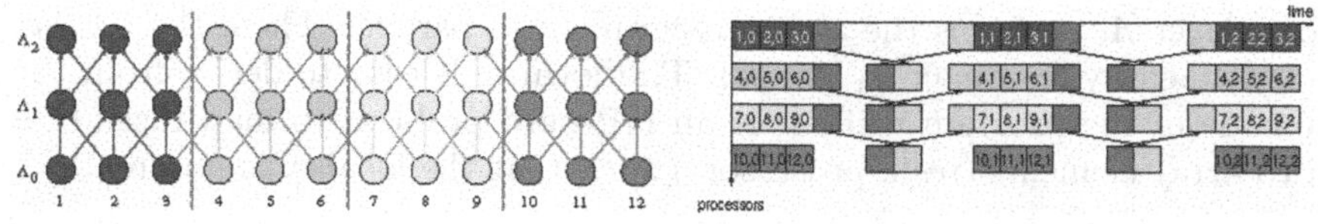

Figure 2. Data Distribution (left) and the Corresponding LogP-Schedule (right).

We denote schedules by calligraphic letters $\mathcal{S}$. The make-span or execution time of $\mathcal{S}$, denoted by $T(\mathcal{S})$ is the smallest time when all tasks are finished. For a task-graph G, $T_{\text{opt}}(G)$ denotes the make-span of an optimal LogP-schedules for G. We rather refer to the sets of processes computed on the same LogP-processor than to their actual sequences of computations, send, and receive operations.

Any oblivious program can be transformed into a semantically equivalent program that can be executed asynchronously on a distributed memory machine, i.e. on the LogP-machine [15].

3 Program Information and Task Graphs

The goal is to minimize the execution time of LogP-schedules and there are quite a few known algorithms. However, the performance of the optimized code or even the applicability of the algorithms heavily depend on structural properties, cf. [9]. They require the task graph to be balanced, symmetric, layered, connected by a certain degree etc. Those properties are difficult to compute for graphs but rather simple for the source code they are generated from.

For simulating a one-dimensional wave, a new value for every simulated point is recalculated in every time step according to the current value of this point(y_0), its two neighbors (y_{-1},y_{+1}), and the value of this point one time step before (y_0'). This update is performed by the function:

$$\Phi(y_o, y_{-1}, y_{+1}, y_o') = 2 \cdot y_0 - y_o' + \Delta_t^2/\Delta_y \cdot 2 \cdot (y_{+1} - 2 * y_0 + y_{-1})$$

The task graph and its schedule are shown in Fig. 2. The task-graph is derived from the following stencil computation:

for $t := 0$ **to** T **do**
    forall $i = 0$ **to** n **do in parallel**
        $a[i] := \Phi(a[i-1], a[i], a[i+1])$

HPF compilers obtain a schedule by block-wise distribution of the array a onto the P processors, i.e. array elements $a[0], \ldots, a[n/P-1]$ are stored on processor P_0, $a[n/P], \ldots, a[2n/P-1]$ on processor P_1 etc (cf. Fig. 2). In general, we can use the idea of block-wise data distribution for all source programs containing a loop whose body is a parallel loop with the assignment $a[j] := \Phi(a[j_1], \ldots, a[j_m])$ This class of programs are called *one-dimensional stencil computations.* It occurs typically for solving partial differential equations in one spatial dimension.

Suppose, layer Λ_i performs the above assignment in parallel. Then, the *locality* of Λ_i is defined by $\delta_i = \max_{k=1}^{m} |j - j_k|$. The locality is easy to derive from the source code of the program while it is an NP-hard problem to map tasks of a graph to array elements (resp. processors) such that the locality is minimal.

4 The Optimization Algorithms

The basic idea for both scheduling techniques is to save communication costs by avoiding many small messages in favor of one large message. However, they require locality of the layers to be bounded by a constant. This information is obtained from the program code. The next two subsections discuss the scheduling algorithms the third subsection compares measurements of run-times.

4.1 Redundant Scheduling

A block-wise schedule communicates in each iteration step. Our goal is to delay communication to the next step: Instead of receiving the result from tasks computed on neighbor processors we compute the tasks themselves on the same processor redundantly. This saves communication time for the price of additional computations. Fig. 3 sketches the scheduling algorithm: We cover the task graph with triangles. Each triangle defines a process computing the tasks corresponding to the covered nodes. If the tasks corresponding to the bottom nodes in the triangles got their data, all other tasks inside can be computed without any communication. As shown in Fig. 3 (top, left), several nodes are covered by different triangles. These nodes are computed redundantly by the respective processes. Fig. 3 (bottom, left) shows the remaining communication between the processes. Actually, the dotted line does not represent communications if we linearly cluster the processes. Finally, we map the processes to processors according to the block-wise scheduling algorithm. Neighbored processes are computed on the same processor, cf. Fig. 3 (top, right), which again reduces the remaining communication, cf. Fig. 3 (bottom, right).

The redundant computations cost additional time. However, due to delaying the communication, we can send larger messages and one larger message is less expensive than many small messages. Hence, there is a trade-off between redundant computation and message bundling. Using the LogP model, we are able to conservatively estimate the execution time for each schedule. Then we find the trade-off point by minimizing this execution time bound. We discuss the time bound for executing the tasks of a processor from one communication to the next communication phase: each process computes $\lceil n/P \rceil$ tasks in each layer. Additionally, it computes tasks that would belong to the neighbor process in a pure block-wise schedule. In Fig. 3 (top, right) a pure block-wise schedule is defined by the vertical dotted lines; the light gray process computes three tasks belonging to its right (dark gray) neighbor in a pure block-wise schedule and vice versa. These redundantly computed tasks correspond to the black nodes and white nodes, respectively. Each process computes say $2k$ tasks redundantly in the first layer, k tasks of the left neighbor and k tasks of the right neighbor.

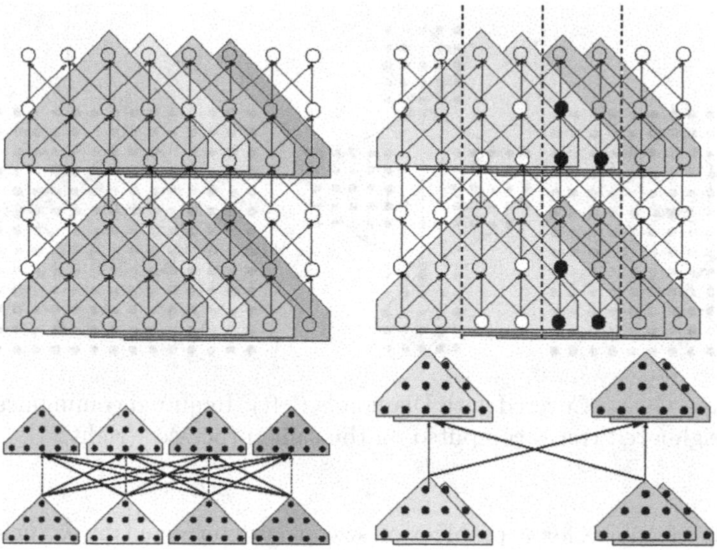

Figure 3. Task Graph Covered with Triangles (top,left). Linear schedule of the triangle processes (bottom, left). Neighbor processes are computed on the same processor (top, right). Final schedule of the task graph (bottom, right).

It is $k = 2$ in Fig. 3. It further computes $2k - 2$ tasks redundantly in the second layer, $2k - 4$ tasks redundantly in the third layer etc. The $(k + 1)$-st layer does not involve redundant computations. Hence, for the first $k+1$ layers, a processor computes $k^2 + k$ tasks redundantly and the computation time of the first $k + 1$ layers is $T_{comp}(n, k, P) \leq ((k+1)\lceil n/P \rceil + k^2 + k) \times c$ where c denotes the weight of one task. The same holds for the next $k+1$ layers and so on. After each $k+1$ layers, each process must communicate with its neighbors. Provided $\lceil n/P \rceil \leq k$, the results of $2(k+1)$ tasks are to communicate, where $k+1$ tasks are computed by the right and $k+1$ tasks by the left neighbor. The communication costs after each $k+1$ layers are $T^{sync}_{comm}(k) \leq 3o(k+1) + 2L(k+1) + \max(g(k+1), o(k+1))$ if synchronous communication is used (send to left neighbor, receive from right neighbor, send to right neighbor, receive form right neighbor). The communication costs after each $k + 1$ layers are

$$T^{async}_{comm}(k) \leq \max(o(k + 1) + L(k + 1), 2\max(g(k + 1), o(k + 1))) + \max(g(k + 1), o(k + 1)) + o(k + 1)$$

if asynchronous communication is used (send to left neighbor, send to right neighbor, receive from right neighbor, receive form left neighbor). Since $o(k+1) \leq \max(g(k + 1), o(k + 1))$ it holds

$$T^{async}_{comm}(k) \leq \max(L(k + 1), g(k + 1), o(k + 1)) + 3\max(g(k + 1), o(k + 1))$$

The computation of t layers is bounded by

$$T_{\text{redundant}}(n, k, t, P) \leq \lceil t/(k + 1) \rceil (T_{comp}(n, k, P) + T_{comm}(k))$$

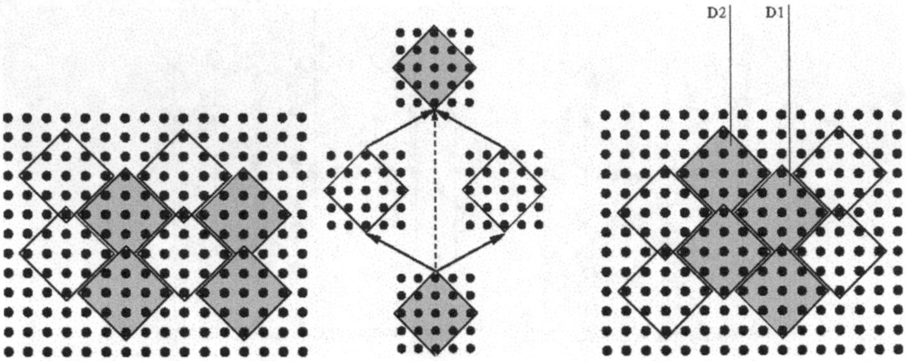

Figure 4. Task Graph Covered with Diamonds (left). Required communications (middle). Neighbored traces computed on the same processor (right).

The schedule of choice for a problem of size n is determined by k such that $T_{\text{redundant}}(n, k, t, P)$ is minimum. This is an easy computation since for all parallel machines that have been modeled as LogP machines the functions $L(x)$, $g(x)$, and $o(x)$ are linear in x. The next subsection discusses an alternative scheduling approach, the next section shows practical results for both approaches.

4.2 Diamond Scheduling

This approach avoids redundancy but still bundles small messages to a single larger messages. It covers the task graph with diamonds that follow communication edges, cf. Fig. 4. Each *trace* of overlapping diamonds defines a process, cf. Fig. 4. Neighbored traces are computed on the same processor if there are more traces than processors; this corresponds to a block-wise scheduling of the traces, cf. Fig. 4 (right).

Let k be the border length of the diamonds. Each diamond covers $2k^2 - 2k + 1$ tasks. Let $P \geq n/(2k-1)$, i.e. there are enough processors to compute the two neighbored traces. Consider a pair of neighbored diamonds $D1, D2$, cf. Fig. 4 (right). Then the diamond schedule proceeds as follows: (i) Receive the message containing the results of the predecessors at the south-east side of diamond $D1$. Altogether, $2k - 2$ results are required accounting with time $o(2k - 2)$. (ii) Compute all but the lowest tasks of diamond $D1$ in time $(2k^2 - 2k) \cdot c$. (iii) Send the results of the tasks at the north east side of $D1$. The message reaches the destination processor at time $o(2k - 2) + L(2k - 2)$. (iv) Receive the message containing the results of the south-west side of diamond $D2$ in time $o(2k - 2)$. (v) Compute the task of diamond $D2$ (except for the lowest) in time $(2k^2 - 2k) \cdot c$. (vi) Send the results of the tasks at the north-west side of $D2$. The message reaches the destination processor at time $o(2k - 2) + L(2k - 2)$. Hence, each pair of diamonds requires time $(4k^2 - 4k) \cdot c + 2o(2k - 2) + 4L(2k - 2)$. For simplicity, suppose that the number t of iterations performed is a multiple of $2k - 1$. Then there are $2k - 1$ such pairs being executed sequentially. The

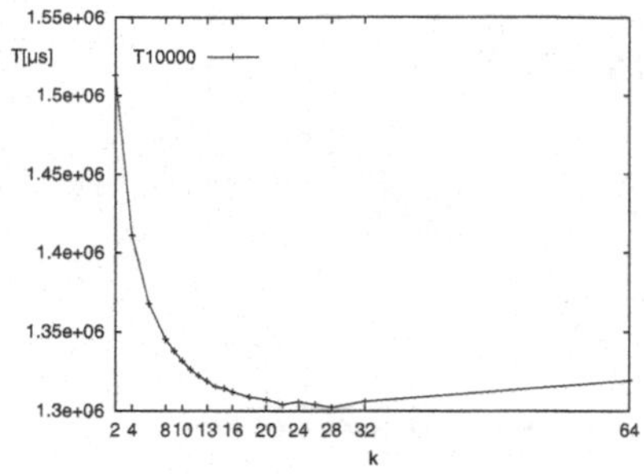

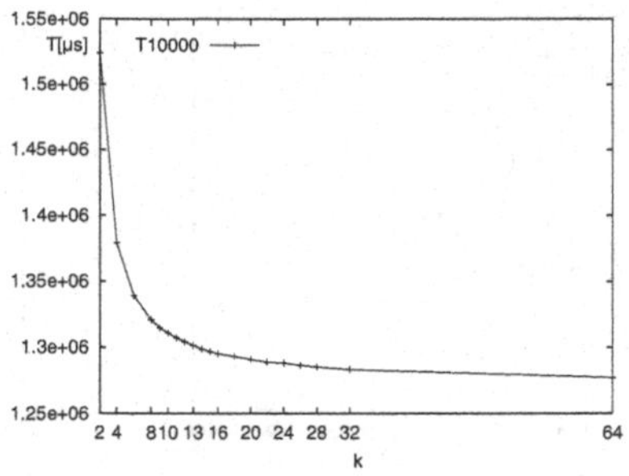

Figure 5. Estimated execution time for 10000 iterations of wave simulation with problem size 10000 on the RS/6000 SP using 8 processors. Redundant (left) and stripe scheduling (right).

lowest pair must account for the lowest task. Note, the last pair still must send messages. Therefore, the total time for the schedule is

$$T_{\text{diam}}(n,k,t) = c + ((4k^2 - 4k) \cdot c + 2o(2k-2) + 4L(2k-2)) \cdot t/(2k-1)$$

This formula can be used to determine an optimal side length of the diamonds.

We now consider the case that $P < n/(2k-1)$, i.e. not enough processors are available. For simplicity, suppose that $n/(2k-1)$ is an integral multiple of P. Then $n/((2k-1) \cdot P)$ neighbored pairs of diamonds, denoted by a *strip*, are computed on the same processor. The messages to be sent and received are only at the borders of such a strip. Hence, they have still the length $2k-2$. The total computation time of the tasks in a strip is $n/((2k-1) \cdot P) \cdot (2k^2-2) \cdot c$. With the same argument as above, we obtain the following execution time for the schedule:

$$T_{\text{strip}}(n,k,t,P) = \left(\frac{n \cdot (2k^2-2) \cdot c}{(2k-1) \cdot P} + 2L(2k-2) + 4o(2k-2)\right) \cdot \frac{t}{2k-1} + \frac{n \cdot c}{(2k-1) \cdot P}$$

The last term comes from the first strip where we have to account for the lowest tasks. This function is monotonously decreasing in k (assuming that L and o are functions linear in message size). Therefore, the execution time is minimized by choosing k as large as possible. Fig. 4.2 uses the measured values for the LogP parameters and determines the optimum values of k in redundant and stripe schedules.

5 Practical Results

Fig. 6 shows that the LogP model is adequate to give quite precise estimations of the programs' execution times.

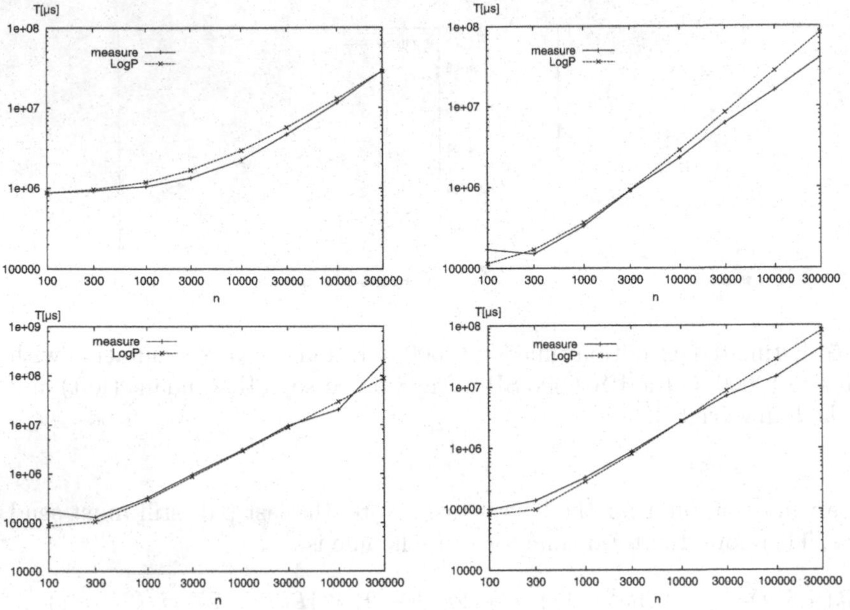

Figure 6. Times for 10000 iterations of wave simulation with $n = 10000, P = 8$ on the RS/6000 SP. Block scheduling (left, top), redundant scheduling (right, top) and diamond scheduling (bottom, left) and stripe scheduling (bottom, right).

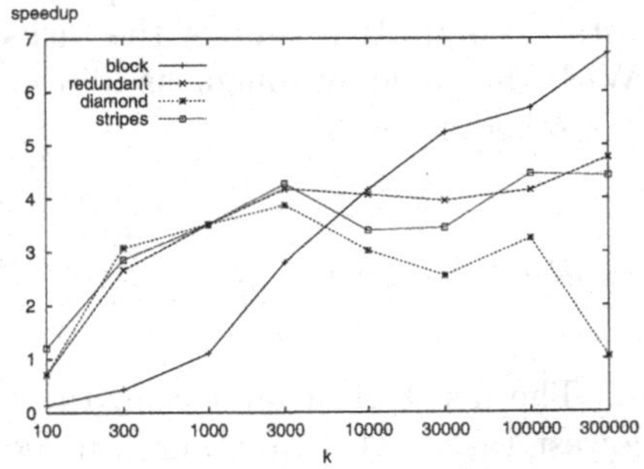

Figure 7. Measured speed-up for 10000 iterations of wave simulation with varying problem size n, RS/6000 SP 8 Processors.

Fig. 7 shows the measured run-times for different scheduling strategies discussed in the previous sections. We compare the block-wise schedule and the redundant schedule with two diamond schedules: "diamond" denotes the schedule where the diamond borders are maximum (as supposed to be optimum from our theoretical observations), "stripes" denotes a schedule with diamond border of 7. For large problems, the latter outperforms the former. This is because for diamond schedule the memory required increases in the length of the diamond sides. Since restricted memory and memory hierarchies are not captured by the LogP model, this behavior is not predicted. It is therefore a target of future work to refine the model.

Fig. 8 varies the number of processors while the problem size is kept quite small. It shows the efficiency of the diamond and block-wise schedules. With increasing problem size, the execution time is always dominated by computation.

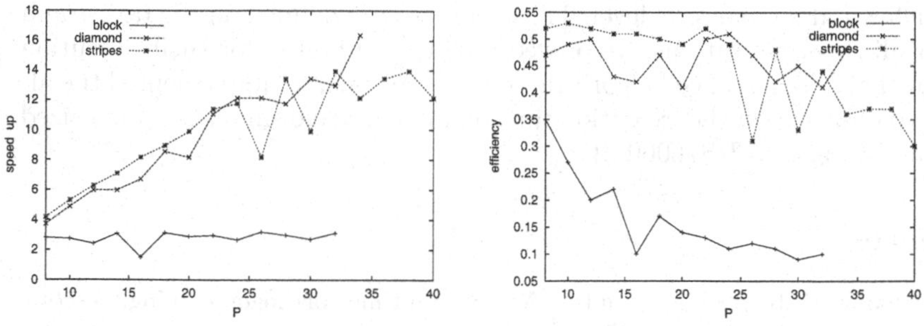

Figure 8. Speed-up and Efficiency for 10000 iterations of wave simulation with problem size 3000 on the RS/6000 SP over a range of processor numbers.

As our optimizations reduce the communication costs, their effects are decreasing as the problems grow. However, for smaller problem size, our optimization outperforms the block-wise schedules.

6 Related Work

Much research has been done in recent years on scheduling task graphs with communication delay (see e.g. [5,6,7,10,11]). Scheduling task graphs on the LogP-machine has been investigated for classes of graphs that arise in special applications like various broadcast problems, summation, or FFT [2]. There are some works investigating scheduling task graphs for the LogP-machine [15,8,12,9]. [12] shows that the computation of a schedule of length at most B is NP-complete even for fork and join trees and $o = g$. They discuss several approximation algorithms for fork and join trees. [8] discusses an optimal scheduling algorithm for some special cases of fork graphs.

Tiling is studied in the context of translation of data-parallel languages as well as in the parallelization context, e.g. [14,13,4]. These works assume source code information (from data flow analysis) to provide the essential data dependencies in the programs. As far as known to the authors, none of these works consider redundant computations for the optimization. The works perform tile shape optimization or tile size optimization (as the present paper). Most papers optimize for simplified machine models that did not prove their adequateness for real parallel machines. [1] uses the BSP as machine model. None of them consider LogP.

7 Conclusion

We demonstrated that the optimization of parallel programs is performed best if source code information is exploited for scheduling heuristics. We showed that loosing either information, the source code or the task graph, may lead to worse results. For the class of parallel algorithms that numerically compute linear partial differential equations, we presented three optimization algorithms: the former may be defined at source code level. The latter two are defined on the task graph level. For all three algorithms, we proved upper time bound for their execution times. For the more advanced algorithms on task graphs, we implemented the algorithms and measured the execution time of a linear wave simulation optimized with them for an IBM RS/6000 SP.

References

1. R. Andonov, S. Rajopadhye, and N. Yanev. Optimal orthogonal tiling. *Lecture Notes in Computer Science*, 1470, 1998.
2. D. Culler, R. Karp, D. Patterson, A. Sahay, K. E. Schauser, E. Santos, R. Subramonian, and T. von Eicken. LogP: Towards a realistic model of parallel computation. In *4th ACM SIGPLAN Symposium on Principles and Practice of Parallel Programming (PPOPP 93)*, pp 1–12, 1993. SIGPLAN Notices (28) 7.
3. B. Di Martino and G. Ianello. Parallelization of non-simultaneous iterative methods for systems of linear equations. In *Parallel Processing: CONPAR 94 – VAPP VI*, volume 854 of *Lecture Notes in Computer Science*, pp 253–264. Springer, 1994.
4. P. Feautrier. Some efficient solutions to the affine scheduling problem. I. one-dimensional time. *International Journal of Parallel Programming*, 21(5):313–347, October 1992.
5. A. Gerasoulis and T. Yang. On the granularity and clustering of directed acyclic task graphs. *IEEE Transactions on Parallel and Distributed Systems*, 4:686–701, June 1993.
6. J.A. Hoogreven, J.K. Lenstra, and B. Veltmann. Three, four, five, six or the complexity of scheduling with communication delays. *Operations Research Letters*, 16:129–137, 1994.
7. H. Jung, L. M. Kirousis, and P. Spirakis. Lower bounds and efficient algorithms for multiprocessor scheduling of directed acyclic graphs with communication delays. *Information and Computation*, 105:94–104, 1993.
8. I. Kort and D. Trystram. Scheduling fork graphs under logp with an unbounded number of processors. In *Europar '98*, pp 940–943, 1998.
9. W. Löwe and W. Zimmermann. Scheduling balanced task-graphs to logp-machines. *Parallel Computing*, 26(9):1083–1108, 2000.
10. C.H. Papadimitriou and M. Yannakakis. Towards an architecture-independent analysis of parallel algorithms. *SIAM Journal on Computing*, 19(2):322 – 328, 1990.
11. J. Siddhiwala and L.-F. Cha. Path-based task replication for scheduling with communication cost. In *Proceedings of the International Conference on Parallel Processing*, volume II, pp 186–190, 1995.
12. J. Verriet. Scheduling out-trees of height 1 in the LogP-Model *Parallel Computing*, 26(9):1065–1082, 2000.
13. M. Wolfe. More iteration space tiling. In ACM, editor, *Proceedings, Supercomputing '89: November 13–17, 1989, Reno, Nevada*, pp 655–664, New York, NY 10036, USA, 1989. ACM Press.
14. Michael Wolfe. Iteration space tiling for memory hierarchies. In Gary Rodrigue, editor, *Proceedings of the 3rd Conference on Parallel Processing for Scientific Computing*, pp 357–361, Philadelphia, PA, USA, December 1989. SIAM Publishers.
15. W. Zimmermann and W. Löwe. An approach to machine-independent parallel programming. In *Parallel Processing: CONPAR 94 – VAPP VI*, volume 854 of *Lecture Notes in Computer Science*, pages 277–288. Springer, 1994.

Event Manipulation for Nondeterministic Shared-Memory Programs

Dieter Kranzlmüller, Rene Kobler, and Jens Volkert

GUP Linz, Johannes Kepler University Linz
Altenbergerstr. 69, A-4040 Linz, Austria/Europe
kranzlmueller@gup.uni-linz.ac.at
http://www.gup.uni-linz.ac.at/

Abstract. Building powerful machines is only one part of high performance computing. Obviously, such supercomputers must be programmed efficiently to obtain the desired performance. This task isdifficult and time consuming due to huge amounts of data being processed and critical anomalies like deadlocks and race conditions.This paper focuses on race conditions in shared-memory programs, which are introduced due to nondeterministic behavior at synchronization or communication operations. Such programs may yield different results, even if the same input data is provided. This complicates testing anddebugging, where techniques for re-executingand controlling the nondeterminism of such programsare needed. Sucha sophisticated technique is event manipulation, whichallows to steer race conditions in parallel programs.While originally applied to message-passing programs, the latest event manipulation approach addresses OpenMP shared-memoryprograms. This paper describes the principal idea of shared-memory event manipulation and demonstrates its application for a simple mutual exclusion example.

1 Introduction

Parallel computing is the major enabling force for high performance computing. Multiple concurrently executing and communicating tasks on massively parallelarchitectures or in clusters of workstations offerperformance beyond the capabilities of sequentialarchitectures. Such parallel systems can be programmed in manydifferent ways according to the user's needs, theapplication's characteristics, or the underlyinghardware architecture. However, especially two programming paradigms share most of today's existingparallel applications, message-passing programmingand shared-memory programming.In message-passing programs, the parallel tasks communicate with each other via dedicated function-calls for sending and receiving messages. Inshared-memory programs, the parallel system offersa single address space, which is used for communication and synchronization. Both paradigms do already exist for a rather long time.Yet, the message-passing approach seems to be stillthe most widely used paradigm, even though it is oftenassumed to be more difficult for in-experienced userthan the shared-memory approach. This persistence may beattributed to the

B. Hertzberger et al. (Eds.): HPCN Europe 2001, LNCS 2110, pp. 283–292, 2001.

existence of standards like the Message-Passing Interface MPI or the Parallel Virtual Machine PVM. Another reason may be the dominance of distributed memory machines, which used to favorthe message-passing approach.Recently it seems, that the shared-memory approach mayfinally be on the rise for broad user acceptance.On the one hand, the finalization of the OpenMP industry standard may be able imitate the success of MPI. OpenMP is a vendor-independent parallel application programming interface (API)to support the implementation of shared-memoryprograms. Instead of being a new programming language,it works in conjunction with either standardFortran or C/C++. It is comprised of a set of compilerdirectives that describe the parallelism in the sourcecode, along with a supporting library of subroutinesavailable to applications.On the other hand, the scalability of shared-memorySMP architectures rises up to 32 processors or evenhigher, and many parallel architectures offerlarge single address spaces via distributed shared-memory.Furthermore, some of the world's biggest machines, like ASCI White [1],are composed from clusters of SMPs, thus combiningthe benefits of both distributed and shared-memory.The proposed approach for programmingASCI White is not pure MPI, but mixed programmingwith a combination of MPI and OpenMP. Based on these observations, we try to shift parts of our testing and debugging environment [6]from message-passing programs to the shared-memoryapproach. In this paper we describe the successfulapplication of event manipulation for shared-memoryOpenMP programs.Event manipulation is a unique technique that has beendeveloped for investigating nondeterminism in message-passing parallel programs. With event manipulation it is possible to evaluate different execution paths of a nondeterministic program for a given set of input data.The original event manipulation method operateson the order of messages at wild card receives.Yet, to apply event manipulation to OpenMP programs, we need to address shared-memory accesses and synchronization.This paper describes the initial results in this area.In the next section, we introduce the eventgraph model as a key formalism to describeparallel program behavior. Based on this model,we describe nondeterministic program behaviorand its effects observed during testing anddebugging in Section 3. Afterwards the theoretical basics ofour event manipulation method and the correspondingartificial replay mechanism for OpenMP programs are presented.Finally, the simplified example ofSection 5 demonstrates theapplication of event manipulation inpractice.

2 The Event Graph - A Parallel Program Model

Describing a parallel program's execution requires some kind ofmodel to express occurring states and their connections. Corresponding to the different needs of program developers and the various possibilities for analyzing a program'sexecution, many different models have already been proposed. One of these models, which is implicitly used in many existing program analysis tools, is the event graph. Formally, such a graph canbe defined as follows [8]:

[1] `http://www.llnl.gov/asci/platforms/white/`

Definition 1. *An event graph is a directed graph $G = (E, \rightarrow)$, where E is the non-empty set of events e_p^i of G, while$\rightarrow$ is a relation connecting events, such that$e_p^i \rightarrow e_q^j$ means, that there is an edge from event $e_p^i$to event e_q^j in G with the "tail" at event $e_p^i$and the "head" at event e_q^j.*

The events in such a graph are state changes occurring duringprogram execution. As a basic definition, we can utilize thefollowing notion [15]:

Definition 2. *An event e_p^i is defined as an action without durationthat takes place at a specific point in time i and changes thestate of a process/thread p.*

Such events are connected by temporal and causalrelations. Temporal relations are established by orderingevents corresponding to their occurrence time. Possible causal relations are expressed whenevents affect other events due to some control or data flow.A widely used formalism in this context isthe happened before relation, which can formally be definedas follows [9]:

Definition 3. *The "happened before" relation denoted as "$\rightarrow$" on aset of events in G is the smallest transitive, irreflexiverelation satisfying the following two conditions for arbitraryevents e_p^i and e_q^j:*

1. *If e_p^i and e_q^j are events on the same thread ($p = q$) and e_p^i occurs before e_q^j, then $e_p^i \rightarrow e_q^j$.*
2. *If e_p^i is the source of a communication or synchronization operation on thread p, and e_q^j is the corresponding target operation on thread q, then $e_p^i \rightarrow e_q^j$.*

With these three definitions, arbitrary parallel programs can bemodeled by defining the set of events and observing theirinstantiations and relations during execution. In [8]an overview of existing approaches for applying the event graph-model to message-passing programs, as well as concretestrategies for exploiting this model during debugging of massively parallel programs is given.In the scope of this work, we are interested in events occurring during execution of shared-memory programs. Corresponding to OpenMP, we can identify two main groups of events, those generatedby compiler directives, and those initiated by dedicated librarycalls. The former set includes parallel control structures likefork and join. The latter consists of operations for communicationand synchronization, like barrier and lock/unlock.In order to obtain such events, the target application needs to beinstrumented, and monitoring is performed during program execution.For obtaining the relations between observed events, differenttechniques have been proposed. Probably the best known mechanismsin this context are scalar logical clocks [9] andvector clocks [2]. The latter are needed for our investigations. After obtaining event data and associated ordering information,the event graph model can be constructed to serve as the basis forintended program analysis activities. In most cases, the event graphwill be visualized with some kind of space-time diagram, which displays the observed events and their relations concerning observation time and associated threads. Some examplespace-time diagrams are given in Section 5.

3 Nondeterminism in Shared-Memory Programs

An important feature of parallel programs is nondeterministic behavior, which describes the possibility of different program executions being observed, although the same input data is provided [8]. Simplified, a program is nondeterministic,if - for a given input - there may be situations where an arbitraryprogramming statement is succeeded by one of two or more follow-upstates. This is called a "race condition".The key characteristic of nondeterministic programs is the freedom ofchoice at particular program states, which may be determined by purechance or unawareness of the complete state of the computing environment. For shared-memory programs, we can identify a rich set of operationsthat permit nondeterminism. Basic operations are accessesto shared variables. We can define a race condition as the process that leads to an indeterminate value, whenever twoor more threads are allowed to simultaneously access a shared-variable [11].Such race conditions are called "data races", as describedin many different research papers [12,3,13].Since race conditions may occur on different levels ofsynchronization, from mutual exclusion toweaker synchronization such as Post/Wait, an additional group of "synchronization races"is often described. While data races are mostly the result ofimproper synchronization, synchronization races areneeded to allow for competition between threads, e.g.to enter a critical section, or to lock a semaphore [17].Data races are usually not intended by the programmer,and therefore represent errors in most cases. They are introduced bythe programmer accidentally or due to laziness.On the other hand, there are many applications for synchronization races, e.g. when modeling reality (Chaos Theory, Monte Carlo Methods) or improving performance (First-In-First-Out queues) [8].For this reason, we will initially stick to synchronizationraces, although we believe, that our method may be appliedto data races as well.In both cases, nondeterminism leads to difficulties duringthe testing and debugging phase, which can be summarizedwith the following three effects:

- The "irreproducibility effect": Nondeterministic programexecutions cannot be repeated at will, since subsequent executionsmay deliver completely different results. This limits the usageof cyclic debugging techniques [17].
- The "completeness problem": Nondeterministic program executionsoffer several results for one set of input data, and obtaining a complete set of results is difficult or impossible, even fornumerous repeated executions [7].
- The "probe effect": Nondeterministic program executionsare affected by observation mechanisms introduced for programanalysis. Monitoring a program affects the probability of choicesat nondeterministic events, thus yielding different results thanwithout monitoring [2].

In the worst case, two scenarios may be observed duringdebugging: On the one hand, existing errors may vanish by preventing certain erroneous computations from occurring.On the other hand, new errors may occur which would neveroccur in the original program due to low probability.Consequently, user's need special support from program analysistools to overcome these problems. Corresponding to the threeeffects described above, different methods have proposed.The irreproducibility effect is addressed by so-called "record&replay" techniques, which provide an equivalent execution of a nondeterministic parallel program basedon some previously observed execution. The solution ofrecord&replay techniques is therefore based on a two-stepapproach. During an initial record phase, the order ofcritical events is stored in trace files, so that thetrace files contain sufficient information to describeall the choices taken at nondeterministic events.These data are used as a constraint during subsequent replay-phases to enforce equivalent program execution. Well-known examples of record&replay mechanisms are described in [10,14,16].In contrast to the irreproducibility effect, thereare only few solutions concerning the completenessproblem. In fact, only the solutions described in[6,5] allow to computeevery possible execution of a nondeterministic program.Both solutions achieve this goal by combinatorial experimentation of possible choices at nondeterministicevents. However, both methods address only message-passingsystems, and there is no equivalent solution providedfor nondeterministic shared-memory programs.Corresponding to the probe effect, a wide variety ofsolutions have been proposed. Unfortunately, most ofthese solutions address only the delay of eventsas generated by the monitoring overhead.The problem of possible event re-ordering isusually ignored, comparable to the completenessproblem described above. In fact, as shown in [7], solving the completeness problemseems to be the only feasible solutions of comprehensively addressing the probe effect.

4 Event Manipulation Approach

In this section, we introduce the event manipulationapproach for nondeterministic shared-memory programs.The original event manipulation approach as implemented for message-passing parallel programs is one oftwo existing solutions for solving the completenessproblem [6]. By converting this technique to shared-memoryprograms, our event manipulation approach offersa unique approach for analyzing nondeterministicbehavior of shared-memory programs.To describe the event manipulation approach for shared-memory programs, we define an event graph as introduced in Section 2. For simplification, we use only a limited set of events,which introduces the problem, that only programssticking to the selected limited set of nondeterministicoperations will be useful for investigations. Programs containing other nondeterministic events may exhibit unknown consequences. However, we intend to implement more and more different event types in thefuture, which should finally lead to a completenondeterminism analysis technique.The events chosen for our initial shared-memoryevent manipulation approach are the lock and unlock operations of OpenMP. These operationsallow flexible implementation of mutual exclusion,because corresponding calls can be placed at arbitrary points in the code. Their definitionin OpenMP is as follows:

```
   void omp_set_lock(omp_lock_t
*semaphore);   void omp_unset_lock(omp_lock_t
*semaphore);
```

The `omp_set_lock` function waits until the simplelock `*semaphore` is available, which means that itis unlocked. This function blocks the calling thread until thespecified lock is available. Then the lock is set and ownershipis granted to the thread executing the function.Its counterpart, the `omp_unset_lock` function releases the simple lock if executed by a thread owning the lock.In addition to these two standard operations, thereare several other interesting operations for nondeterminismanalysis. One example is the `omp_test_lock` function, whichattempts to set a lock but does not block the executionof the thread. Upon return, `omp_test_lock` indicateswhether the lock was set or not. After selecting events of interest, the program is instrumentedand an initial execution is monitored. The generated data can thenbe used to construct an event graph or a visual representationof it, respectively. Based on this graph, the user can perform nondeterminism analysis with the following four steps:

1. Selection of an arbitrary nondeterministic event
2. Evaluation of race candidates corresponding to the selected event
3. Event manipulation by exchanging the selected event with another candidate
4. Artificial replay to enforce the manipulated execution

Step 1 represents the selection of an event e_p^i, for which thefollowing question should be explored: What would have happened, if the nondeterministic choices would have been different from what has been observed? The possible events are all the lock operationsoccurring during the program's execution, because only their instantiationmay be affected by race conditions. During step 2, all possible choices for the selected event e_p^i areevaluated. This set is called the *race_candidates*, which can bedefined as follows:

$$race\_candidates(e_p^i) = \{e_p^i\} \cup \{\min(e_q^j) \mid \neg(e_q^j \rightarrow e_p^i) \forall 0 \leq q < n, p \neq q\} \quad (1)$$

In this definition, the term $\min(e_q^j)$ revealsthe first event on thread q, that satisfies the subsequentcondition. This assumes implicitly, that lock operations are non-overtaking, which is true for OpenMP because a lock operation is blocking until a semaphore is acquired. Additionally, the number of elements in one set of*race_candidates* is always lower equal the numberof threads n, because only one lock operation per threadcan be included.The term $\neg(e_q^j \rightarrow e_p^i)$ describes,that only those events can be considered a race candidate,that did not occur before, which satisfies the followingcondition:

$$\neg(e_q^j \rightarrow e_p^i) \iff e_p^i \rightarrow e_q^j \vee e_p^i \parallel e_q^j \quad (2)$$

Please note, that the set of *race_candidates* contains onlylock operations for one dedicated semaphore s. It is assumed,that there are no lock operations e_q^k for other semaphores $\bar{s}$ before e_q^j, for which the following condition holds:

$$e_p^i \rightarrow e_q^k \wedge e_q^k \rightarrow e_q^j \quad (3)$$

If lock operations concerning another semaphore $\bar{s}$ according to this condition do exist, the corresponding event e_q^j on thread q has to be removed from the set of *race_candidates* due to its dependencyon semaphore $\bar{s}$.After identifying the set of *race_candidates*, the user perform step 3 - event manipulation - by choosing one element, which shouldoccur at the nondeterministic choice instead of the observedevent. This defines a point-of-exchange (*poe*) as follows:

$$poe = \{e_p^i, e_q^j \in race\_candidates\} \tag{4}$$

The consequences of the event manipulation are evaluated during step 4, where the program is re-executed under control of a sophisticated record&replay mechanism. This task is called artificial replay, and consists of three distinct phases:

a Before the *poe*, the program is executed as previously observed.
b At the *poe*, the exchange of events is enforced.
c After the *poe*, the program is executed without control, because it is unknown, how the event manipulation affects the program's behavior.

A basic necessity in this case is to identify a border, where therecord&replay mechanism switches from the previously observedexecution to the uncontrolled execution. This border is calleda *cut*, which is constructed based on the *poe* as the following set of events:

$$cut(e_p^i) = poe \cup \{\min(e_r^k) \mid e_p^i \rightarrow e_r^k \forall 0 \leq r < n, r \neq p \neq q\} \tag{5}$$

After artificial replay is completed, the user can identify theconsequences of the event manipulation. This is usually basedon comparing the results of the previously observed executionwith the results of the artificial replay. In addition, theartificial replay may reveal errors in the program, which were hidden due to nondeterminism. An concrete examplefor the event manipulation approach is described in thenext section.

5 Example

As a practical example for application of the event manipulationapproach, a very simple source code of a nondeterministic OpenMPprogram is chosen. This program contains only nondeterministiclock/unlock operations, which ensure mutual exclusion for a given`shared_variable`.

Example 1. Simple nondeterministic OpenMP program

```
#include <omp.h>    ...
int shared_variable;
omp_lock_t semaphore;   ...
#pragma omp parallel
{       omp_set_lock(&semaphore);
        operation(&shared_variable);
        omp_unset_lock(&semaphore);
}
```

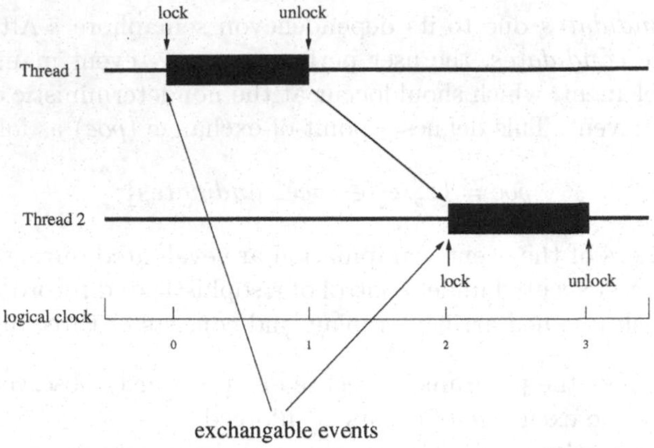

Fig. 1. Observed Program Execution of Two OpenMP Threads.

An example execution of this program with two OpenMP threadsis shown as a space-time diagram in Figure 1.The `lock` operation onthread 1 is performed before the `lock` operation on thread 2, which indicates that thread 1 can modifythe `shared_variable` before thread 2.If the result of this simple program is based on theorder of the operations on the `shared_variable`,the order of the lock operations may change the observed program behavior.Due to the nondeterministic behavior of this program, thefollowing situations may emerge:If the user observes the execution as shown in Figure 1, ...

- ... and this execution reveals incorrect results, debugging is initiated. However, during debugging it cannot be assumed, that the same execution is observed, and thus the error may be irreproducible.
- ... and this execution delivers correct results, and the user assumes that this result is the only possible result, any further testing with the same set of input data may be neglected.
- ... and this execution delivers correct results, and the user knows that there may be other results for the same set of input data, computing these other results may be impossible even for large numbers of tests.

All these situations can be solved with the event manipulationapproach described in this paper. After performing the foursteps introduced in Section 4, the artificialre-execution of the program delivers the execution as shownin Figure 2. As displayed, the order of the lockoperations is reversed. The results of this program run representthe effect of event ordering for this particular set of nondeterministic events.

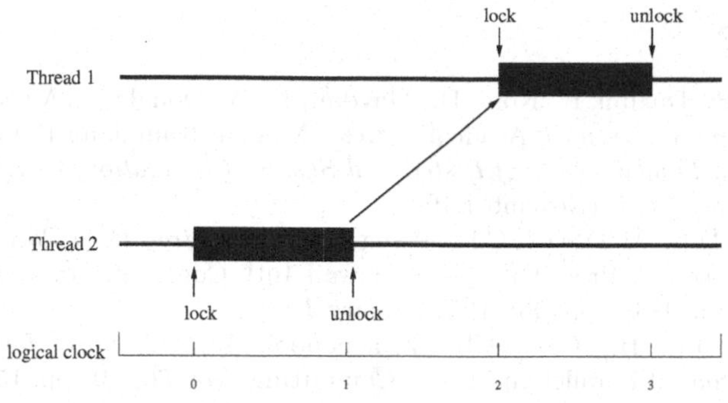

Fig. 2. Program Execution after Event Manipulation.

6 Conclusions and Future Work

Parallel programs are often nondeterministic due to applicationrequirements or performance improvements. Unfortunately, thisintroduces many problems during testing and debugging, thus stimulating many tool developments in this area. A technique for analyzing nondeterministic behavior is event manipulation. In this paper, we have shownits latest revision, which allows its application to shared-memory programs.The current implementation of shared-memory event manipulationrepresents only the first prototype, which vindicates its currentlimitations to lock/unlock operations. At present, we are tryingto integrate other nondeterministic events, especially nonblockinglocks as described in Section 4. This increasesthe number of race conditions, but also the target applicationsfor our approach.Another goal of our future work in this project is automatizationof the event manipulation technique. In [8], automaticnondeterminism analysis for message-passing programs is described.In terms of shared-memory programs, this technique can be implementedif every member of the set of *racing_messages* is automaticallytested with event manipulation. The biggest problem in this contextis certainly the number of possible executions, where correspondingsolutions do not exist at present.

Acknowledgements

Contributions to this work have been made by many people, most notable Michiel Ronsse, ELIS Dept., University Ghent, Belgium, who providedhis knowledge about nondeterminism to bridge the gap between message-passing programs and shared-memory programs.

References

1. Chandra, R., Dagum, L., Kohr, D., Maydan, D., McDonald, J., Menon, R.,*Parallel Programming in OpenMP*,Academic Press, Morgan Kaufmann Publishers (2001).
2. Fidge, C.J., *Fundamentals of Distributed System Observation*,IEEE Software, Vol. 13, No. 6, pp. 77-83 (November 1996).
3. Helmbold, D.P., McDowell, C.E., Wang, J.-Z., *Detecting Data Races by Analyzing Sequential Traces*, Proc. HICCS-24, Hawaii Intl. Conference on System Sciences, Vol.2, Hawaii, USA, pp. 408-417 (January 1991).
4. Hu, Y.C., Lu, H., Cox, A.L., Zwaenepoel, W., *OpenMP for Networks of SMPs*,Journal of Parallel and Distr. Computing, Vol. 60(12), pp. 1512-1530 (Dec. 2000).
5. Kacsuk, P., *Systematic Macrostep Debugging of Message Passing Parallel Programs*, Future Generation Computer Systems, North-Holland, Vol. 16, No. 6, pp. 597-607 (April 2000).
6. Kranzlmüller, D., Grabner, S., Volkert, J.,*Debugging with the MAD Environment*, Parallel Computing, Vol. 23, No. 1–2, pp. 199–217 (Apr. 1997).
7. Kranzlmüller, D., Schaubschläger, Ch., Volkert, J.,*The Completeness Problem of Parallel Program Testing*,Proc. SCI 2000, Orlando, Florida, USA, Vol. VII, pp. 435-440 (July 2000).
8. Kranzlmüller, D.,*Event Graph Analysis for Debugging Massively Parallel Programs*,PhD Thesis, GUP Linz, Joh. Kepler University Linz (September 2000). `http://www.gup.uni-linz.ac.at/~dk/thesis`
9. Lamport, L., *Time, Clocks, and the Ordering of Events in a Distributed System*,Communications of the ACM, pp. 558 - 565 (July 1978).
10. LeBlanc, T.J., Mellor-Crummey, J.M., *Debugging Parallel Programs with Instant Replay*,IEEE Trans. on Computers, Vol. C-36, No. 4, pp. 471-481 (April 1987).
11. Lewis, T.G., *Foundations of Parallel Programming - A Machine-Independent Approach*, IEEE Computer Society Press, Los Alamitos, CA, USA (1993).
12. Miller, B.P., Choi, J.-D., *A Mechanism for Efficient Debugging of Parallel Programs*,Proc. PLDI'88, ACM SIGPLAN'88 Conference on Programming Language Design and Implementation, Atlanta, Georgia, pp. 135-153 (June 1988).
13. Netzer, R.H.B., Miller, B.P., *What are Race Conditions? - Some Issues and Formalizations*, ACM Letters on Programming Languages and Systems, Vol. 1, No. 1, pp. 74-88 (March 1992).
14. Netzer, R.H.B., *Optimal Tracing and Replay for Debugging Shared-Memory Parallel Programs*, Proc. of the 3rd ACM/ONR Workshop on Parallel and Distributed Debugging, San Diego, CA, USA (May 1993).
15. van Rick, M., Tourancheau, B., *The Design of the General Parallel Monitoring System*,Programming Environments for Parallel Computing, IFIP, North Holland, pp. 127-137 (1992).
16. Ronsse, M.A., Levrouw, L.J., *On the Implementation of a Replay Mechanism*, Proc. EuroPar '96, LNCS, Vol. 1123, Lyon, France, pp. 70-73 (August 1996).
17. Ronsse, M.A., De Bosschere, K., Chassin de Kergommeaux, J., *Execution Replay and Debugging*, Proc. AADEBUG 2000, 4th Intl. Workshop on Automated Debugging, Munich, Germany, pp. 5-18 (August 2000).

An Open Distributed Shared Memory System[1]

G. Manis, L. Lymberopoulos, N. Koziris, and G. Papakonstantinou

National Technical University of Athens
Department of Electrical and Computer Engineering
Zografou Campus, Zografou 15773
{manis,llymber,nkoziris,papakon}@cslab.ece.ntua.gr

Abstract. In this paper we present a set of Distributed Shared Memory (DSM) tools. These tools follow the philosophy of Open Implementation. They do not include built-in policies for management of shared data, while the task of definition and implementation of the basic design choices for the manipulation of shared data is shifted from the DSM system to the programmer. One of the main reasons that DSM systems are not widely used today in the development of parallel applications is the fact that no existing DSM system meets the requirements of a wide variety of applications, usually due to inflexible built-in policies. The system described here provides the programmer with a set of tools enabling the customization of the DSM system, according to the needs of any specific application. This set of tools has resulted from the decomposition of existing DSM systems into primitive functions.

1 Introduction

Distributed shared memory, first introduced in 1986 by Li [1,2], combines the advantages of shared memory with those of distributed memory systems. Unfortunately, it combines some of their disadvantages as well, introducing new problems and subsequent fields of research. Approximately one and a half decade later, distributed shared memory is included in every relative textbook and is widely accepted by the scientific and academic communities. Regardless, companies do not seem ready to share this conviction, and rarely invest money in the development of commercial software to support distributed memory, or in the development of applications based on distributed shared memory. The question, then, seems to be 'who is more intuitive'; the companies or the academic community?

Distributed shared memory, as it was proposed in 1986 by Li, was a single paged virtual address space over a LAN with local and remote pages. When a page that was not in the local memory was requested, an operating system fault was activated and the page was transferred from a remote processor. New approaches appeared later, in which only a part of the whole address space was shared, or even a single data object (shared variable), separating distributed memory system to page based (e.g. Mirage [3]) and data object based (e.g. Orca [4,5,6]) systems. In this paper we are interested in shared object based distributed shared memory systems.

---

[1] This work is partially funded by the Greek Secretariat of Research and Technology.

B. Hertzberger et al. (Eds.) : HPCN Europe 2001, LNCS 2110, pp. 293-301, 2001.

In the shared data object based systems, the unit of the shared data is the variable. Shared variables can be accessed by every process participating in the DSM system in a way that is defined by a specific policy that differs from system to system. Part of this policy is the consistency model that is supported by the system.

The most important consistency models as described by Tanenbaum in [7], are the strict consistency model, the sequential consistency model, the causal consistency model and the processor consistency model. Weak, release and entry consistency are also presented in this book. In strict consistency, "any read to a memory location x returns the value stored by the most recent write operation to x". In sequential consistency, "the result of any execution is the same as if the operations of all processors were executed in some sequential order, and the operations of each individual processor appear in this sequence in the order specified by its program". In causal consistency, "writes that are potentially causally related must be seen by all processes in the same order. Concurrent writes may be seen in a different order in different machines". Finally, processor consistency defines that "writes done by a single process are received by all other processes in the order in which they were issued, but writes from different processes may be seen in a different order by different processes". One more question which rises from the above is why so many different consistency models exist.

Apart from the consistency model selected for the DSM system, there are several other design choices which affect overall performance. Issues like granularity (the size of unit of sharing), data location (fixed or dynamic ownership, centralized or distributed scheme), write synchronization policies, etc, are very significant design choices. Since, however, there are no unanimous solutions to decisions related to these issues, the result is that each DSM system must propose something new in order to meet the needs of a specific application. This has led to the development of a variety of distributed shared memory systems.

These systems may be examined based on the type of shared variables they support and the way that these variables are managed. Orca is an object-based system designed for the development of portable parallel applications. Orca is built on top of the Amoeba [8,9] distributed operating system and consists of a language, a compiler and a run-time system. Data objects in Orca can be single copy or replicated. Replicas are located in every processor. Read operations are performed locally, if a local copy exists. Write updates are performed using an update message when the variable is replicated, ensuring a satisfactory level of consistency. The choice of whether a shared object will be single-copy or replicated is up to the compiler, which collects the necessary information to form this decision from the source code. Also, the state of a variable can change from single copy to replicated data and vice versa at run-time. Munin [10,11] is based on release consistency, which is ensured by the use of synchronization variables. "Acquire" and "Release" mechanisms are used to ensure that shared variables are consistent when a critical section is exited. Munin also supports read-only, migratory, write-shared and conventional variables. Unlike Munin, Midway [12] supports entry consistency, in which each shared variable should be associated with a synchronization variable, while consistency is ensured at the entry point of critical sections. In both systems, the programmer is responsible for the activation of the synchronization mechanisms. Other existing DSM systems include Linda [13], Mirage, Clouds [14,15], Mether [16], etc, each one being very different from the other. The last question one may ask is why no DSM system has been accepted as a standard yet, when so many systems have been proposed.

We cannot answer the questions stated above, but they all lead to the conclusion that the field is still open to much research. The approach we introduce in this paper is closer to that OpenTS [17] and of the Munin system than to any of the other mentioned. OpenTS provides a primary interface inspired in the tuple space access primitives and a separate interface that allows the user to adjust the implementation strategies decisions. Munin provides a kind of application-specific consistency by giving the programmer various shared objects to select from, according to the specific application that will be developed. The idea of not providing standard policies for the management of DSM objects, but rather giving the programmer the flexibility to customize the DSM system according to the requirement of the specific application, is the main philosophy behind our proposed set of tools.

In the following, the basic theory and design choices of a new DSM system will be outlined, implementation issues will be described and experimental results will be presented. Some conclusions and plans for future work will also be discussed.

2 System Overview

One of the conclusions reached from the above discussion is that existing DSM systems did not succeed in becoming widely accepted as parallel programming tools. A new system consisting of tools for building customized DSM models according to the needs of specific applications seems an alternative solution to some of the issues stated earlier. The main idea is simple. Rather than implementing a DSM system with built-in fixed policies fitting well to the requirements of a (possibly large) group of applications, but not to the requirements of any application, we move the task and the responsibility of designing, implementing and managing the DSM system to the application programmer. In our layout, the programmer can choose from a set of primitive functions structured in an hierarchical way, according to the management policy that intends to follow.

The set of the proposed tools results from a through examination and assessment of existing DSM systems. Policies, algorithms and type of data shared in existing DSM systems were studied and decomposed into simpler, primitive functions. The common denominator of all these function formed the basis of the proposed set of tools. This set of tools has been designed in a hierarchical way: simple primitive functions are at the bottom of the pyramid, while more complicated functions, which in turn use the lower primitive ones, are at the higher levels. The programmer can choose any function he/she needs from any level of the hierarchy tree.

This system is completely different from the DSM systems presented or mentioned earlier. Munin is the only one whose theory is close to our system, since it supports an application-specific type of consistency. The difference is that Munin concentrates on application-specific data objects rather than application-specific tools in general. Consistency in Munin is ensured by mechanisms provided especially for this reason and which support only release consistency. Both in Munin and in our system, the programmer is responsible for the activation of the consistency synchronization mechanisms, with the addition that in our system the programmer has the option to customize those mechanisms themselves.

System calls supported by this solution are listed below, as well as a short description of each one of them. Calls are assorted in groups of similar operations.

System initialization:
- DSM_Init: initialize the DSM system, starts the necessary daemon processes in every processor
- DSM_Register: called by each process intending to share data through the DSM system

Ownership and Migration:
- DSM_CreateOwner: a new shared variable is created in the local server
- DSM_AreYouTheOwner: ask a DSM server if this server is the owner of the specified variable
- DSM_BroadcastForOwner: sends a broadcast message looking for the owner of the specified variable
- DSM_GetOwnerFromLocalServer: asks the local DSM server about the owner of a shared variable
- DSM_GetOwner: asks the specified DSM server about the owner of a shared variable
- DSM_SetOwner: informs the specified DSM server about the owner of a shared variable
- DSM_UpdateOwner: the owner of a shared variable is updated with a new value for the shared variable
- DSM_GetPossibleOwner: indicates a possible owner for the variable. Some migration models use this option
- DSM_SetPossibleOwner: sets a possible owner for a shared variable

Read-Write operations:
- DSM_ReadValue: reads from a specified server the value of a shared variable stored in this server
- DSM_WriteValue: a new value of the shared variable is stored in the specified server

The replica related operations:
- DSM_CreateReplica: a replica of a shared variable is created in a server
- DSM_DeclareReplicaToOwner: the owner is informed for the creation of a new replica
- DSM_ReadReplica: returns the value of a local replica of the shared variable
- DSM_GetReplicasVector: returns the vector containing the location of all replicas of a shared variable
- DSM_SetReplicasVector: a replicas vector with the location of all replicas is stored in the specified DSM server
- DSM_UpdateLocalReplica: a new value is stored in the local replica of the shared variable

- DSM_SetClosestReplica: informs the DSM server where the closest replica of the variable is located. Some models use this options
- DSM_GetCloserReplica: returns the location of the closest replica of the variable
- DSM_InvalidateLocalValue: invalidates the local replica of the specified shared variable
- DSM_ValidateLocalValue: sets the local replica as a valid copy of the shared variable
- DSM_IsReplicaValid: checks if the local copy of the variable has a valid value
- DSM_InvalidateAllReplicas: invalidates all replicas
- DSM_InvalidateAllReplicasAndReceiveAck: invalidates all replicas and wait for acknowledgements to ensure that there is no valid copy of the replica left.
- DSM_UpdateOwnerAndAllReplicas: updates the owner of a variable and all replicas for the new value of a shared variable

Synchronization mechanisms operations:

- DSM_OpenGlobalSemaphore: initialization of a global semaphore
- DSM_WaitGlobalSemaphore: The P operation on a global semaphore
- DSM_SignalGlobalSemaphore: The V operation on a global semaphore
- DSM_CloseGlobalSemaphore: release the global semaphore
- DSM_CreateGlobalBarrier: initialization of a global barrier
- DSM_Barrier: a process has reached the meeting point

Using the above mechanisms, it becomes relatively simple to implement various DSM models. As an example, as well as for debugging and evaluation purposes, some DSM models have been implemented with this system. Below, we will briefly outline the pseudocode of the implementation of three typical DSM models.

The first one is a DSM model similar to the one supported by Orca: after a write operation, replicas are updated by a simple multicast message (write-update model). In this model the definition of a shared variable is as follows:

```
{
DSM_CreateOwner
}
```

The read operation:

```
{
DSM_ReadValue
if a replica is not stored in the local server
   {
   DSM_BroadcastForOwner
   DSM_CreateReplica
   DSM_ReadValue
   }
}
```

The write operation:

```
{
DSM_BroadcastForOwner
```

```
DSM_UpdateOwnerAndAllReplicas
}
```

The second is a stricter model. After a write operation, the replicas are invalidated. The declaration of the shared variables and the read operation is the same with those described in the above paradigm. The pseudocode for the write operation follows:

```
{
DSM_BroadcastForOwner
DSM_InvalidateAllReplicasAndReceiveAck
DSM_UpdateOwnerAndAllReplicas
}
```

Migration can be implemented as follows:

```
{
DSM_SetOwner
DSM_GetReplicasVector
DSM_SetReplicasVector
}
```

Some of the previous definitions lack a high level of consistency. Consistency is not always essential, but if considered necessary it can be achieved using the global semaphore mechanisms. For example the migration operations can be locked as shown, using a semaphore initialized to zero.

```
{
Wait (S)
DSM_SetOwner
DSM_GetReplicasVector
DSM_SetReplicasVector
Signal (S)
}
```

3 Experimental Results

In its first implementation, the proposed Open DSM system was built on top of the PVM platform. Apart from this implementation, the proposed set of tools will be further implemented in different environments. For evaluation and debugging reasons, several DSM systems were implemented using the tools provided and toy scale applications were developed on top of them. Experimental results have been collected in a cluster of 16 Linux PCs connected with in a FastEthernet LAN, isolated from any external traffic. In the following sub-sections, we will present two examples that prove the flexibility of our system to construct different DSM platforms for the execution of parallel applications.

3.1 Use of the Open DSM System to Resolve the ASP Problem

The all-pair shortest-paths (ASP) problem demands the calculation of the length of the shortest-path between any two nodes of a given weighted graph.

Using the deployed API, we implemented three different DSM platforms to run the parallel application. In the first system, System A, all Read – operations on the shared variables are all forwarded to the processor that is the single owner of this variable. The synchronisation is achieved using barriers between the parallel processes. Both of the other two systems implement the WRITE_UPDATE protocol, where the processor that possesses the variable updates after a Write - operation all the replicas that are held by the other processors of the system. The difference between these two DSM systems is that the first one (System B) uses barriers to achieve synchronisation, whereas the other (System C) uses the deployed global semaphores.of the Open DSM system to resolve the ASP problem

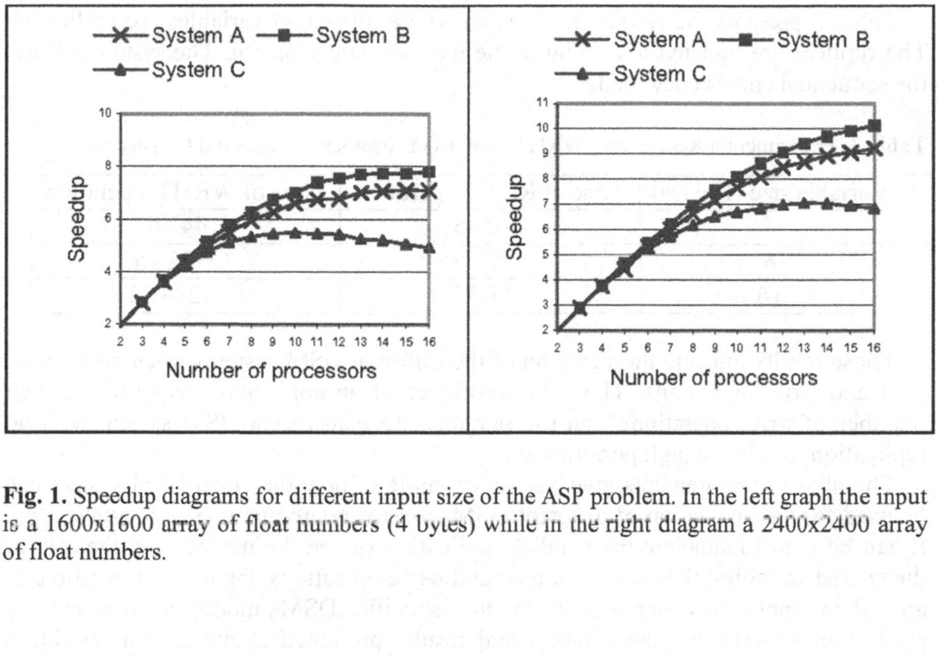

Fig. 1. Speedup diagrams for different input size of the ASP problem. In the left graph the input is a 1600x1600 array of float numbers (4 bytes) while in the right diagram a 2400x2400 array of float numbers.

3.2 Performance of Read – Write Operations in Different DSM Systems Constructed from the Open DSM System

In this test, we implemented some DSM systems with different policies in the management of the shared data. We collected experimental results for read and write operations on shared variables. We used all (16) processors of our Linux cluster. All times are in milliseconds.

Table 1 presents the results in a DSM system where all read and write operations are forwarded to the owner of the shared variable (READ_ONLY protocol).

Table 1. Experimental Results from READ_ONLY Protocol.

Variable size (Kbytes)	READ operation (msec)	WRITE operation (msec)
4	9.06	11.04
8	14.94	19.19
16	25.62	34.00

Table 2 presents the results obtained from a deployed DSM system where all processors keep a replica of the shared variable. The replicas are updated according to the write-update protocol. The system follows the processor consistency model.

Table 2. Experimental Result from WRITE_UPDATE Protocol. Processor Consistency.

Variable size (Kbytes)	Time of READ operation	Time of WRITE operation
4	0.47	53.27
8	1.47	81.79
16	1.79	140.03

Table 3 presents the results in a system where all shared variables are replicated. The replicas are updated according to the write-update protocol. The system follows the sequential consistency model.

Table 3. Experimental Result from WRITE_UPDATE Protocol. Sequential Consistency.

Variable size (Kbytes)	Time of READ operation	Time of WRITE operation
4	0.46	447.60
8	1.46	722.25
16	1.79	1294.32

These results indicate the variation of the different DSM systems when performing read and write operations. Thus, the developer of an application requesting a high mumber of write operations, can use our tools to construct a DSM system with no replication to achieve high performance.

The above measurements give also an example of how the Open DSM system can be used for the evaluation of different DSM systems related to a specific application. It can be used to simulate the parallel application (given the number and the size of the shared variables, the number of read and write operations, the read-write ratio etc) and then apply this application to the specific DSM model to measure its performance. Note that the experimental results presented above do not provide a comparison between the systems constructed with our tools and other stand-alone systems, as the purpose of this first implementation was to investigate the flexibility of our system to deploy relatively easily various DSM platforms.

4 Conclusions and Future Work

In this paper, we present a set of tools, which can be used by the programmer to implement a DSM system according to the need of a specific application. This set of tools has arisen from the decomposition of existing DSM systems into primitive functions. In its first implementation, our system was built on top of the PVM platform, which was extended to provide a synchronization mechanism based on global semaphores. Further deployment has to be done in order to improve the efficiency of these tools, so that the overall performance of the proposed system be increased.

References

1. K. Li and P. Hudak. "Memory Coherence in Shared Virtual Memory Systems". Proceedings of 5th Annual ACM Symposium on Principles of Distributed Computing, pp. 229-239, 1986
2. K. Li and P. Hudak. "Memory Coherence in Shared Virtual Memory Systems". ACM Transactions on Computer Systems, 7(4) pp. 321-359, 1989
3. B. D. Fleisch and G. J. Popek, "Mirage: A Coherent Distributed Shared Memory Design". Proceedings of the 12th ACM Symposium on Operating Systems Principles, pp. 211-223, 1989
4. H. E. Bal, R. Bhoedjang, R. Hofman, C. Jacobs, K. Langendoen and T. Ruhl. "Performance Evaluation of the Orca Shared Object System". ACM Transactions on Computer Systems, 16(1), pp. 1-40, 1998
5. H. E. Bal, M. F. Kaashoek and A. S. Tanenbaum. "A Distributed Implementation of the Shared Data-Object Model". Distributed and Multiprocessor Systems Workshop, pp 1-19, 1989
6. A. S. Tanenbaum and H. E. Bal. "Programming a Distributed System Using Shared Objects". Proceedings of the Second International Symposium on High Performance Distributed Computing, pp. 5-12, 1990
7. A. S. Tanenbaum. "Distributed Operating Systems". Prentice Hall, 1995
8. A. S. Tanenbaum, R. van Renesse, H. Van Staveren, G. J. Sharp, S. J. Mullender, J. Jansen, and G. van Rossum. "Experiences with the AMOEBA Distributed Operating System". Communications of the ACM, vol. 33, no. 12, 1990
9. S. J. Mullender and A. S. Tanenbaum. "The Design of a Capability-based Distributed Operating System". Computer Journal, vol. 29, no 4. pp. 289-300, 1985
10. J. K. Bennet, J. B. Carter and W. Zwaenepoel, "Munin: Distributed Shared Memory Based on Type-Specific Memory Coherence". Proceedings of the 2nd ACM SIGPLAN Symposium on principles and Practice of Parallel Programming, pp. 168-176, 1990
11. J. B. Carter. J. K. Bennet and W. Zwaenepoel, "Implementation and Performance of Munin". Proceedings of the 13th ACM Symposium on Operating Systems Principles, pp. 152-164, 1991. ACM Operating Systems Review, Special Issue 25(5)
12. B. N. Bershad, M. J. Zekauskas and W. A. Sawdon. "The Midway Distributed Shared Memory System". Technical Report CMU-CS-93-119, School of Computer Science, Carnegie Mellon University, 1993
13. L. S. Ahuja et al. "Linda and Friends". IEEE Computer, pp. 26-34. 1986
14. J. M. Bernabeu-Auban, P.H. Hutto, M. Yousef, A. Khalidi, M. Ahamad, W. F. Appelbe, P. Dasgupta, R. J. LeBlanc and U. Ramachandran. "Clouds – A Distributed, Object-Based Operating System Architecture and Kernel Implementation". Proceedings of the EUUG autumn Conference, pp. 25-37, 1988
15. U. Ramachandran, M. Yousef and A. Khalidi, "An Implementation of Distributed Shared Memory". Distributed and Multiprocessor Systems Workshop, pp.21-38, 1998
16. R. G. Minnich and D. J. Farber. "The Mether System : Distributed Shared Memory for SUN OS 4.0". USENIX, pp. 51-68, 1989
17. J. Carreira, G. Silva, K. Langendoen, "Efficient and Portable Parallel Programming: An Open Distributed Shared Memory Implementation". PDCS'97, 1997

Two Layers Distributed Shared Memory

F. Baiardi[1], D. Guerri[2], P. Mori[1], L. Moroni[1], and L. Ricci[1]

[1] Dipartimento di Informatica, Università di Pisa
Corso Italia 40, 56125 - Pisa (Italy)
[2] Synapsis S.r.l., P.zza Dante 19/20, 57124 - Livorno (Italy)
{baiardi,guerri,mori,ricci}@di.unipi.it

Abstract. This paper presents a methodology to design a distributed shared memory by decomposing it into two layers. An application independent layer supplies the basic functionalities to access shared structures and optimizes these functionalities according to the underlying architecture. On top of this layer, that can be seen as an application independent run time support, an application dependent layer defines the most suitable consistency model for the considered class of applications and it implements the most appropriate caching and prefetching strategies for the consistency model. To exemplify this methodology, we introduce DVSA, a package that implements the application independent layer and SHOB, an example of the second layer. SHOB defines a release consistency model for iterative numerical algorithms and it implements the corresponding caching and prefetching strategies. We present some experimental results of the methodology and discuss the performance of a uniform multigrid method developed through SHOB on a massively parallel architecture, the Meiko CS2, and on a cluster of workstations.

1 Introduction

A Distributed Shared Memory, DSM, is a software layer that emulates a shared memory on distributed memory machines. A DSM allows the application developer to focus its interest on the application problems, rather than managing the physical allocation of the data onto the local memories and the coordination of data movement. We introduce a methodology that defines a DSM by composing two layers:

- an application independent layer, that includes the basic functionalities to access shared data and to synchronize the accesses. The functionalities of this application independent layer are optimized according to the concurrent mechanisms supported by the underlying architecture;
- an application dependent layer built on top of the previous one to implement a given consistency model. This layer implements the most effective caching and prefetching strategies for the considered model. This is an application dependent layer because different classes of applications require alternative consistency models.

B. Hertzberger et al. (Eds.): HPCN Europe 2001, LNCS 2110, pp. 302–311, 2001.

To exemplify this methodology, we present DVSA, Distributed Virtual Shared Areas, a package that implements the application independent layer and structures the shared memory as a set of areas. Currently, two DVSA implementations have been developed on, respectively, the Meiko CS2 [4] and a cluster of workstations, COW.

As an example of how the application dependent layer can be developed using the DVSA functions, we present SHOB, SHared OBjects, a package that defines a release consistency model well suited for numerical iterative algorithms. The caching and prefetching strategies supported by SHOB have been designed to exploit at best this consistency model to minimize the cost of accesses to shared data. Another package implemented on top of DVSA is HPF-Share [4], that defines a consistency model to support HPF applications.

The decomposition of the DSM into two layers simplifies the porting of the second layer, and of the corresponding programming tools, onto any architecture where the DVSA package has been implemented.

Notice that the second layer is not implemented by the application developer. Rather, it is the developer of the DSM that implements this layer to offer a given consistency model. The application developer should choose the consistency model and, consequently, the most appropriate second layer for the application of interest.

The rest of the paper is organized as follows. Sect. 2 presents the main DVSA features, sect. 3 presents the SHOB package and its implementation. Sect. 4 presents some experimental results of an uniform multigrid method developed through SHOB.

2 DVSA

Distributed Virtual Shared Areas, DVSA, is a package that provides a shared memory abstraction on distributed memory parallel architectures. Similar abstractions of a shared memory are TreadMarks [1], HIVE [3], Munin [5] and Tempest and Typhoon [10]. The DVSA functionalities are simpler than those of these systems because, while these packages have been developed to support user applications, DVSA is application independent and its goal is the development of application dependent run time supports. For this reason, DVSA offers a wide range of primitives to access areas and to synchronize processes, but it does not impose a given consistency model. In our approach, this model as well as the corresponding caching and prefetching strategies are defined and implemented by a second layer, that depends upon the applications of interest.

DVSA defines the shared memory as a set of areas. An area is a sequence of contiguous memory locations and it is paired with an identifier. Each process can access an area through the corresponding identifier, independently of the actual physical allocation of the area itself. Each area is considered as an atomic entity, and all its locations are either read or updated by each DVSA operation. In order to guarantee the correctness of the accesses, these areas can be managed through the DVSA primitives only, see Tab. 1.

To efficiently execute the package on different architectures, the size of each area may be freely chosen in an architecture dependent range. In fact, the hardware/firmware support of an architecture determines a range of area sizes that the architecture itself can efficiently support. The lower bound of this range is the most critical one from the performance point of view, because it defines the smallest amount of data exchanged when accessing an area. The lower bound has to be chosen according to architectural parameters such as the time to set up a communication and the bandwidth of the interconnection network. For instance, in the Meiko CS2, that includes specialized co-processors for message exchange and a fast interconnection network, this value is lower than that of a COW, that provides little hardware/firmware support for data exchange. On the other hand, to minimize problems such as false sharing, the largest size of an area should be chosen according to the semantic of the application too, according to the data structures recorded into an area.

The DVSA primitives can be partitioned into four subsets: notification, access, synchronization and utility.

The *notification* primitives initialize and terminate the DVSA support. At initialization time, each process declares, through the *Share* primitive, all and only the areas it is willing to share. Both the physical allocation, i.e. the local memory of the processing node, p-node, where the area is mapped, and the size of each area are fixed at this time and cannot be changed at run time. The physical allocation of an area can be chosen either by the user or by the built-in DVSA allocation procedure.

The DVSA *synchronization* primitives support the coordination of processes operating on the same area. A *Lock* primitive issued by a process P delays synchronized accesses from other processes to the area until the corresponding *Unlock* primitive is issued by P. A *Lock(read,A)* primitive delays any update of A, while a *Lock(write,A)* delays any access to A.

The *utility* primitives return information on the areas. The *AreaInfo(A)* primitive returns the size and the local memory where A is stored. The *MyAreas* primitive returns the list of the identifiers of the areas declared by the invoking process P and of the areas allocated in the local memory of P.

Four *access* primitives are available: *Read* and *Write*, to, respectively, read and write an area, *Copy*, to copy an area into another one and *Read_Write*, to read an area before updating it. The synchronization protocols and the kind of termination of each access primitive can be chosen through the parameter m, see table 1. In terms of the synchronization protocol, the DVSA accesses are partitioned into *synchronized* and *non synchronized* while, from the point of view of termination, they are partitioned into *blocking* and *non blocking*.

A *synchronized* access may be delayed when other synchronized accesses on the same area are concurrently executed. An update is delayed if at least one other process is accessing the same area. A synchronized read, instead, is delayed iff another process is updating the same area. Hence, several processes can read the same area simultaneously, but only one process can update it. As soon as a synchronized access on an area is terminated, pending accesses on the same

Table 1. DVSA Primitives.

Notification primitives	
Share(n,L(A_i))	declares a list of n areas $A_0, A_1, ...A_{n-1}$
End()	free all the areas
Access primitives	
Read(m,A,b)	reads the area A into the buffer b
Write(m,A,b)	writes the buffer b into the area A
Copy(m,A,B)	copies the area A into the area B
Read_Write(m,A,b,c)	reads the area A into the buffer b, writes the buffer c into A
Synchronization primitives	
Lock(w,A)	delays synchronized access on A
Unlock(A)	permits synchronized access on A
Test(h)	test if the operation returning h is terminated
Wait(h)	waits till the operation returning h is terminated
Utility primitives	
AreaInfo(A)	returns information on A
Exist(A)	returns 1 if A exists
MyAreas()	returns information about the areas of the process

area are considered. Write accesses have priority over the read ones. Then, if both read and write accesses have been delayed, a write one is executed first. This scheduling algorithm is unfair, but we assume that the average number of delayed accesses is so low that fairness is not an issue. If no write accesses are pending, then all read accesses are concurrently executed. A *non synchronized* access, instead, is immediately executed even if other accesses, synchronized or non synchronized, are in progress on the same area. Hence, this kind of access can be safely used only if either the semantics of the application enables concurrent accesses, or if proper process synchronizations in the application guarantee the mutual exclusion on the area.

A *blocking* access terminates only after the access has been completed. A blocking *Read* terminates as soon as the shared area has been copied into the local buffer, while a blocking *Write* terminates as soon as the local buffer has been copied into the shared area. Instead, *non blocking* accesses terminate immediately, returning an handle *h* that identifies the issued operation. To test the status of a non-blocking access, the DVSA library defines two operations on the handle *h*: *Test(h)*, to test whether the access is terminated, and *Wait(h)*, to wait till the access has been terminated. Non blocking accesses are used to overlap the execution of the application with the DVSA accesses.

As far as a single area is concerned, the DVSA provides a sequential consistency memory model [9]. The next section shows how an alternative consistency model can be implemented on top of DVSA.

3 SHOB

The SHared OBject, SHOB, package is an application oriented support corresponding to the second layer of our DSM system. It implements operations on shared data structures according to a release consistency model that is well suited for numerical iterative algorithms. To minimize the overhead due to remote accesses, i.e. accesses to data stored into a remote memory, SHOB implements caching and prefetching strategies suited to this model. In the following, we describe the implementation of these strategies and their effectiveness for numerical iterative applications. Furthermore, we focus on shared structures that are two dimensional matrices.

3.1 SHOB Primitives

SHOB requires that each process declares, through the primitive *Declare*, all and only the structures that it is going to share. A structure is shared only among the processes that have declared it, i.e. a process that does not declare a structure S cannot access it but avoids any overhead related to S. The primitive *Declare* returns an handle that has to be specified by any process on any p-node to access the structure. In the case of two dimensional matrices, one of two allocation strategies is chosen at declaration time. The strategies differ in the definition of the *allocation unit*, i.e. the set of elements that are allocated in the same DVSA area. The first strategy maps K columns, or rows, into a DVSA area. The second strategy partitions the matrix into KxK square submatrices and maps each submatrix into a DVSA area. The most appropriate strategy depends upon the application behavior. The value of K and the physical allocation of the data, i.e. the p-node where each allocation unit is stored, can be chosen either by the programmer in the declaration or by the SHOB support and cannot be updated at runtime. As shown in Tab. 2, SHOB defines three access primitives to handle a matrix: *read*, *write* and *sync*. The *read* and the *write* primitives work on one element of the shared matrix and they are executed in a non synchronized mode. Hence, the same element can be accessed in parallel by several processes. SHOB does not supply any mechanism to implement mutual exclusive accesses. Moreover, the *write* primitive updates the values of shared data in an asynchronous mode. Any process synchronization exploits a *sync(M)* primitive, that implements a barrier among all the processes sharing M. The barrier can be passed only when all the pending accesses on M have been terminated and all the processes have flushed back their local caches. No consistency

Table 2. SHOB Primitives.

Access primitives
Read(M,i,j,b) reads M[i][j] into the local variable b
Write(M,i,j,b) copies the local variable b into M[i][j]
Sync(M) completes the update of M

problem arises because of the updates, provided that distinct processes updates distinct allocation units of *M*. This corresponds to a release consistency model where the number of operations on M in between two successive invocations of *sync(M)* determines the trade off between data consistency and synchronization overhead.

3.2 SHOB Implementation

The SHOB handle of a shared matrix *M* is implemented as pointer to a structure, replicated in each p-node. The structure records the identifier of the DVSA area containing the first allocation unit of *M*, the adopted allocation strategy the information to implement the caching and prefetching strategies and the identifier of the DVSA area to implement *sync(M)*. The allocation units of a shared matrix are DVSA areas whose identifiers are contiguous. In the case of programmer-defined allocation, the DVSA library manages the physical allocation of the areas according to the programmer directives. When an element of *M* is read or written, the SHOB support computes the identifier of the area where the element is stored, using the identifier of the initial DVSA area and the allocation method of *M*.

We describe now the implementation of a *SHOB read operation* and the caching and prefetching strategies. The first time a process P_h invokes a SHOB read on an element *M[i][j]* of a shared matrix, the DVSA area including *M[i][j]* is copied, through a DVSA read, into the cache C_h in the local memory of P_h. If, later on, P_h invokes a SHOB read on an element *M[k][l]*, $i \neq k$ or $j \neq l$, that belongs to the same area, SHOB returns the value of this element from C_h. As soon as all the elements of the area have been read from C_h, the SHOB support automatically updates C_h by starting a non blocking DVSA read operation on the corresponding area. In this way, all the elements in the area are *prefetched*. Moreover, between two consecutive SHOB read operations on the same element, the value of the cache is always updated through a DVSA read operation. The overhead of these strategies is very low, because most read accesses are overlapped with the application.

The implementation of a *SHOB write operation* is similar to that of a read one. When a process P_h writes *M[i][j]*, if a copy of the area containing this element is not present in its local cache C_h, then P_h copies the area in C_h through a DVSA read and it updates the local copy of *M[i][j]*. The updated value of the area will be copied back through a DVSA write into the corresponding area either when a *sync*(M) is invoked or as soon as all the elements of the area have been updated in C_h. In the latter case, a non blocking DVSA write updates the area in parallel with the user computation.

sync(M) is implemented through DVSA operations on a further DVSA area paired with M. When a *sync*(M) is issued, any copy of an area that includes at least an updated element of M, is copied back into the area, provided that SHOB has not autonomously issued such an update. The update is implemented through a DVSA write. The application is suspended till all pending updates are terminated.

The goal of these caching strategies is to fully exploit any value in the local cache and to access through a simple operation any data in a DVSA area. To take into account updates to the area due to other processes, SHOB updates through a DVSA read any copy of a DVSA area in the local cache as soon as it any data in the copy has been read. However, some updates of an element may be lost, because the corresponding DVSA area may be read just before it is updated by another process. Hence, the consistency model guarantees that only the last update of an area before a *sync(M)* will be seen by the other processes. These strategies try to exploit at best the properties of numerical iterative algorithms, that update each element of a shared data only once for each iteration. Furthermore, due to their iterative behaviour, these algorithms can easily tolerate the delay of some updates. Obviously, to fully exploit these strategies, the allocation units should be chosen according to the application behaviour.

4 Experimental Results

To evaluate on a real application the performances and the effectiveness of the DSM that composes DVSA and SHOB, we have implemented a parallel version of an non adaptive multigrid method. Before discussing the performance of this application, we consider that of the DVSA primitives. Currently, two versions of DVSA have been developed: one on the Meiko CS2 and the other on a COW. The CS2 system is a tightly coupled network of HyperSPARC (100 Mhz) processors with 128 Mbyte of local memory running the Solaris operating system. Processors are interconnected by a multi-stage switched network optimized for high performance inter-processor communications. Each workstation of the COW is a PC with an Intel Pentium II CPU (266 MHz) and 128 Mbytes of local memory and it runs the Linux operative system. The interconnection network is a 100Mbit Fast Ethernet Switch.

In the Meiko CS2, the time to read/write an area ranges from 10 μ sec for a 1 Kbyte local area to 160 μ sec for a 16 Kbyte remote area. In the COW architecture, instead, the times to read/write an area ranges from 2 μ sec for 1 Kbyte local area to 2 msec for a 16 Kbyte remote area. In the CS2 architecture, for the same area, there is a one order of magnitude difference between the times of, respectively, a local access and a remote one. In the COW architecture, instead, there is a two orders of magnitude difference.

To achieve satisfactory performances on such a broad range of architectures, the effectiveness of the caching and prefetching strategies supported by the consistency model of interest is fundamental. Furthermore, the ability of implementing these strategies starting from an architecture independent level, as the one defined by the DVSA, strongly simplifies the implementation.

The next step of our experimentation investigates the performance of a uniform multigrid method on the Meiko CS2 and on the COW. The implementation uses the SHOB library primitives embedded in the C programming language.

4.1 The Performance of a Uniform Multigrid Method

Multigrid methods solve partial differential equations (PDE) by discretizing the domain through a hierarchy of grids. The non adaptive version uses a statically defined hierarchy, [2], where each grid is a discrete representation of the whole domain at a distinct abstraction level. The grids at the higher levels of the hierarchy use a larger number of points than those at the lower levels. The highest level grid is the *finest grid*, while the lowest level one is the *coarsest grid*. Several operators, multigrid operators, are applied to the grid hierarchy in a predefined order, V-cycle. Each operator updates the current value of each point p of each grid g using the values of the neighbors of p. The neighborhood stencil of p depends upon the considered operator and it may include points on the same grid of p or on the grids above or below in the hierarchy [6]. The V-cycle is iteratively applied on the grid hierarchy until the current error on the finest grid is lower than a fixed threshold. The discrete solution of the PDE is represented by the values of the points in the finest grid. We apply the method to the Poisson problem, i.e. the Laplace equation along with the Dirichlet boundary condition:

$$-\frac{d^2u}{dx^2} - \frac{d^2u}{dy^2} = f(x,y) \qquad in \qquad \Omega =]0,1[\times]0,1[$$

$$u = h(x,y) \qquad in \qquad \delta\Omega$$

with $f(x,y) = 0$ and $h(x,y) = 10$.

In the experiments, the finest and the coarsest grids are implemented by, respectively, a 1024x1024 matrix and a 256x256 one. Our implementation represents each grid as a shared matrix that is allocated by mapping K columns into each DVSA area. If n is the number of columns, p the number of processes, we assume that K divides n/p and map onto the local memory of P_h all the areas storing the matrix columns from $(n/p) * h$ to $(n/p) * (h+1) - 1$. To port the application, we only had to change the value of K, the number of columns in a DVSA area.

Because of the neighborhood stencil of the operators, this allocation guarantees that most of the accesses to a shared matrix are served by the local cache. To show this, consider that each P_h applies the multigrid operators to all the columns mapped onto its local memory and that all the elements of a column c are updated before considering the next column. To update an element e in c, P_h needs the values of the neighbors of e in $c-1$ and in $c+1$. When P_h reads the first element of $c-1$ or of $c+1$, SHOB copies the DVSA area containing the corresponding column into the cache of P_h. Hence, P_h finds in its local cache all the values to update all the other elements of c. Moreover, SHOB transparently copies back the values from the local cache to the proper DVSA area as soon as all the elements of the column have been updated. Even if SHOB cannot guarantee that all the updates to an element will be seen by a process reading this element, the solution of the PDE can be reached even if some processes use an out-of-date value. However, the use of out-of-data values results in a larger number of iterations to compute the solution. This does not imply

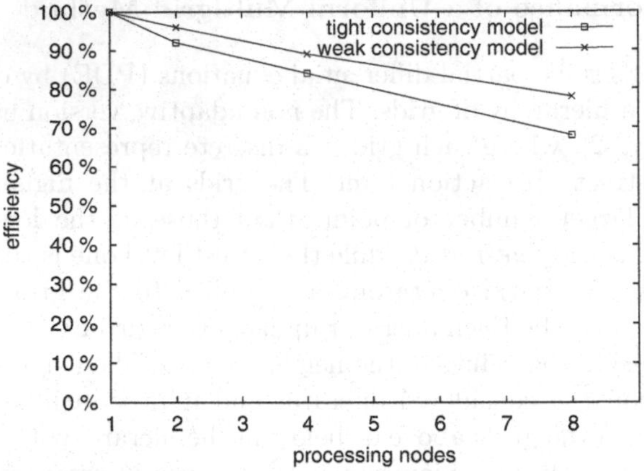

Fig. 1. Efficiency of the Uniform Multigrid Method on the CS2 Architecture.

that more time is required to compute the solution, because the time due to synchronizations may be larger than the one due to the additional iterations. To investigate the execution times of the multigrid method as an increasing degree of inconsistency is allowed, we have considered two versions that define the range of solutions supported by the consistency model of SHOB. In the version with the lowest degree of inconsistency, tight version, the processes issue a $sync(A)$, for each matrix A, after updating a column of A. In the version allowing the largest degree of inconsistency, weak version, the processes issue a $sync(A)$ only at the end of each V-cycle.

The experiments confirm that both versions produce the same numerical results using a different number of V-cycles. The tight version produces the final results in 50 V-cycles, while the weak version requires at most two more V-cycles to produce the same results. Figure 1 shows the experimental results of the CS2 implementation, whereas Fig. 2 shows the results of the COW implementation. In the both platforms, the weak version results in a better performance.

While the best performance has been achieved on the CS2, the performance increase achieved by the weak version is larger in the COW architecture. In the case of the CS2 the difference between the performances of the two version is negligible, whereas in the COW architecture an iteration of the weak version takes 65% of the time of an iteration of the tight version.

These results show the effectiveness of a two layers DSM: it is important to have an application independent layer, that optimizes the basic functionalities to access shared data, according to the tools and the hardware support of the architecture. However, an application dependent layer is important too, because it specializes caching and prefetching strategies to hide the latency to read and update remote data according to the consistency model most appropriate for the class of applications.

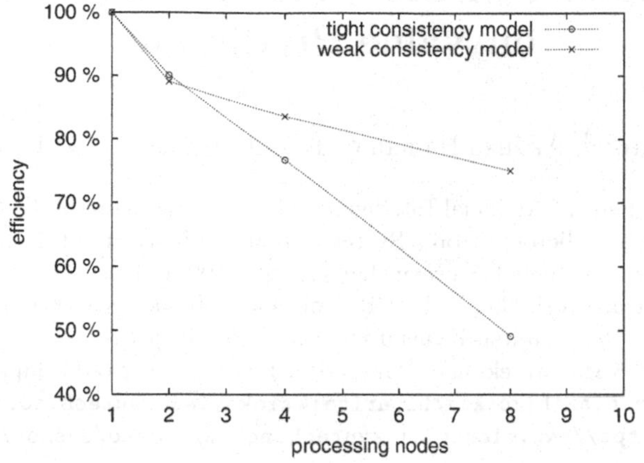

Fig. 2. Efficiency of the Uniform Multigrid Method on the COW Architecture.

References

1. Amza, C., Cox, A.L., Dwarkadas, S., Keleher, P., Lu, H., Rajamony, R., Yu, W., Zwaenepoel, W.: TreadMarks: Shared Memory Computing on Networks of Workstations. IEEE Computer **2**(29) (1996) 18 – 28
2. Baiardi, F., Chiti, S., Mori, P., Ricci L.: Parallelization of Irregular Problems Based on Hierarchical Domain Representation. Proceedings of HPCN 2000: Lecture Notes in Computer Science 1823 (2000) 71 – 80
3. Baiardi, F., Dobloni, G., Mori, P., Ricci L.: Hive: Implementing a virtual distributed shared memory in Java. DAPSYS - 3rd Austrian Hungarian Workshop on Distributed and Parallel Systems Kluwer Press (2000)
4. Baiardi, F., Guerri, D., Mori, P., Moroni, L., Ricci, L.: Evaluation of a Virtual Shared Memory by the Compilation of Data Parallel Loops. 8th Euromicro Workshop on Parallel and Distributed Processing (2000)
5. Bennett, J.K., Carter, J.B., Zwaenepoel, W.: Munin: Distributed shared memory based on type-specific memory coherence. ACM SIGPLAN Symposium on Principles & Practice of Parallel Programming (1990) 168 – 176
6. Briggs, W.: A Multigrid Tutorial. SIAM (1987)
7. Coelho, F., Germain, C., Pazat, J.L.: State of Art in Compiling HPF. In The parallel programming Model: Foundations, HPF Realization and Scientific Application, Lecture Notes in Computer Science 1132 (1996)
8. Gupta, S.K.S., Kaushik, S.D., Sharma, S., Huang, C.H., Sadayappan, P.: Compiling Array Expression for Efficient Execution on Distributed-Memory Machines. Journal of Parallel and Distribute Computing **32** (1996) 155 – 172
9. Lamport, L.: How to Make a Multiprocessor Computer That Correctly Executes Multiprocess Programs. IEEE Trans. on Computers **C-28**(9) (1979) 690 – 691
10. Reinhardt, S.K., Larus, J.R., Wood, D.A.: Tempest and Typhoon: User-level Shared Memory. Int. Symposium om Computer Architecture (1994) 325 – 336
11. Stichnoth, J.M., O'Hallaron, D., Gross, T.R.: Generating Communication for Array Statements: Design, implementation, and evaluation. Journal of Parallel and Distribute Computing **21**(1) (1994) 150 – 159

Influence of Compiler Optimizations on Value Prediction

Toshinori Sato[1,2], Akihiko Hamano[3], Kiichi Sugitani[4], and Itsujiro Arita[1]

[1] Department of Artificial Intelligence, Kyushu Institute of Technology
[2] Center for Microelectronic Systems, Kyushu Institute of Technology
680-4 Kawazu, Iizuka, 820-8502 Japan
[3] Heart at Work Co., Ltd., 3-18 Honmachi, Iizuka, 820-0042 Japan
[4] Fujitsu Kyushu Digital Technology Ltd.,
3-22-8 Hakata-ekimae, Hakata-ku, Fukuoka, 812-0011 Japan
{tsato,akihiko,kiichi,arita}@mickey.ai.kyutech.ac.jp
http://www.mickey.ai.kyutech.ac.jp/~tsato/cosmos/

Abstract. The practice of speculation in resolving data dependences based on value prediction has been studied as a means of extracting more instruction level parallelism. There are many studies on value prediction mechanisms with high predictabilities. However, to the best of our knowledge, the influence of compiler optimizations on value prediction has not been investigated. In this paper we evaluate efficiency of value prediction on several binaries which are compiled with different optimization levels. Detailed simulations reveal that value prediction is still effective for highly optimized binaries.
Keywords: Instruction level parallelism, data speculation, value prediction, optimization level, high-performance compilers.

1 Introduction

In order to improve performance, modern microprocessors rely on exploiting instruction level parallelism (ILP). However, ILP is limited by dependences between instructions. They are classified into three classes — control, name, and data dependences. There are many studies to reduce control and name dependences, but data dependence remains as a major bottleneck limiting ILP. Data speculation based on value prediction is such a technique that addresses the problem by resolving data dependences speculatively. An outcome of an instruction is predicted by value predictors[6,8]. The instruction and its dependent ones can be executed simultaneously, thereby exploiting ILP aggressively. Early works on value prediction mechanisms mainly consider hardware structures, and only their predictabilities and hardware costs are investigated. However, it is obvious that the characteristics of program binaries affect the value predictability, and thus it is not enough to study only hardware construction. For example, someone might expect that sophisticated optimizations obviate value prediction. From these considerations, in this paper we evaluate the relationship between compiler optimization levels and efficiency of value prediction.

B. Hertzberger et al. (Eds.): HPCN Europe 2001, LNCS 2110, pp. 312–321, 2001.

The organization of the rest of this paper is as follows. Section 2 surveys related works. Section 3 describes our evaluation methodology. Section 4 contains simulation results. Finally, Section 5 presents our conclusions.

2 Related Work

Data speculation[6,8] is a technique which executes instructions speculatively using predicted data values. Data dependences are speculatively resolved and thus ILP is increased. There are many studies proposing value prediction mechanisms, some of which achieve prediction accuracy as high as 80%, such as 2-level[20] and context-based[16] predictors. In order to improve prediction accuracy, several hybrid predictors[9,11,20] are proposed. The hybrid predictor is a combination of several value predictors with a selector choosing the probably most accurate one. An another approach to improve value prediction accuracy is using program execution profiles[3,7,13]. Using the profiles, instructions are classified according to the predictability, and a compiler provides the classified information to processors.

High prediction accuracies require considerable hardware cost. Morancho et al. [9], Rychlik et al. [11], and Calder et al. [4] examine capacity constraints of value predictors. Morancho et al. [9] and Rychlik et al.[11] proposed to reduce the hardware cost by classifying instructions based on their value predictability. Instructions which are easily predicted use simpler predictors such as the last-value and the stride predictors, whose hardware cost is low. High-cost predictors such as the 2-level, the hybrid, and the context-based predictors are used only for hard to predict instructions. Calder et al. [4] filter instructions based on their influence on processor performance. Only instructions on critical paths are held in a value predictor, reducing the number of entries of the predictor. Recently, we examined a technique reducing the hardware cost by exploiting narrow width values[14] and partial resolution[15]. Across SPEC programs, over 50% of integer operands are 16 bit or less[1]. That is, high-order bits of value prediction tables are merely utilized. Therefore, we proposed to keep only low-order bits of data values in the tables to reduce their hardware cost. Tag array size can be saved by employing partial resolution, using fewer tag address bits than necessary to uniquely identify every instruction. Full resolution of value predictors is not necessary since they do not have to be correct all the time. On the other hand, Fu et al. [5] and Tullsen et al. [19] remove value prediction hardware completely with the aid of compiler management of values in registers.

However, to the best of our knowledge, research on the influence of compiler optimizations on value prediction has been neglected. Recently, we have evaluated the relationship between compiler optimization levels and value predictability and found that there are meaningful value predictabilities in highly optimized binaries[18]. However, we have not evaluated the influence of compiler optimizations on processor performance. Therefore, in this paper we evaluate the efficiency of value prediction on variety of binaries compiled with different optimization levels and its contribution to processor performance.

3 Evaluation Methodology

In this section, we describe our evaluation methodology by explaining processor model and binaries used in this study.

3.1 Processor Model

We model a realistic 8-way out-of-order execution superscalar processor based on register update unit (RUU)[17] which has 128 entries. Each functional unit can execute any operations. The latency for execution is 1 cycle except in the case of multiplication (4 cycles) and division (12 cycles). A 4-port, non-blocking, 128KB, 32B block, 2-way set-associative L1 data cache is used for data supply. It has a load latency of 1 cycle after the data address is calculated and a miss latency of 6 cycles. It has a backup of an 8MB, 64B block, direct-mapped L2 cache which has a miss latency of 18 cycles for the first word plus 2 cycles for each additional word. No memory operation can execute that follows a store whose data address is unknown. A 128KB, 32B block, 2-way set-associative L1 instruction cache is used for instruction supply and also has the backup of the L2 cache which is shared with the L1 data cache. For control prediction, a 1K-entry 4-way set associative branch target buffer, a 4K-entry gshare-type 2-level adaptive branch predictor, and an 8-entry return address stack are used. The branch predictor is updated at instruction commit stage.

In this paper, we investigate two value prediction mechanisms — last-value predictor[8] and hybrid predictor consisting of the stride and the 2-level predictors [20]. We use direct-mapped tables for these predictors. The former represents simple predictors and the latter does complex ones. The configuration of the hybrid predictor is based on [20]. When a misspeculation occurs, it is necessary to revert processor state to a safe point where the speculation is initiated. We use an instruction reissue mechanism[12] which selectively flushes and reissues misspeculated instructions.

We use two types of simulators for this study. One is a functional simulator for counting value predictability and the other is a timing simulator for evaluating processor performance. We implemented the simulators using the SimpleScalar tool set (ver.3.0a)[2]. The SimpleScalar/PISA instruction set architecture (ISA) is based on MIPS ISA.

3.2 Benchmark Binaries

The binaries evaluated in this study are distributed by University of Michigan. They were compiled by GNU GCC (version 2.7.2.3) and MIRV[10]. For each compiler, three optimization levels, `-O0`, `-O1`, and `-O2`, were performed. Seven programs from eight SPEC95 CINT benchmarks are used for this study. The input files are modified so that evaluation time is practical. Each program is executed to completion. The candidate instructions predicted by the value predictors are register-writing ones, and do not include branch and store instructions. Tables 1 and 2 present total number of instructions executed and that

Table 1. Dynamic Instruction Count (GCC).

program	-O0		-O1		-O2	
099.go	268,225,895	(223,777,492)	146,032,803	(115,9999,78)	134,766,291	(104,239,257)
124.m88ksim	215,027,498	(155,551,958)	124,461,169	(86,167,274)	119,705,428	(81,391,241)
126.gcc	146,724,176	(100,785,825)	105,315,933	(69,649,020)	103,357,196	(67,359,249)
129.compress	77,092,827	(54,636,345)	48,821,478	(32,717,148)	47,719,938	(31,615,607)
130.li	307,778,502	(185,680,534)	208,780,589	(123,022,538)	206,414,110	(121,247,791)
132.ijpeg	18,082,420	(14,009,018)	8,831,646	(6.490,266)	8,606,970	(6,255,132)
134.perl	12,166,065	(7,775,114)	10,641,524	(6,611,315)	10,645,288	(6,607,758)

Table 2. Dynamic Instruction Count (MIRV).

program	-O0		-O1		-O2	
099.go	179,701,392	(142,959,132)	131,900,827	(98,921,730)	132,506,520	(99,186,262)
124.m88ksim	184,814,315	(130,030,485)	122,977,073	(82,888,272)	123,766,329	(85,244,701)
126.gcc	140,356,442	(96,215,914)	115,904,906	(75,804,673)	115,334,584	(86,961,482)
129.compress	69,108,399	(48,019,527)	49,210,033	(32,525,461)	47,700,633	(31,587,802)
130.li	270,793,655	(161,858,905)	207,709,666	(121,168,612)	207,705,881	(121,167,937)
132.ijpeg	12,045,397	(8,705,739)	9,751,916	(7,033,935)	9,995,242	(7,282,418)
134.perl	12,574,740	(8,233,448)	10,489,957	(6,644,109)	10,510,075	(6,660,903)

of instructions touched to predicted (in bracket) for every optimization level in the cases of GCC and MIRV respectively. Please note that the numbers are equivalent between the last-value and the hybrid predictors, since we count them using the functional simulator. In general, MIRV -O0 executes considerably less instructions than GCC -O0, because MIRV has graph coloring allocation and copy propagation in -O0 while GCC has no register allocation in -O0[10]. For the remaining optimization levels, the two compilers have almost equal abilities.

4 Simulation Results

In this section, we present simulation results. We define predictability as the number of instructions that are (correctly and incorrectly) predicted by a value predictor over that of all register-writing instructions. Prediction coverage is defined as the number of instructions correctly predicted over that of all register-writing instructions. Prediction accuracy is the percentage of instructions correctly predicted over all predicted instructions. Therefore, we have the following equation.

$$(Prediction\ coverage) = (Predictability) * (Prediction\ accuracy)$$

It has been found that the dominant factor in performance improvement is not prediction accuracy but prediction coverage[12,19]. Therefore, We use the predictability and the prediction coverage as metrics for evaluation. After that, processor performance is evaluated.

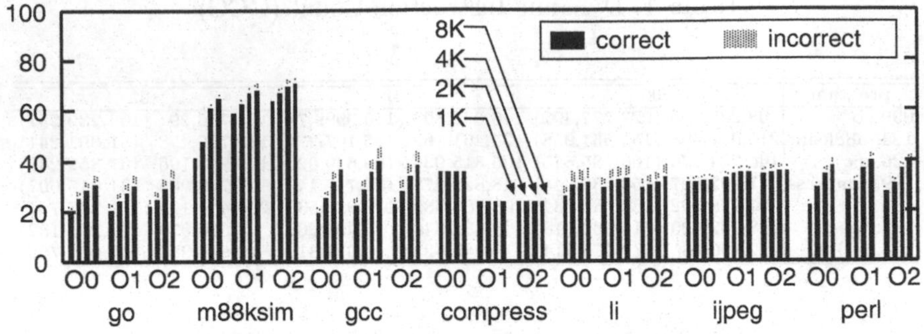

Fig. 1. %Value Predictability (GCC/last).

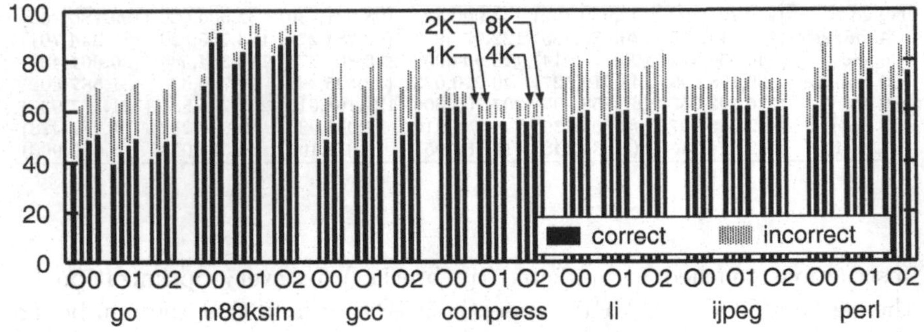

Fig. 2. %Value Predictability (GCC/hybrid).

4.1 Predictability

Figure 1 shows simulation results when the last-value predictor is utilized on GCC binaries. The horizontal and vertical axes denote the optimization levels and the percentage respectively. Each bar, divided into two parts, indicates the predictability and the prediction coverage. The lower part (black) indicates the percentage of the instruction whose data value is correctly predicted. The upper part (gray) indicates the percentage that is mispredicted. That is, the lower part is the prediction coverage and the sum of the two parts is the predictability. For each group, bars from left to right indicate the results when the value predictor has 1024-, 2048-, 4096-, and 8192-entry tables, respectively. We can find the followings. First, in the case of `129.compress`, both the predictability and the prediction coverage are considerably reduced when the optimization level changes from `-O0` to `-O1`. This might mean that relatively high optimizations obviate the value prediction. Second, different from the previous observation, both the predictability and the prediction coverage are slightly improved as the optimization level becomes higher in the cases of the remaining programs. This denies the first investigation. Third, the efficiency of the value prediction changes

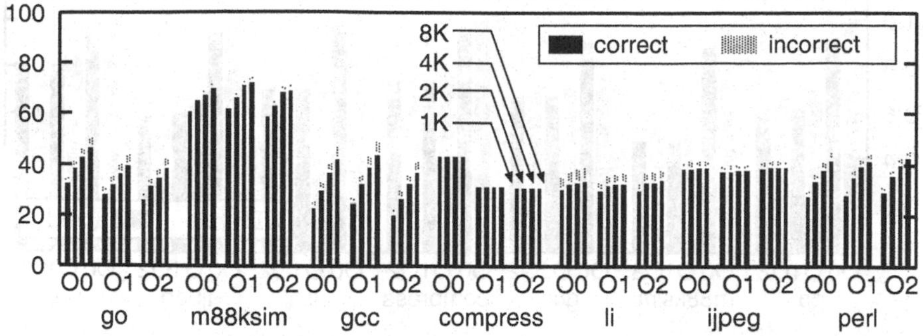

Fig. 3. %Value Predictability (MIRV/last-value).

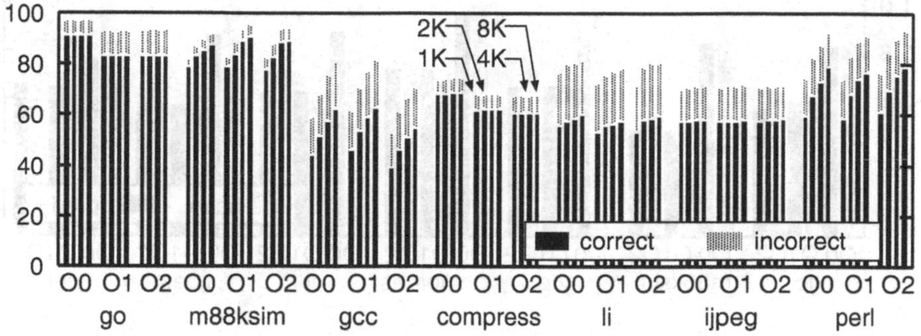

Fig. 4. %Value Predictability (MIRV/hybrid).

little when the optimization level changes. Therefore, we can confirm that the conservative last-value predictor is effective for highly optimized binaries.

Figure 2 shows simulation results, when the hybrid predictor is utilized on GCC binaries. The characteristics similar to the last-value predictor case is observed. However, two programs should be mentioned. In the case of `129.compress`, the difference in the predictability and the prediction coverage between `-O0` and `-O1` are considerably smaller than the last-value predictor case. In the case of `134.perl`, both the predictability and the prediction coverage are slightly reduced as the optimization level becomes higher. However, these observations does not require the different conclusions from the last-value predictor case. In general, both the predictability and the prediction coverage rarely changes when different optimization levels are considered. And thus, the aggressive hybrid predictor is also effective for highly optimized binaries.

Figure 3 shows simulation results, when the last-value predictor is utilized to MIRV binaries. The different characteristics from GCC binaries can be observed. First, we can find that in general both the predictability and the prediction coverage are greater in MIRV than in GCC. This can be observed throughout all programs and all optimization levels. Tables 1 and 2 say that there are not signif-

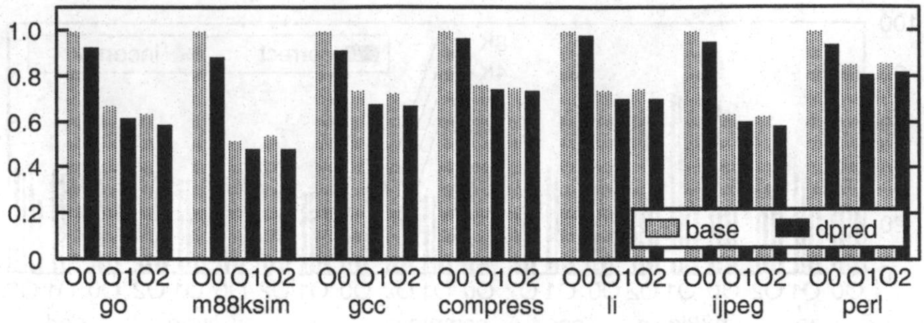

Fig. 5. Execution Cycle Time (GCC/last-value).

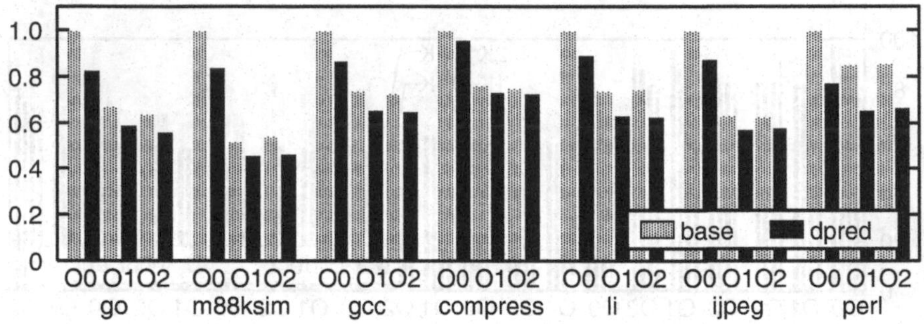

Fig. 6. Execution Cycle Time (GCC/hybrid).

icant difference on dynamic instruction counts between two compilers. Therefore, MIRV can benefit from value prediction more than GCC. Second, it is generally found that both the predictability and the prediction coverage decrease slightly as the optimization level becomes higher. This difference from the GCC case can be found especially in the case of `099.go`. However, the decrease is not as large as that can be found in GCC `129.compress`. Thus, we can conclude that the conservative last-value predictor is still effective for MIRV binaries.

Figure 4 shows simulation results, when the hybrid predictor is utilized to MIRV binaries. The characteristics similar to the last-value predictor case is observed, while `099.go` shows considerably different behavior. Both the predictability and the prediction coverage, which are greater than in the GCC case, are reduced slightly when higher optimization levels are applied. In addition, for `099.go` both the predictability and the prediction coverage are considerably reduced when the optimization level changes from `-O0` to `-O1`. However, we can conclude that the aggressive hybrid predictor is also effective for highly optimized binaries.

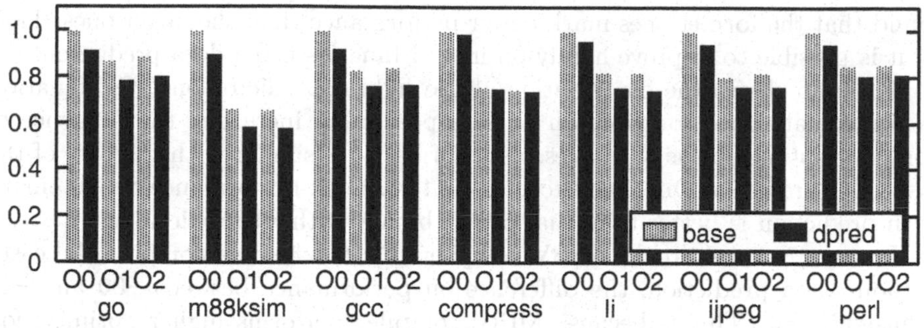

Fig. 7. Execution Cycle Time (MIRV/last value).

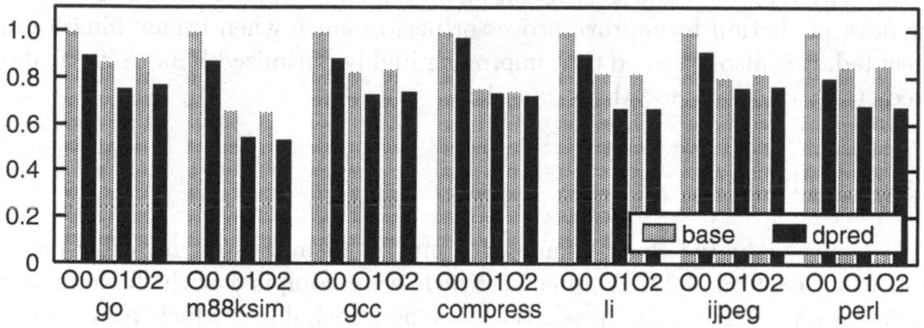

Fig. 8. Execution Cycle Time (MIRV/hybrid).

4.2 Processor Performance

Next, the influence on processor performance is evaluated. We use execution cycle time as a metric for the evaluation. Hence, the smaller the number is the higher the performance is. Every execution cycle time is normalized by the time in the case of `-O0` binary without data prediction. We use 4096-entry direct-mapped tables for this evaluation.

Figure 5 shows the contribution of the last-value predictor on GCC binaries. The horizontal axe indicates program names and optimization levels, and the vertical axe indicates relative processor performances. For each group of two bars, the left bar (gray) is for the case without data prediction and the right one (black) is for the case with prediction. First, when we compare the cases without data prediction, it is easily observed that processor performance is considerably improved as the optimization level is changed from `-O0` to `-O1`. In addition, the difference between performance of `-O0` binaries with data prediction and that of `-O1` binaries without data prediction is still significant. This implies that it is difficult for data prediction to contribute on binaries with conservative compiler optimizations such as legacy binaries. In other words, re-compilation is better than hardware-based optimizations for these binaries. Second, when we compare `-O1` binaries with prediction with `-O2` binaries without prediction. It is easily found that the former ones mark better performance than the latter ones. That is, it is possible to improve highly optimized binaries using data prediction.

Figure 6 shows the contribution of the hybrid predictor on GCC binaries. While the absolute contribution of data prediction increases, the influence of compiler optimizations on processor performance is similar with the case of the last-value predictor. Only one exception is `134.perl`. Performance of `-O0` binary with prediction is better than that of `-O1` binary without prediction.

Figures 7 and 8 are for MIRV compiler. First, when we compare the cases without data prediction, the difference on performance between `-O0` and `-O1` is insignificant. This is because MIRV compiler performs higher optimization with `-O0` option than GCC as shown in Table 2. Nevertheless, the difference between performance of `-O0` binaries with data prediction and that of `-O1` binaries without prediction is still considerable. This confirms that it is difficult for data prediction to improve processor performance when legacy binaries are executed. It is also observed that improving highly optimized binaries using data prediction is possible for MIRV binaries.

5 Conclusion

This paper evaluated the influence of compiler optimizations on value prediction using GCC and MIRV binaries which were compiled with different optimization levels. The value prediction is a new technique, which resolves data dependences speculatively in order to extract more ILP from programs with small parallelism. Previous works on value prediction proposed several tradeoff points between value predictability and hardware complexity. However, to the best of our knowledge, the relationship between compiler optimization levels and contribution of value prediction to processor performance has not been investigated. Therefore, in this paper we have evaluated value predictability of several binaries. From detailed simulation results, we have found that for both GCC and MIRV the value prediction is still effective for binaries compiled with every optimization level.

This paper is one result of our ongoing research in high-performance, low-power, and complexity-effective microprocessor *COSMOS* at Kyushu Institute of Technology. More information is available at our home page.

Acknowledgments

This work is supported in part by the grants from Japan Society for the Promotion of Science (no.12780273) and from Fukuoka Industry, Science & Technology Foundation (no.H12-1).

References

1. Brooks D., Martonosi M.: Dynamically exploiting narrow width operands to improve processor power and performance. 5th Int'l Symp. on High Performance Computer Architecture (1999)
2. Burger D., Austin T.M.: The SimpleScalar tool set, version 2.0. ACM SIGARCH Computer Architecture News, 25(3) (1997)
3. Calder B., Feller P., Eustace A.: Value profiling. 30th Int'l Symp. on Microarchitecture (1997)
4. Calder B., Reinman G., Tullsen D.M.: Selective value prediction. 26th Int'l Sym. on Computer Architecture (1999)
5. Fu C-y., Jennings M.D., Larin S.Y., Conte T.M.: Software-only value speculation scheduling. Technical Report, Dept. of Electrical and Computer Engineering, North Carolina State University (1998)
6. Gabbay F.: Speculative execution based on value prediction. Technical Report #1080, Dept. of Electrical Engineering, Technion (1996)
7. Gabbay F., Mendelson A.: Can program profiling support value prediction? 30th Int'l Symp. on Microarchitecture (1997)
8. Lipasti M.H., Shen J.P.: Exceeding the dataflow limit via value prediction. 29th Int'l Symp. on Microarchitecture (1996)
9. Morancho E., Llaberia J.M., Olive A.: Split last-address predictor. Int'l Conf. on Parallel Architectures and Compilation Techniques (1998)
10. Postiff M., Greene D., Lefurgy C., Helder D., Mudge T.: The MIRV SimpleScalar/PISA compiler. Technical Report CSE-TR-421-00, Dept. of Computer Science, University of Michigan (2000)
11. Rychlik B., Faistl J.W., Krug B.P., Kurland A.Y., Sung J.J., Velev M.N., Shen J.P.: Efficient and accurate value prediction using dynamic classification. Technical Report CMuART-98-01, Dept. of Electrical and Computer Engineering, Carnegie Mellon University (1998)
12. Sato T.: Analyzing overhead of reissued instructions on data speculative processors. Workshop on Performance Analysis and its Impact on Design held in conjunction with 25th Int. Symp. on Computer Architecture (1998)
13. Sato T.: Profile-based selection of load value and address predictors. 2nd Int'l Symp. on High Performance Computing (1999)
14. Sato T., Arita I.: Table size reduction for data value predictors by exploiting narrow width values. 14th Int'l Conf. on Supercomputing (2000)
15. Sato T., Arita I.: Partial resolution in data value predictors. 29th Int'l Conf. on Parallel Processing (2000)
16. Sazeides Y., Smith J.E.: Implementations of context based value predictors. Technical Report TR-ECE-97-8, Dept. of Electrical and Computer Engineering, University of Wisconsin-Madison (1997)
17. Sohi G.S.: Instruction issue logic for high-performance, interruptible, multiple functional unit, pipelined computers. IEEE Trans. Comput., 39(3) (1990)
18. Sugitani K., Hamano A., Sato T., Arita I.: Evaluating effect of optimization level on value predictability. 4th Int'l Conf. on Algorithms and Architectures for Parallel Processing (2000)
19. Tullsen D.M., Seng J.S.: Strageless value prediction using prior register values. 26th Inte'l Symp. on Computer Architecture (1999)
20. Wang K., Franklin M.: Highly accurate data value prediction using hybrid predictors. 30th Int'l Symp. on Microarchitecture (1997)

Experiments with Sequential Prefetching

Sathiamoorthy Manoharan and Chaitanya Reddy Yavasani

Department of Computer Science, University of Auckland, New Zealand

Abstract. Sequential prefetching is a classic method of prefetching cache contents. There are two variants of sequential prefetching. *Prefetch-on-miss* prefetches the next consecutive cache lines following the line that misses in the cache. *Tagged prefetch* prefetches the next consecutive lines of the cache line that is currently being accessed. In this paper, we compare these two variants both analytically and experimentally, and show that while tagged prefetch is better with smaller latencies, prefetch-on-miss is better when the memory latency is large.

1 Introduction

Sequential prefetching is a hardware prefetching scheme that brings data into the cache before the data is expected by the processor [4,3,2,5]. The simplest form of sequential prefetching prefetches line $p + 1$ when line p is accessed. This is also referred to as *one-line-lookahead* or *one-block-lookahead*. One-line-lookahead is not always sufficient to prefetch data early enough to avoid processor stalls, especially when there is not enough computation between the use of successive cache lines. A multi-line-lookahead can solve this problem. The scheme known as *Next-N-line prefetching* prefetches the N successive lines $p + 1$, $p + 2$, ..., $p + N$, when line p is accessed. Here N is said to be the degree of prefetching. One-line-lookahead is a special case of Next-N-line prefetching where N is 1.

In an *adaptive sequential prefetching*, the degree of prefetching N changes dynamically depending on how useful the prefetched data is [1]. Successful use of prefetched data increment N, while wasted prefetches lead to decrementing N. If N reaches 0, prefetching is effectively turned off. It is turned on again if the hardware finds enough misses that could have been avoided if N were 1.

Depending on what type of access causes cache lines to be prefetched, sequential prefetching is classified into two schemes: *prefetch-on-miss* and *tagged prefetch*.

A prefetch-on-miss scheme prefetches successive cache lines when the current line misses in the cache. That is, when line p misses in the cache, prefetches are initiated for lines $p+1$ through to $p+N$. It can reduce the number of misses in a strictly sequential reference stream by a factor of $N+1$, if the memory latency is not large.

A tagged prefetch scheme prefetches successive lines when the current cache line is referenced for the first time. This referencing includes both hits and misses. In this scheme, each cache line has a tag associated with it. The tag is set to zero when the line is loaded, and when the line gets used, the tag is set to one. When a

B. Hertzberger et al. (Eds.): HPCN Europe 2001, LNCS 2110, pp. 322–331, 2001.

cache line's tag changes from zero to one, prefetches are initiated for the next N successive cache lines. Using the transition point of the tag to initiate prefetches makes sure that repeated references to the same cache line do not result in many identical prefetch requests. Tagged prefetch can reduce the number of misses to zero in a strictly sequential reference stream, if the memory latency is not large.

In this paper, we study *prefetch-on-miss* and *tagged prefetch* analytically and experimentally, and show that while tagged prefetch is better with smaller memory latencies, prefetch-on-miss is better when the memory latency is large. To show these results analytically, we use a strictly sequential reference stream. We also include experimental results from running some benchmarks.

The rest of the paper is organized as follows. Section 2 introduces the basis for the analysis and defines the notations that are used subsequently. Section 3 compares the two schemes under one-line-lookahead, and section 4 compares them under two-line-lookahead. Section 5 presents generalized results for multi-line lookahead. Section 6 includes comparison results from running some benchmarks. The final section concludes with a summary.

2 The Analysis Basis

To analyze and compare prefetch-on-miss and tagged prefetch, we use a strictly sequential reference stream. This can be a stream of sequential data (such as an array access with unit stride) or a large basic block of instructions.

Let the number of words per cache line be w, and the size of the sequential reference stream be m cache lines. Assume that the computation cycles spent using a single cache line is C, and the memory latency to access a single word in memory is L.

If no prefetching is used, a sequential reference stream will see a repeating pattern of miss, cache line fill, and computation. The miss takes L cycles to get the first word of the line into the cache, and to get the last word into the cache will require a further w cycles. Then the processor will compute for C cycles without any misses. This process repeats m times for the entire sequential reference stream. Therefore, the total execution time without prefetching, T_0, is given by

$$T_0 = m(L + w + C) \tag{1}$$

We define the speedup due to prefetching as

$$\frac{\text{Execution time without prefetching}}{\text{Execution time with prefetching}}$$

The speedup is achieved by overlapping memory accesses with computation.

3 One-Line-Lookahead

The prefetch-on-miss scheme, when executing a strictly sequential reference stream, will have a cache miss for every other cache line. This is because a prefetch for line $p+1$ is not issued until line p misses. The prefetch-on-miss scheme will therefore see a miss costing $L+w$ cycles and $2C$ computation cycles corresponding to the two lines of cache (one fetched, and the other prefetched). This will repeat $\frac{m}{2}$ times since the size of the sequential stream is m cache lines.

Note that the prefetched cache line would only take an extra w cycles to arrive from memory after the fetched line arrived. The computation cycles performed using a cache line, C, will usually be more than w, so these w cycles will be completely overlapped by computation.

The time T_{1m} taken to complete the execution of the sequential stream using prefetch-on-miss is therefore given by

$$T_{1m} = \frac{m}{2}(L + w + 2C) \tag{2}$$

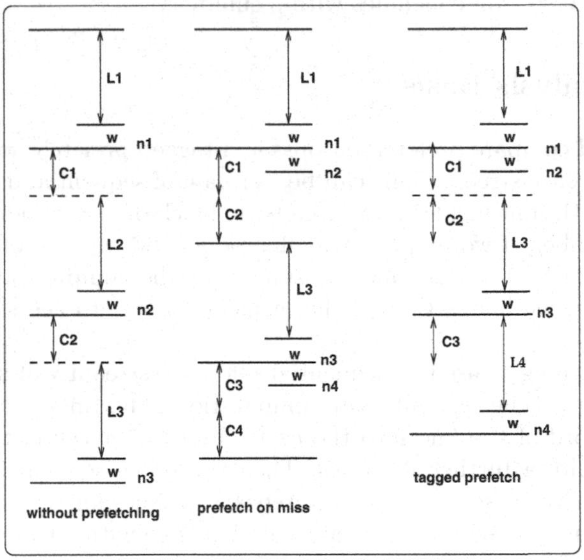

Fig. 1. Execution Timing for One-Line-Lookahead.

See Figure 1. The numbers next to C and L in the figure denote the cache line to which the computation or latency correspond. For example, $C3$ denotes the C computation cycles that corresponds to line 3, and $L3$ denotes the latency for getting line 3 from memory. Also note that nx in the figure denotes the arrival of line x in the cache.

A tagged prefetch scheme overlaps memory access and computation as shown in Figure 1. There is a one-off waiting of $L+w$ cycles to get the first line into the cache. After that, the computation corresponding to line i and the latency for getting line $i+1$ from memory overlap.

The time T_{1t} taken to complete the execution of the sequential stream using tagged prefetch is therefore given by

$$T_{1t} = (L + w + C) + (m - 1)\max(L + w, C) \tag{3}$$

3.1 Speedup Due to Prefetching

Using (1) and (2), the speedup due to prefetch-on-miss, S_{1m}, is given by

$$S_{1m} = \frac{T_0}{T_{1m}} = \frac{m(L + w + C)}{\frac{m}{2}(L + w + 2C)} = 1 + \frac{L + w}{L + w + 2C} \tag{4}$$

This shows that the speedup due to prefetch-on-miss monotonically increases with latency. It gets close to 2 with large latencies.

From (1) and (3), the speedup due to tagged prefetch, S_{1t}, is given by

$$S_{1t} \approx \frac{L + w + C}{\max(L + w, C)} \tag{5}$$

When $L + w < C$, the speedup S_{1t} is $1 + \frac{L+w}{C}$, and when $L + w > C$, it is $1 + \frac{C}{L+w}$. As latency increases, S_{1t} increases until the latency is close to the computation time, and starts to decrease as the latency further increases. See Figure 2 which shows the behavior of both S_{1m} and S_{1t} when w is 32 and C is 100.

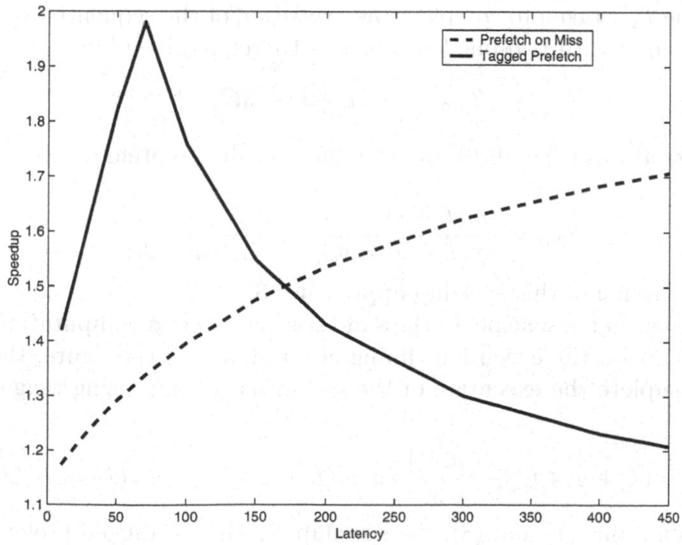

Fig. 2. Speedup for Prefetch-on-Miss and Tagged Prefetch with One-Line Lookahead.

Figure 2 shows that for larger latencies, prefetch-on-miss leads to a better speedup than tagged prefetch.

Prefetch-on-miss scheme gets a cache miss for every other cache line. This is because the next line is not prefetched until a line misses. A tagged prefetch scheme issues prefetches even when a line hits. Therefore a tagged prefetch scheme performs better when the computation is large enough to overlap the prefetch latency. See Figure 1.

At each miss, the prefetch-on-miss streams the prefetch requests after servicing the miss. Servicing the miss costs $L + w$ cycles, but prefetching the next line costs only an extra w cycles. With a tagged prefetch scheme, on the other hand, each prefetch request is initiated separately, so each costs an entire $L + w$ cycles. When C is larger than $L + w$, this latency is fully hidden; consequently, tagged prefetch performs better than prefetch-on-miss. However, when $L + w$ is larger than C (i.e., large latencies), a substantial part of the latency is not hidden. This leads to tagged prefetch performing worse than prefetch-on-miss at large latencies.

4 Two-Line-Lookahead

With two-line lookahead, the prefetch-on-miss scheme will have a cache miss for every third cache line. This is because prefetch requests for lines $p+1$ and $p+2$ are not issued until line p misses. The prefetch-on-miss scheme will therefore see a miss costing $L + w$ cycles and $3C$ computation cycles corresponding to the three lines of cache (one fetched, and the others prefetched). This will repeat $\frac{m}{3}$ times (since the size of the sequential reference stream is m cache lines). See Figure 3.

The time T_{2m} taken to complete the execution of the sequential stream using prefetch-on-miss with two-line lookahead is therefore given by

$$T_{2m} = \frac{m}{3}\,(L + w + 3C) \tag{6}$$

Using equations (1) and (6), the speedup S_{2m} due to prefetch-on-miss is given by

$$S_{2m} = \frac{m(L + w + C)}{\frac{m}{3}\,(L + w + 3C)} = \frac{3(L + w + C)}{L + w + 3C} \tag{7}$$

With large latencies, this speedup approaches 3.

A tagged prefetch scheme overlaps memory access and computation as shown in Figure 3. From the execution timing chart shown in the figure, the time T_{2t} taken to complete the execution of the sequential stream using tagged prefetch is given by

$$T_{2t} = (L + w + C) + \frac{m-1}{3}\,(\max(L + w, C) + \max(L + 2w, 2C)) \tag{8}$$

From equations (1) and (8), the speedup S_{2t} due to tagged prefetch is given by

$$S_{2t} \approx \frac{3(L + w + C)}{\max(L + w, C) + \max(L + 2w, 2C)} \tag{9}$$

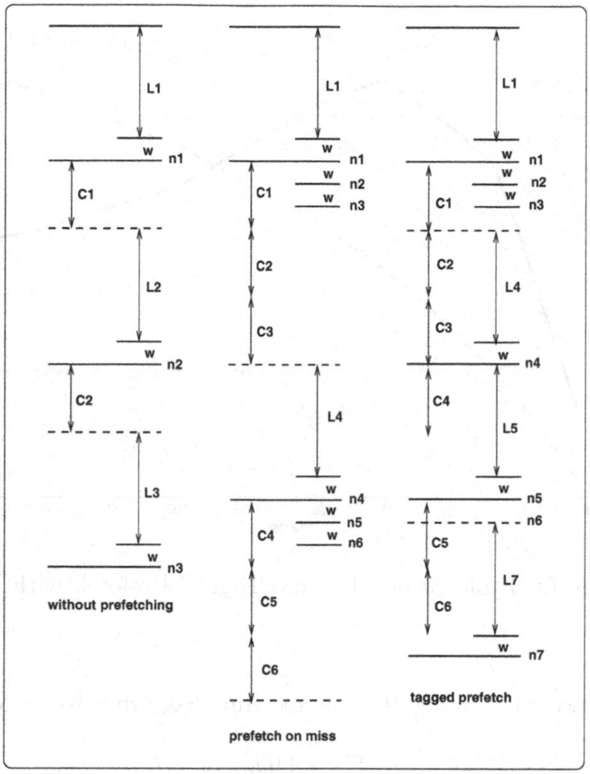

Fig. 3. Execution Timing for Two-Line-Lookahead.

Note that, with large latencies, S_{2t} will approach $\frac{3}{2}$. See Figure 4.

5 Multi-Line-Lookahead: Generalized Analysis

With an N-line lookahead, the prefetch-on-miss scheme will have a cache miss for every $(N+1)^{th}$ cache line. This is because prefetch requests for lines $p+1$ through to $p+N$ are not issued until line p misses. The prefetch-on-miss scheme will therefore see a miss costing $L+w$ cycles and $(N+1)C$ computation cycles corresponding to the $(N+1)$ lines of cache (one fetched, and the others prefetched). This will repeat $\frac{m}{N+1}$ times (since the size of the sequential reference stream is m cache lines).

The time T_{Nm} taken to complete the execution of the sequential stream using prefetch-on-miss with N-line lookahead is therefore given by

$$T_{Nm} = \frac{m}{N+1}(L + w + (N+1)C) \qquad (10)$$

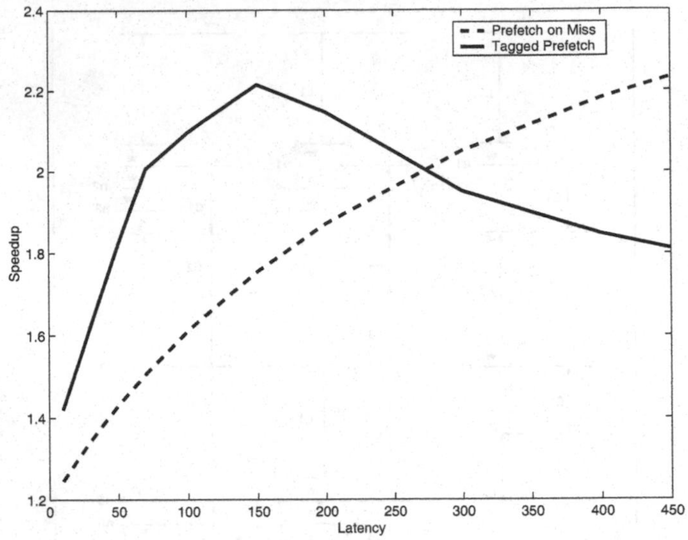

Fig. 4. Speedup for Prefetch-on-Miss and Tagged Prefetch with Two-Line Lookahead.

From equations (1) and (10), the speedup S_{Nm} due to prefetch-on-miss is given by

$$S_{Nm} = \frac{(N+1)(L+w+C)}{L+w+(N+1)C} \tag{11}$$

With large latencies, S_{Nm} approaches $N+1$.

The time T_{Nt} taken to complete the execution of the sequential stream using tagged prefetch with N-line lookahead is given by

$$T_{Nt} = (L+w+C) + \frac{m-1}{N+1}(\max(L+w,C) + \max(L+Nw,NC)) \tag{12}$$

Using equations (1) and (12), the speedup S_{Nt} due to tagged prefetch is given by

$$S_{Nt} \approx \frac{(N+1)(L+w+C)}{\max(L+w,C) + \max(L+Nw,NC)} \tag{13}$$

With large latencies, S_{Nt} approaches $\frac{N+1}{2}$.

5.1 Practical Limitations

Basic blocks in real programs are not big enough to benefit from multi-line lookahead schemes with $N > 2$. Therefore, in practice, only one-line and two-line lookahead schemes will find profitable use.

If the memory system is to return data requests out of order, then buffer space will be required to hold information on the identity of data returned from memory. Also, the address and returned data will have to be suitably tagged. The buffer space and the tag bits required are proportional to N in a multi-line lookahead. This is another practical limitation to using multi-line lookahead with large N values.

6 Experimental Results

This section reports experimental results to verify the analytical results derived in the previous sections. It presents speedup vs. latency graphs for three benchmark applications using sequential prefetching.

The benchmarks are run on a cycle-level simulator of the Alpha architecture. The simulator simulates the EV4 architecture with the exception that it does not simulate out of order execution and super-scalar instruction issue. It simulates the primary instruction and data caches, the secondary unified cache, the write buffer, a prefetch issue buffer, and a memory subsystem. The size of the secondary cache is 256KB, and the prefetch issue buffer has 8 entries. System calls are also simulated, so that any statically-linked binary executable can run on the simulator.

We report results from running three benchmark applications: VARIANCE, INRPROD, SEQSTREAM. VARIANCE and INRPROD use data prefetching, while SEQSTREAM uses instruction prefetching.

VARIANCE computes the variance of all elements in a one-dimensional array of size 2MB. INRPROD computes the inner product of two vectors of size 250KB. SEQSTREAM is a synthetic program which simply initializes an array of 40KB without using a loop. The program is strictly sequential in terms of its instruction access.

Figures 5–7 show how speedup varies with latency for the three benchmarks. Results for one-line-lookahead are marked $N = 1$, and the results for two-line-lookahead are marked $N = 2$ in the figures. The graphs show that tagged prefetch offers better speedups at smaller latencies, while at larger latencies, prefetch-on-miss offers better speedups. This is in agreement with the analytical results derived earlier.

Note that for VARIANCE and INRPROD, the computation time C is large, and therefore the maximum speedup for tagged prefetch is attained only at large latencies (when the latency roughly matches the computation time for a perfect overlap).

7 Summary and Conclusions

Sequential prefetching is a hardware scheme to prefetch cache contents. This paper studied two sequential prefetching schemes, *prefetch-on-miss* and *tagged prefetch*, both analytically and experimentally. It showed that while tagged prefetch is better with smaller memory latencies, prefetch-on-miss is better when the memory latency is large.

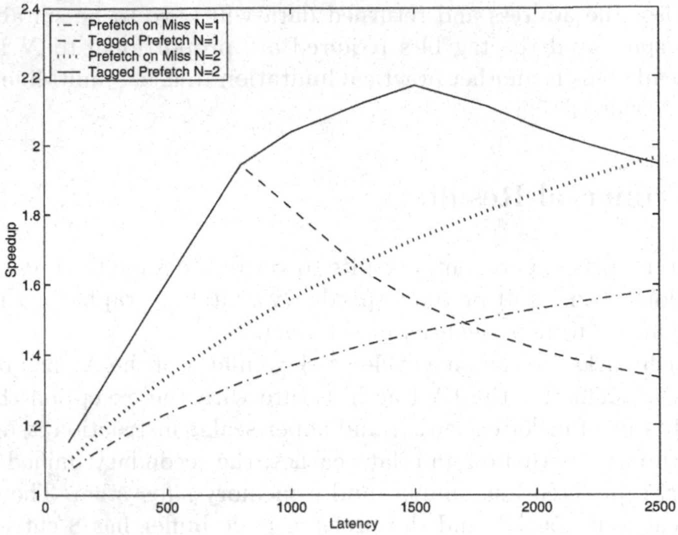

Fig. 5. Speedup vs. Latency for VARIANCE.

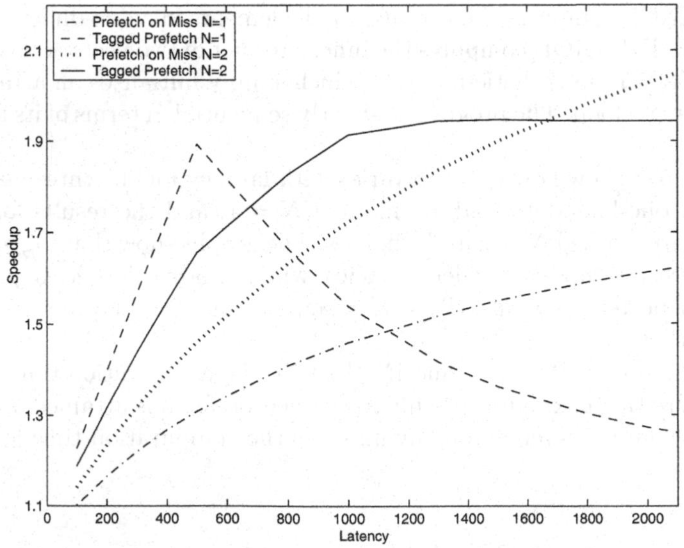

Fig. 6. Speedup vs. Latency for INRPROD.

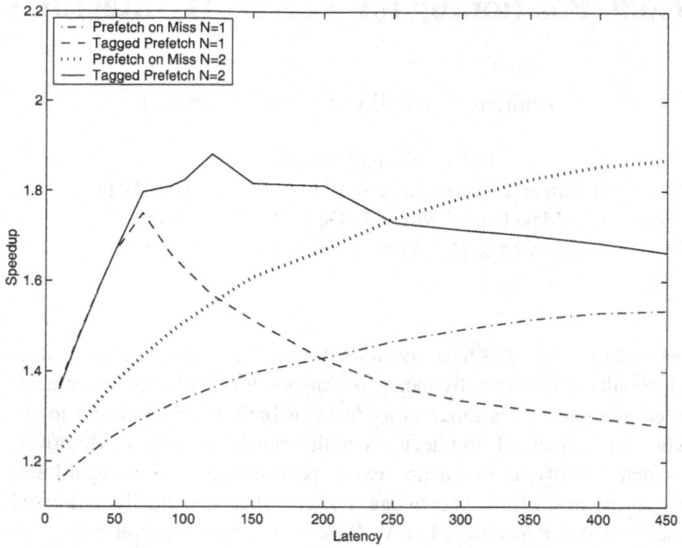

Fig. 7. Speedup vs. Latency for SEQSTREAM.

References

1. Fredrik Dahlgren, Michel Dubois, and Per Stenstrom. Sequential hardware prefetching in shared-memory multiprocessors. *IEEE Transactions on Parallel and Distributed Systems*, 6(7):733–746, July 1995.
2. Nathalie Drach. Hardware implementation issues of data prefetching. In *Proceedings of the International Conference on Supercomputing*, pages 245–254. ACM Press, 1995.
3. Norman P Jouppi. Improving direct-mapped cache performance by the addition of a small fully-associative cache and prefetch buffers. In *Proceedings of the 17th International Conference on Supercomputing*, pages 364–373. IEEE Press, 1990.
4. A J Smith. Cache memories. *ACM Computing Surveys*, 1992.
5. Steven VanderWiel and David Lilja. When caches aren't enough: Data prefetching techniques. *Computer*, pages 23–30, July 1997.

Code Positioning for VLIW Architectures

Andrea G.M. Cilio and Henk Corporaal

Delft University of Technology
Computer Architecture and Digital Techniques Dept.
Mekelweg 4, 2628CD Delft, The Netherlands
{A.Cilio,H.Corporaal}@et.tudelft.nl

Abstract. Several studies have considered reducing instruction cache misses and branch penalty stall cycles by means of various forms of code placement. Most proposed approaches rearrange procedures or basic blocks in order to speed up execution on sequential architectures with branch prediction. Moreover, most works focus mainly on instruction cache performance and disregard execution cycles. To the best of our knowledge, no work has specifically addressed statically scheduled ILP machines like VLIWs, with control-transfer delay slots.
We propose a new code positioning algorithm especially designed for VLIW-style architectures, which allows to trade off tighter schedule for program locality. Our measurements indicate that code positioning, as a result of tighter program schedule and removed unconditional jumps, can significantly reduce the number of execution cycles, by up to 21%, while improving program locality and instruction cache performance.

1 Introduction

Several trends of today's architectures contribute to make the instruction memory system a performance bottleneck. Reduced instruction set architectures have almost doubled [1] the instruction fetch bandwidth requirements. (Instruction-level parallel machines like VLIWs and superscalars further increase the required bandwidth.) The gap between memory and processor speed has incremented the memory cycle latency and consequently the cache miss penalty. Finally, deep pipelines and high ILP levels have increased the penalty which a processor can incur when a control transfer instruction is executed. These trends render several compile-time optimizations, like compile-directed prefetching of code and data, loop transformations for data caches, and code positioning critical for achieving optimal performance. This paper addresses *code layout*, or positioning for VLIW-style architectures. By positioning the code blocks adequately, the following improvements can be obtained:

1. *More effective use of the instruction cache.* The spatial locality of instruction cache accesses can be increased, while reducing the number of conflict misses, thereby improving cache utilization [2][3][4].
2. *Reduction of branch penalty overhead.* For architectures with branch prediction basic blocks can be arranged so as to reduce the branch penalty [5][6][7].
3. *Reduction of unconditional branches.* The basic blocks can be rearranged so as to eliminate frequently executed unconditional jumps [5][6].

B. Hertzberger et al. (Eds.): HPCN Europe 2001, LNCS 2110, pp. 332–343, 2001.

In contrast to architectures with branch prediction, some recent VLIWs machines, like Trimedia TM1000 [8] and the TI C62x family [9] expose the branch penalty to the compiler by introducing several branch delay slots. One of the goals of code positioning for these architectures is then to maximize the number of useful operations that can be imported into delay slots from other basic blocks while minimizing the number of frequently executed delay slots that cannot be sufficiently filled. Application-specific architectures may not even have an instruction cache: the whole program is downloaded to a local, fast-access instruction memory before starting execution. Clearly, for such machines the locality of the program code is irrelevant. This gives a degree of freedom in the selection of the branch direction, which can be exploited for reducing the execution cycle count.

These considerations suggest that the cost model and the goals of code positioning for a statically scheduled, VLIW-style machine with control transfer delay slots may differ from those found by previous works. The purpose of this paper is to explore alternatives to the existing code positioning algorithms specifically suited for this class of machines (with different instruction cache configurations), and to investigate the effects on the program schedule.

This paper is organized into 5 sections. Section 2 reviews previous work on code positioning. Section 3 presents our code positioning algorithm, which is evaluated on a group of benchmarks in Section 4. Section 5 summarizes the results.

2 Related Work

Code positioning can be applied at various levels of granularity. Early work in this direction focused on reducing page faults of virtual memory machines by positioning memory pages. Later, as the instruction cache began to play a critical role in the overall performance, the attention shifted towards finer levels of granularity, namely single procedures, segments of procedures and single basic blocks. Most of the techniques mentioned below, as ours, base their algorithms on profiling information gathered by previous program executions using representative input data.

In [3] Hwu *et al.* present a set of compiler techniques that improve instruction cache performance; among these are *Function layout* (i.e., basic block positioning) and *Global layout*, which tries to arrange whole procedures in a sequential order that minimizes conflict misses. McFarling [10], who also addresses both levels of granularity, proposes a code positioning algorithm that minimizes conflict misses; he also proposes to avoid caching instructions that are not frequently executed. Pettis and Hansen [5] present a refined version of Hwu's algorithm and evaluate also the reduction of branch penalty cycles and executed instructions. Gloy and others [4] build on the work of Pettis by taking into account the *temporal* ordering of executed procedures. Their work is restricted to procedure placement (i.e., global layout). Mendlson *et al.* [2] consider reducing of cache misses due to conflicting blocks in a loop. Their approach uses the concept of *abstract caches* to place a set of blocks of the same loop in a conflict-free manner. Controlled code duplication is used when multiple uses of the same blocks in different loops would conflict with other blocks in the respective loop. Differently from other works, this approach is directed by a static cost model and does not use program profil-

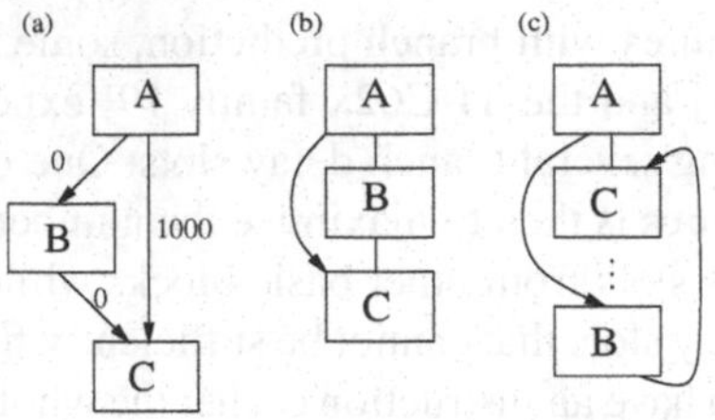

Fig. 1. Example of Code Placement. The numbers next to the edges represent execution counts obtained from profiling.

ing. Calder and Grunwald [6] exclusively address branch cost reduction. This restricted form of the code positioning problem, termed *branch alignment*, does not consider program locality. The authors propose improved models of the branch costs that expose the branch prediction schemes of the underlying machine microarchitecture. Young *et al.* [7] modeled the branch alignment optimization problem as a Directed Traveling Salesman Problem (DTS) and attained near-optimal speedup.

All these works evaluate sequential architectures, as opposed to explicitly programmed ILP machines. Furthermore, all these works have mainly been aimed at reducing the instruction cache misses or the branch penalty. We will show that for VLIW architectures with branch delay slots code positioning has a significant impact on the program schedule and the execution cycle count.

3 Code Positioning Algorithm

Traditional compilers generate code in which segments rarely executed during typical runs are interspersed with frequently used parts. Often, infrequent code is executed to handle exceptional situations. See for example block B in the control flow graph (CFG) of Fig. 1(a). A traditional compiler would place this CFG as shown in Fig. 1(b), thus forcing the processor to take a branch every time block A is entered. Figure 1(c) shows a better code placement of the same CFG. In this case, the branch will never be taken, and the control flow will fall through to block C. Even in architectures (like VLIWs) for which the branch penalty is identical in both paths, the second layout is preferable, because it increases the probability that blocks A and C end up in the same cache line.

We consider the problem of code positioning in the context of explicitly programmed ILP architectures of the VLIW type in which control transfer instructions have architecturally visible delay slots. Delay slots are an architectural feature to hide the penalty of control transfer instructions. The scheduler tries to fill the delay slots with operations always useful, which may logically precede the control transfer or may belong to basic blocks that are dominated by it. In machines with jump delay slots the scope and the algorithm of the instruction scheduler interact with code positioning and affect its effec-

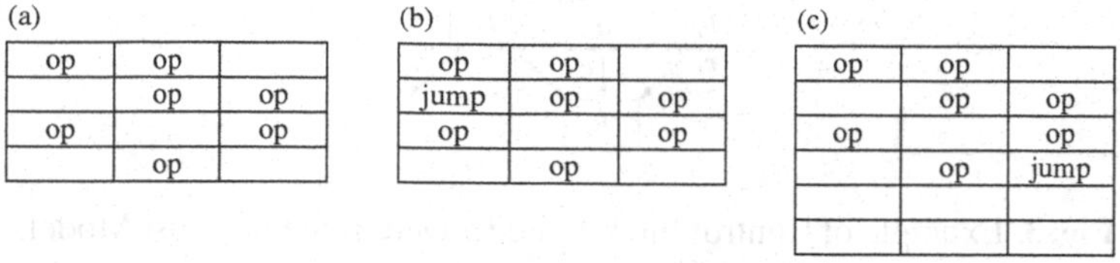

Fig. 2. Effect of adding a jump to a basic block: (a) original, partially scheduled code; (b) jump added with zero cost; (c) jump added with cost proportional to $L = 2$.

tiveness. Our scheduler uses the *region* scheduling scope [11][1] which allows to extract parallelism from several paths and move operation across join points. The scheduler first schedules the current basic block with a list-based algorithm and then tries to import operations from successor basic blocks in order to fill the unused operation slots and possibly reduce the overall path length to the successors.

We implemented code positioning as a pre-scheduling pass because of its simplicity. A post-scheduling pass would require partial rescheduling. Though simpler, a pre-scheduling pass leaves us with the difficulty of finding an accurate cost model to guide positioning. Recent works suggest that to achieve better results in cache miss reduction, a cost model based on cache and basic block size must be used [2][4]. However, this is not possible in a VLIW instruction scheduler with global scope because the block sizes depend on the program scheduling and are very difficult to estimate. Some blocks can be enormously enlarged[2], while other blocks may disappear altogether, absorbed into predecessor blocks owing to operation importing. To make things worse, the exact schedule itself depends on the code layout, as will be shown below. For these reasons, we will use the proximity heuristic to minimize cache misses: blocks that are placed closer have less chances of competing for the same cache line.

Cost Estimates. Differently than for architectures with branch prediction [6][7], the exact cost of branches and unconditional jumps strongly depends on the schedule. For example, the cost is very low if the scheduler succeeds in placing it in an empty operation slot L cycles before the end of the basic block, where L is the number of delay cycles that follow a jump or a branch. See the example in figure 2(a–b); each tile represents an operation slot and "op" indicates that a slot is occupied.

The goal of our algorithm is to arrange the basic blocks of a procedure in an order that minimizes the number of operations executed and maximizes the locality of the instruction fetch. This objective can be reformulated as the problem of deciding for each edge $e = a \rightarrow b$ whether block b must be the sequential successor of block a.

[1] Regions, which correspond to loop bodies, are the most general scheduling scope, in the sense that traces, superblocks and trees are also regions. Our schedule supports speculation and guarding, therefore can achieve an effect equivalent to hyperblock if-conversion.

[2] This can be due, e.g., to the presence of dependencies from long-latency operations, or to the fact that a block is a duplication point for operations imported up along another path.

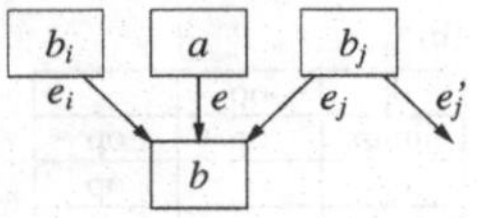

Fig. 3. Example of Control Flow Used to Determine the Cost Model.

This decision is guided by a function which estimates the cost of placing b immediately below a.

Definitions. In order to describe the algorithm, the following definitions are given:

1. $CFG(B, E, s)$ is a directed cyclic graph representing a procedure;
2. nodes $b \in B$ are the basic blocks of the procedure;
3. directed edges $b_i \rightarrow b_j \in E$ $(b_i, b_j \in B)$ are the control flow edges from a source block to a destination block;
4. let $b \in B$, then $pred(b) = \{b_i : \exists b_i \rightarrow b \in E\}$, $succ(b) = \{b_i : \exists b \rightarrow b_i \in E\}$;
5. $\exists! s \in B : pred(s) = \emptyset$; i.e. there is one and only one block (the entry point of the procedure) that does not have predecessors;
6. to each edge $e \in E$ is attached a weight f_e that gives the execution count obtained from the profiling.

We will begin with a cost model that considers only the execution cycles and ignores locality. Let us consider an edge $e = a \rightarrow b$ of an unconditional jump, as shown in fig. 3. If we knew the exact effect on scheduling of making b the sequential successor of a, we would be able to compute the cost of this decision. Given the complexity of our operation-based, global list scheduler [11], it is not possible to predetermine the effect of reordering a block without backtracking. We therefore resort to a cost estimate.

Let $E_u \cup E_c \subset E$ be the set of incoming edges of b; E_u is the set unconditional edges, while E_c is the set of edges which come from a basic block terminating with a conditional branch; f_e is the frequency of execution of the control flow edge e. As a first approximation, we consider only the *local* effects of placing b in the fall-through path of a, i.e., consider only the effects on the schedule of the predecessors of b and ignore the other basic blocks. The general formula of the local cost (estimated number of additional cycles required) of taking b as sequential successor of a is then:

$$SeqCost(e) = \sum_{e_i \in E_u \cup E_c} \phi(e_i) - \phi(e). \quad (1)$$

where $\phi(e)$ is the cycle penalty associated with an edge e when it cannot be selected as fall-through path. The cycle penalty depends on the characteristics of the source and destination blocks of the edge e. For an unconditional edge $\phi(e) = Lf_e \cdot c_u(e)$, where L is the number of delay slots of a control flow operation and $c_u(e)$ depends on how early the scheduler can place the jump operation of e (compare figure 2(b) and 2(c)). For a conditional branch edge the value of $\phi(e)$ is more problematic, because it depends on the placement of the alternative edge:

$$\phi(e) = \begin{cases} 0, & \text{if the other branch edge } e' \text{ is sequential} \\ (L+1) \cdot \min\{f_e, f_{e'}\}, & \text{if neither branch directions is fall-through.} \end{cases} \quad (2)$$

if either branch direction can be selected as fall-trough path then no unconditional jump needs to be inserted. Equation (2) takes the minimum frequency between the two branch edges because if a basic block with an unconditional jump must be inserted, it is always possible to assign the less frequent direction to this jump. Note that it is possible to have branches for which neither direction is fall-through. As noted in [6], in some cases such branch alignment can be preferable.

The expressions for $SeqCost(\cdot)$ and $\phi(\cdot)$ are local approximations: since the decisions taken in other blocks of the procedure affects the scheduling globally, $\phi(\cdot)$ depends on the ordering of all blocks of the scheduling scope. Given a basic block b, its best sequential predecessor is indicated by the edge $e \in pred(b)$ that minimizes $SeqCost(\cdot)$. If we assume that the sum terms in (1) are not affected by the positioning, then the minimization becomes a problem of maximizing the only term that is different in each expression. For unconditional edges the best fall-through path is then the most frequent path, as in the greedy algorithms of the literature [5][6]. This path is also the best choice from the point of view of program locality, given our proximity heuristic. For branch edges $\phi(\cdot)$ completely depends on the placement of the alternative edge. On the one hand, we would like to postpone the decision for less critical branches until we know the placement of one of the two edges and we can use (2) to estimate the local costs. On the other hand, a critical branch edge should be decided first.

This problem motivated the introduction of a *priority function* to determine the order in which edges are selected for placement. Our priority (applied to blocks) ensures that outgoing edges of more critical blocks are selected first, thereby postponing the decision for less critical branch edges until we might have a better estimate for $\phi(\cdot)$. As priority, we take the frequency for the blocks with unique outgoing edge, and the following function for the blocks with outgoing branch edges e, e':

$$prio(b) = \min\{f_e, f_{e'}\}. \quad (3)$$

For the cost of a branch edge when the placement of the alternative edge is unknown we take the following formula, in which $c_b \in [0, 1]$ reflects the probability that the branch needs an additional unconditional jump: $\phi(e) = c_b(L+1)\min\{f_e, f_{e'}\}$.

These cost and priority functions are not adequate if program locality is also to be taken into account. If only the program locality is relevant, a more suitable priority is simply based on the edge execution frequency, therefore a block terminating in a branch has the same priority of its most frequent outgoing edge:

$$prio(b) = \max\{f_e, f_{e'}\}. \quad (4)$$

Similarly, the cost $\phi(\cdot)$ for a branch edge uniquely depends on its frequency, like for unconditional edges: $\phi(e) = Lf_e$.

3.1 Algorithm Description

Our algorithm can be divided into two parts; the first part sorts the basic blocks for later selection according to a priority, the second part selects the basic blocks and places them according to the local cost estimate described above.

First, a priority is assigned to every basic block of a procedure. For blocks with a unique successor, the priority is proportional to the execution count of the jump. Blocks which end with a conditional branch are assigned the following priority, which combines (3) and (4):

$$prio(b) = \alpha \cdot \min\{f_e, f'_e\} + (1 - \alpha) \cdot \max\{f_e, f'_e\}. \tag{5}$$

The parameter $\alpha \in [0, 1]$ allows to balance execution count reduction ($\alpha = 1$) against enhancement of program locality ($\alpha = 0$). The latter choice results in the same selection order of the greedy algorithms found in the literature.

Algorithm 1 places the basic blocks of CFG by assigning a label to every edge $e \in E$. Initially, all edges are labelled as undefined. An edge $e = b_i \rightarrow b_j$, is labelled *SEQ* if b_j is the sequential successor of b_i; vice versa, $label(e)$ = *NSEQ* if b_j does not follow b_i. The actual placement is performed after all edges have been assigned a label and merely involves moving pointers to list elements around; this pass inserts any unconditional jumps and basic blocks necessary to ensure that the code executes correctly. The function BestCandidate estimates the sequential cost with (1) and returns the destination block with smaller cost. The estimation of $\phi(e)$ for a branch edge e combines execution cycle reduction and locality enhancement objectives:

$$\phi(e) = \begin{cases} 0, \text{if only the other edge, } e' \text{ is sequential,} \\ c_b(1 - \alpha)Lf_e + c_b\alpha(L + 1) \cdot \min\{f_e, f_{e'}\}, \text{otherwise.} \end{cases} \tag{6}$$

Note that at this point $\phi(e)$ may be exactly determined for some branch edges because their branch has been already aligned. When both branch directions are known not to be fall-trough we take $c_b = 1$.

Like the greedy algorithm proposed by Pettis and Hansen [5], this algorithm is able to restructure the placement of loop basic blocks so that a loop header appears in the middle of the loop. In addition, this algorithm partially solves the problem indicated in [6], whereby a conditional edge is given precedence over an unconditional edge with the same frequency, without need to exhaustively search several possible orders, simply because unconditional paths can be given priority over conditional edges.

A few additional, practical considerations are in order. For blocks that contain indirect (computed) jumps the positioning of the successors is relevant only for the locality. All successor edges are marked *NSEQ* before starting LayoutBasicBlocks. This improves the placement of other blocks with branches. We must check for cyclic chains of edges marked *SEQ*. If such a chain would be formed by assigning a *SEQ* label, the edge is assigned label *NSEQ* (and the cycle is broken).

4 Evaluation

Our placement algorithm has been evaluated in a series of experiments. This section first describes the method used to perform our measurements and the target machines

Algorithm 1 LayoutBasicBlocks (CFG).

```
AssignLabels(CFG)
for all b ∈ B, sorted by priority(·) do
  if |succ(b)| = 1 ∧ ¬label((b, b1)) = NSEQ then
    label((b, b1)) = SEQ,   label((b', b1)) = NSEQ   ∀b' ∈ pred(b1), b' ≠ b
  else if |succ(b)| = 2 then
    if label((b, b1)) = NSEQ ∧ label((b, b2)) = NSEQ then
      // Nothing to do
    else if label((b, b1)) = NSEQ then
      label((b, b2)) = SEQ,   label((b', b2)) = NSEQ   ∀b' ∈ pred(b2), b' ≠ b
    else if label((b, b2)) = NSEQ then
      label((b, b1)) = SEQ,   label((b', b1)) = NSEQ   ∀b' ∈ pred(b1), b' ≠ b
    else
      b_best = BestCandidate(b1, b2)
      label((b, b_best)) = SEQ,   label((b, b_worst)) = NSEQ
      label((b', b_best)) = NSEQ   ∀b' ∈ pred(b_best), b' ≠ b
    end if
  else
    label((b, b')) = NSEQ   ∀b' ∈ succ(b)
  end if
end for
```

used. Then the benchmark programs are briefly presented. The rest of this section is dedicated to the results of the simulated execution and their analysis. Also, we compared our algorithm with an improved version of Pettis and Hansen's [5] which, as proposed in [6], uses a specific cost model for our architecture.

4.1 Experimental Setup

To perform our measurements we used a cycle-accurate simulator for transport-triggered architectures (TTAs)[11]. TTAs are a class of statically scheduled architectures for which data transports and bypasses of the general-purpose registers are explicitly programmed. For the purpose of this paper, we can consider our test architecture to be comparable to a VLIW. The simulator is part of a software design system that allows to schedule and simulate the execution of programs for a range of machines. The cache performance has been measured by using the *cheetah* cache simulator [12], which can model the behavior of different cache configurations. In this way we are able to evaluate the effect of code placement on the cache performance.

We performed our evaluations on DSP applications and on the Unix programs *compress* and *cjpeg*, the standard JPEG compressor. The DSP programs, taken from [13], are subdivided into audio applications (*arfreq, g722main, music*), and image processing applications (*edge, expand, smooth*). These benchmarks were compiled with *gcc* (ported to our architecture) into sequential code and then scheduled. We profiled and simulated complete programs, including library code.

The simulations have been performed on two different machines, one "low-end" (M1) and one "high-end" (M2). Table 1 summarizes their characteristics. Both ma-

Table 1. Simulated Target Machines. "Long immediates" is the number of long immediates that can be specified each cycle.

	quantity				quantity	
resource	**M1**	**M2**	**unit**	**latency**	**M1**	**M2**
transport busses	3	8	ld/st unit	2	2	3
long immediates	1	2	IALU	1	2	4
integer registers	24	64	multiply	3	1	1
FP registers	16	48	divide	10	1	1
boolean registers	2	4	FPU	3	1	1

chines have 2 cycles of delay for control transfer operations. The transport busses are used for the data traffic between functional units or registers. Two transport busses roughly deliver the same instruction bandwidth of one conventional VLIW operation slot. Unoccupied transport slots contain a no-transport code.

4.2 Experimental Results

Table 2 summarizes the effect of code positioning on the program schedule. The cycle counts do not include the stall cycles due to instruction cache miss. Data memory accesses are assumed to always hit the cache. Each row of the tables refers to a different benchmark. For each machine, the left column below label *cycles* shows the cycles spent to execute the benchmark without code positioning optimization (i.e., using the code layout generated by gcc); the right column gives the cycle count reduction obtained with code positioning. This reduction is the performance improvement to be expected on a machine in which the fetch subsystem never stalls the execution engine. The columns below *size* show the static code size (in instruction words) of each benchmark before code positioning and the size reduction achieved by code positioning. The next two columns show the percentage of control transfer operations *(CT)* executed in the original and in the optimized program. Function calls are not influenced by this optimization and are not included in the counts (their percentage is very low). Although the extent of the speedup varies largely, all benchmarks improve. The programs showing best speedup are *compress*, *smooth* and *expand*; this is partly explained by the high frequency of control flow transfers (16–24% of operations are control transfers).

Our algorithm tries to reduce conflict misses and exploit cache line locality in the instruction cache by ordering frequently executed paths sequentially. This results in a longer sequence of instructions executed between two taken control transfer operations. To evaluate the instruction cache performance, we measured the miss reduction of a number of direct-mapped caches. We simulated caches with 16 instruction-word lines, with sizes ranging from 128 to 8192 instruction words.[3] Figure 4 shows the cache miss count in the original and in the optimized code for a selection of benchmarks. All

[3] Our binary encoder produces instruction words of 12 and 32 bytes for M1 and M2 machines, respectively. Much denser encodings are easily attainable, but we did not investigate this aspect, since it is marginal to our discussion.

Table 2. Effect of Code Placement on the Program Schedule.

benchmark	machine 'M1' exec. cycles orig.	red.%	size orig.	red.%	CT % orig.	opt.	machine 'M2' exec. cycles orig.	red.%	size orig.	red.%	CT % orig.	opt.
compress	22.3M	16.1	6015	7.2	20.4	13.1	18.9M	21.3	5143	9.6	16.2	10.2
cjpeg	4.4M	3.7	13031	0.2	12.3	10.9	3.2M	4.0	9838	1.0	12.1	10.9
arfreq	11.5M	1.8	1541	7.7	8.3	7.9	8.3M	2.4	1352	8.4	7.5	7.1
g722main	19.1M	11.2	7353	8.5	11.9	8.1	12.1M	20.1	5804	10.3	10.1	7.1
music	37.6M	3.4	6494	-0.3	10.2	9.6	24.8M	8.7	5648	-0.6	9.1	8.0
edge	1.0M	11.8	8941	19.3	19.6	16.0	0.8M	17.2	8155	22.5	15.9	12.1
expand	0.6M	16.5	6766	4.5	24.7	19.7	0.5M	21.2	6005	5.9	23.6	18.1
smooth	0.4M	12.5	6049	11.1	20.5	16.7	0.4M	19.8	5377	13.6	17.1	12.6
average	12.1M	9.6	7023	7.3	16.0	12.8	8.6M	14.3	5915	8.8	13.9	10.8

these programs have a rather small footprint (all fit in a 128KB cache). The miss rate reduction is of course relevant only for caches where conflict and capacity misses occur. For *compress*, for example, the reduction is above 50% for all relevant cache sizes. Other benchmarks, like *cjpeg*, show more modest improvement. The case of *arfreq* is interesting; this benchmark shows a slight miss reduction increase for certain cache sizes. This is due to the presence of two very critical loops in different procedures. Depending on the placement of the two procedures, these loops may or may not generate a large number of conflict and capacity cache misses. Such problem can be substantially alleviated by procedure positioning, as shown in [2][4].

The overall execution time is computed by combining the execution cycle count with the stall cycles due to instruction cache miss. Table 3 summarizes the results obtained by simulating the benchmarks on target machine M2 with a cache size of 128 and 1024 instruction words, (4KB and 32KB, respectively). For each cache, the column *cycles* shows the total number of cycles (including stalls). Next columns show the execution cycle speedup and the overall cycle count reduction including stalls. The columns under *stalls* show the fraction of stall cycles in the original and the optimized program. The results show that the speedup obtained is highly dependent on the cache size and so is the effect of the two optimizations: program locality and execution cycle count. For large caches, e.g., execution cycle reduction dominates the achieved speedups.

Alongside our code placement algorithm we implemented and tested the block placement algorithm described by Pettis and Hansen. Our algorithm always yields lower execution cycle counts, although the difference is very small. On average, we measured 14.3–13.5% execution cycle reduction (depending on the value of α) versus 13.2%. The cache performance is also very similar. Pettis and Hansen's algorithm performs slightly better when $\alpha = 1$ is chosen for our algorithm and the cache is small. For low values of α our algorithm always performs better.

We also evaluated the impact of the heuristic parameter α (described in section 3.1). For lack of space, the discussion of the results are omitted. In summary, while most programs, as we anticipated, showed a trade-off between execution cycle count and program locality, the entity of such differences were minimal.

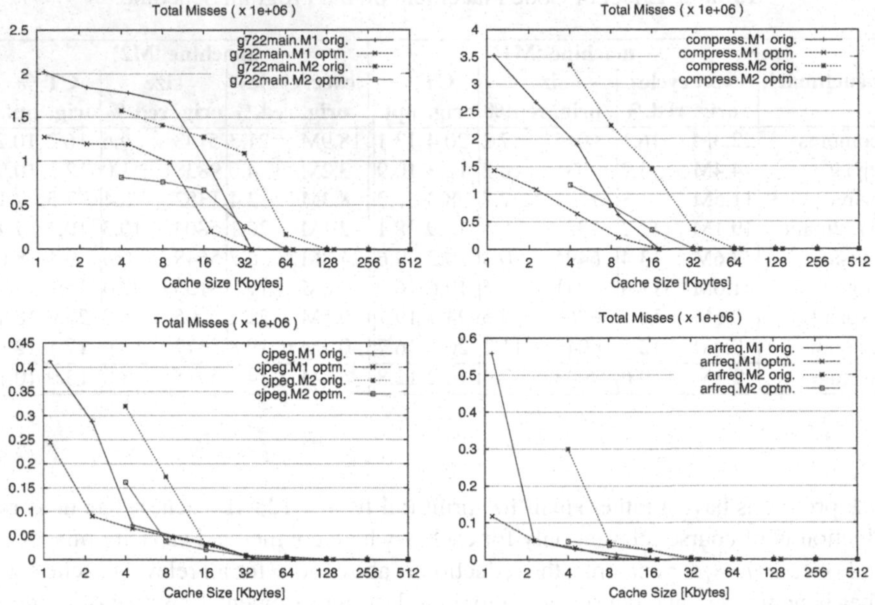

Fig. 4. Miss Reduction with Various Instruction Caches for *g722main*, *compress*, *arfreq* and *cjpeg*.

5 Conclusions

Code placement is an important optimization technique, which has a sizeable effect on horizontally scheduled ILP machines with jump delay slots. In this paper we showed that a placement algorithm can be tuned for VLIW-type architectures by placing higher priority to reducing the unconditional jumps executed. Reducing the number of critical jump delay slots has a positive effect on the program schedule. The measurements confirm that the simple reduction in execution cycle count (up to 21%) is a significant factor of the overall speedup and becomes dominant when the cache size is large. The algorithm by Pettis and Hansen achieves comparable results, indicating that the heuristic parameters have weak influence on the speedup.

References

1. J. W. Davidson and R. A. Vaughan. The effect of instruction set complexity on program size and memory performance. In *ASPLOS-II*, pages 60–64, Palo Alto, CA, 1987.
2. Abraham Mendlson, Shlomit S. Pinter, and Ruth Shtokhamer. Compile time instruction cache optimizations. In *Compiler Construction*, pages 404–418, April 1994.
3. W. W. Hwu and P. P. Chang. Achieving high instruction cache performance with an optimizing compiler. In *ISCA-16*, pages 242–251, Jerusalem, Israel, May 1989.

Table 3. Overall Speedup Obtained with Two Different Cache Sizes, Target Machine 'M2'.

benchmark	128-word cache					1024-word cache				
	cycles	cycle red. %		stalls %		cycles	cycle red. %		stalls %	
	orig.	exec.	total	orig.	optm.	orig.	exec.	total	orig.	optm.
compress	86.0M	21.3	55.6	78.03	61.07	18.9M	21.3	21.3	0.16	0.15
cjpeg	9.6M	4.0	34.3	66.69	51.36	3.3M	4.0	4.4	3.63	3.25
arfreq	14.3M	2.4	36.4	41.89	10.90	8.3M	2.4	2.4	0.01	0.01
g722main	43.6M	20.1	40.6	72.26	62.67	15.5M	20.1	37.6	21.99	0.14
music	108.5M	8.7	52.1	77.14	56.48	25.1M	8.7	9.0	1.16	0.86
edge	3.2M	17.2	37.9	74.50	66.00	1.3M	17.2	43.6	34.33	3.66
expand	1.6M	21.2	47.4	65.63	48.53	0.7M	21.2	37.1	21.77	2.02
smooth	1.5M	19.8	41.5	74.97	65.68	0.5M	19.8	43.4	30.10	0.86
average	33.5M	14.3	43.2	68.9	52.8	9.2M	14.3	24.9	14.1	1.4

4. Nikolas Gloy and Michael D. Smith. Procedure placement using temporal-ordering information. *ACM TOPLAS*, 21(5):977–1027, September 1999.
5. Karl Pettis and Robert C. Hansen. Profile guided code positioning. In *PLDI*, pages 16–27, White Plains, New York, June 1990.
6. Brad Calder and Dirk Grunwald. Reducing branch costs via branch alignment. In *ASPLOS-VI*, pages 242–251, October 1994.
7. Cliff Young, David S. Johnson, David R. Karger, and Michael D. Smith. Near-optimal intraprocedural branch alignment. In *PLDI*, pages 183–193, June 1997.
8. Jan Hoogerbrugge. Instruction scheduling for trimedia. *JILP*, 1(1–2), 1999.
9. Texas Instrument Inc. *TMS320C6000 Programmer's Guide*, 2000.
10. S. McFarling. Program optimization for instruction caches. In *ASPLOS-III*, pages 183–193, May 1989.
11. Jan Hoogerbrugge. *Code Generation for Transport Triggered Architectures*. PhD thesis, Technical University of Delft, February 1996.
12. Rabin Sugumar. *Multi-Configuration Simulation Algorithms for the Evaluation of Computer Architecute Designs*. PhD thesis, University of Michigan, August 1993.
13. Paul M. Embree. *C Algorithms for Real-Time DSP*. Prentice Hall, 1995.

Selective Register Renaming: A Compiler-Driven Approach to Dynamic Register Renamin

N.Zingirian and M.Maresca

DEI - University of Padua, Italy
Dipartimento di Elettronica ed Informatica – University of Padua, Italy

Abstract. Dynamic register renaming is a mechanism present in many high performance microprocessors of latest generation aimed at removing false dependencies from the code. Unfortunately, in many cases, this mechanism keeps busy more registers than the necessary. In this paper we introduce a novel technique, called Selective Register Renaming, in which the compiler helps the processor to save physical registers when the hardware renames registers. The paper explains the principles of this technique and shows its effects on several Livermore Kernel Loops.
Keywords: Dynamic Register Renaming, Instruction Level Parallelism, Loop Parallelization, Register Allocation.

1 Introduction

Register Renaming (RR) consists of removing false dependencies present in the code by allocating new registers to eliminate either Write-after-Read or Write-after-Write hazards[1]. RR can take place either via software[4], during register allocation/assignment at compile time, or via hardware[6], during instruction decode at run time. Since Hardware RR is widely adopted on the current microprocessor generation (e.g., [7]), this paper focuses on the latter approach. This approach consists of modifying the register identifiers contained in the operand fields of the instruction words on the fly at the decoding stage, replacing them with the identifiers of physical registers that are not in use[5].

Unfortunately this approach, while very simple and effective, is not fully efficient. In fact it is well known that this approach frequently leaves registers idle many clock cycles before considering them available again. Excellent works in literature deal with this inefficiency adopting more sophisticated hardware mechanisms, such as the virtual-physical register support[3].

In this paper we propose and evaluate a new scheme, called Selective Register Renaming (SRR). The main idea of SRR is that the processor does not allocate a new physical register whenever it decodes an instruction that writes a results on a logical register, like in the current Hardware RR schemes. On the contrary it decides whether to allocate a new physical register or to use the old physical registers, relying on compiler hints. The paper is organized as follows. Section 2 describes the actions taken by last generation superscalar processors to perform register renaming. Section 3 proposes the SRR approach. Section 4 shows the

B. Hertzberger et al. (Eds.): HPCN Europe 2001, LNCS 2110, pp. 344–352, 2001.

improvement delivered by SRR in terms of register utilization and parallelism exposed on a set of Livermore Kernel Loops. Section 5 provides a discussion of the SRR.

2 Background

In this paper we refer to the Hardware Register Renaming scheme based on a single register file, like the one adopted in the Mips R10000[7] and Alpha 21264[2] processors. We present this scheme by referring to the *Rename* and *Reclaim* functions of Fig. 1.

While decoding instruction I, the processor applies the *Rename* function to the source operand identifiers of I (i.e., the instruction word fields $I.r_1$ and $I.r_2$) and to the destination operand identifier (i.e., the instruction word field $I.w$) specified in the current instruction. The source register logical identifiers are directly replaced with the corresponding physical identifiers retrieved from a table called *Register Alias Table* (**RAT**), as shown in lines 2-3. The renaming of the destination register, on the contrary, consists of several operations (lines 4 to 8). In particular, a new physical register identifier is extracted from a list of free register identifiers (*Free List*, **FL**). If no free registers are available, no renaming can take place and the current instruction (as well as the following ones) stalls until at least one physical register is released. Otherwise, the new physical register identifier extracted from **FL** replaces the logical identifier (line 7) and the **RAT** is updated (line 8). Before updating $I.w$ the processor needs to save the old physical register identifier corresponding to the current logical identifier in an additional field of I, called $I.old$. In fact the old physical register cannot be released at this stage, because it can be accessed by instructions still in execution. The processor will release that physical register when instruction I will be graduated.

```
1.  Rename (I)
2.      I.r1←RAT[I.r1]
3.      I.r2←RAT[I.r2]
4.      while (is_void(FL)) stall
5.      new← dequeue (FL)
6.      I.old←RAT[I.w]
7.      I.w ← new
8.      RAT[I.w]←new
```

```
1.  Reclaim(I)
2.      enqueue (I.old, FL)
```

Fig. 1. Pseudocode of Register Renaming.

When the processor graduates instruction I, it performs the *Reclaim* function (see Fig. 1). An instruction is graduated when all the preceding [1] instructions

[1] The precedence refers to the instruction order as it appears in the code text, regardless the execution order decided by dynamic scheduler

have been already executed and graduated. As a consequence the old physical register identifier saved in *I.old* is surely not used by any other instruction, therefore it can be released, i.e. added to the **FL** (line 2).

We observe that the scheme presented allocates a new physical register for each logical identifier *I.w*, regardless whether *I.old* causes WAW/WAR hazards or not. The point is that, if *I.old* does not cause hazards, the action of associating logical identifier *I.w* to a *new* physical register does not increases the execution efficiency of *I*, but keeps both *I.old* and *new* busy. If more registers are kept busy than the necessary, then the Rename function stalls (line 4) more times than the necessary and introduces execution inefficiency.

3 Selective Register Renaming

In this section we propose a register renaming scheme, called Selective Register Renaming (SRR), that does not allocates new physical registers unconditionally, like the conventional scheme. On the contrary, in some cases, it leaves or the old logical-to-physical mapping unvaried, to reduce the amount of physical registers in use.

The SRR scheme makes the compiler and the hardware cooperate to avoid to allocate new physical registers when it is not necessary.

The following proposition identifies the case in which the compiler can detect that the conventional Hardware Register Renaming allocates an useless new physical register.

Proposition 1. *If the destination register of the current instruction appears as a source register only in the instructions that precede the current instruction in the DAG*[2]*, then this logical register can be mapped on the same old physical register on which it is already mapped, without introducing false dependencies.*

For example, Proposition 1 is true when an optimizing compiler generates an instruction sequence like

```
add r1<- r2,r3
mul r3<-r1,r4
mul r2<-r1,r5
```

The compiler immediately reuses registers `r3` and `r1` because it exactly know the life ranges of all operands. On the contrary, RR allocates two additional registers, although the compiler allocates register efficiently, as shown below:

```
add new1<- r2,r3
mul new2<-new1,r4
mul new3<-new1,r5
```

The two additional physical registers allocated by RR (in particular, `new2` and `new3`) are not necessary because the instruction that writes them cannot be executed, due to true dependencies, before the operand stored in the old physical register dies.

[2] The Direct Acyclic Graph of the true dependencies

The idea of the SRR is to mark the current instruction with an appropriate bit, say $I.b$ when, at compile time, the compiler detects the condition identified by Proposition 1. The decoding logic allocates no new physical registers when it decodes instruction I with bit $I.b$ enabled, and the graduation logic does not free any register when it graduates instructions when bit $I.b$ is enabled.

Figure 2 shows the SRR pseudocode and reports the differences from the RR presented in Figure 1 in boxes. In the SRR scheme, when $I.b$ is enabled, destination register identifiers are processed in the same way as source register identifiers. When an instruction with the b bit enabled graduates, the Reclaim function does not do anything, because the physical register does not evict any old physical register.

```
Rename (I)
    I.r1←RAT[I.r1]
    I.r2←RAT[I.r2]
    if I.b == 0
        while (is_void(FL)) stall
        new← dequeue (FL)
        I.old←RAT[I.w]
        RAT[I.w] ← new
    else
        I.w←RAT[I.w]
```

```
Reclaim(I)
    if I.b == 0
        enqueue (I.old, FL)
```

Fig. 2. Pseudocode of Selective Register Renaming.

4 Experiments

4.1 Evaluation of Register Utilization

In order to estimate the register utilization inefficiency due to the allocation of new physical registers we devised an appropriate metric and measured it from simulation traces. This metric, called "idle register-cycles", aggregately measures how many registers are wasted by each instruction during program execution and how many cycles each of these registers is kept busy. For instance, 12 "register-cycles" correspond to the fact that 6 unnecessary registers have been kept busy along 2 cycle, or, equivalently, 4 unnecessary registers for 3 cycles, or 3 registers for 4 cycles, etc.

The following proposition specifies how we computed the "idle register-cycles" of an instruction.

Proposition 2. *When instruction I is issued, if $I.old$ is not involved in output dependencies or anti-dependencies, then I is said to be* issuable without renaming *and adds c "idle-register cycles" to the total "idle register-cycles" of I, where c is the number of cycles spent between renaming of I and graduation of I.*

The c cycles are counted between renaming and graduation, because, if the current instruction allocates a new unnecessary physical register, the registers is unnecessary not until it is deallocated, but until the previous physical register corresponding to the same logical register is deallocated, i.e. when the current instruction is graduated (see Sect 2).

In loops, being each instruction repeated several times, the total "idle register-cycles" for an instruction in the overall execution is equal to the sum of the "idle register-cycle" measured at each iteration for that instruction. The metric is normalized for convenience over the total "register-cycles" available, that is the product of the number of execution cycles by then number registers available.

In synthesis, given instruction I with one destination operand in the body of a loop of N iterations (so that N instance of I, say $I_1 \ldots I_N$, are executed), being TOT_{cycles} the clock cycles taken by a microprocessor to execute the program, and TOT_{regs} the number of physical registers present in the architecture of the same type as the destination operand, the "idle register cycles" of I, IRC(I) is defined as follows

$$IRC(I) = 100 \times \frac{\sum_{I_i \,|\, issuable\ w/o\ renaming}^{I_N} (cycle_{grad}(I_i) - cycle_{renam}(I_i))}{TOT_{cycles} \times TOT_{regs}} \quad (1)$$

Case Study. We consider the Livermore Kernel 1 as an example. Figure 3 shows its source code. Figure 4 shows the corresponding assembly code of the loop (as generated by GNU C targeted to HP PA-RISC 7100 processor) represented in the true dependency graph. Dark arrows indicate the operands for which Proposition 1 holds. For these operands the compiler does not waste any register and no false dependencies take place. However, if a processor equipped with RR executes the code in Fig. 3 then it wastes a new physical register for each operand pointed by the dark arrow, for each iteration. On the contrary the white arrow indicates the only case of false dependency introduced by the compiler. In this case the hardware register renaming removes such a false dependency.

Table 1 reports the "idle register-cycles" of each instruction obtained by simulating the execution of the Livermore Kernel Loop 1, for different register file sizes. In the simulated processor, the functional unit number, the maximum instruction issue, the instruction queue size and the graduation degree have been set in such a way that none of them causes performance bottleneck, so that physical register limitation is the only bottleneck.

The reader should notice that the sum of the IRCs of all the instructions of the loops exceeds 50%, for all the register file sizes simulated . It means that the inefficiencies of RR strategy in this experiment wastes more than half register resources, regardless the register file size.

Table 1. Percentage of idle register-cycles, for each instruction and some register file sizes.

Instruction	32+32 regs	40+40 regs	48+48 regs	56+56 regs	64+64 regs
add %r22,%r26 → %r19	6.5	3.4	3.0	1.2	1.2
add %r23,%r26 → %r20	6.5	3.4	3.0	1.2	1.2
load (%r19) → %fr9	0.0	0.0	0.0	0.0	0.0
load (%r20) → %fr8	0.0	0.0	0.0	0.0	0.0
fmult %fr10,%fr9 →%fr9	9.3	9.8	10.2	10.5	10.7
fmult %fr11,%fr8 → %fr8	9.3	9.8	10.2	10.5	10.7
fmult %fr9 ,%fr8 → %fr9	9.2	9.8	10.2	10.5	10.7
fload (%r29+%r21),%fr8	0.0	0.0	0.0	0.0	0.0
fmult %fr8, %fr9 → %fr8	9.8	10.3	10.4	10.7	10.7
add %r23, 8, %r23	5.0	2.6	2.4	1.0	1.0
fadd %fr22,%fr8 → %fr8	10.7	10.8	10.8	10.8	10.9
store %fr8 → (%r31+%r21)	-	-	-	-	-
add %r21, 1 → %r21	0.1	0.0	0.0	0.0	0.0
add %r22, 8 → %r22	0.1	0.0	0.0	0.0	0.0

To measure the effects of the SRR scheme, we considered the same kernel. We implemented the SRR scheme in our simulator and measured the IRC. Table 2 reports the simulation results.

```
for ( k = 0 ; k < SPAN ; k++ )
   X[k] = q + Y[k]*( r*Z[k+10] + t*Z[k+11] );
```

Fig. 3. Livermore Kernel 1: Loop Structure.

The reader should notice that SRR forces to zero the IRC of instructions that have the b bit enabled. which are represented in the true dependency DAG of Figure 4 pointed by a black arrow because satisfy Proposition 1. In fact such instructions cannot waste any register-cycle, since they do not require any new phyisical register to rename identifiers.

4.2 Evaluation of Instruction Level Parallelism

In this section we show the effectiveness of SRR in a set of experiments based on the first 13 Livermore Kernels to verify that the reduction of IRC corresponds to the enhancement of the instruction level parallelism. The experiments have been carried as follows. First, we transformed the source code of each Livermore Kernel applying, when necessary, scalar replacement to make memory dependencies explicit and loop unrolling to expose the parallelism of the task. We compiled the resulting code with GNU C targeted to PA-RISC architecture version 1.0 (-`m-mpa-risc-1-0` option, in order to avoid instructions with multiple destination registers, such as *prog fmpyadd* instruction). All the instructions of the code have been set with $b = 0$. Next, we simulated the execution of the kernel loop

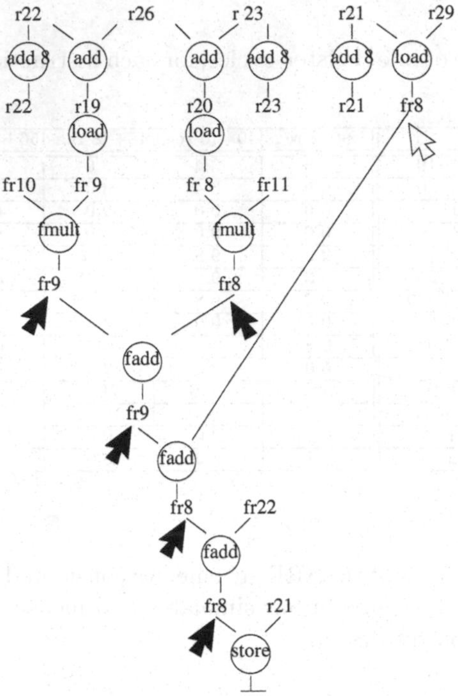

Fig. 4. DAG of Livermore Kernel 1: Instructions pointed by dark arrows satisfy Proposition 1.

Table 2. Percentage of idle register-cycles, for each instruction for different RR schemes, with 48 registers available.

Instruction	RR	SRR
add %r22,%r26 → %r19	2.8	1.3
add %r23,%r26 → %r20	2.8	1.3
load (%r19) → %fr9	0.0	0.0
load (%r20) → %fr8	0.0	0.0
fmult %fr10,%fr9 →%fr9	10.2	0.0
fmult %fr11,%fr8 → %fr8	10.2	0.0
fmult %fr9 ,%fr8 → %fr9	10.2	0.0
fload (%r29+%r21),%fr8	0.0	0.0
fmult %fr8, %fr9 → %fr8	10.3	0.0
add %r23, 8, %r23	0.0	0.0
fadd %fr22,%fr8 → %fr8	10.1	0.0
store %fr8 → (%r31+%r21)	0.0	0.0
add %r21, 1 →	0.0	0.0
add %r22, 8 → %r22	0.0	0.0

body, which is a basic block in the kernels selected, using a simulated processor, with infinite resources, apart from the register file size. The average number of instructions graduated at each cycle and the idle register-cycle are the outcomes of the simulation. Then we set $b = 1$ for the instructions that satisfy Proposition 1 and then we re-executed the code, reporting the results in Table 3.

Tab. 3 shows that for several Kernels SRR delivers a significant speedup, up to the factor 2.20 measured in the Livermore Kernel 7 in case of 64 + 64 registers available. SRR has no effects in the case in which true dependencies or control dependencies create a critical path (Kernels 3 and 11) and in the case in which the condition of Proposition 1 never occurs in the code (Kernels 2,10 and 12).

Table 3. Average number of instructions graduated per cycle using RR and SRR on Livermore Kernels 1-13.

	48 + 48 Regs		56 + 56 Regs		64 + 64 Regs	
Ker	RR	SRR	RR	SRR	RR	SRR
1	7.923	10.163	9.043	12.310	10.163	13.523
2	7.183	7.183	7.613	7.613	11.967	11.967
3	9.607	12.727	11.417	14.210	12.727	16.983
4	7.630	8.450	9.477	11.737	11.660	11.7503
5	3.017	3.017	3.017	3.017	3.017	3.017
7	5.953	10.397	5.953	13.647	6.840	15.063
8	6.590	9.890	7.580	11.703	9.890	13.023
9	5.907	8.797	8.300	8.797	8.797	11.687
10	7.667	7.667	10.110	10.110	10.137	10.137
11	2.863	2.863	2.863	2.863	2.863	2.8633
12	9.717	9.717	10.727	10.727	14.590	14.590
13	6.380	7.327	7.843	8.063	8.063	9.533

5 Conclusions and Future Work: The Precise Interrupts

In this paper we have shown that the cooperation of the compiler and the hardware for register renaming allows increasing the efficiency of the register utilization and, consequently, of the instruction level parallel execution. However a delicate point is still under investigation about the handling of precise interrupts.

In the traditional RR scheme, the set of physical registers currently in use holds not only the current process state but also the history of the previous states. This history is sufficient to undo all the instructions decoded and not yet committed, since the RR scheme never overwrites old values before committing them in order. As a consequence, in event of interrupt, the "in-order state" recovery consists only of restoring the logical-to-physical mapping valid just before the interrupted instruction has been decoded.

Unfortunately, unlike RR, the SRR scheme does not keep track of the full history in the physical registers, because, whenever an instruction having the b

bit enabled writes back a result, it may overwrite that result on a non-committed value, i.e. it may cancel out an history item from the physical register file. As a consequence, to support precise interrupts, the SRR scheme needs to record old values overwritten by instructions having the b bit enabled.

We are investigating a combined approach based on the recovery based both on the values present in the register file and in appropriate fields of the reorder buffer. Qualitatively speaking, we argue that conventional RR makes some history values spend several clock cycles in a costly memory (i.e., the register file) with concurrent access capabilities, even though these values will be accessed only in event of interrupt.

On the contrary the SRR eliminates this inefficiency by allocating the values that can be accessed simultaneously (because their life ranges overlap) on costly register file locations, and allocating the remaining history values on a cheaper structure, such as the reorder buffer entry. The quantitative evaluation of the costs in terms of ports for both solutions is the theme of our future works.

References

1. J.L Hennessy and D.A. Patterson. *Computer Architecture: A Quantitative Approach.* Morgan Kaufmann, San Mateo, CA, second edition, 1996.
2. Richard E. Kessler. The Alpha 21264 microprocessor: Out-of-order execution at 600 MHz. In IEEE, editor, *Hot chips 10: conference record: August 16–18, 1998, Memorial Auditorium, Stanford University, Palo Alto, California*, 1109 Spring Street, Suite 300, Silver Spring, MD 20910, USA, 1998. IEEE Computer Society Press.
3. T. Monreal, A. González, M. Valero, J. González, and V. Viñals. Delaying physical register allocation through virtual-physical registers. In *Proceedings of the 31th Annual International Symposium on Microarchitecture*, pages 186–193, 1999.
4. David A. Padua and Michael J. Wolfe. Advanced compiler optimization for supercomputers. *Communications of the ACM*, 29(12):1184–1201, December 1986.
5. Dezsö Sima. The design space of register renaming techniques. *IEEE Micro*, 20(5):70–83, September/October 2000.
6. Gurindar S. Sohi. Instruction issue logic for high-performance, interruptable, multiple functional unit, pipelined processors. *IEEE Transactions on Computers*, 39(3):349–359, March 1990.
7. Kenneth C. Yeager. The MIPS R10000 superscalar microprocessor — emphasizing concurrency and atency-hiding techniques to efficiently run large, real-world applications. *IEEE Micro*, 16(2):28–40, April 1996.

An Adaptive Space-Sharing Policy for Heterogeneous Parallel Systems

Zhengao Zhou[1] and Sivarama P. Dandamudi[2]

[1] Manugistics Canada, Ottawa, Canada
zzhou@manu.ca

[2] Center for Parallel and Distributed Computing, School of Computer Science
Carleton University, Ottawa, Canada
sivarama@scs.carleton.ca

Abstract. As the distributed-memory parallel systems become heterogeneous in nature, it is important to devise scheduling policies that take node heterogeneity into account. Previous studies on this topic have focused on homogeneous systems. In this paper, we propose a new two-level space-sharing policy for heterogeneous systems. The proposed policy is an adaptive space-sharing policy that takes system load and user requirements into account. We compare the performance of the new policy with a previous policy. The results presented here show that the proposed policy provides substantial performance improvements.

1 Introduction

Distributed-memory parallel systems can be built with off-the-shelf components (processors, networks, memory). As a result, these systems tend to use different processors (different clock rate, cache memory, etc.) leading to node heterogeneity. Over the lifetime of a machine, replacing some of the original processors by improved versions also results in heterogeneity. Thus, either by design or over lifetime, distributed-memory parallel systems tend to be heterogeneous in the sense that all nodes are not the same. For example, the system in our research laboratory has four types of Pentium processors. This paper proposes a space-sharing scheduling policy for heterogeneous distributed-memory systems.

Space-sharing policies divide the set of system processors into disjoint partitions so that each partition can be allocated to a single job. Commercial distributed-memory systems such as IBM SP2, Meiko, Intel iPSC and Paragon and Cray T3D use space-sharing processor allocation policies [3].

Space-sharing policies can be broadly divided into fixed, variable, adaptive, and dynamic policies [5]. In fixed policies, the partition size is fixed on a long term basis and is typically modified by a system reboot. In variable partitioning policies, the partition size is determined at job submission time based on the user request. In adaptive partitioning policies, partition size is determined by the system by taking system load conditions and user requirements into account. Adaptive policies perform better than the other two types of policies. Several adaptive policies have been proposed for distributed-memory systems [6,7,8].

In all three types of policies, once a partition is allocated to a job, the partition size remains fixed for the lifetime of that job. This kind of allocation wastes

B. Hertzberger et al. (Eds.): HPCN Europe 2001, LNCS 2110, pp. 353–362, 2001.

resources if the job parallelism varies during execution due to, for example, different execution phases. Dynamic partitioning eliminates this disadvantage. In dynamic partitioning, processor allocation varies over the lifetime of a job by responding to the changes in parallelism of the job.

Dynamic partitioning has several disadvantages [3]. Dynamic partitioning does not support popular programming styles (such as SPMD and data parallel models). Furthermore, repartitioning overheads, which include switching protection domains, may negate the benefits of dynamic partitioning [9]. In addition, it requires coordination between the application and the operating system. As a result of these drawbacks, dynamic partitioning has been used only to a limited degree [3]. Thus, we focus on adaptive partitioning policies in this paper.

The focus of the policies proposed in the literature is for homogeneous systems. Heterogeneous systems have not received much attention. This paper presents a new policy for heterogeneous systems and evaluates its performance. Our results show that it provides substantial performance improvements over two previously proposed policies.

2 Space-Sharing Policies for Homogeneous Systems

It has been observed that, for better overall performance, job scheduling policies should facilitate sharing the processing power equally among the jobs [5]. This implies that space-sharing policies should attempt to allocate equal partitions. We now describe the space sharing policies proposed by Rosti et al [7] that attempt to generate equal sized partitions while adapting to transient workload. The basic adaptive policy is shown below:

Given: a specific implementation of compute_target_size
target_size ← compute_target_size
while (queue_length > 0) **and** (free_processors > target_size) **do**
    free_processors ← free_processors − target_size
    queue_length ← queue_length − 1
    schedule_a_job (target_size)
od

Rosti et al. have proposed five adaptive policies (AP1 through AP5) and evaluated their performance under various workload and system conditions. The adaptive nature of the policies adjusts to various workloads by computing the partition size of each job at schedule time. On each such partition, a job exclusively runs until completion. These policies differ in how the target size is computed (i.e., in the definition of *compute_target_size* procedure).

2.1 Adaptive Policy AP

The space sharing policy that we have implemented is their AP2 policy. Among the five policies, this is the only policy that reserves a partition for future job arrivals. Such policies are referred to as processor saving policies [8]. This reservation alleviates the performance deterioration associated with non-preemptive first-come/first-served scheduling of jobs. The results in [8] show that processor saving policies provide better performance over their greedy counterparts under various system and workload characteristics. In this policy, which we refer to as the adaptive policy (AP), the partition size PS is computed as

$$PS = max\left(1, \left\lfloor \frac{total\ processors}{queue\ length + 1} + 0.5 \right\rfloor\right) \quad (1)$$

The policy always reserves one partition for future job arrivals (that is why we use (*queue length* + 1) rather than *queue length*). However, from the expression above, we notice that AP tends towards premature queuing. The main reason for this behavior is that this policy considers only the queued jobs to determine the partition size. This policy also contravenes the equal allocation principle mentioned earlier.

2.2 Modified Adaptive Policy MAP

In order to obviate the problems of the AP policy, it has been modified by taking into account the total number of jobs in the system [2]. The MAP policy computes the partition size PS as

$$PS = max\left(1, \left\lfloor \frac{total\ processors}{queue\ length + (f \times number\ schduled\ jobs) + 1} + 0.5 \right\rfloor\right) \quad (2)$$

where $0 \leq f \leq 1$. The parameter f is used to control the contribution of the scheduled jobs to the partition size. Note that we get the original AP policy by setting f to 0. The motivation behind the modification is two fold. One, the goal of equal partition for each job is better realized by taking all the jobs into account (this is the case when $f = 1$). Two, it provides a good heuristic during processor allocations. It has been shown that the modified policy provides appreciable improvement in performance [2]. The amount of improvement obtained is a function of parameter f, system load, and workload.

3 Proposed Policy for Heterogeneous Systems

The new policy HAP proposed here is based on the MAP policy, but adapted to heterogeneous systems. The main problem with the MAP policy is that it partitions at the physical processor level. Thus, if the processors in the system have different processing capacities, this policy still allocates the same number of physical processors to a job based on the partition size calculated from Eq. (2). For example, assume that the system has 3 processors, P1, P2, and P3 with processing capacities of 1, 2, and 3 respectively. (We introduce a unit called the BPU to express the differences among the processors shortly.) If the number of jobs in the waiting queue is 1 and there are no scheduled jobs, the partition size is computed from Eq. (2) as 2. For this example, MAP allocates two physical processors for the job. The allocated processors can be any two of the three physical processors in the system. If the allocated partition consists of P1 and P2, it represents 50% of the processing capacity of the system (as it should be to allocate equal processing power to all jobs). However, the worst-case allocation consists of P2 and P3, which means giving the job more than 80% of the total system capacity (as opposed to 50%).

The new policy proposed here solves the above problem for the heterogeneous distributed-memory systems. A new concept, called Basic Processing Unit (BPU), is introduced to take the processor heterogeneity into account. The physical processor with the lowest processing capacity in the system is considered as representing one BPU. We express the rating of other processors in BPUs.

The proposed policy uses the total BPUs in place of total number of processors to compute the partition size in Eq. (2). That is, partition size is computed as in MAP except that the partition size is given in BPUs as opposed physical processors.

Once the partition size is determined, the next step is to actually partition the free processors into subsets and assign them to each job. In a homogeneous system, it is easy to partition, as all nodes are identical. One can choose any partition whose size is equal to or larger than the partition size PS computed at the first level. In a heterogeneous system, however, it is more complex. We have to deal with two issues. First, how do we partition the system? The partition should be at least partition size PS in terms of BPUs. Among the candidate partitions, our goal is to find a set of physical processors, the sum of whose BPUs has the least difference from the required size (best fit policy).

The second issue we encounter is if, two or more partitions have exactly the same size, which should be selected? This is not a problem in a homogeneous system. Since all processors are similar, it makes no difference which partition is selected. In heterogeneous systems, we do have some concerns in making such decisions. Our policy chooses the partition with the fewest processors due to the following considerations: (i) Communication overhead tends to be smaller with smaller number of processors in a partition. (ii) Better performance is obtained with faster processors (two faster processors is better than four slower processors of equivalent aggregate processing power).

When no partition that meets the partition size requirement is found, jobs will remain in the queue until the system state changes (due to the arrival of a new job or departure of a completed job).

4 Job and Workload Characterization

4.1 Job Structure

One of our major objectives is to get an idea of the performance improvement one can obtain with the new policy. In a sense, the results reported here should be treated as a first step to fully evaluate the performance of this new policy. For this preliminary study, we have selected the fork-and-join job structure that has been used extensively in the literature. A further justification for using this job structure is provided by our previous studies that showed that the type of job structure did not affect the *relative performance* of policies (even though their absolute performance is affected) [1,2].

In algorithms exhibiting fork-and-join type of job structure, work can be decomposed into independent sub-computations. The parallel implementation of these types of algorithms can be done by creating tasks to work on the sub-computations. The fork-and-join (FJ) job structure is reasonable for the class of problems with a solution structure that iterates through a communication phase and a computation phase. In this paper, only a single phase that involves one synchronization point is assumed. Similar job structure has been used in several previous studies (see [1] for references).

4.2 The Workload Model

This section presents an abstract workload model for parallel programs [1]. The abstract model consists of a set of input parameters along with a job structure. The job arrival process is characterized by two parameters: λ and CV_a. The job arrival rate is represented by λ and CV_a represents the coefficient of variation of the inter-arrival times. The other parameters are described next.

Maximum Parallelism (mp). The maximum parallelism of a parallel program refers to the maximum number of tasks that can be run in parallel, if an unlimited number of processors is provided. Since we assume that the tasks in a fork-and-join job can be executed independently, its maximum parallelism is always equal to the number of tasks the job has.

Mean Service Demand (D). There are two ways to model the service demand of a parallel program: correlated or uncorrelated. In the correlated workload, overall service demand of a job is proportional to the number of tasks. This is not true in uncorrelated service demand. We use parameter D to represent the mean service demand of a parallel program.

We have conducted experiments with both correlated and uncorrelated workloads. The results are qualitatively similar. Therefore, for the sake of brevity, we report results for the uncorrelated workload only.

Coefficient of Variation of Service Demand (CV_d). CV_d is the coefficient of variation of service demand of parallel programs. Measurements at some supercomputer centers indicate that the service time coefficient of variation can be very high (as high as 30 in some cases) [4].

Synchronization Cost (syn). This parameter represents the cost (i.e., processor service time required) of performing synchronization.

4.3 Simulation Model

We implemented the simulation model in C using discrete event simulation techniques. We used a two-stage hyper exponential model to generate service times and inter-arrival times with CVs greater than 1 (i.e., 2, 3, $\cdots$, 6). Details on this model can be found in [10]. For the uncorrelated job service demand, we generate D and distribute this to the tasks of the job.

A batch strategy has been used to compute confidence intervals (30 batches were used for the results reported here). We actually used 31 batches but the first batch statistics were ignored as it produces optimistic values due to cold start. This strategy produced 95% confidence interval that is less than 5% of the mean value except when the system utilization is very high.

5 Performance Comparison

This section presents results of the simulation experiments to show the relative performance of HAP and MAP policies for heterogeneous systems. The results are presented for uncorrelated workload. Unless otherwise stated, the following default parameter values are used. Other parameters are $mp = 32$, $D = 16$, $CV_d = 3.5$, $CV_a = 1$, and $syn = 0.1$. All times in our plots are expressed in units of parameter D time units. The syn is set to 0.1 as only a single synchronization phase is to be modeled. The maximum parallelism of a job is varied from 1 to mp, which is 32. The default service time CV is fixed 3.5 as empirical observations at several supercomputer centers indicated this to be a reasonable value [4]. However, these studies have also shown that the service time CV can be very high as well (as high as 30 in some cases). The arrival CV is fixed at 1 (i.e., we assume Poisson arrivals). However, we study the sensitivity of performance to these variances in Section 5.2. The default f value is set to 0.5. Section 5.3 discusses the performance sensitivity to this parameter.

All simulation experiments assume a system with $N = 64$ processors. These 64 physical processors are divided into two classes, Class I and Class II. Class I contains $N_1 = 32$ physical processors with same processing capacity. Class II contains the remaining $N_2 = 32$ physical processors that are two times faster than those in Class I. Thus, each physical processor in Class I has one BPU and each physical processor in Class II has two BPUs. We have also done experiments with different classes of processors. Due to space restriction, these results are not presented here (see [10] for these results).

5.1 AP versus HAP

This section reports the relative performance of HAP and AP policies for homogeneous systems. Note that, for homogeneous systems, HAP reduces to MAP policy. The relative performance of the AP and HAP policies is shown in Figure 1. The x-axis gives the system utilization and the y-axis represents the mean response time. The data in Figure 1 shows that the HAP policy provides performance improvement over AP policy. Also, the performance difference tends to be smaller at low and high system utilizations. At very low system utilizations, both policies use the same partition size as the probability of a job receiving service is small. As the system utilization increases, the number of jobs receiving service increases. This increase causes the HAP policy to use smaller partitions than the AP policy. Smaller partition sizes improve system utilization as it decreases, for example, internal fragmentation. Internal fragmentation refers to the situation in which an application is not able to use all the processors in the partition assigned to it. For example, during a serial phase, smaller partitions cause fewer processors to idle than larger partitions. At high system utilizations, both policies tend to use the same partition size—for example, at very high system loads, each partition may have only one processor under both policies.

5.2 Performance Improvement of HAP Over MAP

The data presented in [10] shows that the HAP policy performs substantially better than the MAP policy at moderate to high system utilizations. Due to

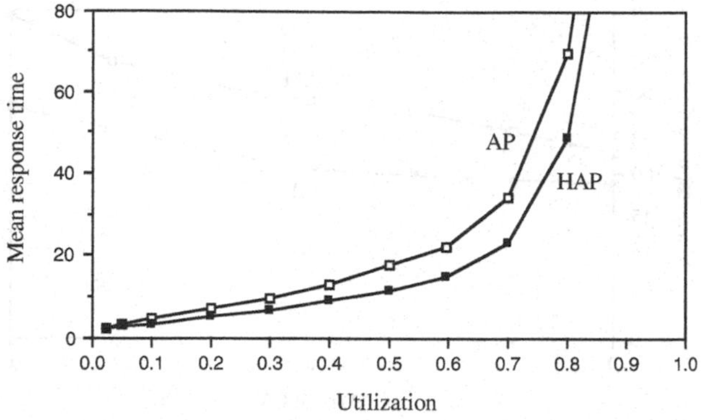

Fig. 1. Performance of the AP and HAP policies in a homogeneous system with 64 processors. All other parameters are set to their default values given at the beginning of Section 5.

space restrictions, we do not present these results here. However, to demonstrate that HAP policy consistently performs better than the MAP policy, the next two sections present the performance of these two policies to variances in inter-arrival and service times.

Performance Sensitivity to Variance in Inter-Arrival Times. The performance sensitivity of the HAP and MAP policies to inter-arrival CV CV_a is shown in Figure 2. The number of physical processors in Class I is 32, f is 0.5, the service demand CV_d is fixed at 1.0, and the system utilization at 70%. The mean response time increases with increasing inter-arrival CV for both HAP and MAP. The HAP policy maintains its performance superiority over MAP policy at this system utilization. Further, as the variance in inter-arrival times increases, the performance difference increases.

It should be noted that increased variance implies clustered arrival of jobs into the system. It also implies longer gaps in job arrivals (to get the same mean arrival value). This increased variance affects performance in two ways. First, clustered arrival of jobs means jobs may have to wait longer to get to the execution stage. This increases the queuing time component of the response time. This causes the response time to increase for both policies. Second, clustered arrival also forces the scheduling policy to assign smaller partitions as clusters of jobs arrive into the system. But, the arrival pattern also includes long gaps (the higher the CV the longer the gaps). Thus, most jobs are stuck with smaller partitions even though some processors are idle due to lack of jobs (because of the long gaps). The overall effect of these factors is to decrease the efficiency of the system by leaving more processors idle. The efficiency decreases as the variance in arrivals increases, resulting in an increase in the response time. Since HAP policy tends to assign smaller partitions, the effect is less pronounced in HAP than in MAP policy.

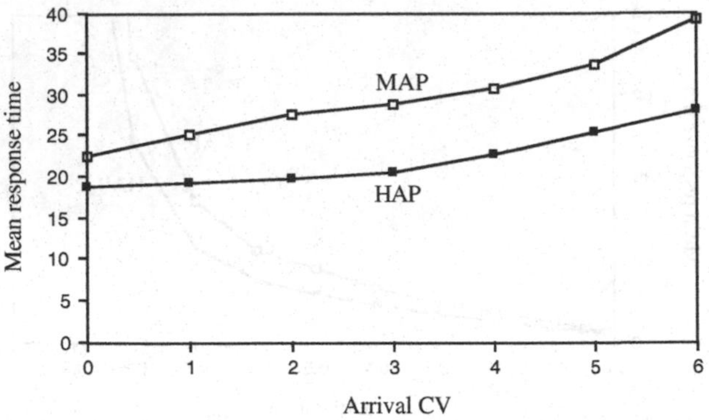

Fig. 2. Performance Sensitivity of MAP and HAP Policies to Inter-Arrival Time Variance.

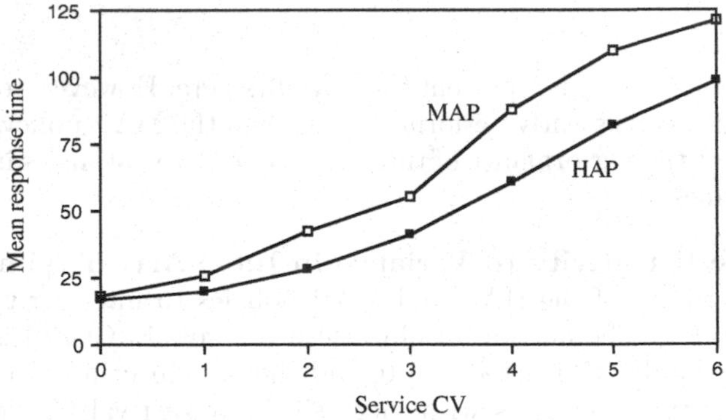

Fig. 3. Performance Sensitivity of MAP and HAP Policies to Service Time Variance.

Performance Sensitivity to Service Time Variance. Performance of the HAP policy to service time CV is shown in Figure 3. For comparison, we have also included the MAP policy. The system utilization is fixed at 70% and the inter-arrival CV at 1 (i.e, Poisson arrivals are used). The data presented in Figure 3 show that the mean response time increases with the coefficient of job service demand variance CV_d for both HAP and MAP policies.

Performance of both policies is more sensitive to variance in service time than in arrival times. This can be seen from the results presented in Figures 2 and 3. This is in contrast to the sensitivity exhibited by MAP in homogeneous systems [2], which shows that the MAP performance is less sensitive to service time CV than to inter-arrival CV.

The reason is that increased CVd implies that there is large number of small service demand jobs and a few large service demand jobs. Thus, as CVd increases,

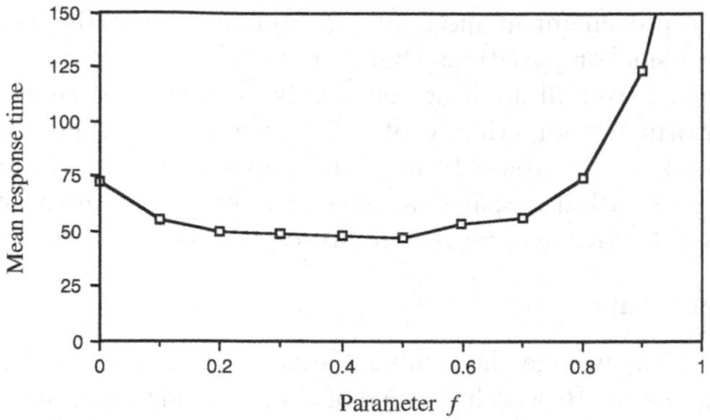

Fig. 4. Performance Sensitivity of the HAP Policy to Parameter *f*.

the service demand of the larger jobs increases even though their number goes down as a fraction of the total jobs. Since HAP allocates fewer physical processors than MAP does, the jobs with large service demand cause the long mean response time. As a result, the service demand variance CVd has more impact than inter-arrival time variance CVa when HAP policy is used.

5.3 Sensitivity to Parameter *f*

The parameter *f* determines the contribution of the scheduled jobs to the partition size. To see the impact of this parameter on the performance, we have conducted experiments for various values of *f* in the range between 0 and 1. Selected results of these experiments are shown in Figure 4. In the data presented in this figure, the number of physical processors in Class I is 32, the system utilization is fixed at 70%, service demand CV_d is fixed at 3.5, and the inter-arrival CV_a is fixed at 1. All other parameters are set to their default values given at the beginning of Section 5.

It should be observed that at low system utilizations, we should assign larger partitions as there are fewer jobs in the system. This means the *f* value should be smaller at these utilizations. As the system utilization increases, the partition size should be reduced by increasing the *f* value. The results presented in Figure 4 suggest the best *f* value is around 0.5. However, performance is only marginally sensitive to a range of *f* values (from 0.2 to 0.7). Our previous results on homogeneous systems have also shown that an *f* value in the range between 0.5 and 0.75 is a reasonable choice [2].

6 Summary

We have proposed a new space-sharing policy HAP for distributed-memory systems to take processor heterogeneity into account. We have also reported performance comparison of the new policy with a previously proposed policy. Due to space restrictions, these results are not presented here. These additional results are available in [10]. Our results show that the new policy provides substantial performance improvement at moderate to high system loads. The HAP policy tends to assign smaller partitions than the MAP policy. Because of this, HAP policy performs better at high system loads. We have presented results that show the performance superiority of HAP over a wide range of variances in inter-arrival and service times. In fact, the performance improvement provided by HAP increases with increasing variance in inter-arrival and service times. We have done several other experiments to study the performance.

Acknowledgements

We gratefully acknowledge the financial support provided by the Natural Sciences and Engineering Research Council of Canada and Carleton University. This work was done while the first author was at the School of Computer Science, Carleton University.

References

1. S. L. Au and S. P. Dandamudi, "The Impact of Program Structure on the Performance of Scheduling Policies in Multiprocessor Systems," *J. Computers and Their Applications,* Vol.3, No.1, April 1996, pp. 17-30.
2. S. P. Dandamudi and H. Yu, "Performance of Adaptive Space Sharing Processor Allocation Policies for Distributed-Memory Multicomputers," *J. Parallel and Distributed Computing*, Vol. 58, 1999, pp. 109-125.
3. D. G. Feitelson and L. Rudolph, "Parallel Job Scheduling: Issues and Approaches," *Job Scheduling Strategies for Parallel Processing,* D. Feitelson, and L. Rudolph (eds.), LNCS Vol. 949, Springer-Verlag, 1995, pp. 1-18.
4. D. G. Feitelson and B. Nitzberg, "Job Characteristics of a Production Parallel Scientific Workload on the NASA Ames iPSC/860," *Job Scheduling Strategies for Parallel Processing,* D. Feitelson, and L. Rudolph (eds.), Vol. 949, Lecture Notes in Computer Science, Springer-Verlag, 1995, pp. 337-360.
5. D. G. Feitelson, L. Rudolph, U. Schwiegelshohn, K. C. Sevcik, and P. Wong, "Theory and Practice in Parallel Job Scheduling," *Job Scheduling Strategies for Parallel Processing,* D. Feitelson, and L. Rudolph (eds.), Vol. 1291, Lecture Notes in Computer Science, Springer-Verlag, 1997, pp. 1-34.
6. D. G. Feitelson and L. Rudolph (editors), *Job Scheduling Strategies for Parallel Processing,* IPPS Workshop Proceedings, Lecture Notes in Computer Science, Springer-Verlag, 1994, 1995, 1996, 1997, and 1998.
7. E. Rosti, E. Smirni, G. Serazzi, L.W. Dowdy, and B.M. Carlson, "Robust Partitioning Policies for Multiprocessor Systems," *Performance Evaluation,* Vol. 19, 1994, pp. 141-165.
8. E. Rosti, E. Smirni, L. W. Dowdy, G. Serrazi, and K. C. Sevcik, "Processor Saving Scheduling Policies for Multiprocessor Systems", *IEEE Trans. Computers*, Vol. 47, No. 2, February 1998.
9. M. S. Squillante, "On the Benefits and Limitations of Dynamic Partitioning in Parallel Computer Systems," *Job Scheduling Strategies for Parallel Processing,* D. Feitelson, and L. Rudolph (eds.), LNCS Vol. 949, Springer-Verlag, 1995, pp. 219-238.
10. Z. Zhou, *An Adaptive Space-Sharing Policy for Heterogeneous Parallel Systems*, M.Sc. Project Report, School of Computer Science, Carleton University, Ottawa, Canada, 1999.

Orthogonal Processor Groups for Message-Passing Programs

Thomas Rauber[1], Robert Reilein[2], and Gudula Rünger[2]

[1] Institut für Informatik, Universität Halle-Wittenberg
06099 Halle(Saale), Germany
rauber@informatik.uni-halle.de
[2] Fakultät für Informatik, Technische Universität Chemnitz
09107 Chemnitz, Germany
{reilein,ruenger}@informatik.tu-chemnitz.de

Abstract. We consider a generalization of the SPMD programming model to orthogonal processor groups. In this model different partitions of the processors into disjoint processor groups can be exploited simultaneously in a single parallel implementation. The parallel programming model is appropriate for grid based applications working in horizontal or vertical directions as well as and for mixed task and data parallel computations[2]. For those applications we propose a systematic development process for message-passing programs using orthogonal processor groups. The development process starts with a specification of tasks indicating horizontal and vertical sections. A mapping to orthogonal processor groups realizes a group SPMD execution model and a final transformation step generates the corresponding message-passing program.

1 Introduction

Parallel machines with distributed memory organization are a popular platform for the implementation of applications from scientific computing because it is now well-understood how to get portable efficient programs. For typical grid-based computation structures of scientific applications, the SPMD programming model usually leads to good efficiency. But there are also applications which can benefit from a parallel programming model that allows more general but still regular communication and dependence patterns.

In this paper, we consider a group SPMD programming model with multiple processor groups in orthogonal directions where different directions are active at different points of the execution time. This programming model is suitable for applications which consist of independent disjoint computations with varying structure. Theoretical investigations have shown that it can be useful to exploit this kind of parallelism pattern in a parallel program with disjoint processor sets which concurrently execute the program in SPMD mode [7,8]. The advantage is that collective communication operations performed on smaller processor groups lead to smaller execution times due to the logarithmic or linear dependence of the communication times on the number of processors. Moreover, the message sizes are usually smaller and different communication operations can often be executed concurrently on disjoint processor groups without interference. Disjoint processor groups can be expressed in the communication library MPI, but the direct coding of orthogonal processor groups may lead to intricate and

B. Hertzberger et al. (Eds.): HPCN Europe 2001, LNCS 2110, pp. 363–372, 2001.

error-prone message-passing programs. Therefore, it seems to be appropriate to provide a way to specify programs with orthogonal processor groups on a more abstract level and to generate the corresponding MPI program by a compiler tool.

The contribution of this paper is to propose a transformation approach to develop appropriate message-passing programs in a group SPMD programming model in several steps. The approach comprises a structuring of the potential parallelism, a mapping onto orthogonal processor groups and a final transformation step into a corresponding MPI program. The structuring is given by a specification program indicating sections of horizontal, vertical, or global interaction structures which are identified by analyzing the potential parallelism of the computations. The mapping onto processor groups leads to a structured parallel program in a group SPMD model with changing or alternating group structure. Different partitions of the processors into disjoint processor groups can be exploited in a single parallel implementation, but at each time of the execution only one partition can be active. The final MPI program realizes the horizontal and vertical sections with group operations. The paper outlines the transformation approach and illustrates the transformation steps for appropriate application programs.

The rest of the paper discusses orthogonal structures and its specification in Sections 2 and shows the mapping onto orthogonal processor groups in Section 3. Runtime tests with orthogonal group structures are given in Section 4. Section 5 discusses related work and Section 6 concludes.

2 Orthogonal Structures of Computations in Scientific Applications

An application has an orthogonal structure of computation and communication with varying directions, if the computations and the data dependencies exhibit regular patterns in a horizontal or vertical direction of the task organization.

Describing Orthogonal Computation Structures. For the organization of the computations and the assignment to processors, we use the following abstraction: A parallel application program is composed of a set $\mathcal{T}$ of n one-processor tasks which are organized in a two-dimensional way. The tasks are numbered with two-dimensional indices in the form T_{ij}, $i = 1, \ldots, n_1, j = 1, \ldots, n_2$, with $n = n_1 \cdot n_2$. Each single task $T \in \mathcal{T}$ consists of a sequence of computations and communication commands. Data needed from other tasks or required by other tasks form a dependence structure between the tasks.

For applications exhibiting orthogonal interaction structures the horizontal and vertical computations and communications are clearly separated, i.e., at each point in time, a task is either involved in operations in the vertical or the horizontal direction, but not both. This can be formalized using an *interaction matrix* which is the adjacency matrix of the dependence graph of a task array $\mathcal{T}$. An interaction matrix $A_{\mathcal{T}}$ for a task array $\mathcal{T}$ has size $n \times n$ with $n = n_1 * n_2$. The row $r = (i-1)n_2 + j$ of matrix $A_{\mathcal{T}}$ shows the interactions of task T_{ij}. The columns of the matrix $A_{\mathcal{T}}$ are also associated with the tasks of $\mathcal{T}$, where column $s = (k-1)n_2 + l$ is associated with task T_{kl}. The matrix $A_{\mathcal{T}}$ has a nonzero entry in $A_{\mathcal{T}}[r, s]$, $r, s = 1, ..., n$, if task T_{ij} associated with row r has an interaction with task T_{kl} associated with column s. If task T_{ij} has no interaction with task T_{kl} then there is no entry in $A_{\mathcal{T}}[r, s]$ for $r = (i-1)n_2 + j$ and $s = (k-1)n_2 + l$.

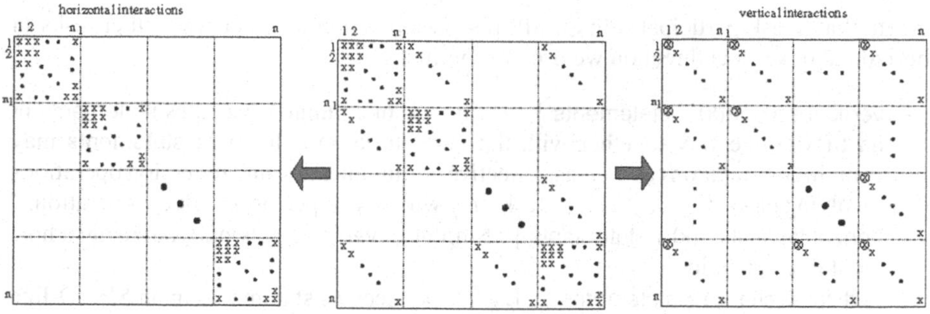

Fig. 1. Interaction matrices of size $n \times n$ for task arrays of size $n_1 \times n_2, n = n_1 n_2$, showing the nonzero pattern for horizontal and vertical (middle), horizontal (left) or vertical (right) dependencies.

Figure 1 shows the pattern of an interaction matrix with horizontal and vertical interactions (in the middle) and two matrices which show the interactions separately. The matrix on the left, depicting horizontal dependencies, consists of n_2 blocks of size $n_1 \times n_1$ where each block indicates the dependencies within one row of the task array $\mathcal{T}$. Since there are no nonzero elements outside those blocks the dependency graph of the corresponding task array consists of n_2 independent subgraphs. The matrix on the right, depicting vertical dependencies, has nonzero entries in the main diagonal and the diagonals with distance n_1. The nonzero elements in all rows with distance n_1 correspond to computations within one column of a task array; e.g. the set of items with circles indicate the pattern for vertical dependencies in the first column of the task array. The nonzero pattern of the illustration in Figure 1 express the maximal set of dependencies for a single vertical or horizontal computation structure which means that an orthogonal program part can have less nonzero elements than depicted but not more or in other positions. To express SPMD computations in horizontal or vertical direction, we can define the interaction matrix in a more general way, such that a nonzero element means that the two tasks associated with the corresponding row and column perform similar computations in an SPMD style.

To exploit orthogonal process structures, an application is not required to have a dependence graph with corresponding interaction matrix for the entire program. Rather, we consider a much larger class of applications for which the orthogonal structures are only given for parts of the program and define orthogonal application structures in the following sense: If an application program can be decomposed into program parts and the program parts have either a dependence graph with only horizontal interactions (Figure 1 left) or only vertical interactions (Figure 1 right), then the application exhibits an orthogonal structure with the potential parallelism suitable for a group SPMD program with varying orthogonal groups. Thus, different program parts may have different interaction matrices. Also program parts with general interaction matrix are allowed but cannot benefit from the specific orthogonal group implementation.

Specification of Orthogonal Structures. The entire task program is specified in a group SPMD style with explicit constructions for horizontal or vertical executions, which we call *horizontal sections* and *vertical sections*, respectively. In horizontal sections, a task T_{ij} has interactions to a set of tasks $\{T_{ij'} | j' = 1, ..., n_2, j' \neq j\}$. In vertical sections, a task T_{ij} has interactions to a set of tasks $\{T_{i'j} | i' = 1, ..., n_1, i' \neq i\}$. To indicate that a task participates in an SPMD-like operation together with other tasks in horizontal or vertical direction we use the commands:

- vertical_section(k) { statements } : Each task in column k executes statements in an SPMD-like way together with the other tasks in column k; statements may contain computations as well as collective communication and reduction operations involving tasks $T_{1k}, \ldots, T_{n_1,k}$. Tasks T_{ij} with $j \neq k$ perform a skip-operation.
- horizontal_section(k) { statements } : Similar to vertical_section(k), but using a horizontal organization.
- vertical_section( ) { statements } : Each task executes statements in an SPMD-like way together with the other tasks in the same column. A task in column k may perform computations as well as collective communication and reduction operations involving the tasks $T_{1k}, \ldots, T_{n_1,k}$ in the same column. Computations in a specific column of the task array are executed in parallel with the other columns in a group SPMD programming model. Thus, vertical_section( ) corresponds to a parallel loop over all columns k where each iteration executes vertical_section(k).
- horizontal_section( ) { statements } : Similar to vertical_section( ), but using a horizontal organization.

The gray parts in the following illustration depict active computation parts in the task array:

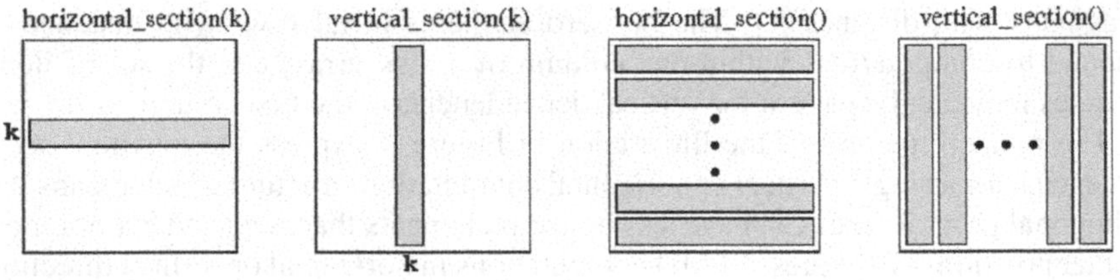

Commands outside of horizontal or vertical sections are executed in SPMD style which also includes task specific operations. The approach can be generalized to more than two dimensions in an analogous way by introducing orthogonal_section() commands specifying subspaces of the task grid in which the communication is performed.

Example: LU Factorization. For an illustration we use the well-known LU factorization for the solution of linear equation systems $Ax = b$ which has been investigated in great detail in the past. The optimal computational structure for the LU decomposition results from a double-cyclic distribution that distributes the rows and columns of the coefficient matrix $A \in R^{n_1 \times n_1}$ cyclically among the processors [7,8]. Each entry a_{ij} of the coefficient matrix is assigned to a task T_{ij}, $i, j = 1, \ldots, n_1$, that is responsible for the computations of this entry in the elimination steps. Each elimination step k, $1 \leq k < n_1$, consists of the following computation phases:

1. vertical_section(k): The tasks in column k of the task array cooperate to determine the global pivot element $a_{rk} = max_{k \leq j \leq n_1} |a_{jk}|$.
2. horizontal_section(): The entries of the pivot row r are distributed in the columns of the task array such that element a_{rj} is distributed to all tasks T_{ij} with $i > r$. Moreover, if $r \neq k$ the tasks T_{kj} and T_{rj} exchange elements a_{kj} and a_{rj}.

3. vertical_section(k): The tasks in column k of the task array computes the elimination factors l_k.
4. horizontal_section(): The tasks T_{jk} in column k distribute the elimination factors l_k in the corresponding rows $j, k+1 \leq j < n_1$.
5. SPMD section: The tasks T_{ij} for $k+1 \leq j < n_1$ compute new values for their associated entries of the coefficient matrix.

3 Mapping to Orthogonal Processor Groups

For the mapping of the tasks to the processors, we assume that the number of processors is smaller than the number of tasks. In order to exploit the potential parallelism of orthogonal computation structures, we map the computations onto a two-dimensional processor grid of size $p_1 \times p_2$ for which we provide two different partitions that exist simultaneously. For the assignment of tasks to processors, we use parameterized mappings similar to parameterized data distributions [3,7] which describe the data distribution for arrays of arbitrary dimension, i.e., a task is assigned to exactly one processor, but each processor might have several tasks assigned to it.

Assignment of Tasks to Processors. A double-cyclic mapping of the two-dimensional task array to the processor grid is specified by block-sizes b_1 and b_2 in each dimension, which determine the number of consecutive rows and columns that each processor obtains of each cyclic block. For a total number of p processors, the distribution of the task array T of size $n_1 \times n_2$ is described by an assignment vector of the form

$$((p_1, b_1), (p_2, b_2)) \tag{1}$$

with $p = p_1 \cdot p_2$ and $1 \leq b_i \leq n_i$ for $i = 1, 2$. For simplicity we assume $n_i/(p_i \cdot b_i) \in \mathbb{N}$. To describe a double-cyclic distribution, we logically arrange the processors in a two–dimensional grid of size $p_1 \times p_2$ by specifying a grid function $\mathcal{G} : P \rightarrow \mathbb{N}^2$ where P is the set of available processors. This defines p_1 row groups R_q, $1 \leq q \leq p_1$, and p_2 column groups C_q, $1 \leq q \leq p_2$,

$$R_q = \{Q \in P \mid \mathcal{G}(Q) = (q, \cdot)\} \qquad C_q = \{Q \in P \mid \mathcal{G}(Q) = (\cdot, q)\}$$

with $|R_q| = p_2$ and $|C_q| = p_1$. The row and column groups build separate orthogonal partitions of the set of processors P, i.e.,

$$\bigcup_{q=1}^{p_1} R_q = \bigcup_{q=1}^{p_2} C_q = P \text{ and } R_q \cap R_{q'} = \emptyset = C_q \cap C_{q'} \text{ for } q \neq q'.$$

The row groups R_q and the column groups C_q are orthogonal processor groups. Using distribution vector (1), row i of the tasks T is assigned to the processors of a single row group $Ro(i) = R_k$ with $k = \left\lfloor \frac{i-1}{b_1} \right\rfloor \bmod p_1 + 1$. Similarly, column j is assigned to the processors of a single column group $Co(j) = C_k$ with $k = \left\lfloor \frac{j-1}{b_2} \right\rfloor \bmod p_2 + 1$. A program mapped to processors refers to the row and column groups by $Ro(i)$ and $Co(i)$, i.e., it uses the original task indices. Thus, after the mapping, the task structure is still visible and the orthogonal processor groups according to the given mapping are known implicitly.

Execution Model for Orthogonal Processor Groups. The mapping of the tasks to the processors defines the computations that each processor has to perform. Computations of tasks that are outside a vertical or horizontal section are executed in SPMD style. Horizontal and vertical sections require the coordination of the participating processors. A vertical_section(k) operation is performed by all processors in column group $Co(k)$. Similarly, a horizontal_section(k) operation is performed by all processors in $Ro(k)$. A vertical_section( ) operation is performed by all processors, but each processor is only exchanging information with the processors in the same column group. A horizontal_section() is executed analogously.

Mapping the tasks to processors may cause that a single processor executes more than one task of a specific row or column of the tasks array. Therefore, the communication operations within a horizontal or vertical section have to be transformed from a row or column oriented communication to a mixed local and global communication on the assigned processor set. For a reduction operation, for example, each processor performs the reduction for its local tasks and then participates in a global communication with the other processors in its row or column group, respectively, assuming an associative reduction operation.

Transformation to MPI Programs. Each orthogonal section is translated separately into a corresponding MPI program fragment. This is possible since the specification of orthogonal sections is compositional. MPI supports the organization of the communication in orthogonal groups by the concept of communicators and group objects. The row and column group for a row i and column j of the task array $\mathcal{T}$ can be obtained by $Ro(i)$ = `RoGroup[i/b1 % p1]` and $Co(j)$ = `CoGroup[j/b2 % p2]`, if block-sizes b_1 and b_2 are used. The row (column) group of an arbitrary processor with a global rank q can be computed efficiently by defining a virtual two-dimensional process topology of size $p_1 \times p_2$ and by using `MPI_Cart_coords()` to obtain the corresponding grid position (x, y). Then `RoGroup[x]` is the corresponding row group and `CoGroup[y]` is the corresponding column group. Based on predefined row and column groups and their corresponding communicators, the definition of horizontal and vertical sections can be translated into executable MPI programs.

Example: Iterated RK Methods. Iterated Runge-Kutta (RK) [5,9] methods are explicit one-step methods for the solution of systems of ordinary differential equations (ODEs) of the form $\frac{dy(x)}{dx} = f(x, y(x)), y(x_0) = y_0, x_0 \leq x \leq x_{end}$ where $y : \mathbb{R} \rightarrow \mathbb{R}^n$ is the unknown solution function and $f : \mathbb{R} \times \mathbb{R}^n \rightarrow \mathbb{R}^n$ is an application-specific function which is nonlinear in the general case; n is the size of the ODE-system. The predefined vector $y_0 \in \mathbb{R}^n$ specifies the initial condition at x_0. An s-stage iterated RK method performs a fixed number m of iterations per time step to compute an approximation of the solution function (denoted by $y_{\kappa+1}^{(m)}$) according to the following computation scheme:

$$
\begin{aligned}
\mu_{(0)}^{l} &= f(y_\kappa) \quad \text{for } l = 1, \ldots, s \\
\mu_{(j)}^{l} &= f(y_\kappa + h_\kappa \sum_{i=1}^{s} a_{li} \mu_{(j-1)}^{i}) \quad \text{for} \quad l = 1, \ldots, s, j = 1, \ldots, m \qquad (2) \\
y_{\kappa+1}^{(m)} &= y_\kappa + h_\kappa \sum_{l=1}^{s} b_l \, \mu_{(m)}^{l}.
\end{aligned}
$$

The advantage of iterated RK methods for parallel execution is that the iteration system (2) of size $s \cdot n$ consists of s independent systems to determine the approximations $\mu^l_{(j)}$, $j = 1, \ldots, m$. These function evaluations can be performed in parallel by s independent, disjoint processor groups $G_1, \ldots, G_s$ in a programming model with mixed task and data parallelism [6]. Group $G_l = \{q_{l,1}, \ldots, q_{l,g_l}\}$ with g_l processors is responsible for the computation of one sub-vector $\mu^l_{(j)}$, $l \in \{1, \ldots, s\}$. Figure 2 illustrates the computations for one time step.

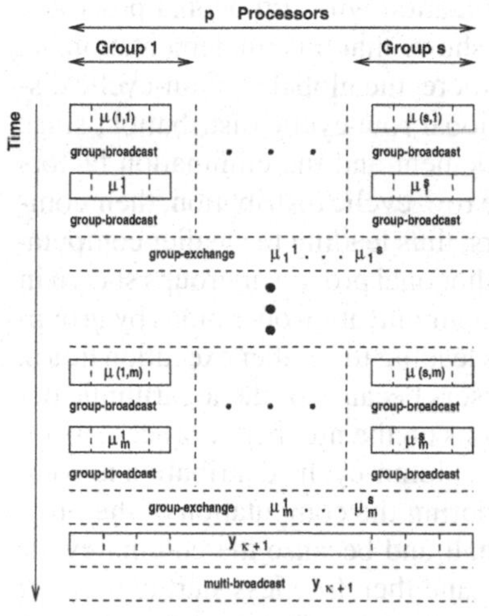

initialize $\mu^l_{(0)} = f(y_\kappa), l = 1, \ldots, s$;
for $j = 1, \ldots, m$ do sequentially {
  forall groups $G_l, l \in \{1, \ldots, s\}$ do in parallel {
    forall processors $q \in G_l$ do in parallel {
      compute $\lceil n/g_l \rceil$ elements of
        $\tilde{\mu}(l, j) = y_\kappa + h_\kappa \sum_{i=1}^{s} a_{li} \mu^i_{(j-1)}$;
      make $\tilde{\mu}(l, j)$ available to all processors of G_l;
      compute $\lceil n/g_l \rceil$ elements of $\mu^l_{(j)} = f(\tilde{\mu}(l, j))$; }}
  forall processors q do
    exchange the computed components of $\mu^l_{(j)}$;
}
forall processors q do {
  compute $\lceil n/p \rceil$ el. of $y_{\kappa+1} = y_\kappa + h_\kappa \sum_{l=1}^{s} b_l \mu^l_{(m)}$;
  compute $\lceil n/p \rceil$ el. of $\hat{y}_{\kappa+1} = y_\kappa + h_\kappa \sum_{l=1}^{s} b_l \mu^l_{(m-1)}$;
}
make $y_{\kappa+1}$ available to all processors;
compute new stepsize;

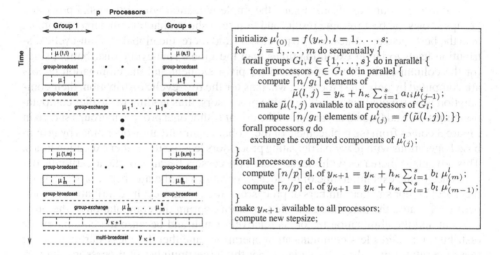

Fig. 2. Parallel Execution of a Time Step of the Iterated RK Method.

For the *special case* that all groups have exactly the same size $g = p/s$ and that p divides n, the implementation can be optimized by using orthogonal groups $Q_1 \ldots, Q_g$ with $|Q_k| = s$ and $Q_k = \{q_{l,k} \in G_l \mid l = 1, \ldots, s\}$.

The iteration steps are performed in the same way as in the general case. But since each processor computes the same number of components, the exchange of the elements of $\mu^1_{(j)}, \ldots, \mu^s_{(j)}$ can be performed more efficiently: Instead of making all components available to each processor, group-multi-broadcast operations can be performed on $Q_1 \ldots, Q_g$ in parallel with each processor $q_{l,k}$ of Q_k contributing n/g components of $\mu^l_{(j)}$, thus making for each processor exactly those components of $\mu^1_{(j)}, \ldots, \mu^s_{(j)}$ available that are needed for the execution of the next iteration.

4 Runtime Tests on Cray T3E

Figure 3 shows speedup results for the parallel LU decomposition on a Cray T3E-1200. The diagrams compare three versions which are based on a row cyclic or column cyclic data distribution on the global group of processors with an implementation which results from the use of orthogonal groups. The orthogonal groups are used for computing the pivot element on a single column group, for making the pivot row available to all processors by parallel executions on all column groups, for computing the elimination factors on a single column group, and for broadcasting the elimination factors in the row groups. Each processor only allocates the elements of the array that it has to compute. This establishes spatial locality when a processor accesses its local elements of one row of the coefficient matrix in storage order. The processor grid is chosen such that there is an equal number of processors in each dimension, i.e., the row and column groups have equal size. All versions result in a good load balance. The block-size in each dimension has been set to 1. Runtime tests have shown that larger block-sizes lead to (slightly) larger execution times.

For a larger number of processors, the implementation with orthogonal processor groups shows the best runtime results and Figure 3 shows that this implementation has also the best speedup values. For 16 processors and more, the global column-cyclic distribution leads to larger execution times than the global row-cyclic distribution, since for the column-cyclic distribution, both the pivot element and the elimination factors are computed by only one processor, whereas for the row-cyclic distribution, their computation is distributed among all available processors, thus leading to smaller computation times. The additional advantage of the use of orthogonal processor groups shown in Figure 3 comes from the replacement of the global communication operations by group-based operations with fewer participating processors leading to smaller execution times. This advantage increases with the number of processors because of the logarithmic dependence of the execution time of broadcast operations on the number of processors on the T3E [7]. For smaller numbers of processors, the column-cyclic distribution is competitive, because the percentage of idle processors during the computation of the pivot element and the elimination factors is relatively small and because the column-cyclic distribution requires less communication operations and therefore less startup time for these operations than the other variants. For the largest number of processors, the percentage difference of the runtime of the version with orthogonal processors groups to the second best version is about 32%.

Figure 4 shows speedup values for different parallel versions of the iterated RK method on a Cray T3E-1200. As basic RK method, a LobattoIIIC6 method with four stages has been used which results in a method of order 6. Each time step performs five iterations. The method has been used for the solution of the Brusselator equation, a time-dependent 2D reaction-diffusion equation. Performing a spatial discretization with a uniform grid with n grid points in each dimension leads to an ODE system of size $2n^2$. The figure shows the results for a 32×32 grid. The resulting right hand side function f of the ODE system is a sparse function, i.e., the evaluation of each component of f depends only on a fixed number of components of the argument vector, thus leading to a linear dependence of the total evaluation time of f on the size of the ODE system. The figure compares a pure data parallel realization of the iterated RK method with a general task-parallel implementation which performs the computations of the approximations of the stage vectors concurrently by different groups of processors and a task-parallel implementation which is optimized by using orthogonal groups. The pure data parallel implementation is in general much faster than the mixed task and data parallel implementation. This is also the case for the implementation with the orthogonal groups for a small number of processors. But the scalability properties are improved by using orthogonal processor groups and for more than 32 processors, this implementation is getting better than the data parallel implementation. For 64 processors, it is twice as fast as the data parallel implementation.

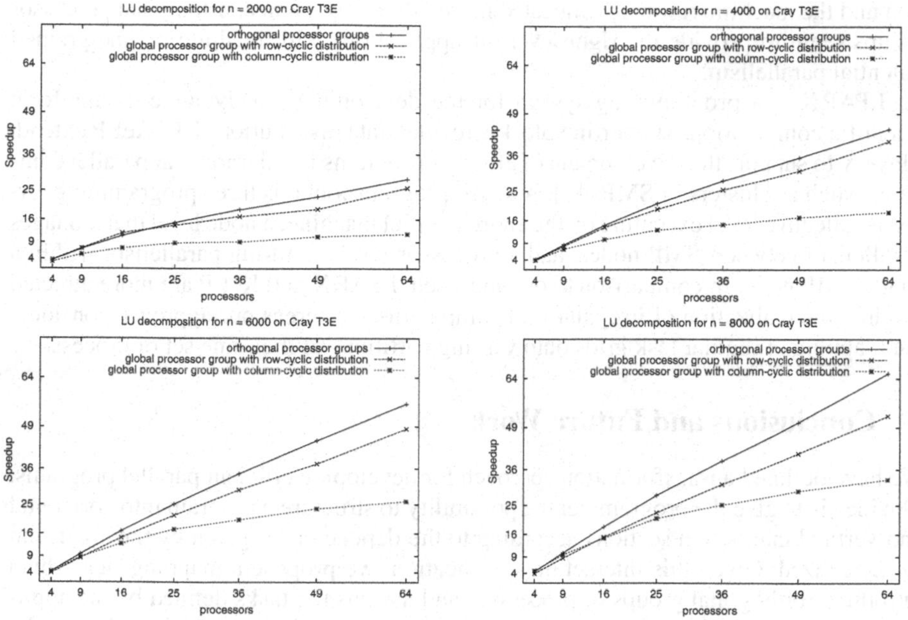

Fig. 3. Speedup Values for the LU Decompositions on a Cray T3E-1200.

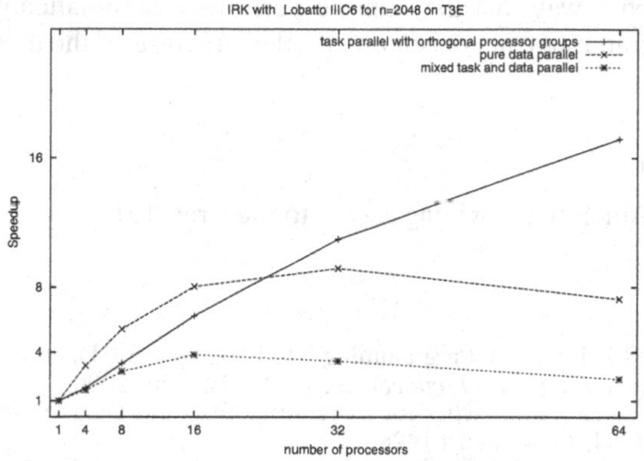

Fig. 4. Speedup Values for the Iterated RK Method on a Cray T3E-1200.

5 Comparison to Related Work

Many environments for scientific computing are extensions to the HPF data parallel language. An example is HPJava[10] which adopts the data distribution concepts of HPF but uses a high level SPMD programming model with a fixed number of logical control threads and includes collective communication operations The concept of processor groups is supported in the sense that global data distributed over one process group can be defined and that the program execution control can choose one of the process groups to be active. In contrast, our approach provides processor groups which can work simultaneously and, thus, can exploit the potential parallelism of the application and the machine resources allocated more efficiently. Hence, orthogonal processor groups seem to provide the right level for applications with medium or fine-grained potential parallelism.

LPARX is a programming system for the development of dynamic, nonuniform scientific computations supporting block-irregular data distributions [4]. KeLP extends LPARX to support the development of efficient programs for hierarchical parallel computers such as clusters of SMPs [1]. A KeLP program contains three programming levels: a collective level executing on the entire parallel machine, a node level that manages parallelism between SMP nodes, and a processor level capturing parallelism within a single SMP node. In comparison to our approach, LPARX and KeLP are more directed towards the realization of irregular grid computations whereas our approach considers the mapping of regular task grids onto varying partitions of the same set of processors.

6 Conclusions and Future Work

We have outlined a transformation approach for developing efficient parallel programs. The idea is to give the programmer the possibility to structure a program into horizontal and vertical interaction sections according to the dependencies given by the algorithm to be realized. Given this interaction specification, we propose a mapping step which introduces orthogonal groups of processors and assigns the tasks defined by the application algorithm to the processors according to a parameterized distribution that can be chosen by the programmer. The mapping and the translation to the final MPI program are designed in such a way that they can be performed automatically by a compiler system, so the programmer can concentrate on the structure of the algorithm to be implemented.

Acknowledgement

We thank the NIC Jülich for providing access to the Cray T3E.

References

1. S.B. Baden and S.J. Fink. A Programming Methodology for Dual-Tier Multicomputers. *IEEE Transactions on Software Engineering*, 26(3):212–226, 2000.
2. H. Bal and M. Haines. Approaches for Integrating Task and Data Parallelism. *IEEE Concurrency*, 6(3):74–84, July-August 1998.
3. A. Dierstein, R. Hayer, and T. Rauber. The ADDAP System on the iPSC/860: Automatic Data Distribution and Parallelization. *JPDC*, 32(1):1–10, 1996.
4. S.R. Kohn and S.B. Baden. Irregular Coarse-Grain Data Parallelism under LPARX. *Scientific Programming*, 5:185–201, 1995.
5. T. Rauber and G. Rünger. Parallel Execution of Embedded and Iterated Runge–Kutta Methods. *Concurrency: Practice and Experience*, 11(7):367–385, 1999.
6. T. Rauber and G. Rünger. A Transformation Approach to Derive Efficient Parallel Implementations. *IEEE Transactions on Software Engineering*, 26(4):315–339, 2000.
7. T. Rauber and G. Rünger. Deriving Array Distributions by Optimization Techniques. *Journal of Supercomputing*, 15:271–293, 2000.
8. E. van de Velde. Data Redistribution and Concurrency. *Parallel Computing*, 16:125–138, 1990.
9. P.J. van der Houwen and B.P. Sommeijer. Parallel Iteration of high–order Runge–Kutta Methods with stepsize control. *J. Comp. Applied Mathematics*, 29:111–127, 1990.
10. G. Zhang, B. Carpenter, G.Fox, X. Li, and Y. Wen. A high level SPMD programming model: HPspmd and its Java language binding. Technical report, NPAC at Syracuse Univ., 1998.

Scheduling Task Graphs on Arbitrary Processor Architectures Considering Contention

Oliver Sinnen and Leonel Sousa

Universidade Técnica de Lisboa, IST / INESC-ID
Rua Alves Redol 9, 1000 Lisboa, Portugal
{Oliver.Sinnen,las}@inesc.pt

Abstract. F or the efficient utilisation of a parallel system a task must be divided into sub-task and these m ustbe assigned to the system's resources. The temporal and spatial assignment of sub-task to the computing and communication resources of a parallel system is known as the scheduling problem.
This article proposes a new scheduling heuristic, called Simple Contention Scheduling (SCS), for arbitrary processor architectures with the consideration of link conten tion. The properties of the algorithm are discussed and it is compared to other existing algorithms (DLS and BSA) for arbitrary processor architectures with the consideration of conten tion. The comparison is done for a large set of random graphs and different architectures. Experimental results show that, while having the same or lower complexity, SCS produces in most cases better results than the other algorithms.

1 Introduction

T o execute a program on a parallel system, it must be divided into subtasks which are then scheduled to the target system's resources under the consideration of their precedence-constraints. The manner of how the temporal and spatial mapping of the sub-tasks to the systems resource is done significantly determines the efficency of the parallel program. The problem of finding an efficient utilisation of the system resources is known as the scheduling problem.

F or the purpose of static scheduling, an algorithm or program is represented as a graph [1]. The nodes represent the sub-tasks, that is the computation, and the edges the communication and precedence-constraints of the program. The most popular graph model used for the scheduling of programs is the Directed A cyclic Graph (DAG) [2]. A DA G is a directed and acyclic graph $G = (V, E)$, where V is a set of nodes (vertices) and E is a set of directed edges. A weigh t $w(n_i)$ assigned to a node n_i represents its computation costs and a weigh t c_{ij} assigned to an edge e_{ij} represents its communication costs. The indices ij denote that the edge is directed form node n_i to node n_j.

The scheduling of a D AG (or task graph) in its general form is a NP-hard problem [3], i.e. an optimal solution cannot be calculated in polynomial time (unless $NP = P$). Sc heduling algorithms are therefore based on heuristics that

B. Hertzberger et al. (Eds.): HPCN Europe 2001, LNCS 2110, pp. 373–382, 2001.

try to produce near optimal solutions. Many algorithms have been proposed in the past, e.g. [2,4,5,6,7,8,9], following different approaches. Early scheduling algorithms do not take communication into account [6], but due to the increasing gap between computation and communication performance of parallel systems, the consideration of the communication became more important and is included in the scheduling algorithms recently proposed. Most of these algorithms assume the target system as a homogenous system with fully connected processors. Moreover contention for communication resources is neglected.

Very few algorithms model the target system as an arbitrary processor network and incorporate contention in the scheduling heuristic [4,7,9]. In these algorithms, the topology of the parallel target system is represented as an undirected graph $G = (V, E)$, where V is a set of vertices and E is a set of undirected edges. A vertex P_i represents the processor i and an undirected edge L_{ij} represents a communication link between the incident processors P_i and P_j. Most scheduling algorithms assume a dedicated communication system. In [10], Macey and Zomaya showed that the consideration of link contention is significant to produce an accurate and efficient schedule.

This article proposes a new scheduling heuristic for arbitrary processor networks with the consideration of link contention. The proposed algorithm is based on the traditional list scheduling enhanced for the consideration of link contention. The new heuristic is compared with the existing algorithms BSA [7] and DLS [9] and its benefits as well as its shortcomings are discussed.

The rest of this paper is organised as follows. In the next section definitions and terms for scheduling algorithms are introduced and the two heuristics for the comparison are discussed. Section 3 presents the proposed algorithm and Section 4 compares this algorithm with the two other algorithms based on random task graphs. This article finishes with the conclusions in Section 5.

2 Scheduling Algorithms

In the classic list scheduling formulation, a priority is assigned to each node, and a node is stored in a priority queue as soon as it is ready. Nodes that have all immediate predecessors queued are referred to as being ready. In each step the node with the highest priority is taken from the queue and scheduled to the target system. All the algorithms considered in this paper employ some kind of list scheduling to establish the (initial) scheduling order of the nodes.

Two terms often used for the calculation of the nodes' priorities are the *bottom-level* and the *top-level* of a node. The bottom-level of a node is the longest path beginning with the node and the top-level is the longest path reaching the node. The length of a path is defined as the sum of the weights of its nodes and edges. A level is called *static* when the length is calculated based only on the node weights.

If a node n_i is scheduled to a processor P, we denote its starting time on this processor as $ST(n_i, P)$ and its finish time as $FT(n_i, P)$. After all nodes of the DAG have been scheduled to the target system the schedule length is defined as

$max_i\{FT(n_i, P)\}$ over all processors. The data ready time $DRT(n_i, P)$ of a node n_i is defined as the time at which the last communication from its parent nodes finishes. The node from which the last arriving message is sent is denoted the Very Important Node $VIP(n_i)$. For a valid schedule, $ST(n_i, P) \geq DRT(n_i, P)$ must be true for all nodes.

The aim of all scheduling algorithms is to minimise the schedule length without violating the precedence-constraints among the tasks.

2.1 Dynamic Level Scheduling (DLS)

The Dynamic Level Scheduling (DLS) algorithm was proposed by Sih and Lee in [9] for interconnection-constrained heterogeneous processor architectures (in this paper we consider the homogeneous version). The DLS algorithm is a list scheduling algorithm with dynamic priorities that accounts for interconnection contention. It achieves the contention sensitive scheduling by treating interconnection links, like the processors, as shared resources. A message is therfore scheduled to an interconnection link and occupies that link exclusively for the period of communication.

Like other list scheduling algorithms, the DLS assigns a priority to every node. However, the priority can change during the scheduling depending on the utilisation of the available resources. The priority (or level) is therefore dynamic, as it changes in each scheduling step. The dynamic level DL of a node is defined as

$$DL(n_i, P_j) = SL(n_i) - max(DRT(n_i, P_j), RT(P_j))$$

where $SL(n_i)$ is the static level of node n_i, which is the *static* bottom-level, and $RT(P_j)$ is the ready time of processor P_j, i.e. the finish time of the last node scheduled on P_j. The second term in the equation above is the earliest starting time on the processor P_j.

The algorithm can be described as follows:

1. Find the ready node-processor pair with the highest DL;
2. Schedule the selected node on the selected processor and the incoming messages on the communication links, according to an architecture dependent routing;
3. Repeat step 1 and 2 until all nodes are scheduled.

Due to the high complexity of the first step, and the fact that an initial node selection results in almost the same behaviour, the first step can be streamlined by first selecting a node and then a processor [9]. The node is selected as $max_i\{SL(n_i)+max_j\{c_{ji}\}\}$ thus, the node with the highest sum of static bottom-level and maximum incoming communication is selected. The selected node has its highest dynamic level on the processor that allows its earliest starting time.

The complexity of determining the priority for initial node selection is $O(V+E)$, based on a Depth First Search (DFS). With the utilisation of a Heap as the priority queue for the ready nodes, the complexity for inserting and removing nodes and for determining new ready nodes is $O(Vlg(V)+E)$. Finding the processor with the earliest starting time for a node has a complexity of $O(P(V + E \cdot O(Routing)))$ for all nodes. Every node is validated on every processor by checking the arrival time of the incoming communication edges, which are routed and scheduled to the interconnection links. Therefore, the complexity of the DLS algorithm for homogeneous processors is $O(Vlg(V) + P(V + E \cdot O(Routing)))$, where $O(Routing)$ is the complexity of the routing..

2.2 Bubble Scheduling and Allocation (BSA)

The Bubble Scheduling and Allocation (BSA) [7] algorithm, proposed by Kwok and Ahmad, uses a scheduling approach different from list scheduling. The first part of the algorithm establishes a serial order among the nodes, which is similar to the attribution of priorities in list scheduling. The serial order of the nodes is based on the order of the critical path nodes. The BSA algorithm distinguishes between critical path nodes (CPN), in-branch nodes (IBN), which are nodes that have a path to a CPN, and out-branch nodes (OBN), which are all nodes that are neither CPNs nor IBNs. The serial order is created by ordering the nodes according to the order of the CPNs, but all IBNs of a CPN are inserted recursiv elybefore that CPN, with largest communication first. The OBN are appended to that list in topological order. The ordered nodes are then scheduled ("injected") to the pivot processor. The pivot processor is the processor with the highest degree, i.e. the largest number of adjacent processors. For the main part of the BSA algorithm a processor list is created in a breadth-first search order, using the pivot processor as the starting processor.

Having the nodes "injected" on the pivot processor and a processor list, the main part of the algorithm starts. In each phase of the algorithm the next processor of the processor list is selected as the pivot processor. Every node that is scheduled on this pivot processor is examined for migration to an adjacent processor. A migration is considered if the starting time of the node is later than its data arrival time. In this case the node might start earlier on another processor and therefore improve the total schedule length. A node is also considered for migration if its VIP-node is not scheduled on the pivot, i.e. the largest communication to the node is done over the interconnection network. A node migrates to an adjacent processor if the starting time improves or, in the case that the starting time stays the same, if the VIP-node is scheduled on an adjacent processor. The node is "bubbled up" when the node migrates to another processor, since it starts there at an earlier time. Nodes that were scheduled after the migrated node on the pivot processor can also "bubble up", occupying the freed time slot.

A node can be scheduled to a processor when there is a sufficient large slot, i.e. a slot larger than the execution time of the node, between two nodes already scheduled on that processor. Additionally, the starting time of a node must be larger than its DRT. This is in contrast to most scheduling algorithms, where nodes are scheduled after the last node on the processor. Messages are scheduled in the same way to links. The earliest starting time of the message is the finishing time of the node from which the communication edge emerges, or, for a ulti hop communication, the arrival time of the message from a previous link.

The BSA algorithm uses an iterative routing, as messages are only rerouted on the link between the two processors involved in a migration of a node.

A crucial part of the algorithm is the updating of the start times for the nodes and messages when a node migrates to another processor. Even though, the migration is only done when the start time of the node decreases, the messages that leave the node might arrive later than before the migration. This implies that the *DRT* of a successor node might become larger than its *ST*, which would violate the precedence-constraints. Unfortunately, in [7] nothing is stated about how this update is realised, except that it has a complexity of $O(E)$. In this article, we therefore assumed that a node is only migrated if the new *DRT*s of the direct successors are lower than their *ST*s.

Kwok and Ahmad give a total complexity of $O(P^2EV)$ for the BSA algorithm.

3 The Proposed Algorithm

The algorithm we propose belongs to the class of list scheduling heuristics and is called Simple Contention Scheduling (SCS). The name reflects the main properties of this algorithm: it is simple in both the algorithm itself and its complexity and it accounts for contention on intercommunication links.

Being a list scheduling algorithm, the first phase of the heuristic is to assign priorities to the nodes. The main measure for the node priority is its bottom-level. The bottom level has several properties beneficial for list scheduling. The node with the highest bottom-level is the first node on the critical path of the graph. This node should be scheduled first to allow the fastest possible execution of the critical path nodes. After the first node has been scheduled to the target system, the node with the highest bottom-level of the remaining unscheduled graph is the first node of the critical path of this remaining graph. So, by choosing always the node with the highest bottom-level, we choose the first node of the dominant sequence [11], that is the dynamic critical path, of the remaining unscheduled graph. Another advantageous property is that during the scheduling the bottom-level of the nodes does not change, thus it has to be calculated only once for every node. The zeroing of communication costs between two nodes scheduled to the same processor has only effect on the top-level of the node.

Ordering the nodes by its bottom-levels in descending order, automatically establishes an order according to the precedence-constraints of the graph. We formulate this in the following theorem.

Theorem 1. *Let $G = (V, E)$ be a directed acyclic graph, where V is a set of nodes (vertices) and E is a set of directed edges. Let $w(n_i) > 0$ be the weight of node n_i and $c_{ij} > 0$ the weight of the edge e_{ij}, where the indices ij denote that the edge is directed form node n_i to node n_j. The bottom-level $bl(n_i)$ of node n_i is defined as the maximum sum of node and edge weights of all path starting with n_i (including n_i).*

If the nodes $n_i \in V$ of G are sorted by its bottom-levels $bl(n_i)$ in descending order the resulting order is a topological order [12] of the nodes.

Proof. (Proof by contradiction) We assume there is a node n_i that has an outgoing edge e_{ij} to node n_j and that n_j has a higher $bl(n_j)$ than the $bl(n_i)$ of n_i. So n_j appears before n_i in the ordering of the nodes by its bottom levels. Since there is an edge from n_i to n_j, there is a path p going through n_j and beginning with n_i. As the bottom-level is defined as the maximum sum of node and edge weights of a path, the sum of path p is $bl(n_j) + c_{ij} + w(n_i)$, which is greater than $bl(n_j)$. This is a contradiction.

The bottom-level can be calculate with a DFS [12] in $O(V + E)$.

The first step of SCS is to calculate the bottom-level and to generate an ordered list of nodes accordingly. For nodes with equal bottom-level the one with the higher top-level comes first. A node with a higher top-level is part of a more dominant sequence of the entire graph and should be scheduled earlier. Ties are broken randomly. The top-level can also be calculated with a basic graph algorithm (e. g. DFS) in $O(V + E)$.

As the schedule order can be established a priori, there is no need to take care of ready nodes during the actual scheduling. The DLS algorithm has to keep track of the ready nodes to avoid the violation of precedence-constraints. The main part of SCS iterates over the ordered node list and schedules a node to the processor with the earliest finish time. Using the finish time instead of the starting time, takes the processor and interconnection speed into account. This allows SCS to be applied to heterogeneous target systems.

To achieve the algorithms awareness of contention, interconnection links between the processors are treated as resources like the processors [9]. Communication is scheduled on the links as it is done for the computation on the processors. The route for the communication is determined by the static routing algorithm of the target architecture. A new node or edge is added to the appropriate resource at the end of the already scheduled nodes or edges. When the main part of SCS is looking for the processor with the smallest finish time for a node, its incoming edges and the node are temporarily scheduled on the considered processor and the employed interconnection links. This is necessary to determine the earliest finish time, since the *DRT* of the node depends on the contention on the links. After the best processor has been found, the edges and nodes are scheduled permanently and the previous step is repeated with the next node of the list.

The Simple Contention Scheduling (SCS) is summarised in the following.

1. Calculate the bottom and top level of the nodes.
2. Generate an ordered list of nodes according to the nodes' bottom-level. For nodes with equal bottom-level, the node with the higher top level comes first. Ties are broken randomly.
3. Schedule the nodes in list order to the processor that allows the earliest *finish* time. Schedule the incoming edges of a node on the communication links of the target system.

The complexity of the first step is $O(V + E)$ as discussed above. Ordering the nodes has a complexity of $O(V lg(V))$, using for example Mergesort [12]. In the main part of SCS the finish time of every node is evaluated together with its incoming edges P times, once for every processor. For every scheduling of an edge the communication route must be determined. Thus, the complexity of step 3 is $O(P(V + E \cdot O(Routing))$. Therefore the complexity of the overall algorithm is $O(Vlg(V) + P(V + E \cdot O(Routing))$. For common regular architectures [3] and for sparse irregular architectures the routing complexity is lower or equal $O(P)$. The complexity of SCS is thus identical to that of DLS and lower than BSA's by the factor of the number of nodes V.

4 Experimental Comparison

The discussed algorithms have been implemented and used to schedule various task graphs to different architectures in order to compare the obtained schedule lengths, i.e. the time when the last processor finishes. We choose, as it is done in literature [8,9,11], random graphs for the comparison as they are not inclined to one particular algorithm and include different algorithm structures.

The task graphs were generated in the following manner. The graph size is varied from 50 to 500 nodes in steps of 50 and the average number of edges per node is chosen as 2, 3 or 5. The computation costs are chosen randomly from a linear distribution around 1, so the average node weight is 1. To achieve several communication to computation ratios (CCR), edge weights are chosen dependent on the average number of edges per node. The communication to computation ratio is defined as the sum of all edge weights divided by the sum of all node weights. For each of the three different average edges per node numbers, we consider CCRs of 0.1, 1, and 10.

The generated random graphs were scheduled to two different architectures, one with few interconnection links, a ring architecture, and the other with many links, a fully connected system. The number of processors for both architectures is chosen to 8. Figure 1 shows the schedule lengths obtained for the various algorithms drawn over the number of nodes. For all displayed figures the average number of edges per node is 2; the results for 3 and 5 edges per node showed the same relations between the scheduling algorithms. On the left side, the diagrams a), c) and e) depict the schedule results for the ring architecture and on the right side the diagrams b), d) and f) depict the results for the fully connected architecture. The three different results for each architecture were obtained for the three different CCRs, namely 0.1, 1 and 10.

First SCS and DLS are examined. Being two algorithms with a common base, both perform equal for low communication (CCR 0.1, a and b). For medium communication SCS yields slightly better results (up to 3%) on the fully connected architecture (CCR 1, d) and equal results on the ring. This can be explained with the difference in the attribution of node properties between the two algorithms. The static level of the DLS algorithm does not include communication costs leading to less accurate properties for the nodes. This behaviour is confirmed for high communication (CCR 10, f), where SCS produces results up to 20 % better than DLS on the fully connected architecture. Overall, SCS pro-

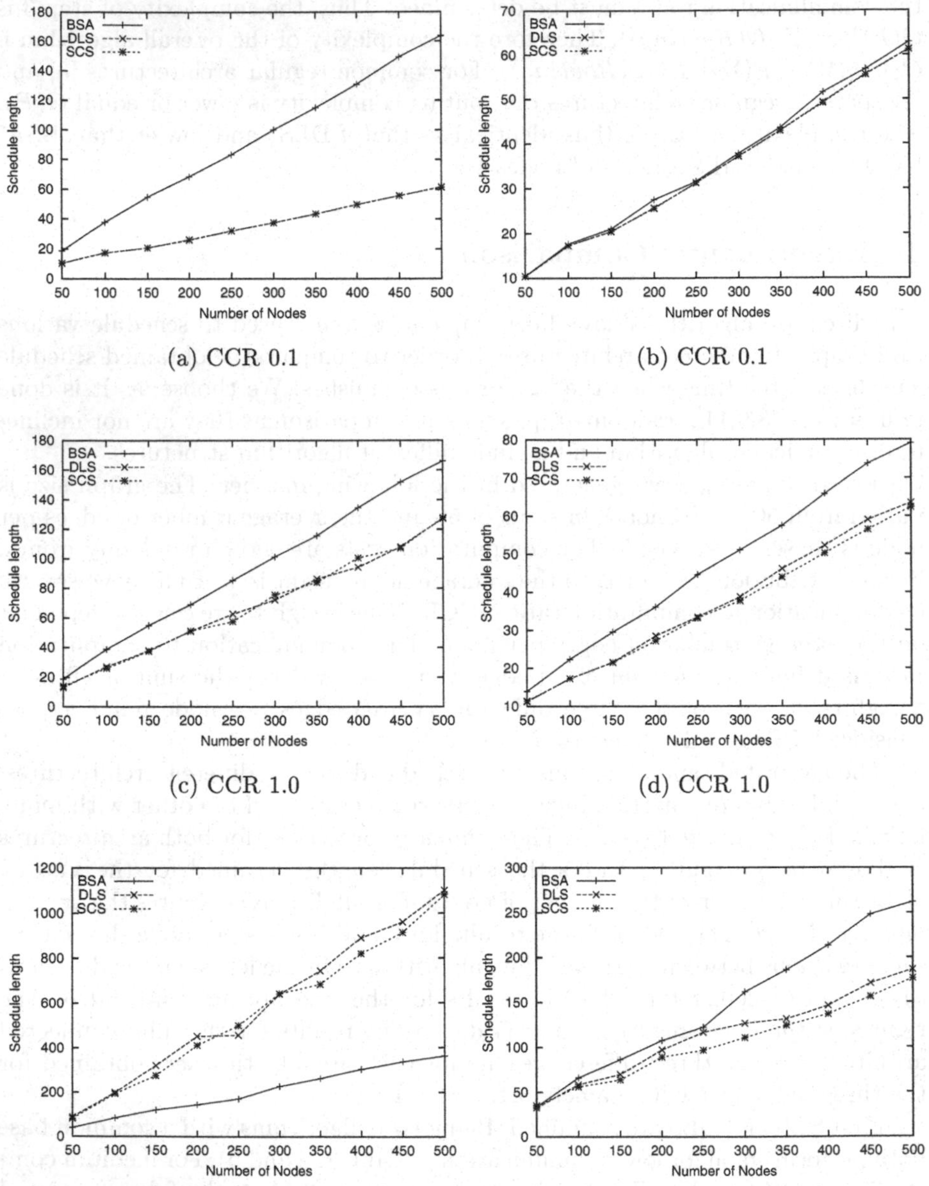

(a) CCR 0.1 (b) CCR 0.1

(c) CCR 1.0 (d) CCR 1.0

(e) CCR 10.0 (f) CCR 10.0

Fig. 1. Sc heduling results on ring architecture (left)and fully connected arc hitecture (right) with 8 processors.

duces better schedules than DLS for higher communication and equal results in the other cases. We now discuss the BSA algorithm compared to SCS and DLS.

The ring architecture serves very well to observe the behaviour of the algorithms. For low communication (CCR 0.1, a) the BSA algorithm performs much worse than DLS and SCS; in fact, the schedule length of BSA is more than the double of SCS's and DLS's schedule length. BSA starts with a sequential schedule on one processor and tries then to improve the schedule. As each processor in the ring has only two neighbours, which are considered in each step of the algorithm, the load cannot be balanced efficiently. For medium communication (CCR 1, c) BSA produces much better results, but still worse than DLS and SCS (about 20%-30%). Here, the significance of the communication becomes more important and BSA only migrates a node if it is beneficial, thus keeping communication low. For high communication (CCR 10, e) both DLS and SCS produce results that are worse than the sequential schedule[1]. Only the BSA algorithm, starting with the sequential schedule and only migrating when it is beneficial, can produce schedules with a length below the sequential. However, it is arguable if a program with such high communication will be run on a parallel machine with 8 processors given a speedup of less than 20% (e.g. 17% for 500 nodes).

For the fully connected architecture the situation is different. Since this architecture has much more intercommunication links than the ring, communication does not play such an important role as in the previous case. The SCS and the DLS algorithm give better results than BSA in all cases. In comparison with the ring architecture, one would expect that the difference between BSA and SCS is even bigger for low communication (CCR 0.1, b). This is, however, not the case, the difference to SCS is almost negligible. A possible explanation is, that in each step of BSA all processors are considered for migration, since each processor is a direct neighbour of every other processor. Thus the load can be balanced better as for the ring architecture. With the increase of the communication the difference between BSA and SCS increases (CCR 1, d and CCR 10, f). Interesting is, that this behaviour is the opposite of the behaviour for the ring. Due to the fully connected architecture the effect of the communication is partly absorbed by the high number of communication links. The migration of one node at a time (BSA) might not be beneficial, but the migration of several nodes together would be.

5 Conclusion

In this article we proposed the new scheduling heuristic SCS for arbitrary processor networks with the consideration of link contention. The algorithm is a list scheduling heuristic and has a simple structure. The algorithm's complexity is lower than the BSA's and is equal to that of the DLS heuristic.

---

[1] The schedule length of the sequential schedule corresponds to the number of nodes as the average node weight is 1 and the communication costs between nodes on the same processor is zero.

For experimental evaluation, we compared SCS with the two algorithm BSA and DLS. The experiments showed that SCS is in all cases equal or superior to DLS due to a more appropriate choice of node priorities. Graphs with low and medium communication were also always best scheduled by SCS. In the case of high communication, the BSA algorithm delivered better results than the SCS when the communication becomes very dominate owing to a small number of links. In fact, SCS produced results worse than the sequential schedule which, however, can be easily detected and substituted by the sequential schedule. Furthermore, the small speedup yielded by BSA in these cases makes a parallelisation at least questionable.

Future work should improve SCS behaviour for high communication and few links without significantly effecting the results for low and medium communication.

References

1. Oliver Sinnen and Leonel Sousa. A comparative analysis of graph models to develop parallelising tools. In *Proc. of 8th IASTED Int'l Conference on Applied Informatics (AI'2000)*, pages 832–838, Innsbruck, Austria, February 2000.
2. V. Sarkar. *Partitionning and Scheduling Parallel Programs for Execution on Multiprocessors* MIT Press, Cambridge MA, 1989.
3. Michel Cosnard and Denis Trystram. *Parallel Algorithms and Achitectures.* Int. Thomson Computer Press, London, UK, 1995.
4. H. El-Rewini and T. G. Lewis. Scheduling parallel program tasks onto arbitrary target machines. *Journal of Parallel and Distributed Computing*, 9(2):138–153, June 1990.
5. A. Gerasoulis and T. Yang. A comparison of clustering heuristics for scheduling DAGs on multiprocessors. *Journal of Parallel and Distributed Computing*, 16(4):276–291, December 1992.
6. H. Kasahara and S. Narita. Practical multiprocessor scheduling algorithms for efficient parallel processing. *IEEE Transactions on Computers*, C-33:1023–1029, November 1984.
7. Y. Kwok and I. Ahmad. Bubble Scheduling: A quasi dynamic algorithm for static allocation of tasks to parallel architectures. In *Proc. of Symposium on Parallel and Distributed Processing(SPDP)*, pages 36–43, Dallas, Texas, USA, October 1995.
8. Y. Kwok and I. Ahmad. Benchmarking the task graph scheduling algorithms. In *Proc. of Int. Par. Processing Symposium/Symposium on Par. and Distributed Processing (IPPS/SPDP-98)*, pages 531–537, Orlando, Florida, USA, April 1998.
9. Gilbert C. Sih and Edward A. Lee. A compile-time scheduling heuristic for interconnection-constrained heterogeneous processor architectures. *IEEE Transactions on Parallel and Distributed Systems*, 4(2):175–186, February 1993.
10. B. S. Macey and A. Y. Zomaya. A performance evaluation of CP list scheduling heuristics for communication intensive task graphs. In *Parallel Processing Symposium, 1998. Proc. of IPPS/SPDP 1998*, pages 538 –541, 1998.
11. T. Yang and A. Gerasoulis. DSC: scheduling parallel tasks on an unbounded number of processors. *IEEE Transactions on Parallel and Distributed Systems*, 5(9):951 – 967, September 1994.
12. Thomas H. Cormen, Charles E. Leiserson, and Ronald L. Rivest. *Introduction to Algorithms* MIT Press, Cambrige, Massachusetts, 1990.

PIO: Parallel I/O System for Massively Parallel Processors

Taisuke Boku[1], Masazumi Matsubara[1*], and Ken'ichi Itakura[2**]

[1] Institute of Information Sciences and Electronics, University of Tsukuba
[2] Center for Computational Physics, University of Tsukuba

Abstract. In this paper, we propose a parallel I/O system utilizing parallel commodity network attached to multiple I/O processors of parallel processing systems. I/O requests from user application are automatically distributed to parallel I/O channels to achieve the best utilization of parallel network, and the system provides highly scalable performance according to the number of channels. We also provides an API for easy user-level programming and a set of utilities for high-speed data transfer and real-time parallelized visualization based on it.

1 Introduction

Recent performance progress on MPPs (Massively Parallel Processors) naturally requires very high-performance I/O system to support large amount of data input and output. These requirements include not only high-speed data transfer from disk to disk but also real-time data visualization with high-performance graphic workstations. On traditional MPP systems, however, the output data are stored into the local disk at first, then transferred to outer environment. In this step, only I/O processors are involved to data transfer, and the performance of the whole system is bounded by these I/O processors and the network attached to them.

Recently, the surrounding environments of such an MPP system are constructed with multiprocessor workstations or clusters, which are equipped with multiple high-speed processors and multi-port network interfaces to provide powerful computational feature and data transfer bandwidth. In order to utilize these resources' power as well as the computational capability of MPP system, it is desired to eliminate the bottleneck of network performance between them.

One of the solutions is to introduce parallel network connections among them to improve absolute data transfer bandwidth, and utilize them as the I/O channel between parallel systems. There have been several researches so far for such high performance parallel I/O systems. HPSS[1] provides very high bandwidth of storage-to-storage data transfer with a special file transfer protocol named PFS. Beowulf[3] cluster utilizes more than one network interface to improve the I/O bandwidth between nodes. In [2], Sang introduces heterogeneous and multiple

* Fujitsu Laboratories Ltd., from April 2001.
** Japan Atomic Energy Research Institute, from April 2001.

B. Hertzberger et al. (Eds.): HPCN Europe 2001, LNCS 2110, pp. 383–392, 2001.

network connection and keeps load-balancing on data transfer. MPI-2[4] specifies MPI-IO API to perform parallel I/O functions on parallel processes. However, these researches have the following limitation:

– Targeting storage-to-storage data transfer to keep load-balancing and static analysis on required bandwidth. (HPSS, Sang's system)
– Omitted well-balanced data transfer load on parallel network interfaces for dynamic parallel processes. (Beowulf)
– No usage and treatment on parallel network connections. (MPI-IO)

In [5], we have proposed the concept and basic design of a parallel I/O system named PIO, equipped with the following features:

– Based on commodity network to achieve very cost-effective system, and
– Useful and easy user-level API to hide the system complexity.

Now we have completed to implement our prototype system on various parallel processing systems including actual MPP, parallel workstation and cluster system. We also have introduced the following features:

– Dynamic load balancing for not only storage-to-storage access but also process-to-storage or process-to-process communication, and
– High portability available on various types of parallel system architecture.

In this paper, we describe the design and implementation of PIO, and its basic performance including dynamic load balancing feature.

2 Experimental Environment

Most of MPPs in the past have been equipped relatively small number of interfaces for the external network even though they have very powerful internal network and high bandwidth to local disk inside the system. For instance, CP-PACS (Computational Physics by Parallel Array Computer System) [6] MPP system, which is a research prototype system of Hitachi SR-2201, is equipped with a single HIPPI channel even it has 128 I/O processors to access the same number of RAID-5 disk units simultaneously. This situation causes a poor I/O bandwidth to communicate with outer environments to behave as front-end system of MPPs. This problem is not solved by enhancing the channel speed of a single network because all data from thousands of processing units are concentrated to a single I/O processor. Therefore, it is desired to equip a large number of I/O processors with multiple network interfaces.

Here, we introduce a large number of commodity network connections like Ethernet or ATM instead of utilizing a small number of high-speed network like HIPPI, for effective use of I/O processors equipped with MPPs. This strategy goals to the high-performance and low-cost external I/O channel.

However, it is very hard for user applications to utilize these parallel network resources directly because they require complicated I/O programming. Moreover, when considering dynamic load-balancing among parallel channels, it is almost impossible to program them with system-independent portability. Currently, generic operating systems also cannot provide such a feature. Therefore, a poweful and flexible user-level system to utilize these high-bandwidth parallel network as an external I/O channel is strongly required.

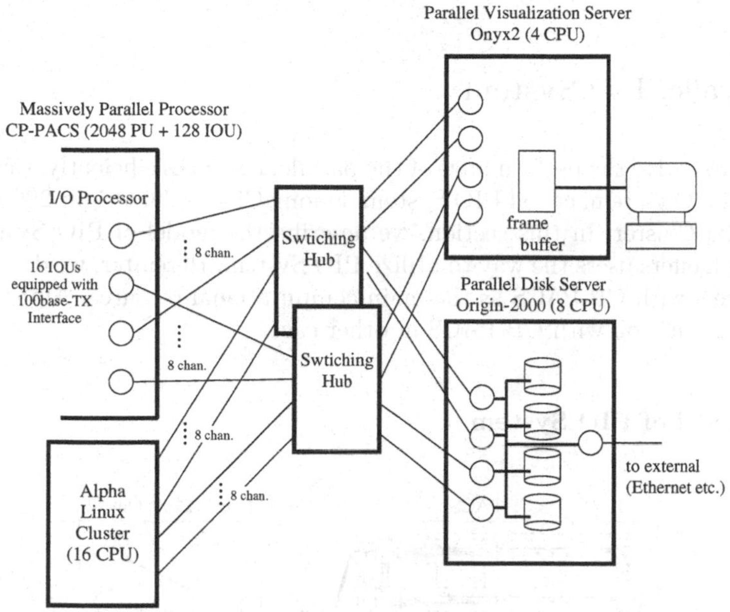

Fig. 1. Experimental Environment for Parallel I/O System.

To build a prototype I/O system based on the above concept, we have constructed an experimental environment with a real MPP system CP-PACS, surrounding it with parallel workstations SGI Origin-2000 and Onyx2 as well as Alpha CPU based cluster. We have selected 100base-TX Ethernet as a basic network for commodity-based parallel network. We design and implement the parallel network controlling system with ordinary TCP/IP protocol which allows to use other commodity network like Gigabit Ethernet according to the availability and cost.

Fig.1 describes the experimental environment on which we have implemented a parallel I/O system. CP-PACS, which is an MPP for large scale scientific calculations developed in University of Tsukuba, consists of 2048 PUs (Processing Units) and 128 IOUs (I/O Units) including 16 special IOUs equipped with 100base-TX Ethernet interface. Any PU can perform TCP/IP communication with outside resources through them. The external systems of CP-PACS are SGI Origin-2000 as a file server with 8 CPUs, Onyx2 as a visualization server with 4 CPUs, and Alpha-CPU Linux cluster with 16 CPUs. Both SGI machines have four 100base-TX Ethernet interfaces, and each CPU of Alpha cluster has a 100base-TX Ethernet interface. Then, there are totally 40 Ethernet cables connected to Switching HUBs. We designed the network controlling system not to limit the number of HUBs for scalability of the system. Therefore, even 48 ports of HUB is available today, we divide all network links from each machine into two classes, and connect each halves to one HUB as shown in the figure. Actually, there is no limitation to the number of HUBs (or network groups) in our system.

3 Parallel I/O System

In order to derive the performance of the parallel networks efficiently, we develop a parallel I/O system, called PIO System, among CP-PACS, Origin-2000, Onyx2 and Alpha Cluster. In this section, we describe the model of PIO System and API which offers users the way to utilize PIO System. Hereafter, we describe our PIO system with CP-PACS as the main computational resource. Alpha Cluster plays the same roll with CP-PACS in other case.

3.1 Model of PIO System

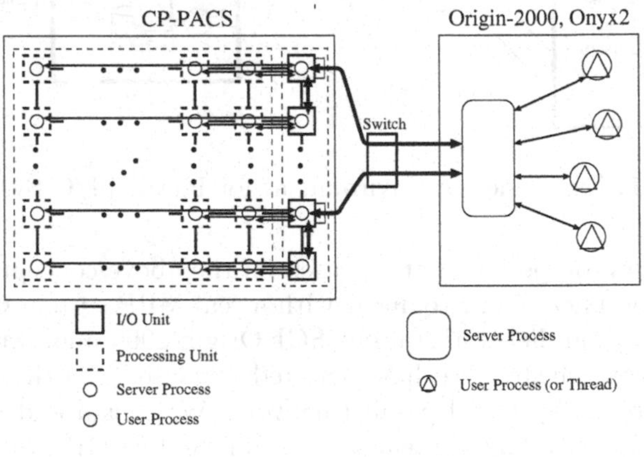

Fig. 2. Model of Parallel I/O System.

Fig.2 shows the model of PIO System we have built on the experimental environment shown in Fig.1. The data produced by one side of user processes are transferred to the other side through server processes. Server processes of CP-PACS are running on IOUs and handle data with user processes which are created on PUs in their own territory. The responsible IOU for a PU is determined according to its location for basic static load balancing. When a stream of data is transferred between a user process and its server process, it is passed with very fast inter-node communication protocol called "Remote DMA transfer" [6] which is a raw data transfer mechanism on CP-PACS. Organized from multiple server processes, PIO System can distribute I/O requests fairly.

On the other hand, there is only one server process on Origin-2000 (or Onyx2) because it has a shared memory architecture and the server process can be implemented with multiple threads in order to handle multiple data streams simultaneously to/from user processes (or threads) and CP-PACS's server processes. User and server processes communicate via shared memory with high bandwidth.

The primary functions of a server process are:

- to manage all data transfer among machines,
- to hide the redundant information about multiple network connections from user processes,
- to preserve the data to/from user processes (buffering), and
- to distribute the workload among parallel channels.

First, if each of user processes must manage data transfer between itself and the other processes, the application program becomes very complex, then it will be heavily burden for users. Therefore, we attempt to reduce the user's burden by undertaking this work at server processes. Similarly, it will be helpful for users to write a program by hiding the redundant information. The third function is namely to do data buffering on server processes. By getting proper size of data buffer, a user process quickly returns from send operation as soon as all of data have transferred to the server process. About the last function, we will describe later in Section 3.3.

Currently, our PIO System is implemented on TCP/IP protocol stack. Therefore, the peak performance per channel is equal to or less than that of TCP/IP on 100base-TX. Since the system is equipped with a large number and amount of buffers in several data transfer stages, the latency for peer-to-peer communication of one message is relatively large compared with raw TCP/IP transfer. However, the ordinary usage of this system is to handle large amount of data produced by one side (CP-PACS) to another (Origin-2000, etc.) with high throughput, then such a latency is not important.

3.2 API

API (Application Program Interface) of PIO System provides an interface for easy programming to bind user and server processes. It consists of a group of library functions as follows:

- Functions to initialize (or finalize) PIO System
  `PIO_Init()`, `PIO_Exit()`
- Functions to input/output data
  `PIO_Send()`, `PIO_Recv()`
- Functions to get various information
  `PIO_Getappinfo()`, `PIO_Gethostinfo()`, `PIO_Getconinfo()`
- Function to select communication partner
  `PIO_Addpartner()`

Usually, it is necessary to make multiple programs for two or more machines because PIO System is implemented among MPPs and other external systems. Then, we unify the API for any platform, which makes it easy for users to write and debug the program. It also increases the portability of PIO-based applications on various platforms.

We also aim to offer users a simple and flexible interface. As mentioned in the previous section, server processes hide redundant information from user processes by default. However, the user process also can require more detailed information for well-balanced I/O control for the application with explicit function calls.

3.3 Communication Load Balancing

The most important issue to exploit high bandwidth with parallel network connections is how to distribute the communication load among them. PIO provides three types of load balancing feature, and users can select one of them in their application programs:

USER: The application program specifies the channel to be used for the communication on each message at sending time. All network properties including the number of available channels, the number of processes, etc. can be retrieved via appropriate functions in PIO-API, and the program can control the distribution of traffic freely.

STATIC: In most of programs written in SPMD-manner in large scale simulations, the amount of data produced by each processor tends to be equal naturally. A simple example is the spatial domain decomposition method, which is the most common parallel programming style for wide variety of scientific problems. Since the responsible PIO server process is determined based on the location of client user process fairly, it is easy and natural to distribute all traffics fairly to parallel channels. This is the default option of PIO communication.

DYNAMIC: PIO servers try to distribute traffics dynamically according to current load of all parallel channels. When the user program generates imbalanced amount of data among parallel processes and the user cannot estimate nor control them, such a dynamic load balancing is necessary.

In general, it is hard to balance the load on an MPP with distributed memory architecture like CP-PACS, because all server processes do not share the traffic information and the cost of information exchange is very high for fine grain traffic control. We have introduced an efficient method for this utilizing the other side of front-end workstations. These machines are shared-memory multiprocessors, and all server threads can share the information on current traffic condition. Therefore, it is easy to notice any requested information to their partner servers on MPP side. We have implemented this information exchanging on three-way data transfer protocol (*sending-request*, *buffer-allocation* and *sending-data*) to minimize the controlling message to be exchanged.

3.4 Supporting Multiple Platforms

Since we are strongly conscious of both distributed and shared memory architecture of parallel processing platforms of PIO, it is designed to be implemented wide variety of parallel architectures including ordinary MPPs, parallel workstations and clusters. Moreover, PIO can be implemented heterogeneous cluster system with single processor and SMP node. For instance in our environment, Origin-2000 and Onyx2 are usually used as individual servers for file service and visualization, but they also can be configured as a cluster of SMP nodes. PIO also supports a network grouping feature to divide multiple network ports into several groups, which allows to connect ports in different group into different network switch. This feature is very important for the system scalability. We can increase the number of network interfaces keeping the maximum number of I/O ports of a switch. Utilizing this feature as well as flexible configuration mentioned above, PIO System provides a scalability to support real MPP systems.

4 Performance Evaluation

4.1 Basic Data Transfer

Fig.3(a) shows the data transfer throughput based on round-trip data exchange (*ping-pong*) between CP-PACS and Origin-2000 with various message sizes. Here, data labeled "TCP" and "PIO" represent the throughput of raw TCP/IP and PIO, respectively. Since CP-PACS achieves only 2.5 MB/s of throughput on single channel TCP/IP[1], the absolute performance is not high[2]. However, it shows the scalability of parallel channel usage is quite well up to 16 channels.

Since intermediate buffers on PIO System increase data transfer latency, the performance of PIO in *ping-pong* is much lower than that of TCP/IP especially for short messages. However, for large scale scientific applications, the average message size is quite long, and the buffering overhead is only 10 to 15%.

On the other hand, unidirectional data transfer is the primitive usage of PIO as I/O system. Fig.3(b) shows the response time on each PU to send a message with raw TCP/IP and PIO. Here, the high-speed data transfer between user and server processes both in CP-PACS and Origin-2000 achieves very low response time for user processes compared with raw TCP/IP.

4.2 Dynamic Load Balancing

To evaluate the dynamic load-balancing feature on PIO, we measured the elapse time to complete a large amount of unidirectional data transfer with artificially

[1] Origin-2000 can achieve over 90% of theoretical peak throughput on TCP/IP. Therefore the number of network ports (CP-PACS=16, Origin-2000=4) is balanced.

[2] Especially with 8KB and 256KB message sizes, the throughput is drastically degraded. It is caused by data transfer protocol switching and the specific size of buffer on the operating system on CP-PACS.

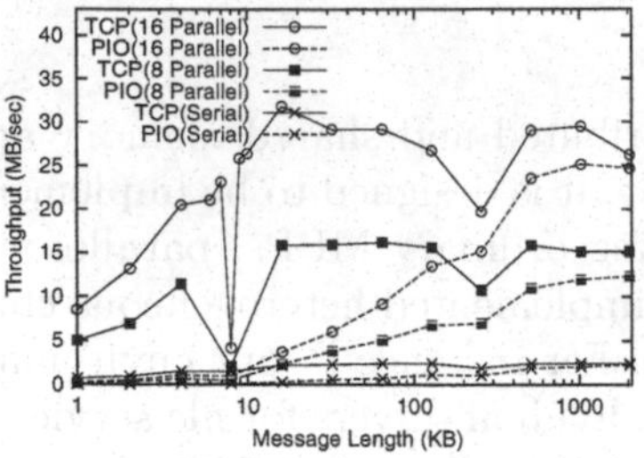

(a) Throughput on Ping-Pong Transfer

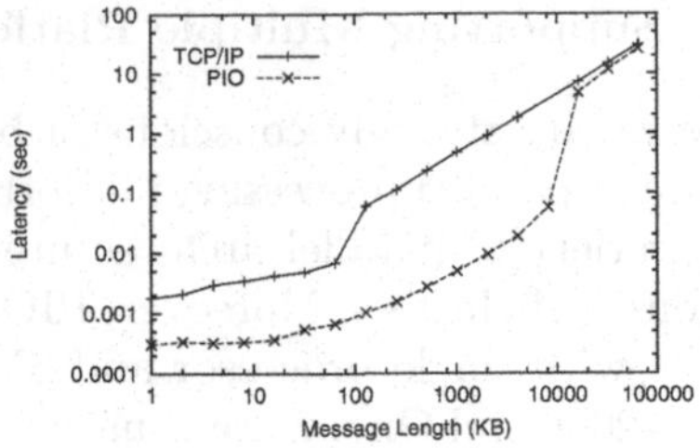

(b) Response Time for Unidirectional Transfer.

Fig. 3. Basic Data Transfer Performance of PIO System.

imbalanced data output by parallel user processes. Various length of messages are sent from CP-PACS where the maximum message size is 16 times as longer as the minimum one. Results are shown in Fig.4. USER shows the result of explicit user-level programming to flatten the data distribution among all parallel channels to minimize the elapse time. STATIC and DYNAMIC show the results of system default (no load-balancing) and dynamic load-balancing feature, respectively. Theoretically, STATIC requires 2.13 times longer elapse time than USER under the given condition.

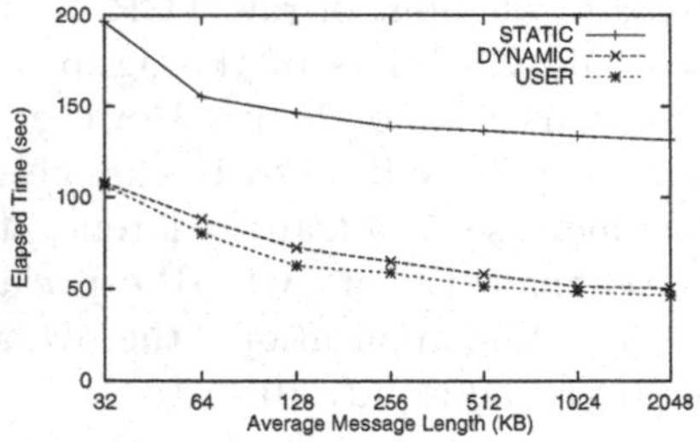

(a) Load Balance Effect on PIO.

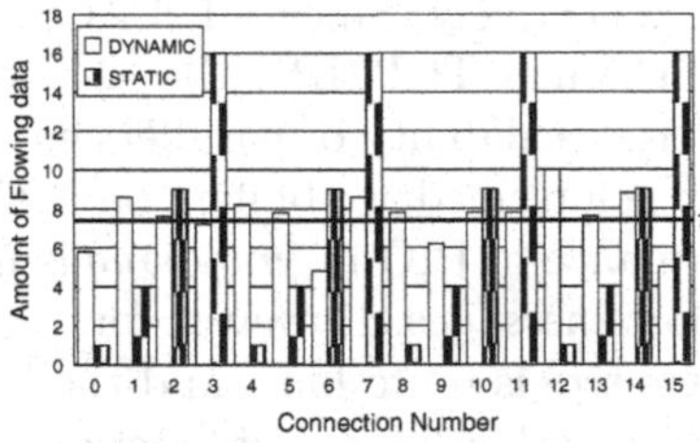

(b) Actual Load among 16 Channels.

Fig. 4. Dynamic Load Balancing Feature of PIO.

Fig.4(a) shows the elapse time to transfer 512MB data in total with various average message sizes. The figure shows the dynamic load-balancing feature achieves almost the same performance as the optimal user programming while the system default takes 2 to 2.5 times longer time, which is close to theoretical time. There is a slight difference between USER and DYNAMIC caused by load-balancing overhead, but it is negligible. Fig.4(b) shows the actual data traffic on all 16 parallel channels (average message size = 512KB). The horizontal line at "7.5" shows the ideal data amount achieved with USER. We can see DYNAMIC distributes the data traffic fairly among all channels with a small fraction of error.

All the data show the dynamic load balancing feature of PIO System works quite well. However, when it is guaranteed by the user program that the amount of all data produced by parallel processes are the same, it is the best to use system default static load-balancing to reduce the overhead.

5 PAVEMENT – A PIO Application

Several applications to utilize PIO System are now under construction. PAVEMENT (Parallel I/O and Visualization Environment) is a high-performance parallel data exchange and visualization system. CP-PACS and Onyx2 are connected via PIO, and user processes on CP-PACS directly output the numerical results to Onyx2 in order to produce real-time graphic image as a result of numerical computation. We are developing a general purpose parallelized visualizer based on AVS/Express[7]. Since AVS/Express consists of a large number of individual modules connected in object oriented manner, it is easy to parallelize several modules which make the bottlenecks in the system.

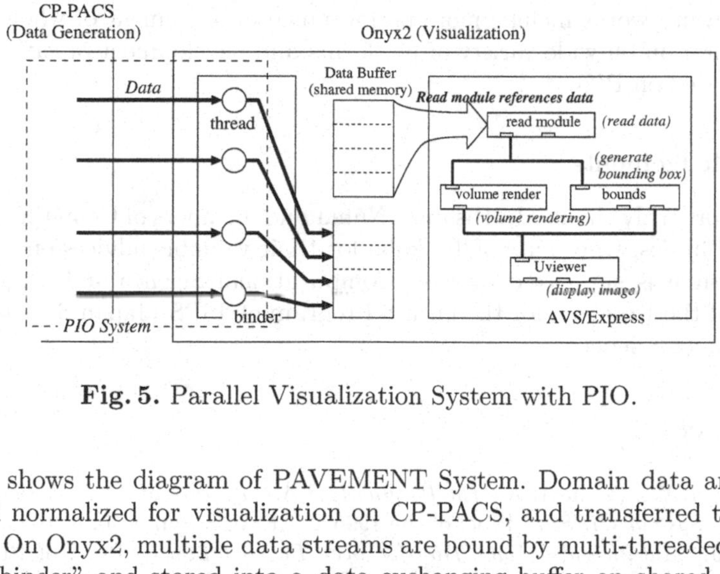

Fig. 5. Parallel Visualization System with PIO.

Fig.5 shows the diagram of PAVEMENT System. Domain data are generated and normalized for visualization on CP-PACS, and transferred to Onyx2 via PIO. On Onyx2, multiple data streams are bound by multi-threaded process named "binder" and stored into a data exchanging buffer on shared memory. They are read by a dedicated data input module implemented on AVS/Express, then processed by parallelized volume rendering module. Except these two special modules, we utilize the basic system modules of AVS/Express to minimize the development costof PAVEMENT. Finally, the 3-D volume renderred image is generated in very high-speed compared with the original sequential version. Here, PIO provides a wide bandwidth of communication channel among two parallel processing systems. We have confirmed this system works correctly in high performance with 2048 PUs on CP-PACS. Since PIO-API is well-designed for easy programming, the program of "binder" is quite simple while it provides high bandwidth of parallel data manipulation. We are also developing a parallel file system on user-level with PIO. It is based on the management of a large amount of file pieces to construct a logical image of a large file. Each file piece is generated by parallel process on CP-PACS, then they are transferred through PIO to Origin-2000 file server. It is a natural way to generate, transfer and manage these pieces completely in parallel.

6 Conclusions and Future Works

We have developed a parallel I/O system named PIO for massively parallel processors based on parallel connections with commodity network. The prototype system with 16 channels connect various types of parallel processing platforms including a real MPP and state-of-the-art parallel workstations and clusters. The scalability in usage of parallel networks is quite well and it provides very cost effective data transfer system. The dynamic load-balancing feature to flatten the data traffic distribution is realized based on co-working distributed memory MPP and shared memory parallel systems. A simple API provides easy programming for application users.

Our future works include more performance improvement of whole system, implementation on wide variety of platforms and development of various applications based on PIO.

Acknowledgements

The authors truly thank Mr. Hisataka Numa and members of Center for Computational Physics, University of Tsukuba for their valuable advices on this work. This research is supported by the "Computational Science and Engineering" Project of the Research for the Future Program of JSPS (Japan Society for the Promotion of Science).

References

1. Watson, R.W., Coyne, R.A.: *The Parallel I/O Architecture of the High-Performance Storage System (HPSS)* , Proc. of the 1995 IEEE MSS Symposium, 1995.
2. Sang, J.: *Multithreaded Data Transfer over Parallel Links* , Proc. of PDPTA'99, 1999.
3. Sterling, T., Becker, D., Savarese, D., et.al.: *BEOWULF : A Parallel Workstation for Scientific Computation* , Proc. of Int. Conf. on Parallel Processing 1995, 1995.
4. *MPI-2: Extensions to the Message-Passing Interface* , `http://www.mpi-forum.org/`
5. Matsubara, M., Numa, H., Boku, T.: *Commodity Network based Parallel I/O System for Massively Parallel Processors* , Proc. of PDPTA'99, 1999.
6. Boku, T., Itakura, K., Nakamura, H., Nakazawa, K.: *CP-PACS: A massively parallel processor for large scale scientific calculations*, Proc. of ICS'97, 1997.
7. Advanced Visual Systems Inc.: *AVS/Express*, `http://www.avs.com/software/soft_t/avsxps.html`

Fast "Short" Messages on a Linux Cluster

M. Danelutto and A. Rampini

Dept. of Computer Science – University of Pisa – Italy

Abstract. We discuss an experiment aimed at lowering the operating system related overheads when performing small size message communications on a Beowulf class Linux PC cluster. The experiment consists in adding a small number of new system calls to the Linux kernel allowing user code to send/receive messages to/from remote processes. The system calls have been implemented using the standard kernel module mechanism provided by Linux. Those new system calls allow small size messages to be exchanged between cluster nodes with times that are 10 to 15% smaller than those achieved using standard TCP/IP communications.

1 Introduction

Clusters are gaining more and more importance in the parallel computing world. On the one side, parallel architectures that can be roughly classified as clusters appear on the top positions of the `top500` list [1]. These architectures mainly differ from Beowulf [2] clusters because they adopt interconnection networks not belonging to the *commodity-off-the-shelf* class. Such interconnection networks provide impressive bandwidth and very low communication latencies at a cost which represents a significant fraction of the whole machine cost. On the other hand Beowulf class clusters promise supercomputer performances at a reasonable cost. They usually adopt Ethernet based interconnection networks delivering reasonable bandwidth but also very high communication latency times.

We are interested in investigating ways to improve the performance of parallel applications on Beowulf class computers (our group is active in the definition of structured [3,4] parallel programming environments suitable for parallel cluster programming based on the skeleton concept [5]). One of the key factors affecting parallel application performances on clusters is communication cost, both in terms of bandwidth and in terms of latency. In particular, the efficiency of interprocessor communications appears to be crucial with respect to performance in at least two distinct cases:

- in interprocessor communications involving "large" messages used to keep the distributed computation state consistent. These communications take substantial advantages of any improvement in interconnection network bandwidth.
- in interprocessor communications involving "small" (possibly zero-length) messages used to keep synchronized the distributed network of processes

B. Hertzberger et al. (Eds.): HPCN Europe 2001, LNCS 2110, pp. 393–402, 2001.

making up the parallel application. These communications take significant advantage of any improvement in interconnection network latency.

To minimize the time spent in communications in a cluster environment, message aggregation techniques are often used. Message aggregation consists in grouping different messages directed to the same destination processing element in such a way that they can be sent in a single message, rather than using multiple messages. The technique is obviously aimed at reducing the *global* overhead involved in message transmission and demonstrated good results [6]. However, the aggregation technique introduces small overheads in the delivering of a single message. In order to effectively transmit the message the run time support must wait that other messages for the destination node become available (or that a short timer expires, signaling that there are no further messages to be combined) and then must "pack" the different messages in a single, larger message to be transmitted to the destination node. This is a price worth to be paid in case the messages involved are those keeping the distributed computation state consistent (i.e. those actually distributing data among the processors). But the (small) overhead involved in message aggregation can have bad effects on the parallel program performance when the messages involved are "pure synchronization" messages. As an example, if the messages are used to implement a barrier, any overhead added to the basic transmission mechanism can result in a global slowdown of the parallel application.

In this work, we want to present the results of an experiment we performed aimed at lowering the overheads involved in a single communication between processes running on a Beowulf cluster based on Linux Pentium machines. In particular, we tried to optimize the communication times required by pure synchronization messages, i.e. by those messages involving the exchange of a very small amount (possibly zero bytes) of information. Such kind of communication is heavily used in parallel computations. The cost paid to deliver a pure synchronization message mostly comes from network latency, i.e. from the time spent in setting up the hardware to deliver the message. Bandwidth is not crucial, as the size of the message to be transmitted is negligible, and a single packet is enough to transfer the whole information required between the sender and the receiver nodes. Therefore, we developed a very small set of communication primitives implemented in such a way that the software overheads involved in the communication are the smallest possible. During our experiments we only take into account Beowulf class clusters equipped with plain Fast Ethernet NICs as this kind of NICs is by far the cheapest network hardware commonly found in Beowulf clusters.

Rather than trying to intervene on the NIC drivers (which is the technique adopted in other projects, such as the Gamma one [7]) we adopted an approach that minimizes the operating system overheads involved in a communication, i.e. the overheads involved in setting up parameters, in copying data from user space to kernel space, etc. Basically, we introduced a small number of new system calls in the Linux Kernel. These system calls implement the basic communication primitives used in parallel programming on PC/WS clusters: send and receive plus functions handling process names. Each syscall completely handles the whole communication required. Therefore the user code just issues a single syscall to implement a remote communication. The number of new system calls introduced is very limited, therefore we can assume that slots are available in the current and future implementations of the Linux kernel. If a larger number of syscalls is needed, a different mechanism should be used, e.g. syscall multiplexing in the kernel space.

By implementing the communication primitives at the kernel level, we can directly interact with the NIC (Network Interface Card) drivers. Therefore all our syscalls use Ethernet protocols to perform physical communications. This requires some additional code (message acknowledging, basically) to ensure reliability in the communications, however.

Our syscalls are implemented using the Linux kernel module mechanism [8,9]. Such mechanism allows portions of code to be dynamically (even "on-demand", if the case) linked to the *running* kernel code.

2 Communication Primitives

Rather than implementing a variety of communication primitives (as it happens in state-of-the-art communication libraries such as MPI [10]) we choose to implement exactly two communication primitives: an asynchronous send (`cls_send`) and a synchronous receive (`cls_receive`). This choice has been made in the perspective of implementing a communication layer that can be used to replace (in part or as a whole) the lower communication layer used in common implementations of MPI, such as `mpich` [11].

Our send primitive returns control to the caller user process as soon as the message has been physically copied to the destination node buffers, even in case no destination process has issued a matching receive syscall. Our receive primitive, on the other hand, is blocking, in that the primitive does not return control to the caller process until a matching send primitive has been issued and the message has actually been transferred. Furthermore, the send primitive is symmetric (i.e. we can only specify a single destination process in the send parameters), while the receive primitive is not: we can either specify a process as the effective sender of the message we want to receive or we can specify that we want to receive a message from any sender process, in FIFO order.

In addition to these two primitives, we implemented a couple of *naming/service* primitives: a `cls_getpid` allowing a process to obtain a valid identifier that can be used for communications and a `cls_exit` that allows a process to release its identifier as well as all the allocated resources before termination. (The identifiers returned by `cls_getpid` happen to be in the range $[0, n-1]$, where n is the total number of processes involved in the communications.)

With these basic primitives most of the usual communication primitives can be implemented. P4, which is one of the communication packages used to run MPICH [11] on clusters using TCP/IP interconnection networks, provides exactly these primitives plus a synchronous send one (the latter one can be emulated by a couple of send/receive calls using our library).

3 Communication Primitives in Kernel Space

In order to implement the primitives discussed in Sec. 2 we introduced some new system calls directly in the Linux kernel. The system calls (`sys_clsgetpid`, `sys_clsexit`, `sys_clssend` and `sys_clsreceive`) implement process registration, process termination[1], message send and message receive primitives directly at the kernel level. User program can issue direct calls to the syscalls or they can issue calls to a wrapping library implementing the `cls_xxx` primitives. The library primitives only perform additional sanity checks before actually calling the corresponding syscalls. No memory copy is performed while traversing these syscall wrapping functions.

The advantages achieved by the kernel level implementation of the communication primitives are twofold: on the one hand, the overheads due to switch between user mode and kernel mode are minimized, as the whole communication takes place within a single system call; on the other hand, by working in kernel mode we have direct access to NIC drivers. Therefore, we can access the network device at the lowest admissible level[2] and therefore we don't pay the extra overheads due, as an example, to the TCP/IP stack protocol implementation.

The process leading to the addition of new system calls in an operating system kernel is usually quite complex. Code for the new system calls has to be developed and then the whole kernel must be rebuild (linked) in order to make the new calls available to the user code. The Linux operating system[3] provides a simpler way to insert new features into a *running* kernel: dynamic module loading. Linux dynamic kernel module loading mechanism allows code to be dynamically linked to the kernel space, while the kernel is running. By issuing an `insmod` command (as superuser), an object code can be loaded into the kernel space and the relevant kernel functions/data structures become accessible from that code. Furthermore, the loaded code can register functions as system calls. As soon as the communication syscalls are no longer needed, an `rmmod` command can be used to unload the kernel module implementing the syscalls from the kernel space, thus freeing up all the resources used to implement our communication layer in the kernel space.

We exploited this mechanism to implement our communication system calls. We observed that in the current kernel version[4] there where some free syscall slots (i.e. entries in the kernel syscall table). Therefore we wrote a kernel module implementing the communication primitives of Sec. 2 as syscalls and we used these free slots to register our communication syscalls.

---

[1] Process leaving the set of processes performing communications, actually.
[2] Excluding the driver level, which is beyond our goals, as stated in the Introduction.
[3] This is our target OS, as we address Beowulf class clusters.
[4] Linux kernel version 2.2.13, when we performed the experiments.

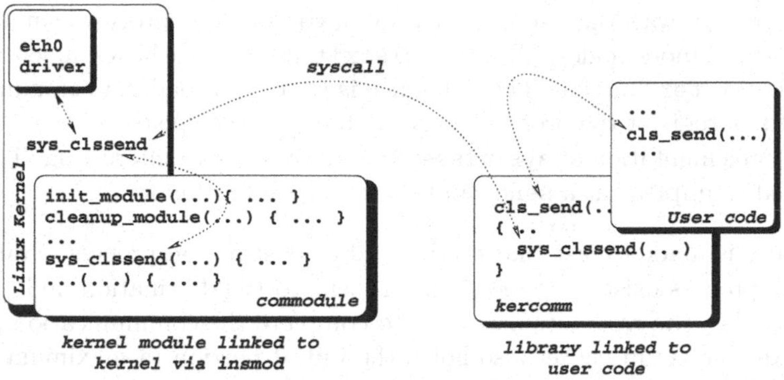

Fig. 1. Software General Structure.

4 Implementation Details

The overall structure of the implemented software is depicted in Figure 1. It consists of two different software packages: `kercom` and `commodule`.

`kercom` represents the library code that must be linked to the user code in order to use our communication primitives; it provides the wrapping functions `cls_XXX` called from user code to invoke the communication syscalls. The C code for the `kercom` module is very small (1 hundred lines of code, rough), as it just adds some sanity checks to the plain syscall. As an example, the following code is the one implementing the function `cls_send` wrapping the syscall `sys_clssend`:

```
int cls_send (int dest, void* buff, int buffsz)
{
 if ((dest<0) || (dest>(CLUSTER_SIZE*MAX_PROC_X_HOST))) return ENOPROC;
 if (buff == NULL)                                      return EINVADDR;
 if ((buffsz>msg_sz) || (buffsz<0))                     return EMSGBIG;
 return sys_clssend(dest, buff, buffsz);
}
```

`commodule` is the dynamic kernel module. With less than 1.5 Klines of C code, it implements all the functions needed to register the new syscalls into the kernel and to implement the send and receive syscalls. In particular, `commodule` provides the `init_module` and `cleanup_module` functions, that are called by the kernel when the module is loaded by an `insmod` and unloaded by an `rmmod` command, a `rx_packet` function, which is the function called to handle incoming packets, the functions `sys_clsgetpid`, `sys_clsexit`, `sys_clssend` and `sys_clsreceive`, which are those registered in the syscall table in the `init_module` function, and a set of auxiliary functions `eth_xxx` called from within a `rx_packet` function to handle different types of incoming packets (ACKs, transmission requests, process registration/de-registration requests, etc.).

The `sys_clsregister` function updates the module internal data structures and broadcasts to all hosts in the cluster a message announcing the new process, in such a way that communication involving the process can be initiated on the remote nodes. The `sys_clsexit` function behaves much like the `sys_clsregister` function. The difference is that the broadcast communication is used to de-register the process id on the hosts of the cluster.

The implementation of `sys_clssend` function is more interesting. The steps performed to implement a `send` can be summarized as follows:

1. a check is made to test that the id used as destination process is valid
2. if the process exists, a *setup* packet is sent to the destination node, in such a way that the resources necessary to complete the communication are prepared. The setup packet also holds the initial portion (a maximum of 1478 bytes, i.e. the Ethernet MTU minus our header size) of the data to be transmitted.
3. if the setup succeeds, the data packets are sent to the destination node. In case of short messages (messages whose size, including the headers, is less than network MTU) a single setup/data packet is sent, otherwise more than a single data packet must be sent[5]. Each data packet is to be acknowledged, as the protocol used for transmission is plain Ethernet protocol. However, we do not acknowledge any packet separately, as this leads to poor performance. Instead, the sender node explicitly requests ACKs relative to all the sent packets when all the data packets have actually been sent. The receiving node, at this point, sends all the ACKs relative to the received packets within a single packet. In case some ACKs are missing, the sender node only re-sends the non-ACKed packets, and then requires the ACKs for these packets only.
4. if the data transmission succeeds, the resources allocated are freed.

Most of the code written to implement `commodule` uses the typical mechanisms used in kernel coding. As an example:

– the check on the existence of the destination process in the `sys_clssend` is performed with interrupts disabled, in such a way that the access to the process data structures can be performed in a safe way[6]
– the physical transmission of the packet is in charge of the Ethernet NIC driver. Therefore the "transmission" code in the `sys_clssend` is just a call to a `dev_queue_xmit` function. This function is the one provided by the Linux kernel to submit an Ethernet packet to be transmitted to the Ethernet device driver.
– incoming packets are routed to the `rx_packet` function as its address is passed to the kernel function `dev_add_pack` that basically registers an handler for incoming packets.

[5] This because the syscalls actually allow communications to be performed with message sizes up to 90Kbytes.

[6] With interrupts enabled, another process may issue a `sys_clsexit`. call, for instance, thus requiring access to the same data structures used to check the existence of the destination process.

The physical communication layer used in our syscalls is the Ethernet communication layer. This means that hosts in the cluster are identified via their MAC address[7], rather than their INET address. The MAC addresses of the hosts in the cluster[8] must be known before compiling the `commodule`. Therefore, the addition of the node to the system requires recompiling the `commodule`. As the `commodules` are dynamically linked to the kernel, recompiling the module requires a `rmmod` of the existing version of the module and a subsequent `insmod` of the new module. This implies that all the running applications using the library must be temporary stopped. However, the addition of a (some) node(s) to a cluster is a rare event and the price can be accepted.

5 Experimental Results

We performed a set of experiments with our communication primitives on `Backus`. `Backus` is a Beowulf class cluster with 16 processing elements. Each processing element (128Mbyte of RAM, 2Gbyte of user disk space) has a 3Com Fast Ethernet NIC. The nodes are interconnected by a 3Com SuperStackII 100Mbit switch. 10 of the nodes have 233Mhz Pentium II processors and the other 6 nodes have a 400Mhz Celeron processor. `Backus` is the cluster we developed at our Department within the Italian Research Council MOSAICO project.

In the experiments, we run a process per processing element continuously performing communications with the other processes. Some of the experiments run for times reaching the 3-4 hours. Within these experiments all the communications take place correctly and we have no problems with the kernel or the user code. This is the first good result. Although the Linux interface to kernel modules is quite simple, system level programming can lead to dangerous situations even in consequence of simple programming errors, possibly leading to a machine crash. Furthermore, the tests have been performed while the cluster was used by other people, using standard parallel tools such as MPI, PVM or plain Unix/Socket tools. In all cases all the standard Unix services of the cluster continued to operate without problems. In the whole period of the test (a couple of weeks) the machine has been never rebooted, nor it has been after the end of the tests.

Some of the experiments were aimed at measuring the latency and the bandwidth provided by the communication primitives we implemented. The results achieved by these experiments have been compared with the numbers obtained by using plain TCP/IP communications[9]. Concerning bandwidth, our kernel

[7] Each MAC address is a 6 byte number, guaranteed to be unique in the world, identifying each NIC.

[8] Actually, the MAC addresses of their Ethernet NICs.

[9] I.e. we rewrote the same programs used for the experiments in such a way that Linux TCP/IP sockets where used to perform interprocessor communications, instead of our communication primitives. As most of the existing communication libraries (MPI and PVM, for example) use the TCP/IP socket layer to implement interprocessor communications, this turns out to be a significant test.

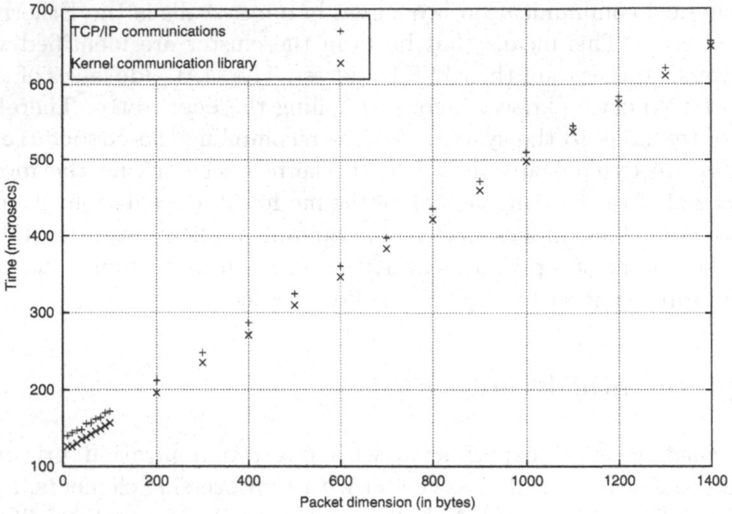

Fig. 2. Comparing Kernel Communication Times with TCP/IP.

implementation of communication primitives achieved a maximum bandwidth of 9.8 Mbyte/sec. Fast Ethernet has a peak bandwidth of 12.5Mbytes/sec. Using TCP/IP communications we achieved a peak bandwidth of 10.74 Mbytes per second. These results should be read as follows: our experiment was mainly aimed at providing fast communications in case of short messages. The communication primitives we implemented support messages whose size are in the range $[1, 90K]$ bytes. Peak performance is achieved with messages whose size is around 10Kbytes. TCP/IP peak performance is achieved with messages whose size is around 90Kbytes. Therefore we can conclude that TCP/IP stack implementation under Linux is very efficient in transferring large messages, which is a known result[10]. The fact that our library only achieved a fraction (near 80%) of the peak bandwidth of the interconnection network is due to the emphasis given to the fast small size message transmission (i.e. on latency minimization) rather than on overall bandwidth (we have not performed any kind of optimization in long message transmission protocol, actually).

Concerning latency, our primitives provided a latency of 64μsecs compared to the one of TCP/IP that was of 76μsecs. Again, these results are relative to the transmission of messages whose size was in the range $[1, 90K]$ bytes. The measured latencies are far away from latencies measured in implementations that modify the NIC drivers (e.g. Gamma latency is around 13μsecs against an hardware lower bound of 7μseconds [7]) but still we demonstrate an 18% improvement with respect to the TCP/IP measured latency.

[10] TCP/IP stack has been rewritten in Linux 2.2 and everybody acknowledges that the new implementation is definitely very efficient.

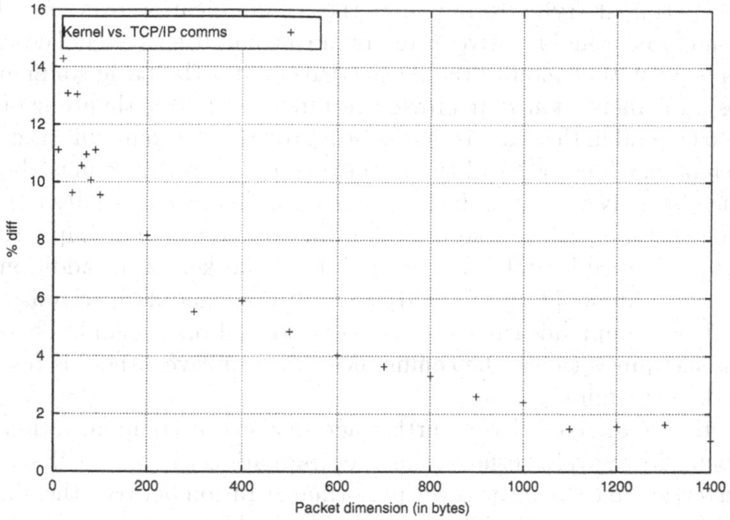

Fig. 3. Percentage difference in times (Kernel vs. TCP/IP communications, packet size smaller than Ethernet MTU).

The most significant result is that on small size messages (those used in synchronizations, in the range of [1, 100] bytes): our implementation of communication primitives demonstrates a raw 15% improvement with respect to the TCP/IP times. Although 15% cannot be considered in absolute a substantial improvement, it can be considered satisfactory in our perspective as this improvement has been reached just using software and on definitely cheap, Beowulf class NIC hardware. Figure 2 plots the *completion time* for such communications (Kernel implementation vs. TCP/IP) and Figure 3 plots the relative differences between those times. The numbers measured demonstrate that the approach is definitely a valid one.

Last but not least, we must remark that the whole implementation has been developed in a couple of man-months. The clear kernel interface provided by Linux demonstrated to be fundamental to achieve this results in such a short time. Also, the amount of documentation available on the Linux kernel due to the fact that Linux is Open Source software, has been a great advantage in the design and implementation of the kernel communication module.

6 Conclusions

We implemented an essential communication primitive set directly in the Linux kernel. By using these communication primitives we achieved better communication times than TCP/IP in short message communications on a Beowulf class cluster with Fast Ethernet interconnection network. At the moment, we know no other experiment trying to minimize the communication overheads by integrating communicating primitives directly in the operating system kernel. Other experiments exist that modify the device drivers and the basic communication primitives implementation that provide results even better than our ones. But the price to be paid in that case is that in order to port the communication library on a different machine, with different network cards, a driver must be rewritten for the new network cards. Instead, our implementation is fully portable on any machine running a Linux kernel 2.2 or higher. You should simply recompile the kernel module code and load the module in the kernel. In addition to this big advantage, the adoption of the dynamic kernel module mechanism proper of Linux allows communication syscalls to be loaded on demand. Therefore, in case no user requires to use the communication primitives, there is no effect at all on the kernel running.

Currently, we are considering further activity on our communication library: new mechanisms to avoid kernel module recompilation when new PEs are added to the cluster, evaluation of the price/performance ration between the library and other hw/sw systems based on more expensive (and more performant) hardware (e.g. Myrinet) and, last but not least, we plan to investigate the possibility to integrate our communication primitives in the `mpich` ADI in such a way that it can be used when small size messages are to be transmitted.

References

1. Top500.org. Top500 supercomputer sites. `http://www.top500.org`.
2. Beowulf.org. The beowulf project. `http://www.beowulf.org`.
3. B. Bacci, M. Danelutto, S. Pelagatti, and M. Vanneschi. SkIE: a heterogeneous environment for HPC applications. *Parallel Computing*, 25:1827–1852, December 1999.
4. M. Danelutto. Task farm computations in java. In Buback, Afsarmanesh, Williams, and Hertzberger, editors, *High Performance Computing and Networking*, LNCS, No. 1823, pages 385–394. Springer Verlag, May 2000.
5. M. Cole. *Algorithmic Skeletons: Structured Management of Parallel Computations*. Research Monographs in Parallel and Distributed Computing. Pitman, 1989.
6. Congduc Pham and Carsten Albrecth. Tuning message aggregation on high performance clusters for efficient parallel simulations. *Parallel Processing Letters*, 9(4):521–532, 1999.
7. G. Ciaccio and G. Chiola. GAMMA and MPI/GAMMA on Gigabit Ethernet. In *Proceedings of 7th EuroPVM-MPI*, LNCS, No. 1908. Springer Verlag, September 2000. Balatonfured, Hungary.
8. M. Beck, H. Bohme, M. Dziadska, U. Kunitz, R. Magnus, and D. Verworner. *LINUX Kernel Internals*. Addison Wesley, 1998.
9. D. A. Rusling. The Linux Kernel. version 0.8-3, `http://www.linuxdoc.org`.
10. M. Snir, S. W. Otto, and J. Dongarra. *MPI – The complete reference. Volume 2: The MPI core*. MIT Press, Cambridge, 1998.
11. Argonne National Laboratory. Mpich home page. `http://www-unix.mcs.anl.gov/mpi/mpich/index.html`.

Track IV

Computational Science

Efficient Monte Carlo Linear Solver with Chain Reduction and Optimization Using PLFG

Maria Isabel Casas Villalba[1] and Chih Jeng Kenneth Tan[2]

[1] Norkom Technologies Ltd.
Norkom House, 43 Upper Mount Street
Dublin 2, Ireland
isabel.casas@norkom.com
[2] School of Computer Science
The Queen's University of Belfast
Belfast BT7 1NN, United Kingdom
cjtan@acm.org

Abstract. In this paper, we show a Monte Carlo linear solver with chain reduction and optimization, coupled with PLFG, a parallel pseudo-random generator. PLFG, designed for MIMD architectures, is highly scalable and with the default parameters chosen, it provides an astronomical period of at least $2^{29}\left(2^{23209}-1\right)$. Numerical experiment results show that Monte Carlo method with chain optimization and reduction gives much better estimates of the solution vector.
Keywords: Monte Carlo Method, Pseudo-random Number Generator, Lagged Fibonnaci Generator, Parallel Computation, Linear Solver.

1 Introduction

It is well known that we can obtain statistical estimates for the components of the solution vector of systems of linear algebraic equations (SLAE) by performing random sampling of a certain random variable whose mathematical expectation is the desired solution [12, 17].

Classical methods such as Gauss-Jordan or non-pivoting Gaussian Elimination methods require $O(n^3)$ steps for a $n \times n$ square matrix [2]. In contrast, Monte Carlo methods require only $O(NT)$ steps to find an element of the inverse matrix, and $O(nNT)$ steps for computing the full solution vector, where N is the number of chains and T is the chain length, both quantities independent of n and bounded [1]. Even though Monte Carlo methods do not yield better solutions than direct or iterative numerical methods for solving SLAE, they are more efficient for large n and are highly parallelizable.

However, Monte Carlo methods require pseudo-random number generator (PRNG) of high quality, high speed and long period. The pseudo-random numbers used on the each of the processors and between the processors has to be statistically independent. Also, the period of the PRNG has to be at least the same length as the number of pseudo-random numbers required in the simulation. If there are k steps in the simulation but the PRNG used has a period

B. Hertzberger et al. (Eds.): HPCN Europe 2001, LNCS 2110, pp. 405–414, 2001.

$\frac{k}{l} < k$, then the same $\frac{k}{l}$ points are being sampled l times, rather than sampling on all the k points. Simulation results obtained under such conditions will be erroneous. It is unfortunate that many researchers tend to overlook this critical aspect of Monte Carlo methods in practice.

In this paper, we show the Monte Carlo method for solution of SLAE with chain reduction and optimization obtained using PLFG, a parallel PRNG based on the lagged-Fibonacci algorithm. First, we outline stochastic methods for solving SLAE. Following, we introduce the concept of optimal and almost optimal transition frequency function. We continue by describing the technique for chain reduction and optimization. Next, we show the lagged Fibonacci algorithm and PLFG. Finally, we present the results of tests conducted to compare the accuracies of the solutions obtained with and without chain reduction and optimization, obtained using PLFG.

2 Stochastic Methods for Solving SLAE

Consider a matrix $A \in \mathbb{R}^{n \times n}$ and a vector $x \in \mathbb{R}^{n \times 1}$. Further, A can be considered as a linear operator $A\left[\mathbb{R}^{n \times 1} \rightarrow \mathbb{R}^{n \times 1}\right]$, so that the linear transformation

$$Ax \in \mathbb{R}^{n \times 1} \tag{1}$$

defines a new vector in $\mathbb{R}^{n \times 1}$.

The linear transformation in Equation 1 is used in iterative Monte Carlo algorithms, and the linear transformation in Equation 1 is also known as the iteration. This algebraic transform plays a fundamental role in iterative Monte Carlo algorithms.

In the problem of solving systems of linear algebraic equations, the linear transformation in Equation 1 defines a new vector $b \in \mathbb{R}^{n \times 1}$:

$$Ax = b, \tag{2}$$

where A and b are known, and the unknown solution vector x is to be solved for.

We introduce a matrix L, such that $(I - L) = A$, where I is the identity matrix. Hence, the Equation 2 can be written as

$$x = Lx + b\,. \tag{3}$$

It is obvious that a numerical approximation to x can be obtained using the first-order stationary linear iterative method for Equation 3,

$$x^{(k+1)} = Lx^{(k)} + b \tag{4}$$

which converges if $\|L\| < 1$. Assuming that $x^0 \equiv 0$, the von Neumann series can be written as

$$x^{(k+1)} = (I + L + L^2 + \ldots + L^k)b = \sum_{m=0}^{k} L^m f \text{ where } L^0 \equiv I\,. \tag{5}$$

Since $\|L\| < 1$ is assumed, then

$$\lim_{k\to\infty} x^{(k)} = \lim_{k\to\infty} \sum L^m b = (I-L)^{-1}b = A^{-1}b = x \tag{6}$$

holds [11, 6].

3 Markov Process

We suppose that $\{s_1, s_2, \ldots, s_n\}$ is a finite discrete Markov chains with n states. At each discrete time $t = 0, 1, \ldots, N$, a chain S of length T is generated: $k_0 \to k_1 \to \ldots \to k_j \to \ldots \to k_T$ with $k_j \in \{s_1, s_2, \ldots, s_n\}$ for $j = 1, \ldots, T$.

We define $\mathbf{P}\,[k_0 = s_\alpha] = p_\alpha$ as the probability that the chain starts in state s_α, and $\mathbf{P}\,[k_j = s_\beta | k_{j-1} = s_\alpha] = p_{\alpha\beta}$ as the transition probability to state s_β from state s_α for $\alpha = 1, \ldots, n$ and $\beta = 1, \ldots, n$.

The probabilities $p_{\alpha\beta}$ thus define the transition matrix P. The distribution $(p_1, \ldots, p_n)^T$ is said to be acceptable to vector h, and similarly that the distribution $p_{\alpha\beta}$ is acceptable to L, if [12]

$$\begin{cases} p_\alpha > 0 \text{ when } h_\alpha \neq 0 \\ p_\alpha \geq 0 \text{ when } h_\alpha = 0 \end{cases} \text{and} \begin{cases} p_{\alpha\beta} > 0 \text{ when } l_{\alpha\beta} \neq 0 \\ p_{\alpha\beta} \geq 0 \text{ when } l_{\alpha\beta} = 0 \end{cases} \tag{7}$$

We also define

$$W_j = W_{j-1}\frac{l_{k_{j-1}k_j}}{p_{k_{j-1}k_j}}, W_0 \equiv 1 \tag{8}$$

and the random variable

$$\eta_T(h) = \frac{h_{k_0}}{p_{k_0}} \sum_{j=0}^{T} W_j b_{k_j}\,. \tag{9}$$

It can be shown that the limit of $\mathbf{M}\,[\eta_T(h)]$, the mathematical expectation of $\eta_T(h)$, is [11]

$$\mathbf{M}\,[\eta_T(h)] = \left\langle h, \sum_{m=1}^{T} L^m b \right\rangle = \left\langle h, x^{(T+1)} \right\rangle \Rightarrow \lim_{T\to\infty} \mathbf{M}\,[\eta_T\,(h)] = \langle h, x \rangle \tag{10}$$

Knowing this, we can find an unbiased estimator of $\mathbf{M}\,[\eta_\infty\,(h)]$ in the form

$$\theta_N = \frac{1}{N}\sum_{m=1}^{N} \eta_\infty(h) \tag{11}$$

Next, we consider the functions $h \equiv h^j = (0, 0, \ldots, 1, \ldots, 0)$, where $h_i^j = \delta_i^j$ is the Kronecka delta. Then,

$$\langle h, x \rangle = \sum_{i=0}^{n} h_i^j x_i = x_j. \tag{12}$$

It follows that an approximation to x can be obtained by calculating the average for each component of every Markov chain

$$x_j \approx \frac{1}{N} \sum_{m=1}^{N} \theta_T^m \left[h^j\right] . \tag{13}$$

In summary, we generate N independent Markov chains of length T and calculate $\eta_T(h)$ for each path. Finally, we estimate the j-th component of x as the average of every j-th component of each chain.

4 Minimal Probable Error

The Monte Carlo method provides a probabilistic error bound which depends on the chains simulated. The value of this error is on average [12, 1],

$$r_N \approx 0.6745 \sigma(\theta) N^{-\frac{1}{2}} \tag{14}$$

where $\sigma(\theta)$ is the standard deviation of the estimator θ. Therefore, if the number of Markov chains N increases the error bound decreases. Also the error bound decreases if the the standard deviation of the estimator decreases. The latter option more feasible since to reduce the error by $O(n)$, the number of paths N would have to be increased by $O\left(n^2\right)$.

This leads to the definition of almost optimal transition frequency for Monte Carlo methods. The idea is to find a transition matrix P that minimize the second order moment of the estimator.

5 Almost Optimal Transition Frequency Function

Define the vectors ψ and $\hat{\psi}$,

$$\psi_\alpha = \left(\sum_\beta \frac{l_{\alpha\beta}^2 b_\beta^2}{p_{\alpha\beta}} \right)^{1/2} \tag{15}$$

and

$$\hat{\psi} = \sum_\beta |l_{\alpha\beta} b_\beta| \quad \forall \alpha = 1, 2, \ldots, n. \tag{16}$$

As shown in [5], the probabilities that minimize the second moment $\mathbf{M}\left[\theta^2\right]$ are

$$\hat{p}_i = c_0 b_{\alpha_0} \prod_{j=1}^{i} |l_{\alpha_{j-1}\alpha_j}| b_{\alpha_i} \tag{17}$$

and

$$\hat{p}_{k,i} = c_0 |h_{\alpha_0}| \prod_{j=1}^{i} |l_{\alpha_{j-1}\alpha_j}| [b_{\alpha_i} L^{(k-i)} b_{k_i}]^{1/2}, \tag{18}$$

where $L^{k-1} = \sum_{\alpha_{i+1}} \cdots \sum_{\alpha_k} l_{\alpha_i \alpha_{i+1}} \cdots l_{\alpha_{k-1}\alpha_k} b_{\alpha_k}$ and $c_0 = \left(\sum_\beta |b_\beta \hat{\psi}_\beta|\right)^{-1}$. The almost optimal frequency is $\tilde{p}_i = c_0 |h_{\alpha_0}| \prod_{j=1}^{i} |l_{\alpha_{j-1}\alpha_j}|$. Thus, the Monte Carlo method is more efficient using optimal or almost optimal transition frequency functions. Therefore, $p_{\alpha\beta}$ is chosen proportional to the $|l_{\alpha\beta}|$.

6 Parameters Estimation

We choose the transition matrix P with elements $p_{\alpha\beta} = \frac{|l_{\alpha\beta}|}{\sum_\beta |l_{\alpha\beta}|}$ for $\alpha, \beta = 1, \ldots, n$. In practice the length of the Markov chain must be finite, and is terminated when $|W_j b_{k_j}| < \delta$, for some small value δ [12]. Since

$$|W_j b_{k_j}| = \left| \frac{l_{\alpha_0\alpha_1} \cdots l_{\alpha_{j-1}\alpha_j}}{\frac{|l_{\alpha_0\alpha_1}|}{\|L\|} \cdots \frac{|l_{\alpha_{j-1}\alpha_j}|}{\|L\|}} \right| |b_{k_j}| = \|L\|^i \|b\| < \delta \,, \tag{19}$$

it follows that

$$T = j \leq \frac{\log\left(\frac{\delta}{\|b\|}\right)}{\log \|L\|} \tag{20}$$

and

$$D\left[\eta_T(h)\right] \leq M\left[\eta_T^2\right] = \frac{\|b\|^2}{(1 - \|L\|)^2} \leq \frac{1}{(1 - \|L\|)^2} \,. \tag{21}$$

According to the Central Limit Theorem,

$$N \geq \left(\frac{0.6745}{\epsilon}\right)^2 \frac{1}{(1 - \|L\|)^2} \tag{22}$$

is a lower bound on N.

In addition, since in Monte Carlo method, we generate N independent Markov chains of length T and calculate $\eta_T(h)$ for each path, an optimal number of chains for each row of L, $N_i < N$, which reduces the number of chains, while increasing efficiency, can be calculated as

$$N_i = \left(\frac{0.6745}{\epsilon}\right)^2 \frac{\|b\|}{(1 - D_i)^2} \text{ where } D_i = \sum_j^n |l_{ij}| \,. \tag{23}$$

7 Lagged Fibonacci Generators

In general, lagged Fibonacci generators (LFGs) are of the form

$$x_i = (x_{i-p_1} \odot x_{i-p_2}) \bmod M \tag{24}$$

where x_i is the pseudo-random number to be output, and $\odot$ is a binary operation performed with the operands x_{i-p_1} and x_{1-p_2}. We call $p_1, p_2, p_1 > p_2$ the lag values. The operand $\odot$ commonly used are addition (or subtraction), multiplication or bitwise exclusive OR (XOR). M is typically a large integer value or 1 if x_i is a floating point number. When using XOR operation, mod$M$ is dropped. It is obvious that we need an array of length p_1 to store the previous p_1 values in the sequence.

XOR operations give the worst pseudo-random numbers, in terms of their randomness properties [8, 3, 16]. Additive LFGs were more popular than their multiplicative counterpart because multiplication operations were considered to be slower, even though the superior properties of multiplicative LFGs has been noted in [8]. Tests comparing operation execution times have shown that, with current processors and compilers, multiplication, addition and subtraction operations are of similar speeds. Thus, multiplicative operations should be favored.

Parameters p_1, p_2 and M should be chosen with care to obtain long period and good randomness properties. [4] suggested lag values be greater than 1279. Large p_1 and p_2 also improves randomness since smaller lags leads to intra-stream higher correlation [8, 3, 4]. In LFGs, the key purpose of M is to limit the output to the data type range.

Initializing the lag table before any pseudo-random numbers are generated with the LFG is also of critical importance. The initial values have to be statistically independent. To obtain these values, another PRNG is often used.

With $M = 2^e$, where e is the total number of bits, additive LFG have a period $2^{e-1}(2^{p_1} - 1)$. The period of multiplicative LFG, however, is shorter than that of additive LFG: $2^{e-3}(2^{p_1} - 1)$. This shorter period should not pose as a problem for multiplicative LFGs, if p_1 is large.

Since the LFG shown above uses two past random numbers to generate the next, it is said to be a "two-tap LFG". Empirical tests have shown that LFGs with more taps, three-tap LFG or four-tap LFG for example, give better results in statistical tests, at the minimal extra time cost of accessing the lag table and performing the $\odot$ operation.

If a multitap additive LFG is used, M can be chosen to be the largest prime number that fits in the data type [7]. It can be proven using the theory of finite fields that random numbers generated in such a manner will be a good source of random numbers. For a more thorough treatment, see [7].

8 The PLFG Parallel Pseudo-random Number Generator

Here we outline the PLFG algorithm. A more complete discussion may be found in [15].

Some of the well known techniques for parallelizing pseudo-random number generators include leap frog, sequence splitting, independent sequences, shuffling leap frog, maximal period sequence splitting and parameterization [4, 18, 10, 9]. PLFG is an LFG parallelized by independent sequences. Aside from being straight forward, parallelization by independent streams is also very efficient, and highly recommended [13].

The Mersenne Twister 19937 (MT19937), a sequential PRNG having the Mersenne prime period, $2^{19937} - 1$, is used to seed the lag tables. `PLFG` is practically highly scalable since with MT19937, the upper limit of the number of `PLFG` processors is $\frac{2^{19937}-1}{23209}$, which is not really a limit at all!

`PLFG` was designed to be a two-tap LFG-based generator, and can be extended to have more taps. We chose the lag table parameters to be $p_1 = 23209$, $p_2 = 9739$, recommended in [7]. Using a round-robin algorithm, the lag table size in `PLFG` is minimized. The pseudo-random numbers generated by `PLFG` are of ANSI Standard C type `unsigned long`. With $p_1 = 23209$, on machines with 32-bit `unsigned long`, this translates to a memory requirement of 92836 bytes, which is insignificant on today's machines.

On 32-bit machines, `unsigned long` is typically 32-bit. However, it is a 64-bit data type on some 64-bit machines. The implied variability in the period of the pseudo-random number sequence generated by `PLFG` is indeed a feature: when machines with wider word size become available, the period of `PLFG` automatically expands! The period remains to be $2^{29}\left(2^{23209} - 1\right)$ on 32-bit machines, but increases to $2^{61}\left(2^{23209} - 1\right)$ on 64-bit machines. The data type size differences may be a concern however, when using `PLFG` on heterogeneous workstation clusters.

Coupling the long period of MT19937 with the long period of each of the independent sequences in `PLFG`, the probability that the sequences overlaping is minimal. The quality of the pseudo-random numbers generated by `PLFG` has also been confirmed to be comparable to that of more commonly used pseudo-random number generators [15].

`PLFG` has been tested to generate 3.9841×10^6 and 1.6287×10^6 pseudo-random numbers per processor on a DEC Alpha 21164 (EV56) 500 MHz cluster and on a shared-memory Intel Pentium Pro 200 MHz machine, respectively, faster than some parallel PRNGs, and on par with others [15].

9 Numerical Tests

Parallel solution for Monte Carlo with almost optimal transition frequency function and chain reduction and optimization was implemented under Message Passing Interface, using a mix of master-slave and single program multiple data stream approach. L and b are broadcasted to all processors at startup, and the slaves are assigned, by the master node, rows to compute. This way, minimal communication is involved. The slave nodes are used efficiently since if the N_i is smaller for any particular row, computations will terminate sooner, and a new row can be assigned to the slave node once the master node receives the result of the computation for the previous row.

The numerical tests were done on a DEC Alpha XP1000 cluster, with 667MHz Alpha EV67 processors, communicating over switched-Fast Ethernet. Dense balanced matrices which have nearly equal sums of elements for each row were used. Table 1 shows the root mean square (RMS) error of the product of x, computed

Table 1. Numerical Experiment Results.

Matrix size	Norm	Time (s)	RMS error	No. chains	Chain reduction and optimization
10×10	0.739	3.982	0.168	348447	Yes
10×10	0.739	7.639	0.173	670340	No
10×10	0.739	3.164	0.189	348447	Yes
10×10	0.739	5.344	0.184	670340	No
100×100	0.437	0.234	0.0626	892290	Yes
100×100	0.437	0.392	0.0640	1433300	No
1000×1000	0.502	0.090	0.0948	50837	Yes
1000×1000	0.502	0.135	0.0924	183000	No
1000×1000	0.640	5.561	0.1404	6203298	Yes
1000×1000	0.640	31.676	0.1407	35119000	No
1000×1000	0.707	12.884	0.1671	14318054	Yes
1000×1000	0.707	47.839	0.1670	52947000	No

using the parameters $\epsilon = 0.01$ and $\delta = 0.01$, with matrix A, along with the average execution times and the total number of chains.

The improved results obtained by Monte Carlo method with chain reduction and optimization is clear. Monte Carlo method with chain reduction and optimization require 3 – 4 times less chains to reach the same, or even better, precision.

10 Conclusion

Monte Carlo methods are well suited for large problems where other solutions methods are impractical or impossible for computational reasons. Where massive parallelism is available and low precision of the solution is acceptable, Monte Carlo algorithms is favorable, particularly when $n \gg N$. In addition, when $\|L\|$ is considerably smaller than 1, Monte Carlo methods are very effective. The method of chain reduction and optimization presented in this paper can be coupled well with the Relaxed Monte Carlo Method [14] which reduce $\|L\| < 1$ to be a desired value, significantly smaller than 1. Despite acceptable low precision, it is often beneficial to have better estimates of the solution, however. As seen in the numerical tests results, Monte Carlo method with chain reduction and optimization yields more accurate estimates, in a much shorter amount of time.

Acknowledgment

Computing resources for this project was provided in part by the MACI Project at the University of Calgary, Canada, to the second author, through J. A. Rod Blais.

The first author would like to thank her employers at Norkom Technologies, Ireland for relieving her from her regular duties, allowing her to work on this research project.

We would also like to thank Vassil Alexandrov, from the High Performance Computing Center, The University of Reading, UK, for the fruitful discussions.

References

[1] ALEXANDROV, V. N. Efficient Parallel Monte Carlo Methods for Matrix Computations. *Mathematics and Computers in Simulation 47* (1998).

[2] BERTSEKAS, D.P., AND TSITSIKLIS, J.N. *Parallel and Distributed Computation: Numerical Methods.* Athena Scientific, 1997.

[3] CODDINGTON, P.D. Analysis of Random Number Generators Using Monte Carlo Simulation. *International Journal of Modern Physics C5* (1994).

[4] CODDINGTON, P.D. Random Number Generators for Parallel Computers. *National HPCC Software Exchange Review*, 1.1 (1997).

[5] DIMOV, I.T. Minimization of the Probable Error for some Monte Carlo Methods. In *Mathematical Modelling and Scientific Computations* (1991), I. T. Dimov, A. S. Andreev, S. M. Markov, and S. Ullrich, Eds., Publication House of the Bulgarian Academy of Science, pp. 159 – 170.

[6] FADDEEV, D.K., AND FADDEEVA, V.N. *Computational Methods of Linear Algebra.* Nauka, Moscow, 1960. (In Russian.).

[7] KNUTH, D. E. *The Art of Computer Programming, Volume II: Seminumerical Algorithms*, 3 ed. Addison Wesley Longman Higher Education, 1998.

[8] MARSAGLIA, G. A Current View of Random Number Generators. In *Computing Science and Statistics: Proceedings of the XVI Symposium on the Interface* (1984).

[9] MASCAGNI, M., CEPERLEY, D., AND SRINIVASAN, A. SPRNG: A Scalable Library for Pseudorandom Number Generation. In *Proceedings of the Third International Conference on Monte Carlo and Quasi Monte Carlo Methods in Scientific Computing* (1999), J. Spanier, Ed., Springer Verlag.

[10] MASCAGNI, M., CUCCARO, S.A., PRYVOR, D.V., AND ROBINSON, M.L. A Fast, High Quality, and Reproducible Parallel Lagged-Fibonacci Pseudorandom Number Generator. *Journal of Computational Physics 119* (1995), 211 – 219.

[11] RUBINSTEIN, R.Y. *Simulation and the Monte Carlo Method.* John Wiley and Sons, 1981.

[12] SOBOL', I.M. *Monte Carlo Numerical Methods.* Moscow, Nauka, 1973. (In Russian.).

[13] SRINIVASAN, A., CEPERLEY, D., AND MASCAGNI, M. Testing Parallel Random Number Generators. In *Proceedings of the Third International Conference on Monte Carlo and Quasi-Monte Carlo Methods in Scientific Computing* (1998).

[14] TAN, C.J.K., AND ALEXANDROV, V. Relaxed Monte Carlo Linear Solver. In *Recent Advances in Computational Science, Proceedings of the 2001 International Conference on Computational Science* (2001), V. N. Alexandrov, J. J. Dongarra, and C. J. K. Tan, Eds., vol. 2073 of *Lecture Notes in Computer Science*, Springer-Verlag, pp. 1275 – 1283.

[15] TAN, C.J.K., AND BLAIS, J.A.R. PLFG: A Highly Scalable Parallel Pseudorandom Number Generator for Monte Carlo Simulations. In *High Performance Computing and Networking, Proceedings of the 8th International Conference on*

High Performance Computing and Networking Europe (2000), M. Bubak, H. Afsarmanesh, R. Williams, and B. Hertzberger, Eds., vol. 1823 of *Lecture Notes in Computer Science*, Springer Verlag, pp. 127 – 135.

[16] Vattulainen, I., Ala-Nissila, T., and Kankaala, K. Physical Models as Tests of Randomness. *Physics Review E52* (1995).

[17] Westlake, J. R. *A Handbook of Numerical Matrix Inversion and Solution of Linear Equations.* John Wiley and Sons, 1968.

[18] Williams, K.P., and Williams, S.A. Implementation of an efficient and powerful parallel pseudo-random number generator. In *Proceedings of the Second European PVM Users' Group Meeting* (1995).

On-Line Tool Support for Parallel Applications

Marian Bubak[1,2], Włodzimierz Funika[1],
Bartosz Baliś[1], and Roland Wismüller[3]

[1] Institute of Computer Science, AGH, al. Mickiewicza 30
30-059 Kraków, Poland
[2] Academic Computer Centre – CYFRONET, Nawojki 11
30-950 Kraków, Poland
[3] LRR-TUM – Technische Universität München
D-80290 München, Germany
{bubak,funika,balis}@uci.agh.edu.pl, wismuell@in.tum.de
phone: (+48 12) 617 39 64, *fax*: (+48 12) 633 80 54 *phone*: (+49 89) 289-28243

Abstract. This paper presents the recent development of the environment of on-line tools for parallel programming support, based on a universal monitoring system, the OCM, which is built in compliance with the OMIS specification. Issues covered include enhancements needed both at the monitoring level and at the user interface level in order to achieve full tool support for message-passing parallel applications, and to enable interoperability of tools. We focus on the evolution of the environment towards support for performance analysis of MPI applications, and interoperability of two tools: the PATOP performance analyzer and the DETOP debugger. We also outline perspectives for further research to extend the environment's capabilities to support other parallel programming paradigms.
Keywords: Parallel programming, monitoring systems, interoperability.

1 Introduction

A number of run-time tools were developed to support the development of parallel message-passing applications [1]. These tools are intended for performance analysis, visualization, debugging, etc. Most of the tools currently available follow the off-line trace based approach, which means that the information about an application's execution is gathered while it is running but it is analysed by a tool after the application has terminated. Representative examples of such tools are Vampir, Pablo, and ParaGraph. The on-line tools that work simultaneously with the application are much fewer.

On-line tools are composed of the GUI for interaction with users and the *monitoring system*, which gathers information about applications and allows for controlling them. If we separate the GUI from the monitoring system and design a standardised protocol for communication between tools and monitoring systems, we can benefit from this in several ways. Firstly, we can develop a single monitoring system on which all future tools can be based. Secondly, we separate the tool development from the monitoring system development which makes

B. Hertzberger et al. (Eds.): HPCN Europe 2001, LNCS 2110, pp. 415–424, 2001.

easier to port tools to other platforms and extend their capabilities. Thirdly, tools based on a single monitoring system are enabled to cooperate with each other, which opens new avenues of *interoperability* of tools.

Among the initiatives to define universal interfaces between tools and monitoring systems, there are DAMS [5], and DPCL [11]. However, DAMS does currently not address performance monitoring, while DPCL does not offer direct support for message passing environments like PVM [6] and MPI [10]. The *On-line Monitoring Interface Specification* (OMIS) [8] is designed to provide support for a wide range of parallel paradigms and different kinds of tools. OMIS provides services which are divided into three categories: *information services* for retrieving information about a program's execution, *manipulation services* for controlling the execution and changing state of observed objects, and *event services* for detecting specified events, and triggering a list of actions whenever relevant events occur.

Based on the compliance with the OMIS specification and its implementation, OCM[1], a performance analysis tool, PATOP, is enabled to function in message passing programming environments, interoperate with other on-line monitoring based, OMIS-compliant tools like the DETOP debugger [15], etc. The OCM, originally developed to monitor PVM applications, has been enhanced to support development of MPI applications. The next natural step was to adapt the mentioned tools to the MPI programming model.

In this paper, we follow the issues concerning the enhancement of the tool functionality and relevant changes to the tools, monitoring system and provide an outlook at the interoperability of OCM-based tools as well.

2 OMIS as Basis for Building a Tool Environment

OMIS is intended to meet the goal of building a standardised interface between application development support tools and parallel programming environments, which addresses the issues of versatility, portability, extensibility and independence of the underlying parallel environment. OMIS is not restricted to a single kind of tools, especially it supports both debuggers and performance analysers [3,4]. This implies that a monitoring system compliant with OMIS needs to be capable to control processes from the very beginning of their execution (crucial for debugging) as well as to monitor the parallel programming library calls efficiently (crucial for performance analysis).

At the beginning, the use of OMIS for parallel programming was rather constrained. On one hand, an OMIS compliant monitoring system, OCM was available to monitor PVM programs only, on the other hand, there already existed a number of powerful tools, most of which did not support parallel applications using PVM or other popular message passing mechanisms. The design we will report on below thus had two goals: to extend the applicability of the OCM to other parallel programming facilities, especially MPI, and to extend the OMIS-based tool environment, specifically w.r.t. performance analysis.

[1] OCM stands for the OMIS Compliant Monitoring system.

The OCM is a distributed monitoring system which consists of multiple *local monitors*, one per node of the target system, and one central component, the *Node Distribution Unit* (NDU). Tools submit their requests to the NDU, which splits the request into sub-requests for local monitors. Local monitors execute the requests and send the replies back to the NDU, which then assembles them into one reply for the tool. The local monitor processes control the application processes via operating system interfaces (`ptrace` and `/proc`) and communicate with additional parts of the monitoring system contained in instrumented libraries, using a shared memory segment.

The ever growing use of MPI motivated us to port and enhance the OCM's functionality so it could also support the development of MPI applications. When porting, our main concern was to preserve the user interface designed for PVM as much as possible. This would ensure that the user could easily switch between PVM and MPI applications. One of the main points in the design of the OCM concerns the way the user can handle both the execution of an application with the OCM and the start-up of all of the OCM's components. Since the usage of OCM-based tools should be as near to sequential on-line tools as possible, this implies that one can start a tool without having to run a program beforehand. Thus, the tool can start the OCM, which next starts the application. As opposed to PVM, in MPI the hosts to be monitored are not known prior to the start of an application. Thus, the start-up procedure must progress in a sequence reversed w.r.t. PVM: first, an application is created, then the OCM's local monitors get started on the nodes relevant to the application. To create the processes of an MPI application, the OCM has been enhanced by an MPI specific service.

The most significant data the monitoring system must know about an application to be monitored is: on which hosts the application is running, and which processes on these hosts belong to the application. So, *how the necessary data can be obtained?* For MPI applications using `mpich` [7], the data concerning the application can be retrieved during its startup in the following way: Once the application is started with a proper command line flag passed to `mpirun`, the latter will start a single monitor process. A procedure following the initial step is used to obtain information from an internal data structure with the UNIX `ptrace` system call or `/proc` file system, when the processes created by the master process are stopped by an endless loop inside `MPI_Init()`. So, it is the first (master) local monitor that starts up the MPI program and distributes explicitly the information about the running program among the components of the OCM. The above information on the application gets also accessible for tools through an MPI specific information service provided by the OCM. For a detailed description of the start-up procedure, the reader is referred to [2].

For performance analysis goals, the OCM gathers information about MPI function calls via instrumentation of the parallel library, what consists in building a wrapper for each library function. To ease making wrapper modules, a kit of universal tools for automatic generation of the individual wrappers has been developed.

3 Operation of PATOP on Top of the OCM for MPI

In general, performance measurement tools for parallel applications are intended to help in finding and fixing bottlenecks by carrying out various measurements [9]. In order to provide *on-line* performance analysis facilities for PVM and MPI applications, an existing tool, PATOP [15], was used. Skipping details, (see [3]) we outline the functionality of the tool and the changes to the OCM and PATOP necessary to make them work together for PVM (very shortly) and MPI applications. PATOP has originally been developed for the on-line performance analysis of applications running under PARIX on Parsytec systems. The tool provides a reasonably rich set of available performance measurements and display diagrams at various levels of detail. The integration of PATOP with the OCM involved an additional layer, a high level routine library, ULIBS. In the case of PATOP, the task of ULIBS is to transform high-level measurement specifications from PATOP to low-level requests acceptable by the OCM and to transform back the results. Both the OCM and ULIBS needed to be extended with modules specific to performance analysis.

3.1 Gathering Performance Data with the OCM

In order to be able to provide performance analysis of a parallel program, tools need to acquire information about the program such as the amount of time spent on waiting for a message from another process, or the message size. The OCM provides all needed support for obtaining such information. To make this data available to the performance analysis tools, we can use on of the three principal approaches: *direct communication* — immediate sending data on having detected an interesting event, *tracing* — storing information in a trace buffer whenever a relevant event occurs, *counters/integrators* — summarising the most important information on each event in counters and integrators for `integer` and `float` values, respectively. The mechanism for the counters and integrators is provided by an extension to the OCM, PAEXT, in which two new types of OMIS objects are defined, *counters* and *integrators*.

One of the most important issues in performance data acquisition is efficiency, i.e. the cost of gaining the data. While the direct communication approach is the most flexible, its overhead can be unacceptable — due to an excessive communication and frequent context switches between a tool and a monitoring system. The tracing approach introduces the concept of *local storing*, in which data is first stored locally, in the context of application processes, and is only sent to a tool when required. Counters and integrators are data structures for processing performance data. Combined with local storing, this mechanism provides a significant saving of resources usage, therefore it is the least intrusive and yet a less informative approach.

The involvement of ULIBS, which had been developed prior to the OCM, as an intermediate abstraction level implied its modification to comply with the OMIS interface. Now, ULIBS provides the following functionality for tools: *asynchronous* requests and callbacks, configuration and current state *info* (nodes, processes), *starting/stopping* processes, *debugger-oriented* mechanisms like symbol tables, variable info, breakpoints, single stepping. Whereas the interface of ULIBS basically remained unchanged w.r.t. to the original version, its implementation had to be completely redesigned for the use with the OCM.

3.2 Adaptation of the Tool Environment to MPI

Extending PATOP with a capability to measure performance of MPI applications implied further extensions in other components of the environment. Fig. 1 shows the general architecture of the environment with MPI-specific parts and performance analysis modules exposed.

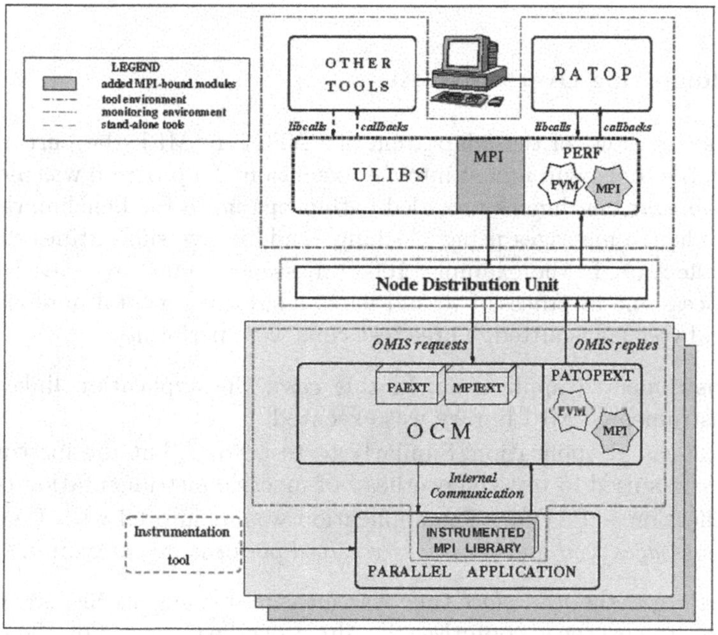

Fig. 1. Architecture of the Monitoring Environment.

Adapting PATOP to MPI required some new modules to be added to the environment. The OCM was extended with a new MPI-bound extension, MPIEXT. This extension handles the start-up of the MPI application under control of the OCM. The PATOPEXT module of the OCM was extended with an MPI-specific part. The instrumented MPI library is intended to be linked to a parallel application. A new module for MPI was added to ULIBS, while the PERF module within ULIBS was extended with an MPI-specific part. This part contains the performance measurements definitions. These definitions are requests to the monitoring system for monitoring parallel library calls. In its turn, the PERF module needed to be extended by the measurement definitions which depend on the MPI communication subroutines semantics.

In case of MPI collective communication, in practice, it is not always possible to distinguish between the sender and the receiver. Therefore, two new measurements have been added to PATOP, i.e. those related to the time and volume of a collective communication call. These measurements are expressed in terms of the OMIS requests similarly to those related to point-to-point communication. Since collective communication always involves a *communicator*, connected with a group of processes, it would be convenient to be able to apply a global measurement to a collection of processes involved, instead of selecting the processes one-by-one. Therefore, the tool environment was enhanced to provide information on communicators created within an application, which is gathered by means of appropriate instrumentation of the MPI communicator-bound subroutines.

3.3 Monitoring Overhead Test

In the closing stage of the adaptation of PATOP to MPI, the perturbation induced by the tool environment into the execution of a program was measured by using *ping-pong* benchmark provided within `mpich`. In the benchmark, two processes exchange messages using blocking send/receive subroutines `MPI_Send()` and `MPI_Recv()`. In the example, 10000 messages, 10000 bytes each were sent from process one to process two and back. That gives a total number of nearly 200 Megabytes transmitted. Three test runs were performed:

1. Uninstrumented application. In this case, the application linked with an uninstrumented MPI library was executed.
2. Instrumented application. Similarly to test No. 1, but the instrumented library was used to test the overhead of inactive instrumentation code.
3. Application + PATOP. The application was monitored with PATOP using the *messages sent* and *messages received* performance measurements.

In each test, the *wall clock* time was measured using the `MPI_Wtime` subroutine. Thus, the overhead comprises the direct overhead caused by the instrumentation and indirect overhead caused by the monitoring processes and PATOP. The table below shows a summary of the results obtained on two SUN workstations (300 MHz UltraSparc II), connected via fast Ethernet (100 MBit/s). The indicated error interval is $\pm\sigma$, i.e. the values' standard deviation.

Test	Mean exec. time [s]	Abs. overhead per event [μs]	Relative overhead [%]
Uninstrumented application	41.77 ± 0.19	–	–
Instrumented application	41.62 ± 0.14	–	–
Application + PATOP	43.54 ± 0.32	88.5 ± 18.5	4.2 ± 0.89

Within the limits of statistical variation the execution times of an uninstrumented application and an application with inactive instrumentation code are the same[2]. On the other hand, the overhead of full monitoring is more significant - the application was running approximately 4.2% slower. However, by executing PATOP and the OCM's NDU process on a separate node, this overhead could be further reduced.

4 Interoperability within the Tool Environment

The term *interoperability*, in the context of monitoring, mainly refers to on-line tools, and means their capability to run concurrently when applied to the same application [14]. Moreover, a cooperation between tools is supposed to be possible to provide additional functionality to the tool environment. For example, when applied to a long-time running parallel application, PATOP can be used to monitor and visualize the application execution. When in PATOP's performance displays one detects an unexpected behavior of the application, DETOP could be helpful to suspend the application's execution and to check proper variables, and afterwards, to resume the application's execution.

The first basic requirement for interoperability concerns the capability of different tools to run concurrently. In case of tools coming from different vendors, *structural conflicts* between different portions of the monitoring systems may occur, which may even prevent tools from running concurrently. As multiple tools may request an operation on a single object at the same time, there must be provided an infrastructure to handle multiple requests. If tools do not form a monolithic environment, the interoperability of tools based on different monitoring systems is hardly attainable due to likely conflicts, e.g. those on exclusively accessible objects [13]. Further problems may occur at the user level and manifest in *logical conflicts*. For example, if a debugger and a analyzer work concurrently, and a process is stopped by the former, the latter might not show it on its performance displays unless it is notified of the event. This could result in inconsistent performance visualization, what relates to *consistency problems*.

Originally, the OCM was designed to provide some coordination in accessing shared objects by multiple tools [14]. This concerns the mutual exclusivity of requests referring to a single object, the distribution mechanism of requests operating on more than one node, locking requests for their execution time.

To provide a management of the environment, we have developed a new OCM-oriented tool, OCTET (OCM-based Tool Environment top-level Tool), which is intended to work on top of the tool environment. OCTET performs two tasks: the first one is to start-up the tool environment, while the second task is to provide tools with some information to resolve consistency problems. For setting up parameters like the name of a parallel environment (PVM or MPI), paths to the application and tools' executables, number of processes to be run (in case of MPI only), the tool provides a simple user interface.

[2] In fact, the mean execution time for the instrumented application by chance even was slightly smaller than the mean execution time of the uninstrumented application.

On the startup of the tool environment, OCTET starts the OCM, application and tools[3]. The tools obtain a list of the application processes' tokens, attach to each of the application processes, and finally, get information on the environment to enable their possible interactions.

The cooperation of PATOP and DETOP can reveal incorrect behaviour which occurs in two cases, depending on what tool started an application. If this was PATOP, which kept retrieving performance data from the OCM, after the application had been suspended by DETOP, this could not comply with the user's expectations that PATOP hangs up monitoring when the application is being suspended. A solution may be in allowing PATOP to be notified on DETOP's activities. On the other hand, if the application is started by DETOP, PATOP may not start performance monitoring. Again, a notification that the application processes have changed their states is necessary.

Fortunately, based on the OMIS specification it is possible to "program" an action of PATOP on each event, which is related to a change in the state of a thread, with conditional requests to the OCM [8]. The process' identifier is actually passed to the appropriate callback function mechanism, which is activated on every occurrence of the event. The callback function performs appropriate actions to stop or resume the measurements.

Questions may arise about interactions of the tools:

1. A tool, e.g. PATOP can program reactions to various scenarios of tools' cooperation. *So, how can PATOP learn of the actual configuration of the tool environment?*
2. *Which part of the software should be responsible for sending the above requests?* When answering we should bear in mind that inserting an additional code into a tool would affect the principle of the independence of tools.

To address the problems, each tool was provided with a specific library, in which every probable scenario of tools' cooperation would be handled [14]. The library is implemented as part of ULIBS and designed so that it does not affect the implementation of tools involved. With this implementation, the interoperability modules within the tools have to be provided with information on what tools are running within the environment. This information can be provided by the OCTET tool. For example, if OCTET knows that it would run two tools, e.g. PATOP and DETOP, it can inform each tool that the other tool is running. This information would actually be processed within ULIBS and passed to the interoperability module, what results in issuing proper conditional requests mentioned above.

5 Concluding Remarks

Our work was focused on extending the range of parallel environments as well as on extending the set of on-line tools available to the user. The OCM has been extended, on the one hand, to support a set of general purpose performance

[3] Currently, in case of an MPI application, this itself is started prior to the OCM.

analysis requests, provided by PAEXT, on the other hand, it was possible to add a tool-specific requests (e.g. those for PATOP, embedded in PATOPEXT). In its turn, a tool, i.e. PATOP, was modified to be capable to support MPI, by adding measurements related to MPI-specific communication.

One of the main concerns within this work was interoperability of on-line tools for parallel programming support, which is a key feature for building a powerful, easy-to-adapt tool environment. The OCM monitoring system provides mechanisms that are sufficient to meet these requirements. Tools adapted to the OCM are enabled to run concurrently and operate on the same object, based on introduction of a high level tool. Due to this, a definition of tools' interactions, which leads to effective tool cooperation, is possible without intrusion into their implementation.

Future work will be concentrated on further extending of the range of parallel environments supported by the tools and on the problem of direct interactions between the tools already involved and those to be added. As for the new environments, our main goal is providing support for OpenMP applications. Special attention will be paid to a well defined naming policy for loops and barriers, based on information from the compiler. Further development in the context of interoperability will be focused on extending the role of OCTET in "programming" tools' interactions.

Acknowledgements

This work has been carried out within the Polish-German collaboration and supported, in part, by KBN under grant 8 T11C 006 15.

References

1. Browne, S.: Cross-Platform Parallel Debugging and Performance Tools. In: Alexandrov, V., Dongarra, J., (eds.): Recent Advances in Parallel Virtual Machine and Message Passing Interface, Proc. 5th European PVM/MPI Users' Group Meeting, Liverpool, UK, September 7-9, 1998, Lecture Notes in Computer Science **1497**, Springer, 1998, pp. 257-264.
2. Bubak, M., Funika, W., Gembarowski, R., and Wismüller, R.: OMIS–Compliant Monitoring System for MPI Applications. In: R. Wyrzykowski, B. Mochnacki, H. Piech, J. Szopa (Eds.), PPAM'99 - The 3th International Conference on Parallel Processing and Applied Mathematics, Kazimierz Dolny, Poland, 14 - 17 September 1999, pp. 378-386, IMiI Czestochowa (1999).
3. Bubak, M., Funika, W., Iskra, K., Maruszewski, R., and Wismüller, R.: Enhancing the Functionality of Performance Measurement Tools for Message Passing Applications. In: Dongarra, J., Luque, E., Margalef, T., (Eds.), Recent Advances in Parallel Virtual Machine and Message Passing Interface. Proceedings of 6th European PVM/MPI Users' Group Meeting, Barcelona, Spain, September 1999, Lecture Notes in Computer Science **1697**, Springer, 1999. pp. 67-74.
4. Bubak, M., Funika, W., Młynarczyk, G., Sowa, K., and Wismüller, R.: Symbol Table Management in an HPF Debugger. In: Sloot, P., Bubak, M., Hoekstra, A.,

Hertzberger, B., (eds.): *Proc. Int. Conf. High Performance Computing and Networking*, Amsterdam, April 12-14, 1999, 1278-1281, Lecture Notes in Computer Science **1593**, Springer, 1999.

5. Cunha, J., Lourenço, Vieira, J., Moscão, B., and Pereira, D.: A Framework to Support Parallel and Distributed Debugging. In: Sloot, P., Bubak, M., Hertzberger, B., (eds.): *Proc. Int. Conf. High Performance Computing and Networking*, Amsterdam, April 21-23, 1998, 708-717, Lecture Notes in Computer Science **1401**, Springer, 1998.
6. Geist, A., et al.: PVM: Parallel Virtual Machine. A Users' Guide and Tutorial for Networked Parallel Computing. MIT Press, Cambridge, Massachusetts (1994)
7. Gropp, W., Lusk, E.: User's Guide for `mpich`, a Portable Implementation of MPI. ANL/MCS-TM-ANL-96/6, 1996.
8. Ludwig, T., Wismüller, R., Sunderam, V., and Bode, A.: OMIS – On-line Monitoring Interface Specification (Version 2.0). Shaker Verlag, Aachen, vol. 9, LRR-TUM Research Report Series, (1997) `http://wwwbode.in.tum.de/~omis/OMIS/Version-2.0/version-2.0.ps.gz`
9. Miller, B.P., Callaghan, M.D., Cargille, J.M., Hollingsworth, J.K., Irvin, R.B., Karavanic, K.L., Kunchithapadam, K., and Newhall, T.: The Paradyn Parallel Performance Measurement Tool, *IEEE Computer*, vol. 28, No. 11, November, 1995, pp. 37-46
10. MPI: A Message Passing Interface Standard. In: Int. Journal of Supercomputer Applications, **8** (1994); Message Passing Interface Forum: MPI-2: Extensions to the Message Passing Interface, July 12, (1997) `http://www.mpi-forum.org/docs/`
11. Pase, D. Dynamic Probe Class Library: tutorial and reference guide, Version 0.1. Technical Report, IBM Corp., Poughkeepsie, NY, June 1998. `http://www.ptools.org/projects/dpcl/tutref.ps`
12. Shende, S.,Malony, A.D., Cuny, J., Lindlan, K., Beckman, P., and Karmesin, S.: Portable Profiling and Tracing for Parallel Scientific Applications using C++.In: Proceedings of SPDT'98: ACM SIGMETRICS Symposium on Parallel and Distributed Tools, pp. 134-145, Aug. 1998.
13. Trinitis, J., Sunderam, V., Ludwig, T., and Wismüller, R.: Interoperability Support in Distributed On-line Monitoring Systems. In: M. Bubak, H. Afsarmanesh, R. Williams, and B. Hertzberger, editors, High Performance Computing and Networking, 8th International Conference, HPCN Europe 2000, volume 1823 of Lecture Notes in Computer Science, Amsterdam, The Netherlands, May 2000. Springer.
14. Wismüller, R.: Interoperability Support in the Distributed Monitoring System OCM. In R. Wyrzykowski et al., editor, Proc. 3rd International Conference on Parallel Processing and Applied Mathematics - PPAM'99, pages 77-91, Kazimierz Dolny, Poland, September 1999, Technical University of Czestochowa, Poland.
15. Wismüller, R., Oberhuber, M., Krammer, J. and Hansen, O.: Interactive Debugging and Performance Analysis of Massively Parallel Applications. *Parallel Computing*, **22**(3), (1996), 415-442 `http://wwwbode.in.tum.de/~wismuell/pub/pc95.ps.gz`
16. Wismüller, R., Trinitis, J., and Ludwig T.: OCM - A Monitoring System for Interoperable Tools. In: Proceedings of the 2nd SIGMETRICS Symposium on Parallel and Distributed Tools SPDT'98, Welches, OR, USA, August 1998.

Cluster Computation for Flood Simulations[1]

Ladislav Hluchy, Giang T. Nguyen, Ladislav Halada, Viet D. Tran

Institute of Informatics, Slovak Academy of Sciences
Dubravska cesta 9, 842 37 Bratislava, Slovakia
giang.ui@savba.sk

Abstract. Simulation of the water flood problems often leads to solving of large sparse systems of partial differential equations. For such systems, the numerical method is very CPU-time consuming. Therefore, the parallel simulation is essential for water flood study with satisfactory accuracy. In this paper, we present some experimental results of parallel numerical solutions done on Linux clusters. Our measurements indicate that Linux cluster can provide satisfactory power for parallel numerical solutions, especially for large problems. We also provide experimental results done on a SGI Origin2000 machine for comparison with Linux clusters.

1 Introduction

Over the past few years, floods have caused widespread damages in the world. Most of the continents were heavily threatened. Therefore, modeling and simulation of flood in order to forecast and to make necessary prevention is very important. The kernel of the problem flood simulation is a flood numerical modeling, which requires an appropriate physical model and robust numerical schemes for a good representation of reality.

Flood simulation systems such as FLDWAV [1], MIKE21 [2], SMS [3] consist of a graphical user interface for pre- and post-processing and computational modules. The graphical user interface is used for reading terrain maps, defining initial conditions, boundary condition and other parameters necessary for the simulation. All the data are passed to the computational modules that perform the simulation based on mathematical models of water flow and their numerical solutions. The output data from computational modules (water level, velocities) are sent back to the graphical user interface for visualization, animation and further processing.

The computational modules are the only computation-expensive part of flood simulation systems. They typically take several days of CPU-time for simulation of large models. For critical situations, e.g. when a coming flood is simulated in order to predict which area will be threatened and to make necessary prevention, such a long

[1] This work is supported by EU 5FP ANFAS IST-1999-11676 RTD and the Slovak Scientific Grant Agency within Research Project No. 2/7186/20

B. Hertzberger et al. (Eds.) : HPCN Europe 2001, LNCS 2110, pp. 425-434, 2001.

time is unacceptable. Therefore, using HPCN platforms to reduce the computational time [4] of flood simulation is imperative.

Numerical methods for flood simulations are generally based on Finite Difference or Finite Elements space discretisation. Both methods lead to solve large-size problems; therefore, the use of HPCN platforms for modeling is very important [10]. Using HPCN platforms can drastically reduce the computational time, which allows simulating larger problem and consequently provides more reliable results. Such HPCN based software could be the backbone of a flood alert system.

The contribution of this paper is to present parallel numerical methods for flood simulations and to study behaviors of a parallel programming models for distributed memory cluster computation [12], also for distributed shared memory machine [11] with a suitable application. Moreover, the implementation of the cyclic reduction algorithm is interesting from the parallel point of view. The algorithm provides enough parallelism and data distribution for parallel computations. Our experiments of large size flood simulations achieve better performance on Linux-cluster than on the SGI Origin machine using the same number of processors.

The rest of the paper is organized as follows. Section 2 gives a brief introduction to the mathematical model of flood simulation problems. Section 3 presents numerical methods to be used. Section 4 is concerned with possibility of parallel implementation, a brief view into our Linux-cluster, numerical experiments and measurements. Section 5 summarizes with some conclusions.

2 Mathematical Models and Motivations

A review of the literature shows that many 2D depth-averaged numerical models have been developed and applied to free-surface flow problems. Although a depth-averaged model may suffer from some limitations on physical interpretation, e.g. invalid governing equations near bore with sharp curvature; the computed results may still be used for a simulation model to predict the evolution of a river flood.

These partial differential equations are of non-linear type of variables of ζ, p, q, where ζ is surface elevation, p, q discharges [1] [2] [3]. These variables are most convenient for defining appropriate boundary conditions. However, the use of this formulation is, however, limited for flows with restricted Reynolds numbers. For convection-dominated open channel flows such as flow affected by tidal flow and dam-break flow, numerical difficulties and non-adequate simulated flow may occur due to inadequate treatment of the non-linear convective terms.

However, the usefulness of such models has been demonstrated on a number of examples, especially where the flow paths on the floodplain can be ill-defined due to the man made structures. One of the 2D depth averaged flow models is MIKE21 [2].

It is known [2] that MIKE21 solves mass equation and momentum equation by finite difference method in the space-time domain using so-called Alternating Direction Implicit (ADI) technique. The equation matrices that result for each individual grid line are resolved by Double Sweep algorithm, i.e. equations are solved in one-dimensional sweeps, alternating between x and y directions. In the x-sweep, the mass equation and momentum equation are solved taking ζ from n to $(n+1/2)$ and p

from n to $(n+1)$ in time space. The variable n indicates time steps in time space. For the term involving q, the two levels of old and known values are used, i.e. $(n-1/2)$ and $(n+1/2)$. In the y-sweep the mass equation and momentum equation are solved taking ζ from $(n+1/2)$ to $(n+1)$ and q from $(n+1/2)$ to $(n+3/2)$ while terms in p use the values just calculated in the x-sweep at n and $(n+1)$. The mass and momentum equation thus expressed in a one-dimensional sweep for sequence of grid points in a line lead to a following coupled linear systems of equations:

$$A_{j,k} \cdot p_{j-1,k}^{(n+1)} + B_{j,k} \cdot \zeta_{j,k}^{(n+1/2)} + C_{j,k} \cdot p_{j,k}^{(n+1)} = D_{j,k} \qquad \text{for } j = 1,2\ldots,J \qquad (1)$$

$$a_{j,k} \cdot \zeta_{j,k}^{(n+1/2)} + b_{j,k} \cdot p_{j,k}^{(n+1)} + c_{j,k} \cdot \zeta_{j+1,k}^{(n+1/2)} = d_{j,k} \qquad \text{for } j = 1,2\ldots,J$$

Thus, equations (1) yield the tridiagonal system of linear equations $Ax = y$, with *tridiagonal matrix A*. In generally, if such lines are K, we need to compute such system K-times with different coefficients of matrix A and vector y. Naturally, this part of computation is very time consuming and is suitable for parallelization.

3 Numerical Solutions

We consider a linear system $Ax = y$ of size n, $A \in R^{n \times n}$ and a given vector $y \in R^n$. For a tridiagonal matrix A, we denote the nonzero value of the matrix as follows:

$$A = \begin{pmatrix} b_1 & c_1 & & & & 0 \\ a_2 & b_2 & c_2 & & & \\ & a_3 & b_3 & c_3 & & \\ & & \ddots & \ddots & \ddots & \\ & & & a_{n-1} & b_{n-1} & c_{n-1} \\ 0 & & & & a_n & b_n \end{pmatrix} \qquad (2)$$

The computation of the Gaussian elimination for a tridiagonal matrix is linear in n but it does not provide a possibility for parallel computation. The Gaussian elimination is a purely sequential algorithm. In the next section, we describe an algorithm, which is suitable for parallel computation [8].

3.1 Recursive Doubling

Recursive doubling [5] is performed as follows: we write matrix A as $A = (B + I + C)$, where diagonal$(A) = I$, B is the strictly lower diagonal part, C is the strictly upper diagonal part. After multiplying matrix A by the matrix $(-B + I - C)$ we obtain a matrix with three nonzero diagonals again, but now the nonzero diagonal entries are in positions (i, j) with $|i - j| = 2$, i.e. their distance to the main diagonal is doubled. We describe the first step in more detail for three neighboring equation $i-1, i, i+1$:

$$\begin{aligned} a_{i-1}x_{i-2} + b_{i-1}x_{i-1} + c_{i-1}x_i &= y_{i-1} \\ a_i x_{i-1} + b_i x_i + c_i x_{i+1} &= y_i \\ a_{i+1}x_i + b_{i+1}x_{i+1} + c_{i+1}x_{i+2} &= y_{i+1} \end{aligned} \tag{3}$$

Equation *(i-1)* is used to eliminate x_{i-1} from the *i*-th equation and the equation *(i+1)* is used to eliminate x_{i+1} from the *i*-th equation. The new equation is:

$$a_i^{(1)}x_{i-2} + b_i^{(1)}x_i + c_i^{(1)}x_{i+2} = y_i^{(1)} \tag{4}$$

with coefficients:

$$\begin{aligned} \alpha_i^{(1)} &= -a_i/b_{i-1} \\ \beta_i^{(1)} &= -c_i/b_{i+1} \\ a_i^{(1)} &= \alpha_i^{(1)}a_{i-1} \\ b_i^{(1)} &= b_i + \alpha_i^{(1)}c_{i-1} + \beta_i^{(1)}a_{i+1} \\ c_i^{(1)} &= \beta_i^{(1)}c_{i+1} \\ y_i^{(1)} &= y_i + \alpha_i^{(1)}y_{i-1} + \beta_i^{(1)}y_{i+1} \end{aligned} \tag{5}$$

After $N=\lfloor log_2(n) \rfloor$ steps, there is only one main diagonal left and we can compute

$$x_i^{(N)} = y_i^{(N)}/b_i^{(N)} \tag{6}$$

for $i=1,\ldots, n$.

3.2 Cyclic Reduction

The recursive doubling algorithm offers a high degree of potential parallelism that is archived by computation redundancy. The *cyclic reduction* [5] algorithm is a modification of the recursive doubling algorithm, which avoids computation redundancy by computing only needed values $a_i^{(k)}$, $b_i^{(k)}$, $c_i^{(k)}$, $y_i^{(k)}$. The entry in position *(2i, 2i-1)* is eliminated using row *(2i-1)* and the entry *(2i, 2i+1)* is eliminated using row *(2i+1)*. This leads to fill-in in positions *(2i, 2i-2)* and *(2i, 2i+2)*. This also eliminates all odd numbered unknowns and all even numbered unknowns are coupled to each other only.

The algorithm runs in two steps:

Step 1: For $k = 1, \ldots, \lfloor log_2(n) \rfloor$

Compute $a_i^{(k)}$, $b_i^{(k)}$, $c_i^{(k)}$, $y_i^{(k)}$ with $i = 2^k, \ldots, n$ and step 2^k by the form (5).

In step $k = \lfloor log_2(n) \rfloor$ there is only one equation for $i = 2^N$ with $N = \lfloor log_2(n) \rfloor$

Step 2: For $k = \lfloor log_2(n) \rfloor, \ldots, 0$ compute x_i by the formulation:

$$x_i = \frac{y_i^{(k)} - a_i^{(k)}.x_{i-2^k} - c_i^{(k)}.x_{i+2^k}}{b_i^{(k)}} \tag{7}$$

Fig. 1 shows illustration for the elimination phase of recursive doubling (a) and cyclic reduction (b) for a matrix with size $n=8$ (or $nxn = 8x8$). The positions of non-zero elements in i-th step are marked as *(i)*. The main diagonal always has non-zero values in all steps. It is easy to see that the matrix in Fig. 1 has non-zero main diagonal and diagonals marked by (0) at start time. After first step, the main diagonal and diagonals, which are marked (1), have non-zero values. Elements in (0) diagonals of the matrix have zero values, etc. The elimination phase of recursive doubling algorithm finishes when only the main diagonal is left. The elimination phase of cyclic doubling algorithms finishes when we can compute the value of x_8.

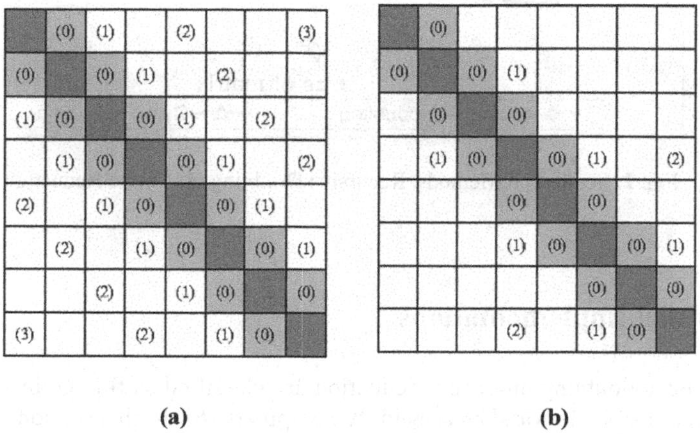

Fig. 1. Recursive Doubling (a) and Cyclic Reduction (b) for n=8.

Both recursive doubling and cyclic reduction algorithms can be generalized to banded matrices by applying the described computational steps for elements to blocks and using matrix operations instead of operations on single elements. In this article, we do not discuss this direction in more details.

3.3 Comparison of Methods

As we describe above, recursive doubling gives a large possibility for parallel computation but does a large amount of redundancy operations. Cyclic reduction computes only the necessary entries for x_N with $N = \lfloor log_2(n) \rfloor$ but still provides potential parallelism. The sequential computational runtime of the recursive doubling algorithm is $O(n \cdot log_2 n)$ that is larger than the Gaussian elimination. The sequential computational runtime of the cyclic reduction algorithm is $O(n)$ as the runtime of the Gaussian elimination. The advantage of both recursive doubling and cyclic reduction is a large possibility for parallel computation [5] [7]. The computation times of recursive doubling and cyclic reduction are shown in Fig. 2.

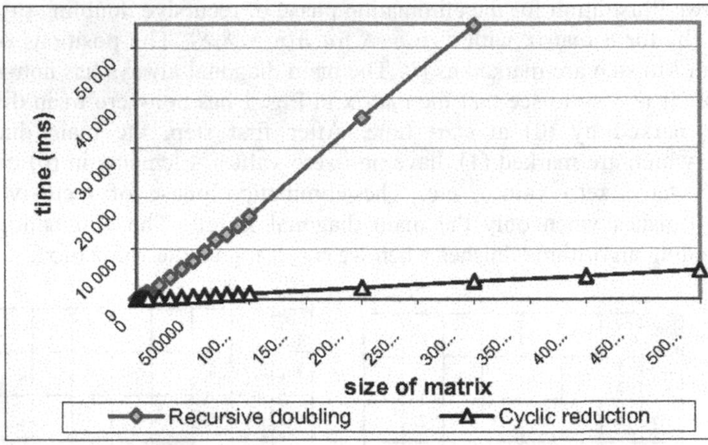

Fig. 2. Sequential Methods: Recursive Doubling vs. Cyclic Reduction.

4 Parallel Implementations

Both recursive doubling and cyclic reduction are classified as fine-grained algorithms [5], which are suitable for shared memory computers [6]. In this section, we describe results of our experiment on our PC Linux cluster and on SGI Origin2000 and. The cyclic reduction algorithm was chosen because of its possibility for parallelism and better computational runtime than the recursive doubling algorithm.

4.1 A Linux Cluster and the SGI Origin2000

Our Linux cluster consists of eight Pentium-III nodes. All computers are interconnected with 100Mb/s Ethernet network, through 16-port switch. Each computing node has 512 MB memory and the whole system memory is 4GB. The Mandrake Linux operating system is installed for the cluster. For comparison, we run our experiment on distributed shared memory SGI Origin2000 [11], which consists of 4 dual computer nodes, where all 8 processors are R10000, 250 MHz, with two 4MB cache memories. The whole system has 2GB of memory. The communication bandwidth between two processors in one node differs from the communication bandwidth between two processors located in different nodes. For reasons of administrative work, we do not use the whole system for our experiments but only six processors of each system.

For our experiments, we use MPI library Mpich [9] implemented for Linux and SGI Origin2000. On SGI Origin2000, parallel applications run under the LSF batch system.

4.2 Parallelization Method

The cyclic reduction is a row-oriented algorithm. The computation of coefficients of the matrix $A^{(k)}$ is performed by computing coefficients $a_i^{(k)}$, $b_i^{(k)}$, $c_i^{(k)}$, $y_i^{(k)}$. In our parallel implementation, if a row is distributed to processor p, then processor p computes all coefficients for the given row. Each processor deals with its own data, e.g. certain number of rows. The data distribution reduces the interaction between processors as much as possible. Communications are required only for bounded values and only if it is necessary. The parallel algorithm is performed as follows:

We assume the matrix size is n and the number of processors is p.

Step 1: For $k = 1, \ldots, \lfloor log_2(n) \rfloor$

Each processor P_j $(1 \leq j \leq p)$ sends four data values to processor P_{prev} and/or P_{next} if they need values located in P_j.

Processor P_j computes coefficient $a_i^{(k)}$, $b_i^{(k)}$, $c_i^{(k)}$, $y_i^{(k)}$ of $\lfloor n/p \rfloor$ rows of the matrix with step size 2^k using form (5).

If P_j needs values from other processor, it receives four data values from each P_{prev} and/or P_{next} to finish its computation,

where $prev = j - 2^{k+1}$
$next = j+2^{k+1}$

Step2: For $k = \lfloor log_2(n) \rfloor, \ldots, 0$

Each processor P_j $(1 \leq j \leq p)$ sends four data values to processor P_{prev} and/or P_{next} if they need values located in P_j.

Processor P_j computes $x_i^{(k)}$ of $\lfloor n/p \rfloor$ rows of the matrix with step size 2^k using form (7).

If P_j needs values from other processor, it receives four data values from each P_{prev} and/or P_{next} to finish its computation,

where $prev = j - 2^{k+1}$
$next = j+2^{k+1}$

4.3 Experiments and Measurements

Processors	n=4000	n=10^5	n=5.10^5	n=10^6	n=5.10^6
1	2,2	59,7	777,4	1 554,5	7 776,9
2	1,3	30,3	563,3	1 207,8	6 040,5
3	1,4	20,2	279,1	781,9	4 052,9
4	1,5	15,0	202,9	701,6	3 588,6
5	1,4	12,6	165,3	487,9	2 878,9
6	2,2	11,3	138,0	387,4	2 347,2

Table 1. Execution Time (ms) on SGI Origin2000.

We have tested our MPI program on two different machines: Linux cluster and SGI Origin2000 with different input tridiagonal matrices. The message-passing program shows good results on both machines in different ways.

On the shared memory SGI Origin2000, for small systems of less than 4000 equations, the execution time per processor is too small in comparison with the communication time (Table 1). On the distributed system Linux cluster, the communication time is larger and the small speedup is reached for systems with more than 40000 equations, i.e. 10 times larger than on SGI Origin2000 (Table 2).

Processors	n=40000	$n=10^5$	$n=5.10^5$	$n=10^6$	$n=5.10^6$
1	55,3	142,1	719,6	1 442,9	7 047,9
2	28,8	75,5	366,4	739,3	3 771,8
3	21,9	68,3	269,8	526,0	2 518,3
4	30,5	50,0	213,6	407,7	1 975,0
5	30,5	67,4	220,4	342,9	1 627,7
6	42,0	41,2	163,6	286,7	1 334,6

Table 2. Execution Time (ms) on Linux Cluster.

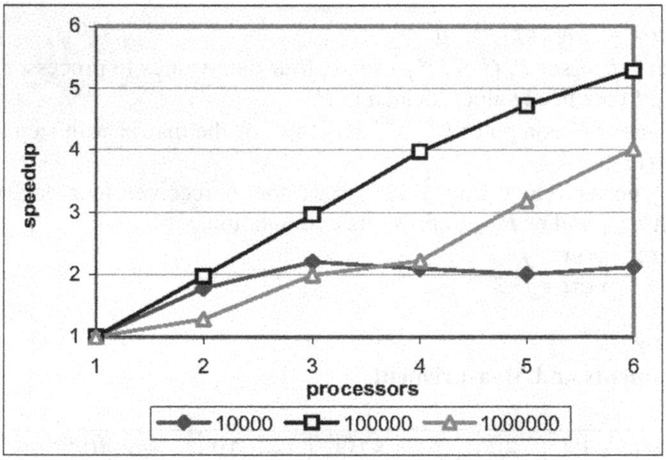

Fig. 3. Speedup on SGI Origin2000.

For systems with 5.10^4 to 5.10^5 equations speedups on SGI Origin2000 with six processors are quite good: from 4 to 5 approximately. The speedups go down when size of systems increases. The result is stable around 3.0 to 3.2 for systems with more than 10^6 equations (Fig. 3). We consider such results cause distributed shared memory effects and cache performance.

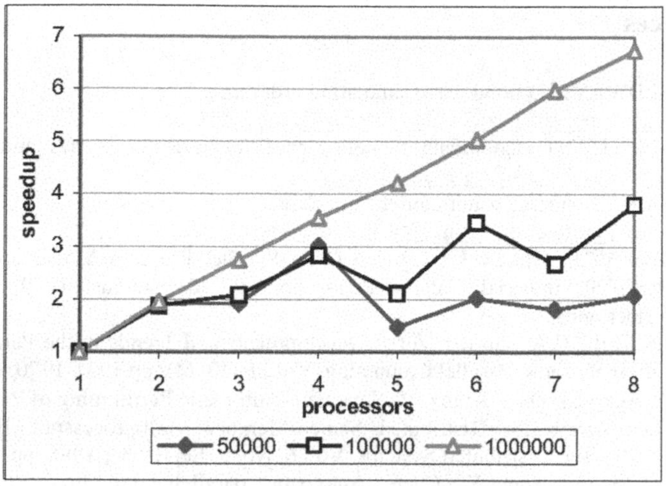

Fig. 4. Speedup on Linux Cluster.

On the Linux cluster speedups grow quite linearly in comparison with the number of processors. With six processors and the size of systems 10^4, the speedup is around 3.5. For larger system with more than 10^6 equations, the speedup values are stable around 5.2 (Fig. 4).

5 Conclusion and Future Work

In this paper, we have shown that the cyclic reduction method can be efficiently implemented for distributed memory systems, especially for Linux-clusters. A comparison with a thread implementation shows that the Linux-cluster leads to better results with more reliable runtime behavior for large systems of partial differential equations than on distributed shared memory machines. Otherwise, SGI Origin2000 shows better results for smaller systems of equations, for which the algorithm needs more communications per given computation rows than large systems.

This work is a part of the EU fifth Framework Programme ANFAS (datA fusioN for Flood Analysis and decision Support) IST-1999-11676 RTD and is supported by the Slovak Scientific Grant Agency within Research Project No. 2/7186/20.

References

1. FLDWAV: 1D flood wave simulation program http://hsp.nws.noaa.gov/oh/hrl/rvrmech/rvrmain.htm
2. Mike21: 2D engineering modeling tool for rivers, estuaries and coastal waters http://www.dhisoftware.com/mike21/
3. SMS: 2D surface water modeling package http://www.bossintl.com/html/sms_overview.html
4. Tran V.D., Hluchy L., Nguyen G.T.: Parallel Program Model and Environment, ParCo'99, Imperial College Press, pp. 697-704, August 1999, TU Delft, The Netherlands.
5. I.S. Duff, H.A. van der Vorst: Developments and Trends in the Parallel Solution of Linear Systems, Parallel Computing, Vol 25 (13-14), pp.1931-1970, 1999.
6. A. Agarwal, D. A. Kranz, V. Natarajan: Automatic Partitioning of Parallel Loops and Data Arrays for Distributed Shared-Memory Multiprocessors, IEEE Trans. on Parallel and Distributed Systems, Vol. 6, No. 9, September 1995, pp. 943 962.
7. G. M. Megson, X. Chen, Automatic parallelization for a class of regular computations. World Scientific, 1997.
8. T. L. Freeman, C. Phillips: Parallel Numerical Algorithms. Prentice Hall 1992.
9. MPICH - A Portable Implementation of MPI http://www-unix.mcs.anl.gov/mpi/mpich/
10. Selim G. Akl: Parallel Computation Models and Methods, Prentice Hall 1997.
11. Origin2000 Architecture http://techpubs.sgi.com/library/manuals/3000/007-3511-001/html/O2000Tuning.1.html
12. Pfister G.F.: In Search of Clusters, Second Edition. Prentice Hall PTR, ISBN 0-13-899709-8, 1998.

Advanced Library Support for Irregular and Out-of-Core Parallel Computing

Peter Brezany[1], Marian Bubak[2,3], Maciej Malawski[2], and Katarzyna Zając[2]

[1] Institute for Software Science, University of Vienna
Liechtensteinstrasse 22, A-1090 Vienna, Austria
brezany@par.univie.ac.at
[2] Institute of Computer Science, AGH, al. Mickiewicza 30
30-059 Kraków, Poland
[3] Academic Computer Centre – CYFRONET, Nawojki 1
30-950 Kraków, Poland
{bubak,malawski,kzajac}@uci.agh.edu.pl

Abstract. Large scale irregular applications involve data arrays and other data structures that are too large to fit in main memory and hence reside on disks; such applications are called *out-of-core* applications. We present a design and implementation of `lip`, a new parallel library that is intended to efficiently support parallel execution of irregular out-of-core codes on parallel systems. We discuss the `lip` structure and application interface, as well as an application that has been implemented using that interface. The library has been implemented on top of MPI/MPI-IO and, therefore, is fully portable. We introduce preliminary performance results from a template CFD codes to demonstrate the efficacy of the presented techniques.

1 Introduction

Parallelising irregular codes for distributed-memory systems is a challenging problem and is of great importance. Application areas in which irregular codes are found include unstructured multigrid fluid dynamic solvers, molecular dynamics codes and diagonal or polynomial preconditioned iterative linear solvers, etc. In such codes, access patterns to major data arrays are not known until runtime, what requires preprocessing in order to determine the data access patterns, and, consequently, to find what data must be communicated and where it is located. Furthermore, the data arrays may be dynamically partitioned and redistributed in order to minimise communication and balance load. It is necessary to support these techniques by an appropriate runtime library. For example, the CHAOS [13] library is a well-known system which has supported the handling of irregular computations on the Intel iPSC/860 and Paragon systems.

Large scale irregular applications involve large arrays and other data structures. Runtime preprocessing and automatic partitioning provided for these applications results in construction of large data structures which increase the memory usage of the program substantially. Consequently, a parallel program may quickly run out of memory. Therefore, some data structures must be stored

B. Hertzberger et al. (Eds.): HPCN Europe 2001, LNCS 2110, pp. 435–444, 2001.

on disks and fetched during the execution of the program. Such applications are called *out-of-core (OOC) applications.* The performance of an OOC program strongly depends on how fast the processors, the program runs on, can access data on disks. However, it is difficult for programmers to manage the memory-disk interface explicitly, and moreover, programmer-inserted I/O significantly reduces portability. Virtual memory is one possible approach to maintaining disk storage which has been provided by many vendors. However, it has been observed [2] that the performance of scientific applications that rely on virtual memory is generally poor due to frequent paging in and out of data. Therefore, the development of appropriate parallelisation methods that are based on portable and optimised libraries is an important research issue.

Parallelising irregular OOC programs is a relatively new topic and there has been little research in this area. The report on the PASSION project [7] very briefly outlines the functionality of the runtime support provided for OOC irregular problems. Brezany et al. [2,3] proposed OOC extensions to the CHAOS library and experimentally implemented several software modules on the Intel Paragon system. So far, more research work has addressed parallelising regular OOC applications. Nieplocha and Foster [11] developed a parallel library optimising accesses to two-dimensional out-of-core arrays. Bordawekar, Choudhary, and Thakur [14] worked on compiler methods for out-of-core HPF regular programs. Cormen and Colvin worked on a compiler for out-of-core C*, called ViC* [8]. Paleczny, Kennedy, and Koelbel [12] proposed a compiler support and programmer I/O directives which provide information to the compiler about data tiles for OOC regular programs.

In this paper we describe an advanced library called `lip` to support out-of-core computing on distributed-memory architectures. This paper enhances ideas outlined in [5,6]. Particularly, implementation of data partitioning algorithm was added to the set of functions of the library. The performance studies of the full implementation of the library using synthetic and real benchmarks on a workstation cluster system are presented as well.

In the following sections of this paper we describe the definition of the high level parallelisation strategy for irregular OOC problems, an appropriate library interface supporting this strategy and the full implementation and performance of `lip`.

2 Design Goals

In out-of-core computations, disk storage is treated as an other level in the memory hierarchy, below cache, local memory, and (in a parallel computer) remote memories. Another important requirement is portability of libraries supporting parallel OOC computations across a wide range of target parallel platforms. Accordingly, `lip` is designed to meet several goals:

- design on top of the MPI and MPI-IO standard,
- efficiently handle transfer of irregular array sections between local and remote memories and disks,
- allow applications to explicitly control file layout and parallelism in file access,
- be flexible enough to support a wide variety of interfaces and policies, implemented in libraries built on top of it in the future,
- provide support for both in-core and OOC irregular codes, because it might be inconvenient to use other libraries to handle in-core code parts as they may occur in real OOC applications,
- allow easy and efficient implementation of high-level libraries.
- scale to many processors,
- minimise memory and performance overhead.

3 Library Structure and Application Interface

In this section, we explain the basic model of data storage, communication, computation, and I/O used by the library and give an overview of the library.

3.1 Data Storage, Communication, Computation, and I/O

The main idea of the `lip` design is presented in Fig. 1. In the SPMD (Single Program Multiple Data) programming model, the distribution of each array determines for each processor an associated set of local elements called the local segment. Each OOC array is stored in a single file called Out-of-Core Array File (OAF)[1].

The portion of the local segment which is in main memory is called the In Core Local Array (ICLA). All computations are performed on data in the ICLA. During the course of computation on a processor, the sections of the operand arrays are fetched from OAFs into the ICLAs of the processor, the new values are computed and the sections updated are stored back into OAFs, if necessary. On each processor, the computation is performed in stages where each stage operates on only so large parts of the arrays which can fit in the local memory. The computation performed in a stage corresponds to the execution of a *i-section* of loop iterations.

Each data-parallel loop is transformed into four main phases: *work distributor*, *partitioner*, *inspector*, and *executor*. The work distributor determines how to spread the work (iterations) among the available processors - on each processor, it computes the *execution set*, i.e. the set iterations to be executed on this processor, and splits the execution set into a set of i–sections. Sections of out-of-core arrays associated with each i–section can fit in the main memory of the processor. The partitioning phase is optional. This phase essentially involves the use of a partitioner which analyses the loop to determine a new data and work distribution, to minimise interprocessor communication and enhance locality and load balance. The description of the new data distribution computed

[1] This corresponds to the global placement model [1].

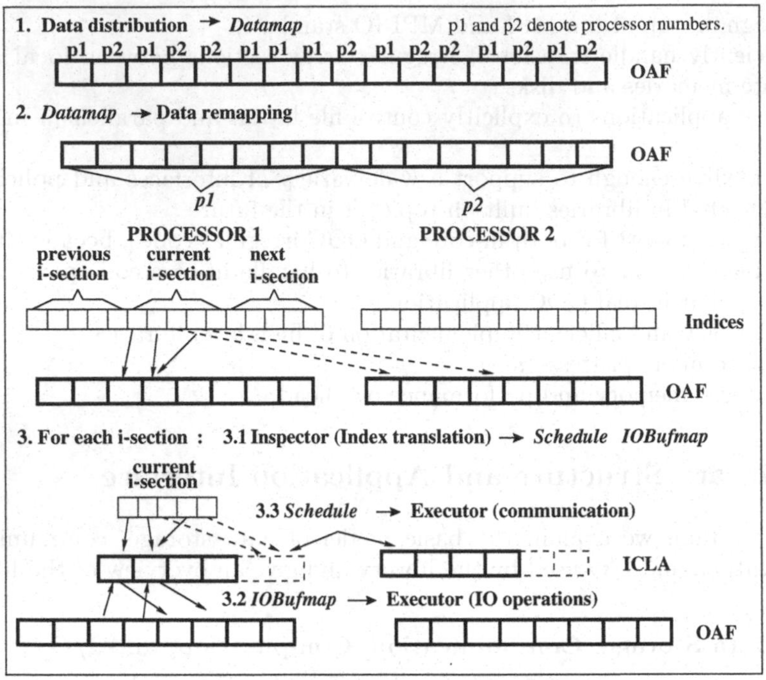

Fig. 1. Inspector/Executor Technique for Out-of-Core Computing.

by the partitioner is stored in a data structure called *Datamap* (Step 1 in Fig. 1). The data arrays and index arrays are remapped to obtain the new distributions determined by the partitioner (Step 2). The array redistribution results in restructuring of OAFs. For each i–section, the *inspector* performs the dynamic loop analysis of the loop to determine *communication schedules* [9] and *I/O mapping structures* called *IOBufmap* to establish an appropriate addressing scheme for access to local elements and copies of non-local elements on each processor (Step 3.1). The *executor* performs I/O operations and communication (Steps 3.2 and 3.3) according to the structures determined in the inspector, and, on each processor, executes the actual computations for all iterations of the i–section.

3.2 Library Design

The `lip` library is built using MPI/MPI-IO procedures[2] that appear at two layers:

1. *Application Layer.* A set of procedures has been designed to support dynamic data distributions and to automatically generate send and receive messages by capturing communication patterns at runtime - this support for in-core

[2] However, they could be very easily adapted to the native communication and I/O primitives of the target system.

computations has been also provided by the CHAOS library. We have included into the system a number of procedures to support the development of irregular OOC applications: generation of I/O schedules, data exchange between processors and local and remote files, index translation for copies of OOC data, disk oriented data partitioning, etc. The application programmer only inserts calls to the procedures of this layer.

2. *Optimisation Layer.* The `lip` optimisation routines can be divided into two main categories based on their functionality:
   (a) Optimising the transfer between memory hierarchy levels. Irregular accesses to arrays exhibit *temporal locality* in many applications. `lip` utilises this feature by reusing the data already fetched to the local or remote memory instead of reading them again from disk or communicate between processors. To achieve this goal, the storage allocated on individual processors to *ICLA* is managed like a *cache*.
   (b) Efficient file access method. A simple way of reading an irregular array section is to read each its element individually (it is called the *direct read method*) which issues a large number of I/O calls and allows low granularity of data transfer. It is possible to use a much more efficient method called *data sieving* [1] to read a regular section of the array that includes the original irregular section. The required data elements are then filtered out from the I/O buffer.

3.3 Application Interface

The `lip`'s interface is primarily intended to allow easy implementation of applications and libraries. These libraries will provide a higher-level functionality needed by some user programmers and developers of parallelising compilers.

Partitioning, Data Distribution, and Redistribution. At present, `lip` supports coordinate-based partitioner using Hilbert's curve [10] to determine what part of data should reside on the same node. The library supports both in-core and the OOC version of the partitioner. A partitioner is activated using the `LIP_create_hilbert_distribution()` call. The resulting data mapping is stored in a special object called `Datamap`. The remapping function `LIP_remap_ooc()` can be used to redistribute the data.

Index Translation. The procedures `LIP_Localize()` and `LIP_OOC_localize()` are used for index translation. The former translates globally numbered indices into locally numbered counterparts. It also updates the communication schedule so it could be used to transmit the non-local data. The latter function transforms the indices which point to the data residing on disk into indices pointing to the memory data buffer. The results of mapping memory contents into disk is stored in IO mapping structure.

IODatamap Manipulation. `lip` includes I/O buffer mapping objects which store information about the mapping of the memory buffer onto the corresponding OAF file. There are also functions, e.g. `LIP_IOBufmap_get_datatype()`, to obtain the MPI derived data types according to the information stored in the I/O

```
MPI_Init( /* ... */ );
LIP_Setup( /* ... */ );

  /* Generate index array describing
   * relationships between user data */

  /* Generate irregular distribution*/
LIP_create_hilbert_distribution(/* ... */);

  /* Data remapping */
LIP_remap_ooc(/*...*/);

  /* Perform irregular computation
   * on the data */
for ( /* ... */ )
{
    /* read i-section's indices
     * from a file */

    /* Inspector Phase (computes
     * optimized communication
     * and disk-memory mapping
     * pattern) */
    LIP_Localize( /* ... */ );
    LIP_OOC_Localize(/* ... */);

    /* Create MPI derived datatypes
     * for moving data between
     * the memory and the disk */
    LIP_IObufmap_get_datatype(/*...*/);

    /*Executor Phase (performs
      communication and computation)*/

      /* Exchange data between disk and
       * memory in order to store the data
       * in the memory that will be needed
       * in this i-section */

      MPI_File_write( /* ... */ );
      MPI_File_read( /* ... */ );

      /* gather non-local irregular data */
      LIP_Gather( /* ... */ );

      /* perform computation on data */
      for ( i = /* ... */ )
      {
        k = edge[i];
        y[k] = f( x[k] );
      }
    /* scatter non-local
     * irregular data (results) */
    LIP_Scatter( /* ... */ );
}

/* Get MPI derived datatypes for moving
 * the data obtained in the last
 * iteration from the memory
 * to the disk */
LIP_IObufmap_get_datatype( /* ... */ );

/* Store the data on a disk */
MPI_File_write( /* ... */ );

LIP_Exit( /* ... */ );
MPI_Finalize( /* ... */ );
```

Fig. 2. The Use of `lip` in a Simple Parallel C Program with Out-of-Core Data Array.

buffer mapping object which, in turn, may be used to perform data movement by MPI-IO functions.

Schedule Manipulation and Communication. The `lip` library includes group of functions used to create and manipulate a schedule structure that stores patterns used for communication.

The library includes two general communication functions: `LIP_Gather()` and `LIP_Scatter()`. They perform collective communication among nodes before and after the computational phase. They support numerous communication patterns and data copy/update schemes and use the schedule object to obtain information about the source and destination of the data.

Parallelisation Example

The parallelisation scheme applied to a typical irregular out–of–core problem is outlined in Fig. 2. After the data distribution step, the index translation is performed. Then each processor reads data from a file (I/O phase) and sends the data needed by other processors (communication phase). In the next step, calculation is performed using the data residing in the processor's local memory buffer. After the computational phase, the data are scattered to their owners and written to disk.

4 Performance Evaluation

4.1 `lip` vs. Virtual Memory

To test the `lip` performance, we used an irregular loop performing operations on data stored in vertices of the directed graph as shown in Fig. 3. The structure of the corresponding parallel code is similar to that depicted in Fig. 2.

The ratio of the data size to virtual and real memory sizes was constant (1300000 vertices per processor), so the number of all data elements was proportional to the number of nodes. In addition, the size of the application grew with the size of the schedule structure and consequently the number of nodes. This increase did not effect performance of ooc lip functions as the data were divided into chunks (16 in our example) small enough to fit in main memory. As can be seen in Fig 4, the execution time does not increase with the number of data elements.

However, the tests that use virtual memory show significant decrease of performance with the increasing size of the application. The reason for it are frequent IO operations during the calculation phase due to more general algorithms used for the implementation of disk data accesses.

To make the computational phase more time consuming, the loop depicted in Fig. 3 was executed 10^2 times.

```
for(i=0;i<edge_counter;i++)
  {
     x2[edge2[i]]+=x1[edge1[i]]/n_succ;
     x2[edge1[i]]-=x1[edge1[i]];
}
```

Fig. 3. Irregular Loop Used for Measurements

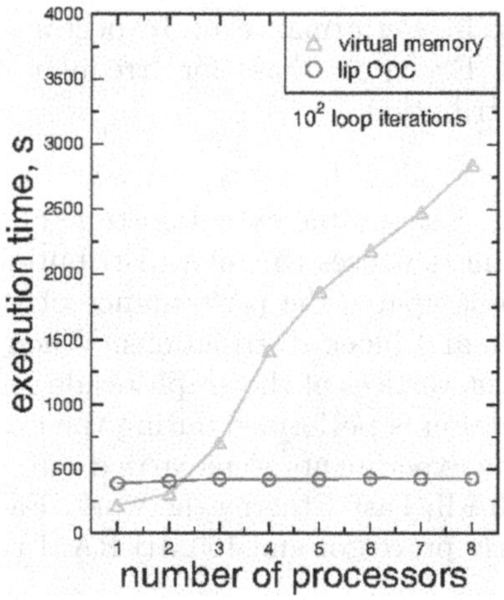

Fig. 4. Performance Comparison.

The experiments were carried out on a cluster of 8 LINUX workstations connected by 100Mb Fast Ethernet with Intel Celeron 600 MHz processor, 64 MB RAM and 256 MB swap size .

4.2 Out-Of-Core Code Coupled to the Hilbert Partitioner

To test the coordinate-based partitioner, we assigned the Cartesian coordinates to the vertices of the graph used in the previous exmple, so the distance between adjacent vertices was not greater then five percent of the distance between farthest vertices.

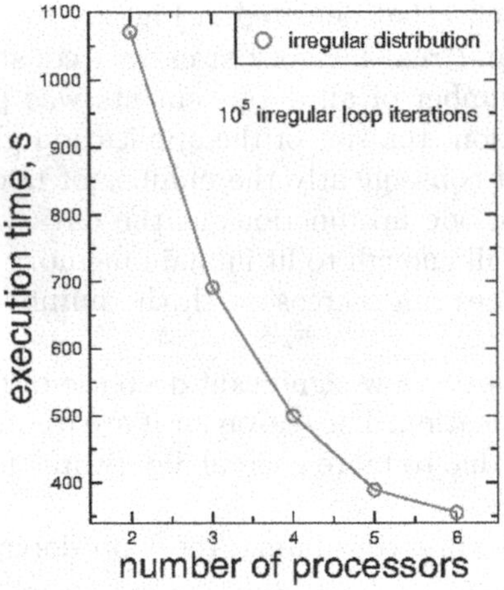

Fig. 5. Performance of Inspector and Executor Phase for Irregular Distribution.

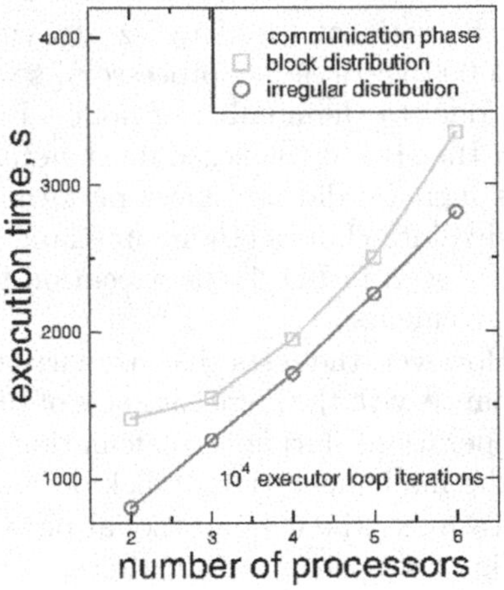

Fig. 6. Performance of Communication Phase.

Fig. 5 shows the execution time of the inspector and executor phases of the program that uses irregular distribution for the graph including 12000 vertices. Fig. 6 compares the performance of the communication phase when using irregular and block distributions. When the irregular distribution is used, more adjacent vertices of the graph reside on the same processor; therefore, less communication is performed during the executor phase.

The experiments were carried out on a cluster of 6 workstations connected by 100 Mb Fast Ethernet network. Each node was SUN SPARCstation 4 with 110MHz processor and 160MB RAM running Solaris 8.

4.3 Parallelisation of the AVL FIRE Benchmark

FIRE [4] is a general purpose computational fluid dynamics (CFD) software developed at AVL company in Graz, Austria. We have parallelised GCCG benchmark solver from FIRE using the `lip` library. The parallelisation scheme depicted in Fig. 2 was used with the irregular code whose extract is shown in Fig. 7.

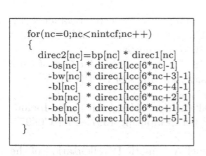

```
for(nc=0;nc<nintcf;nc++)
{
   direc2[nc]=bp[nc] * direc1[nc]
      -bs[nc]  * direc1[lcc[6*nc]-1]
      -bw[nc]  * direc1[lcc[6*nc+3]-1]
      -bl[nc]  * direc1[lcc[6*nc+4]-1]
      -bn[nc]  * direc1[lcc[6*nc+2]-1]
      -be[nc]  * direc1[lcc[6*nc+1]-1]
      -bh[nc]  * direc1[lcc[6*nc+5]-1];
}
```

Fig. 7. Irregular Loop from the FIRE Benchmark Solver.

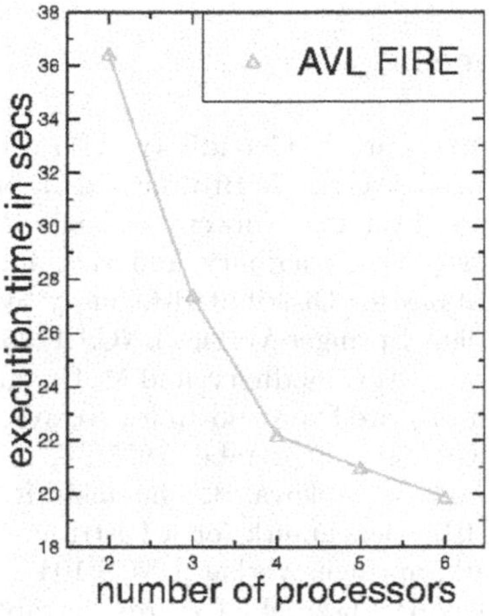

Fig. 8. The Solver Performance.

The tests were applied to the dataset consisting of 13845 cells (size of `direc2` array). The data was block-wise distributed and all arrays were located *in-core*. The tests were performed on a cluster of 6 workstations connected by 100 Mb Fast Ethernet network. Each node was SUN Ultra 10 with UltraSPARC-IIi 440MHz processor and 256MB RAM running Solaris 8. The results of the tests are shown in Fig. 8.

5 Summary

Based on several studies of parallelisation of in-core and out-of-core irregular applications for distributed-memory systems, we have designed a new parallel library, `lip`, that is intended to provide high performance for a wide variety of applications. It is based on the inspector and executor approach and on the MPI–I/O concepts.

While `lip` has been so far evaluated on relatively simple benchmark codes, in the near future, its performance will be tuned and evaluated on large-scale real applications and its stability will be improved. Future work will further focus on performance optimisation, and, finally, the library will be available on the web.

Acknowledgements

We are grateful to Agnieszka Wierzbowska and Piotr Łuszczek for their contribution. This research is being carried out as part of the research project "Aurora" supported by the Austrian Research Foundation. It was done in the framework of the Polish-Austrian collaboration and it was supported in part by the KBN grant 8 T11C 006 15.

References

1. R. Bordawekar, A. Choudhary. Issues in Compiling I/O Intensive Problems. In: R. Jain, J. Werth, J. Browne (eds.) Input/Output in Parallel and Distributed Computer Systems, Kluwer Academic Publ., 1996, pp. 69–96.
2. P. Brezany, A. Choudhary, and M. Dang. Parallelization of Irregular Out-of-Core Applications for Distributed-Memory Systems. Proceedings of HCPN 97, Vienna, April 1997, Springer-Verlag, LNCS 1225.
3. P. Brezany, A. Choudhary, and M. Dang. Parallelization of Irregular Codes Including Out-of-Core Data and Index Arrays, "Parallel Computing 1997 - PARCO'97", Bonn, Germany, September 1997.
4. P. Brezany, V. Sipkova, B. Chapman, R. Griemel Automatic Parallelization of the AVL FIRE Benchmark for a Distributed-Memory System In: PARA'95, Lyngby, Denmark, Springer Verlag, LNCS 1041, pp. 50-60.
5. M. Bubak, P. Łuszczek, Towards Portable Runtime Support for Irregular and Out-of-Core Computations. In: J. Dongarra, E. Luque, T. Margalef, (Eds.), *Proceedings of 6th European PVM/MPI Users' Group Meeting*, Barcelona, Spain, September 1999, LNCS 1697, Springer-Verlag Berlin-Heidelberg, 1999, pp. 59-66.
6. M. Bubak, P. Łuszczek, M. Malawski, K. Zając, Irregular and Out-of-Core Computing on SGI Origin 2000 in: Proceedings of 1st SGI Users Conference, Cracow, Poland, October 2000, ACC Cyfronet UMM, 2000, pp. 185-193.
7. A. Choudhary, et al. PASSION: Parallel and Scalable Software for Input-Output. *CRPC-TR94483*, Rice University, Houston, 1994.
8. T. H. Cormen and A. Colvin. ViC*: A Preprocessor for Virtual-Memory C*. TR: PCS-TR94-243, Dept. of Computer Science, Dartmouth College, Nov. 1994.
9. C. Koelbel, P. Mehrotra, J. Saltz, and S. Berryman. Parallel Loops on Distributed Machines. In *Proceedings of the 5th Distributed Memory Computing Conference*, Charleston, pages 1097–1119, IEEE Comp. Soc. Press, April 1990.
10. C.-W. Ou and S. Ranka SPRINT: Scalable Partitioning, Refinement, and Incremental partitioning Techniques. `http://www.npac.syr.edu/projects/~pcrc/maryland/`.
11. J. Nieplocha and I. Foster. Disk Resident Arrays: An Array-Oriented I/O Library for Out-Of-Core Computations. In *Proc. of the Sixth Symposium on the Frontiers of Massively Parallel Computation*, IEEE Press, October 1996, pp. 196–204.
12. M. Paleczny, K. Kennedy, and C. Koelbel. Compiler Support for Out-of-Core Arrays on Parallel Machines. In *Proceedings of the 7th Symposium on the Frontiers of Massively Parallel Computation, McLean, VA*, pages 110-118, February 1995.
13. R. Ponnusamy et al. CHAOS runtime library. Techn. report, University of Maryland, May 1994.
14. R. Thakur, R. Bordawekar, and A. Choudhary. Compiler and Runtime Support for Out-of-Core HPF Programs. In *Proceedings of the 1994 ACM International Conference on Supercomputing*, pages 382–391, Manchester, July 1994.

A Parallel ADI Method for Linear and Non-linear Equations

I.V. Schevtschenko

Rostov State University
Department of Informatics and Computer Experiment
34/1 Communistichesky avenue, Apt. 111
344091, Rostov-on-Don, Russia
ishevtch@uic.rnd.runnet.ru

Abstract. The key goal of the current paper is to implement a parallel alternating-direction implicit, or ADI, method for solving linear and non-linear equations describing gravitational flow of ground water and realize it on a distributed-memory MIMD-computer under the MPI message-passing system. Aside from that, the paper represents a comparison of the parallel algorithms for solving aforementioned equations and their evaluation in terms of relative efficiency and speedup. The obtained results show that for reasonably large discretization grids the parallel ADI method, both for the linear and non-linear equations, is effective enough on a large number of processors, and $\lim_{M\to\infty}\frac{C_c^A}{C_c^B} = 1.05$, where C_c^A and C_c^B are computational complexity of the algorithms for solving non-linear and linear equations respectively.

1 Introduction

The past decades a substantial progress has been made in mathematical description of water flow and pollutant transport processes. Today's mathematical models are available to predict admixtures migration in subterranean waters under various conditions of the process. Approximation of such models generates large linear systems of algebraic or differential equations which demand utilizing modern supercomputers proposing powerful computational resources to solve large problems in various fields of science. In particular, by solving the equation of pollutant transport in subterranean waters it is necessary to know a ground water table described by the balance mass equation. In this paper it is considered a parallel solution of the linear and non-linear balance mass equations for approximation of which the finite difference method is used. The solution of the problem is found with the aid of the ADI method, in particular, with using Peaceman-Rachford difference scheme [8]. We exploit inherent parallelism of the difference scheme to apply it to distributed-memory MIMD-computers.

An outline of the paper is as follows. Section 2 introduces to general formulation and numerical approximation of the original mathematical model, represents results on accuracy and stability of the used difference scheme and substantiates applicability of the conjugate gradient (CG) method [1] to solving systems of

B. Hertzberger et al. (Eds.): HPCN Europe 2001, LNCS 2110, pp. 445–453, 2001.

linear algebraic equations (SLAEs) generated by Peaceman-Rachford difference scheme. Section 3 is represented by a parallel realization of the ADI method both for the linear and non-linear equations. We compare the parallel algorithms and evaluate them in terms of relative efficiency and speedup. Finally, in section 4, we give our conclusions.

2 General Formulation and Numerical Approximation

One of the existing non-linear models describing gravitational flow of ground waters in an isotropic inhomogeneous element of a water-bearing stratum Ω, in according to [2], can be represented in the form

$$a^* \frac{\partial h}{\partial t} = \frac{\partial}{\partial x}\left(h \frac{\partial h}{\partial x}\right) + \frac{\partial}{\partial y}\left(h \frac{\partial h}{\partial y}\right) + \frac{1}{k}\left(v_{(x)} + v_{(y)}\right). \quad (1)$$

Here $h(x, y, t)$ represents a ground water table, k is a filtrational coefficient, $v_{(x)}$ and $v_{(y)}$ are filtration velocities from below and from above of the water-bearing stratum respectively. Parameter $a^* > 0$ depends on physical characteristics of the water-bearing stratum.

Sometimes it is necessary to simplify equations we solve. As a rule, such a simplification allows us to solve equations faster, if we exploit numerical methods, or even obtain an analytical solution. The linear model of equation (1) can be expressed as

$$b^* \frac{\partial h}{\partial t} = \Delta h + \frac{1}{k h_{av}}\left(v_{(x)} + v_{(y)}\right), \quad (2)$$

where h_{av} is a mean value of the ground water table, parameter $b^* > 0$ depends on h_{av} and physical characteristics of the water-bearing stratum.

The initial condition and Dirichlet boundary value problem for equations (1) and (2) are

$$h(x, y, t = 0) = h_0(x, y), \quad h|_{\partial\Omega} = f(x, y).$$

Here $h_0(x, y)$ is a given function and $f(x, y)$ is a function prescribed on the boundary of the considered field.

The approximation of the original problem is based on Peaceman-Rachford difference scheme, where along with prescribed grid functions $h(x, y, t)$, $h(x, y, t+\tau)$ an intermediate function $h(x, y, t + \frac{\tau}{2})$ is introduced. Thus, passing (from n time layer to $n+1$ time layer) is performed in two stages with steps 0.5τ, where τ is a time step.

Let us introduce a grid of size $M \times N$ in a simply connected area $\Omega = [0, a] \times [0, b]$ with nodes $x_i = i\Delta x$, $y_j = j\Delta y$, where $i = 1, 2, \ldots M$, $j = 1, 2, \ldots N$, $\Delta x = \frac{a}{M}$, $\Delta y = \frac{b}{N}$. In order to write the difference approximation for equations (1) and (2) we denote $h = h^n$, $\bar{h} = h^{n+\frac{1}{2}}$, $\hat{h} = h^{n+1}$, $h_{\bar{x}} = \frac{h_{ij} - h_{i-1,j}}{\Delta x}$, $h_{\underline{x}} = \frac{h_{i+1,j} + h_{ij}}{2}$, $h_{\tilde{x}} = \frac{h_{i-1,j} + h_{ij}}{2}$, $h_{\bar{y}} = \frac{h_{ij} - h_{i,j-1}}{\Delta y}$, $h_{\underline{y}} = \frac{h_{i,j+1} + h_{ij}}{2}$, $h_{\tilde{y}} = \frac{h_{i,j-1} + h_{ij}}{2}$. The difference approximation for equations (1) and (2) can then be represented as

$$\begin{cases} a^* \dfrac{\bar{h}_{ij} - h_{ij}}{0.5\tau} = \dfrac{h_{\underline{x}} \bar{h}_x - h_{\tilde{\bar{x}}} \bar{h}_{\bar{x}}}{\Delta x} + \dfrac{h_{\underline{y}} h_y - h_{\tilde{\bar{y}}} h_{\bar{y}}}{\Delta y} + \dfrac{v_{(x)ij} + v_{(y)ij}}{k}, \\ a^* \dfrac{\hat{h}_{ij} - \bar{h}_{ij}}{0.5\tau} = \dfrac{h_{\underline{x}} \bar{h}_x - h_{\tilde{\bar{x}}} \bar{h}_{\bar{x}}}{\Delta x} + \dfrac{\bar{h}_{\underline{y}} \hat{h}_y - \bar{h}_{\tilde{\bar{y}}} \hat{h}_{\bar{y}}}{\Delta y} + \dfrac{v_{(x)ij} + v_{(y)ij}}{k}. \end{cases} \tag{3}$$

and

$$\begin{cases} b^* \dfrac{\bar{h}_{ij} - h_{ij}}{0.5\tau} = \dfrac{\bar{h}_{i-1,j} - 2\bar{h}_{ij} + \bar{h}_{i+1,j}}{\Delta x^2} + \dfrac{h_{i,j-1} - 2h_{ij} + h_{i,j+1}}{\Delta y^2} + \dfrac{v_{(x)ij} + v_{(y)ij}}{k h_{av}}, \\ b^* \dfrac{\hat{h}_{ij} - \bar{h}_{ij}}{0.5\tau} = \dfrac{\bar{h}_{i-1,j} - 2\bar{h}_{ij} + \bar{h}_{i+1,j}}{\Delta x^2} + \dfrac{\hat{h}_{i,j-1} - 2\hat{h}_{ij} + \hat{h}_{i,j+1}}{\Delta y^2} + \dfrac{v_{(x)ij} + v_{(y)ij}}{k h_{av}}. \end{cases} \tag{4}$$

The initial condition and Dirichlet boundary value problem can be approximated as

$$h|_{t=0} = h_{(0)ij}, \quad h|_{\partial\Omega} = f_{ij}. \tag{5}$$

By addressing stability investigation of the difference equations (3) and (4) we formulate the following lemmas

Lemma 1. *Peaceman-Rachford difference scheme for equation (1) with Dirichlet's boundary conditions at $a^* > 0$ is stable.*

Lemma 2. *Peaceman-Rachford difference scheme for equation (2) with Dirichlet's boundary conditions at $b^* > 0$ is stable.*

Concerning the difference schemes (3) and (4) it can be noted that they have second approximation order [9] both in time and in space.

By using natural regulating of unknown values in the computed field let us reduce difference problems (3),(5) and (4),(5) to the necessity of solving SLAEs $A_k^{(3)} u_k^{(3)} = f_k^{(3)}, A_k^{(4)} u_k^{(4)} = f_k^{(4)}, k = 1, 2$ with special matrices. The coefficient matrices $A_k^{(3)}, A_k^{(4)}, k = 1, 2$ are not constant here.

The obtained SLAEs have been scaled and solved with CG method [9] afterwards. Here the scaling means that the elements of the coefficient matrices and (Right Hand Side) RHSs: $A_k^{(3)} = (a_{ij})_k^{MN}, f_k^{(3)}, k = 1, 2$ and $A_k^{(4)} = (a_{ij})_k^{MN}, f_k^{(4)}, k = 1, 2$ have the following form

$$\hat{a}_{ij} = \frac{a_{ij}}{\sqrt{a_{ii} a_{jj}}}, \ \hat{f}_i = \frac{f_i}{a_{ii}}, \ i, j = 1, 2, \ldots, MN.$$

The selection of the CG method is based on its acceptable calculation time [5] in comparison with the simple iteration method, the Seidel method, the minimal residual method and the steepest descent method.

To proceed, we note that from previous lemmas it can be infered the appropriateness of using CG method since

$$A_k^{(3)} = \left(A_k^{(3)}\right)^T, A_k^{(4)} = \left(A_k^{(4)}\right)^T, A_k^{(3)} > 0, A_k^{(4)} > 0, k = 1, 2.$$

3 Algorithm Parallel Scheme

Before passage to the description of the parallel algorithms we would like to say a few words about the library with the help of which the parallel algorithms have been realized and the computational platform on which they have been run.

The implementation of the parallel algorithms was carried out using C and MPI message passing system. Relative to the paradigm of message passing it can be noted that it is used widely on certain classes of parallel machines, especially those with distributed memory. One of representatives of this conception is MPI (Message Passing Interface) [3], [4], [6], [7]. The MPI standard is intended for supporting parallel applications working in terms of the message passing system and allowed to use its functions in C/C++ and Fortran 77/90.

All the computational experiments took place on a nCube 2S, a MIMD-computer, the interconnection network of which is represented as a hypercube. The number of processing elements (PEs) is 2^d, $d \leq 13$. These PEs are unified into a hypercube communication structure with maximum length of a communication line equaled d. A d-dimensional hypercube connects each of 2^d PEs to d other PEs. Such a communication scheme allows to transmit messages fast enough irrespective of computational process since each PE has a communication coprocessor aside from a computational processor. At our disposal we have the described system with 64 PEs.

The parallel algorithm for solving equations (3), (4), as mentioned above, is based on inherent parallelism which is suggested by Peaceman-Rachford difference scheme. The ADI method gives an opportunity to exploit any method to solve SLAEs obtained in $n + \frac{1}{2}$ and $n + 1$ time layers. In this paper we use the CG method. Along with it, application of Peaceman-Rachford difference scheme to equations (1) and (2) allows to find numerical solution of the SLAEs in each time layer independently, i.e. irrespective of communication process. The main communication loading lies on connection between two time layers. Thus, one simulation step of the algorithm to be executed requires two interchanges of data at passage to $n + \frac{1}{2}$ and to $n + 1$ time layers.

As mentioned before, the ADI method generates two SLAEs with special matrices. One of those matrices obtained in $n+\frac{1}{2}$ time layer is a band tridiagonal matrix and consequently can be transformed by means of permutation of rows to a block tridiagonal matrix, while the second one, obtained in $n+1$ time layer, is a block tridiagonal matrix originally.

Taking into account aforesaid one simulation step of the parallel algorithm, let us denote it **A**, for solving equation (3) with the SLAEs $A_k^{(3)} X_k^{(3)} = B_k^{(3)}$, $k = 1, 2$ can be represented in the following manner

1. Compute $B_1^{(3)}$ in $n + \frac{1}{2}$ time layer.
2. Make permutation of vectors $X_1^{(0)}$, $B_1^{(3)}$, where $X_1^{(0)}$ is an initial guess for CG method in $n + \frac{1}{2}$ time layer.
3. Solve equation $A_1^{(3)} X_1^{(3)} = B_1^{(3)}$ in $n + \frac{1}{2}$ time layer with CG method.
4. Compute $B_2^{(3)}$ in $n + 1$ time layer partially, i.e. without the last item of the second equation (3).

5. Make permutation of vectors $X_2^{(0)} = X_1^{(3)}$ and $B_2^{(3)}$, where $X_2^{(0)}$ is an initial guess for CG method in $n+1$ time layer.
6. Compute the missing item so as the computation of $B_2^{(3)}$ in $n+1$ time layer has been completed.
7. Solve equation $A_2^{(3)} X_2^{(3)} = B_2^{(3)}$ in $n+1$ time layer with CG method.

Let us consider the described algorithm in more detail. Suppose, we have p PEs and it has to solve a system of size $M \times N$. We proceed from the assumption that $\left\{\frac{M}{p}\right\} = 0$ and $\left\{\frac{N}{p}\right\} = 0$, where $\{x\}$ is a fractional part of number x, i.e. vectors $X_k^{(0)}$, $B_k^{(3)}$, $k = 1, 2$ are distributed uniformly.

First step of algorithm **A** is well-understood while the second one claims more attention. Let vectors $X_1^{(0)}$, $B_1^{(3)}$ be matrices distributed in the rowwise manner (these matrices consist of elements of corresponding vectors), then to solve equation $A_1^{(3)} X_1^{(3)} = B_1^{(3)}$ in $n + \frac{1}{2}$ time layer with CG method in parallel we need to transpose the matrix corresponding to vector $B_1^{(3)} = \left\{b_1^{(3)}, b_2^{(3)}, \dots, b_{MN}^{(3)}\right\}$

$$\begin{pmatrix} b_1^{(3)} & b_2^{(3)} & \dots & b_N^{(3)} \\ b_{N+1}^{(3)} & b_{N+2}^{(3)} & \dots & b_{2N}^{(3)} \\ \dots & \dots & \dots & \dots \\ b_{(M-1)N+1}^{(3)} & b_{(M-1)N+2}^{(3)} & \dots & b_{MN}^{(3)} \end{pmatrix} \to \begin{pmatrix} b_1^{(3)} & b_{N+1}^{(3)} & \dots & b_{(M-1)N+1}^{(3)} \\ b_2^{(3)} & b_{N+2}^{(3)} & \dots & b_{(M-1)N+1}^{(3)} \\ \dots & \dots & \dots & \dots \\ b_N^{(3)} & b_{2N}^{(3)} & \dots & b_{MN}^{(3)} \end{pmatrix}$$

and the matrix corresponding to vector $X_1^{(0)} = \left\{x_1^{(0)}, x_2^{(0)}, \dots, x_{MN}^{(0)}\right\}$

$$\begin{pmatrix} x_1^{(0)} & x_2^{(0)} & \dots & x_N^{(0)} \\ x_{N+1}^{(0)} & x_{N+2}^{(0)} & \dots & x_{2N}^{(0)} \\ \dots & \dots & \dots & \dots \\ x_{(M-1)N+1}^{(0)} & x_{(M-1)N+2}^{(0)} & \dots & x_{MN}^{(0)} \end{pmatrix} \to \begin{pmatrix} x_1^{(0)} & x_{N+1}^{(0)} & \dots & x_{(M-1)N+1}^{(0)} \\ x_2^{(0)} & x_{N+2}^{(0)} & \dots & x_{(M-1)N+1}^{(0)} \\ \dots & \dots & \dots & \dots \\ x_N^{(0)} & x_{2N}^{(0)} & \dots & x_{MN}^{(0)} \end{pmatrix}.$$

Of course, such a transposition requires transmission of some sub-matrices of size $\frac{M}{p} \times \frac{N}{p}$ to the corresponding PEs. Thus, the number of send/receive operations $C_{s/r}^A$ and the amount of transmitted data C_t^A are

$$C_{s/r}^A = 2p(p-1), \quad C_t^A = 2M\left(N - \frac{N}{p}\right).$$

Further, in accordance with the algorithm **A**, to avoid extra communications we compute vector $B_2^{(3)}$ partially and then permute vectors $X_2^{(0)} = X_1^{(3)}$ and $B_2^{(3)}$ as above. Afterwards, we complete the computation of vector $B_2^{(3)}$ (its missing item) and solve equation $A_2^{(3)} X_2^{(3)} = B_2^{(3)}$ in $n+1$ time layer with CG method in parallel. By resuming aforesaid one simulation step of algorithm **A** to be run requires

$$C_{sr}^A = 4p(p-1), \quad C_t^A = 4M\left(N - \frac{N}{p}\right),$$

$$C_c^A = \frac{N}{p}\left((19m-6)\left(I_{CG^A}^{n+\frac{1}{2}} + I_{CG^A}^{n+1}\right) + 46M - \frac{8M}{p} + 14\right) + 16M + 4.$$

Here $I_{CG^A}^{n+\frac{1}{2}}$ and $I_{CG^A}^{n+1}$ represent the number of iterations of CG method in solving system (3) in $n+\frac{1}{2}$ and $n+1$ time layers, and C_c^A is a computational complexity of the algorithm.

The parallel algorithm, let us denote it **B**, for solving equation (4) with the SLAEs $A_k^{(4)} X_k^{(4)} = B_k^{(4)}$, $k = 1, 2$ is realized similarly

1. Compute the value $h_{av} = \frac{1}{MN} \sum_{i=1}^{M} \sum_{j=1}^{N} h_{ij}$.
2. Compute $B_1^{(4)}$ in $n + \frac{1}{2}$ time layer.
3. Make permutation of vectors $\tilde{X}_1^{(0)}$, $B_1^{(4)}$, where $\tilde{X}_1^{(0)}$ is an initial guess for CG method in $n + \frac{1}{2}$ time layer.
4. Solve equation $A_1^{(4)} X_1^{(4)} = B_1^{(4)}$ in $n + \frac{1}{2}$ time layer with CG method.
5. Compute $B_2^{(4)}$ in $n + 1$ time layer partially, i.e. without the last item of the second equation (4).
6. Make permutation of vectors $\tilde{X}_2^{(0)} = X_1^{(4)}$ and $B_2^{(4)}$, where $\tilde{X}_2^{(0)}$ is an initial guess for CG method in $n + 1$ time layer.
7. Compute the missing item so as the computation of $B_2^{(4)}$ in $n+1$ time layer has been completed.
8. Solve equation $A_2^{(4)} X_2^{(4)} = B_2^{(4)}$ in $n + 1$ time layer with CG method.

One difference with regard to algorithm **A** is the computation of the value h_{av} which involves extra communication: to gather all the values h_{ij} from all the PEs, sum and broadcast them to all the PEs afterwards. As above, by denoting the number of send/receive operations, the amount of transmitted data and computational complexity of the algorithm by $C_{s/r}^B$, C_t^B and C_c^B we get that one step of algorithm **B** to be executed requires

$$C_{s/r}^B = 2p + 4p(p-1), \; C_t^B = 2p + 4M\left(N - \frac{N}{p}\right),$$

$$C_c^B = \frac{N}{p}\left((18m-3)\left(I_{CG^B}^{n+\frac{1}{2}} + I_{CG^B}^{n+1}\right) + 47M - 4\right) + 12M + 31,$$

where $I_{CG^B}^{n+\frac{1}{2}}$ and $I_{CG^B}^{n+1}$ are the number of iterations of CG method in solving system (4) in $n + \frac{1}{2}$ and $n + 1$ time layers.

As it can be seen from above, algorithm **A** is 1.05 time slower than algorithm **B** with regard to computational complexity for

$$\lim_{M\to\infty} \frac{C_c^A}{C_c^B} = 1.05, \; N = I_{CG^A}^{n+\frac{1}{2}} = I_{CG^A}^{n+1} = I_{CG^B}^{n+\frac{1}{2}} = I_{CG^B}^{n+1} = M, \; p = 1,$$

but we cannot say the same about overall communication time of the algorithms because of their difference in the communication schemes.

As a result, the ratio $D = \frac{T_p^A}{T_p^B}$ (the values T_p^A, T_p^B are computational time of algorithms **A** and **B** respectively) may be a little lower or higher than 1.05 for there is a slight difference in communication schemes and program realizations of both algorithms (in some places of the programs cycles or memory-to-memory copy operations may be different and so on). Besides, D will be decreasing as the number of available PEs increases, and the number of iterations is constant. That is confirmed by Fig. 1 below. Such diminution is explained by the increasing of negative influence of the global sum operation the communication time of which will be increasing as the number of PEs grows.

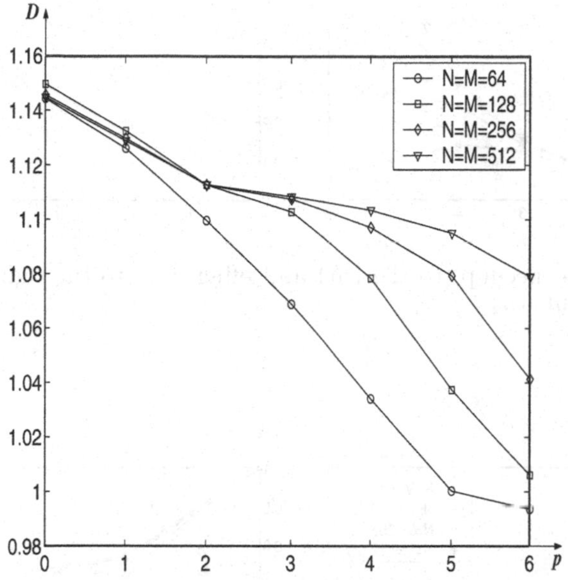

Fig. 1. The ratio $D = \frac{T_p^A}{T_p^B}$ $\left(I_{CG^A}^{n+\frac{1}{2}} = I_{CG^A}^{n+1} = I_{CG^B}^{n+\frac{1}{2}} = I_{CG^B}^{n+1} = 1\right)$ for various grid sizes and processors number.

Apart from it, ratio $D \to 1.05 \pm \varepsilon$ (ε is an error depending on program realization) as the number of available PEs is constant, and the iteration number increases. Such state of affairs connects with insignificant influence of the global sum operation in increasing the number of iterations.

At this we finish the description of the algorithms and consider some test experiments all of which are given for one simulation step of the algorithms and for $I_{CG^A}^{n+\frac{1}{2}} = I_{CG^A}^{n+1} = I_{CG^B}^{n+\frac{1}{2}} = I_{CG^B}^{n+1} = 1$. The horizontal axis, in all the pictures, is $2^p, p = 0, 1, \ldots, 6$.

By following [10] let us consider relative efficiency and speedup

$$S_p = \frac{T_1}{T_p}, \quad E_p = \frac{S_p}{p},$$

where T_p is a time to run a parallel algorithm on a computer with p PEs ($p > 1$), T_1 is a time to run a sequential algorithm on one PE of the same computer.

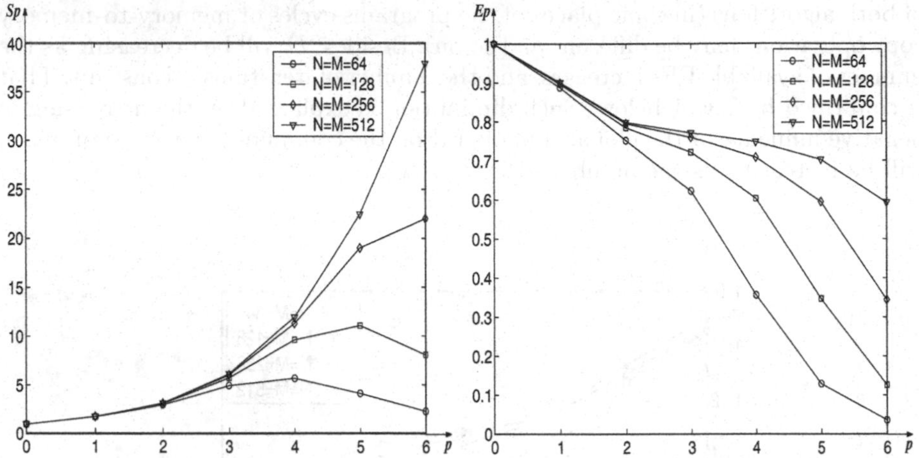

Fig. 2. Relative speedup (to the left) and efficiency (to the right) of algorithm **A** at various grid sizes.

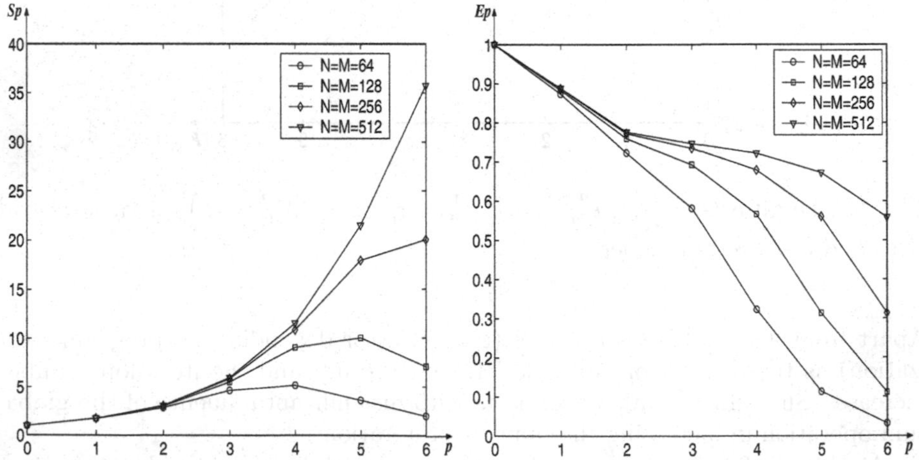

Fig. 3. Relative speedup (to the left) and efficiency (to the right) of algorithm **B** at various grid sizes.

As we can see from Fig.2 and Fig.3 relative speedup and efficiency are satisfactory even for the grid of size $N = M = 512$.

4 Conclusion

In the paper we have obtained the parallel ADI method for the non-linear equation with computational complexity almost equaled $\left(\lim_{M\to\infty} \frac{C_c^A}{C_c^B} = 1.05\right)$ to the computational complexity of the parallel ADI method for the linear equation. That will allow to find solutions of equations (1) and (2) in the same time if $I_{CG^A}^{n+\frac{1}{2}} = I_{CG^A}^{n+1} = I_{CG^B}^{n+\frac{1}{2}} = I_{CG^B}^{n+1} = N$.

Generally speaking, for real problems solved by the described algorithms, the algorithm **B** will be faster than algorithm **A** since, as a rule, the number of iterations required for one simulation step of algorithm **B** is smaller than for one simulation step of algorithm **A**, but this difference in overall computational times of the algorithms is minimal.

References

1. Barrett, R., Berry, M., Chan, T.F., Demmel, J., Donato, J.M., Dongarra, J., Eijkhout, V., Pozo, R., Romine, C., Henk Van der Vorst: Templates for the Solution of Linear Systems: Building Blocks for Iterative Methods, `http://www.netlib.org/templates/Templates.html`
2. Bear, J., Zaslavsky, D., Irmay, S.: Physical principles of water percolation and seepage. UNESCO, (1968)
3. Foster, I.: Designing and Building Parallel Programs: Concepts and Tools for Parallel Software Engineering, Addison-Wesley Pub. Co., (1995)
4. Gropp, W., Lusk, E., Skjellum, A., Rajeev, T.: Using MPI: Portable Parallel Programming With the Message-Passing Interface, Mit press, (1999)
5. Krukier L.A., Schevtschenko I.V.: Modeling Gravitational Flow of Ground Water. Proceedings of the Eighth All-Russian Conference on Modern Problems of Mathematical Modeling, Durso, Russia, September 6-12, RSU Press, (1999), 125-130
6. *MPI: A Message-Passing Interface Standard, Message Passing Interface Forum*, (1994)
7. Pacheco, P.: Parallel Programming With MPI, Morgan Kaufamnn Publishers, (1996)
8. Peaceman, D., and Rachford, J. H.H.: The numerical solution of parabolic and elliptic differential equations. J. Soc. Indust. Appl. Math., No.3 (1955), 28-41
9. Samarskii, A.A., and Goolin, A.V.: Numerical Methods, Main Editorial Bord for Physical and Mathematical Literature, (1989)
10. Wallach, Y.: Alternating Sequential/Parallel Processing, Springer-Verlag, (1982)

Study of the Parallel Block One-Sided Jacobi Method*

E.M. Daoudi, A. Lakhouaja, and H. Outada

University of Mohamed First, Faculty of Sciences
Department of Mathematics and Computer Science
LaRI Laboratory, 60 000 Oujda, Morocco
{mdaoudi,lakhouaja,outada}@sciences.univ-oujda.ac.ma

Abstract. In this paper, we study the parallelization of the one-sided Jacobi method for computing the eigenvalues and the eigenvectors of a real and symmetric matrix. We use a technique to overlap the communications by the computations in order to decrease the global communication time. We also extend the obtained results to the block version for using the level-3 BLAS.
Keywords: Eigenvalue problem; Jacobi method; Parallel algorithm.

1 Introduction

The parallelization of the one-sided Jacobi method [1] for computing the eigenvalues and the eigenvectors of a real and symmetric matrix A of size n, needs less communications than the classical Jacobi method ("*two-sided Jacobi method*" [6]). It needs only local communications (column translation) between processors in contrast to the two-sided Jacobi which needs, in addition, global communications (rotations broadcast) [2,3]. In this work, we study, on a distributed memory architecture composed of p processors denoted by P_m with $0 \leq m \leq p-1$, the parallelization of the one-sided Jacobi method using a technique to overlap the communications by the computations in order to decrease the global communication time. We also present a parallel block version for using level-3 BLAS.

We consider a distributed memory architecture composed of p processors denoted by P_m with $0 \leq m \leq p-1$. The cost of communicating L data between two neighbor processors is modeled by $\beta + L\tau$, where β is the start-up time and τ is the time to transmit one data.

This paper is organized as follow: in section 2, we present the sequential algorithm, section 3 is devoted to the parallelization. In section 4, we present a block parallel version.

2 Sequential Algorithm

The basic idea of the Jacobi method [6] for computing the eigenvalues and eigenvectors of a real symmetric matrix A of size n, consists in constructing a sequence

* This work is supported by the European program INCO-DC, "DAPPI" Project.

B. Hertzberger et al. (Eds.): HPCN Europe 2001, LNCS 2110, pp. 454–463, 2001.

of symmetric matrices $\{A^{(k)}\}$, which converges, according to the Frobenius norm, to a diagonal matrix D which diagonal elements are the eigenvalues of A, by means of:

$$A^{(k+1)} = {J^{(k)}}^\top A^{(k)} J^{(k)} \quad \text{for } k = 1, 2, \cdots \tag{1}$$

where $A^{(1)} = A$ and $J^{(k)} = J(i,j)$, for $1 \leq j < i \leq n$, is a Jacobi rotation, in the (i,j) plane, chosen in order to annihilate $a_{i,j}^{(k)} = a_{j,i}^{(k)}$. It is completely determined by $a_{i,i}^{(k)}$, $a_{j,j}^{(k)}$ and $a_{i,j}^{(k)}$ and each product ${J^{(k)}}^\top A^{(k)} J^{(k)}$ only modifies the columns and rows i and j of $A^{(k)}$. By symmetry, only the columns or the rows are updated, which requires $6n flops$.

The eigenvectors are obtained by constructing the sequence of matrices $\{U^{(k)}\}$, which converges to a matrix which contains the eigenvectors, by the relation:

$$U^{(k)} = \prod_{i=1}^{k-1} J^{(i)}$$

A cyclic Jacobi method consists in applying successive sweeps until convergence is reached, where a sweep is a set of $n(n-1)/2$ distinct rotations which annihilate all non-diagonal elements of the lower triangular part of A. The different ways of choosing (i,j) define an ordering [5,7]. On figure 1, we present the steps of one sweep for the odd-even ordering [5] which is used in our work, where the indices (i,j) indicate the rotation $J(i,j)$. Each sweep needs n steps.

$$\begin{aligned}
&\textit{Step } 1 : (1,2), (3,4), (5,6)\\
&\textit{Step } 2 : 2, (1,4), (3,6), 5\\
&\textit{Step } 3 : (2,4), (1,6), (3,5)\\
&\textit{Step } 4 : 4, (2,6), (1,5), 3\\
&\textit{Step } 5 : (4,6), (2,5), (1,3)\\
&\textit{Step } 6 : 6, (4,5), (2,3), 1
\end{aligned}$$

Fig. 1. One Sweep of the Odd-Even Ordering for $n = 6$.

The idea of the one-sided Jacobi consists in transforming the relation (1) in the following manner:

$$\begin{aligned}
A^{(k)} &= {J^{(k-1)}}^\top A^{(k-1)} J^{(k-1)}\\
&= (\prod_{i=1}^{k-1} J^{(i)})^\top A (\prod_{i=1}^{k-1} J^{(i)})\\
&= {U^{(k)}}^\top \bar{A}^{(k)}
\end{aligned}$$

with $\bar{A}^{(k)} = A \prod_{i=1}^{k-1} J^{(i)}$ and $U^{(k)} = \prod_{i=1}^{k-1} J^{(i)}$.

Then, instead of constructing the sequences $\{A^{(k)}\}$ and $\{U^{(k)}\}$, we construct the sequences $\{\bar{A}^{(k)}\}$ and $\{U^{(k)}\}$, with:

$$\begin{cases} \bar{A}^{(1)} = A \text{ and } \bar{A}^{(k+1)} = \bar{A}^{(k)} J^{(k)} \text{ for } k \geq 1 \\ U^{(1)} = I_n \text{ and } U^{(k+1)} = U^{(k)} J^{(k)} \text{ for } k \geq 1 \end{cases}$$

where I_n is the unitary matrix of size n.

The determination of $J^{(k)} = J(i,j)$, needs the knowledge of $a_{i,i}^{(k)}$, $a_{j,j}^{(k)}$ and $a_{i,j}^{(k)}$. Since we only know $\bar{A}^{(k)} = (\bar{A}_1^{(k)}, \cdots, \bar{A}_n^{(k)})$ and $U^{(k)} = (U_1^{(k)}, \cdots, U_n^{(k)})$, these coefficients can be computed by using the following relation:

$$a_{m,l}^{(k)} = (U_m^{(k)})^\top \bar{A}_l^{(k)} \quad \text{for all} \quad 1 \leq m, l \leq n$$

which requires $6n flops$.

At each step k of the algorithm, we compute one rotation, which needs the computation of the parameters c and s and the updates of two columns of $\bar{A}^{(k)}$ and two columns of $U^{(k)}$. Then the computational time for the sequential one-sided Jacobi method, by sweep, is $\frac{n(n-1)}{2}(18n + cte)\omega$ (ω is the cost of an arithmetic operation performed using the level-1 BLAS).

3 Study of the Parallelization

Since each rotation $J(i,j)$, for $1 \leq j < i \leq n$, only modifies the columns and rows i and j of A, and since:

$$J(i,j)J(i',j') = J(i',j')J(i,j) \text{ if } i,j \neq i',j'$$

we deduce that $J(i,j)$ and $J(i',j')$ are disjoint, then at most $\frac{n}{2}$ rotations can be computed simultaneously.

We consider a column distribution of A (figure 2), which consists in assigning at each processor, $\frac{n}{p}$ consecutive columns of A and $\frac{n}{p}$ consecutive columns of U.

For the parallelization:

p 1: each processor computes all possible rotations and updates its local data:

Since each processor P_m, for $0 \leq m \leq p-1$, contains $\frac{n}{p}$ consecutive columns of A noted $A_{m1(j)}$, $A_{m2(j)}$, and $\frac{n}{p}$ consecutive columns of U noted $U_{m1(j)}$ and $U_{m2(j)}$ for $1 \leq j \leq \frac{n}{2p}$,

- it computes, during the first step, the rotations $J(m1(i), m2(j))$ for $1 \leq i,j \leq \frac{n}{2p}$, the rotations $J(m1(i), m1(j))$ for $1 \leq i < j \leq \frac{n}{2p}$ and the rotations $J(m2(i), m2(j))$ for $1 \leq i < j \leq \frac{n}{2p}$, (figure 2(a)) which correspond to $\frac{\frac{n}{p}(\frac{n}{p}-1)}{2}$ rotations.

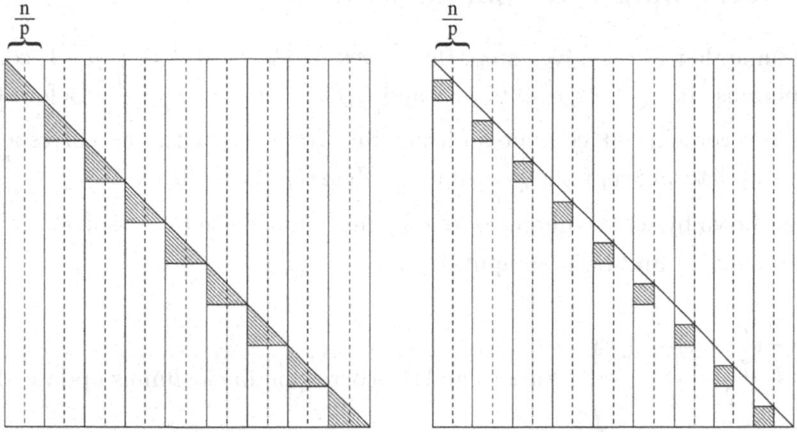

Fig. 2. Computation of the Rotations.

- it computes, during the other steps, the rotations $J(m1(i), m2(j))$ for $1 \leq i, j \leq \frac{n}{2p}$ which correspond to $(\frac{n}{2p})^2$ rotations (figure 2(b)), since a rotation of the form $J(m1(i), m1(j))$ or $J(m2(i), m2(j))$ is computed in the first step.

We deduce that the parallel computation time, by sweep, is:

$$T_{comp} = \frac{n(n-1)}{2p}(18n + cte)\omega$$

p 2: exchanges columns with other processors in order to complete a sweep according to the chosen ordering:
Consider the case where the communications are not overlapped. At each step, each processor communicate only with its two neighbors in order to send and receive $\frac{n}{2p}$ columns of A and $\frac{n}{2p}$ columns of U, each of size n. Then the communication time, T_{comm}, by sweep is:

$$T_{comm} = 2p(\beta + \frac{n^2}{p}\tau)$$

The best strategy is that which overlap efficiently the communications by the computations in a efficiency manner which depends on the machine parameters, τ, β, ω and the size of the matrix.

In the following section, we give an algorithm which permits to overlap the communications by the computations.

4 Overlapping the Communications

We assume that at one given step, the processor P_m for $0 \leq m \leq p-1$, contains the columns $\bar{A}_{m1(j)}$, $\bar{A}_{m2(j)}$, $U_{m1(j)}$ and $U_{m2(j)}$ for $1 \leq j \leq \frac{n}{2p}$. Before translating the columns, it computes all possible rotations with the corresponding column updates, except the $\frac{n}{2p}$ rotations $J(m1(j), m2(j))$, for $1 \leq j \leq \frac{n}{2p}$ (rotations shown by white squares in the figure 3). Since the processor P_m contains $\frac{n}{p}$ columns of A and U, it computes:

- $\left(\frac{\frac{n}{p}(\frac{n}{p}-1)}{2} - \frac{n}{2p}\right)$ rotations and the corresponding columns update, during the first step (rotations shown by the black squares in the figure 3(a)).
- $\left((\frac{n}{2p})^2 - \frac{n}{2p}\right)$ rotations and the corresponding columns update, during the other steps (figure 3(b)).

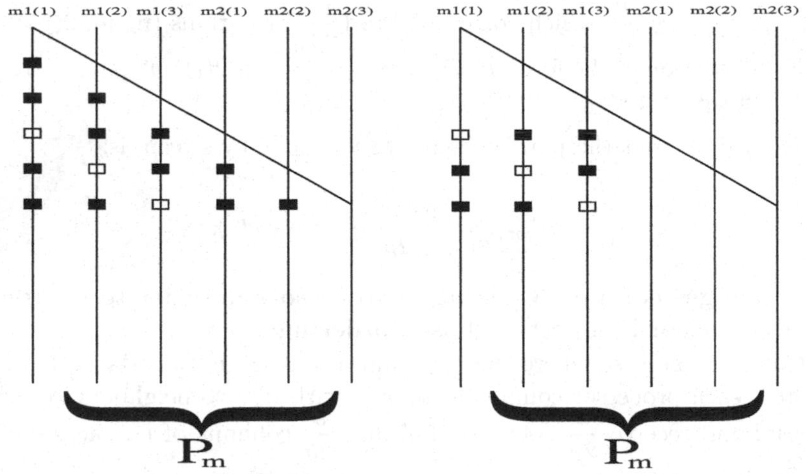

(a) step 1 (b) other steps

Fig. 3. Computation of the Rotations.

The translation will be performed during the computation of the remaining rotations $J(m1(j), m2(j))$, for $1 \leq j \leq \frac{n}{2p}$, in order to overlap the communication by the computation. The strategy is as follow:

For $1 \leq j \leq \frac{n}{2p}$, each processor:

- computes $J(m1(j), m2(j))$ and updates $\bar{A}_{m2(j)}$ and $U_{m2(j)}$, then, simultaneously:
  - updates $\bar{A}_{m1(j)}$ and $U_{m1(j)}$
  - sends $\bar{A}_{m2(j)}$ and $U_{m2(j)}$ to its neighbor processor.

The update of one column needs $3n\omega$ and the computation of one rotation needs $6n\omega$, while the communication of two columns needs $\beta + 2n\tau$.

Lemma 1. *If $6n\omega \geq \beta + 2n\tau$, all the communications can be overlapped by the computations.*

5 Experimental Results

In figure 4, we give the execution time by sweep of the implementations done under the PVM environment on the parallel distributed memory machine TN310. In the sequential case, because of memory limitations, we have only given the tests for matrix sizes less than 512 . In order to improve the technique used by decomposing the columns in blocks of optimal sizes [4].

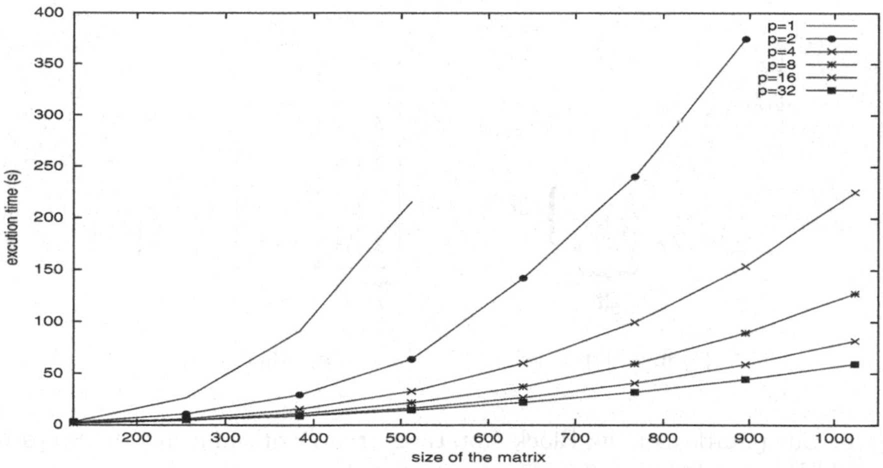

Fig. 4. Execution time in Seconds.

Remark: For our system (TN310), $3\omega - \tau < 0$, ($\beta = 0.002s$, $\tau = 22\mu s$ and $\omega = 0.25\mu s$) which implies that for this strategy we can not overlap all the communications. To do this, we have used a decomposition of the matrix by blocks (section 6).

6 Block Version

In this section, we present a block version which allows exploiting the level-3 BLAS. To do this, the matrix A is decomposed into k^2 square blocks $A_{i,j}$ ($k = \frac{n}{r}$), each one of size r, as follow:

$$A = \begin{pmatrix} A_{1,1} & \cdots & A_{1,k} \\ \vdots & \ddots & \vdots \\ A_{k,1} & \cdots & A_{k,k} \end{pmatrix}$$

Each block rotation $J(i,j)$ is computed from the block $\begin{pmatrix} A_{i,i} & A_{i,j} \\ A_{j,i} & A_{j,j} \end{pmatrix}$, and consists in applying:

- during the first step, $\frac{2r(2r-1)}{2}$ elementary rotations, if $j = i + 1$ (diagonal blocks), which forms one sweep of the sequential Jacobi algorithm applied to a square block of size $2r$ (figure 5(a)).
- during the other steps, r^2 elementary rotations, if $j > i + 1$ (subdiagonal blocks), to the block $A_{j,i}$ (figure 5(b)).

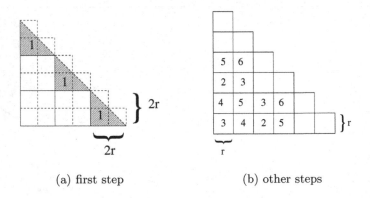

(a) first step (b) other steps

Fig. 5. Computation of the Block Rotations (the Digits Indicate the Steps of the Odd-Even Ordering).

This block decomposition of the matrix A can be exploited for using the level-3 BLAS, since the elementary rotations can be accumulated in a square matrix $J_{i,j}$ of size $2r$ and then, this matrix can be used to update the corresponding columns of the matrices $\bar{A}$ and U, using level-3 BLAS as follow:

$$(\bar{A}_i \ \bar{A}_j) = (\bar{A}_i \ \bar{A}_j)\, J_{ij}$$

$$(U_i \, U_j) = (U_i \, U_j) \, J_{ij}$$

with $\bar{A}$ and U decomposed in rectangular blocks of size $n \times r$ ($\bar{A} = (\bar{A}_1, \cdots, \bar{A}_k)$ and $U = (U_1, \cdots, U_k)$).

The determination of $J_{i,j}$, needs the knowledge of $A_{i,i}$, $A_{j,j}$ and $A_{i,j}$; since only the coefficients of $\bar{A}$ and U are known, these parameters can be computed by using the following relation:

$$A_{ij} = U_i^\top \bar{A}_j \qquad \text{for} \quad i, j = 1, \cdots, k$$

6.1 Parallelization

At each processor we assign $\frac{k}{p}$ consecutive blocks of $\bar{A}$ and U. At each step, each processor executes the two following phases:

i. computes $\dfrac{\frac{k}{p}(\frac{k}{p} - 1)}{2} = \dfrac{k(k-p)}{2p^2}$ block rotations during the first step and $(\frac{k}{2p})^2$ during the other steps and updates the corresponding columns.

ii. translates $\frac{k}{2p}$ block columns each of size $n \times r$

Computation

- The computation of A_{ii}, A_{ij}, and A_{jj}, needs $6nr^2 flops$
- The computation of the block rotation needs $O(r^3) flops$.
- The updates of $\bar{A}$ and U needs $16r^2 n flops$.

We deduce that the computational time by sweep is:

$$\begin{aligned} T_{comp} &= T_{comp}(step\ 1) + (2p-1) T_{comp}(step\ i) \quad i \neq 1 \\ &= (\frac{k(k-p)}{2p^2} + (2p-1)(\frac{k}{2p})^2)(22nr^2\omega_3 + O(r^3)\omega) \\ &= (\frac{11}{2}\frac{n^3}{p^2} + 11\frac{n^3}{p} - 11\frac{n^2 r}{p})\omega_3 + (\frac{n^2}{p^2}O(r) + \frac{n^2}{p}O(r) + \frac{n}{p}O(r^2))\omega \end{aligned}$$

Communication

At each step, each processor communicates only with its two neighbors in order to send and receive $\frac{k}{2p}$ columns of $\bar{A}$ and $\frac{k}{2p}$ columns of U, each of size $n \times r$. Then the communication time, by sweep is:

$$\begin{aligned} T_{comm} &= 2p(\beta + 2\frac{n^2}{2p}\tau) \\ &= 2p\beta + 2n^2\tau \end{aligned}$$

6.2 Overlapping the Communications

We adopt the same technique as for the point case. We suppose that at one given step, the processor P_m, for $0 \leq m \leq p-1$, contains the block columns $\bar{A}_{m1(j)}$, $\bar{A}_{m2(j)}$, $U_{m1(j)}$ and $U_{m2(j)}$ for $0 \leq j \leq \frac{k}{2p}$. Before translating the columns, it computes, all possible rotations with the corresponding columns updates, except the $\frac{k}{2p}$ rotations $J(m1(j), m2(j))$, for $1 \leq j \leq \frac{k}{2p}$.

For the computation of the rotations $J(m1(j), m2(j))$, $1 \leq j \leq \frac{k}{2p}$, the updates of the corresponding columns and the translation, we adopt the following strategy:

For $1 \leq j \leq \frac{k}{2p}$, each processor:

- computes $J(m1(j), m2(j))$ and updates $\bar{A}_{m2(j)}$ and $U_{m2(j)}$ and simultaneously:
  - updates $\bar{A}_{m1(j)}$ and $U_{m1(j)}$
  - sends $\bar{A}_{m2(j)}$ and $U_{m2(j)}$ to its neighbor processor.

The updates of one block columns needs $(4nr^2 - nr)\omega_3$, while the communication of two block columns, each of size $n \times r$ needs $\beta + 2nr\tau$.

Lemma 2. *If $(8nr^2 - 2nr)\omega_3 \geq \beta + 2nr\tau$, i.e. $2nr((4r-1)\omega_3 - \tau) \geq \beta$, all the communications can be overlapped by the computations.*

6.3 Experimental Results

In figure 6, we give the execution times of the simulations using PVM environment done on a PC-Linux. The tests are not done on the TN310 since the level-3 BLAS libraries are not implemented on the TN310.

7 Conclusion

In this paper, we have proposed two techniques for overlapping of the communications by the computations in the parallelization of the point and block one-sided Jacobi methods. Our future work is to generalize these techniques.

References

1. J. Cuenca, D. Giménez, *Implementation of parallel one-sided block Jacobi methods for the symmetric eigenvalue problem,* ParCo'99, Delft (1999).
2. E.M. Daoudi, A. Lakhouaja, *Exploiting the symmetry in the parallelization of the Jacobi method,* Parallel Computing 23 (1997) 137-151.
3. E.M. Daoudi, A. Lakhouaja, *New parallel block distribution for Jacobi method,* Research Report, Faculty of Sciences, LaRI, Oujda, 2001.

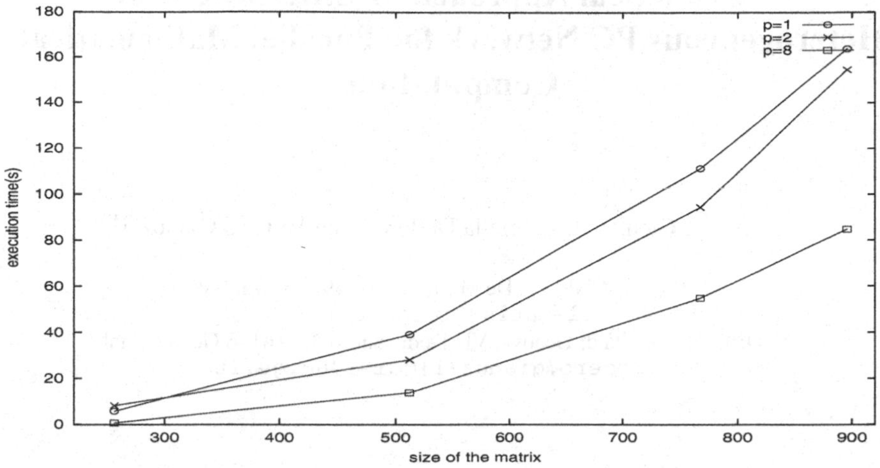

Fig. 6. Execution Times in Seconds for the Simulations.

4. F. Desprez, *Procdures de bases pour le calcul scientifique sur machines parallles mmoire distribue,* PhD thesis, Institut National Polytechnique de Grenoble, France (1994).
5. P.J. Eberlein, *On using the Jacobi method on the hypercube,* In Heath, M.T.(Ed.).Proc. of the Second Conference on Hypercube Multiprocessors (1987) 605-611.
6. G.H. Golub, C.F. Van Loan, *Matrix computation,* Johns Hopkins University Press, 2nd edition (1989).
7. F. T. Luck, H. Park, *On parallel Jacobi orderings,* SIAM J.Sci Stat. Comp. 10 (1989) 18-26.

A Practical Approach to Efficient Use of Heterogeneous PC Network for Parallel Mathematical Computation

Andrea Clematis[1], Gabriella Dodero[2], and Vittoria Gianuzzi[2]

[1] IMA -CNR Via De Marini 6, 16149 Genova, Italy
clematis@ima.ge.cnr.it
[2] DISI, Università di Genova, Via Dodecaneso 35, 16146 Genova, Italy
{dodero,gianuzzi}@disi.unige.it

Abstract. This paper presents an experience in use of a parallel mathematical library, ScaLAPACK, on a network composed by heterogeneous workstations. The good performance results have been obtained by means of a distributed programming environment which is able to dynamically evaluate available computing power at each workstation and to distribute accordingly the set of parallel processes.

1 Introduction

Networks of workstations (NoWs), mainly based on low-cost, PC-based systems, are becoming increasingly performing both for computational power and for communication speed: the interest arises from large scale computations which can hardly be solved on a local resource like a single workstation.

Thus, we considered the possibility of providing an easy support to adapt library functions, originally developed for parallel machines, to cluster of Workstations, minimizing the efforts required for such adaptation. In particular, the integration of the run-time allocation system PINCO, developed at DISI (University) and IMA (CNR) of Genova [5] with the mathematical library ScaLAPACK will be discussed.

As we shall see in the next section, ScaLAPACK received great attention mainly on homogeneous processor clusters, completely devoted to parallel computation. However, as the experience has shown, the progress in computer technology is very fast, and a workstation can become obsolete in a very short time. It is usual to have laboratories composed by architecturally similar workstations, incrementally growing with new hosts, with higher speed and wider memory size, as soon as it is possible to upgrade at least a part of them. At DISI, as an example, we have 40-50 Intel PC-based workstations with different configurations, all running some Linux implementation. To exploit the computational power of such a not completely homogeneous cluster we needed (and developed) a distributed programming environment with a specialized competence in scaling and locating parallel applications, able to maintain load balanc-

B. Hertzberger et al. (Eds.) : HPCN Europe 2001, LNCS 2110, pp. 464-473, 2001.

ing among heterogeneous workstations, in order to obtain the maximum efficiency yet preserving the simplicity of using existing code. This system will be presented in Section 3.

Section 4 describes our experiment in executing a typical library function, which solves the symmetric eigenvalues problem, on an heterogeneous workstation cluster, showing actual execution timings. Comparisons with related works and conclusive remarks are presented in Section 5.

2 ScaLAPACK on Homogeneous Clusters

ScaLAPACK, Scalable Linear Algebra PACKage, or Scalable LAPACK, is a public-domain portable library of high-performance linear algebra routines for distributed-memory message-passing MIMD computers and networks of workstations supporting PVM and/or MPI communication platforms. It has been developed by several institutions: Oak Ridge National Laboratory, Rice University, University of California at Berkeley, University of California at Los Angeles, University of Illinois and University of Tennessee, Knoxville [7]. It contains routines for solving systems of linear equations, least squares problems, and eigenvalue problems, and also handles many associated computations such as matrix factorisations or estimating condition numbers. It is currently written mainly in Fortran 77 in a Single Program Multiple Data (SPMD) style.

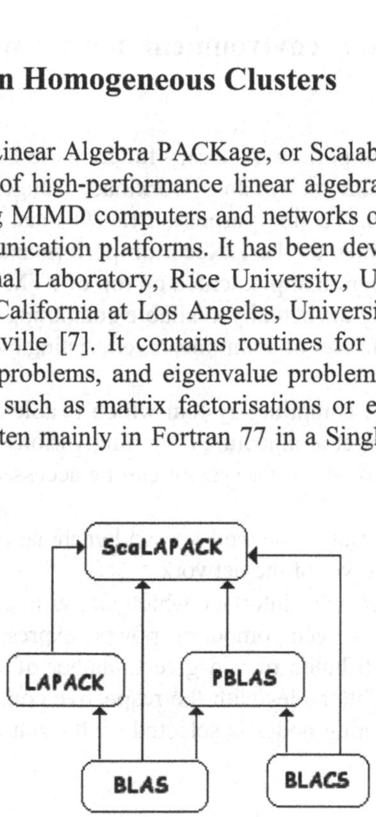

Fig. 1. ScaLAPACK Hierarchical View.

ScaLAPACK has been incrementally built starting from two standard libraries called the BLAS, or Basic Linear Algebra Subprograms (executed on a single processor), and BLACS, or Basic Linear Algebra Communication Subprograms, providing the needed communication routines. Machine dependencies are limited to these two libraries. At a higher level, PBLAS (Parallel BLAS) library provides the fundamental building blocks of the ScaLAPACK: it is a distributed-memory versions of the BLAS (see Figure 1).

It is possible to download the source of the library for many parallel machines and for Unix based workstation clusters. In the ScaLAPACK Guide [10] there are some suggestions to achieve high performance on NoWs. For example, it is required that all processors should be similar in architecture and performance, since ScaLAPACK is limited by the slowest processor, and only one process must be started on each processor. The last requirement follows immediately from the fact that, if the net is homogeneous, it makes no sense to have more than one process per processor, since that would only increase the communication and context switch times.

3 The Programming envIronment for Network of COmputers (PINCO)

PINCO aims to support the porting of library functions to heterogeneous networks of workstations: adapting them to dynamic workload changes requires only minimal source code modifications. To this purpose, PINCO must be able to monitor each node's workload, to guarantee the best execution performance. The run-time environment is a C distributed program implemented on top of PVM [12], which continuously monitors the net, evaluating the actually available computing power.

PINCO has also a "static" set of components, consisting on:

- A simple but effective compiling system which is able to automatically generate code for any set of target architectures, running various UNIX-like OS. The only assumption is that a common file system can be accessed by all nodes in the network.
- A set of programs running some well-known benchmarks, in order to evaluate the total raw computing power of the network.
- An application programmer's interface which allows to a process to start a parallel execution; to request a given computing power, expressed in normalized unit of measure, possibly distributed over a given number of workstations. The system returns a list of available nodes with the respective computing power. Allocation of processes to computing nodes is selected by the starting process and executed by the system.

Considering that PINCO run time environment is made by heterogeneous nodes and that the code distribution is decided in a dynamic way, charging the user with the task of recompiling code for each different target architecture he/she wants to include in the parallel machine, and placing each copy of executable code in the proper directory, constitutes a tedious task, especially when the number of heterogeneous architectures is greater than two: PINCO may save the user a lot of time and many possible mistakes by means of its compilation environment.

The user has only to set two parameters: a variable which indicates the target application (the PINCO_ROOT environment variable in Unix systems), and a configuration file which contains the names of selected target architectures.

This situation allows implementers to tune their parallel code by experiencing with different allocation strategies. However PINCO provides by default a simple scheduler which performs automatic allocation of tasks to processors without any optimization algorithms. In our first experiences this simple round-robin scheduler is sufficient, as we shall see, to achieve performance gains.

3.1 Evaluating System Resources

For an efficient management of computing resources available on the local area network, PINCO must be able to evaluate resources. Consider first how to evaluate total computing power of the parallel system as the sum of heterogeneous resources. This point could be very subtle: different machines may be more or less suited for different applications (e.g. matrix multiplication and sorting algorithms), and many factors may determine the actual efficiency of a single node in processing a selected application (CPU speed, memory hierarchies, specialized coprocessors and the like).

PINCO approach to this problem has been experimentally derived by collecting data about a sample target network and a set of suitable benchmarks. Running the benchmarks on the different nodes, it is possible to get performance figures for the whole network. These figures are not strictly defined because the different nodes may change their relative performance selecting a different benchmark. In any case it is possible to get an approximate value of relative speeds of the different machines, which may be used in most cases as suitable indication, and normalize such numbers among them.

PINCO is designed to run on shared systems, hence we have to consider that actual computing power at each node will be only a fraction of the statically computed power of the same node. Let us call s_i the static computing power at node i. The following simple formula:

$$ds_i = (1-\sigma_i)\, s_i$$

correlates the dynamic relative speed ds_i with the static relative speed s_i of node i. The parameter σ_i expresses the load factor of node i. In Unix based systems, a simple but effective way of evaluating load factors is to use the percentage of CPU usage as provided by `ps` system call. Actually this is the choice currently implemented inside PINCO.

3.2 Resource and Task Manager

The execution of parallel processes on the network is initiated by a user command (`PINCO_init`) which is issued on some machine, which shall be called the host node in the following. The command also starts automatically one process, called the local daemon, on each node of the network. All further interactions shall take place among the machines in user transparent way.

Allocation of parallel tasks is distributed across the network, yet maintaining a centralized information collector, called PINCO Resource and Task Manager (PRTM). It is a process running on the host node, where information sent by the various machines in the network is kept, and used to take decisions about process allocation to the network. At present allocation decisions result from negotiation with the user process, since the choice suggested by PRTM is evaluated and partially modified, if needed, by the user. As far as experience in this decision making process is collected, it is expected to include "smart" allocation policies in PRTM, thus making user negotiation no longer needed for most applications.

Information to be collected on each node i of the network can be classified as:

- static information (s_i): initialized as collected from off-line benchmarks;
- dynamic information (σ_i): periodically updated by local daemons.

Local daemons, running on each network computer, compute the percentage of CPU free every 5 seconds by issuing a local system call, such as the command `ps` from a shell. This value is sent to PRTM at startup, and it is sent again only if different for more than 10% from the previous one. When PRTM receives a message from node x, it updates the x-th field in the `load_table` (where dynamic and static network load parameters are kept) by computing ds_X from the above formula.

3.3 Application Program Interface

Allocation requests from user processes are based on a data structure called `grant_record`. It is filled by PRTM and passed to the user process by means of a two phase protocol:

- in the first phase, the user process issues a request expressed by the call `err = PINCO_grant (requested_power, & grant_record);`
- the user process may check compliance of the proposed allocation to possible additional constraints and then issues the activation request: `err = PINCO_spawn ( & executable_file, & grant_record);`

The function `PINCO_grant` fills the following data structure, in accordance with values actually stored in the `load_table`:

```
struct grant_record
{ float grant_power;  /* granted power */
  int cluster_dim; /* host number */
  node * topology;
} grant_record;
```

where `node` is a list of `cluster_dim` pairs (`processor_name, local_granted_power`), such that the sum of all `local_granted_power` fields, as stored in `grant_power`, matches the request or default approximates it if it is impossible.

If the user only wishes to experience a parallel execution, without bothering with sophisticated allocation mechanisms, a simpler call to `PINCO_spawn` is sufficient, following the format:

```
err = PINCO_spawn(requested_power,&grant_record,
                  &executable_file);
```

4 PINCO Application to ScaLAPCK Library

In order to use a ScaLAPACK library function, four basic steps are required to the programmer:

- Initialise the process grid: each process is assigned an identifier inside a bi-dimensional matrix (row and column indexes) and a communication context is built around the grid to allow safe communications.
- Distribute the matrix on the process grid: each global matrix to be distributed across the process grid must be assigned an *array descriptor* to this aim.
- Call ScaLAPACK routine.
- Release the process grid.

It is thus important to understand the data distributions model implemented by ScaLAPCK, since the application data partitioning within the hierarchical memory of a concurrent computer is critical in determining the performance and scalability of the parallel code.

4.1 ScaLAPACK Data Distribution

The library uses a two-dimensional block-partitioned algorithm with cyclic distribution, in order to minimise the frequency of data movement between different levels of the memory hierarchy, thereby reducing the fixed startup cost incurred each time a message is communicated.
A global matrix *(MxN)* is partitioned into *NB* equal sized blocks, where each block is a submatrix *(M/(kp) x N/(tq))*. Each block is then assigned using a cyclic distribution to each process. One-dimensional block row or column distributions (or stripwise distributions) can be obtained as well, considering $k=1/p$ or $t=1/q$.

This kind of distribution is used in ScaLAPACK to minimize data transfer between processes, however, it has also been proved to be useful to balance the computational time. Load balancing can be obtained only when each process computation is well balanced, independently from the density of the matrix. In the general case, simple stripwise data partitioning is not convenient, since at each step some processors can be underloaded. Static two dimensional block distibution seems the best suited for generic computations, so that each process might evaluate a data set equally distributed, with the same probability of having elements inside the matrix [9,6]. It is a user's responsibility to decide the grain of such a scattering.

4.2 The Experiments Performed on Solving Generalized Eigenproblems and on LU Factorization Routines

PINCO allows the run-time monitoring of the network, in order to have a precise view of the actual computational power at each workstation. Thus, it would be possible to adapt process distribution to a dynamically varying availability of computational power, even if other users are working on the same workstations. Since in this case the workload may change, dynamic data distribution (migration) could be needed. However, migration may prove to be impractical due to the communication and control overhead. Presently we are studying a different approach, that is, to extract the user's profiles in order to foresee "steady windows" in which to run parts of the parallel program. Results presented in the following have been obtained considering the cluster completely devoted to parallel execution, a situation which typically occurs outside working hours, when no interactive users are sitting at the workstation's keyboard.

Changes performed to ScaLAPACK to allow the use of PINCO are limited to a BLACS routine, *Blacs_setup*, where PINCO is started calling *pinco_init*, and the call to *pvm_spawn* (that distributes the processes round robin on the processors) has been substituted with *pinco_spawn*. The allocation algorithm included in PINCO performs automatically the process distribution, accordingly with the computational power of each workstation. PINCO proved in such a way, its ease of use.

Presently, only a few tests (those provided with ScaLAPACK library itself) have been performed, and a careful analysis is needed to completely evaluate the gain obtainable using PINCO on other non homogeneous networks. As an example, we show in the following, results obtained executing *xdgsep*, that tests PDSYGVX, the ScaLAPACK routine for solving generalized symmetric definite eigenproblems on double precision data (reduced to a standard symmetric eigenvalue problem using the ScaLAPACK Cholesky factorization of B), and the routine performing the LU factorization.

PDSYGVX assumes that all global data has been distributed to the processes with a one-dimensional or two-dimensional block-cyclic data distribution. This last distribution has been chosen. The set of target network includes the hosts listed in Table 1: the comparative computational power has been obtained using the result of the PINCO benchmark, as described in Section 3, and all of them run Linux.

Table1: The Network.

Name	**Architecture**	**Normalised Computational Power**
lyra	Pentium 200 MMX	36
aquarius	K6-500	68
begato	Pentium 2 350	380
zaffiro	Pentium 200 MMX	29
aries	Pentium 3 350	538
smeraldo	Pentium 2 300	295
scorpio	Pentium Pro 200	91

Table 2 shows the results obtained by averaging 5 test runs using the original ScaLAPACK library, with a process grid *PxQ* (5x2), and a global matrix *NxN* (100x100) of randomly chosen values, considering the last 5 workstations listed in Table 1. The matrix is considered already partitioned on the grid and 2 processes are started on each host. *Wall* is the time in seconds from the beginning to the end of the computation, while *CPU* is the computational plus the synchronization time. Thus, *Wall-CPU* gives the time needed for communications. Finally, *chk* is the scaled residual which is compared with a given threshold to test the correctness of the result. The average is shown because we experienced time variations, due to the different initial data distribution and to the possibility of having some active system processes, for example the update daemon, even when no interactive user is working.

The second row shows the average of 5 runs obtained with ScaLAPACK with PINCO. `PINCO_spawn` allocation algorithm selected only 4 workstations to do the work, avoiding zaffiro, which is considered too slow with respect to the others. This choice also impacts on the CPU time that decreases due to the smaller synchronization times. The 10 processes have been distributed as follows: the master and 3 slaves on begato, 1 slave on scorpio, 2 on smeraldo, 3 on aries. The latter, aries, is relative underloaded: it could have 1 more process than begato, however, due to the choice of begato as console, this solution was not possible.

Table2: Tests with the Original ScaLAPACK (row 1) and with PINCO (row 2).

N	Wall	CPU	chk
100	15.04	0.44	0.92
100	7.65	0.32	0.99

Table 3 shows the results of 3 tests performed using `xdlu` program of ScaLAPACK test suite (average values on 5 runs), that implements the real double precision LU factorization. The 3 tests consider 6 workstations, partitioning the program in 15, 20 and 50 processes respectively. Results are given with couples of row: the first one is obtained by considering ScaLAPACK as it is, the second row of each test, by considering ScaLAPACK plus PINCO.

Table3: Test Sequence with Different Number of Processes (LU Factorization).

Test	N	P	Q	Wall	CPU
Test1	100	5	3	1.627	0.064
	100	5	3	1.551	0.056
Test2	100	5	4	2.219	0.057
	100	5	4	1.999	0.053
Test3	100	10	5	4.376	0.064
	100	10	5	2.797	0.056

Processes have been spawned as shown in Table 4 (begato is always the master host):

In both experiments we used procedures included in the ScaLAPACK test suite, mainly studied in order to test communication functions, and the times related to this part are clearly more important than the execution times. In every test, ScaLAPACK with PINCO shows a better performance than ScaLAPACK alone, even if processes have been allocated on Workstations using the "trivial" non-optimized scheduler included in PINCO. Better results are expected when the allocation is decided after a closer look at the particular communication paths of the routines.

Table4: Processes Allocation in the LU Factorization Tests.

Test	begato	zaffiro	scorpio	smeraldo	aquarius	lyra
Test1	3	3	3	2	2	2
	7	-	1	5	1	-
Test2	4	4	3	3	3	3
	9	-	1	6	1	1
Test3	9	9	8	8	8	8
	22	2	5	16	2	3

5 Conclusions

The paper has described an experience on an heterogeneous network of workstations, which has been used as a parallel computer in order to run a typical mathematical library routine, taken from ScaLAPACK. The experience has been eased by means of a dedicated distributed environment, which the authors have been developing for a while.

An interesting use of ScaLAPACK on homogeneous clusters is outlined in the Beowulf project [1]. Beowulf clustering is defined as off the shelf computer systems running any one of several open-source Unix-like operating systems, Linux in particular, networked together to function as a parallel processing super-computer. The computers, or nodes, are linked together and controlled by a master node. Many nodes, up to thousands, have been networked together to form very powerful computational resources. Even until now Beowulf has been mainly used in scientific environments, Scyld Computing Corporation [11] is now providing the infrastructure for commercial production systems for a wider diffusion.

Despite the many researchers working on workstation clusters, there are few environments that support an easy-to-use access to heterogeneous distributed systems, where mathematical libraries are available, with transparent access to resources. Usually, either they have a complex user interface, or they have limitations in architectural support. The goal of providing and supporting a Parallel Library for arbitrary networks of workstations, including the heterogeneous ones, is still far from being achieved.

Closest to PINCO is the DAME system [4] (Data Migration Environment) which aims to dynamically balance the workload of SPMD regular computations with the irregular data partitioning model: one process is started for each processor, but an irregular distribution of data is adopted, inversely proportional to the computational power of each processor. Instead, PINCO adopts the solution of the regular data partitioning, allocating more than one process per processor. DAME methodology allows to reduce the communication workload, but requires that programs are written in a data decomposition-indepent way, and cannot be applied on parallel library that are based on a regular data distribution. Remarkable are various studies and proposals on this topic, among the other we cite [2, 3] comparing the ScaLAPACK original implementation with a different data distribution mainly based on a static tiling approach. We notice that one of the most relevant result of the work on static tiling is to outline a load balancing strategy for heterogeneous network of workstations which is asymptotically optimal, and which could be implemented in the PINCO framework using a process multiplexing approach.

References

1. http://www.beowulf.org
2. P. Boulet et al., “Algorithmic issues on heterogeneous computing platform”, *Parallel Processing Letters* 9(2), pp. 197-213, 1999.
3. P. Boulet, J. Dongarra, Y. Robert, F. Vivien, "Static tiling for heterogeneous computing platforms", Parallel Computing, 25 (1999) pp. 547-568
4. M. Cermele, M. Colajanni, G. Necci, “Dynamic load balancing of distributed SPMD computations with explicit message-passing”, *Proc. 6th Heterogeneous Computing Workshop*, pp. 2-16, 1997.
5. A. Clematis G. Dodero, V. Gianuzzi, “A resource management tool for heterogeneous networks”, *Proc. 7th Euromicro Workshop on Parallel and Distributed Processing*, IEEE Comp. Soc. Press, pp. 367-373, 1999.
6. G. Dodero, V. Gianuzzi, M. Moscati, M. Corvi , “A Scalable Parallel Algorithm for Matching Pursuit Signal Decomposition”, *Proc. HPCN'98 Conference*, Lect. Notes in Comp. Sciences, pp. 458-466, 1998.
7. J.J. Dongarra, D.W. Walker, “Software libraries for linear algebra computations on high performance computers”, *SIAM Review*, 37(2), pp. 151-180, 1995.
8. http://www.gridcomputing.com
9. R. v.Hanxleden, Scott L. Ridgway, “Load balancing on message passing architectures”, *Journal of Parall. Distrib. Computing*, 13, pp. 312—324, 1991.
10. http://www.netlib.org/scalapack/slug/index.html
11. http://www.scyld.com
12. V.S. Sunderam, G.A. Geist, J. Dongarra, R. Manchek, “The PVM concurrent system: evolution, experiences, and trends”, *Parallel Computing*, Vol.20(4), pp. 531-546, 1994.

An Hierarchical MPI Communication Model for the Parallelized Solution of Multiple Integrals

Peter Friedel[1], Jörg Bergmann[2], Stephan Seidl[3], and Wolfgang E. Nagel[3]

[1] Institute of Polymer Research Dresden, Hohe Str. 6, 01069 Dresden
Peter.Friedel@zhr.tu-dresden.de
WWW home page: http://rcswww.urz.tu-dresden.de/~friedel
[2] Ludwig-Renn-Allee 14, 01217 Dresden, Germany
[3] Center for High Performance Computing, Dresden University of Technology
Zellescher Weg 12, 01069 Dresden, Germany

Abstract. For the modeling of polymer meso-structures, the spinodal points can be obtained by random phase approximations. The necessary number of these spinodal points to describe a phase diagram can significantly be reduced, if the usual sampling-point method is replaced by a Newton iteration, utilizing all the transiently computed data. This has the consequence that the simple inner parallelism of the problem gets lost, i.e. the possibility to compute a high number of independent sampling-points in parallel. On the other hand, the overall CPU time requirement is rather drastic and parallelism seems to be the only way to achieve acceptable turn-around times. Hence, there is no other way than to parallelize the objective functions of the Newton iteration, mainly consisting of a handy set of multiple integrals, which have to be numerically solved using step-width adaption.
Computing multiple integrals with step-width adaption in parallel results in nested parallelism, which is difficult to implement. A lot of applications need the concept of nested parallelism. In many implementations, OpenMP does not support nested parallelism at the moment, and for MPI, the granularity has to be sufficient. As a case study, the present paper describes the parallelization and the communication model for the solution of multiple integrals. The results show adequate parallel efficiency values for executions up to 256 processes. The parallel integral solver code is modularized in the sense that it can easily be applied to any other integrands.
Keywords: Parallel numerical solution, multiple integrals, step-width adaption, army communication model.

1 Introduction

In polymer chemistry, the modeling of meso-structures allows a better understanding of material properties. The goal here is to compose new polymers with tailored characteristics.

For the phase diagrams, one theoretical approach is to obtain spinodal points by random phase approximations[1]. Furthermore, the necessary number of these

B. Hertzberger et al. (Eds.): HPCN Europe 2001, LNCS 2110, pp. 474–482, 2001.

spinodal points to describe a phase diagram can significantly be reduced, if the usual sampling-point method is replaced by a special Newton iteration, utilizing all the transiently computed data. While the sampling-point method allows easy computation of spinodal points in parallel, a Newton algorithm serializes the appropriate calculations. In other words, the simple opportunity to parallelize the code gets lost and the gain of point number reduction does not fully compensate that. Therefore, due to CPU time, parallelism is needed. The only possibility is to parallelize the objective functions of the Newton iteration. These functions consist of $n(n+1)/2$ multiple integrals, where n is the number of components of the polymer. This number of integrals is by far too small to be successfully parallelized with respect to the integrals. Hence, what remains is the possibility of parallelization of the multiple integrals themselves.

Traditional solutions for numerical integration are sequential algorithms and use a moderate set of processors calculating an equidistant function field[3]. On the other hand, algorithms using step-width adaption are much more effective, depending on integral kernel curvatures[4,5]. They are well-discussed for simple processor machines and use a recursive adaptive algorithm. One of the goals of this work was to implement an analogous procedure for massively parallel computers like CRAY T3E, applying an iterative process over recursive data structures (see also Wirth[6]). The corresponding routines can be used for solving multiple integrals of unlimited depth by applying an hierarchical MPI communicator structure.

2 Physical and Mathematical Background

2.1 A Motivating Example from Polymer Physics

A motivating example is taken from polymer physics (see van Kampen[2] and Goldberg[7]). Polymer physical details of the used integral kernel $p(q, s, s', L)$ can be read in [2,8,9,10]. We are rather interested in the integration procedure, so we will use the following triple integral

$$\langle P(q) \rangle = \int_0^\infty dN \int_0^L ds \int_0^L ds' \, p(q, s, s', L) \tag{1}$$

as our test example, using relevant physical values for L and q[8].

2.2 The Seven-Point Simpson Rule

For a single integral, the seven-point Simpson equation

$$I(x) = \int_a^b \mathrm{F}(x)\, dx = Q\,(\,h,\, \mathrm{F}(x_0) \ldots \mathrm{F}(x_6)\,) + \frac{h^8 \cdot \mathrm{F}^{(8)}(\varepsilon)}{1567641600}, \tag{2}$$

is an excellent numerical approximation (with $h = b - a$). This eqn. (2) can be written as described in[4,5]:

$$Q\,(h,\,\mathrm{F}(x_0)\ldots\mathrm{F}(x_6)\,) = \frac{h}{840}\cdot(\,41\;\;\mathrm{F}(x_0) + 216\,\mathrm{F}(x_1) + 27\,\mathrm{F}(x_2) + 272\,\mathrm{F}(x_3) + 27\;\;\mathrm{F}(x_4) + 216\,\mathrm{F}(x_5) + 41\,\mathrm{F}(x_6)\,). \quad (3)$$

Eqn. (3) can also be applied for multiple integrals on every integration level, hence, $\mathrm{F}(x)$ may describe a further inner integral.

2.3 Recursively Determined Adaptive Step-Widths

Let $\mathrm{F}(x)$ be the integral kernel $p(x, s, s', L)$ from eqn. (1), with s, s', L as additional arguments combined inside $arglist_{\mathrm{F}}$. The shown integration function in algorithm (4) describes the splitting of integration area into two equal parts and the computing of the corresponding values for $\mathrm{F}(x)$. Depending on an arbitrary given precision ε, either the function returns the sum of the $left$ and $right$ part of the local integration area δx by using $\delta x_0 = (a - b)/6$ as starting value (see eqn. (2)), or the step-width adaption is performed by a recursive function call. f_{amax} describes the maximum of an absolute value of the function $F(x)$ with $f_{\mathrm{amax},0} = 0$ at the beginning of the integral calculation.

3 Parallelization and Communication Model

Parallelization requires a change of the recursive function call into an iterative procedure over a recursive data structure. In our parallelization model, we decide between communication and calculation nodes. The communication node performs the management over the recursive data structure. It sends the jobs to and receives the computed results from all other calculation nodes. This separation between calculation and communication nodes is an important advantage of the presented approach. In addition to this, it can be emphasized for more complicated integrations.

Due to load balancing reasons, it was necessary to develop different models depending on the number of available nodes. If there is more than one node available, the numerical integration can be performed with the so-called simple parallelization model. This means the calculation of the most outer integral is carried out in parallel and the calculation of the more inner integrals in sequential mode. The complex parallelization model, which was developed for more than 15 nodes, is also capable of calculating deeper integral levels and uses more than one communicator in a communicator tree for message passing.

3.1 Simple Parallelization Model

In this model we have one master and the other nodes are slaves (Fig. 1a). The master manages the subdivision of the integrals. If there are leaves on the

Algorithm (4) :

numerical function INTEGRAL $(x_0, y_0, x_2, y_2, x_4, y_4, x_6, y_6,$
$x_8, y_8, x_{10}, y_{10}, x_{12}, y_{12},$
$formerIntegral, f_{\mathrm{amax}}, \delta x,$
$\varepsilon, \mathrm{F}, arglist_{\mathrm{F}})$

inout f_{amax};

begin

local $x_1, x_3, x_5, x_7, x_9, x_{11}, y_1, y_3, y_5, y_7, y_9, y_{11}, i, left, right;$

for $i := 1\,(+2)\; i < 12$ **do**

$x_i := \frac{1}{2}(x_{i-1} + x_{i+1});\quad y_i := \mathrm{F}(x_i, arglist_{\mathrm{F}});$

$f_{\mathrm{amax}} := max(f_{\mathrm{amax}}, |y_i|);$

$left := \mathrm{SEVENPOINT}\;(\frac{1}{2}\delta x, y_0, y_1, y_2, y_3, y_4, y_5, y_6);$

$right := \mathrm{SEVENPOINT}\;(\frac{1}{2}\delta x, y_6, y_7, y_8, y_9, y_{10}, y_{11}, y_{12});$

if $|\,left + right - formerIntegral\,| < \varepsilon\, f_{\mathrm{amax}}\, \delta x$

then return $(left + right);$

else return (INTEGRAL $(x_0, y_0, x_1, y_1, x_2, y_2, x_3, y_3, x_4, y_4,$
$x_5, y_5, x_6, y_6, left, f_{\mathrm{amax}},$
$\frac{1}{2}\delta x, \varepsilon, \mathrm{F}, arglist_{\mathrm{F}})$

\+ INTEGRAL $(x_6, y_6, x_7, y_7, x_8, y_8, x_9, y_9, x_{10}, y_{10},$
$x_{11}, y_{11}, x_{12}, y_{12}, right, f_{\mathrm{amax}},$
$\frac{1}{2}\delta x, \varepsilon, \mathrm{F}, arglist_{\mathrm{F}}));$

end □ (4)

dynamic data tree, where the precision ε is not reached at a particular tree level, the tree will be extended at this position by the master node. This gives an appropriate number of jobs for precisioning. If one slave is free of work, the master gives one of these jobs to it. Otherwise, if precision ϵ is reached at all, the master node collects the results and gives them back to the calling routine.

3.2 Complex Parallelization Model with Communicator Trees

Numerical calculation of multiple integrals suggests a more complex communication model, because all levels lower than the outer one can also be calculated in parallel mode.

Parallel calculation of the outer two levels of multiple integrals leads to a master-sub-master-slave model, including a hierarchical communication structure with communicator trees.

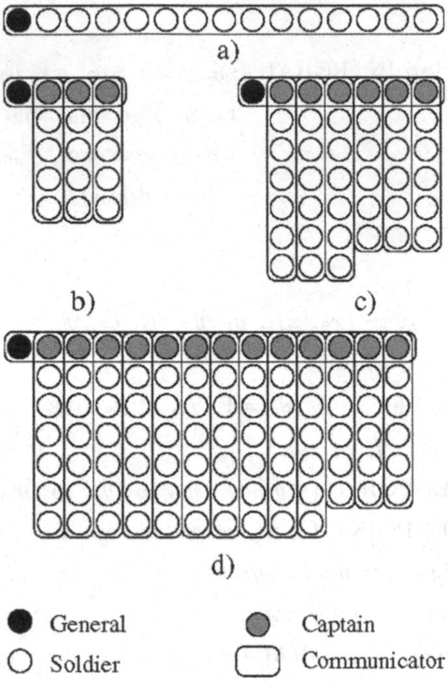

Fig. 1. Communication models with one and two communicator levels.

Because of its similarity with the structure of the army, we call this model the "army model" (Fig. 1). The communicator tree contains one "General". The General gives instructions to its "Captains", the General and all the Captains are members of the top-level communicator (outer integral calculation). Every Captain is a member of the top-level communicator and is also a member of one sub-communicator (second integral calculation). The difference between the available number of nodes and the number of Captains, advanced by one, gives the number of "soldiers". Every soldier node calculates all sub-levels of integration inside its assigned integration interval, working in sequential mode. It receives its instructions from *its* Captain alone and sends the results back to it.

Initially, the number of Captains is chosen as the integer part of the square root of available number of nodes. A system with up to 12 Captains can be managed by the General, which depends on the seven-point Simpson rule. Due to the same reason, one can also assume that every Captain is able to handle 12 soldiers. This efficiently works as long as the integral kernel is heavy enough. That means, that every soldier is very busy with calculations. Otherwise, if the integral kernel is a "light" one (as in the case of our test kernel), then the soldiers rapidly calculate their sampling points, so that the communication cannot follow appropriately. With our test kernel, using the tracing tool VAMPIR[11], we found out that one Captain should only manage the communication for up to six soldiers. This is the reason for the rather low, but relatively constant parallel efficiency of our army model up to 83 nodes (see Fig. 2). On the other hand, using a heavy integral kernel, one would be able to employ one General, 12 Captains and 144 soldiers as the theoretical limit of the two-level army model. Naturally, the heavier the integral kernel, the better the appropriate calculation-to-communication ratio will be.

In Fig. 1, one can see some different communication models for our "light" integral kernel. Fig. 1a) shows the upper limit scheme for the simple communication model (15 nodes). In Fig. 1b), the "smallest" army model can be seen (16 nodes) and Fig. 1c) demonstrates a model with 40 nodes. Fig. 1d) shows the two-level army model near the upper limit for constant efficiency. Using many more nodes, the efficency decreases in our example (see Fig. 2).

Note that a three-level communication model may be implemented in the same manner as the described two-level model. Having a three-level integration problem as may come from polymer physics, one can construct an army model of one General, 12 Captains, 144 Lieutenants and 1728 soldiers which works with a sufficiently high parallel efficiency.

4 Results and Discussion

4.1 Parallel Efficiency

We performed calculations to obtain the parallel efficiency $E(n)$ as the ratio of t_{seq} to $n \cdot t_{par}$, which depends on the number n of used nodes and t_{seq} and t_{par} as the consumed time in sequential and parallel mode respectively.

It can be seen in Fig (2) (solid line), that at 15 nodes, the efficiency of the simple parallelization model is much higher (0.84) than the efficiency of the army model (0.68 on CRAY T3E-600 and 0.59 at CRAY T3E-1200, because of their equal networks but their different CPU clock rates, dashed lines).

If there are more than 36 nodes available for calculation, the army model is used with the modification that the number of Captains is increased to the sixth part of the available nodes. Every sub-communicator now contains six soldiers. By applying this model, we can obtain an almost constant efficiency for a maximum of 83 nodes on CRAY T3E-1200 (Fig. 2). This is an army structure with one General, 12 Captains and 70 soldiers. For more than 83 nodes, the General splits the integral on the top level into 12 equidistant parts, reasoned by the seven-point Simpson rule, with one part for each Captain.

In addition to this, it should be noted that it seems to be possible to hold efficiency almost constant by applying this method to deeper integral levels and the efficiency should increase by calculating heavier integral kernels than the used test case.

4.2 Communication Rates

The procedures for the three integral levels (see eqn. (1)) and the numerical integration were traced[11], by using the army model with 43 nodes (one General, six Captains and 36 soldiers) on CRAY T3E-600.

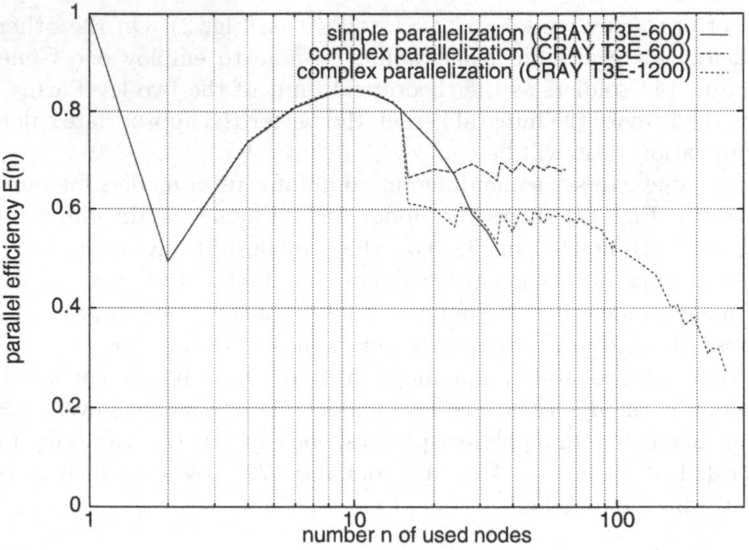

Fig. 2. Scaling behavior for numerical integration, calculated on CRAY T3E-600 and CRAY T3E-1200 using different communication models.

By using a double buffering mode, we achieved a high communication speed. This means that the "chairman" of every communicator, either the General or one of the Captains, gives the jobs to the other members of its communicator when starting integral calculation. These members are now working. If all members are busy, the chairman gives further jobs to the communication cloud, which causes the following effect. No soldier has to wait to give the results back to and get a further job from its Captain. The communication time in the used MPI procedure "*MPI_Sendrecv_replace*" between soldiers and Captain shrinks from approximately 200 μs, without double buffering mode, down to 30 μs with double buffering in our example.

4.3 Calculation Speed

Another aspect can be shown to underline the efficiency of our model. One can easily imagine that the soldiers, nodes 7 to 42 here, have to do the majority of the calculation work.

We chose only node no. 7 and only a middle-part of an integral calculation as an example to show the possible calculation speed in MFLOPS (million floating-point operations per second), using VAMPIR[11] as the visualization tool for traced algorithms.

81 MFLOPS show that node no. 7, as an example, is fairly busy.

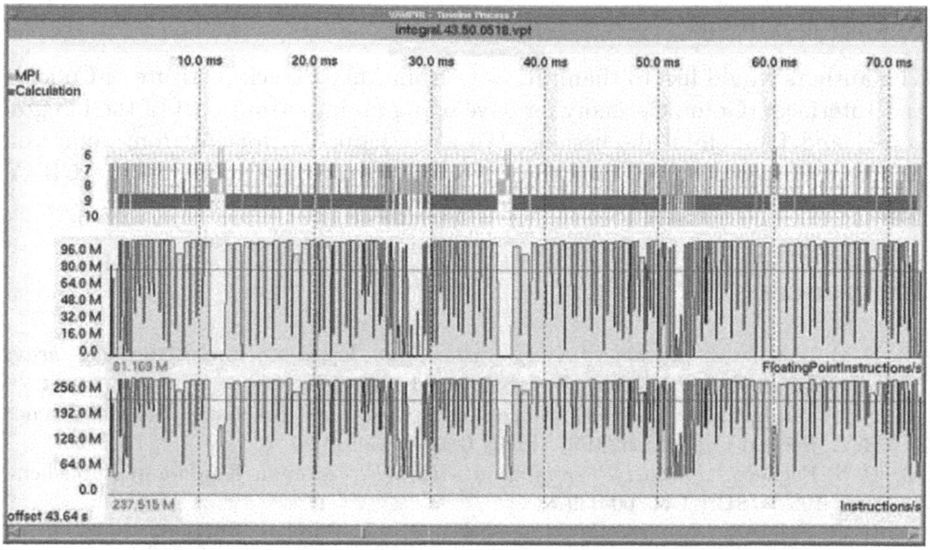

Fig. 3. Middle-part of the node-7-soldier VAMPIR time lines for the call hierarchy, the floating-point operations per second and the executed instructions per second, obtained at CRAY T3E-600.

5 Summary and Conclusions

We presented an hierarchical communication model for the parallel numerical solution of multiple integrals with step-width adaption. We applied this method to a triple integral with an integral kernel from polymer physics.

Parallelization was performed by two different communication models using MPI (Message Passing Interface). Firstly, the outer integration was parallelized with a maximal parallel efficiency of 0.84 using ten nodes. Secondly, two outer levels of integration were included to distribute the calculation work to the available nodes using a two-level communication model. This model was called the army model as a result of its similarity to the command structure of the army. The two-level communication model led to a parallel efficiency of 0.68 with 64 nodes, which remained constant up to 83 nodes. Deeper General-Captain-Lieutenant- ... -soldier models would be useful if more than three levels of integration had to be calculated and/or the integrands were much heavier, e.g. in problems from theoretical polymer physics. This is one of the most important advantages of our chosen communication model.

To summarize, the paper shows that nested parallelization can be performed with up to 256 nodes on T3E-600/1200, even though it is done at the lowest computation level. Furthermore, the code is modularized, so that the *Physics* can be written without knowing what the *Computer Science* does with the data.

Acknowledgments

The authors would like to thank R. Netz from Max-Planck-Institute of Colloids and Interfaces, Golm, Germany for developing an important part of the integral kernel and for helpful discussions. We also thank F. Hoßfeld from John von Neumann Institute for Computing (NIC) for providing access to the CRAY T3E-1200.

References

1. P. G. de Gennes. *Theory of X-ray scattering by liquid macromolecules with heavy atom labels J. de Physique (Paris)* 1979. **31**, 235-238.
2. N. G. van Kampen. *Stochastic processes in physics and chemistry*, Elsevier Science B.V., Second impression: 1997. ISBN 0-444-89349-0.
3. P. S. Pacheco. *Parallel Programming with MPI*, Morgan Kaufmann Publishers, Inc., 1997. ISBN 1-55860-339-5.
4. H. R. Schwarz. *Numerische Mathematik.* B. G. Teubner Stuttgart, 1993. ISBN 3-419-22960-9.
5. C. W. Ueberhuber. *Numerical Integration on Advanced Computer Systems.* Springer-Verl. Berlin, Heidelberg New York, 1993. ISBN 3-540-58410-2.
6. N. Wirth. *Algorithmen und Datenstrukturen*, B. G. Teubner Stuttgart, 1995. ISBN 3-519-12250-2.
7. A. Yu. Goldberg and A. R. Khokhlov. *Statistical Physics of Macromolecules*, AIP Press New York, 1994. ISBN 1-56396-071-0.
8. R. R. Netz. *Random-Phase Approximation for Semi-Dilute Semiflexible Polyelectrolyte Solutions European Polymer Journal.* (subm.).
9. G. Z. Schultz. *Z. Phys. Chem., Abt. B*, 1939. **43**, 25.
10. B. H. Zimm. *J. Chem. Phys.*, 1948. **16**, 1099.
11. *VAMPIRtrace - MPI Profiling library*, Pallas GmbH, Hermlheimer Str. 10., D-50321 Brühl, Germany. e-mail: info@pallas.com, http://www.pallas.de.

Impact of Data Distribution on Performance of Irregular Reductions on Multithreaded Architectures*

Gary Zoppetti, Gagan Agrawal, and Rishi Kumar

Department of Computer and Information Sciences
University of Delaware, Newark DE 19716
{zoppetti,agrawal,kumar}@eecis.udel.edu

1 Introduction

Computations from many scientific and engineering domains use irregular meshes and/or sparse matrices. The codes expressing these computations involve *irregular reductions*. The main characteristics of irregular reduction loops are 1) elements of left-hand-side arrays may be incremented in multiple iterations of the loop, but only using associative and commutative operations (these arrays are called *reduction arrays*), 2) there are no loop carried dependencies, except on elements of reduction arrays, and 3) one or more arrays are accessed using indirection arrays.

It is very challenging to efficiently parallelize codes involving irregular reductions, especially on large parallel machines. Because of accesses through indirection arrays, communication and locality are hard to manage. Not only is the total communication volume large, but the communication requirements typically cannot be determined at compile time. It is also hard to efficiently allocate space for non-local elements. Particularly, there are no effective solutions for parallelization of *adaptive* irregular reductions. In an adaptive irregular reduction, the elements of the indirection arrays are modified after every few iterations. This significantly increases the overhead associated with partitioning and runtime preprocessing routines [7,11], which have been critical for achieving locality, communication efficiency, and effective buffer management.

Recently, there has been much interest in multithreaded architectures. A multiprocessor based upon a multithreaded architecture supports multiple threads of execution on each processor. These architectures also support low-cost thread initiation, low-overhead communication, and efficient communication and synchronization between threads on different processors. Multithreaded architectures are considered a promising medium for scalable parallelization of irregular applications, where the frequent communication and synchronization make parallelization hard on conventional parallel machines.

We have developed an execution strategy for irregular reductions on a multithreaded architecture. The key idea in our execution model is that the frequency

* This research was supported by NSF grant CCR-9808522. Agrawal and Kumar were also supported by NSF CAREER award ACI-9733520.

B. Hertzberger et al. (Eds.): HPCN Europe 2001, LNCS 2110, pp. 483–492, 2001.

and volume of communication is independent of the contents of the indirection arrays. Thus, unlike other approaches to scalable parallelization of irregular reductions, our approach does not require mesh partitioning [1], array renumbering [6], or a high-cost inspector that itself requires communication between processors [7]. The performance depends upon the architecture's ability to support low-cost communication and overlap communication and computation, and is largely independent of the problem partitioning. Thus, the same performance can be obtained on adaptive problems, without paying the high overhead of partitioning frequently.

Though the frequency and volume of communication is independent of the data distribution, other factors are not. These factors are locality, load balance, and buffer space requirements. This paper evaluates the impact of data distribution on the performance of irregular reductions, which use our execution strategy, on a multithreaded architecture. We used two scientific kernels involving irregular reductions (`euler` and `moldyn`) for our experiments. We compare the performance achieved using block, block-cyclic, and cyclic data distributions, which are three common distributions used in data parallel languages like HPF [5]. Our results show that a block-cyclic distribution achieves a good balance between locality and load balance, and hence obtains the best performance. For `euler`, a cyclic distribution performs the same as a block-cyclic distribution, whereas the performance of a block distribution is significantly worse. For `moldyn`, a block-cyclic distribution is significantly better than both cyclic and block.

The rest of the paper is organized as follows. Our execution strategy, along with an example, is presented in Section 2. The experimental evaluation of impact of data distribution is presented in Section 3. We conclude in Section 4.

2 Execution Strategy

In this section, we give an example of a code that performs an irregular reduction. We then describe the execution strategy we propose for this class of applications.

2.1 Irregular Reduction Loops

```
Real    X(num_nodes), Y(num_edges) ;        ! data arrays
Integer IA(num_edges,2) ;                   ! indirection array

for(i = 0; i < num_edges; i++) {
        X(IA(i,1)) = X(IA(i,1)) + Y(i) ;
        X(IA(i,2)) = X(IA(i,2)) - Y(i) ;
}
```

Fig. 1. A simple loop involving indirection

Figure 1 shows a simple loop that performs an irregular reduction. In iteration i of the loop, the code makes two indirect references to array X using $IA(i, 1)$ and $IA(i, 2)$. We say that array X is accessed using two *indirection array sections*, denoted by $IA([1, num\_edges, 1], 1)$ and $IA([1, num\_edges, 1], 2)$.

2.2 Execution Strategy for Multithreaded Architectures

```
{* On processor proc_id: *}

   {* length of the portion of array X updated in each phase *}
   reduc_length = num_nodes / (k * num_procs) ;

   {* runtime preprocessing on each processor * }
   LIGHTINSPECTOR(num_nodes,num_procs,num_edges, IA[[1,num_edges_local,1],1],
          IA[[1,num_edges_local,1],2], &IA1_out, &IA2_out, &IB1_out, &IB2_out,
          &loop1_pt, &loop2_pt) ;

   for(phaseno = 0 ; phaseno < k * num_procs ; phaseno++) {

      {* portion of array X updated in each phase *}
      reduc_portion = (k * proc_id + phaseno) mod (k * num_procs) ;

      Receive X[reduc_portion * reduc_length: (reduc_portion + 1) * reduc_length - 1: 1]
         from processor proc_id - 1 ;

      {* main execution loop *}
      for(i = loop1_pt[phaseno] ; i < loop1_pt[phaseno + 1] ; i++) {
         X[IA1_out[i]] = X[IA1_out[i]] + Y[i] ;
         X[IA2_out[i]] = X[IA2_out[i]] - Y[i] ;
      }

      {* second loop to copy from buffer *}
      for(i = loop2_pt[phaseno] ; i < loop2_pt[phaseno + 1] ; i++) {
         dest = IB1_out[i] ;
         source = IB2_out[i] ;
         X[dest] = X[dest] + X[source] ;
      }

      Send X[reduc_portion * reduc_length: (reduc_portion + 1) * reduc_length - 1: 1]
         to processor proc_id + 1 ;

   }
```

Fig. 2. Strategy for Executing Irregular Reductions on a Multithreaded Architecture. The number of phases per processor is $k * num\_procs$.

Recently, there has been much interest in multithreaded architectures. A multiprocessor based upon a multithreaded architecture supports multiple threads of execution on each processor. These architectures also support low-cost thread initiation, low-overhead communication, and efficient communication and synchronization between threads on different processors. Multithreaded architectures are considered very effective for irregular and communication intensive applications. Therefore, it is natural to consider multithreaded architectures for the efficient execution of irregular reductions.

We now present our proposed execution strategy for irregular reductions on a multithreaded architecture. The execution strategy for the loop shown in Figure 1 is given in Figure 2. The key idea in this strategy is to make the communication independent of the data and work distribution and the values in the indirection arrays. Though the total volume and frequency of communication is high, our premise is that sophisticated support for overlapping communication and computation will enable high performance.

Initially, iterations and the data are divided between the processors. Each processor is responsible for executing *num_edges_local* iterations. When the code is run on *num_procs* processors, each processor executes the loop in $k{*}num\_procs$ phases. k is a small constant, and the most obvious choices of k are 2 and 4. In each phase, only a portion of the reduction array X is available locally on each processor. On processor 0, the first $num\_nodes/(k * num\_procs)$ elements are available locally during the 0th phase, the next $num\_nodes/(k * num\_procs)$ elements are available locally during the 1st phase, and so on.

The runtime routine LIGHTINSPECTOR is used to divide the iterations assigned to a processor into different phases. We now explain the functionality of this routine.

The phase to which an iteration is assigned depends upon the elements of the reduction array X updated in that iteration, which in turn depends upon the contents of the indirection arrays. Consider iteration i and the values of $IA(i,1)$ and $IA(i,2)$. Let $X(IA(i,1))$ belong to the *reduc_portion* for phase $r1$, and let $X(IA(i,2)$ belong to the *reduc_portion* for phase $r2$. The following three possibilities exist:

- $r1 = r2$. In this case, iteration i is obviously assigned to phase $r1$. Since both $X(IA(i,1))$ and $X(IA(i,2)$ are within the range of X owned by this processor during this phase, both of them are updated during the execution of the main loop.
- $r1 < r2$. In this case, iteration i is assigned to phase $r1$. The element $X(IA(i,1))$ is owned by the processor during phase $r1$ and is updated. However, the element $X(IA(i,2)$ is not owned by the processor during this phase. The length of the array X is extended to create a remote buffer location. The value of $IA(i,2)$ is reassigned to point to this buffer location. During phase $r2$, an iteration of the second inner loop (shown in Figure 2) increments the value of $X(IA(i,2)$ using the value in the buffer location. The arrays *loop2_pt*, *IB1_out*, and *IB2_out* are used for managing the second loop.

– $r1 > r2$. In this case, iteration i is assigned to phase $r2$. The rest of the processing is analogous to the previous case.

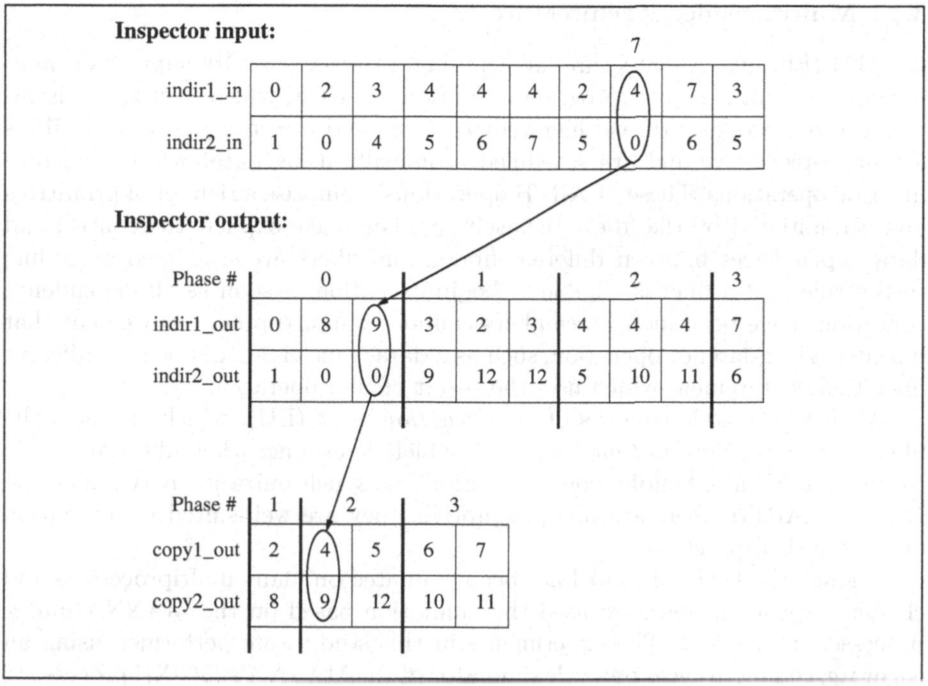

Fig. 3. Inspector Example: The number of processors (*num_procs*) is 2 and the value of k is 2.

Figure 3 illustrates the input and output of the LIGHTINSPECTOR routine running on $num\_procs = 2$ processors with $k = 2$. The sample mesh contains 8 nodes and 20 edges. Each processor therefore owns 10 edges. Since there are $k * num\_procs = 4$ phases per processor, each processor owns 2 nodes per phase. For each reduction array, each processor must maintain $2 * k = 4$ local buffers and 1 remote buffer. Phases 0, 1, and 2 are assigned 3 edges while phase 3 is assigned 1 edge. The inspector places the edges in the new indirection arrays and modifies their values to point to local or remote buffer locations. Consider the 7th edge in the example. Since node 0 is owned locally during phase 0, this edge is assigned to the 0th phase. Node 4 is owned in a future phase, namely phase 2. This value must be changed to point to a buffer location, which in this case is 9.

3 Experimental Results

In this section, we describe the multithreaded architecture used for evaluating our techniques and the experimental results we have obtained.

3.1 Multithreaded Architecture

EARTH (Efficient Architecture for Running THreads) [3,9,10] supports a *non-preemptive* multithreaded program execution model in which a program is divided into a two-level thread hierarchy of *fibers* and *threaded procedures*. Fibers are non-preemptive and are scheduled atomically using dataflow-like synchronization operations. These "EARTH operations" comprise a rich set of primitives and are initiated by the fibers themselves. They make explicit the control and data dependences between different fibers, and fibers are scheduled according to the rule that a fiber is eligible to begin execution as soon as all dependence conditions have been met. Since fibers cannot be interrupted, a statement that initiates a long-latency operation, such as a data transfer, should be in a different fiber than a statement which uses the result of that operation.

An EARTH node consists of an *Execution Unit* (EU), which executes the fibers, and a *Synchronization Unit* (SU), which determines when fibers are ready to run. The SU also handles communication and synchronization between nodes. Because EARTH fibers are non-preemptive, they are well-suited for execution on off-the-shelf processors.

Though the EARTH model has been emulated on many multiprocessors and clusters of workstations, we used the emulation based on the MANNA multiprocessor from GMD. The experiments in this study were performed using an accurate, complete cycle-by-cycle simulator of the MANNA's i860XP processors, system bus, memory system and the interconnection network. We obtained our results using the simulator's `manna-dual` mode, where the 2 i860XP processors per node function as the EU and SU, respectively.

3.2 Benchmarks and Experiment Design

Our goal in this study was to evaluate the impact of data distribution on the performance of irregular reductions. Data parallel languages like HPF have typically used 3 types of data distributions: block, cyclic, and block-cyclic [5]. In a *block* distribution, *num_edges*/*num_procs* consecutive iterations are assigned to each processor. In a cyclic distribution, elements of the arrays are distributed among the processors in a round-robin fashion. In a block-cyclic distribution with a block size *bs*, blocks of size *bs* are mapped to the processors in a cyclic fashion.

We present experimental results from two scientific kernels involving irregular reductions: `euler` and `moldyn`. `euler` is derived from a Computational Fluid Dynamics (CFD) code [2] and `moldyn` is derived from a molecular dynamics code [4]. We used two datasets for each of these codes. The two meshes used for `euler` had 2,800 nodes and 17,377 edges, and 9,428 nodes and 59,863 edges, respectively. `moldyn` was evaluated using problems with 2,916 molecules and 26,244 interactions, and 10,976 molecules and 65,856 interactions, respectively.

The sequential versions of these codes were timed on one i860XP processor, and then the multithreaded version produced by the compiler was executed on 2, 4, 8, 16, and 32 nodes. All measurements were performed using 100 iterations of the time-step loop. For parallel versions, the inspector was executed once.

One factor in using our strategy is the choice of the parameter k. The number of phases executed on each processor is $k * num\_procs$. A larger value of k allows better overlap of communication and computation and increases the ability to tolerate load imbalance. However, it also results in higher threading overheads and loss of locality. In our initial experiments, we used $k = 1, 2$, and 4, and found that $k = 2$ generally outperforms $k = 1$ and $k = 4$. Therefore, all the results presented in this paper use $k = 2$.

For each benchmark and dataset, we present performance data from four versions: a block distribution; a block-cyclic distribution using a larger block size ($bs = 8$ for the 2k datasets and $bs = 16$ for the 10k datasets); a block-cyclic distribution using a smaller block size ($bs = 2$ for the 2k datasets and $bs = 4$ for the 10k datasets); and a cyclic distribution. These four strategies will be referred to as `block`, `bc(8|16)`, `bc(2|4)`, and `cyclic`, respectively.

3.3 Results

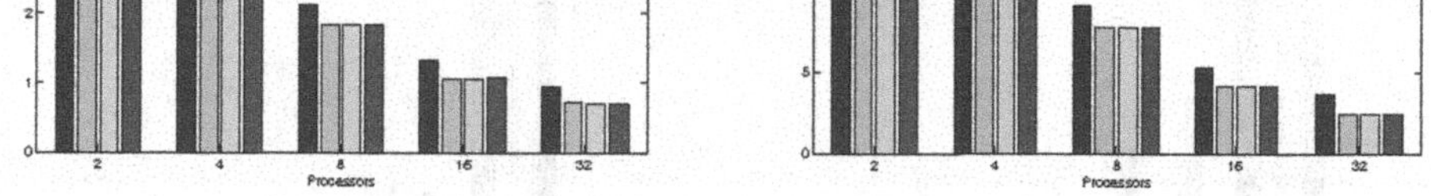

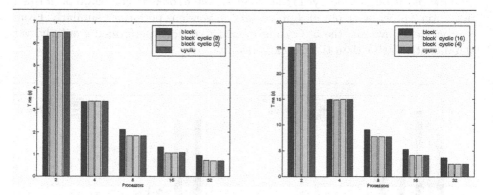

Fig. 4. Parallel Performance of `euler`: 2K Mesh (left) and 10K Mesh (right).

Figure 4 shows the execution times for `euler` on its 2K and 10K meshes. The sequential execution time on the 2K mesh was 7.85 seconds. The speedups of the 2 processor versions were 1.24 with the `block` strategy, and 1.21 with the `bc(8)`, `bc(2)`, and `cyclic` strategies. The relative speedups[1] in going from 2 to

[1] All relative speedups were computed against the best 2 processor version, which in this case was the `block` version.

32 processors were 6.78 with the `block` strategy, 8.87 with the `bc(8)` strategy, 9.02 with the `bc(2)` strategy, and finally, 9.22 with the `cyclic` strategy.

Two main observations from these results follow. First, the block-cyclic and cyclic distributions performed nearly identically. This shows that any locality gains from using a larger block size were insignificant or were offset by a load imbalance.

Second, in comparing the block and cyclic distributions, the block distribution performed better by 3% on 2 processors and by 0.7% on 4 processors, while the cyclic distribution performed better on 8 or more processors. The cyclic version was faster by 14% on 8 processors, 20% on 16 processors, and 26% on 32 processors. We carefully analyzed the number of iterations assigned to each phase on all processors. A block distribution resulted in a significant load imbalance, whereas cyclic or block-cyclic distributions did not. We believe this is the main reason the block-cyclic and cyclic distributions performed significantly better. On smaller configurations, a reasonably good load balance was possible with either distribution. Though we could not measure cache miss rates, we believe that the block distribution performs marginally better on small configurations because of better locality.

The performance on the 10K mesh followed the same trend. The sequential execution time was 29.07 seconds. The speedups of the 2 processor versions were 1.16 with the `block` strategy, 1.13 with the `bc(16)` and `bc(4)` strategies, and finally, 1.12 with the `cyclic` strategy. The relative speedups in going from 2 to 32 processors were 8.05 with the `block` strategy, 11.84 with the `bc(16)` strategy, and 12.00 with the `bc(4)` and `cyclic` strategies. The `block` version again performed better on 2 processors, but at 4 processors all versions performed similarly. From 8 processors onwards, the `bc(x)` and `cyclic` versions performed similarly, yet significantly better than the `block` version.

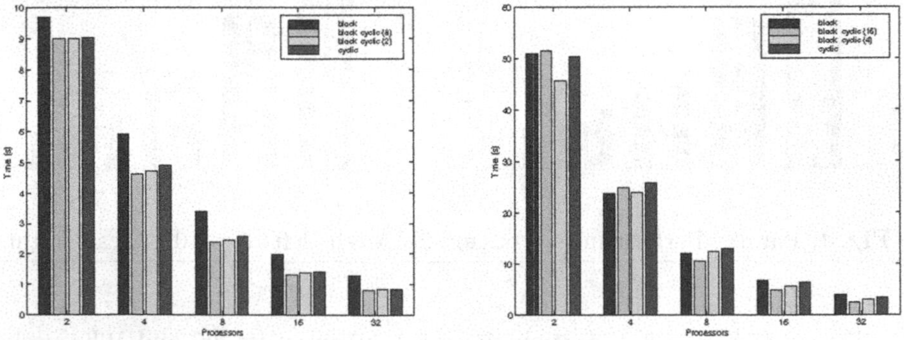

Fig. 5. Parallel Performance of `moldyn`: 2K Mesh (left) and 10K Mesh (right).

Figure 5 shows the execution times for `moldyn` on its 2K and 10K datasets. The sequential execution time on the 2K dataset was 10.80 seconds. The speedups of the 2 processor versions were 1.11 with the `block` strategy, and 1.20 with the `bc(8)`, `bc(2)`, and `cyclic` strategies. The relative speedups in going from 2 to 32 processors were 7.05 with the `block` strategy, 11.17 with the `bc(8)` strategy, 10.82 with the `bc(2)` strategy, and finally, 10.55 with the `cyclic` strategy. The `bc(8)` version performed the best, although the `bc(2)` version differed by at most 3%. The `block` version was consistently slower by 7% to 37%. This is because the 2K dataset resulted in a severe load imbalance using a block distribution, even on small configurations.

The sequential execution time on the 10K dataset was 28.98 seconds. For this dataset, the two processor versions showed significant slowdowns compared to the sequential version. The "speedups" of the 2 processor versions were 0.57 with the `block` strategy, 0.56 with the `bc(16)` strategy, 0.63 with the `bc(4)` strategy, and finally, 0.58 with the `cyclic` strategy. We believe that the loss of locality is the main reason for the performance on the 2 processor versions. The level of performance degradation is dataset dependent, and turns out to be much more significant on this dataset, as compared to the two meshes used for `euler` and the 2K dataset used for `moldyn`. The relative speedups in going from 2 to 32 processors were quite good. They were 11.78 with the `block` strategy, 18.54 with the `bc(16)` strategy, 15.51 with the `bc(4)` strategy, and finally, 13.84 with the `cyclic` strategy. Like the 2K results for `moldyn`, a block-cyclic version gave the best performance on all processors (on 4 processors, the `block` version was similar). On 2 and 4 processors, the `bc(4)` version was the better performer while on 8 or more processors, the `bc(16)` version performed best. We attribute this to the following: as the number of processors increases, a larger block size improves locality but not at the expense of creating a load imbalance. The load balance achieved with the cyclic distribution is not very good for this dataset. As a result, the `block` version performed similarly to the `cyclic` version.

4 Summary

In this paper, we have presented an experimental study to evaluate the impact of data distribution on the performance of irregular reductions executing on a multithreaded architecture. The execution strategy we proposed for such an architecture made the volume and frequency of communication independent of the data distribution. However, the choice of data distribution influenced locality and load balance, and thus the overall performance. The block distribution maintained better locality, and often performed the best on a small number of processors. However, it did not achieve good load balance on a larger number of processors, resulting in low performance. The cyclic distribution achieved a good load balance, but resulted in a significant loss of locality on some datasets. The block-cyclic distribution achieved a good balance between locality and load balance, and had the best overall performance on a large number of processors.

References

1. S.T. Barnard and H. Simon. A fast multilevel implementation of recursive spectral bisection for partitioning unstructured problems. Technical Report RNR-92-033, NAS Systems Division, NASA Ames Research Center, November 1992.
2. R. Das, D. J. Mavriplis, J. Saltz, S. Gupta, and R. Ponnusamy. The design and implementation of a parallel unstructured Euler solver using software primitives. *AIAA Journal*, 32(3):489–496, March 1994.
3. Herbert H. J. Hum, Olivier Maquelin, Kevin B. Theobald, Xinmin Tian, Guang R. Gao, and Laurie J. Hendren. A study of the EARTH-MANNA multithreaded system. *International Journal of Parallel Programming*, 24(4):319–347, August 1996.
4. Yuan-Shin Hwang, Raja Das, Joel H. Saltz, Milan Hodoscek, and Bernard R. Brooks. Parallelizing molecular dynamics programs for distributed memory machines. *IEEE Computational Science & Engineering*, 2(2):18–29, Summer 1995. Also available as University of Maryland Technical Report CS-TR-3374 and UMIACS-TR-94-125.
5. C. Koelbel, D. Loveman, R. Schreiber, G. Steele, Jr., and M. Zosel. *The High Performance Fortran Handbook.* MIT Press, 1994.
6. S.S. Mukherjee, S.D. Sharma, M.D. Hill, J.R. Larus, A. Rogers, and J. Saltz. Efficient support for irregular applications on distributed-memory machines. In *Proceedings of the Fifth ACM SIGPLAN Symposium on Principles & Practice of Parallel Programming (PPOPP)*, pages 68–79. ACM Press, July 1995. ACM SIGPLAN Notices, Vol. 30, No. 8.
7. R. Ponnusamy, J. Saltz, A. Choudhary, Y.-S. Hwang, and G. Fox. Runtime support and compilation methods for user-specified irregular data distributions. *IEEE Transactions on Parallel and Distributed Systems*, 6(8):815–831, August 1995.
8. Joel Saltz, Kathleen Crowley, Ravi Mirchandaney, and Harry Berryman. Run-time scheduling and execution of loops on message passing machines. *Journal of Parallel and Distributed Computing*, 8(4):303–312, April 1990.
9. Kevin B. Theobald, José Nelson Amaral, Gerd Heber, Olivier Maquelin, Xinan Tang, and Guang R. Gao. Overview of the Threaded-C language. CAPSL Technical Memo 19, Department of Electrical and Computer Engineering, University of Delaware, Newark, Delaware, March 1998. In `ftp://ftp.capsl.udel.edu/pub/doc/memos`.
10. Kevin Bryan Theobald. *EARTH: An Efficient Architecture for Running Threads.* PhD thesis, McGill University, Montréal, Québec, May 1999.
11. Janet Wu, Raja Das, Joel Saltz, Harry Berryman, and Seema Hiranandani. Distributed memory compiler design for sparse problems. *IEEE Transactions on Computers*, 44(6):737–753, June 1995.

Implementing and Benchmarking Derived Datatypes in Metacomputing

Edgar Gabriel, Michael Resch, and Roland Rühle

High Performance Computing Center Stuttgart
Allmandring 30, D-70550 Stuttgart, Germany
gabriel@hlrs.de

Abstract. Flexible data structures have become a common tool of programming also in the field of engineering simulation and scientific simulation in the last years. Standard programming languages like Fortran and C or C++ allow to specify user defined datatypes for such structures. For parallel programming this leads to a special problem when it comes to exchanging data between processors. Regular data structures occupy contiguous space in memory and can thus be easily transferred to other processes when necessary. Irregular data structures, however, are more difficult to handle and the costs for communicating them may be rather high. MPI (Message Passing Interface) provides so called "derived datatypes" to overcome that problem, and for many systems these derived datatypes have been implemented efficiently. However, when running MPI on a cluster of systems in wide-area networks, such optimized implementations are not yet available and the overhead for communicating them may be substantial. The purpose of this paper is to show how this problem can be overcome by considering both the nature of the derived datatype and the cluster of systems used. We present an optimized implementation and show some results for clusters of supercomputers.

1 Introduction

MPI (Message-Passing Interface) has become the standard library for communication in distributed memory parallel systems. It provides the user with different abstraction layers for the communication that allow to express different levels of complexity of communication necessary for an application. The base level is the exchange of a chunk of memory between two processes. In the simplest case, such a contiguous piece of memory is copied from one processors' buffer to another one's. The programmer only has to specify the start address of the memory block and its size. For regular data structures like matrices and vectors which are common in many legacy codes this perfectly reflects the users' needs.

However, using more complex and irregular data structures this low level approach is no longer feasible. MPI therefore provides a higher level of abstraction called "derived datatype". The derived datatypes routines give the user the possibility to describe the data structures to MPI. The user then does not have to take care of packing the right data into the right buffer at the sending process. And there is no need for unpacking them at the receiving side. Giving

B. Hertzberger et al. (Eds.): HPCN Europe 2001, LNCS 2110, pp. 493–502, 2001.

a generic description of a new datatype and making this description available to the MPI library the application can use this datatype in any point-to-point communication as well as for several collective operations.

For MPI implementations, the handling of such derived datatypes is not trivial on a single system. One has to introduce a very general framework to support them and has to make sure, the right data is transferred in each communication. Things become even more complicated when moving from the homogeneous environment of a single system (Massively Parallel Processors or Parallel Vector Processors) to the heterogeneous environment of hierarchically clustered system, as can be found in grid environments.

For quite a long time, the optimization of handling of derived datatypes didn't have a high priority for MPI implementors. Nevertheless, some work has been done in this field recently. This reaches from the optimization of loops for the Pack/Unpack operations [4], [5], to strengthening the type matching rules of MPI [8] and to the development of accurate benchmarks for derived datatypes [6]. Most remarkably, the FT-MPI project [7] introduces three different packing techniques for derived datatypes, depending on the datatype and user settings. All these approaches are quite interesting and have shown good results. However, further work is necessary for optimization in that field.

The purpose of this paper is to present an optimized implementation of MPI's derived datatypes for hierarchically clustered systems. This is part of a full implementation of MPI for clusters of MPPs and PVPs called PACX-MPI [3]. Experiments with this library have shown that an efficient implementation of derived datatypes is crucial for the performance of the communication.

The structure of the paper is as follows: In chapter 2 the basic concept of PACX-MPI is introduced which sets the framework for the following chapters. General issues concerning the implementation of derived datatypes are discussed in chapter 3. This general discussion leads to the specific method chosen in PACX-MPI which is described in more detail in chapter 4. In order to verify the performance of our work we did a number of communication tests. The results of are shown in chapter 5.

2 Concept of PACX-MPI

PACX-MPI is an implementation of the message-passing standard MPI which aims to support the coupling of high performance computing systems distributed in a grid. The characteristics of such clustered systems from the communication point of view are at one hand low communication latencies (in the order of microseconds) and high bandwidth (in the range of several hundred Megabytes / second) for communication between MPI processes on the same host (referred to as *internal communication* throughout the rest of this paper). On the other hand, the library and the simulation have to deal with high communication latencies (in the range of tens of milliseconds) and small bandwidth (ranging from a few Kilobytes / second to a few Megabytes/second) for communication between MPI processes on different hosts, called *external communication.*

Taking the characteristics of clustered system into account, PACX-MPI relies on three main concepts:

1. Two level hierarchy: in clustered systems, a message-passing library has to deal with two different levels of quality of communication. Therefore the library uses two completely independent layers, one to handle operations inside a single system, and one to handle operations between different systems.
2. Usage of the optimized vendor MPI library: Internal operations are handled using the vendor MPI environment on each system. This allows to fully exploit the capacity of the underlying communication subsystem in a portable manner.
3. Usage of communication daemons: on each system in the metacomputer two daemons take care of communication between systems. This allows to bundle communication and to avoid to have thousands of open connections between processes. In addition it allows to handle security issues centrally. The daemon nodes are implemented as additional, local MPI processes. Therefore, no additional TCP-communication between the application nodes and the daemons is necessary, which would needlessly increase the communication latency.

An additional list of features, like optimized point-to-point operations for each communication layer, optimized collective operations for hierarchically clustered systems [3], and optionally data compression for the communication between different machines to reduce the size of the transferred data, makes PACX-MPI well suited for heterogeneous, clustered systems. All these features have allowed PACX-MPI to be used successfully in a number of metacomputing projects linking resources in Japan, the US and Europe [1,2].

3 Analysis of MPI's Derived Datatypes

A derived datatype in MPI can be described as a collection of blocks of variables. Each block holds a certain number of items of the same basic datatype (e.g. integers). The size of each block may vary. Throughout this paper we will use N as the argument for the Number of blocks and l for the length of each block in bytes. This simplification does not limit the scope of our investigation too much but allows to set up a simplified theoretical model that helps to better understand the performance issues involved. All the findings can be generalized for more complex derived datatypes.

In the following we give a short overview of possible implementation methods for derived datatypes in MPI. Basically there are two options: the Pack/Unpack Model and the the Direct Communication Model. Additionally, we will have a closer look at two implementation possibilities, which are bound to some specific hardware characteristics.

3.1 Pack/Unpack Model

In this model, the structure specified for the derived datatype by the user is first copied into a contiguous memory segment (see left part of figure 1). This contiguous memory block is then sent to the destination processes, where the data is copied again from the contiguous memory segment into the user structure. This model therefore involves N memcpy operation on each side. Thus, the overall execution time $t_{pack/unpack}$ for a point-to-point operation for the datatype described above is:

$$t_{pack/unpack} = 2 \cdot N \cdot t_{memcpy}(l) + t_{transfer}(N \cdot l) \quad (1)$$

where N is the number of blocks of the derived datatype, l is the length of each block of the derived datatype, $t_{memcpy}(l)$ is the time for copying a chunk of data of length l out of or into a buffer and $t_{transfer}(N \cdot l)$ is the actual transfer time between two processes for a message of size $N \cdot l$. In general the costs for sending a derived datatype in this model are dominated by the actual transfer costs and by the costs for the 2 memcpy operations.

3.2 Shared Memory Model

On shared memory systems point-to-point communication is often implemented in two steps. First the sender copies the data into a shared memory segment which can be accessed by both processes involved. Second the data are copied from that segment into the memory of the receiving process.

For derived datatypes this two-hop communication can easily be used to avoid the additional local memcpy operations: the sender copies all elements of the derived datatypes into the shared-memory segment, and the receiver can copy the data from the shared memory segment directly into the data structures of the user (see right part of figure 1).

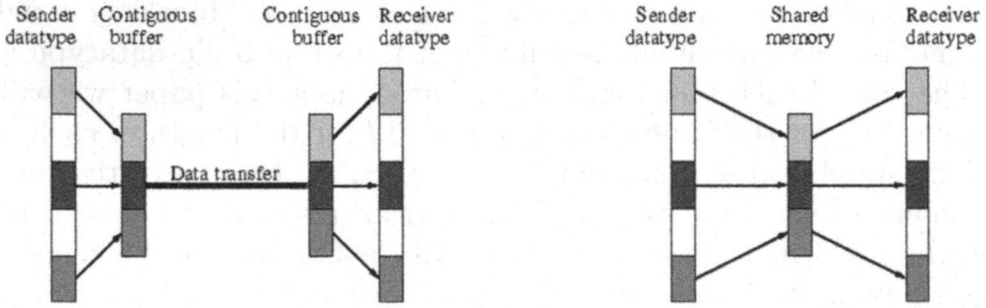

Fig. 1. Concept of the Pack/Unpack Model (left) and the Shared Memory Model (right).

3.3 Direct Communication Model

This model avoids all memcpy operations necessary for the Pack/Unpack Model by sending each element of the derived datatype separately to the destination node. The total time for execution $t_{directcomm}$ of a point-to-point operation is therefore:

$$t_{directcomm} \leq N * t_{transfer}(l) \tag{2}$$

The $\leq$ indicates that because of pipelining effects some parts of communications may overlap which may lead to a lower total amount of time spent in the communication process for the derived datatype.

Modified Direct Communication Model. This model is basically identical to the Direct Communication Model above. The major difference is, that on some systems, the MPI library is built on top of another communication library, which may be much faster. The Cray T3E implementation of MPI with a latency of $15\mu sec$ is for example based on the Cray shmem-library, which in fact has a communication latency of only about $2-3\mu sec$. Thus, the sending of N messages in this model does not need N MPI messages, but N shmem_put or shmem_get operations, which can be executed much faster. The total time for the Direct Communication Model is then reduced by the increase in speed for the transfer of a block of data of length l.

Equations (1) and (2) indicate the dependency of all models on the following parameters:

1. The derived datatype itself: important characteristics are the number of blocks and the length of each block in the derived datatype.
2. The memory subsytem of the machine: there is no way to simply characterize the performance of memcpy operations without going into machine-specific details. Some indicators for the memory subsystem could be the startup-time for a memcpy operation and the effective bandwidth of memcpy operations.
3. The communication subsystem of the machine: again, there is no way to simply characterize the behavior of the communication subsystem. One could use the communication latency and the communication bandwidth as clues for the quality of the communication subsystem.

For a deep analysis of all these factors a more detailed specification of the mathematical model would have to be done. For the practical investigations in this paper simple models for modelling communication and memcpy operations are enough.

4 A Grid-Enabled Implementation of Derived Datatypes

In Metacomputing, the first two parameters mentioned above remain unchanged for every operation, while the third factor depends on the location of the destination processes. Inside a single system the communcation may be very fast while between two different systems, the communication may be rather slow. Therefore, communication speed depends on which communication subsystem can be used to send the message to the destination node. Depending on the location of the destination processes, different models might have to be used to achieve optimal performance:

The first reason is, that one single model may not be available on both communication subsystems. For example, there is no way to make the Shared Memory Model run across different machines.

The second reason is, that since the execution time of a point-to-point operation is strongly influenced by the communication subsystems, different methods may be optimal for different destination nodes. Lets consider a datatype consisting of a small number of blocks N and small length per block, such that $t_{transfer}(l) \approx t_{latency}$ and $t_{transfer}(N \cdot l) \approx t_{latency}$, with $t_{latency}$ being the latency of the communication subsystem. In this case, equation (1) for the Pack/Unpack Model becomes

$$t_{pack/unpack} \approx 2 \cdot N \cdot t_{memcpy}(l) + t_{latency}.$$

For the Direct Communication Model equation (2) reduces to:

$$t_{directcomm} \approx N \cdot t_{latency}.$$

We can see that $t_{pack/unpack}$ is smaller than $t_{directcomm}$ if

$$t_{memcpy}(l) < t_{latency} * \frac{N-1}{2N}. \tag{3}$$

Equation (3) shows, that depending on $t_{latency}$ the Pack/Unpack Model may or may not be faster than the Direct Communication Model for the same derived datatype. These findings have now to be considered for metacomputing where the costs for a transfer may be substantially higher than the costs for a local memcpy operation.

4.1 The Implementation of Derived Datatypes in PACX-MPI

The implementation of derived datatypes in PACX-MPI takes care of the hierarchical communication system given in clustered wide-area systems, by providing different ways of handling derived datatypes, depending on the location of the destination process.

For internal communication, PACX-MPI uses the methods provided by the vendor MPI library. Since these libraries are usually highly optimized, they can implement and use the model which suits their system best, like the shared memory model on SMP's or other optimization possibilities.

For external operations, PACX-MPI has implemented the Pack/Unpack Model, since avoiding several wide-area latencies has turned out to be critical for the communication performance.

To describe a derived datatype, PACX-MPI uses a flat tree. Thus, even if the user defines recursive data types these are reduced by PACX-MPI to basic datatypes of MPI. Additionally, PACX-MPI has implemented an algorithm to detect, whether two blocks form a contiguous segment in memory. Two blocks can be merged to a single one if both blocks consist of the same basic datatype and the distance between them is exactly the number of elements in block i times the extent of the datatype of block i. The conditions expressed in formulas are:

$$d(i) = d(i+1) \tag{4}$$

$$n_e(i) \cdot extent(i) = displ(i+1) - displ(i) \tag{5}$$

with $d(i)$ being the basic datatype used in block i, $extent(i)$ being the extent of the basic datatype of block i, $displ(i)$ being the absolute displacement to a reference point in the memory of the block i and $n_e(i)$ being the number of basic datatypes in block i of type $d(i)$.

5 Tests and Results

We evaluate two platforms in this chapter and compare their implementation of the derived datatypes to the implementation of PACX-MPI. The platforms on which we do our experiments are the Cray T3E and the NEC SX-5. We perform ping-pong tests between two MPI processes using different derived datatypes. The executed test are:

Byte-test: For this test we send a stream of bytes without any structure that may cause an overhead to the communication. This is the most basic test. The results of all other tests will be compared to this one, since the deviations from the byte-test gives a measure for the quality of the implementation of the derived datatypes.

Vector-test 1: The datatype in this test was setup using MPI_Type_vector. The derived datatype consists of two integers. To make sure that no additional overhead occurs there is no gap between the two subsequent blocks when sending a message.

Vector-test 2: The datatype in this test was also setup using MPI_Type_vector. The datatype again consists of two integers, but this time we introduce a gap of two integers between two subsequent elements of this datatype.

Struct-test: For this test we used a datatype used in a real application. It is created using MPI_Type_struct and consists of 92 elements, which can be reduced to three blocks consisting of 11 KByte of characters, 61 32-Bit integers and 56 double, using the block-merging technique presented above. This datatype shows how important the block-merging technique presented in the chapter before can be. An MPI implementation not having this technique will need 92 memory operations to copy it into a contiguous buffer, while PACX-MPI recognizes, that just three memory operations are enough.

5.1 Results on the Cray T3E

The results achieved on the Cray T3E are presented in figure 2. We show the overhead measured for the handling of the different derived datatypes. It was determined by comparing the execution time for the byte-test with that of the derived datatypes for the same amount of data.

The results for the Cray MPI implementation show, that the overhead is very small for the vector-1 test. But for the vector-2 test and for the struct test, the overhead of the MPI implementation is tremendous and can reach more than 80 ms for the vector-2 test. This is about 40 times higher than sending the same amount of data in a contiguous message.

The behaviour can be explained with the implementation of derived datatypes on the Cray T3E. Like on the T3D, the MPI library distinguishes between regularly strided datatypes and unregular datatypes. Regular datatypes are using the Modified Direct Communication Model, while unregular datatypes are using the Pack/Unpack Model.

PACX-MPI on the same platform has a very small overhead for the vector-1 test and the vector-2 test. The overhead in both cases is in the range of 0.1 ms. For the struct test, PACX-MPI has an overhead of 3 ms. Compared to sending the same amount of data as a byte-stream, which takes approximately 23 ms with PACX-MPI, this means an overhead of only approximately 13 percent.

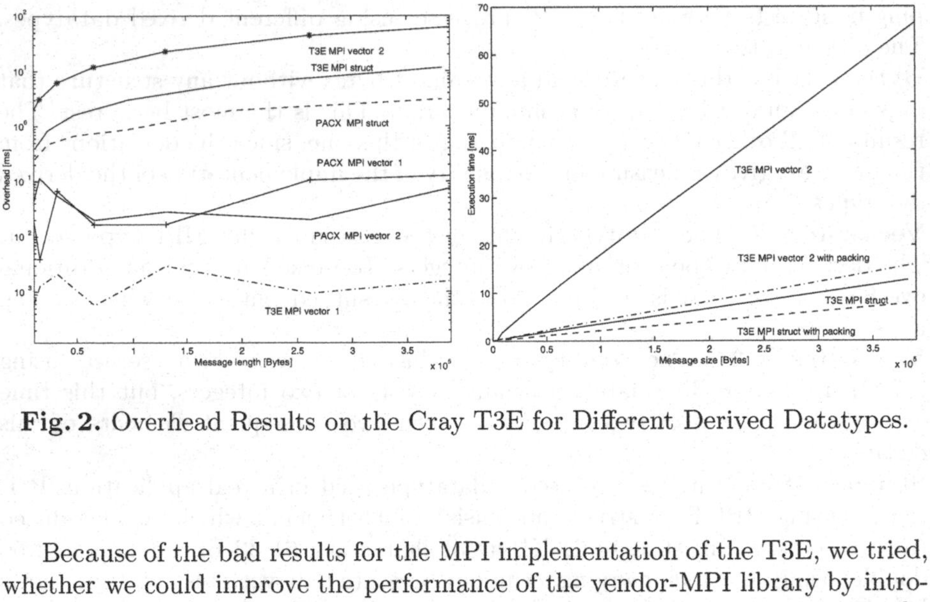

Fig. 2. Overhead Results on the Cray T3E for Different Derived Datatypes.

Because of the bad results for the MPI implementation of the T3E, we tried, whether we could improve the performance of the vendor-MPI library by introducing a pack/unpack step before sending the message to the destination node. The result of this test is presented in the right part figure 2. While the overhead for the struct-test has slightly been reduced, it has been reduced dramatically for the vector-2 test. By introducing this pack/unpack step the execution time could be reduced from 89 ms to approximately 21 ms for the biggest configuration, sending a total amount of 512 KByte of data. Obviously, condition 3 does not hold on the T3E for this datatype. Therefore the usage of the Pack/Unpack Model would achieve better results.

5.2 Results on the NEC SX5

The results achieved on the NEC-SX5 are shown in figure 3. For PACX-MPI, the results are quite similar as on the Cray T3E. There is a small overhead for the vector-1 and vector-2 test, and there is an overhead of up to 1.3 ms for the struct test.

The difference between the T3E and the SX-5 is mainly the quality of the vendor MPI library. On the SX-5, none of the tested derived datatypes show a significant overhead compared to sending only MPI_Bytes. Even the struct-test, which had some performance problems in previous versions is now handled in an optimal manner, such that an explicit packing and unpacking of the data is no longer necessary and even harmful.

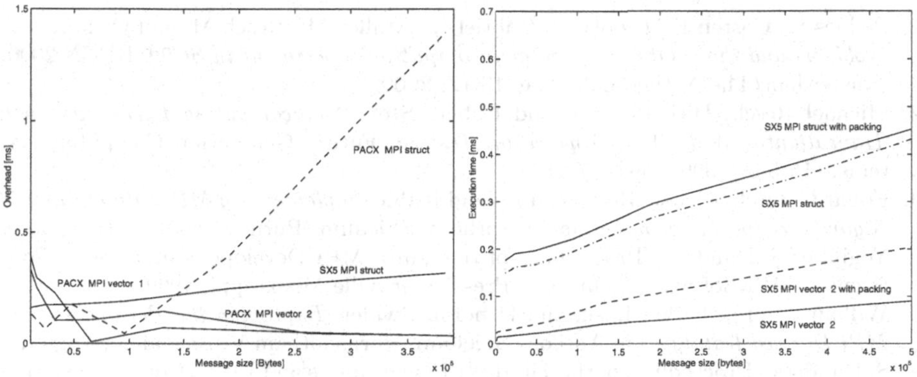

Fig. 3. Overhead Results on the NEC SX-5 for Different Derived Datatypes.

The NEC SX-5 is a good example for a platform, where the usage of a single model for the derived datatype in a wide area cluster could lead to a performance degradation. Figure 3 indicates, that the usage of the Pack/Unpack Model for communication between MPI processes on the same host would introduce superfluous memcpy operations, which would decrease the communication performance with derived datatypes on the SX-5.

6 Summary

We have presented in this paper an implementation of the Message Passing Interface (MPI) for distributed wide-area systems. The focus of this paper was on the implementation of the derived datatypes of MPI. We pointed out the need for different methods of handling derived datatypes within one library, depending on the characteristics of the underlying communication system. PACX-MPI decided to use the vendor-MPI methods to handle derived datatypes for internal operations, since they are best optimized to the given hardware. For external operations, PACX-MPI uses the Pack/Unpack Model, since avoiding multiple wide-area latencies is, at least for small messages, crucial to achieve good performance.

The limitations of this implementation have turned out to be the vendor-MPI library. On some platforms, like the NEC SX-5, the assumption, that the vendor MPI library knows best how to handle user defined datatypes, is true. On these platforms, PACX-MPI performs as good as the vendor MPI. On other platforms, like the Cray T3E, the handling of derived datatypes has turned out to be not optimal. In this case we have to replace their methods by our own ones.

References

1. Pickles S., Costen F., Brooke J., Gabriel E., Müller M., Resch M. and Ord S., *The problems and the solutions of the metacomputing experiment in SC99*, HPCN'2000, Amsterdam/The Netherlands, May 10-12, 2000.
2. Michael Resch, Dirk Rantzau and Robert Stoy, *Metacomputing Experience in a Transatlantic Wide Area Application Testbed*, Future Generation Computer Systems, (15)5-6 (1999), pp. 807-816.
3. Edgar Gabriel, Michael Resch and Roland Rühle *Implementing MPI with Optimized Algorithms for Metacomputing* in Anthony Skjellum, Purushotham V. Bangalore, Yoginder S. Dandass, 'Proceedings of the Third MPI Developer's and User's Conference', MPI Software Technology Press, Starkville Mississippi, 1999.
4. William Gropp, Ewing Lusk, and Deborah Swider *Improving the Performance of MPI Derived Datatypes* in Anthony Skjellum, Purushotham V. Bangalore, Yoginder S. Dandass, 'Proceedings of the Third MPI Developer's and User's Conference', MPI Software Technology Press, Starkville Mississippi, 1999.
5. J.L.Träff, R. Hempel, H. Ritzdorf, and F. Zimmermann *Flattening on the fly: efficient handling of MPI derived datatypes* in Jack Dongarra, Emilio Luque, Tomas Margalef (Eds.) 'Recent Advances in Parallel Virtual Machine and Message Passing Interface', pp 109-116, Springer, 1999.
6. Ralf Reussner, Jesper Larsson Träff, Gunnar Hunzelmann *A Benchmark for MPI Derived Datatypes* in Jack Dongarra, Peter Kacsuk, Norbert Podhorszki (Eds.) 'Recent Advances in Parallel Virtual Machine and Message Passing Interface', pp 10-17, Springer, 2000.
7. Graham E Fagg and Jack J. Dongarra *FT-MPI: Fault Tolerant MPI, Supporting Dynamic Applications in a Dynamic World* in Jack Dongarra, Peter Kacsuk, Norbert Podhorszki (Eds.) 'Recent Advances in Parallel Virtual Machine and Message Passing Interface', pp 346-353, Springer, 2000.
8. William D. Gropp *Runtime Checking of Datatype Signatures in MPI* in Jack Dongarra, Peter Kacsuk, Norbert Podhorszki (Eds.) 'Recent Advances in Parallel Virtual Machine and Message Passing Interface', pp 160-167, Springer, 2000.

MPC++ Performance for Commodity Clustering

Yoshiaki Sakae[1] and Satoshi Matsuoka[2]

[1] Tokyo Institute of Technology, Japan
[2] Tokyo Institute of Technology/JST, Japan
{sakae,matsu}@is.titech.ac.jp

Abstract. In order to verify the viability finer-grained parallel language MPC++, which had originally been developed for Myrinet-specific environments, we performed ports on top of different breeds of MPI, to be executed on two networks of large performance/cost difference, as well as porting NPB 2.3 apps to test ease of expressiveness of parallel programs. Results were positive, (a) the port of the NPB 2.3 apps were effortless, (b) small penalty of additional MPI layer was negligible for NPB applications, and (c) for large data sets, MPC++/MPI on the 100Base-T network was competitive to both the C+MPI on Myrinet, and the original implementation of MPC++ on PM/Myrinet.

1 Introduction

Although commodity "Beowulf" clusters are becoming widespread, programming on such with languages with the class of parallel languages that provide finer-grained multi-threading and fast message passing at the language level has been believed to require expensive messaging hardware with low latency and high bandwidth. Examples are parallel object-oriented language MPC++[1], *independently* developed in the past at the Real World Computing Partnership (RWCP), Split-C[2](UC-Berkeley) and Charm++[3](UIUC). In particular, the original MPC++ assumes a specialized user-level messaging library PM[4] on top of fast and relatively expensive networks, such as Myrinet[5].

We claim that such languages still lacks systematic studies to identify (1) whether they embody sufficient expressive power to easily describe traditional parallel programs, (2) how much performance one expects to maintain/sacrifice by using commodity software/hardware, especially commodity networking layer (including software), for their implementation, and (3) the degree of scalability compared to dedicated software/hardware implementations.

In order to verify the viability of such parallel languages, we took MPC++ which originally required specialized software/hardware layers, in a portable way on top of different breeds of MPI, to be executed on two networks of substantial performance/cost differences, namely, Myrinet and 100Base-T Ethernet. We then investigated whether some NAS Parallel Benchmark (NPB)-2.3 applications (CG, IS) can be ported "naturally" on top of MPC++, to be benchmarked in

B. Hertzberger et al. (Eds.): HPCN Europe 2001, LNCS 2110, pp. 503–512, 2001.

such a environment. Finally we performed detailed analysis of the overhead of MPI layer and compared its performance to original implementation of MPC++ on more "expensive" platforms. We have found that, (1) portings of programs from C+MPI were a matter of few hours, including debugging, (2) benchmark apps ported to MPC++ on MPI performs and scales well, compared both to a) the original application written with C+MPI as well as b) the application ported to run on MPC++, but run on the original implementation that employed the fast user-level communication library PM on Myrinet, and (3) benchmark apps on MPC++ on MPI on commodity 100Base-T network ran and scaled competitively, so long as the problem size was large, and the local computation/communication ratio increased as we scaled the problem larger. The results indicate that languages such as MPC++ could be made to run efficiently on a commodity clusters when running 'standard' parallel programs.

2 MPC++ and Port to MPI

The language features of MPC++ v.2.0 Level 0[1] include object-oriented features of C++, finer-grained, user-level multi-threading, fast remote method invocation, remote memory read/write, synchronizing data structures, etc., basically embodying the features of so-called "concurrent object-oriented programming languages". Program code is distributed identically to all physical processors and a process for the program runs on each processor. Each process has several finer-grained threads of control which are not preemptable. A program may locally or remotely invoke a *function instance* with its own thread of control, using the `invoke` and `ainvoke` template functions. Invoking a function instance will involve creation of a new thread and the execution of the function. The original thread invoking the function instance could either block until the end of the invoked function instance execution, or could continue asynchronously in parallel. All variables are processor-local. In order to access variables of remote processors, a *global pointer* must be employed which provides remote variable read-write transparently at the language level.

The original MPC++ implementation was tightly coupled with the underlying PM communication library[4], which in turn was originally tightly coupled with Myrinet. This provided for both low-latency and high-bandwidth communication, as well as finer-grained multithreading via fast user-level communication handling.

In order to achieve portability on commodity clusters, we segregated MPC++ from PM and the custom threading layer, as seen in (Fig. 1). We centralized and re-defined the communication layer API so that various communication libraries could be employed; for the purpose of the paper we have employed MPI, but other communication libraries such as VIA[6] could be used. We reimplemented the threads as standard user-level threads using standard techniques, gaining portability at the expense of efficiency of context switching and synchronization. The resulting artifact, MPC++ on MPI, could be run on a commodity clustering environment without any special software, driver, or hardware installation.

Although there have been middle-tier libraries such as Nexus[7] and Madeleine[8] that provide common infrastructure for communication and multithreading, the point of this particular research is what happens if such features are provided integrally at the language level, rather being employed as a library . Providing features at the language level does have its pros and cons; one could either eliminate the overhead by using language-level semantics, but one could also decrease efficiency by constraining a user to the particular programming model and implementation thereof. For example, the implementation of CC++[9] on Nexus largely assumes coarser-grained user programs than MPC++.

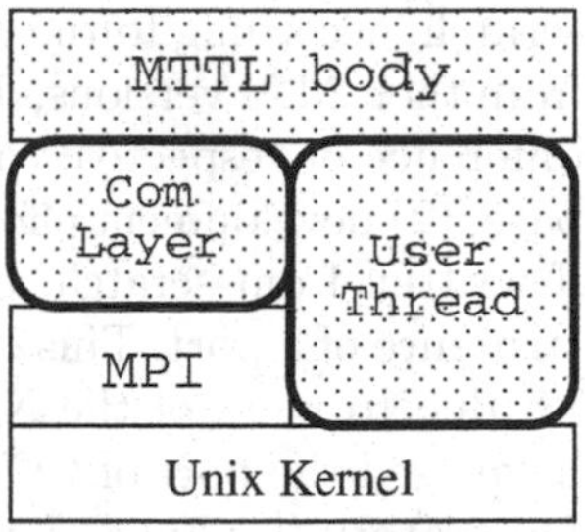

Fig. 1. Structure of MPC++/MPI.

3 Evaluation Procedure

Because of the commodity platform, the overhead includes the a) extra software overhead imposed by the MPI software layer, and b) overhead of the underlying communication layer on which MPI is implemented, such as the TCP/IP, including both software and hardware layers. Threading efficiency is sacrificed as mentioned earlier. Also, in the current implementation, collective communication such as "Reduction" or "Barrier" is implemented at the MPC++ level, and currently does not utilize the collective communication features of the underlying MPI. Although this was done for portability reasons this could turn out to be less efficient due to extra software handling overhead.

In order to investigate (1) whether MPC++ allows easy expressiveness of traditional SPMD style parallel programs, (2) how much performance one retains/loses by using commodity software/hardware, and (3) the degree of scalability compared to dedicated software/hardware implementations, we first created the portable implementation of MPC++ as mentioned in Section 2. Then, we ported the CG and IS from NPB-2.3 benchmarks onto MPC++ to identify whether MPC++ could readily express these applications. Next, we performed comprehensive benchmarks of the applications, varying the followings:

- number of processors (1–32),
- problem size (class A and B),
- messaging layer (MPI vs. Native PM),
- different incarnations of MPI (LAM[10] vs. MPICH[11]),
- low-level software messaging layer (PM vs. TCP/IP)
- underlying hardware messaging layer (Myrinet vs. 100Base-T Ethernet).

3.1 Porting NPB CG and IS to MPC++

The NPB-2.3 benchmarks are written in parallel SPMD-style. All are in Fortran+MPI, except for IS which is written in C+MPI.

We had initially considered rewriting the NPB benchmarks in MPC++ from scratch, according to the NPB-2.3 specification[12]. However, we decided to port from their MPI versions, due to: (1) it would be difficult to precisely quantify the difference of respective communication layers, due to drastically different code base, (2) development effort will be substantial, and (3) it was deemed that, even if written from scratch, there is a good change that it will resemble the code structure of a port. This decision still restricted us to the use of NPB to CG and IS, due to most of the NPB programs being written with Fortran+MPI. (We have used the port of CG to C + threads by Yoshio Tanaka at Electrotechnical Laboratory, Tsukuba, Japan.)

Here are the general strategies employed to port CG and IS from their C+MPI versions to the MPC++ versions:

- *The main control thread on node 0 distribute work to slave threads on other nodes.* MPI program written in SPMD style is executed equally on each compute node. Thus, on parallelization by dividing the task to each node, one must make conditionals according to the node number of each node. For example, serialized region must be guarded by a `if` statement to check for execution on node 0. For MPC++, on the other hand, even if the program is written in SPMD style, its main thread will be executed only on node 0 and the other nodes will each wait for a divided task to be assigned from node 0 via remote invocations. Hence we must modify the MPI program so that the control thread on node 0 performs such distribution to slave threads on other nodes on initialization of a parallel region.
- *Translation of message send/receive pair to remote read/write.* In a SPMD MPI program, the programmer usually pairs the `MPI_Send` on the send side and `MPI_Recv` on the receive side manually, and will be careful to avoid any deadlocks since all the nodes will be executing the (possibly blocking) `MPI_Recv`. On the other hand, in MPC++ the programmer can perform one sided access to remote memory with a global pointer, if the remote address is known earlier. Here, one must be careful with synchronization and updating so that correct data will be read by the receiver. This is usually achieved by preceding the reads/writes with a global barrier.
- Translation of other, simpler MPI primitives to MPC++. These include operations such as collective operations.

Although it is not clear whether all SPMD program in this manner, in practice, we have found it rather straightforward to port C NPB programs into MPC++, using the above strategies. The specifics of each program is as below:

Port of CG. CG measures the time elapsed in solving an unstructured sparse linear system with the conjugate gradient method. For direct comparison with an MPC++ version, we used the port of the original Fortran+MPI CG by Tanaka which parallelized the program with C+Threads into C+MPI, and then subsequently ported it to MPC++. The modification from C+MPI into MPC++ took approximately 2 hours, and the total modified lines of code are about 60 out of 800 lines.

Port of IS. IS is an integer sorting program. Its kernel loop consists of node internal histogram calculation, total histogram gathering, data re-distribution, ranked internal sorting, and subsequent validation. Three of these phases need to communicate: total histogram gathering, exchanging number of data to be re-distributed, and data re-distribution. These are implemented using MPI_Reduce, MPI_Alltoall and MPI_Alltoallv in the original C+MPI version. Due to the simplicity of the data structures, we were able to do the port in approximately 1.5 hours. The total lines of code modified are about 50 out of 1000 lines.

4 Performance Evaluation

Based on the ported codes, we conduct performance evaluations on a major subset of the combinations as described in Section 3, as shown in Table 1. Some combinations are simply not available (e.g., MPC++ on LAM on PM). Here, Native-PM denote the original incarnation of PM on Myrinet, while PM-Ether[13] denote the port of PM onto Ethernet. MPICH-PM[14] is MPICH with PM as the underlying communication layer. Any combination of MPC++ with underlying MPI are the portable implementation in Section 2, while MPC++ directly coupled with PM (MPC++/PM-Ether and MPC++/Native-PM) is the original MPC++ implementation where PM is hardwired.

Table 1. Evaluated Data.

	Ethernet	Myrinet
CG	Fortran/LAM	Fortran/MPICH-PM/Native-PM
	C++/LAM	C++/MPICH-PM/Native-PM
	MPC++/LAM	MPC++/MPICH-PM/Native-PM
	MPC++/PM-Ether	MPC++/Native-PM
	MPC++/MPICH-PM/PM-Ether	
IS	C/LAM	C/MPICH-PM/Native-PM
	MPC++/LAM	MPC++/MPICH-PM/Native-PM
		MPC++/Native-PM

4.1 Evaluation Environment

As an evaluation platform, we employ a 32-node portion of our Presto I experimental PC cluster interconnected with 3 networks, i.e., dual independent full 32-port switches (Planex FHSW-3232NW) for the 100Base-T Ethernets, and with Myricom M2M-OCT-SW8 switch × 2 for the Myricom Myrinet (M2M-PCI64A-21). Each node has a single Pentium II 350Mhz processor with 256MB of SDRAM. The operating system is Linux 2.2.14, augmented with the RWCP SCore 3.0 as the clustering environment for Myrinet. Only one 100Base-T network (Intel EtherExpress Pro) was used. We employed the pgcc (Pentium gcc) compiler for all programs with the switches `-O6 -mcpu=i686 -malign-double -fstrength-reduce -funroll-loops -fexpensive-optimizations`.

4.2 Benchmark Results of CG and IS

The graphs Fig. 2, Fig. 3 show the number of nodes vs. elapsed time for CG for different combinations described above. Overall they exhibit reasonable scalability up to 16 nodes for class A and 32 nodes for class B, for all combinations. For class A, as node size increases (up to 32) the working-set size per node becomes quite small and the communication time becomes dominant for Ethernet, as we will verify later. As a result, there is no room for further speedup for node increase, both for C+MPI and MPC++. Fortran+LAM still shows speedup at 32, and this might be due to the relatively coarse-grained communication as compared to the C version (communication granularity for the two programs differ). For Myrinet, on the other hand, we still observe scalability at 32 nodes for all combinations. For Class B, since computation increases with $\mathcal{O}(n^2)$ while communication increases only with $\mathcal{O}(\log(n))$, communication time is not as dominant. Here, all systems except for MPC++/PM-Ether, shows scalability up to and possibly beyond 32 nodes [1].

We note that, even for Class A, as long as we are executing within the number of nodes where a particular combination is exhibiting scalability (up to 16–32 nodes), there is little significant difference in execution time, usually within 10-20%, and even at worst well within the factor of 2.

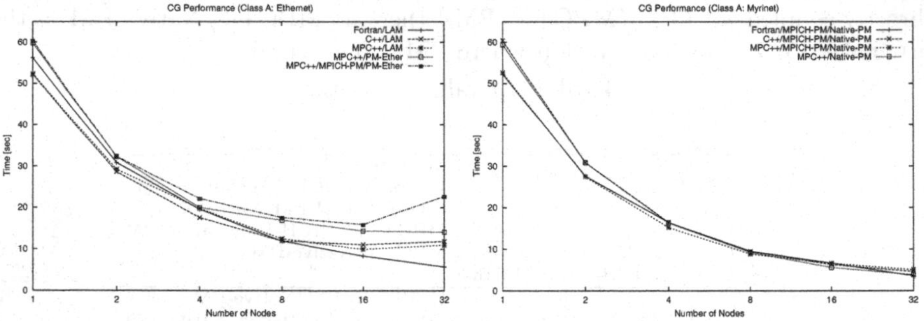

Fig. 2. CG Performance (Class A).

The graphs Fig. 4, Fig. 5 show the number of nodes vs. elapsed time for IS, again for different combinations described above. Since IS involves finer-grained communication compared to CG, communication overhead increases significantly along with the number of nodes. Still, they show some degree of speedup up to 16 nodes for class A and 32 nodes for class B, for all combinations. For Class A, Myrinet provides good scalability even at 32 nodes, irrespective of the intermediate software layer—as a matter of a fact The C/MPICH-PM/Native-PM combination shows almost the same performance as the MPC++/MPICH-PM/Native-PM version, and in fact is superior to the MPC++/Native-PM version. By contrast scalability can only be achieved up to 16 nodes for Ethernet, and moreover, the speedup is not as stable compared to Myrinet. For Class B,

[1] Fig. 3 dose not include MPC++/LAM graph, because of its mysterious behavior on Class B Ethernet.

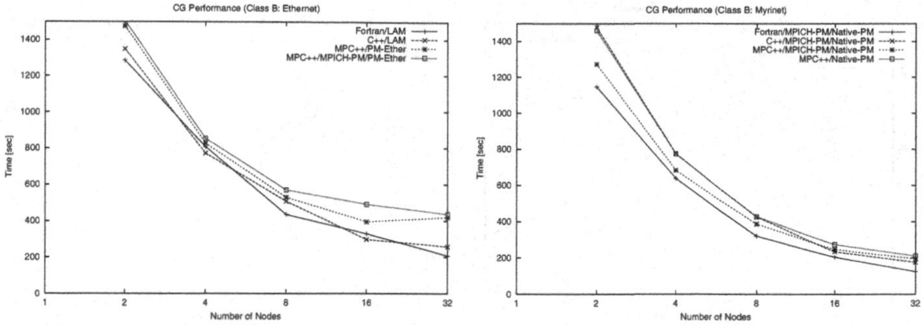

Fig. 3. CG Performance (Class B).

because of a larger working set we obtain much better scalability for Ethernet. One interesting note is that MPC++/LAM continues to scale whereas it levels off for C/LAM. There are several potential reasons for this, but they will require further detailed profiling to determine.

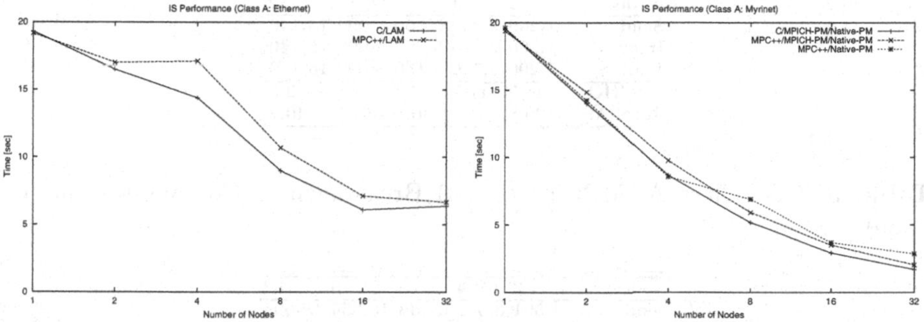

Fig. 4. IS Performance (Class A).

Although scalability was somewhat limited beyond 16 nodes for class A, for (more realistic) class B both benchmarks exhibited competitive performance and scaled well under both 100Base-T and Myrinet. These observations support the viability (albeit somewhat preliminarily) of portable MPC++/MPI implementation to execute well not only on dedicated platforms, but also on everyday commodity platforms.

4.3 Details of CG (Class A)

We analyzed the breakdowns of communication, to investigate the rather subtle performance difference between the original NPB and the MPC++ CG code. Table 2 (C/LAM) and Table 3 (MPC++/LAM) show the results. The numbers indicate the average number of respective MPI operations per node (collective communication is tallied as count of one for each node), "AVG" indicates the average message length per message, and "Total" denotes the average of total number of bytes sent as messages per node. Figure 6 shows the breakdown of time

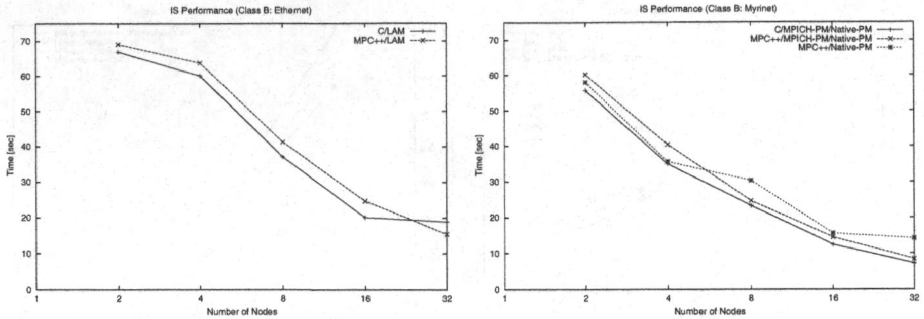

Fig. 5. IS Performance (Class B).

spent on communication/computation for CG for C/LAM and MPC++/LAM. (Possible overlap of communication/computation are not taken into account.)

Table 2. CG (Class A: C/LAM) Breakdown of Communication (per node).

Nodes	2	4	8	16	32
Send	1,200	2,790	5,160	9,090	16,140
Irecv	1,200	2,790	5,160	9,090	16,140
Wait	1,200	2,790	5,160	9,090	16,140
AVG(KB)	18.2	11.7	7.4	4.5	2.6
Total(MB)	20.8	31.3	36.5	39.1	40.4

Table 3. CG (Class A: MPC++/LAM) Breakdown of Communication (per node).

Nodes	2	4	8	16	32
Send	1,591	3,572	6,333	10,654	18,095
Isend	390	1,170	2,730	0	0
Recv	1,981	4,742	9,063	10,654	18,095
Iprobe	1,598	3,575	6,334	10,654	18095
Test	390	1170	2,730	0	0
AVG(KB)	10.80	6.77	4.14	3.87	2.37
Total(MB)	20.9	31.4	36.6	39.4	40.8

In comparison, both send almost equivalent amount of data, but the MPC++/LAM average message size is smaller, resulting in greater number of messages. This is attributable to small control messages, as well as the artifact of converting from C+MPI to MPC++ code. Still, the effect of this is largely negligible as we have seen. As a small note the number of Isend's dropping to 0 is due to the change in communication strategy in MPC++/MPI when message size becomes small.

Figure 6 shows that communication time largely becomes dominant. This is despite that the total size of messages sent per node decreases considerably—rather, the number of communications per node increases almost linearly as the number of nodes increase. Thus, the performance penalty is largely due to the communication latency in this case, for both C++ and MPC++ versions, and the superior low-latency characteristics of PM over Myrinet, not the bandwidth, is likely the cause of superior scalability over 100Base-T Ethernet.

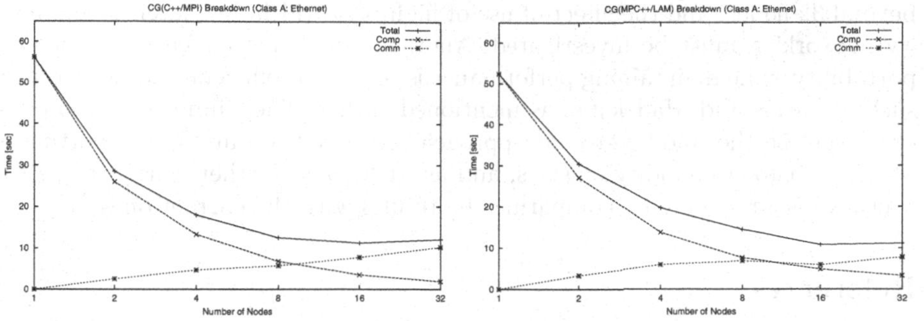

Fig. 6. CG Breakdown (Class A).

We have performed similar analysis for IS, but we omit the details for brevity. In a nutshell, the MPI operations and the number of messages sent greatly differ between the original IS and the MPC++ version, due to the difference in the collective operations—the original IS uses the MPI collective operations, while the MPC++ version uses its own language directive, which is in turn implemented in terms of point-to-point MPI communication.

5 Conclusion

We have performed detailed analysis of viability of MPC++ on commodity clustering environments. Compared to the original MPC++ which assumed a specialized user-level messaging library PM on top Myrinet, a portable implementation could sit on top of various messaging layers such as MPI, which could in turn might sacrifice performance in various ways. We ported NPB-2.3 CG and IS into MPC++, respecting its natural programming style (e.g., remote read/write through global pointers instead of MPI send/receive), and verified that the efforts involved is quite small, i.e. approximately 60 out of 800 lines for CG and about 50 out of 1000 lines for IS, each taking only a few hours. This is an (albeit indirect) evidence that one could program parallel applications that resemble those in NPB naturally in MPC++. We then performed performance analysis of running on different combinations of applications, the programming language, the underlying software messaging layers, and networking hardwares. The results show that for larger, realistic data sets (class B), the portable implementation of MPC++ on MPI and commodity clustering hardware using 100Base-T Ethernet scales quite competitively to both the original NPB code and the versions running on Myrinet. For smaller data sets, we have performed more detailed analysis to examine the source of the overhead.

Still, many future work remain. First of all, our results are restricted to a few benchmarks. Secondly, we need to perform more finer-grained analysis of scalability, especially the profiling of different parts of code to investigate what usage pattern in the algorithm results in what kind/size/frequency of communication, affecting the overall performance. Next, we need to perform further detailed analysis of the communication characteristics. In particular, scalability beyond 32 nodes, and the effect of use of high-bandwidth, low-latency commodity networking must be investigated. Another interesting endeavor to increase portability while maintaining performance is to employ efficient middle-tier layer such as Nexus and Madeleine, as mentioned earlier. Their implementation will effectively be the middle-ground approach compared to our 'direct' porting of MPC++ onto commodity clusters, and as such could further clarify where the overhead is for scaling by comparing the results with the current ones.

References

[1] Yutaka Ishikawa. Multi Thread Template Library – MPC++ Version 2.0 Level 0 Document –. Technical Report 012, Tsukuba Research Center, Real World Computing Partnership, September 1996.
[2] Stephan S Luna. Implementing an Efficient Portable Global Memory Layer on Distributed Memory Multiprocessors. Technical report, UCB, May 1994.
[3] L. V. Kalé and S. Krishnan. CHARM++: A Portable Concurrent Object Oriented System Based on C++. In *Proceedings of OOPSLA'93*, pages 91–108. ACM Press, September 1993.
[4] Hiroshi Tezuka, Atsushi Hori, Yutaka Ishikawa, and Mitsuhisa Sato. PM: An Operating System Coordinated High Performance Communication Library. In *HPCN'97*, pages 708–717. LNCS, April 1997.
[5] `http://www.myri.com/`.
[6] Compaq Computer Corp., Intel Corporation, and Microsoft Corporation. *Virtual Interface Architecture Specification*, draft revision 1.0 edition, December 1997.
[7] I. Foster, C. Kesselman, and S. Tuecke. The Nexus approach to integrating multithreading and communication. *Journal on Parallel and Distributed Computing*, pages 37:70–82, 1996.
[8] Luc Boug, Jean-Franois Mhaut, and Raymond Namyst. Madeleine: An efficient and portable communication interface for RPC-based multithreaded environments (revised version). Technical report, LIP, ENS Lyon, December 1999.
[9] K. Mani Chandy and Carl Kesselman. CC++: A Declarative Concurrent Object Oriented Programming Notation. Technical Report CS-92-01, California Institute of Technology, September 1992.
[10] Gregory D. Burns, Raja B. Daoud, and James R. Vaigl. LAM: An Open Cluster Environment for MPI. In *Supercomputing Symposium'94*, June 1994.
[11] W. Gropp, E. Lusk, N. Doss, and A. Skjellum. A high-performance, portable implementation of the MPI message passing interface standard. *Parallel Computing*, 22(6):789–828, September 1996.
[12] D. Bailey, E. Barszcz, J. Barton, D. Browning, R. Carter, L. Dagum, R. Fatoohi, S. Fineberg, P. Frederickson, T. Lasinski, R. Schreiber, H. Simon, V. Venkatakrishnan, and S. Weeratunga. THE NAS PARALLEL BENCHMARKS. Technical Report 007, RNR, 1994.
[13] Shinji Sumimoto, Hiroshi Tezuka, Atsushi Hori, Hiroshi Harada, Toshiyuki Takahashi, and Yutaka Ishikawa. High Performance Communication using a Commodity Network for Cluster System. In *HPDC 2000*, August 2000.
[14] Francis O'Carroll, Hiroshi Tezuka, Atsushi Hori, and Yutaka Ishikawa. The Design and Implementation of Zero Copy MPI Using Commodity Hardware with a High Performance Network. In *ACM SIGARCH ICS'98*, pages 243–250, July 1998.

Dynamic Instrumentation and Performance Prediction of Application Execution

A.M. Alkindi , D.J. Kerbyson, and G.R. Nudd

High Performance Systems Laboratory, Department of Computer Science
University of Warwick, Coventry CV4 7AL, UK
{ahmed,djke,grn}@dcs.warwick.ac.uk

Abstract. This paper presents a new technique that enhances the process and the methodology used in a performance prediction analysis. An automatic dynamic instrumentation methodology is added to Warwick's Performance Analysis and Characterization Environment PACE [1]. The automation process has eliminated the need to manually obtain application information and data. The Dynamic instrumentation has given PACE the ability to extract and utilize data that were hidden and unobtainable prior to execution. We give two examples to illustrate our methodology. While it was impossible to perform the analysis using the original method due to lack of essential information, the new technique successfully enabled PACE to conduct the prediction analysis in a dynamic environment. The results show that with the automated dynamic instrumentation, the performance prediction analysis of dynamic application execution is possible and the results obtained are reliable. We believe that the technique implemented here could eventually be used in other performance prediction tool-sets, and therefore enhance the ways in which the performance of systems and applications is analysed and predicted.

Keywords: Dynamic instrumentation, performance optimization, performance analysis, modelling, PACE.

1 Introduction

In this age of Information Power GRID [2], a significant amount of work is being undertaken to ease the process of performance prediction. A number of valuable toolsets for performance optimisation through prediction exist. Falcon [3], Paradyn/Dynamist [4], Pablo [5] are being used to assist in performance prediction, tuning and identification of bottlenecks in applications. These tool-sets typically include various features to allow predicting with great accuracy the performance of applications. Some also allow instrumentation at various levels, data collection, and interrogation through visualisation modules. Warwick's Performance Analysis and Characterization Environment PACE is concerned with the prediction analysis and is central to our work in this paper.

In this work, we introduce a new methodology that enhances the performance prediction process in PACE. The new technique utilizes dynamic instrumentation by inserting special sensors in an application source code prior to its first execution. This automated insertion benefits the prediction process in two ways. First, the automation

B. Hertzberger et al. (Eds.) : HPCN Europe 2001, LNCS 2110, pp. 513-523, 2001.

would make it easier for the user to conduct as many studies as he wishes and without the need to manually obtain application information and data. Second, it would allow for the gathering of valuable run-time information from an application execution. This run-time (on-the-fly) information is then used to give PACE the accurate parameters it needs for application performance analysis and prediction. The results of this work show that the process of application performance prediction is greatly enhanced, and the cost of this process is reduced.

An important aspect of PACE is its ability to conduct performance prediction analysis of a given application, and to produce valuable information for the user [6]. However, one analysis aspect this tool-set fails to address is the complete run-time analysis. That is, no information is gathered or analysed while the system is executing. This prevents the user from getting real-time information (such as data, loop sizes, or conditional statements/computations) regarding the system. Up to now, this limitation has been slightly overcome through manual procedures in which specific data are manually inserted into the analysis process. However, these manual procedures have been limited to loops and conditional statements and not data. Furthermore, when the application execution is dynamic and is data-dependent, then the user of PACE cannot pass the correct information to it for a reliable analysis.

Beside our work, several studies have been conducted in order to address this particular problem. Some are introducing the use of historical performance data [7] also known as profiling, while others are implementing an instrumentation procedure through a language for dynamic program instrumentation [8]. However, current tools have yet to produce an accurate dynamic, on-the-fly performance prediction of a given application. For example, while Dynamist is a powerful performance analysis tool, that can give a variety of useful information about the program being executed, it has a number of problems. It produces huge overhead for the user because the analysis is done on the executable, therefore large system resources are required in order to conduct a stable analysis. Furthermore, the analysis must be conducted on the same platform that is being analysed. Another tool, Falcon, supports instrumentation, tuning and steering. It also applies the sensors technique on applications. However, it does not provide modelling and it relies on the availability of shared memory between application and local monitoring threads. None of these systems, however, provide or try to acquire information with regards to loop/conditional statements or data dependencies in the given application in order to enhance the accuracy of the prediction analysis. Our work, as indicated earlier, introduces a dynamic instrumentation technique, which we believe will tackle the problem stated above and improve on solutions provided by available performance prediction packages. We believe that the methodology presented here may pave the way for the development of future generations of performance prediction tools, which currently lack the ability to analyse or predict an application performance on a given platform dynamically and on-the-fly.

Many publications of previous work on the PACE tool exists, which we do not have space to reference properly. However, we will give a brief introduction to this performance prediction tool-set in the next section. A good tutorial on PACE can be found in our web site http://www.dcs.warwick.ac.uk/~hpsg/pace_top.htm.

2 PACE: Performance Analysis and Characterisation Environment

PACE is a modelling toolset for high performance and distributed applications. It includes tools for model definition, model creation, evaluation, and performance analysis. It uses associative objects organised in a layered framework as a basis for representing each of a system's components. An overview of the model organisation and creation is presented in the following subsections.

2.1 Model Components

Many existing techniques, particularly for the analysis of serial machines, use Software Performance Engineering (SPE) methodologies [9], to provide a representation of the whole system in terms of two modular components, namely a software execution model and a system model. However, for high performance computing systems, the organisation of models must be extended to take concurrency into account. The layered framework is an extension of SPE for the characterisation of parallel and distributed systems. It supports the development of several types of models: software, parallelisation (mapping), and system (hardware).
The functions of the layers are:

Application Layer – describes an application in terms of a sequence of subtasks. It acts as the entry point to the performance study, and includes an interface that can be used to modify parameters of a performance scenario (a user might perform an analysis many times with different parameters).

Application Subtask Layer – describes the codes within an application that can be executed in parallel.

Parallel Template Layer – describes the parallel characteristics of subtasks in terms of expected computation-communication interactions between processors.

Hardware Layer – collects system specification parameters, micro-benchmark results, statistical models, analytical models, and heuristics that characterise the communication and computation abilities of a particular system.

In the layered framework, a performance model is built up from a number of separate objects. Each object is of one of the following types: application, subtask, parallel template, and hardware. A key feature of the object organization is the independent representation of computation, parallelisation, and hardware. Each software object (application, subtask, or parallel template) is composed of an internal structure, options, and an interface that can be used by other objects to modify its behaviour. The main aim of these objects is to describe the system resources required by the application, which are modelled in the hardware object. Each hardware object is subdivided into many smaller component hardware models, each describing the behaviour of individual parts of the hardware system. For example, the memory, the CPU, and the communication system are considered as separate component models [10].

2.2 Model Creation

PACE users can employ a workload definition language CHIP³S to describe the characteristics of the application. CHIP³S is an application modelling language that supports multiple levels of workload abstractions [11]. When application source code is available, the Application Characterization Tool (ACT) can semi-automatically create CHIP³S workload descriptions. ACT performs a static analysis of the code to produce the control flow of the application, operation counts in terms of SUIF language operations, and the communication structure. This process is illustrated in Fig.1. SUIF, Stanford University Intermediate Format [12], is an intermediate presentation of the compiled code that combines the advantages of both high level and assembly language. ACT cannot determine dynamic related performance aspects of the application such as data dependent parameters. These parameters can be obtained either by profiling or with user support.

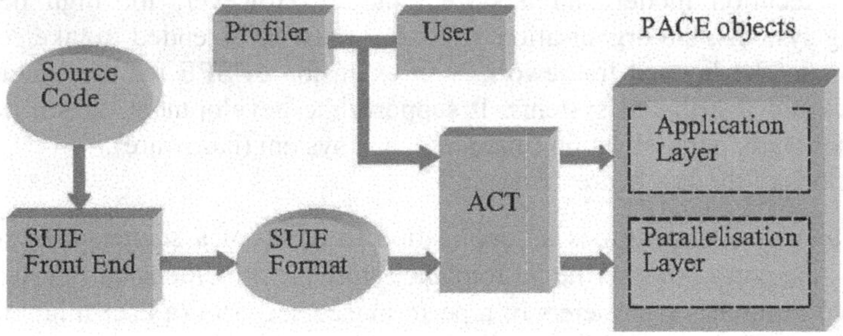

Fig. 1. Model Creation Process with ACT.

The CHIP³S objects adhere to the layered framework. A compiler translates CHIP³S scripts to C code that is linked with an evaluation engine and the hardware models. The final output is a binary file, which can be executed rapidly. The user determines the system/application configuration and the type of output that is required as command line arguments. The model binary performs all the necessary model evaluations and produces the requested results. PACE includes an option to generate predicted traces that can then further be analysed by visualization tools (e.g. PABLO) [13].

3 Dynamic Instrumentation of Application Execution

3.1 Overview

In this work, we introduce a new approach to gather maximum run-time information for the performance prediction analysis process. Our goal is to acquire application data dynamically, while at the same time not to lose the ability to analyse these data due to their incorrectness, or resources overhead. The run-time (on-the-fly) procedure would give the user an acceptable view of what is really happening inside the application. This makes analysing and predicting the performance of this application more accurate, and therefore gives us a greater chance for better optimisation. Our aim has also been to enhance performance prediction methodology without losing the tool's original functionality. We believe that any on-the-fly performance prediction tool-set should be easy to use, which would make it a system for any user, and not just for performance experts.

3.2 An Automated Profiling Technique for PACE

The long and sometimes inefficient procedures mentioned above and described in [6] and [13] are summarized in the following steps:

- Execute an ACT command on the required application.
- Manually, collect the code produced from the output of the previous run and insert it into subtask object and then execute the modified object.
- Execute and link the remaining subtask objects to produce the desired model.

These long and painstaking steps must be executed separately and manually. Fig.2 shows the original manual process of creating an executable model in PACE. For an application to be analysed, and its performance to be precisely predicted, the user must interfere and insert the output from ACT to produce a reliable subtask, which is then linked with other subtasks to produce the executable model.

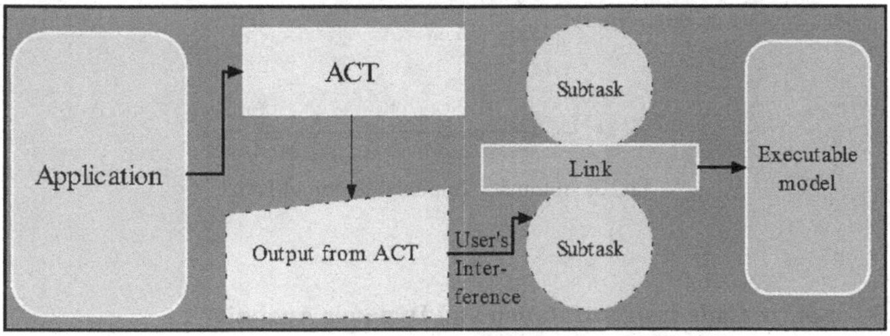

Fig. 2. Manual Implementation of PACE.

Table 1 shows loops representation in an ACT format from a matrix multiplication program. When the analysis requires hundreds or even thousands of runs, this interference is going to be an inefficient way to create an executable model.

Our automation solution creates a filtering routine that takes as input the output of ACT and filters out the necessary code (see fig.3). The routine inserts this code inside the subtask object without user interference. This automation enhances the process described earlier. A simple batch program could now run ACT, execute and link all the required subtask objects, and finally produce the executable model. This automation process reduces the cost for conducting multiple analyses on multiple or distributed systems.

Table 1. Loops Representation in an ACT Format from a Matrix Multiplication Program.

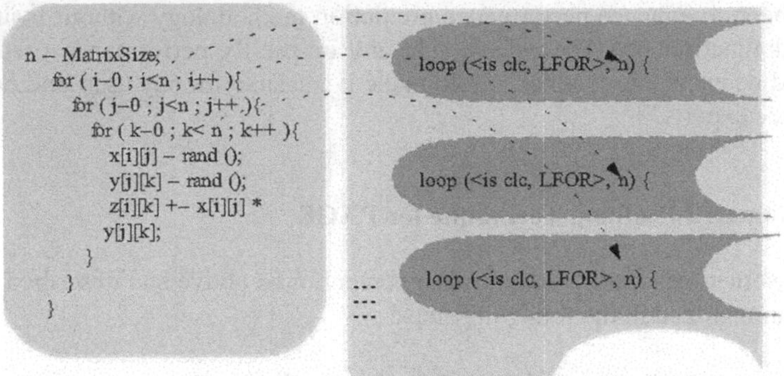
n – MatrixSize;
for (i–0 ; i<n ; i++){
for (j–0 ; j<n ; j++){
for (k–0 ; k< n ; k++){
x[i][j] – rand ();
y[j][k] – rand ();
z[i][k] +– x[i][j] *
y[j][k];
}
}
}
loop (<is clc, LFOR>, n) {
loop (<is clc, LFOR>, n) {
loop (<is clc, LFOR>, n) {

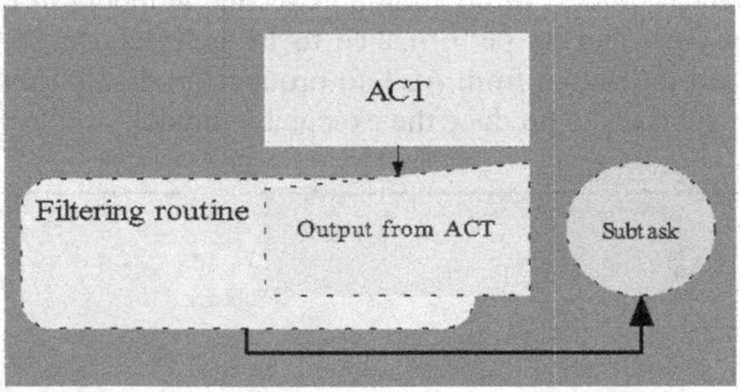
ACT
Filtering routine
Output from ACT
Subtask

Fig. 3. Automatic Implementation of PACE.

3.3 Source Code Instrumentation and Dynamic Analysis

For PACE to be able to function in a dynamic environment with a reliable outcome, we introduce the source code instrumentation and dynamic analysis methodology. The original method requires a user to input application information (which might consist of parameters like loop sizes and conditional statements probabilities) to PACE when an analysis is performed. While the user might correctly guess the information in small applications with few loops or conditional statements using small sized data, it is going to be harder when big applications with huge loops or confusing conditional statements are the target of analysis. Trying to guess, or manually figuring out, the sizes of application loops and the probabilities of its conditional statements are not helpful in getting an accurate application performance prediction.

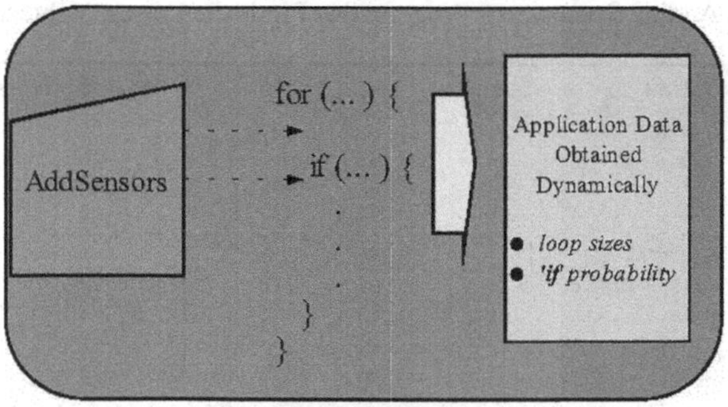

Fig. 4. Dynamic Instrumentation of an Application.

Our solution to this problem is to use an analyser, AddSensors, that inserts special sensors into the required application. The sensors, as shown in fig.4, reside inside fundamental statements (loops and conditional statements in the case of our examples) and are triggered when the application is first executed. These sensors gather run-time information and data for the user, and pass them to the subtask using the new automated procedure.

We illustrate this methodology on two examples, a matrix multiplication program, and a program to find the smallest value in an array. Tables 2 and 3 below show the result of executing the AddSensors routine on both programs. The advantage of this method is that it enables the user to have PACE gather information automatically and dynamically. With this information conducting a performance prediction analysis using PACE is now possible and the results are reliable.

Table 2. Adding Sensors to a Matrix Multiplication Program.

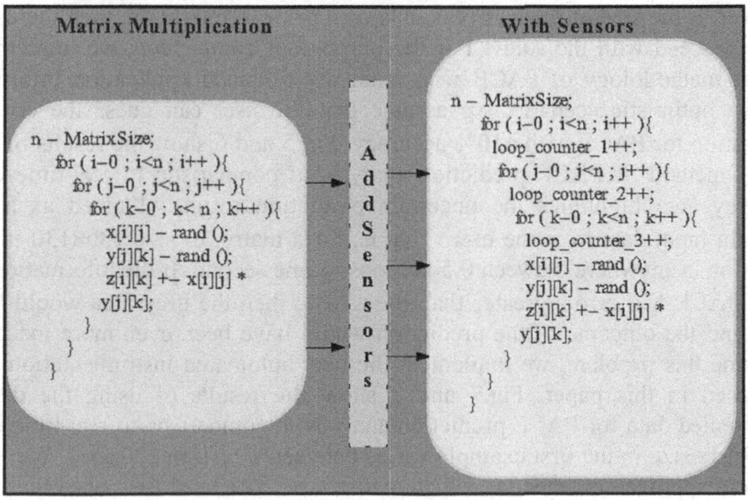

Table 3. Adding Sensors to a Program that Finds the Lowest Value in an Array.

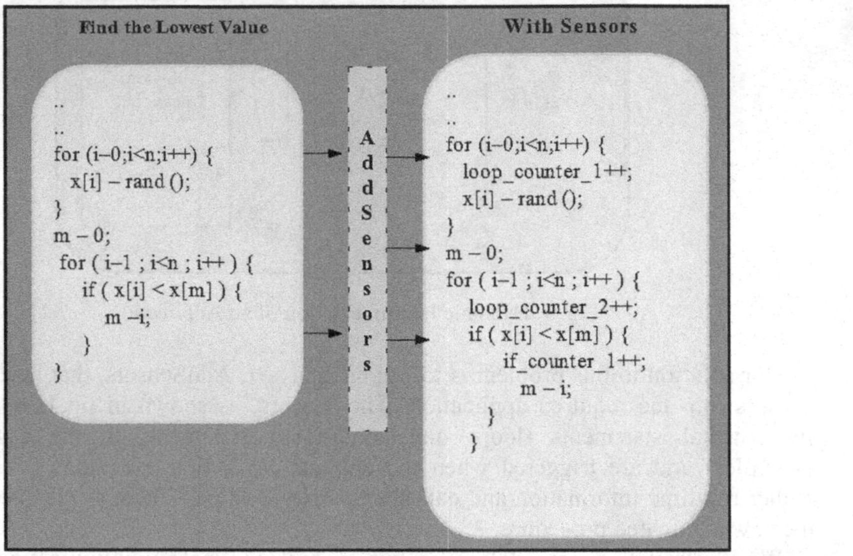

Without dynamically obtained data, a user is not able to conduct performing prediction analysis by a tool-set like PACE. This is especially true when the analysis is performed in a dynamic environment where application execution is data-dependent. To illustrate this difficulty, we ran the prediction analysis for both example programs on a Sun Ultra10 machine using the original method of PACE. In both examples, there are data and information that PACE would require before utilizing its prediction capabilities. For example, the user should pass the loop size as well as the conditional statement probability in the matrix multiplication routine. However, if the size of the matrix is unknown prior to execution, then the user can no longer proceed with the study. For the purpose of comparison, we implemented the original methodology of PACE with manually obtained application information. In the best optimistic scenario, we assume that the user can guess the correct input information for PACE with ±10% accuracy. Fig.5 and 6 show the results of using this manual method of PACE prediction analysis in comparison to real-time execution. The grey area highlights the uncertain prediction results obtained as a result of uncertain input data from the user. That is, for a matrix of size 130×130 in fig.5, the prediction is anywhere between 0.5 seconds to one second. If the information that was fed to PACE had errors greater than the ±10%, then the grey area would have been larger and the outcome of the prediction would have been even more inaccurate. To overcome this problem, we implement the new automated instrumentation technique introduced in this paper. Fig.7 and 8 show the results of using the dynamically instrumented data for PACE prediction analysis in comparison to real-time execution. The matrix size in the first example varies between 10x10 to 200x200. We implement a random approach of selecting which matrix size to be used first when the program is executed. The reason behind this is so that the user gets unpredictable information when manually analyzing the source code, which makes him/her rely solely on the dynamic instrumentation technique. The same rule is applied for the array sizes in the second example.

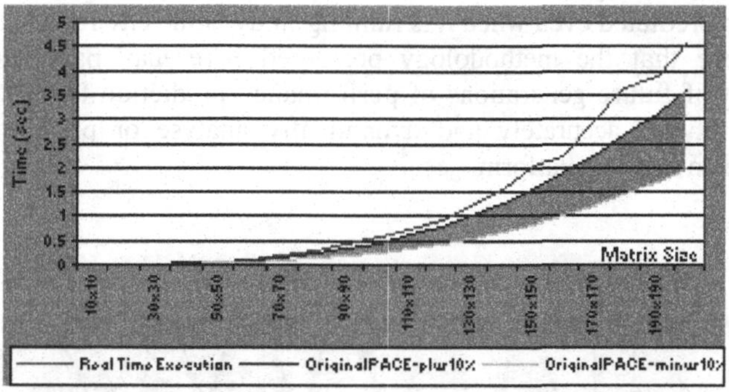

Fig. 5. Matrix Multiplication Performance Prediction with Uncertain Prediction Times.

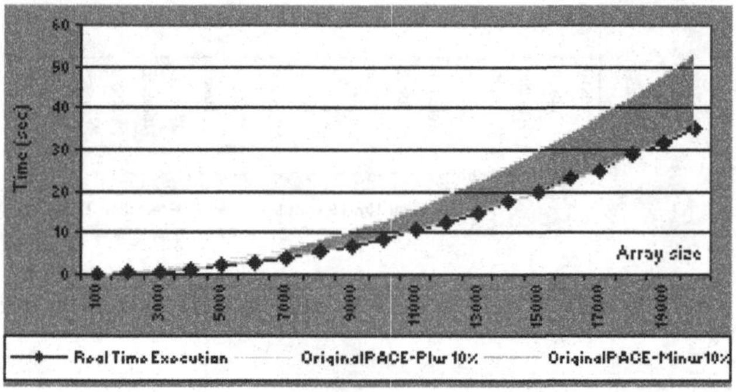

Fig. 6. Uncertain Prediction Times for a Program to find the Lowest Value in an Array.

4 Analysis and Conclusion

In this work, we show that the new automation and instrumentation technique for conducting performance prediction research using PACE is cheaper, faster, and more accurate than the original approach of PACE. This is true, because the user can now perform a large amount of analysis for many applications without the need for him/her to interfere. The automation process eliminates the need of manually obtaining ACT output and inserting it to an application object for a later use. The dynamic instrumentation gives PACE the ability to extract data that were hidden and unobtainable prior to execution. These data are essential for producing reliable prediction results from PACE. The user can now be satisfied that correct information was gathered for the application in question and that PACE is performing prediction analysis using reliable data.

We have used two examples, a matrix multiplication program, and a program to find the lowest value in an array, to illustrate our methodology. The results show that with the automated dynamic instrumentation, an application execution can now be analysed and predicted even when it is running in dynamic environment.

We believe that the methodology presented here may pave the way for the development of future generations of performance prediction tools, which currently lack the ability to accurately and dynamically analyse or predict an application performance on a given platform.

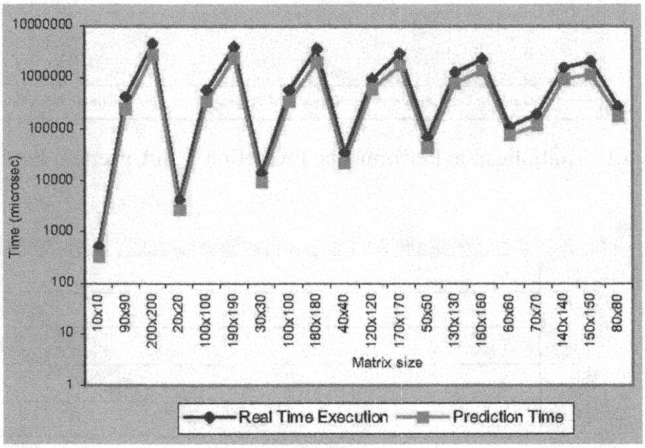

Fig. 7 Matrix Multiplication Performance Prediction Using Dynamic Instrumentation vs. Real Time Execution.

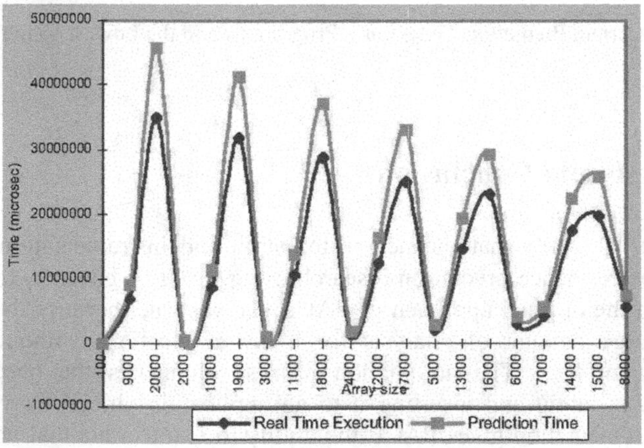

Fig. 8. Dynamically Instrumented Prediction for a Program to Find the Lowest Value in an Array.

Acknowledgments

This work is funded in part by DARPA contract N66001-97-C-8530, awarded under the Performance Technology Initiative administered by NOSC. Mr. Alkindi is supported on a scholarship by the Sultan Qaboos University, Oman.

References

1. Nudd, G.R., Papaefstathiou, E., et.al., A layered Approach to the Characterization of Parallel Systems for Performance Prediction, in Proc. of Performance Evaluation of Parallel Systems, Warwick (1993) 26-34.
2. Foster, I., Kesselman, C.: The Grid, Morgan Kaufmann, (1998).
3. Gu, W., Eisenhauer, G., Schwan, K.: On-line Monitoring and Steering of Parallel Programs, Concurrency: Practice and Experience 10 9 (1998) 699-736.
4. Miller, B.P., Callaghan, M.D., Cargille, J.M., Hollingsworth, J.K., Irvin, R.B., Karavanic, K.L., Kunchithapadam, K., Newhall, T.: The Paradyn Parallel Performance Measurement Tools, IEEE Computer 28 11 (1995) 37-46.
5. DeRose, L., Zhang, Y., Reed, D.A.: SvPablo: A Multi-Language Performance Analysis System, in Proc. 10th Int. Conf. on Computer Performance, Spain (1998) 352-355.
6. Papaefstathiou, E., Kerbyson, D.J., Nudd, G.R., Atherton, T.J.: An overview of the CHIP3S performance prediction toolset for parallel systems, in: 8th ISCA Int. Conf. on Parallel and Distributed Computing Systems, Florida (1995) 527-533.
7. Karen L. Karavanic and Barton P. Miller, Improving Online Performance Diagnosis by the Use of Historical Performance Data, SC'99, Portland, Oregon (USA) November 1999.
8. Jeffrey K. Hollingsworth, Barton P. Miller, Marcelo J. R. Gonçalves, Oscar Naim, Zhichen Xu and Ling Zheng, MDL: A Language and Compiler for Dynamic Program Instrumentation, International Conference on Parallel Architectures and Compilation Techniques San Francisco, California, November 1997.
9. Smith, C.U.: Performance Engineering of Software Systems, Addison Wesley (1990).
10. Harper, J.S., Kerbyson, D.J., Nudd, G.R.: Analytical Modeling of Set-Associative Cache Behavior, IEEE Transactions on Computers 48 10 (1999) 1009-1024.
11. Papaefstathiou, E., Kerbyson, D.J., Nudd, G.R., Atherton, T.J.: An overview of the CHIP3S performance prediction toolset for parallel systems, in: 8th ISCA Int. Conf. on Parallel and Distributed Computing Systems, Florida (1995) 527-533.
12. Wilson, R., French, R., Wilson, C., et.al.: An Overview of the SUIF Compiler System, Technical Report, Computer Systems Lab Stanford University (1993).
13. Kerbyson, D.J., Papaefstathiou, E., Harper, J.S., Perry, S.C., Nudd, G.R.: Is Predictive Tracing Too Late for HPC Users?, in High Performance Computing, Kluwer Academic, March 1999, 57-67.

Improving Load Balancing in a Parallel Cluster Environment Using Mobile Agents

M.A.R. Dantas and F.M. Lopes

Department of Computer Science, University of Brasilia
Brasilia 70910-970, Brazil
mardantas@computer.org

Abstract. In this article we present an improvement on the load balancing of a parallel cluster environment, considering the MPI parallel programming paradigm and employing a mobile agent system. Our approach is to apply the mobile agent technology to provide a better scheduling, which could represent in a cluster configuration an enhancement on the load balancing. MPI in cluster of heterogeneous machines could lead parallel programmers to obtain frustrated results, mainly because of the lack of an even distribution of the workload in the cluster. As a result, before submitting a MPI application to a cluster, we use the *Aglets* mobile agent package to acquire a more precise information of machines' workload. Therefore, with a more precise knowledge of the load (and characteristics) in each machine, we are ready to gather lightweight workstations to form a cluster. Our empirical results indicates that it is possible to spend less elapsed time when considering the execution of a parallel application using the agent approach in comparison to an ordinary MPI environment.

Keywords: Mobile Agents, Message Passing Interface (MPI), Load Balancing, Scheduling, Cluster Computing.

1 Introduction

Personal computers clustered together either physically or virtually represent interesting hardware platforms to meet the growing computational demand of many organizations. Several software packages have been designed to improve the utilization of cluster environments, usually focusing either on throughput or parallelism of tasks. The main target of the packages in the first category (i.e. throughput-oriented) is to improve the distribution of multiple, independent tasks. In contrast, parallel software environments are designed to enhance the execution of concurrent tasks, achieving reasonable parallel speedup. In practice, this means that all tasks should complete at the same time for optimal performance. Examples of packages developed to enhance the throughput of tasks are Condor [1], LSF [2] and Load Lever [3]. In contrast, MPI [4], PVM [5], ThreadMarks [6], Mirage+ [7], Shasta [8] are examples of message-passing and distributed shared memory packages intended to improve the parallelism of the tasks.

B. Hertzberger et al. (Eds.): HPCN Europe 2001, LNCS 2110, pp. 524–531, 2001.

Although parallel software packages, such as MPI, represent an enhancement for application programmers to use cluster environments as a specialised parallel computer, these environments suffer from the absence of an improved policy for load balacing. In other words, computation on workstation clusters when using MPI software environment might result in low performance, because there is no elaborate scheduling mechanism to efficiently submit processes to those environments. Consequently, although a cluster could have available resources (e.g. lightly load workstations and memory), these parameters are not completely considered and the parallel execution might be inefficient.

As we have noticed before [9,10], the imbalance of the workstations should be tackled to enhance substantially the performance of applications on a high-performance environments. Therefore, in this paper we present some empirical results in the use of a mobile agents system which can be customized to work together with MPI parallel applications in a workstation cluster environment. In the search of a mobile agents system suitable to a heterogeneous cluster configuration, we decided to use the *Aglets* package (mainly because this software has proved to have necessary portability for heterogeneous configurations).

The article is organised as followos. In section 2 we show our cluster environment considering the architecture of the machines and software packages. In the following section (section 3) we draw a comparison related to mobile agents systems. In section 4 we show our experiments and results. Finally, section 5 states our conclusions.

2 Cluster Environment

It is important to mention that the cluster which we are considering in this article is an heterogeneous parallel environment. The configuration is represented by six loosely-coupled IBM personal computers (without monitor, keyboard or mouse), two IBM RS/6000 workstations and two SUN Sparcstation 10. All machines interconnect by a 100 Mbps switch. The degree of heterogeneity of this cluster is high, i.e. we have different architectures and operating systems. Table 1 shows more characteristics of the cluster environment.

Table 1. Cluster Environment.

Processors type	IA 32 - SPARC - PowerPC
Memory (MB)	32 - 64 - 128
Operating Systems	Linux - Solaris 7 - Aix 4.1
Aglets Version	Aglets1.1b3
Switch	10-100 Mbps

Table 2. Mobile Agent Systems.

Agent	Company/ Organisation	Programming Language
Agent Tcl	Dartmouth College	Tcl/Tk
Aglets	IBM	Java
Ara	University of Kaiserlautern	Tcl and C
Concordia	Mitsubishi	Java
D'Agent	Dartmouth College	Java
MOA	Open Group	Java
Mole	Stuttgart University	Java
Odyssey	General Magic	Telescript
Tacoma	Tromso University/ Cornell University	Tcl/Tk / C
Telescript	General Magic	Java
Voyager	ObjectSpace	Java

3 Mobile Agents Systems

Mobile agents systems are broadly characterized by a programming language (e.g. Tcl/Tk and Java), the communication among agents (examples are *now-type messaging, future-type messaging and one-way-type messaging* [11]) and the support to a variety of user applications.

We have adopted a classic method to find a suitable mobile agent system to execute in our cluster. First, we did a search in [12,13,14,11,15,16,17,18,19,20,21], [22,23,24] to know more about these systems. In a second phase, we have installed and configured some of these (free) available packages. Before deciding for a specific package, we have experimented the systems (an example was the *D'Agent* system). Our experience, in this second phase, indicates that some mobile systems are not completely ready to support different operation systems. This is an essential aspect to take in consideration especially when a heterogeneous cluster is the parallel environment. In addition, some of the systems are not easy to configure.

After analysing the available systems, we decided to use the *Aglets* mobile agents package. This package has presented no difficults to install and configure. More important, the *Aglets* works absolutely how we expected in a heterogeneous cluster environment.

Table 2 shows a general view of some mobile agents systems available. As we have mention before, these systems have (a) different implementation facilities, such as programming languages (e.g. C, Java Tcl/Tk), (b) operating systems support (e.g. Linux, Solaris, Aix) and (c) degree of completeness for a mobile agent system.

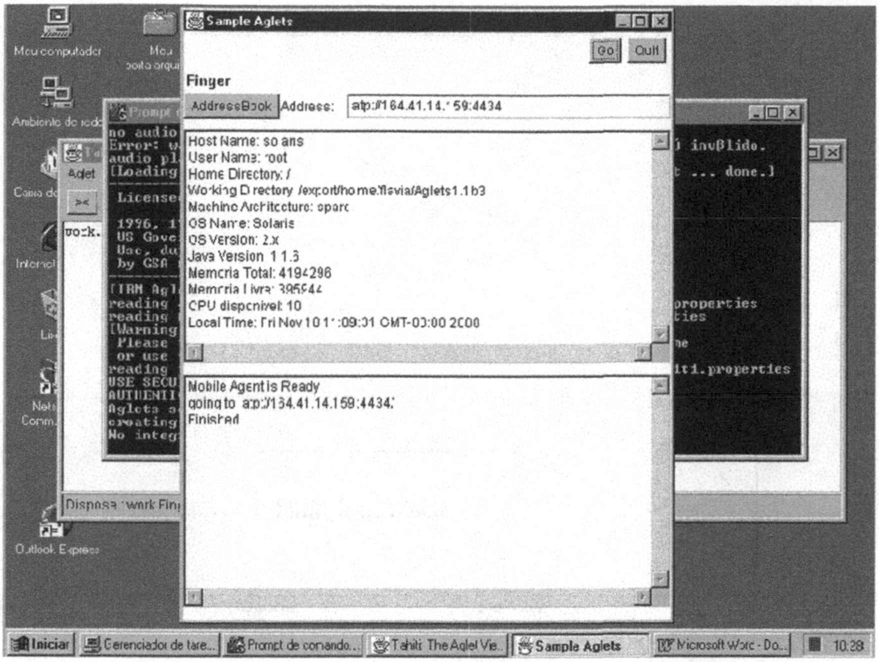

Fig. 1. The Windows Interface of the Environment.

4 Experimental Results

4.1 Introduction

In this section, we show results from our experiments using the *Aglets* mobile agents system. It is interesting to mention that our goals were to verify how the package works in a heterogeneous environment, to check the possibility to combine functions of the system with other environments such as MPI and finally to analyse the performance of the results in comparison to an ordinary MPI cluster configuration.

We have used as parallel application a matrix multiplication algorithm. This application was chosen because the granularity of this application is ideal to our tests (it is a coarse grain algorithm when compare to others algorithms, for example an *fft* algorithm). In addition, it is revelant to remember that the interoperability of the cluster was our main concern and we could increase the size of the problem to verify the performance. In other words, increasing the size of the matrices we were able to check whether we still have profit with the modified MPI cluster environment, or not.

In addition to what we have explained about the MPI parallel application, it is important to mention that our approach to form the cluster was to use first the mobile agents system. Figure 1 illustrates an example of the Windows interface of the *Aglets* environment. Each operating system has an similiar interface. The main advantage behind the graphical interface of the *Aglets* system, was

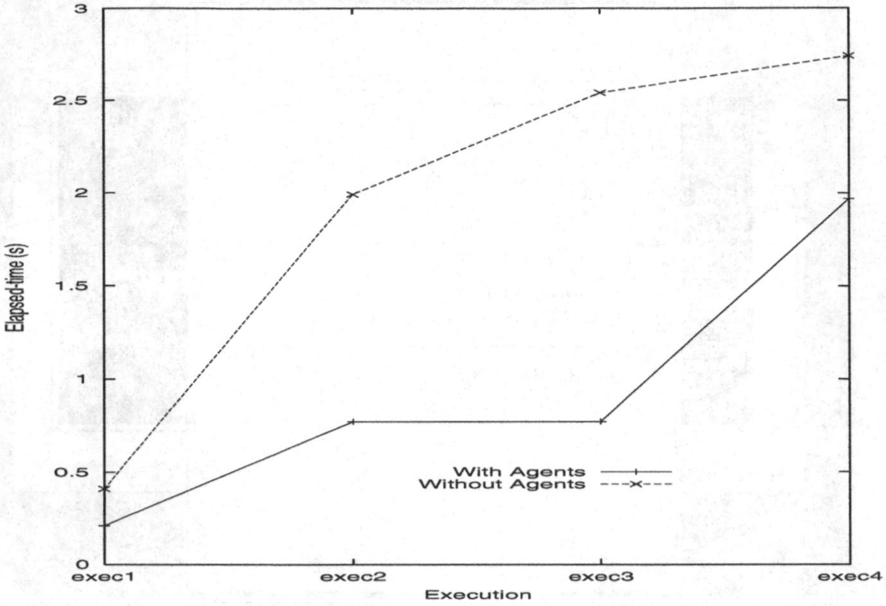

Fig. 2. A 150 x 150 Matrix Multiplication.

the introduction, through the programming language, some queries to obtain characteristics of a machine. Thus, before submitting any MPI parallel application we were able to verify characteristics of all machines and then decide which workstations were candidate to form the cluster. Aspects such as memory usage, available CPU, operating system version and machine architecture are some topics which we have employed to accept (or not) a workstation to execute a parallel application.

4.2 Experiments

The results presented in this section represent ten times the execution of the parallel application. We decided to submitt first the parallel application on a *quite* environment (i.e. nobody was using the cluster facilities). The second submission (and the third in figure 2) represents an ordinary diarly cluster use.

In our experiments, as we mention before, we have used a matrix multiplication parallel application, with different granularity to verify the performance of the environment. We understand that the importance of our research, at this stage, was to verify how the mobile package works on a heterogeneous environment and check the possibility to combine functions of the system with other environments such the MPI.

Figure 2 presents the execution cases of 150x150 matrix multiplication with the agent approach and without agent approach. In the first classe of execution (*with Agent*), the *Aglets* environment was used to form the cluster. On the other

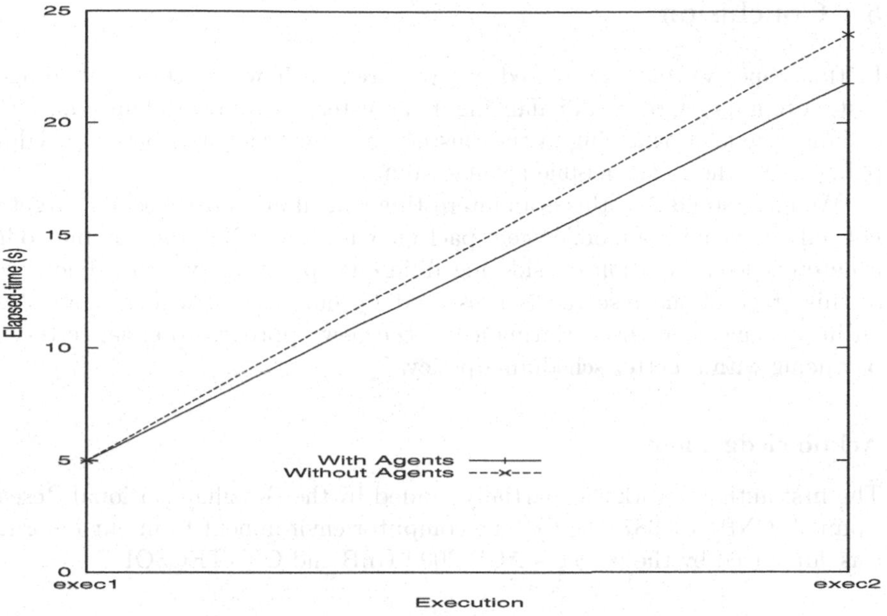

Fig. 3. A 300 x 300 Matrix Multiplication.

hand, the second approach (*without Agent* was characterised by the use of the MPI environment in an ordinary way *mpirun*). In this figure four execution tests are shown (i.e. *exec1, exec2, exec3* and *exec4*). The first test (*exec1*) represents the two classes of execution (i.e. with and without Agent) considering that the cluster was *quite*. The second and the third executions, for both classes, the cluster was in ordinary use. Finally, the fourth execution was characterised by a heavy use of the parallel cluster. From this figure is clear to observe that our approach has reached a successful performance in comparison to the ordinary MPI environment. More important to note, is the interval between execution 2 and execution 3. During typical hours of the cluster usage, the MPI parallel application has almost the same amount of elapsed time. On the other hand, in the same interval the conventional MPI environment has significant different results. The last execution (*exec4*) also demonstrates that the use of mobile agent system has advantages.

In figure 3 we have the a similar pattern of the figure 2. Now executing an 300 x 300 matrix multiplication parallel application. In this figure, we are illustrating the two classes of executions (i.e. with and with the agent approach) and the two cases of workload, *quite* and *heavyweight*. This figure shows that for a *quite* the two classes of executions were almost similar, mainly because all workstations were lightweight. However, the case when the cluster was *hevayweight* the agent approach spent less elapsed time on the execution of the parallel application.

5 Conclusion

In this paper we have presented our research on how to use a mobile agents system to improve the load balancing in a cluster environment. Our approach to enhance the load balancing in the cluster was to implement a better scheduling policy using the *Aglets* mobile agent system.

We understand that this is an interesting contribution because our target was not only to combine a mobile agent package with the MPI paradigm, but to have a heterogeneous solution considering different operating systems. The success of this stage of the research is presented by our results, which demonstrates an improvement on the performance of the new approach to enhance the load balancing with a better scheduling policy.

Acknowledgement

The first author's work was partially funded by the Brazilian National Research Council- CNPq (300874/00-6). The computer environment from our laboratory was supported by the projects ACP 2000-UnB and CIC-TECSOFT.

References

1. M. L. Michael J. Litzkow and M. W. Mutka, "Condor - A Hunter of Idle Workstations," *Proceedings of IEEE 8th International Conference on Distributed Computing Systems*, pp. 104–111, 1988.
2. J. Suplick, "An Analysis of Load Balancing Technology," January 1994.
3. I. B. M. Corporation, *IBM LoadLever : User's Guide.* Kingston, NY: IBM, September 1993.
4. MPI-Forum, "MPI: A Message-Passing Interface Standard," *International Journal of Supercomputer Application*, vol. 8, no. 3-4, 1994.
5. A. B. Al Geist, Jack Dongarra and V. Sunderam, *PVM : User's Guide and Reference Manual (Version 3.3).* Oak Ridge, USA: Oak Ridge National Laboratory Technical Report ORNL/TM-12187, May 1994.
6. J. R. C. et al, "Project Zeus," *IEEE Network*, pp. 20–30, 1993.
7. N. J. B. Fleish, R. Hyde, "Mirage+: a kernel implementation of distributed shared memory on a network of workstations," *Software: Practuce and Experience*, pp. 1–21, March 1994.
8. C. T. D. Scales, K. Gharachorloo, "Shasta: A Low-Overhead Software-Only Approach for Supporting Fine Grain Shared Memory," *Aplos-96*, 1996.
9. M. Dantas and E. Zaluska, "Improving Load Balancing in an MPI Environment with Resource Management," *Lecture Notes in Computer Science, Proceedings of the HPCN Europe 1996, Brussels, Belgium*, pp. 959–960, April 1996.
10. M. Dantas and E. Zaluska, "Efficient scheduling of mpi applications on network of workstations," *Future Generation Computer Systems*, vol. 13, pp. 489–499, May 1998.
11. D. B. Lange and M. Oshima, "Programming and deploying java mobile agents with aglets," November 1998.
12. D. College, "D'agent," `http://agent.cs.dartmouth.edu/software/agent2.0/download.html`.

13. D. Johansen, R. van Renesse, and F. B. Shneider, "An introduction to the tacoma distributed system - version 1.0," June 1995.
14. R. G. et al, "Mobile agents: The next generation in distributed computing," *IEEE - Proceedings of the 2nd AIZU International Simposium on Parallel Algorithms / Architecture Synthesis (pAs'97)*, pp. 8–24, 1997.
15. G. Magic, "General magic," `http://www.genmagic.com/telescript`.
16. J. White, "Mobile agent white paper," `http://www.genmagic.com/agents/Whitepaper/whitepaper.html`, 1996.
17. IBM, "Aglets home page," `http://www.trl.ibm.co.jp/aglets/index.html`.
18. H. Peine, "An introduction to mobile agent programming and the ara system," 1996.
19. W. L. Dejan S. Millojicic and D. Chauhan, "Mobile objects and agents (moa)," `http://www.opengroup.org`.
20. B. Pierce, "Mobile agent computing: A white paper," `http://www.cis.upenn.edu/~bcpierce/courses/629/papers/Concordia-WhitePaper.html`, 1997.
21. C. University, "Concordia mobile agent site," `http://www.meitca.com/HSL/Projects/Concordia`.
22. D. M. et al, "Masif - the OMG mobile agent system interoperability facility," September 1998.
23. J. Baumann, F. Hohl, K.Rothermel, and M. Straber, "Mole - concepts of mobile agent system," *IPVR (Institute for Parallel and Distributed High-Performance Computers - University of Stuttgart*, no. 1997/15, 1997.
24. G. Glass, "Voyager. overview of voyager: Objectspace's produtct family for state-of-the-art distributed computing," 1999.

Track V

Posters

Simulation and 3D Visualization of Bioremediation Interventions in Polluted Soils

M.C.Baracca, G.Clai, and P.Ornelli

ENEA HPCN Via Martiri di Montesole 4 40129-Bologna, Italy
{baracca,ornelli}@bologna.enea.it

Abstract. The in-situ bioremediation technique, based on the indigenous bacteria capability to degrade organic contaminants, can be viewed as a way to accelerate and amplify natural phenomena. This paper is focused on the results of real interventions simulations, based on the use of high-performance computers, and on their 3D visualization.

1 Introduction

The in-situ bioremediation techniques are based upon the capability of indigenous microorganisms to degrade the organic contaminants in polluted soils. The stimulation of indigenous bacteria is performed providing them with appropriate nutrients, in aqueous solutions, in order to create the soil conditions which favour their growth and their metabolic activity. The aim of the Esprit-HPCN COLOMBO Project was the definition of a methodology useful to approach the degradation interventions, integrating computational simulations and 3D visualization with experimental results and real interventions design. This paper is focused on the results of the computational simulations, based on the use of high-performance computers, and on the 3D graphical investigation of the simulation outcome.

2 The Bioremediation Simulation

The bioremediation dynamical model [1] simulates the fluid flows inside the soil, with diffusion and transport of pollutants and their degradation due to the metabolic activity of the indigenous bacteria. The model has a "layered" structure: the fluid dynamical layer refers up to three different phases (water, pollutant, air); the chemical layer describes the physical-chemical dynamics of the solutes; the biological layer describes biomass growth and those phenomena where microorganisms play a role, e.g. contaminant tranportation. The cellular automata [2] modelling has been chosen since the global behaviour of the bioremediation system arise from the collective effect of locally interacting phenomena. This technique is appropriate for modelling biodegradation in real systems since it allows to adopt an intermediate scale for the cell size, rather than the micoscopic scale. Each cell represents a portion of soil characterized by local values of the system variables (the substates). The implemented software model leads us from the lab tests to the real fields operations, allowing the evaluation of the main bioremediation variables: pollutant, oxygen and hydrogen

B. Hertzberger et al. (Eds.) : HPCN Europe 2001, LNCS 2110, pp. 535-538, 2001.

peroxide concentrations, biodegradation rate and intervention duration. The model predictions fit well with the field test results, so that in future applications an estimate of the intervention costs will be possible. The code was tested on two real fields located in Germany at the US-Depot in Germersheim and at the Deutsche Bahn in Frankfurt: a real intervention simulation requires a very large number of iterations, running from several hours to some days, depending on whether soil conditions are saturated or unsaturated. The locality of the considered phenomena (bacteria and pollutants interact gradually over time and only within their neighbourhood area) benefits from parallel computers through the domain decomposition technique.The benchmarks on Cray T3E-900 [3], CS-2 Meiko and Beowulf cluster, have established that the code execution scales well with the number of available processors, as long as the size of the per processor data is more than half of the exchanged boundary data. In the pictures, the benchmark results for the large model (256x128x13 cells) on the Cray platform, performed by the Edinburgh Parallel Computing Center researchers, are provided toghether with analogous benchmark results for the medium model (128x96x13 cells) on a Beowulf cluster and Cs-2 Meiko machine.

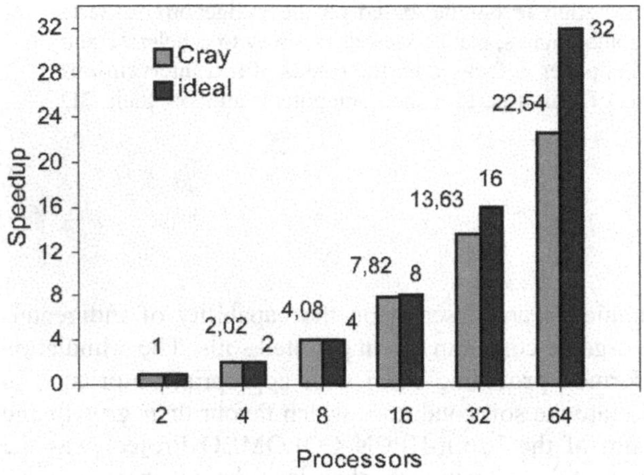

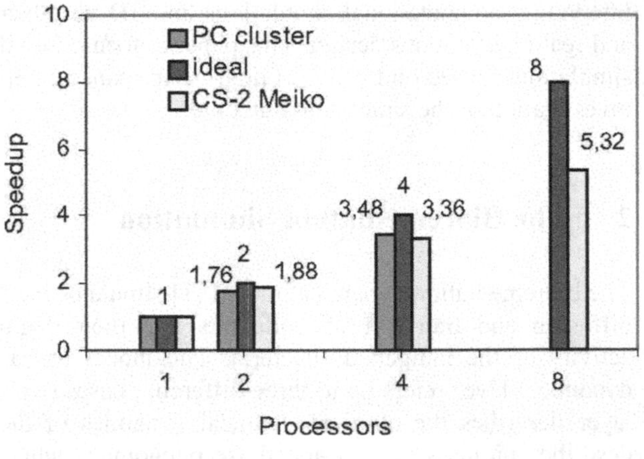

3 The 3D-Visualization Application

A bioremediation intervention simulation can take great advantage from sophisticated 3D visualization, since the dynamics of different physical and chemical processes are better observed in graphical form. The interplay between the visualization and the simulation software, besides constant monitoring of intermediate states, enables the user to adjust on the fly the bioremediation intervention strategy, without the complete re-execution of the simulation. On the other hand, the presentation in graphical form

of the bioremediation process predictions is crucial for proving the advantage and the reliability of the interventions computational design.
The visualization application was first developed on SGI Onyx2 by means of AVS/Express [4], a commercial package running on several platforms of industrial interest, then it was succesfully ported on a Beowulf cluster at the end-user site.

3.1 The Visualization Methods

The methods of visualization have been chosen to give evidence to the more relevant features and more interesting aspects of the biodegradation phenomena [5]. Each substate is interpolated and can be visualized by means of **orthoslices**, **isosurfaces** and **volumes** in a three dimensional Cartesian space representing the soil portion in terms of cells. The geometrical visualization modalities can be selected interactively and it is possible to rotate, to shift and to zoom the 3-D picture.
Each **orthoslice** module subsets a cells array by extracting one slice plane perpendicular to one of the Cartesian axis: the orthoslices can be moved in the range

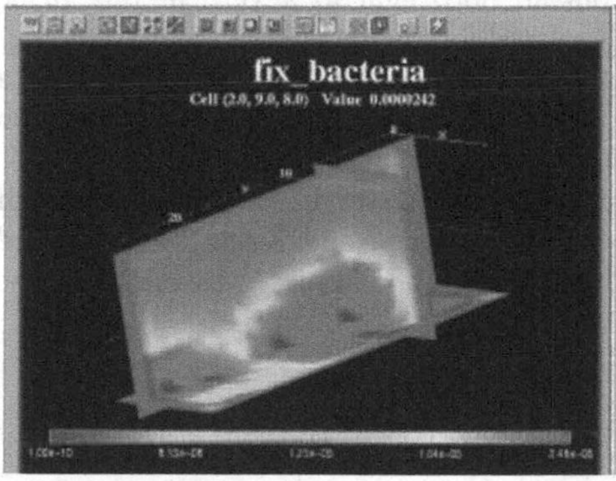

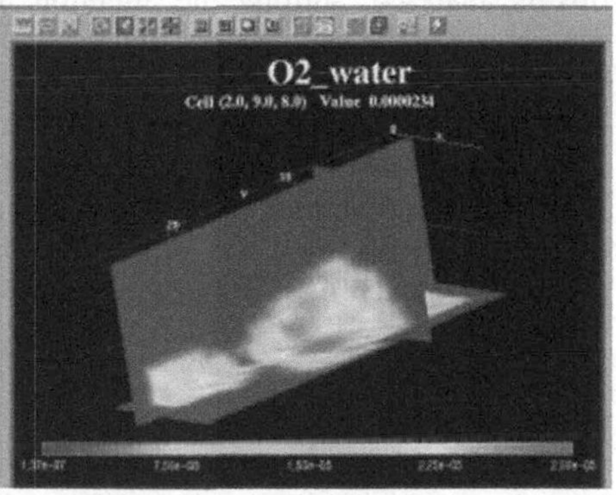

from 0 to the axis maximum dimension. It is possible, with a mouse click, to retrieve the substate value and the coordinates of a cell by means of the **probe** feature, cutting out the interpolated data produced by the visualization process. The above pictures, referred to the Frankfurt test, show the maximum of the biomass growth and the related oxigen concentration, by means of orthoslices.

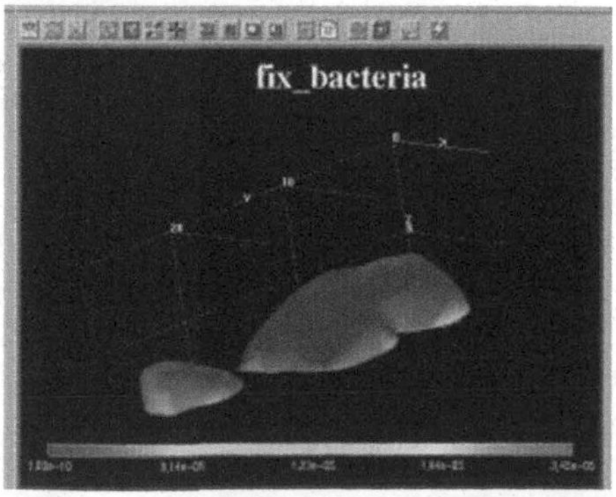

In order to outline correlated substates or threshold effects, the **isosurface** module, that creates a surface with a given constant value level, offers the option to map on a substate isosurface the cell values of an other substate. On the right, the correlation between the biomass and oxigen concentration is shown: chosen a specific value of the oxigen concentration, the related isosurface shape is created while the corresponding bacteria distribution is mapped onto it.

Vectorial fields visualization has been provided, in order to show in an effective way the air, water and pollutant fluxes.

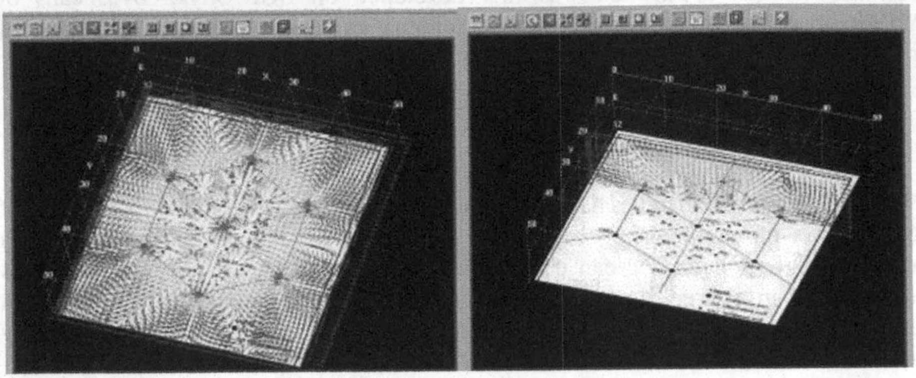

The above pictures refer to the real field test case located in Germersheim: the water flux is analyzed, shifting the attention from an horizontal to a vertical flux plane superposed to the wells distrbution design, realized for the remediation intervention.
The capability to produce 3D animations, visualizing the evolution of a bioremediation process, is crucial for demonstrating the reliability of the visualization application as a powerful tool for planning and designing actual interventions.
Therefore, the **temporal evolution** of the automaton substates, based on a sequence of data files saved during the simulation, has been provided and it can be performed according to each one of the implemented geometrical modalities, as well as for the vectorial fields.

References

1. M.Villani, M.Mazzanti, R.Serra, M.Andretta, S.Di Gregorio "Simulation model implementation description", Deliverable D11 of COLOMBO Project, July 1999.
2. J.von Neumann "Theory of Self Reproducing Automata", Univ.Illinois Press, Champaign, Ill.,1966.
3. K.Kavoussanakis et al., "CAMELot Implementation and User Guide", Deliverable D9 of COLOMBO Project, September 2000.
4. Advanced Visual System Inc. "Using AVS/Express", July 1998.
5. M.C.Baracca, G.Clai, P. Ornelli "Pre/Post Processor Description", Deliverable D12 of COLOMBO Project, July 1998.

Resource Planning in Converged Networks

T.T. Mai Hoang and Werner Zorn

Institute for Operating and Dialog Systems, University of Karlsruhe
Am Fasanengarten 5 - 76128 Karlsruhe, Germany
{hoangmai,zorn}@ira.uka.de

Abstract. Network resource planning aims to determine cost-effective resources needed to carry current volumes of traffic. Determining where and when to add bandwidth or expand routing or switching capacity are the most fundamental resource planning decisions. This paper describes this problem in converged networks and an model-based approach as an efficient solution for it. Computational tests of the approach with different backbones ware carried.

1 The Resource Planning Model

Converged networking is an emerging technology thrust that integrates voice, video and data traffic on a single network. In comparison with traditional data network planning approaches which only take one class of best-effort traffic into consideration, the planning of an converged network should consider all types of traffic as well as the converged network protocols, such as, RSVP, MPLS, OSPF, Constraint-based routing and the resource allocation [1, 4]. The resource planning problem for such networks is formulated in our approach as follows: **Given are** a fixed IP-based converged backbone, the capacities and buffering of its components, the node locations and the link topology, K classes of real-time traffic by means of traffic parameters and QoS requirements, one class of "best-effort"traffic by means of given demand metric and packet delay requirements. **To determine** the cost-effective resource needed for a given traffic demand and QoS requirement. This problem is conceptualized on one hand as an modeling of the converged-based backbone with a graph $G = (V, E)$ in which V is a set of backbone nodes and E is a set of backbone edges, and on the other hand as an simulation of the traffic mapping in respect of network dynamics. The goal of the traffic mapping is to find an optimal network throughput for a given traffic and QoS requirements so that the cost of the bandwidth needed will be minimal. This traffic mapping consists of two steps, the real-time traffic mapping and the best-effort traffic mapping.

The real-time traffic mapping is an analytical simulation of its transmission over real IP-based converged networks. This mapping consists of 3 steps: characterizing the real-time traffic with Tspec and Rspec [3], selecting the path for projecting the flow and allocating the resource on the selected path. The path selection is done via an constraint-based routing algorithm developed within our approach. For an selected path Φ, the optimal resource allocation is formulated as the following optimization task:

B. Hertzberger et al. (Eds.) : HPCN Europe 2001, LNCS 2110, pp. 539-542, 2001.

To optimize
$$F = \sum_{i \in \Phi} \frac{S_i}{S_i - R_i} + F_0 \tag{1}$$

Subject to
$$R_i \le S_i \text{ and } d_{reqd} \le d_0 \tag{2}$$

Whereby, F_0 is the constants calculated using the traffic parameters and the spare capacities at the nodes. S_i is the spare capacity at the node i belong Φ. d_0 is the upper bound of the desired queue delay bound d_{reqd}. To find are the resource R_i to be allocated at nodes $i \in \Phi$ so that the cost function F will be minimal. This optimization task is a non linear combinatorial optimization problem that is solved using the lagrangean procedure and the convexity characteristic of the cost function. The real-time traffic mapping is implemented as an genetic algorithm (GA1). **The "best-effort" traffic mapping** is an analytical simulation of the best-effort traffic transmission over packet networks. This transmission can be seen as the approximation of the OSPF Border Gateway Protocol (BGP). This traffic mapping is solved using flow deviation"[2] concept, which is implemented in our approach as an genetic algorithm (GA2).

2 Implementation and Test of the Model

The model described above is implemented as an genetic algorithm (GA) consisting of two sub genetic algorithms (GA1 and GA2) in Java. The architecture of the prototype is described in figure 1 below. The prototype consists of 4 main modules: an external interface to NetFlow or network management system, an common GUI, an common data and a NAP. The network planning and upgrading algorithms are implemented within NAP module.

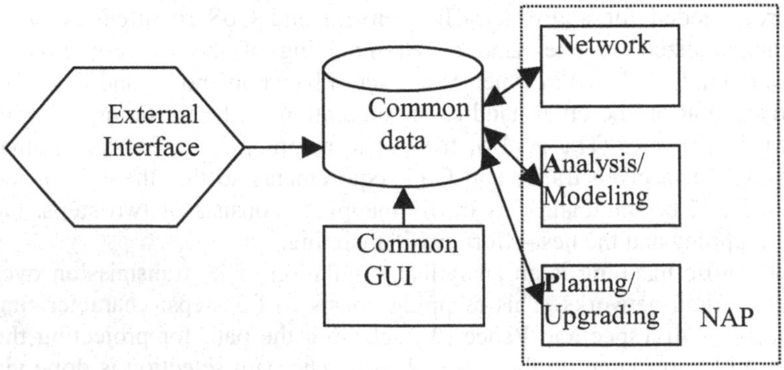

Fig. 1. The Architecture of Resource Planning Prototype.

Our prototype is successful applied for resource planning and upgrading in IP-based converged network infrastructures. The computation was carried out on an Intel

Pentium Processor 333 under Window NT. Fig. 2 shows the runtime of the whole genetic algorithm for capacity planning. The runtime depends on the complexity of the backbone infrastructure and the number of real-time and best-effort traffic demands, the number of generic strings per generation and the total number of evolution cycles. The diagram in figure 2 shows the evolution of the network costs for a typical run of the genetic algorithm for a network with 42 nodes, 106 links, 3600 real-time flows and 98 best-effort traffic demands. In GA1, the number of chromosomes in each generation is between 3 to 24 (equal to the number of all paths connecting two nodes in the network) and the chromosome length is 151. In GA2, these values are 4 and 106. For every generation, the cost values are shown. Starting from a first generation with the usually high cost values, it can be observed that the costs decrease gradually. It indicates that the principle Reproduction/Selection" works in the area of capacity planning, and that the crossover of two good solutions has a high probability of creating another good solution [5].

Network name	Node number	Link number	Real-time traffic number	Best-effort traffic number	Runtime in seconds
NW 1	9	24	276	72	0.7
NW 2	16	48	368	240	1.0
NW 3	25	80	460	600	1.5
NW 4	36	120	552	1260	2.1
NW 5	49	168	644	2352	3.2
NW 6	64	224	736	4032	13.0
NW 7	100	180	920	9900	28.6
NW 8	121	220	1012	14520	58,2
NW 9	144	264	1104	20529	107.6
NW 10	169	312	1196	28392	191,4

Table 1: The Runtime of the Genetic Algorithm.

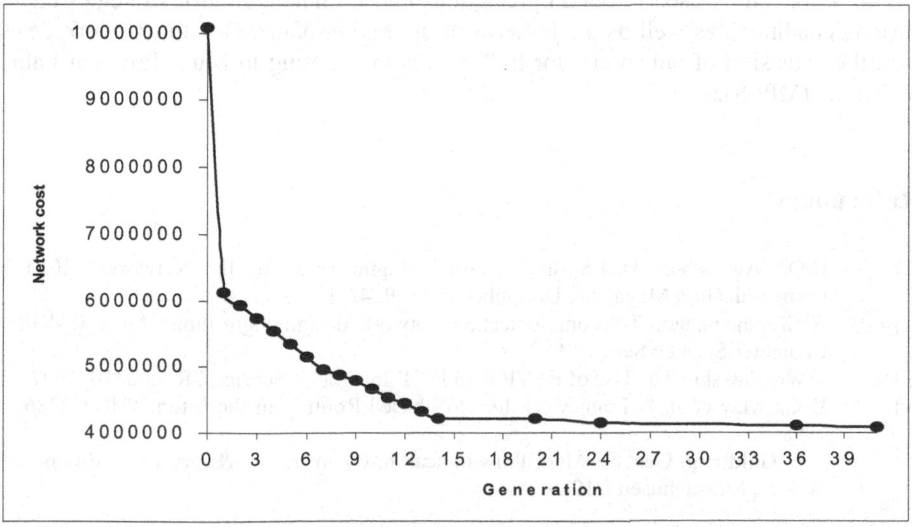

Fig. 2. The Cost Evolution of the Genetic Algorithm for Capacity Planing.

Table 2 show the converged network features which are considered in our model.

features of converged networks	**Considered (Yes/No)**
Integrated Services	yes
Differentiated Services	no
MPLS	yes
Survivability (refer to link/node failure)	yes
Reliability	yes
For real-time traffic End-to-end delay	yes
End-to-end jitter	no
End-to-end blocking probability	no
End-to-end Packet loss probability	yes
Delay at each intermediate node	yes
Packet loss probability at each intermediate node	yes
For best-effort traffic Total end-to-end delay	yes

Table 2. The Converged Network Features within the Resource Planning Model.

Conclusion

This paper describes an model and its prototype used for resource planning in IP-based converged networks supporting the Integrated Services (IntServ), RSVP, the Constraint-based Routing (CR) for real-time and the OSPF BGP for the best-effort traffic, and the resource allocation .

Our work can be used to obtain predictions and formulate control strategies under various conditions as well as guide network upgrading plans. Our future work deals with the extension of our model for traffic mapping relating to Multi Protocol Label Switching (MPLS).

References

[1] D.O. Awduche. MPLS and Traffic Engineering in IP Networks." IEEE Communication Magazine, December 1999, P. 42-47.

[2] A. Kershenbaum. Telecommunication network design algorithms." McGraw-Kill Computer Science Series, 1993.

[3] J. Wroclawski. "The Use of RSVP with IETF Integrated Services," RFC 2210, 1997.

[4] E. Crawlay et al. 'A Framework for QoS-based Routing in the Internet." RFC2386, 1998.

[5] D.E. Goldberg. Genetic Algorithms in search," Optimization &Learning, addition-Wesley, Massachusetts, 1989.

Parallelization of the STEM-II Air Quality Model

J.C. Mouriño[1], D.E. Singh[2], M.J. Martín[1], J.M. Eiroa[1], F.F. Rivera[2], R. Doallo[1], and J.D. Bruguera[2]

[1] Depto. de Electrónica y Sistemas, Univ. A Coruña
jmourino@des.fi.udc.es
[2] Depto. de Electrónica y Computación, Univ. Santiago de Compostela
david@dec.usc.es

Abstract. STEM-II is an Eulerian numerical model to simulate the behavior of pollutant factors in the air. In this paper the computational requirements of the program in terms of memory storage and execution times are analyzed. The results of this analysis are conclusive as regards to the need of using parallel processing to achieve reasonable execution times. Then, the improvements achieved after the parallelization of the code on a distributed memory multiprocessor using the MPI standard message passing library are shown.

1 Introduction

STEM-II (Sulphur Transport Eulerian Model 2) [1] is an Eulerian numerical model to simulate the behavior of pollutant factors in the air. This model is being applied to the Power Plant of As Pontes (A Coruña, Spain) to study the relationships between the emissions, the atmospheric transport, the chemical transformations, the elimination processes, the resultant distribution of the pollutants in the air, and the deposition patterns.

The high computational load of the model forces to execute it on high-performance computing systems, with the aim of obtaining the results with appropriate response times and with the reliability for the industrial exploitation where it is intended to be applied. The purpose of this work is the parallelization of the code corresponding to the STEM-II model, following the message passing programming paradigm. Specifically, we have used for the implementation of the parallel code the MPI (Message Passing Interface) library [3].

The work is organized as follows: In Section 2 an analysis of the sequential code is performed allowing us to make decisions on where to focus our parallelization efforts. Section 3 describes the parallelization of the code. In Section 4 the results of executing our code on the Fujitsu AP3000 multicomputer are shown. Conclusions and main contributions are finally framed within section 5.

2 Sequential Code Analysis

Four main nested loops can be identified in the code of our program: the outermost one is the temporal loop (loop_t); the other three loops cover the three dimensional simulated space (loop_x, loop_y and loop_z). For each iteration of loop_t

B. Hertzberger et al. (Eds.): HPCN Europe 2001, LNCS 2110, pp. 543–546, 2001.

we have two main modules which include the three spatial loops: the horizontal transport module and the vertical transport module (`vertlq`). Besides, `vertlq` is divided into three modules: cloud parameters module (`asmm`), vertical transport module (`vertcl`) and, the most important, the chemical reaction module (`rxn`). The program starts reading and initializing the data independent of time. Each sixty iterations of the temporal loop read and initialize the data dependent of time. When loop_t finishes the totals accumulated in time are computed and printed to the result files. The resulting code consists of more than 130 Fortran 77 files, and approximately 15500 lines of code.

The execution of the program in a Sun Enterprise 250 server with a 296 MHz UltraSparc microprocessor has shown the following results: `vertlq` module takes 240 seconds per iteration which means an 88% of the total time. Besides, STEM-II uses a large number of multidimensional arrays to store physical information (concentration of species, temperature, etc.). The memory used by the program is proportional to the number of entries of the simulated space. In our case, 175794 Kb of memory are required for storing data. This great amount of required resources, both of memory storage and execution time, makes the simulation process be non-viable for the standard workstations. For these reasons, this model is an excellent candidate for its parallel implementation on a multicomputer.

3 Parallelization of the Code

For the parallelization of the program we have focused on the most time-consuming module, `vertlq`. A SPMD parallel programming paradigm was used. The data matrices are distributed by the root processor using the `MPI_SCATTERV` function. Then, each one of the processors execute the corresponding portion of the parallel code and finally, the data matrices are gathered using the `MPI_GATHERV` function to proceed with the sequential portion of the code.

Table 1. Data Dependences.

Module		Dependences	Loops that can be parallelized
`vertlq`	`rxn`	None	x, y, z
	`asmm`	z	x, y
	`vertcl`	z	x, y
Horizontal Transport		x, y	z

The lack of dependences among different loop iterations will determine the chances of code parallelization. Main modules of the program, indicating on which loop they present dependences and which loops can be parallelized, are shown in Table 1. Note that the most-time consuming module (`vertlq`) can be parallelized both in the x-axis and in the y-axis, however the parallelization of the whole program is not easy due to the dependences in the x and y axis that the horizontal transport module presents.

4 Experimental Results

The parallel code was evaluated on the Fujitsu AP3000 system [2]. The available AP3000 system consist of 16 nodes, having 4 of them two processors (20 UltraSparc processors at 300 MHz were then available)

Figures 1 and 2 show the execution times and speedups of the `vertlq` module for up to 16 processors, parallelized by the x and y axis, respectively. The execution time is larger in the first iterations due to the cost of the data loading process. In order to ignore this effect, speedups have been calculated considering the average of the execution times of iterations 6 to 10.

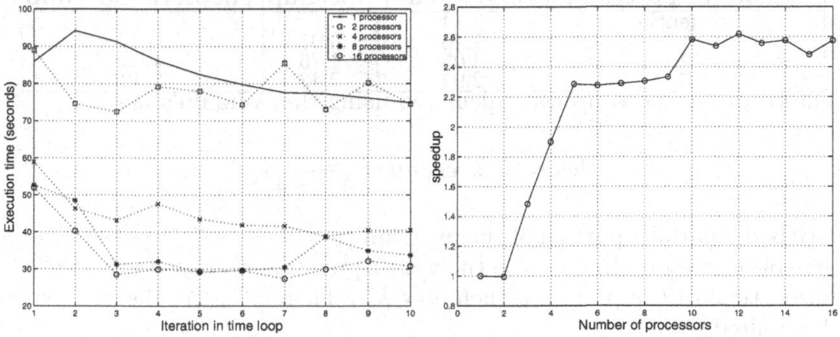

Fig. 1. Execution Times and Speedup of `vertlq` Parallelized on the x Axis.

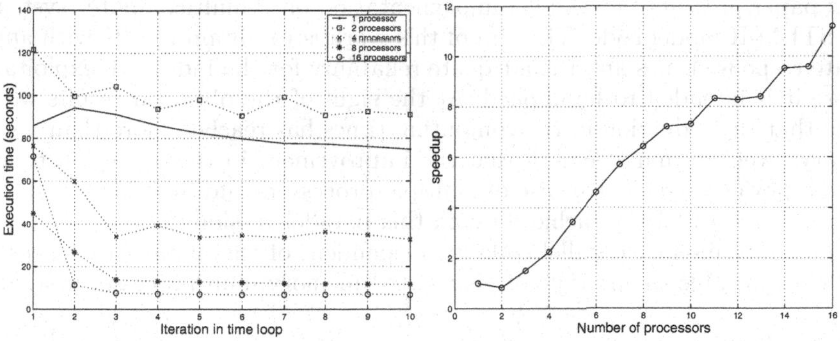

Fig. 2. Execution Times and Speedup of `vertlq` Parallelized on the y Axis.

Note that the results of the parallelization according to the x axis are not as good as could have been expected. Data accesses in the distribution and gathering phases does not occur to consecutive memory positions. This is not the same in the parallelization of the y axis due to the fact that, parallelizing the outer-most loop, we obtain a better exploitation of the memory hierarchy of the system. As can be seen in Figure 2, a good scalability of the parallel code is obtained, reaching a speedup of 11 for 16 processors. The achieved speedups for the parallel version on the y axis are due to both, the parallelization itself and the greater data locality (each processor accesses to a smaller data volume, taking better advantage of the memory hierarchy).

Table 2 shows the obtained speedup corresponding to the whole program execution. A number of 300 iterations (5 real time hours) was performed using the parallelization on the y axis and biprocessor nodes. As can be observed, two processors would be enough to obtain a simulation in real time. As the number of processors increases, the execution time is reduced, obtaining reasonable speedups.

Table 2. Achieved Speedups in the whole of the Program.

Number of processors	Time (hours)	Speedup	Theoretic Speedup
Sequential	7.14		
1	7.40	0.97	
2	4.05	1.76	1.79
3	2.95	2.42	2.42

The theoretic speed-up was calculated using the Amdahl's law:

$$theoretic\_speedup = \frac{1}{1 - c + \frac{c}{p}} \tag{1}$$

where c is the parallel portion of the program ($c = \frac{parallel\_section\_time}{total\_time}$) and p the number of processors. In our case the `vertlq` module represents the 88% of the execution time of the program, then $c = \frac{88}{100}$. Therefore, the theoretic speedup will be limited to 8.

5 Conclusions

This paper presents the parallel implementation on a multicomputer system of the STEM-II model code. The aim of this work is to obtain results with appropriate response times and the adequate reliability for the industrial exploitation where it is intended to be applied. At the sight of the obtained results we can state that the reduction in the computing times has reached more than a satisfactory level. We must remark that the improvement in the execution times is related both to the efficient use of multiple processors and, to a large extent, to the greater data locality achieved with this parallel version.

The implemented parallel code is independent of the number of processors, and it is portable to any type of distributed memory multiprocessor system.

Acknowledgements

This work was supported by the project ref. 1FD97-0118-C02-02.

References

1. Carmichael, G.R., Peters, L.K., Saylor, R.D. *The STEM-II regional scale acid deposition and photochemical oxidant model - I. An overview of model development and applications.* Atmospheric Environment, **25A**, 10, pp. 2077-2090, 1991.
2. H. Ishihata, M. Takahashi & H. Sato. *Hardware of AP3000 Scalar Parallel Server.* Fujitsu Sci. Tech. Journal, **33**(1), pp. 24-30, 1997.
3. P.S. Pacheco. *Parallel Programming with MPI.* Morgan Kaufman Publishers, Inc., 1997.

Simulating Parallel Architectures with BSPlab

Lasse Natvig

Dept. of Computer and Information Science
Norwegian University of Science and Technology
Gløshaugen, N-7491 Trondheim, Norway
Lasse.Natvig@idi.ntnu.no

Abstract. BSPlab is a simulation environment for studying parallel architectures. It offers the BSPlib parallel programming library and is based on Bulk Synchronous Parallel (BSP) computing [1], [2]. BSPlab contains a set of high-level performance models of parallel architectures. It can be used as a tool for architectural level design space exploration of BSP computers. The paper introduces BSP, BSPlib and the architectural models in BSPlab. A sample experiment, the status and future plans of the project are outlined.

1 Introduction

Although it has been claimed an ongoing convergence in parallel computer architecture [3] — understanding the interplay between hardware and software in parallel computing is still difficult. This understanding is crucial both for developing efficient parallel applications on contemporary supercomputers and for designing competitive high-performance parallel computers. An important way of studying this HW/SW interplay is by simulation. The BSPlab environment was developed for use in research and teaching of parallel computer architectures as well as for performance studies of parallel applications.

BSPlab offers the *BSPlib* library for parallel programming. BSPlib is based on the BSP (Bulk Synchronous Parallel) model [4]. Libraries such as PVM and MPI have a larger user community, but we believe the simplicity of BSPlib and the generality of BSP makes it more suitable for developing efficient applications that are portable to a large variety of parallel computer architectures. BSPlab contains architectural performance models for evaluating the efficiency of such portable software.

The paper starts by introducing the BSP model, the BSPlab environment and its architectural models. This is followed by an outline of a sample experiment and a summary of the status and future plans of the BSPlab project.

2 A Short Introduction to BSP and BSPlab

Leslie Valiant proposed the Bulk Synchronous Parallel (BSP) model in 1990 [1]. It is a theoretical framework outlining how parallel computations can be organised in a way

B. Hertzberger et al. (Eds.) : HPCN Europe 2001, LNCS 2110, pp. 547-550, 2001.

that bridges the gap between the needs of the programmers and the hardware offered by the computer architects. The model is defined as the combination of three attributes: *i)* a number of *components* performing processing and/or memory functions, *ii)* a *router* that delivers messages point to point between pairs of components, and *iii)* a *synchronisation facility* that is able to synchronise all or a subset of the components at regular intervals. A computation is described as a sequence of *supersteps*. During a superstep, the components perform computations asynchronously, and the synchronisation facility guarantees that all components have finished the current superstep before they proceed to the next one.

The performance of BSP computers can be characterised by as few as four parameters. Informally these are processor speed (s), number of processors (p), synchronisation cost (l) and global computation/communication balance (g). The BSP parameters are central in algorithm analysis for BSP applications, and make it possible to develop parallel programs with *portable* efficiency.

BSPlab is an environment for experimenting with BSP programs on different parallel architectures. Applications, benchmarks or specialised performance testing programs are written in C or C++ using BSPlib. The user may select among predefined parallel architecture models or might define her own architecture. The programs are debugged and executed in the BSPlab environment to achieve the measures specified by the user to support the current foci of the performance study. A typical goal can be a better understanding of the interplay between the selected BSP-architecture and the parallel program. In short, BSPlab can be regarded as an experimental tool for architectural level *design space exploration* of BSP computers.

BSPlab uses the standard Microsoft Developer Studio for the C++ programs. A consequence is that it has a good programming environment that also can be used for developing BSPlib applications that are targeted for real parallel computers. It was developed at NTNU in Trondheim, Norway, as the diploma work of Dybdahl and Uthus [5]. It is built upon the process-oriented discrete event simulation package C++SIM [6]. BSPlab is available from `www.idi.ntnu.no/bsplab`.

2.1 Architectural Models in BSPlab

When BSPlab starts execution, it loads an architecture definition file that selects an architectural model and gives the values of its parameters. BSPlab currently includes models of two abstract architectures and three real architectures.

- **Null machine.** This abstract machine models an ideal parallel computer where both communication and synchronisation is free. However, the computation time used by the processors can be modelled in several ways.
- **Simple machine.** The simple machine model extends the null machine by offering simple means of representing the time used for communication and synchronisation.
- **Distributed shared memory (DSM).** This architecture is functionally like a multiprocessor with the processors connected through a single common bus to a shared memory. The *logically* shared memory is *physically* distributed among the processors. Each processor can access data in own local memory, in shared memory stored

locally, or in remote shared memory. The parameters for this architecture are modelling the speed of the common bus and the method of synchronising the processors.

- **Tightly coupled multiprocessor (TCM).** This is a *family* of architectures where the processors are interconnected through one of a set of different network topologies. Parameters are available for specifying topology, network size, the communication between the processor and the network, and inter-processor communication.
- **Network of workstations (NOW).** This model represents a set of workstations interconnected in a local area network (LAN). The Ethernet is chosen as network standard and the model contains 10 parameters that specify the BSPlab Ethernet implementation. In addition, it has parameters for specifying the simulation of the network traffic generated by other applications using the same network.

3 Status and Example Experiment

During development, BSPlab was tested on the test programs following the BSPlib standard and a few simple BSP applications. Lilleaas did a thorough testing of BSPlab where he defined 25 different BSP architectures and tested these on 18 different test programs. Very few problems were found [7].

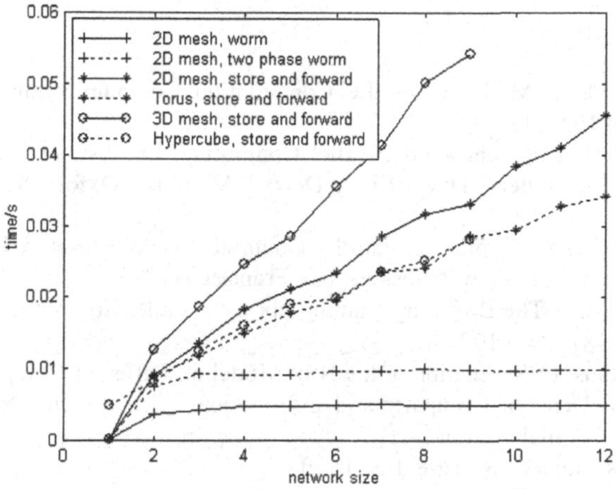

Fig. 1. Performance of 6 Different Message Passing Implementations (from [7]).

Figure 1 summarises the performance that Lilleaas found running a message-passing test on some of the topologies in the BSPlab TCM model. 2D mesh, 2D torus, 3D mesh and hypercube are tested with wormhole routing and store and forward message passing. The test program does 100 iterations letting a processor send a large message to a random destination and measures the message latency. The time used is plotted as function of the network size with values up to 12. Here the network size is the diameter of the network, which equals the maximum number of hops a message must travel. As expected, the time used by wormhole routing is nearly independent of network size while store and forward has a time consumption that is nearly linear in the number of hops. Further details and many other experiments were documented by Lilleaas [14].

4 Concluding Remarks

In the use of BSPlab we have experienced the importance of not varying to many parameters in the same experiment. The interplay between HW and SW in parallel applications is complicated and a systematic approach starting with simple and understandable experiments is recommended.

Several student projects have been done on testing BSPlab, and we have seen that it is working well. Validation of the performance models in BSPlab is still going on. In this context, we have realised that it is a good tool for demonstrating well known aspects of relations between algorithms, applications and architectures. We are therefore planning to use it as a central part in the teaching of a new course in parallel computer architecture. A step by step text introducing parallel computing aspects and performance effects using BSPlab as demonstration tool is under development. This might also be useful in teaching of BSP programming and parallelism in general. In addition, BSPlab can be used as a research tool in the field of portable and efficient software. Its source code is available from its website and the author appreciates comments on its use and possible improvements.

References

1. Valiant, L. A Bridging Model for Parallel Computation. Communications of the ACM, Vol. 33, No. 8 (1990) 103–111
2. McColl, W. F., Bulk Synchronous Parallel Computing, In Abstract Machine Models for Highly Parallel Computers, Davy, J.R. , Dew, P.M. (eds), Oxford Science Publications (1995) 41–63
3. Culler, D., Singh, J.P., Gupta, A.: Parallel Computer Architecture: A Hardware/Software Approach. Morgan Kaufmann Publishers, San Fransisco (1999)
4. Hill, J. et al.: BSPlib: The BSP programming library. Parallel Computing, Elsevier Science, Vol. 24 (14) (1998) 1947–1980 (http://www.bsp-worldwide.org)
5. Dybdahl, H., Uthus, I.: Simulation of the BSP Model on Different Computer Architectures, Diploma Thesis, Dept. of Computer and Information Science (IDI), NTNU, Trondheim, Norway (1997). (Available from http://www.idi.ntnu.no/bsplab)
6. C++SIM User's Guide. Release 1.5. Draft Ver. 1.0, Computing Laboratory, Univ. of Newcastle upon Tyne, UK (1994) (Available from http://cxxsim.ncl.ac.uk/)
7. Lilleaas, E.: Evaluation of BSPlab. Project Report, Dept. of Computer and Information Science (IDI), NTNU, Trondheim, Norway (1998)

A Blocking Algorithm for FFT on Cache-Based Processors

Daisuke Takahashi

Department of Information and Computer Sciences, Saitama University
255 Shimo-Okubo, Urawa-shi, Saitama 338-8570, Japan
daisuke@ics.saitama-u.ac.jp

Abstract. In this paper, we propose a blocking algorithm for computing large one-dimensional fast Fourier transform (FFT) on cache-based processors. Our proposed FFT algorithm is based on the six-step FFT algorithm. We show that the block six-step FFT algorithm improves performance by effectively utilizing the cache memory. Performance results of one-dimensional FFTs on the Sun Ultra 10 and PentiumIII PC are reported. We succeeded in obtaining performance of about 108 MFLOPS on the Sun Ultra 10 (UltraSPARC-IIi 333 MHz) and about 247 MFLOPS on the 1 GHz PentiumIII PC for 2^{20}-point FFT.

1 Introduction

The fast Fourier transform (FFT) is an algorithm widely used today in science and engineering. FFT algorithms have been well studied [1,2,3,4]. Many FFT algorithms work well for data sets that fit into a cache. When a problem exceeds the cache size, however, the performance of these FFT algorithms decreases dramatically. One goal for large FFTs is to minimize the number of cache misses.

In this paper, we propose a blocking algorithm for computing large one-dimensional FFTs on cache-based processors. Our proposed FFT algorithm is based on the six-step FFT algorithm [1,2]. The six-step FFT algorithm requires two multicolumn FFTs and three data transpositions. The three transpose steps typically are the chief bottlenecks in cache-based processors.

Some previously presented six-step FFT algorithms [2,4] separate the multicolumn FFTs from the transpositions. Taking the opposite approach, we combine the multicolumn FFTs and transpositions to reduce the number of cache misses, and we modify the original six-step FFT algorithm to reuse data in the cache memory. We have implemented the proposed FFT algorithm on cache-based processors, the Sun Ultra 10 and the PentiumIII PC, and in this paper we report the performance results.

2 A Block Six-Step FFT Algorithm

The discrete Fourier transform (DFT) is given by

$$y_k = \sum_{j=0}^{n-1} x_j \omega_n^{jk}, \quad 0 \le k \le n-1, \tag{1}$$

B. Hertzberger et al. (Eds.): HPCN Europe 2001, LNCS 2110, pp. 551–554, 2001.

where $\omega_n = e^{-2\pi i/n}$ and $i = \sqrt{-1}$.

If n has factors n_1 and n_2 ($n = n_1 \times n_2$), then the indices j and k can be expressed as:

$$j = j_1 + j_2 n_1, \quad k = k_2 + k_1 n_2. \tag{2}$$

We can define x and y as two-dimensional arrays (in Fortran notation):

$$x_j = x(j_1, j_2), \quad 0 \le j_1 \le n_1 - 1, \quad 0 \le j_2 \le n_2 - 1, \tag{3}$$

$$y_k = y(k_2, k_1), \quad 0 \le k_1 \le n_1 - 1, \quad 0 \le k_2 \le n_2 - 1. \tag{4}$$

Substituting the indices j and k in equation (1) with those in equation (2), and using the relation of $n = n_1 \times n_2$, we can derive the following equation:

$$y(k_2, k_1) = \sum_{j_1=0}^{n_1-1} \sum_{j_2=0}^{n_2-1} x(j_1, j_2) \omega_{n_2}^{j_2 k_2} \omega_{n_1 n_2}^{j_1 k_2} \omega_{n_1}^{j_1 k_1}. \tag{5}$$

This derivation leads to the six-step FFT algorithm [1,2]. We combine the multicolumn FFTs and transpositions to reduce the number of cache misses, and we modify the original six-step FFT algorithm to reuse data in the cache memory. We assume in the following that $n = n_1 n_2$ and that n_b is the block size, and assume that each processor has a multi-level cache memory. A block six-step FFT algorithm can be stated as follows.

1. Consider the data in main memory as an $n_1 \times n_2$ complex matrix. Fetch and transpose the data n_b rows at a time into an $n_2 \times n_b$ matrix. The $n_2 \times n_b$ array fits into the L2 cache.
2. For each of n_b columns, perform n_b individual n_2-point multicolumn FFTs on the $n_2 \times n_b$ array in the L2 cache. Each column FFT fits into the L1 data cache.
3. Multiply the resulting data in each of the $n_2 \times n_b$ complex matrices by the twiddle factors. Then transpose each of the resulting $n_2 \times n_b$ matrices, and return the resulting n_b rows to the same locations in the main memory from which they were fetched.
4. Perform n_2 individual n_1-point multicolumn FFTs on the $n_1 \times n_2$ array. Each column FFT fits into the L1 data cache.
5. Transpose and store the resulting data on an $n_2 \times n_1$ complex matrix.

Fig. 1 gives the pseudo-code for this block six-step FFT algorithm. Here the twiddle factors $\omega_{n_1 n_2}^{j_1 k_2}$ are stored in array `U`, and the array `WORK` is the work array. The parameters `NB` and `NP` are the blocking parameter and padding parameter, respectively. If we do not require an ordered transform, the $n_2 \times n_1$ complex matrix (the array `Y` in Fig. 1) and the transposition in step 5 can be eliminated.

3 Performance Results

To evaluate the proposed FFT algorithm, we compared its performance against that of the proposed FFT algorithm and the FFT library of FFTW (version

```
COMPLEX*16 X(N1,N2),Y(N2,N1),U(N1,N2),WORK(N2+NP,NB)
DO II=1,N1,NB
  DO JJ=1,N2,NB
    DO I=II,II+NB-1
      DO J=JJ,JJ+NB-1
        WORK(J,I-II+1)=X(I,J)
      END DO
    END DO
  END DO
  DO I=1,NB
    CALL IN_CACHE_FFT(WORK(1,I),N2)
  END DO
  DO J=1,N2
    DO I=II,II+NB-1
      X(I,J)=WORK(J,I-II+1)*U(I,J)
    END DO
  END DO
END DO
DO JJ=1,N2,NB
  DO J=JJ,JJ+NB-1
    CALL IN_CACHE_FFT(X(1,J),N1)
  END DO
  DO I=1,N1
    DO J=JJ,JJ+NB-1
      Y(J,I)=X(I,J)
    END DO
  END DO
END DO
```

Fig. 1. A Block Six-Step FFT Algorithm.

2.1.3) [3] which is known as one of the fastest FFT libraries for many processors. We averaged the elapsed times obtained from 10 executions of complex forward FFTs. The FFTs were performed on double-precision complex data, and the table for twiddle factors was prepared in advance.

The Sun Ultra 10 Model 333 (UltraSPARC-IIi 333 MHz, SunOs 5.6) and the PentiumIII PC (Coppermine 1 GHz, Intel i840, Linux 2.2.16) were used. All routines were written in Fortran. For the proposed FFT algorithm, the compiler used on the Sun Ultra 10 was Sun WorkShop Compilers Fortran 5.0, while the compiler used on the PentiumIII PC was g77 version 2.95.2. For the FFTW, the compiler used was Sun WorkShop Compilers C 5.0 on the Sun Ultra 10, while for the PentiumIII PC the compiler was gcc version 2.95.2.

Table 1 compares the proposed FFT algorithm and the FFTW in terms of their run times and MFLOPS. The column headed by n shows the number of points of FFTs. The next eight columns contain the average elapsed time in seconds and the average execution performance in MFLOPS. The MFLOPS values are each based on $5n \log_2 n$ for a transform of size $n = 2^m$.

With the Sun Ultra 10, for $n \leq 2^{17}$ the proposed FFT algorithm is slower than the FFTW, whereas for $n \geq 2^{18}$ the proposed FFT algorithm is faster than the FFTW. This is because the L2 cache size of the UltraSPARC-IIi is 2 MB and it holds a low cache-miss ratio even for a larger problem size on the FFTW.

With the PentiumIII PC, on the other hand, for $n \geq 2^{16}$ the proposed FFT algorithm is faster than the FFTW. This is because the L2 cache size of the

Table 1. Performance of One-Dimensional FFTs on Cache-Based Processors.

n	Sun Ultra 10 (UltraSPARC-IIi 333 MHz)				PentiumIII PC (1 GHz, Intel i840)			
	New FFT		FFTW		New FFT		FFTW	
	Time	MFLOPS	Time	MFLOPS	Time	MFLOPS	Time	MFLOPS
2^{12}	0.00143	171.53	0.00103	239.09	0.00066	371.45	0.00038	645.28
2^{13}	0.00297	179.58	0.00215	248.14	0.00157	339.73	0.00154	346.08
2^{14}	0.00908	126.27	0.00472	243.05	0.00458	250.41	0.00423	270.98
2^{15}	0.02122	115.81	0.01317	186.55	0.01027	239.22	0.00990	248.33
2^{16}	0.04526	115.83	0.03818	137.32	0.02145	244.48	0.02146	244.32
2^{17}	0.09707	114.77	0.09510	117.16	0.04434	251.29	0.04765	233.82
2^{18}	0.22981	102.66	0.25147	93.82	0.09063	260.34	0.10605	222.46
2^{19}	0.46862	106.28	0.52714	94.49	0.19250	258.74	0.24583	202.61
2^{20}	0.96896	108.22	1.10718	94.71	0.42469	246.91	0.54097	193.83

PentiumIII is 256 KB and the cache-miss ratio is high for a larger problem size on the FFTW. The performance of the proposed FFT algorithm remains at a high level even for a larger problem size, owing to cache blocking. These results clearly indicate that for larger problem sizes the proposed FFT algorithm is superior to the FFTW.

4 Conclusion

In this paper, we proposed a blocking algorithm for computing large one-dimensional FFTs on cache-based processors. We reduced the number of cache misses for the original six-step FFT algorithm. The proposed FFT algorithm is most advantageous with processors that have a considerable gap between the speed of the cache memory and that of the main memory.

We succeeded in obtaining performance of about 108 MFLOPS on the Sun Ultra 10 (UltraSPARC-IIi 333 MHz) and about 247 MFLOPS on the 1 GHz PentiumIII PC for 2^{20}-point FFT. These performance results demonstrate that the proposed FFT algorithm utilizes cache memory effectively.

References

1. D. H. Bailey, "FFTs in external or hierarchical memory," *The Journal of Supercomputing*, vol. 4, pp. 23–35, 1990.
2. C. Van Loan, *Computational Frameworks for the Fast Fourier Transform.* SIAM Press, Philadelphia, PA, 1992.
3. M. Frigo and S. G. Johnson, "The fastest Fourier transform in the west." Technical Report MIT-LCS-TR-728, MIT Lab for Computer Science, 1997.
4. K. R. Wadleigh, "High performance FFT algorithms for cache-coherent multiprocessors," *The International Journal of High Performance Computing Applications*, vol. 13, pp. 163–171, 1999.

Monte Carlo Simulations of a Biaxial Liquid Crystal Model Using the Condor Processing System

C. Chiccoli[1], P. Pasini[1], F. Semeria[1], and C. Zannoni[2]

[1)] Istituto Nazionale di Fisica Nucleare, Sezione di Bologna
Via Irnerio 46, 40126 Bologna, Italy
{chiccoli,pasini,semeria}@bo.infn.it
[2)] Dipartimento di Chimica Fisica e Inorganica, Universitàdi Bologna
Viale Risorgimento 4, 40136 Bologna, Italy
Claudio.Zannoni@cineca.it

Abstract. We study a lattice system of biaxial particles interacting with a second rank anisotropic potential by means of Monte Carlo simulations over a wide distributed network. We use the Condor processing system installed on the Italian Nuclear Physics Institute computer network. We have done calculations for a large number of different values of molecular biaxiality and we have determined a phase diagram for the system that we compare with previous simulations. The results of this work seems to be very promising and will allow us to use the Condor system for our large scale simulation studies.

1 Introduction

Computer simulations are a useful tool for investigating many fields of physics and are currently widely used in condensed matter research. We are interested in studying states of condensed matter intermediate between solids and liquids. These states are indicated by the somewhat contradictory name of liquid crystals (LC) and consist of various phases with different molecular organizations [1]. The main characteristic of liquid crystals at molecular level is that they possess orientational order, together with a translational mobility similar to that of liquids in nematic phases and reduced in other, so called smectic, types. A theoretical investigation of LC can be undertaken, as for any other complex fluids, by means of approximate theories or by performing numerical experiments on models. The Monte Carlo method, one of the foremost simulation techniques [2], commonly used in studying phase transitions and critical phenomena, plays an important role also in the investigation of liquid crystals [3]. One of the most important approaches deals with lattice models [3] where the molecules, or tightly ordered cluster of molecules, represented by three dimensional unit vectors ("spins") are considered to have a fixed position at the lattice sites. The spins possess full rotational freedom, subject to a certain intermolecular potential, so that this restriction does not affect their long range orientational ordering. The main advantage in using lattice models is the great number of particles which can be treated in comparison with off-lattice systems. A detailed investigation of these models requires, however, a very significant amount of

B. Hertzberger et al. (Eds.) : HPCN Europe 2001, LNCS 2110, pp. 555-560, 2001.

computing power which can imply using parallel computing, typically when studying large lattices, or employing distributed resources for smaller lattices but with many different values of parameters corresponding to different physical conditions.

Here we wish to present a distributed approach for studying a biaxial liquid crystal model by means of the software CONDOR [4] developed at the Computer Science Department at the University of Wisconsin-Madison and implemented on the network of the Italian National Institute for Nuclear Physics. The paper is organized as follows: first we briefly summarize the main features of the Monte Carlo simulation model; then we describe the CONDOR software and finally we discuss how we have performed the simulations of the biaxial liquid crystal system.

2 The Biaxial Liquid Crystal Lattice Model

The prototype lattice model for modelling nematic liquid crystals formed of uniaxial molecules was devised many years ago by Lebwohl and Lasher (LL) [5] and is the simplest one with the correct symmetry for nematics (in particular the potential is invariant for an head-tail flip of the molecules). The LL interaction tends to bring molecules parallel to one another and effectively models whatever underlying intermolecular interaction either attractive or repulsive that does that. While in this model, as in the large majority of theoretical calculations of liquid crystals, the mesogenic molecules are assumed to be cylindrically symmetric, it is important to recall that nematogenic molecules are invariably non cylindrically symmetric and that a much more realistic approximation is to treat them at least as biaxial objects. A simple lattice model of a biaxial system is defined by the second rank attractive pair potential [6]:

$$U(\omega_{ij}) = -\varepsilon_{ij} \{ P_2(\cos\beta_{ij}) + 2\lambda [R_{02}^2(\omega_{ij}) + R_{20}^2(\omega_{ij})] + 4\lambda^2 R_{22}^2(\omega_{ij}) \} \qquad (1)$$

where ε_{ij} is a positive constant, ε, for nearest neighbour molecules i and j and zero otherwise, P_2 is the second Legendre polynomial. $\omega \equiv (\alpha, \beta, \gamma)$ is the set of Euler angles specifying the orientation of a molecule and R_{mn}^L are symmetrized combinations of Wigner functions [7]. The parameter λ takes into account the deviation from cylindrical molecular symmetry: when λ is zero, the biaxial potential reduces to the Lebwohl - Lasher P_2 one , while for λ different from zero the particles tend to align not only their major axis, but also their short axis. In this latter case and varying the temperature then there is the presence of different orientational phases (see Fig. 1), as shown by Luckhurst and Romano for $\lambda=0.2$ [8] and by us on a L x L x L cubic lattice for a fairly large set of biaxialities [9].

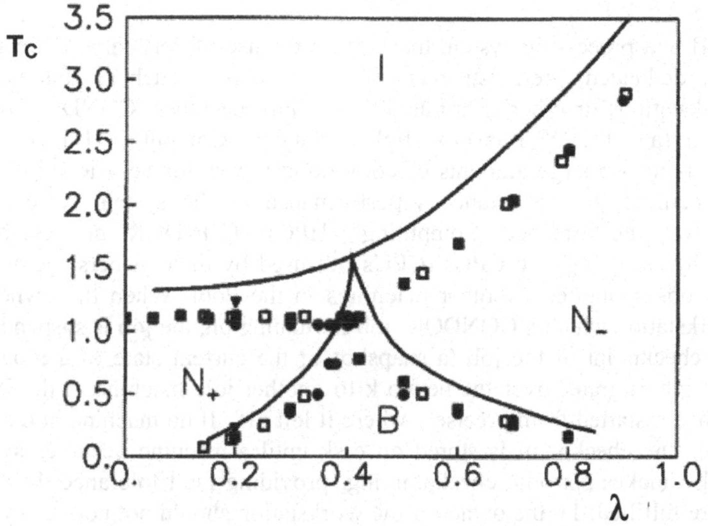

Fig. 1. Phase diagram showing the transition temperature T_C (in reduced units $T^* = kT/\varepsilon$) vs. λ as obtained by MC simulations (symbols) and Mean Field Theory (lines) from Ref. [9]. N_+, N, B represent regions of uniaxial and biaxial nematic phases.

The Monte Carlo simulation are performed using a standard Metropolis algorithm [10] with periodic boundary conditions for the lattice updates. The configuration of the system is given by a set of N trebles of unitary vectors $\mathbf{u}_{ij}$, i=1,2,3, where N is the number of particles. A new configuration is generated by moving a particle at random and we call a set of N attempted moves a cycle. To change the particle orientation we firstly choose at random an integer number $k \in [0,1,2]$ to identify the rotation axis. Then the orientation of the chosen particles is changed by generating a new uniformly distributed random value of the rotation angle. The maximum angular jump is chosen so as to maintain a rejection ratio not too far from 0.5. In this preliminary work we have used at least 30000 equilibration cycles far from the transition and 40000 in the pseudo critical regions. Apart from equilibration, production runs were also of varying length, according to the distance from the transition. Close to a phase change typical sequences of 20 kcycles have been used to produce the averages. We routinely determine a number of observables. In particular we calculate the heat capacity by differentiating the energy against temperature and the full set of second rank order parameters which are essential to define the different type of ordering in the different phases [9].

3 The CONDOR Processing System

CONDOR [4] is a processing system that allows the use of very large collections of available non-dedicated, pre-existing computing resources, such as (but not only) personal workstations or other distributed ownership machines. CONDOR provides an environment (a CONDOR Pool) for High Throughput Computing (HTC). The key idea of HTC is to use large amounts of computing power for very lengthy periods, with no concern in the instantaneous performance of the system typical of the traditional High Performance Computing (HPC). CONDOR creates a HTC environment by assigning idle CPUs (CPUs not used by their owners) belonging to the Pool, to jobs submitted by other machines in the Pool. When the owner starts using the workstation which a CONDOR job is running on, the job is suspended, and eventually a checkpoint of the job (a snapshot of the current state of a program) is done and the job migrates over the network to another idle machine in the Pool, on which the job is restarted from precisely where it left off. If no machine in the Pool is available, then the checkpoint is stored on disk until a machine becomes available. CONDOR also makes periodic checkpointing, providing fault tolerance. In this way two results are fulfilled: 1) the owner of the workstation should not notice any impact on the use of the workstation itself, 2) the job, migrating from one machine to another, restarting from the last checkpoint, will eventually come to the end of its execution. The Italian National Institute for Nuclear Physics (INFN) has developed a wide academic computer network linking its 24 sites since 1982. In 1998 the INFN project of building a CONDOR Pool over the INFN network was started, using already existing machines in many INFN sites. The Pool has been used by many INFN groups. In 1999 the equivalent of 38 years of CPU have been deployed. The Pool has presently more than 200 Unix machines, mainly Compaq Alpha and Linux PC.

4 LC Simulations Using the INFN CONDOR Pool

We have started to use the INFN CONDOR Pool for Monte Carlo simulation of the Liquid Crystal model formed by biaxial molecules to test if this processing system can be used for further studies in our research field. As can be seen from equation 1) the intermolecular potential depends on the biaxiality parameter λ, which has to be varied to study in detail the model. Of course for each value of λ an independent simulation has to be performed over a wide range of temperatures. This problem is then in principle well suited for the use of CONDOR because each simulation can be submitted on different computers. In this way we can send a CONDOR job for each value of the parameter, performing many simulations in parallel. For each value of the parameter λ we start the simulation with a low temperature from a totally aligned configuration of the molecules. For each of the subsequent temperatures the starting configuration was the last one of the previous temperature. We thus need a system for taking into account these jobs' dependencies (the job for a temperature must only start when the run of the previous job is completed). This mechanism is provided by CONDOR with a tool called DAGMan (Directed Acyclic Graph Manager). In this way we can submit a job for each temperature and for each value of the parameter λ, but only one job for each value of the parameter λ (corresponding to a particular temperature) is executing at any time. We have made the simulation for 15 values of the parameter λ and about 20 values of the temperature for each of them. So the total number of jobs submitted was about 300. CONDOR took care of the sequencing and the management (checkpointing, migration etc.) of all these jobs. We have performed the simulations of a 40 x 40 x 40 lattice system and we have been able to reproduce the complete study performed some years ago [9] in just two weeks. In Fig.2 we report as an example a plot of the orientational order $<P_2>=<3\cos^2\beta-1>/2$, where β is the angle between long molecular axis and preferred ordering direction, as a function of reduced temperature obtained from our CONDOR runs at various λ.

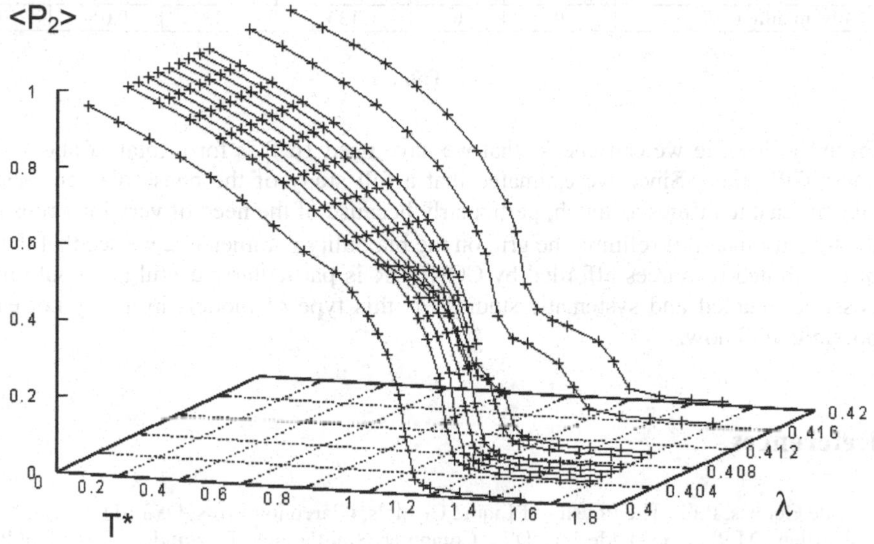

Fig. 2. A summary plot of the orientational order $<P_2>$ for models of various biaxiality λ as a function of the reduced transition temperature $T^* = kT/\varepsilon$.

In practice we have submitted our jobs to a CONDOR Pool of 70 Linux PCs but imposing the condition that the processor speed should be at least 200 MIPS. We report in Table 1 an excerpt of the log file for the simulation of the case λ =0.3 where the machines used can be identified, together with the values of some performance parameters, such as the CPU Usage.

Host/Job	Wall Time (hours)	Good Time (hours)	CPU Usage (hours)	Avg Alloc (minutes)	Avg Lost (minutes)	Goodput	Util.
p3d450.bo.infn.it	44	44	43	872	0	100.0%	97
chandra.bo.infn.it	47	47	47	565	0	100.0%	99
Pcmzz.bo.infn.it	19	19	19	564	0	100.0%	99
Pceng2.bo.infn.it	50	50	50	745	0	100.0%	99
linux1.ba.infn.it	0	0	0	0	0	0.0 %	0
to414xl.to.infn.it	27	27	26	536	0	100.0%	99
to44xl.to.infn.it	13	13	13	763	0	100.0%	99
pcl3c.bo.infn.it	12	12	12	705	0	100.0%	99
pcglob.bo.infn.it	22	22	22	658	0	100.0%	99
Pceng4.bo.infn.it	13	13	13	756	0	100.0%	99
Pceng3.bo.infn.it	13	13	13	751	0	100.0%	99
to40xl.to.infn.it	2	0	0	133	133	0.0%	0

Table 1

From the log file we can check that we have used 16 PCs for a total of about 1700 hours CPU time. Since we estimated that a full study of the phase diagram, should take at least ten times as much, particularly because of the need of very long runs near the transition and of refining the grid on the biaxiality parameter λ, we see that the use of distributed resources afforded by CONDOR is particularly useful and could make possible detailed and systematic studies of this type of models in a way not easily possible until now.

References

1. de Gennes, P.G., The Physics of Liquid Crystals, Clarendon Press, Oxford (1972).
2. Allen, M.P. and Tildesley, D.J., Computer Simulation of Liquids, Clarendon Press, Oxford (1987).
3. Pasini, P. and Zannoni, C, (eds.) Advances in the Computer Simulations of Liquid Crystals, Kluwer, Dordrecht, (2000).
4. http://www.cs.wisc.edu/condor/
5. Lebwohl, P.A. and Lasher, G., Phys. Rev. A 6, (1972) 426.
6. Luckhurst, G.R, Zannoni, C., Nordio, P.L. and Segre, U., Mol. Phys., 30, (1975) 1345.
7. Rose, M.E., Elementary Theory of Angular Momentum, Wiley, N.Y. (1957).
8. Luckhurst, G.R, and Romano, S., Mol. Phys. 40, (1980) 129.
9. Biscarini, F., Chiccoli, C., Pasini, P., Semeria F., and Zannoni, C., Phys. Rev. Lett. 75, (1995) 1803.
10. Metropolis, N., Rosenbluth, A.W., Rosenbluth, M.N, Teller, A.H. and Teller, E., J. Chem. Phys. 21, (1953) 1087.

Generic Approach to the Design of Simulation-Visualization Complexes

Elena V. Zudilova

Integration Technologies Department, Corning Scientific Center
4 Birzhevaya Linia, St. Petersburg 199034, Russia
ZudilovaEV@rning.com

Abstract. Simulation-visualization complexes combine tools for numerical simulation and data visual representation. They facilitate together the research process of investigating a phenomenon by decreasing necessary time and cost resources. The paper is devoted to the process of design and development of such kind of complexes. It introduces the approach that permits to combine simulation and visualization compounds together and represents at the same time a minimum of user discomfort related with the increasing functionality of a final complex.

1 Introduction

The paper presents the approach of how to combine the features of simulation and visualization tools in the framework of one generic exploration complex. This generic exploration environment will provide computation, presentation and interaction features wherever each of them is necessary for exploration of investigated objects or phenomena. Simulation and visualization processes are briefly described in section 2 of the paper. Section 3 is devoted to the integration of simulation and visualization compounds and shows what kind of feedback exists between them. It also contains an example of generic simulation-visualization complex - IntelliSuite™ CAD for MEMS.

2 Simulation-Visualization Processes

Supercomputing applications aimed to numerical simulation often possess a zero-dimensional (0-D) user interface. Command line here is usually standard interface solution [6]. User may only vary simulation parameters, i.e. initial conditions, criterions, scale, etc. Only people participated in the development of this software or specially pre-trained persons can effectively interact with the computational processes and analyze output data.

B. Hertzberger et al. (Eds.) : HPCN Europe 2001, LNCS 2110, pp. 561-564, 2001.

If data generated by numerical simulation software is large and complicated, then the best way to analyze it is to present this data visually. Moreover, there are special information systems based on numerical simulation methods, where visualization can be considered as the only solution of representing output results, such as weather forecasting, medical diagnostics and different types of monitoring.

Scientific visualization means simulating research conditions in two or three coordinates through establishing choreography or movement for the 2D/3D-objects step by step, setting up lights and applying textures [5].

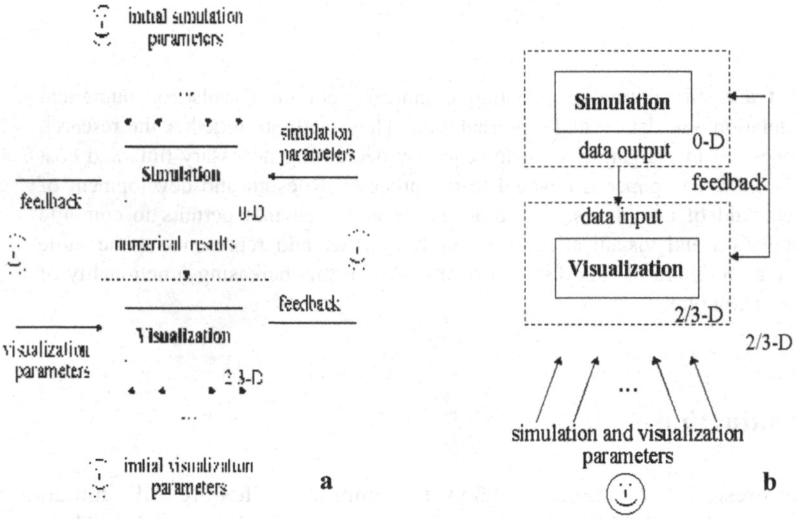

Fig. 1. Traditional feedback scheme between simulation and visualization software in if they are used as separate systems (a) or combined in one generic exploration complex (b).

There are many software applications to work with video, graphics and animation that can be used for the visualization of objects and phenomena either existing or being simulated. In ideal case this software includes integrated systems of analyses, processing and data imaging based on the classical representation as graphs with the complicated data visualizations as 2D surfaces or 3D objects. This functionality is available by so called 2D or 3D (2/3-D) graphical user interfaces [4].

3 Interaction between Simulation and Visualization Compounds

The main advantage of combination of simulation and visualization features in the framework of one exploration complex is the significant decrease of time resources necessary for the conduction of the experimental cycle by minimization of the number of feedback processes. [3] Fig. 1 illustrates how the situation changes when separate simulation and visualization software systems are combined into one generic complex.

Simulation-visualization complexes can be static or dynamic [1]. The static complexes deal with time independent data. Generated once, the following data does not change. So the represented visual image is always the same, only visualization parameters can be varied for better observation. As for dynamic complexes, the interaction between simulation and visualization compounds is more complicated. Numerical data is generated by simulation software periodically and the visualization results are also updated on the screen with the same frequency. So the dynamic complexes usually need more computational resources than static ones.

There are also two different approaches to implement generic exploration complexes. A complex can be implemented as distributed software system where numerical simulation is conducted on a separate computer and visualization process is also generated on a separate graphical workstation or server. There is another approach when it is implemented as standalone generic simulation-visualization software located on one computer system possessing necessary computational and visualization resources. [2]

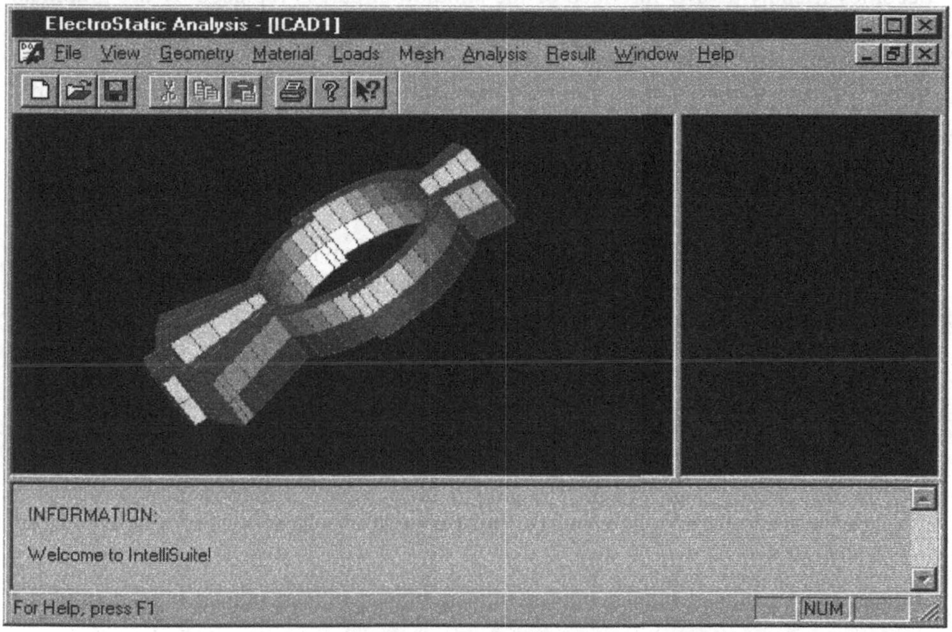

Fig. 2. 3D Interface of IntelliSuite CAD for MEMS.

IntelliSuite CAD for MEMS (Micro Electro Mechanical Systems) can be considered as an example of static simulation-visualization complex implemented as a standalone application. MEMS is a microfabrication technology which exploits the existing microelectronics infrastructure to create complex machines with micron feature sizes.

Fig. 2 represents the 3D interface of the following exploration complex. It provides the ability to simulate different classes of electrostatically induced MEMS devices with high accuracy and then to obtain the graphical presentation of the appearance of each simulated device. ACIS 6.2 kernel visualization library has been used as basis for the implementation of visualization feature of IntelliSuite CAD for MEMS.

4 Conclusion

Integration of two interrelated software compounds of different purposes in a generic complex leads to necessity of design and development of a generic graphical interface that provides a user the ability to interact with each component and to influence on feedback between them. This interface permits to minimize the exploration cycle as a whole, as the features of each compound are now available to user via the same environment. And, moreover, the user interaction with the simulation compound becomes friendly as 0-D interface of the simulation part is replaced by the graphical 2/3-D user interface of the entire generic simulation-visualization complex.

The main aim of the generic interface is to facilitate the work of users so that increasing functionality of simulation-visualization complex does not lead to rapid increasing of user discomfort while working with it.

Complex interface solution for a generic simulation-visualization complex should be based on at least three main usability criterions: consistency at all levels, informative feedback and design simplicity.

The ideal variant is even to provide the generic environment with adaptive user interface that permits a concrete user to have at the top level the features that he is usually deals with because of his experience or preferences.

References

1. Belleman R.G., Sloot P.M.A: The Design of dynamic Exploration Environments for Computational Steering Simulations, Proceedings of the 1st SGI Users' Conference, pp. 57-72 (2000).
2. Bogdanov A.V., Stankova E.N., Zudilova E.V. Visualization Environment for 3D Modeling of Numerical Simulation Results, Proceedings of the 1st SGI Users' Conference, pp. 487-494 (2000).
3. Bogdanov A.V., Zudilova E.V., Zatevakhin M.A., Shamonin D.P.: 3D Visualization of Convective Cloud Simulation. Proc. The 10th International Symposium on Flow Visualization, pp. 118-124 (2000).
4. Gorbachev Y., Zudilova E: 3D-Visualization for Presenting Results of Numerical Simulation. Proceedings of HPCN Europe - 99, pp. 1250-1253 (1999).
5. Zudilova E.: Using Animation for Presenting the Results of Numerical Simulation. Proceedings of International Conference of Information Systems Analysis and Synthesis) / Conference of the Systemics, Cybernetics and Informatics, Orlando, Florida, USA, pp. 643-647 (1999).
6. Zudilova E., Shamonin D.: Creating DEMO Presentation on the base of visualization Model. Proceedings of HPCN Europe - 2000, pp. 460-467 (2000).

On Parallel Programming Language Caper

Sergey R. Vartanov

The Center of Mathematical Modeling and Computation Technologies of Slavonic University, H. Emin, 123, Yerevan, Armenia
svartanov@mail.ru

Abstract. *Caper* is a parallel programming language, which supports declarative parallel computations and control of all architectures by Flynn [1]. *Caper* has a self-organization and asynchronous events processing programming means. Represented language has various variables with different scope, time of creation and survival time. Besides, *Caper* has so called "controlled variables" or variables with statuses, which allow to regulate usage of variables by different parallel processes. *Caper* based on virtual machines system, including own parallel virtual machine. This property allows to create programs not depended on multitasking management of operating systems.

1 Introduction

Caper is a parallel programming language based on the following principles:
- the possibility of calculations in the terms of the main parallel models;
- parallel performing the program structure components without using the parallelizing means of operating systems;
- the possibility of program self-organization during a computing process: dynamic compiling and performing a source code and individual commands, dynamic composing the running code by means of loading and removing object modules;
- controlling the computing process based on different classes events;
- interpreting the object-oriented programming for parallel calculations.

Caper is based on the virtual machine (CVM), which supports both sequential and parallel programs performing. *Caper* commands provide synchronous and asynchronous starts of program procedures with using *Caper* own model of pseudo-parallelism, called as "command-by-command"[2] and the familiar model with time quantums. *Caper* allows to start multiple parallel procedures with common data, a single procedure with multiple data, multiple parallel procedures with multiple data.

Caper is portable. It is independent of what calculation model (sequential or parallel) is used on the computing set. From the viewpoint of an operating system a program in *Caper* can be considered as a single task without subtasks.

From the very beginning the possibility of program self-organization during running was put in *Caper* to adapt of a loaded and running program to calculation

B. Hertzberger et al. (Eds.) : HPCN Europe 2001, LNCS 2110, pp. 565-568, 2001.

needs. *Caper* allows dynamically change a single command, or fragments of a running program, load and remove separate object modules, compile a source code and perform it.

All enumerated possibilities of the language machine and self-organization means allow to transport source codes, object modules and even fragments of the performed code with data to different computers united by network or any other means.

Caper provides possibility to describe various events and to assign the procedures of asynchronous sequential or parallel processing them. The language machine distinguishes classes and subclasses of events. The language is provided with multiple facilities of events managing: to define events, freeze, defreeze, remove and initialize them.

All *Caper*'s variables are polymorphic and differ in the scope, creating way and time. Besides usual variables *Caper* has such resource as managing variables. They are characterized by the states, which can be changed at the different moments of calculations. This allows to regulate access to the variables and initialize calculations when the variables acquire certain states.

2 Program Structure, Data, and Control

A program in *Caper* is an aggregate of so called blocks. Block is a named set of different data types: blocks of commands, constants, text, image and array. Blocks of commands are general logical executable units of programs in *Caper* including start, interaction and termination of parallel processes. Any command is either a sequence of arithmetical or logical expressions or a control statement. *Caper* expressions are constructed from arithmetic, comparing, logical, binary, assignment and some other operators. A block is a general aggregate of executable commands and different data.

Source code compilation result is the *Caper*'s object module or an executable file. *Caper* compiler creates CVM byte-code in different regimes, which can be selected by *Caper*'s compiler commands. In particular, compiler can to create so called "critical fragments"– program fragments, which monopolize resources of CVM.

Caper doesn't permit direct access to computer memory. Variables typification has internal supporting by CVM.

Caper has so called "controlled variables", or *Places*. They are global variables with status, which regulates the access to variable contents.

Caper supports dynamic and static multidimensional arrays with the length violation control, as well as so called "regions"– virtual arrays, which are defined as subarrays of created arrays, and filearrays, which support work with files as with arrays. The strings of *Caper* are whole units of data. At the same time, any string is an array of bytes (symbols).

Collection is a variables group, which must be described and created dynamically or statically. Collection descriptions can be redefined, masked or deleted during calculations time.

3 Caper Virtual Machine

Caper virtual machine can start in three different ways:

- starting object code implanted in *Caper* executable code.
- loading a source code file (source code module);
- loading an object module.

If it's source code module, then *Caper* compiles this code and create the first executable module and block (every module is represented by corresponding block). This module is named automatically as MAIN.

If it's object module, then this module will be loaded and named as MAIN automatically. Implanted module and block has the MAIN name, too.

In fact, *Caper* machine consists of three submachines (a sequential machine, a parallel machine and events machine).

Caper parallel machine executes command-by-command each of started parallel processes. The switching from a parallel process to other CVM carry out after every command of executed blocks or in special points, which defined by program commands or compiler commands.

CVM has three regimes: PRIMARY, when CVM dominates over operating system (OS), SECONDARY, when CVM activity depends on OS, and SOLE, when OS is suppressed by CVM to monopolize computer resources.

Caper has powerful means for asynchronous events processing. All events are divided into the following groups: logical events, program events, virtual machine events, operating system events.

All types of events are supported by collection of functions united by a common style of setting, "freezing," "defreezing" and deleting events.

All asynchronous events are accepted by *Caper* events machine, which starts an events processing block at the special control moment called as "a virtual machine step." If an event processing block was set, then the machine calls this block in the current parallel branch or initiates a new parallel branch for this block.

Caper has a lot of facilities to support parallel processes control and interactions. In first, *Caper* has abilities to activate, deactivate or broken parallel processes.

Stopped processes can remain not activated (in fact, all other processes can be terminated; e.g. there will be no active process that could activate stopped ones). In this case the program will be terminated and all stopped processes will be deleted by *Caper* machine (just as local and private variables, input parameters and others).

Caper allows controlling the moment of switching over to the next parallel process. Besides using public resources (variables, *places*, blocks) interaction of parallel and sequential processes by means of local variables and parameters setting and receiving is allowed. All means are grouped by functions of reading and writing local variables and parameters.

CVM supports a set of events, which can asynchronously inform about parallel computation states: whether the parallel process was started, stopped, terminated and so on.

4 Programming Experience in Caper

Caper was used for calculations with a few thousands CVM's parallel processes in a single processor (Pentium I). There are very natural and effective realizations of such components as multiple simultaneous animations (20-120 items), parallel reading and processing files, parallel searching in memory and in databases, etc. (in [3] a few algorithms of image processing were described; their real parallelizing for 8 computers in network was made in *Caper*-2 and represented in [2]).

Caper is in progress now.

References

1. Flynn, M.: Ultraspeed Computing Systems. IEEE Trans.Comp., Vol. 54, No. 12. (1966) 311-320
2. Vartanov, S.: The CAPER Programming Language. Preprint 97-5. National Academy of Sciences of Ukraine, Glushkov Institute of Cybernetics. Kiev (1997)
3. Vartanov, S., Agayan, S.: Representation of Image Processing Algorithms by Systems of Cellular Transformations. Mathematical Questions of Cybernetics and Computing Technique, vol.15, Yerevan. (1988) 5-19

Increased Efficiency of Parallel Calculations of Fragment Molecular Orbitals by Using Fine-Grained Parallelization on a HITACHI SR8000 Supercomputer

Yuichi Inadomi[1*], Tatsuya Nakano[2], Kazuo Kitaura[3], and Umpei Nagashima[4]

[1] Fuji Research Institute Corporation, 2-3 Kanda Nishikicho Chiyoda-ku, Tokyo 101-8443, Japan

[2] Division of Chem-Bio Informatics, National Institute of Health Science 1-18-1 Kamiyoga, Setagaya-ku, Tokyo, 158-8501, Japan

[3] Department of Chemistry, College of Integrated Arts and Science Osaka Prefecture University, 1-1 Gakuen-cho, Sakai, Osaka 599-8531, Japan

[4] Advanced Industrial Science and Technology, 1-1-4 Higashi Tsukuba, Ibaraki 305-8566, Japan

Abstract. A program for the fragment molecular orbital (FMO) calculation was developed that uses fine-grained parallelization to determine the electronic structure of large molecular systems such as proteins in practicable time. Using this program on a HITACHI SR8000 supercomputer, we found that the method improved the most time-consuming step, which is the calculation of molecular integrals, by a factor of 7. The improvement in the ERI eveluation was proportional to the number of processor units used in the calculations. Despite this improvement, clock time was not dramatically shortened because communication time in an HITACHI SR8000 is relatively slow.

1 Introduction

A fragment molecular orbital (FMO) method[1,2,3] was previously developed to obtain the electronic structure of large molecules such as proteins. This method "divides" a molecule into small fragments and then calculates the molecular orbitals for these fragments and fragment pairs to evaluate the total energy and the charge distribution of the molecule. In this FMO method, parallel executions of these MO calculations can be done easily because the MO calculations for each fragment and fragment pair can be carried out independently. These parallel calculations are called "large-grained parallelization". The ABINIT-MP program[3] used in this study allows large-grained parallel FMO calculation by using a message-passing interface (MPI) [4]. When a HITACHI SR8000 supercomputer is used for FMO calculations, the large-grained parallelization efficiency of this method exceeds 90%[5]. In these FMO calculations, the generation of the

* Corresponding author. e-mail: inadomi@nair.go.jp.

B. Hertzberger et al. (Eds.): HPCN Europe 2001, LNCS 2110, pp. 569–572, 2001.

electron repulsion integrals (ERIs) is still the rate-determining step. Therefore, the fast ERI evaluation is needed for fast FMO calculation. There are two way to shorten the ERI calculation time; one is using the algorithm for the fast ERI calculation, the other is the parallelization of ERI evaluation. For the fast ERI evaluation, Obara et al.[6] and Head-Gordon et al.[7] developed algorithms referred as the vertical recurrence relation (VRR) and the horizontal recurrence relation (HRR), respectively. Although ERIs can be evaluated using only VRR, Head-Gordon et al.[7] showed that using the combination of VRR and HRR in calculating ERI reduces the computation cost. In addition, the VRR and HRR involve simple addition and multiplication, thus making their implementations and vectorizations is relatively easy. Accordingly, we selected the combination of the VRR and HRR as the algorithm for the ERI calculation. Furthermore, we developed the program for parallel computation of the ERIs, referred to as "fine-grained parallelization". We implemented new ERI calculation program and embedded it into ABINIT-MP (henceforth denoted this program as "modified ABINIT-MP").

To evaluate the effectiveness of this program, we used the program on a HITACHI 8000 supercomputer to calculate ERIs for a test molecule. As the results, the computation time of the ERI calculation is shortened by a factor 7. On the other hand, fine-grained parallelization efficiency is very low since the communication cost becomes relatively large to the computation cost due to the shortening of ERI calculation.

2 Results and Discussion

In this study, we used a polypeptide, 15 glycine $(Gly)_{15}$ with the HF/STO-3G[8,9] level, as a target molecules.

Our results show that the efficiency (CPU time) of the ERI computations using modified ABINIT-MP was improved by a factor of about $6 \sim 7$ compared with that using original ABINIT-MP whose ERI calculation program was implemented with only VRR. This improvement also decreased the clock time by 1/3.

Our results also show that the efficiency of the fine-grained parallelization on SR8000 is very low. Here, we discuss the the efficiency of the fine-grained parallelization in the FMO calculation. Assuming that the fine-grained parallelization efficiency of FMO calculations is p and assuming that the cost of the calculation using only one processor is T and the communication cost between 2 processors is L, the total calculation cost (W_{cal}) using n processors is

$$W_{\text{cal}} = T\left((1-p) + \frac{p}{n}\right) . \tag{1}$$

This is Amdahl's law. Because communication cost is mainly due to the reduction process in this calculation, the communication cost (W_{com}) is

$$W_{\text{com}} = L \log_2 n . \tag{2}$$

Table 1. MPI communication cost and ERI computation cost per processor when fine-grained parallelization is used.

number of processors per fragment (pair)	1	2	4	8	16
ERI calculation(Sec.)	72.6	36.6	25.6	12.4	7.9
(ratio)[a]	(1.00)	(0.50)	(0.35)	(0.17)	(0.11)
communication(Sec.)	—	608.8	1269.9	1394.9	1787.0
(ratio)[b]	(—)	(1.00)	(2.09)	(2.29)	(2.94)

a; ratio of computation cost to that with only one processor
b; ratio of the communication cost to that between two processor

Thus, the total cost ((W_{tot}) is

$$W_{\mathrm{tot}} = T\left((1-p) + \frac{p}{n}\right) + L\log_2 n \ . \tag{3}$$

Here, when a parameter q is defined as the ratio of T to $L\,(T/L)$, efficiency (P_{eff}) becomes

$$P_{\mathrm{eff}} = (1-p) + \frac{p}{n} + \frac{\log_2 n}{q} \ . \tag{4}$$

Table 1 lists the execution times for the ERI calculations using 1~16 processor(s) per fragment(pair). High parallelization efficiency was achieved ($p \approx 0.9$) in the ERI calculations using modified ABINIT-MP. Table 1 also lists the MPI communication time per processor unit for the entire calculation using multi processors for a fragment(pair). This table shows that the communication time per processor for the ERI calculations using fine-grained parallelization is similar or better than the theoretical time obtained by using Eq.(2).

Figure 1 shows P_{eff} versus q, for p=0.9. This figure shows that fine-grained parallelization is not effective if q is relatively small (i.e., < 3). In contrast, the parallelization using a few processors in the fragment(pair) computation is effective when $q > 3$. Because q is ca. 1.6 for a HITACHI SR8000 supercomputer, little advantage will be gained by using the fine-grained parallelization on this computer.

3 Conclusions

We implemented an algorithm that uses fine-grained parallelization for fast calculation of ERIs. Computation time for the ERI calculation using this algorithm decreased by 1/6~1/7, and the parallelization efficiency exceeded 90%. However, when a HITACHI SR8000 supercomputer is used, such improved computation time and efficiency by fine-grained parallelization can only be achieved by increasing the ratio of the computation time to the communication time (q) to more than 3. Therefore, when a computer whose communication speed is relatively slow is used for FMO calculations, the fragment size should be large.

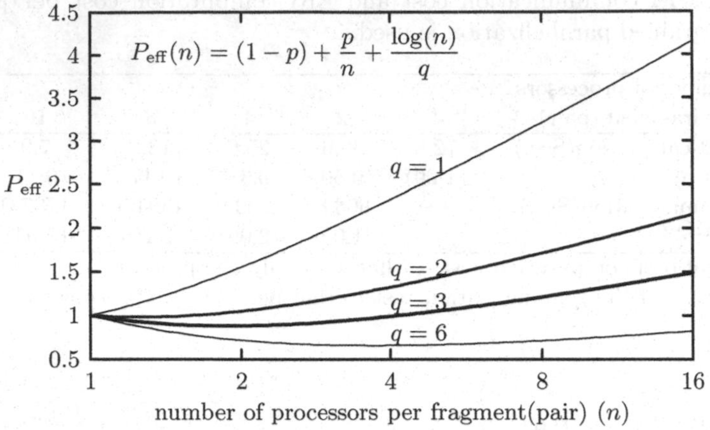

Fig. 1. Expected Efficiency Using Fine-Grained Parallelization for Various q.

This work was supported by the research project, Platform Architecture for Embedded High-Performance Computing. All numerical FMO calculations were carried out using the ABINIT-MP program on the HITACHI SR8000 super computer at the Tsukuba Advanced Computing Center of the Agency of Industrial Science and Technology.

References

1. K. Kitaura, T. Sawai, T. Asada, T. Nakano, and M. Uebayasi, Chem. Phys. Lett. **312** (1999) 319.
2. K. Kitaura, E. Ikeo, T. Asada, T. Nakano, and M. Uebayasi, Chem. Phys. Lett. **313** (1999) 701.
3. T. Nakano, T. Kaminuma, T. Sato, Y. Akiyama, M. Uebayasi, and K. Kitaura, Chem. Phys. Lett. **318** (2000) 614.
4. MPI: A Message-Passing Interface Standard, MPI forum, 1995.
5. T. Sato, Y. Akiyama, T. Nakano, M. Uebayashi, and K. Kitaura, IPSJ Trans. on High Performance Computing Systems **41** (2000) 104.
6. S. Obara and A. Saika, J.Chem.Phys. **84** (1986) 3963.
7. M. Head-Gordon and J. A. Pople, J.Chem.Phys. **89** (1988) 5777.
8. W. J. Hehre, R. F. Stewart, and J. A. Pople, J. Chem. Phys. **51** (1969) 2657.
9. W. J. Hehre, R. Ditchfield, R. F. Stewart, and J. A. Pople, J. Chem. Phys. **52** (1970) 2769.

Customer-Centered Models for Web-Sites and Intra-Nets

Reinhard Riedl

Department of Computer Science, University of Zurich

Abstract. We describe how content structures in Web-sites and Intra-nets can be be linked with customer behavior represented in the log files of Web-servers. More precisely, we present a framework for the design of user-centered Markov models of Web-domains.

1 Practical Experience with Internet Event Mining

This paper discusses *Internet event mining* for the purpose of building customer-centered models of Web-sites and Intra-nets, which combine knowledge about content structures and measurements of user behavior.

1.1 Heaps' and Zipf's Laws

We have analyzed log-files of the following four servers

- the search engine of a major international provider of financial products [4]
- the web-server of a marketing site of another international company
- the web-server of a web-based theater magazine
- the web-server of an academic department

In all four cases, we have detected similar statistical patterns as they are known for natural languages and for database traces [6], namely Heaps' Law ([3]) and both versions of Zipf's Law ([12]). Compare [7].

Heap's Law states that the number of the set of words in a natural language text of size x grows with x^α for some $0 < \alpha < 1$. Hereby, originally Heap observed the exponent 0.5. The function mapping texts of a given size to the average number of words appearing in the texts can be generalized to a mapping from client server traces of a given size and type to the average number of request types appearing in the given traces. That mapping may be considered as a formal workload representation. That representation can further be generalized to all kinds of accesses to resources. Our analyses show, that if we apply that mapping to the four scenarios, Heap's Law holds, although with varying exponents > 0.5.

Zipf's law is already folklore. It describes the distribution of the most popular words in a text. According to Zipf's original observations they decrease with order n^{-1}. Another folklore version of Zipf's Law states that the cardinalities of sets of words appearing exactly n times in a text decrease with n^{-1}.

B. Hertzberger et al. (Eds.): HPCN Europe 2001, LNCS 2110, pp. 573–578, 2001.

Again, generalizations to workload representations for client/server computing or other accesses to resources in computing are possible. See e.g. [6] for corresponding measurements for high-performance transaction processing, which provided coefficients larger than 1. In the four scenarios depicted above we have made observations comparable to Zipf's Laws, but again with varying coefficients.

These findings led us to develop stochastic models for Intra-nets and Web-sites, which we shall present in this paper. Note, that a confirmative discussion of related research results may be found in [1], while a critical discussion of related research results may be found in [11]. We conjecture, that Zipf's Law and Heap's Law will hold for a large variety of traces, but not for all, and that the coefficients represent intrinsic workload properties, which characterize the workload and which have a significant impact on performance data when the workload is processed by a specific type of machine, or a specific system of machines, respectively.

2 Internet Event Mining

Internet event mining is concerned with the deduction of knowledge from the observation of events in some C/S driven part of the Internet. Web-servers are of special interest: They provide a logging of request events for free, plus they record the context of requests, wherever that context is provided by the browser. For example, if a surfer accesses a web-page upon a recommendation of a search-engine, the original search request is usually found in the log-file of the web-server providing the web-page. However, context information does not suffice to identify individual surfing behavior. Proxy and cache transparencies must be broken up, and recorded data must be 'cleaned' and stored appropriately. Afterwards queries on the stored data enable us to answer questions of the type *What is the probability that a session starting with access to page x and ending with access to page y "contains" a particular set of pages?* Such information may be quite useful. For example, our analysis of the marketing site revealed, web-accesses of a competitor and it showed the distribution of footprint lengths on the session level. For more systematic investigations we have drawn up a formal modeling framework for Web-sites (and for Intra-nets in general). The basic idea hereby is quite simple. We generate a Markov model from the measurements and then we apply classical stopping techniques for Markov processes. Accessed data objects are understood as states and the subsequent access of another object is understood as a state transition. Or, on the more abstract level, the attribute values of accessed data objects are considered as states, and the subsequent access of another object with a different (or the same attribute value) is understood as a state transition.

2.1 Prerequisites

Let $\mathcal{D}$ denote the set of requests for data objects, i.e. static and dynamic web-pages, plus a symbol ∞ which jointly represents the beginning or the end of a session. And let $\mathcal{A}$ denote a set of attribute values characterizing the content, meaning, and relevance of a data object. A session is defined as a sequence of page accesses with the GET operation, which are initiated by the same person, such that the time between two subsequent accesses is smaller than a given threshold. We assume that the data in the log-file have been partitioned into sessions $S_i := \{s_{i_1}, s_{i_2}..., s_{i_{n_i}}\}$, $1 \leq i \leq N$, where each s_{i_j} identifies an accessed page. There are various ways to achieve such a partitioning. We first have to perform a user tracking, which enables us to partition the resulting log-file into sessions for individual users with time threshold infinity, and then we compare each two accesses subsequent in time to create sessions with a given finite threshold. We have applied various techniques and our experience can be summarized as follows

Technique	client support	server support	complexity	precision
Java-Script	+	-	++	-
Cookies	+	+	+	+
Hidden Fields	-	+	++	+
unique URI	-	+	++	+
SSL	+	+	+	++
Authentication	+	+	+	+++

We assume that the contents of the web-pages have been classified and an attribute value has been assigned to each web-page. Given a partitioning into sessions and an attribute mapping $A : \mathcal{D} \mapsto \mathcal{A}$, we can define imaginary sessions $S^A := \{a_{i_1} := \mathcal{A}(s_{i_1}), a_{i_2}..., a_{i_{n_i}}\}$, $1 \leq i \leq |\mathcal{A}|$ consisting of imaginary accesses to attribute values. Finally, we assume that sets $\mathcal{T} \subseteq \mathcal{D}$, $\mathcal{T}^A \subseteq \mathcal{A}$ of terminal requests and terminal attributes has been identified, and that value functions $F^{\mathcal{T}} : \mathcal{T} \mapsto \mathcal{R}$, $F^{\mathcal{A}} : \mathcal{T}^{\mathcal{A}} \mapsto \mathcal{R}$ are given, which assign a real number to each terminal object or attribute value. For example, that valuation may describes the benefit for the site owner created when a web-surfer eventually accesses a data objects.

2.2 The Markov Models

We both count the pairs $(i, j) \in \mathcal{D} \times \mathcal{D}$ of subsequent accesses to pages i and j and subsequent imaginary accesses to the pairs $(a, b) \in \mathcal{A} \times \mathcal{A}$ in these sessions, and divide the resulting numbers by the numbers of accesses to page i, and to attribute a, respectively. This defines the transition probabilities for two Markov chains X_k and Y_k:

$$p_{ij} := \frac{\sum_k |\{l|i = s_{k_l}, j = s_{k_l+1}\}}{\sum_k \{l|i = s_{k_l}\}}, \; p^A_{uv} := \frac{\sum_k |\{l|u = a_{k_l}, v = b_{k_l+1}\}}{\sum_k \{l|u = a_{k_l}\}},$$

for all data objects i, j and for all attribute values u, v. Further, we define the stopping time $\tau : \Omega \mapsto \mathcal{N}$, $\tau(\omega) := \min_n X_n(\omega) \in \mathcal{T}$, where $\mathcal{N}$ denotes the natural numbers. Now we can define the real and the imaginary stopped profile

$$\mathcal{P}_{\mathcal{T}}(i) := (P\{X_{\mathcal{T}} = t|X_0 = i\})_{t \in \mathcal{T}}, \; \mathcal{P}^A_{\mathcal{T}}(a) := (P\{Y_{\mathcal{T}} = t|X_0 = a\})_{t \in \mathcal{T}^A},$$

$\forall a \in \mathcal{A}, i \in \mathcal{D}$, the corresponding valuation functions $\tilde{F}^{\mathcal{T}} : \mathcal{D} \mapsto \mathcal{R}$ and $\tilde{F}^{\mathcal{A}} : \mathcal{A} \mapsto \mathcal{R}$, which extend $F^{\mathcal{T}}$ and $F^{\mathcal{A}}$, respectively:

$$\tilde{F}^{\mathcal{T}}(.) : \sum_{j \in \mathcal{D}} \mathcal{P}_T(.)(j) F^{\mathcal{T}}(j), \; \tilde{F}^{\mathcal{A}}(b) : \sum_{a \in \dashv} \mathcal{P}_T^{A}(b)(a) F^{\mathcal{A}}(a),$$

the real and the imaginary link differentials $\mathcal{L}_T$ and $\mathcal{L}_T^A$ as well as the corresponding link valuation functions $\tilde{F}_L$ and $\tilde{F}_L^A$

$$\mathcal{L}_T(i,j) := \mathcal{P}_T(j) - \mathcal{P}_T(i), \; \mathcal{L}_T^A(a,b) := \mathcal{P}_T^A(b) - \mathcal{P}_T^A(a)$$

$$\tilde{F}_L^{\mathcal{T}}(i,j) := \tilde{F}^{\mathcal{T}}(j) - \tilde{F}^{\mathcal{T}}(i), \; \tilde{F}_L^{\mathcal{A}}(a,b) := \tilde{F}^{\mathcal{A}}(b) - \tilde{F}^{\mathcal{A}}(a).$$

2.3 Applications

Our Markov model represents the user behavior on two levels, namely the concrete level of page impressions, and the abstract level of predicates. The transition probabilities on the level of page impressions may seen as *fuzzy virtual links*, while the transition probabilities on the abstract level may be seen as *fuzzy market rules* about user behavior, which abstract from single pages to classes of pages with the same attribute values. If the valuation function reflects the economic value an activity, associated with an impression on a terminal place, then the extended valuation function approximates the expected value under the condition, that a user accesses an object. Moreover, $\tilde{F}_L^{\mathcal{T}}$ tells us, whether this expected value is likely to increase, when a user follows this link, or it is likely to decrease. That creates a geometry on the web pages and on the set of attributes, which yields a new form of user feedback for the marketing department, when marketing goals are represented by attribute values and/or a valuation function. The marketing department can perform experimental redesign of its site and observe how the virtual, user-centered geometry of the site is changed, that is whether their goals are better achieved. Thus, it can be used to implement an adaptive, user-behavior-reactive design of Web-sites, which has clear advantages to a one-guess design by abstract design rules. See also [9] and [8]. The same type of experimental redesign works for Intra-net re-engineering. Further the constructed geometry may be used by Internet agents to create rules for navigation, which imitate human surfing behavior.

2.4 Modeling without Markov Processes

The above modeling is also possible without the explicit introduction of Markov processes. Whether we actually use that term or not is not important. The main idea is the valuing of content objects based on a priori values for target objects and the monitoring of user behavior. That assignment process may be better understood if one thinks about it in terms of Markov processes, and this helps when one tries to set up simulation experiments, but the wording and the adoption of the concept of a stochastic process is not essential.

2.5 n-Markov Chains and Further Discussion of Markov Models

Our design of the stochastic model implicitly relies on a probability space consisting of ordered pairs of content objects, while the natural probability space for analyzing Heap's and Zipf's Laws consists of single objects. On the other hand, if we integrate the referring context i.e, information about the last web-page accessed before the current object was accessed, into our picture, the natural model for the probability space is ordered triples, and the corresponding stochastic process is a 1-Markov process. Analogously, we could construct n-Markov models by integrating an n-step history and thus approximate real world behavior. However, then the statistical relevance of data will eventually be lost.

We have represented the outside world by a single object, but it may be worthwhile to distinguish between different domains of the environment. For example we might distinguish between search engines, a set of selected domains, the rest of the universe, and ∞. Unfortunately, not all browsers provide information, where a users comes from. Therefore, we can only extrapolate probabilities for the corresponding transitions from a subset of log-data. Further, web-sites may be modeled as graphs, where nodes correspond to pages and edges correspond to links and we can represent user-behavior with colored full graphs, including nodes which represent the environment and edges between all nodes, whereby colors represent transition probabilities. If we have enough log-data to create statistically relevant n-Markov chains, we may base our modeling on hyper-graphs with hyper-edges connecting $n + 2$ nodes.

Models related to ours are used in some spiders to calculate the relevance of a Web-page as the limit (equilibrium) distribution of a random walk on the link structure. The difference between these models and ours is that we measure user behavior, while the cited models assume equal likelihoods for all links on the same page. These likelihoods could be chosen according to normalized probabilities p^A_{uv} or hybrid abstract transition probabilities, which result from a joint classification of pages (for the source state) and links (for the target state) of a transition.

2.6 Building Innovative Retrieval Tools

The conditional probabilities arising from the stopped Markov processes may be given the interpretation of inference values, as they are known from information retrieval. (Compare [5] and [10] as well as [2] for a survey.) The analogy may be exploited in order to combine the knowledge of a retrieval system, i.e. its inference values, with the knowledge represented by log-files, i.e. the knowledge represented by stopped Markov processes, by gluing the two corresponding inference networks together. If the retrieval system is described as a four level inference network, then the result of the glueing can be described as a five level inference network: 1. target or terminal objects, 2. objects in the retrieval base, 3. attributes of objects, 4. query profiles, and 5. user need(s). That construction materializes in observations that somebody first uses a search-engine, and then tries to improve the results by surfing for a more accurate information. When the indicated inference networks are combined, the search engine may learn from the behavior of its users without explicit feedback from the user.

3 Conclusions

We have described a modeling framework which provides us with user-centered models of Web-sites, based on the information contained in extended log-files of Web-sites and on content classifications. Thereby, we combine measured user-behavior with content descriptions for Web-pages, and with value functions for Web-pages. Content descriptions may be derived from meta-data or parsed full text, while the value function reflects the value of a page view of a particular page by a customer. As a result we obtain functions which desribe the likelihood that a page will guide the customer to a particular other page and which are the links which support this guidance positevely. It is well-known that direct links to all destinations are not necessarily the optimal solution. Our modeling facilitates the quantitative analysis and the comparison of link structures of marketing sites. Equally, it can be employed for Intranet engineering. Our analysis of a large Intra-net has shown that one of the major problems there is that no information about the usage of new tools is fed into the system re-engineering cycles and into the capacity planning. Our modeling approach enables the calculation of that missing information from log-files.

References

1. R. Baeza-Yates, B. Ribeiro-Neto, Modern Information Retrieval, Addison Wesley 1999;
2. Belkin, N.J., and Croft, W.C., Information Filtering and Information Retrieval: Two sides of the Same Coin, Communications of the ACM, Vol 35, No 2, 1992;
3. J. Heaps, Information Retrieval - Computational and Theoretical Aspects, Academic Press 1978
4. C. Lueg and R. Riedl How Information Technology Could Benefit from Modern Approaches to Knowledge Management, 3rd International Conference of Practical Applications of Knowledge Management, PAKM 2000, Basel 2000
5. J. Pearl, Probabilistic Reasoning in Intelligent Systems: Networks of Plausible Inference, Morgan Kaufmann, 1998
6. R. Riedl, The Impact of Workload on Simulation Results for Distributed Transaction Processing, HPCN Europe '99, Amsterdam 1999
7. R. Riedl: Need for Trace Benchmarks, in "Performance Evaluations with Realistic Applications", edited by R. Eigenmann (SPEC), MIT-Press (to appear)
8. R. Riedl: Meaning and Relevance, to appear in Proceedings of the Fourth International Conference on Cognition Technology: Instruments of Mind, Warwick 2001
9. R. Riedl: Report-based Web-Marketing and Intranet Reengineering, to appear in Proceedings of e-business and e-work, Venice 2001
10. H. Turtle, W.B. Croft: Inference networks for document retrieval, Proc. 13th Ann. Int. ACM SIGIR Conference, Brussels 1990
11. D. Wolfram, Applying informetric characteristics of databases to IR system file design, Part I: Informetric models, Information Processing and Management, Vol. 28, No. 1, 1992.
12. G. K. Zipf Human Behavior and the Principle of Least Effort: An Introduction to Human Ecology, Addison-Wesley, Reading, MA, 1949

A Prototype for a Distributed Image Retrieval System

Odej Kao

Department of Computer Science, Technical University of Clausthal
Julius-Albert-Strasse 4, D-38678 Clausthal-Zellerfeld, Germany
okao@informatik.tu-clausthal.de

Abstract. This paper gives an overview over different techniques necessary for the realisation of distributed image retrieval systems by considering a running prototype CAIRO as an example. CAIRO uses the performance capabilities of a cluster architecture for the realisation of a high-quality, object-based image analysis and search. Integrated scheduling algorithms balance the workload over the available nodes and minimise the communication within the cluster.

1 Introduction

Image management systems – usually divided into pattern recognition systems and image databases – are major components of general multimedia databases. The importance of image databases allowing a search for a *number of images* similar to a given sample image rose enormously in recent years. One of the reasons is the spreading of digital technology and multimedia applications producing Petabytes of pictorial material per year.

The image database presented here is called CAIRO[1] (*Cluster Architecture for Image Retrieval and Organisation*). It is a distributed system for image storage and retrieval combining standard indexing and searching methods with an efficient processing on a cluster architecture. The data is distributed among several nodes and subsequently processed in parallel. In following the main components for the creation and parallel execution of queries, for the extraction, organisation, and update of features and images are discussed.

The web-based user interface provides a number of visual querying methods such as Query-by-Pictorial-Example/Sketch, Mosaic, Browsing, Selection-in-given-Standards, etc. The iconic presentation of the retrieval results allows a refinement of the initial search and further approximation of the desired image set. The user interface depicts also all available algorithms for feature extraction and comparison.

An object-relational database system stores a part of the information derived from the raw data and manages the technical and the world-oriented attributes as well as the existing algorithms for feature extraction and comparison. The information on the size and number of pixels is for example vital for the image

[1] `http://www.in.tu-clausthal.de/cairo/`.

B. Hertzberger et al. (Eds.): HPCN Europe 2001, LNCS 2110, pp. 579–582, 2001.

partitioning and for the dynamic re-distribution strategies. The raw data and the downscaled thumbnails are stored as BLOBs (*Binary Large Objects*) and used for visualising of the query results.

CAIRO offers a large set of algorithms for feature extraction and comparison. Several features such as colour moments and wavelet coefficients are extracted a priori and stored in the index structures such as VA files and VP trees. The low-level attributes can be combined and weighted in different ways resulting into advanced features, which depict the image content on a higher abstraction level. The similarity degree of a query image and the target images equals to the distance between the corresponding features, which is measured using adaptations of well-known metrics or specialised similarity functions. Acceptable system response times are achieved, because no further processing of the image raw data is necessary during the retrieval process resulting into immense reduction of computing time. The easy integration in existing database systems is a further advantage of this approach.

Extraction of simple features, however, often results in disadvantageous reduction of the image content. Important details are not sufficiently considered during the retrieval, thus an object-based, precise search is not possible. Therefore, image retrieval with dynamic feature extraction is necessary, where the user selects manually certain region of interest and uses it as a starting point. Other regions of the query image and the object background are not regarded, so that a detail search can be performed. The template matching approach gives an example for this operation: the region of interest is represented by a minimal bounding rectangle and subsequently correlated with all images in the database. Thereby, the rectangle slides according to a chosen step size over the image. The similarity with the searched region is computed for each new position for example by subtraction of the corresponding pixel colour values. Distortions caused by rotation and deviations regarding the size, colours, etc. have to be considered. A survey of image retrieval methods can be found among many others in [1].

2 Parallel Components

The dynamic feature extraction increases the computational complexity for the query processing significantly. CAIRO exploits therefore the parallelism provided by Beowulf clusters [2,3]. These architectures have an advantage that each node has an own I/O subsystem, thus the transfer effort is shared by a number of nodes. Moreover, the reasonable price per node enables the creation of large systems with the required storage and processing capacity.

Based on their functionality the nodes are subdivided into:

- Query stations host the web-based user interfaces for the access to the database and visualisation of the retrieval results.
- Master node controls the cluster, receives the query requests and broadcasts the algorithms with the search parameters to the computing nodes. Furthermore, it acts as a redundant storage server and unifies the sub results of the computing nodes in a final ranking.

– Computing nodes perform the image processing and comparisons. Each of these nodes contains a disjunctive subset of the existing images and executes all operations with the data stored on the local devices. The computed results are sent to the master node.

Figure 1 shows a schematic of the cluster architecture for the distributed image retrieval system CAIRO.

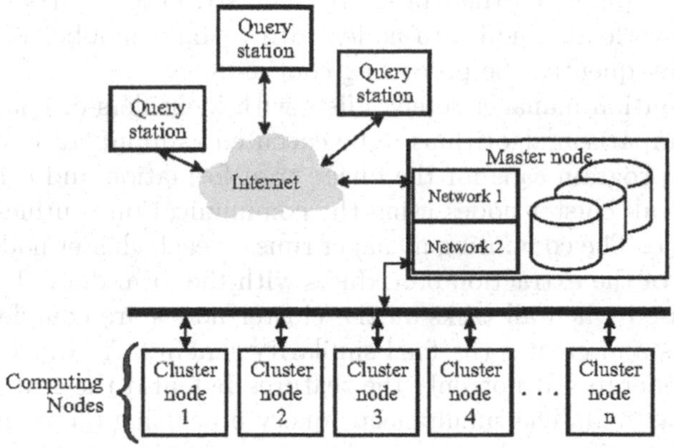

Fig. 1. Cluster Architecture for the Distributed Image Retrieval System CAIRO.

For the initial distribution of the images across the cluster a content independent partitioning is selected: the compressed size of the images stored on the local devices is nearly the same for all nodes. If a new image has to be inserted, the partition with the minimal storage size is determined and the image is sent to the corresponding node. This 'equal-size' distribution leads to similar processing times of all computing nodes and minimises the intra-node communication. The retrieval operations are independent from one another, thus a large performance gain by simultaneous execution can be obtained.

Components called transaction, distribution, computation, and result manager are necessary for the parallel execution of the retrieval operations. They are based on the well-known parallel libraries PVM and MPI [4,5]. The functionality of the transaction manager encompasses the analysis of the transformations to be executed and the determination of operation rank order. Opposed to a conventional database management system, the data is usually only read, so that no access conflicts need to be resolved.

The order of operations should be set in a way that the time for the processing and the presenting of the system response is minimised, and all suitable images have been considered. The most interesting case is given by an evaluation of the a priori extracted features in the first phase. Thereby, a list of potential hits is constructed and forwarded to the distribution manager. This combination limits the set of images to be considered during the time-intensive, dynamic retrieval and achieves the fastest system response time. On the other hand the even image distribution is distorted resulting into varying processing times of the nodes. In this case a dynamic re-distribution of images across the cluster for balancing the workload is necessary. It can be proven, that this data placement problem is NP complete. Therefore, a heuristic strategy called LTF (*Largest Task First*) is implemented [6]. The images on the local devices are sorted in descending order according to their size, thus large images with long compute times are analysed first and are mainly processed on the local node. As soon as some of the nodes have completed their tasks, the re-distribution starts by transferring images from overloaded nodes to nodes with a small number of images to be processed. Subsequently, the processing continues.

The distribution manager receives lists with identifiers of images and of the extraction/comparison algorithms to be executed as input. It analyses the lists, generates the program calls for the image transformation and comparison, and sends these to all cluster nodes using the communication routines of the active virtual machine. The computing manager runs on each cluster node and controls the execution of the extraction procedures with the local data. The result manager is initialised when all tasks on the cluster nodes are completed. It unifies the sub results and creates the final similarity ranking. A large communication overhead is generated if not only the features but also the raw data needs to be compared as well. The simultaneous query processing on all available nodes and the minimisation of the intra-node communication enable a nearly linear speedup, which is confirmed by exhaustive performance measurements.

3 Conclusions

Many well-suited algorithms for image retrieval and detail search are not often applied, as they require immense computational and memory resources. The integration of parallel methods in the organisation and retrieval of images enables the use of such alternative approaches in existing applications. This paper demonstrates this idea by considering an efficient distributed image retrieval system called CAIRO as an example. Thereby, basic image processing, scheduling and data placement methods as well as the cluster architecture are depicted.

References

1. J.P. Eakins, M. Graham. Content-based image retrieval. Technical Report JTAP-039, University of Northumbria at Newcastle, October 1999.
2. G. F. Pfister. *In Search of Clusters*. Prentice Hall, 2. edition, 1998.
3. D.F. Savarese, T. Sterling. Beowulf. *High Performance Cluster Computing - Architectures and Systems*, pp 625–645, Prentice Hall, 1999.
4. Parallel Virtual Mashine: `http://www.epm.ornl.gov/pvm/`.
5. Massage Passing Interface: `http://www.mpi-forum.org/`.
6. O. Kao, G. Steinert, F. Drews. Scheduling Aspects for Image Retrieval in Cluster-Based Image Databases. *Proceedings of the IEEE/ACM Symposium on Cluster Computing (CCGrid 2001)*, to be published.

Data-Object Oriented Design for Distributed Shared Memory[1]

G. Manis

University of Ioannina, Department of Computer Science
P.O. Box 1186, GR-45110 Ioannina, Greece
manis@cslab.ece.ntua.gr

Abstract. This paper describes a design alternative for organizing shared (data) objects in a distributed memory system. Shared objects are organized based on concepts derived from the fields of object-oriented programming and applied to the philosophy of a distributed shared memory system. Inheritance is implemented through an object hierarchy in which each object inherits information from the ancestor objects, located higher in the hierarchy tree. All other object oriented programming concepts are preserved or adapted to the philosophy of the proposed system. To the best of our knowledge, no similar DSM systems have been proposed by now.

1 Introduction

Research in the field of distributed shared memory (DSM) has been very active in the last decade. Many different DSM implementations are available today, while there is always a chapter that is dedicated to DSM in every textbook relative to distributed operating systems. Existing DSM systems can be categorized as object based and page based. In object based systems the shared units are objects (e.g. Orca [1]), while in page based ones the shared unit is a page in memory [2]. The model proposed in this paper belongs in the first category but it is very different from any other object based DSM system; it is the only one that uses inheritance as a means of sharing information. These ideas derive from the field of object oriented design and are applied to the data- objects of an object oriented system. Hence, we will characterize this design as a "data-object oriented" design.

Object oriented design is described by its use of objects, classes and inheritance. Inheritance is the key in object oriented design; the systems that do not support inheritance are characterized as "object based" rather than "object oriented." One question is if it is possible to implement inheritance without classes. The answer is located in the definition of an object hierarchy [3]: an object may inherit information from another object, the latter forwarding the invocations it cannot serve, to the former. This procedure stops when an object is finally able to respond to the initial request, or if there is no other parent object in the hierarchy tree. In latter case, the

[1] This work has been partially funded by the Greek Secretariat of Research and Technology

B. Hertzberger et al. (Eds.) : HPCN Europe 2001, LNCS 2110, pp. 583-586, 2001.

invocation fails. In the rest of this short presentation, the basic concepts of the system will be outlined, and some implementation issues will be discussed.

2 Basic Concepts

In many systems (e.g. Orca [1]) objects represent shared data structures that can be accessed by any process participating in the DSM system. In Chorus [4] objects are passive entities, while in Emerald [5] objects can be passive or active. An Amoeba [6] object is something more general: it can be data, process or another operating system resource. All the above are different conceptions of the idea "object."

The object is a data type that can be accessed only by well-defined operations. In this system, objects are data structures that are accessed strictly through a set of DSM operations (read, write, migrate etc.). Objects in this system are not instantiated by classes. They follow the concepts of "delegation" [3]. In this philosophy, inheritance is expressed through a hierarchy of objects, as presented in the introduction.

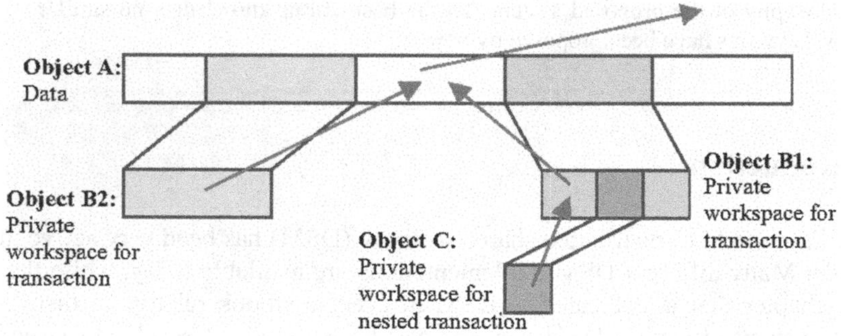

Fig. 1. A transaction system. Object C (nested transaction) inherits information from object B1 (the main transaction). Objects B1 and B2 inherit data from object A (Data). A request to a private workspace will be served using the data located in this workspace if this is possible, or it will be forwarded to the parent objects. In this example the user "sees" the whole workspace, without worrying about other transactions.

An example is presented in figure 1. In this example the data are stored in object A. Data can be modified only through transaction processes. Suppose there are two concurrent transactions acting on the same data. Each transaction creates a private workspace in which all writes are performed until the transaction commits or aborts. Object B1 and B2 inherit information from object A. Objects B1 and B2 reply to requests related to data located in the private space, and forwards the rest of the requests to the real workspace. Each transaction "sees" its virtual workspace, without affecting or being affected by other concurrent transactions.

The whole idea is simple. Next, we will briefly discuss how this design keeps up with the main concepts of object-oriented programming: abstraction, encapsulation, polymorphism and inheritance.

According to the concept of abstraction, the object appears for the rest of the world as a simple and coherent entity. The fact that the information is decomposed and located in different places does not affect the user.

In the case of encapsulation, things are more complicated. Shared data are accessed through a well-defined interface: the available DSM operations. This is compatible with the concept of encapsulation, especially because this is done through messages: for each DSM operation a message is created and sent to the proper object. However, the lack of user-defined methods and the restrain of access to DSM operations only, narrows the ability of the programmer to hide the implementation details. Since we limit the design to data only, the problem is concentrated on the fact that the programmer cannot change the name of the shared variables or the fields of the shared structures.

Three of the main concepts of polymorphism are data overriding, method overriding and method overloading. What is interesting in this system is what happens with overriding of data, rather than method overriding or overloading. In the example of figure 1, private workspaces B1 and B2 overrides data stored in object A while data in object C overrides some of the data in object B2.

The role of inheritance and how it is supported by this system have been discussed above. Objects inherit information from other objects located higher in the hierarchy tree. Information on the parents is located inside each object and not in a class, since there are no classes in this model. This characteristic allows the building of different hierarchy trees with the same objects, according to each specific application. In addition, it is possible to change the hierarchy tree, even during run-time.

3 Implementation Issues

The main operations of a typical DSM system are the creation, access, location, migration of shared data, as well as creation and access for read replicas. In the proposed system there is one more fundamental operation: inheritance. Let's have a look at some implementation issues related to these operations.

A read or write request that cannot be performed by the invoked object is sent to the parent object. This process continues until one object is able to reply or there are no other objects higher in the hierarchy tree, in which case the invocation fails.

Many DSM systems support read replicas. Read replicas are read-only copies of the original objects, and are located at the processors from which invocations are performed frequently. When local copies are kept in many processors, the overall system performance is usually improved. In this system, read requests are addressed to local copies, if a local copy for the object exists in the processor, while all write requests are addressed directly to the original object. The original object keeps track of all read-only copies of the object. Upon receiving a write request, the object updates itself and brings the read-only copies up to date. Updating the original first object and then the replicas allows a high degree of memory consistency (sequential consistency).

Access to all objects (original or read replicas) is possible through a location server. All objects are registered in this server, which provides the necessary information for their access. Requests for the location of an object should include the name of the object and whether the object is to be accessed by a read or a write operation. When a write operation is performed, the location server returns the location of the original object. If the request is related to a read operation, the location server returns the address of the local copy of the object, if a local object exists in the specific processor, otherwise, the location server returns the location of the original object.

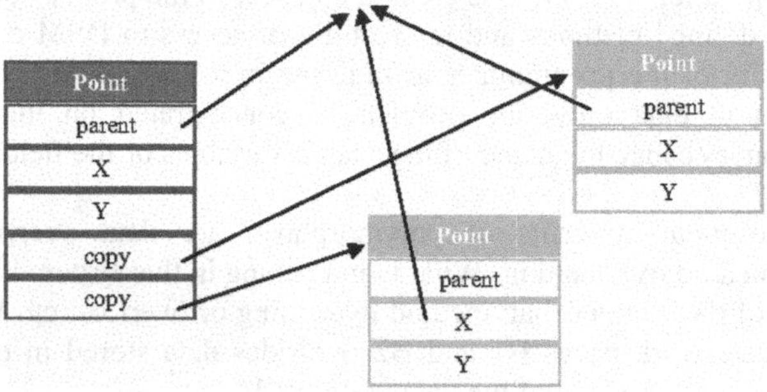

Fig. 2. The structure of an object. The left object is the original. The last two fields of the original object are pointers to the read-only replicas. The rest of the fields are the same. All three has the same parent.

The location server also performs the main operations for object migration. Objects are located by their names. When an object migrates, it copies itself to the new location and informs the location server. The location server re-directs all read and write requests to the new location. The location server addresses from now on, all new read and write requests to the new location of the object. The structure of an object is depicted in fig. 2

References

1. H. E. Bal, M. F. Kaashoek, A. S. Tanenbaum, Experience wit Distributed Programming in Orca," *Proc. IEEE Int. Conf. On Computer Languages,* pp. 79-89, 1990
2. K. Li, P. Hudak, Memory Coherence in Shared Virtual Memory Systems," *ACM Trans. On Computer Systems*, vol 7, no 4, pp. 321-359, 1989
3. H. Lieberman, Using Prototypical Objects to Implement Shared Behavior in Object Oriented Systems," *Proc. Conf. Object Oriented Programming Stsems, Languages and Applications; reprinted in Signal Notices, 21(11)*, pp 214-223, 1986
4. M. Rozier, U. Abrossimov, F. Armand, I. Boule, M. Gien, M. Guillemont, F. Herrmann, C. Kaiser, S Langlois, P. Leonard, W. Neuhauser, Overview of the Chorus Distributed Operating System," *Tech. Report CS/TR-90-25-1*, Chorus Systemes
5. E. Juil, H. Levy, N. Hutchinson, A. Black, Fine-Grained Mobility in the Emerald System," *ACM Trans. on Computer Systems,* vol 6, no 1, pp. 109-133, 1988
6. A. S. Tanenbaum, R. van Renesse, H. van Staveren, G. Sharp, S. Mullender, J. Jansen, G. van Rossum, Experiences with the Amoeba Distributed Operating System," *Communications ACM,* vol 33, no 12, pp. 46-63, 1990

Using Virtual User Account System for Managing Users Account in Polish National Cluster

M.Kupczyk, M.Lawenda, N.Meyer, and P.Wolniewicz

Poznań Supercomputing and Networking Center (PSNC)
ul. Noskowskiego 10, PL-61-704 Poznań, Poland
{miron,lawenda,meyer,pawelw}@man.poznan.pl

Abstract. In this paper we present the configuration of the Polish national cluster using the Virtual Users Account System. In order to give users a possibility to access computer resources we configured a set of LSF queues that sent jobs to other sites. To avoid the problem with managing users account on remote supercomputers we are using special middleware that allows to run and manage users jobs without the necessity of creating account for all users.

1 Introduction

Most often the computing power given to the users in one computing centre is not enough for them. Therefore it is inevitable to provide the users with a possibility to take advantage of the resources belonging to all national supercomputer centres at the same time. Nowadays, owing to the optical technologies, it is possible, because the time of access to remote machines is comparable with the time of accessing the local machines. Based on the Polish Optical Network [1] we developed a national HPC cluster.

The simplest solution to create the cluster is the usage of a job processing system that manages jobs. It is a very good solution for local site conditions: for systems installed in one place, where there is a constant or slow change in the number of users and computing systems. While connecting distributed, geographically distant systems, belonging to different institutions, we were faced with the problem of managing user accounts of the whole nation-wide structure. The problems are due to policies of user management, which are different in each centre and due to the problems with maintaining user account coherency on all machines. We are using the Virtual Users Account System [3], developed in Poznan Supercomputing and Networking Center, that simplifies user accounts management in the distributed environment.

2 Polish National Cluster

In Poland there are over a dozen supercomputers located in several supercomputing centres. Most of them are SGI systems (Challenge, Power Challenge, Origin2000, Onyx2, Origin3000) but there are also IBM's, Cray's, HP's. All Polish supercomputers are used mainly to perform chemical, physical, mathematical and

B. Hertzberger et al. (Eds.) : HPCN Europe 2001, LNCS 2110, pp. 587-590, 2001.

engineering computations. The most popular applications are Gaussian98, Gamess, MSI, Abaqus, Fidap, Matlab, Maple, Amber. A large part of CPU time is also used by users` own developed applications.
Most Polish supercomputing centres agreed to form the Polish national supercomputer cluster. It includes Gdańsk, Poznań, Wrocław, Kraków and Łódź. Additionally, there are some more Polish centres, which are going to join the national cluster as clients and to have the possibility to access computing resources from the other centres. All these centres are connected by the ATM network with a throughput equal to 34 or 155 Mb/s. There is a 10 Mb/s channel dedicated to the multicluster computation. The current stage of this cluster construction is built on LSF queuing system (Load Sharing Facility).

3 Virtual User Account System

The Virtual User Account System (VUS) is a system allowing a load balance of machines installed in different computing sites without overhead related to creating and maintaining additional user accounts. Instead of using real users' account, a set of generic account is used. This system can co-operate with any queuing system. In connection with queuing systems, it allows a better usage of computing resources by sending jobs to currently less utilised machine.

The Virtual Users Account System has the following benefits in comparison with a typical queuing system:

1. **Simplified user accounts administration.** The user does not have to have accounts on remote systems to use them. The pool of virtual users' account is used.
2. **Co-operation with other queuing systems.** System can be configured in such a way that the jobs can be sent also to other queuing systems.
3. **Full accounting information.** Information regarding used CPU time is collected after completing the task and sent to the virtual users' server. Thus, for every user, full information is stored about the used CPU on separated systems and globally.
4. **Automatic file transfer.** There is no need to share files via NFS on all systems connected into a cluster. Files are transferred automatically.
5. **Easy authorisation.** There is no need to define an authorised set of users to a queue in each centre. Access to the queue can be granted by adding a user to the queue access list in the central server.

4 Cluster Implementation

The first phase of building the Polish National Cluster has been finished. In the second phase we plan to move toward grid environment with more grid middleware and grid applications. Currently the cluster consists of SGI systems connected by LSF software. LSF queuing system has been used to integrate some of the computational resources available in POL-34 network. Its main goal is to manage the distribution of system resources, paying special attention to the processor time, operating memory, disk space, etc. Because of using the Virtual Users Account System, users can send jobs to other supercomputer centres without having account on the remote machines.

Table 4.1 lists the supercomputers included in the cluster and the list of applications running on each of them.

Table 4.1. Supercomputing Centres Involved in the National Cluster Configuration.

Centre	Cluster Name	Hardware Platform	WWW address	Application Type
Gdańsk	gdansk	SGI Power Challenge XL, 8CPU, 2GB Origin 2000, 24 CPU, 16 GB RAM	www.task.gda.pl	users own, Matlab
Łódź	lodz	SGI Power Challenge L, 6 CPU, 256 MB RAM	ck-sg.p.lodz.pl	users own, Gamess, MSI
Poznań	poznan	SGI Power Challenge XL, 12 CPU r8k, 1GB RAM Origin3200c, 32 CPU r12k, 8 GB RAM Cray SV1, 8 CPU, 16 GB RAM	www.man.poznan.pl	users own, Gaussian, Gamess, Abaqus, MSI
Wrocław	wroclaw	Onyx, 4 CPU r10k, 1280MB Origin 2000, 32 CPU r10k, 8 GB RAM	info.wcss.wroc.pl	Gaussian GAMESS, MSI, abaqus, users own
Kraków	krakow	Origin 2000, 128 CPU r10k, 16 GB RAM	www.cyf-kr.edu.pl	Gaussian, GAMESS, MSI, users own

The queues configuration was deliberately designed to let the users take advantage of the interactive applications, especially in the graphical environment. Apart from many queues dedicated to the tasks defined by the users, separate queues for the third party applications were created, e.g. for Gaussian98 or MATLAB. The same solution is planned for MSI and ABAQUS as well. Our cluster is heterogeneous and this causes problems with binaries compatibility. Application queues require only submitting data files so it is not important on which architecture jobs run. But when users submit jobs to general queues their have to supply binaries, which are transferred to destination system. Users can define an architecture or operating system as resource requirements. By default the jobs start on systems with the same architecture as the submitting host.

Application queues allow running only specific applications. It is realised by means of so called job starters. LSF allows defining a program, which will be run before the user script is processed. For example, as a job starter for the Gaussian queue, g98 is defined. Users cannot submit jobs to a queuing system directly. It is realised by a set of scripts but users still can specify parameters and queuing system limits.

Not all supercomputers in the Polish centres have installed LSF. But our system allows sending jobs to the systems with other queuing systems. We configured queues that send jobs to NQE queuing system on Cray in PSNC. LSF itself has modules responsible for communication with NQE, but it does not allow transferring files. We also tested our system in communication between LSF and LoadLeveler installed on IBM SP2 and our tests were successful. But because all Polish SP2 are a little bit outdated, we did not decide to include them into the cluster until their upgrade.

Because a job can be run on a remote machine, it is important to transfer all necessary files from the local machine to the remote machine and get the results back. Virtual users' account system allows defining files necessary for the application. For example, for Gaussian98 system have to transfer .inp file to the remote machine and get back .log and .chk file. Files are transferred automatically if during the submission of a job the user specified the type of application. For users' own application the user is responsible for specifying which files should be transferred.

Communication between machines from different centres is conducted on a public network, where data is not safe. Users do not quite like to send their jobs and result unencrypted, therefore a system that transmits the data remotely must encrypt all data. We are using SSL for all communications. SSL certificates are also used to ensure that no one wants to intrude upon the cluster.

In every centre one Virtual User System daemon should be running, and it would be responsible for gathering VUS information from machines and storing it in the database. VUS daemon is also responsible for resolving the real names of users, when job monitoring application is invoked. Users can monitor a state of their jobs, regardless of the fact that the jobs are running on a remote machine and on the generic account. VUS daemon can be duplicated for reliability reasons. Currently we are using Oracle database for storing user mappings, rights and accounting, but because not all the computing centres have it, we plan to use simpler free database.

5 Conclusions

In this paper we outlined the current configuration of the Polish National Cluster that consists of several supercomputers from different centres connected by a fast network. We are using our Virtual Users Account System that saves our administrators much work connected with the users account management. Because of that system we can dynamically change the configuration of the cluster, add new machines, change destination of the queues transparently for users. The users did not have to care about applying for an account on all Polish supercomputers. They just run their jobs and receive data and the system takes care of transmitting and running the job. And now jobs from the overloaded sites are sent to currently less utilised sites that, when improved, mean the utilisation of all supercomputers. Although users can now run a job on different systems we still have full account information.

References

[1] J. Rychlewski, J. Węglarz, S. Starzak, M. Stroiński, *PIONIER – Polish Optical Internet,* conference material ISTHMUS 2000, Poznan, 2000

[2] W. Dymaczewski, N. Meyer, M. Stroiński, P. Wolniewicz, *Virtual User Account System for distributed batch processing*, HPC 99, Amsterdam, April 1999

Performance Evaluation of XTP and TCP Transport Protocols for Reliable Multicast Communications

M.A.R. Dantas and G. Jardini

Department of Computer Science, University of Brasilia
Brasilia 70919-970, Brazil
mardantas@computer.org

Abstract. This article presents a performance evaluation of XTP (Xpress Transport Protocol) and TCP (Transmission Control Protocol) transport protocols for reliable multicast communications. Considering high performance distributed parallel environments, such as workstation clusters, multicast is an interesting function to be observed. On this basis, our research work presents empirical results of reliable multicast communications, which indicates an enhanced performance of the XTP lightweight transport protocol for high performance computing in comparison to the TCP protocol.
Keywords: XTP, TCP, Transport Protocols, Reliable Multicast, Cluster Computing.

1 Introduction

Cluster environments represent interesting hardware platforms to meet the growing computational demand of many organizations. Several software packages have been designed to improve the utilization of cluster environments, usually focusing either on throughput or parallelism of tasks. The main target of the packages in the first category (i.e. throughput-oriented) is to improve the distribution of multiple, independent tasks. In contrast, parallel software environments are designed to enhance the execution of concurrent tasks, achieving reasonable parallel speedup. In practice, this means that all tasks should complete at the same time for optimal performance. Examples of packages developed to enhance the throughput of tasks are Condor [1] and LSF [2]. In contrast, MPI [3] and ThreadMarks [4] are examples of message-passing and distributed shared memory packages intended to improve the parallelism of the tasks.

Although throughput-oriented and parallel software packages represent an enhancement for application programmers on cluster environments, the transport protocol usually employed, *Transmission Control Protocol* (TCP) or it User Datagram Protocol (UDP), do not provide an efficient collective communications among processors and waste a large amount of bandwidth of the network.

The paper is organised as follows. In section 2 parallel cluster environment considered in the article is presented, and we state some characteristics of the

B. Hertzberger et al. (Eds.): HPCN Europe 2001, LNCS 2110, pp. 591–594, 2001.

XTP protocol. Our experimental results are shown in section 3. Finally, in section 4 we present our conclusions.

2 Cluster Environment and XTP Protocol

The configuration considered in the article is represented by a cluster of eight IA 32 processors, 350 MHz, memory with 32 MB and interconnect by a 100 Mbps switch. Only one machine has all peripheral items (monitor, keyboard and mouse) because it works as a monitor of the cluster (i.e. a machine from where we install all software packages and some management actions are taken). Tle Linux operating system and XTPSandia 4.0 were our basic software environment.

The cluster was designed to execute only local parallel applications and it was not connected to main campus network. All experiments were executed in a quite workload environment (i.e. the cluster was dedicated to our experiments whithout any other user or application).

The XTP protocol is a new transport protocol approach which brings new features (when compared to conventional protocols such as TCP and UDP) to be employed in high performance computing environments. Several functions of the protocol were designed to provide more flexibility to the execution of distributed applications, examples are [5,6,7] :

- *multicast and management of multicast group*;
- *support to priority*;
- *transmission control rate*;
- *selective retransmission*;

In our experiments have used the XTPSandia [5].The XTPSandia is an XTP implementation which employs concepts of object oriented approach. The package uses a meta transport library called *MTL*[5]. This library is a collection of classes written in C++ from where it is possible to obtain the transport protocol implementation. The idea behind the XTPSandia is a *daemon* executing into the user space. Any application should requires to the *daemon*, which is responsible in turn to serve the application. It is also interesting to mention that the XTP-Sandia has a feature that allows an ordinary user to customize the protocol to a large number of different operating systems such as Linux, Solaris and AIX.

3 Experimental Results

We have used on our experiments two application programs : *xfile* and *ftp*. The *xfile* application, supported by the XTP protocol, provides services similar to TCP when considering a file transfer application. In the same fashion as the TCP, the *xfile* application allows a reliable exchange of data. On the other hand, because *ftp* does not have a native *multicast*, every time when a file transfer is required to *n processors* it is necessary *n connections* to each processor (in our experiments we represent this fact as *multicast 1:n*). Therefore, in our examples we are presenting the elapsed-time and throughput necessary to the transmission of a file to each workstation. In other words, we decided not to use an average measurement.

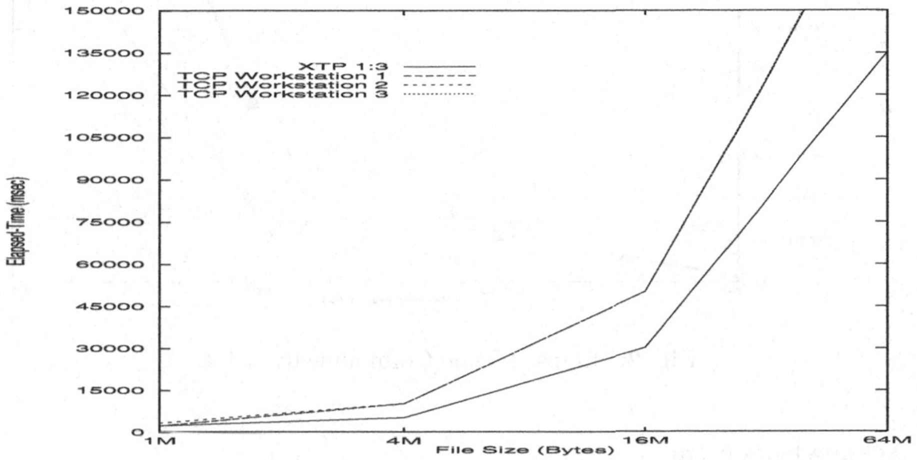

Fig. 1. Elapsed-Time Communication 1:3.

Our experiments presented in figures 1 and 2 show the improvement of the XTP environment. The main explanation for the enhanced performance of the XTP can be understand by how the multicast works. In the *ftp* environment each connection is competing for the use of the network, thus increasing the number of processors to delivery a file less performance is expected. On the other hand, the XTP native multicast function does not require n connections to n processors.

4 Conclusion

In this paper we have presented some experimental results from a lightweight transport protocol, called *XTP*. The existing support for multicast in old fashion transport protocol, such as TCP and UDP, is a bottleneck which should be tackle, especially when considering cluster of loosely-coupled machines as a parallel environment.

The *XTPSandia* implementation was used to compare the use of this lightweight transport protocol to some existing services of the TCP and UDP protocols. The results from our experiments indicate an efficient performance of the XTPSandia transport protocol. Important features have also been verified during our experiments such as (1) the facility to install the package at user space without requiring special privileges to use the operating system kernel, (2) the multicast communication, (3) scalability of the multicast transmissions, (4) good performance when errors were added to the transmission and finally (5) the flexibility to customize metrics of the package.

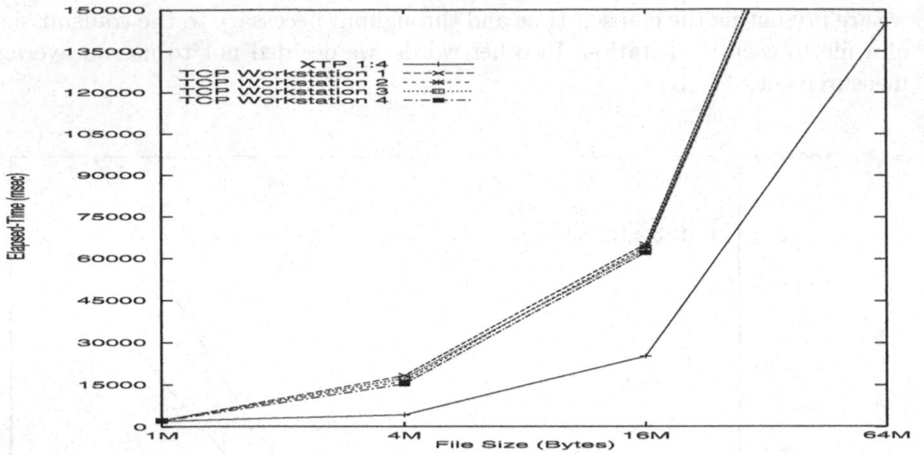

Fig. 2. Elapsed-Time Communication 1:4.

Acknowledgement

The first author's work was partially funded by the Brazilian National Research Council- CNPq (300874/00-6) and Finatec(Fundacao de Empreendimentos Cientificos e Tecnologicos). The Computer environment from the LAICO laboratory was supported by the PROJECT INEG/CNPq-UnB.

References

1. M. L. Michael J. Litzkow and M. W. Mutka, "Condor - A Hunter of Idle Workstations," *Proceedings of IEEE 8th International Conference on Distributed Computing Systems*, pp. 104–111, 1988.
2. J. Suplick, "An Analysis of Load Balancing Technology," January 1994.
3. MPI-Forum, "MPI: A Message-Passing Interface Standard," *International Journal of Supercomputer Application*, vol. 8, no. 3-4, 1994.
4. J. R. C. et al, "Project Zeus," *IEEE Network*, pp. 20–30, 1993.
5. Sandia, "Xpress transport protocol 4.0 specification," `http://www.ca.sandia.gov/xtp`, 1995
6. A. Weaver, "The Xpress Transport Protocol Cluster Computing," High-Performance Cluster Computing, pp. 307–353, 1999.
7. M.L.M.A.R Dantas and M. Rodrigues, "Analysis of a Leightweight Transport Protocol for High-Performance Computing," *Proceedings of the 14th Canadian International Symposium in HPCS, Victoria*, June 2000.

Parallelization of the Continuous Global Optimization Problem with Inequality Constraints by Using Interval Arithmetic*

Abdeljalil Benyoub and El Mostafa Daoudi

Research Laboratory in Computer Science
Department of Mathematics and Computer Science
Faculty of Sciences, University of Mohammed First
Oujda, Morocco
{benyoub,mdaoudi}@sciences.univ-oujda.ac.ma

Abstract. In this work, we study on distributed memory architecture, the parallelization of the continuous global optimization problem, based on interval arithmetic, with inequality constraints. Since this algorithm is dynamic and irregular, we propose, in particular, some techniques taking into account the load balancing problem.
Keywords: Continuous global optimization problem, interval arithmetic, Branch-And-Bound, Hansen's algorithm, parallelization, load balancing.

1 Introduction

The continuous global optimization problem with inequality constraints is well known and can be formulated as follows:

$$\begin{cases} \text{Minimize } f(x), \quad x \in S \subset \mathbb{R}^n. \\ \textit{Subject to constraints}: \\ c_i(x) \leq 0, \quad i = 1, \cdots, m, \quad m \in \mathbb{N}. \end{cases}$$

Where the objective function f and the constraints c_i, $i = 1 \cdots m$, are continuously differentiable real functions defined over the domain $S = [a_1, b_1] \times \cdots \times [a_n, b_n]$, où $(a_i, b_i)_{1 \leq i \leq n} \in \mathbb{R}^2$. Let $f^* = \min_{x \in S} f(x)$ be the minimum of f on S and $P^* = \{x^* \mid f(x^*) = f^*\}$ the set of feasible global minimizers x^* of f on S. Solving the same problem by using interval arithmetic consists in finding subboxes X^* which bound the feasible point x^* and a real interval $[\underline{f}, \overline{f}]$ which bounds the global minimum f^*.

The main idea behind interval Branch-And-Bound methods consists in the definition of [3]: (i) an inclusion function F to compute bounds for f, (ii) an inclusion functions $(C_i)_{1 \leq i \leq n}$ for the inequalities constraints $(c_i)_{1 \leq i \leq n}$, (iii) an interval vector (box) covering the domain S and (iv) some criteria which permit to reject boxes where the global minimum of f can be guaranteed not to lie.

* This work is supported by the European program INCO-DC, "DAPPI" Project.

Several works in the literature, based on the interval Branch-And-Bound methods are developed for the Global Optimization Problem and its parallelization [2,4,3,6,5,7,9,10]. During the parallel implementation some problems occur concerning the load balancing and the memory space. In several works [2,3,5,7], two basic approaches were proposed (centralized and decentralized approaches) for the parallelization of this problem.

This work deals with the parallelization of the Hansen's algorithm [6], in particular we propose a technique in order to improve the load balancing. In Section 2, the principle of Hansen's sequential algorithm is presented. Section 3 is devoted to present a first parallel algorithm which does not take into account the load balancing. In Section 4, we propose a second parallel version with a dynamic load balancing.

2 Hansen's Sequential Algorithm

This algorithm is based on the interval arithmetic and consists in making an exhaustive search in a box. If this box is not rejected, during the different elimination phases, then it will be reduced by applying the reduction step, during which several sub-boxes, of lower sizes, will be generated. The sub-boxes of too small sizes (lower than a given tolerance) will be stored in a final list. On the other hand the boxes with large sizes (upper than the given tolerance) will be subdivided into several new sub-boxes with lower sizes and will be stored in a working list in order to be treated later. After that, the bounds over the global minimum are then computed. At the end of the execution (the working list becomes empty), the global minimum is localized inside a real interval with small size and the final list is composed of certainly feasible or indeterminate small sub-boxes containing the global minimizers.

3 First Parallel Version

It is based on the decentralized approach where the working list WL is distributed among the $(p-1)$ processors (slave processors). But, on the processor P_0 (master processor), we only store the lower and the upper bound values $\underline{f_k}$ and $\overline{f_k}$ computed locally in each processor P_k, $1 \leq k \leq p-1$.

Presentation of the Parallel Algorithm: First, the master processor begins by partitioning and distributing the initial box among the slave processors P_k, $k = 1, \cdots, p-1$. Then, each processor P_k computes a first local upper bound $\overline{f_k}$ of the global minimum and sends it to the master processor, this one computes the best global first upper bound $\overline{f}$ for the global minimum and broadcasts it to all processors. After that, each processor P_k applies the Hansen's sequential algorithm on its local working list. During the parallel execution, the value of $\overline{f}$ can be computed according to two different strategies. We present in the following the steps of the parallel algorithm.

Step 1: P_0 partitions and distributes the initial box $X^{(0)}$ among the $(p-1)$ processors.

Step 2: Each slave processor P_k:

- Computes a first local upper bound $\overline{f_k}$ by using one of the methods described in [1,6,10].
- Sends $\overline{f_k}$ to the processor P_0.

Step 3 : P_0 computes $\overline{f} = \min_{1 \leq k \leq p-1} (\overline{f_k})$ and then broadcasts $\overline{f}$ to processors P_k.

Step 4: Each processor P_k, $1 \leq k \leq p-1$, applies the sequential algorithm for the optimization on its local working list. For updating the global bounds ($\overline{f}$ and $\underline{f}$), we distinguish two strategies:

First strategy: each slave processor P_k computes its local final upper and lower bounds $\underline{f_k}$, $\overline{f_k}$ and its local final list Fin_List_k, then it sends them to the master. When the master processor receives all data from all processors P_k, it computes $\overline{f} = \min_k(\overline{f}_k)$ and $\underline{f} = \min_k(\underline{f}_k)$, then eliminates from each list Fin_List_k, all sub-boxes X such that $F^L(X) > \overline{f}$, the remaining sub-boxes contain the global minimizers with $f^* \in [\underline{f}, \overline{f}]$. One advantage of this strategy is that a few communications between the slave processors and the master are needed for computing the first global upper bound and the final results.

Second Strategy: consists in updating dynamically $\overline{f}$. When each slave processor P_k computes a new local value of $\overline{f_k}$, it sends it to the master processor and becomes idle waiting a new global value of $\overline{f}$ from the master. When the master receives $\overline{f_k}$ from all slave processors P_k, it computes the new global value of the upper bound $\overline{f} = \min_k(\overline{f}_k)$ and then broadcasts it to all slave processors. After receiving the value of $\overline{f}$, each processor P_k eliminates from its local working list each sub-box X that verifies $F^L(X) > \overline{f}$, then it continues its work with the new local upper bound $\overline{f_k} = \overline{f}$. If a slave processor becomes idle i.e. its working list is empty, then it must continue updating its final list by using the value of $\overline{f}$ sent by the master processor. By doing this updating, global minimizers will be assuredly located inside slave processors. This strategy needs more communications and synchronizations, but on the other hand, it reduces the computation time, since the elimination step is performed with the global upper bound $\overline{f} = \min_k(\overline{f}_k)$ and not with the local upper bound. Also, since the value of $\overline{f}$ is known by all processors, the updating of the final lists can be done locally in each slave processors.

Remark: The two strategies remain inefficient since they do not take into account the load balancing strategy. When a local working list becomes empty then the corresponding processor becomes idle.

4 Parallel Version with the Load Balancing Strategy

The idea consists in making a dynamic and uniform redistribution of the boxes among all processors after an estimation of the number of the remaining boxes in the local working lists. The first strategy is to balance the load (estimation of the number of boxes and the uniform redistribution) when a new box, from a working list, is selected in order to be treated. But, with a simple analysis, we deduce that the number of synchronizations needed to balance the load becomes very large, therefore this can increase the global execution time. The second strategy is to balance the load when a working list becomes empty. The redistribution, of the boxes among all processors, avoids that the corresponding processor becomes idle. This strategy needs less synchronizations than the first strategy.**The Algorithm :**

Step 1: P_0 partitions and distributes the initial box $X^{(0)}$ among the $(p-1)$ processors.

Step 2: Each P_k applies the sequential Hansen's algorithm on its local working list WL_k while this one is not empty. If WL_k becomes empty ($WL_k = \emptyset$) then:

• a synchronization is established, in order to estimate the total number of the remaining boxes among the slave processors.

• a uniform redistribution of the remaining boxes among all slave processors is performed.

Step 3: the program ends when all local lists become empty.

Updating the Global Upper Bound: For updating the global upper bound $\overline{f}$, we distinguish two possibilities:

(a) The value of $\overline{f}_k$ is dynamically communicated while the local list WL_k is still not empty. This possibility needs more communications, but on the other hand, it requires less computations since the dynamic exchange of the values of $\overline{f_k}$ presents the advantage to update dynamically the global upper bound $\overline{f}$ which is used to eliminate the sub-boxes verifying $F^L(X) > \overline{f}$.

(b) The value of $\overline{f}_k$ is communicated during the synchronization phase (step 2). This possibility needs less communications but it needs more computations since the update of the global upper bound $\overline{f}$ is delayed until a synchronization is established (a local list becomes empty).

5 Conclusions

In this work, we have presented and discussed several approaches for the parallelization of Hansen's algorithm for the continuous global optimization with inequality constraints by using interval arithmetic. The parallelization of this kind of algorithms needs an efficient strategy for load balancing to be integrated. Since the computation time for treating a box is not known a-priori, we have adopted a strategy for the load balancing based on a dynamic and uniform redistribution. The real implementations of our approaches are in progress. We think that we will obtain a good results since our algorithm takes into account:
- the reduction of the number of synchronizations (a synchronization is only established when a local working list becomes empty).
- the improvement of the load balance: as soon as a processor becomes idle, a uniform redistribution a the remaining boxes is performed.

Acknowledgments

We would like to thank Pr. Nathalie Revol for here valuable helpful.

References

1. Benyoub A., Daoudi E.M.: An Algorithm for Finding an Approximate Feasible Point by Using Interval Arithmetic, LaRI Laboratory, Dept. Maths&CS. Faculty for Science, Univ. Mohammed First, Oujda, Morocco, 2000.
2. Berner S.: A Parallel Method for Verified Global Optimization, Scientific Computing and Validated Numerics: Proc. Int. Sym. on Sc. Computing; G. Alefeld, A. Frommer, B. Lang editors, Akademie Verlag GmbH, Berlin, Germany, 1996.
3. Casado L. G., Garcia I.: Work Load Balance approaches for Branch-and-Bound Algorithms on Distributed Systems, Department of Computer Architecture and Electronics, Almeria University, Almeria, Spain.
4. Denneulin Y., Mehaut J. F., Planquelle B., Revol N.: Parallelization of Continuous Verified Global Optimization, Presented at the 19th Conference TC7 on System Modeling and Optimization, Cambridge, United Kingdom, July 1999.
5. Eriksson J., Lindstöm P.: A Parallel Interval Method Implementation for Global Optimization using Dynamic Load Balancing, Reliable Computing, 1, 1995, 77-91.
6. Hansen E.: Global Optimization Using Interval Analysis, Marcel Decker, New York, USA, 1992.
7. Leclerc A.P.: Efficient and Reliable Global Optimization, PhD Thesis, The Ohio State University, USA, 1992.
8. Leclerc A.P.: Parallel Interval Global Optimization and its Implementation in C++, Interval Computations, 3, 1993, 148-163.
9. Messine F.: Méthodes d'Optimisation Globale basées sur l'Analyse d'Intervalle pour la Résolution de Problèmes avec Contraintes, Thèse, Université Paul Sabatier, IRIT-Toulouse, France, Septembre, 1997.
10. Ratschek H., Rokne J.:New Computer Methods for Global Optimization, Ellis Horwood Limited, Market Cooper Street, Chichester, West Sussex, PO19 1EB, England, 1988.

Track VI

Workshops

A Distributed Platform with Features for Supporting Dynamic and Mobile Resources*, **

Serge Chaumette and Asier Ugarte

LaBRI, Université Bordeaux 1, 351 Cours de la Libération
33405 Talence Cedex, France
{Serge.Chaumette,Asier.Ugarte}@labri.u-bordeaux.fr
http://www.labri.u-bordeaux.fr/

Abstract. Networks of workstations are now common places. They are more and more often used by the parallel computing community to achieve large computations. These platforms have a lower cost than dedicated machines but still provide very good results. At the software level, a middleware is required to make them usable. In this paper we present Jem, the middleware that we have developed at the Laboratoire Bordelais de Recherche en Informatique – LaBRI –. We put the stress on two features – dynamic naming service and smart proxies - that make it easy for Jem to support dynamic and mobile resources.

1 Introduction

The domain of parallelism has evolved a great deal in the last few years. Some problems that required expensive high-performance machines can now be solved on high speed networks of workstations. These configurations have a lower cost but still provide very good results.

This new context raises new problems, the main difficulty of which is to provide support for an application with increasing demand to make an optimal usage of a set of resources, often heterogeneous, connected by a network. This cannot be reasonably achieved without dedicated software support, a middleware. In this paper we present Jem, the middleware that we have developed to cope with this problem. We insist on some of the features – dynamic naming service and smart proxies - that are integrated into Jem and that make it possible to support dynamic and mobile resources. Of course, there are some applications, for instance in the fields of chemistry, physics and weather forecast, that still require supercomputers specifically designed for intensive computation. These are specialized applications that we do not directly address in this paper, although Jem can of course be used to manage them.

---

* Java and all Java-based marks are trademarks or registered trademarks of Sun Microsystems, Inc. in the United States and other countries. The authors are independent of Sun Microsystems, Inc.

** This work is partly supported by the Université Bordeaux I, the Region Aquitaine and the CNRS.

B. Hertzberger et al. (Eds.): HPCN Europe 2001, LNCS 2110, pp. 603–612, 2001.

2 Related Work

The evolution presented in section 1 leads to the need for sharing resources and the services they provide, i.e. to communicate between resources to offer high level-services. When the number of shared resources becomes important, and when these resources are heterogeneous and distributed in a wide-area network, this is sometimes referred to as the *wide-area computing problem* [9]. Wide-area heterogeneous distributed systems can be considered from two points of view, according to the emphasis they put on the features they provide either with respect to computing or service sharing. Thus, we define two approaches [5]:

- **Computation Centric.** This type of systems focus on the computational features of resources. They use the fact that the resources are connected by a network to make an optimal usage of them, but the network is not on the first level. This is a quantitative approach, the stress is put on computation.
- **Network Centric.** In this approach the network is the key aspect. It is not the nature of resources that is the most important, but the fact that the network makes them usable from everywhere in a seamless manner. This approach introduces the notion of community. A resource can join, i.e. become a member of a community. Thereafter it can use all the resources of the community it belongs to. Thus, the cooperation between resources is the main goal. This is a qualitative approach, the stress is put on the network.

In the following sections we present some of the projects that are related to our own project or that inspired us – we had to choose just a few among the many environments that are being developed worldwide –.

2.1 Network Centric Approach

There are a number of environments that implement to a certain extent a network centric approach. For instance, it basically corresponds to what was available in old UNIX-like operating systems where all devices could be directly accessed via the `/dev` directory. For instance, to print a file one could use a command like `cat foo > /dev/printer`. This is no longer the case with UNIX systems that are available now, mainly because device drivers have become more and more sophisticated. Furthermore, the development of the network has made the operation of hardware even more complex. Nevertheless, there are environments that try to cope with the problem that way. We present here some of those that we consider the most significant in the domain.

Probably one of the most representative software environments when talking about the network centric approach is Jini[15][1]. Jini is a software architecture designed and implemented by Sun. It mainly relies on Java™[1] and on the RMI[14] (Remote Method Invocation) framework. Jini is an implementation of the "plug and participate" concept defined by Sun Microsystems. The basic idea

[1] See also Aladdin [20], Java Spaces [16], TSpaces [12] or Active Objects [3].

is that a new piece of hardware (or software) can join a network of resources, say it is present and both offer its services to other resources and use features provided by other resources. The key aspect of this architecture is a lookup service that registers the code of a proxy written in Java; this code migrates to the client when required. This proxy is then free to effectively implement the given service as it wants to, by running local code, or by invoking operations on a remote server.

Inferno[13] is another environment that implements the network centric approach. It can be compared to Java because it relies on a virtual machine called Dis that is alike the Java Virtual Machine. The associated language called Limbo makes it possible to abstract from physical considerations by writing a module that hides the low level details. Every resource is seen as a file that can be accessed by means of read and write operations. One of the main differences with Java is that the distributed aspect of resources is built within the system, whereas Jini is built on top of Java.

Another environment that could be considered as a member of this category is Windows NT. This operating system provides the user/programmer with an abstraction of the resources of the machine he/she is working with, thus offering some sort of abstraction of the services featured by the resources. One of the main differences between Windows NT and Inferno is that an application using Inferno does not need to know if the resources it handles are local or remote: the network is hidden and remote calls are transparent.

2.2 Computation Centric Approach

The other trend is that of computation centric approaches. Most of them basically rely on a combination of technologies, the most representative being: CORBA[19] for making services available over the network; Java to provide code migration; MPI[10], PVM[8], RMI or other communication libraries to provide communication between the software components that are made available to users over the network.

Globus and Legion are two of the most representative environments that implement this approach. The aim of Globus [6] is to provide homogeneous services over a wide-area network of computers. A layer of basic services is built on top of the operating system services available at individual nodes. This layer is then used to provide high level homogeneous services and environments to the programmers. The main characteristics of the Legion [9] environment are the same. Both systems are built on top of distributed heterogeneous networks and try to make an optimal usage of a set of heterogeneous resources. The fundamental difference is that in Legion everything is an object.

The other strategy of computation centric approaches is to offer an environment to control heterogeneous resources instead of trying to make them homogeneous. Netsolve [4] implements such a strategy to control computational resources distributed across the network. It is based on a client/server environment that furthermore provides agents that are in charge of identifying the most appropriate services to answer the requests of clients. Many other systems that could be compared to Netsolve are implemented in various projects around the world. They basically offer the same kind of services. This is why the Netsolve effort is important in order to provide the community with a single widely adopted and widely available environment that offers the most commonly required features.

2.3 Other Approaches – Global Computing

It is often difficult to classify the available platforms in one of the two previous categories, because they offer some functionalities that belong to both. For instance, there is a new trend in the way wide-area networks of personal computers are used to do computation. This is referred to as *global computing.* A project that is representative of this trend is SETI@home [11], which is carried out at the Berkeley University. It provides computational support to a scientific experimentation that uses Internet-connected computers to Search for Extraterrestrial Intelligence (SETI). The computational power of the platform is the result of the execution of the multiple instances of the screen saver that the project distributes and that is run by the people willing to participate.

3 The Jem Platform

3.1 Main Goals of the Platform

The main difficulty when willing to cope with the wide-area computing problem is to provide a software system that makes the usage of such an architecture as simple as possible. This kind of middleware has to offer three main features (some of the platforms presented in section 2 can be used to deal, at least partially, with some of these problems):

- a shared address space. We need a way to access the resources independently of their location, therefore they must be mapped into memory. A mechanism must also be provided to allow a fair sharing of these resources between applications, and between users;
- remote control of the execution support. This is mandatory both to achieve the deployment of applications and to manage the different parts of the application once it has been started. This is even required for simple features – simple from a user perspective – such as detecting the termination of an application within the distributed environment;
- support for parallel execution. This is required in order to use the computational resources in an optimal way. For instance we want to be able to take advantage of the parallelism that is made apparent by distributing the application; this should be achieved without any direct intervention of the programmer.

3.2 The Platform

Jem is a middleware designed to take advantage of a *universe of heterogeneous entities*, either software or hardware, connected by a network, currently the Internet. Real world entities are rarely isolated, therefore the entities that make the universe of Jem maintain *relationships between each other.* We consider one-sided relations where a set of entities depend on a unique other entity. This leads to a *tree-like hierarchy.* For instance a machine has several processors; a processor can have several processes; a process has several threads; a machine possibly has several users logged in; etc. An entity can be indirectly related with another one through other intermediate entities. For instance, a machine can be composed of one processor within which a process is being executed by several threads.

The entities that compose the universe of Jem are managed by a software layer called a **resource**. A resource is a software component that encapsulates a hardware or software entity to make **uniform** the way it is viewed and managed. For instance, a physical printer device can be managed through the use of a resource; we will talk about *printer resource* to refer to the printer management software layer. Inside Jem everything is a resource and all the resources are managed the same way, through the same interface, independently of their effective nature. To integrate a new entity into Jem, it must be *adapted* to that interface. We use the adapter design pattern as defined in [7] and as used for instance by CORBA. The adapter pattern defines an object adapter as an object that adapts the interface of a given object to the interface expected by the caller. The adapter object delegates the requests coming from the caller to the adapted object and the caller can use this object without being aware of its effective interface. A Jem resource is a pair <*rawdata, adapter*>, where *rawdata* is any data, and where *adapter* is the object that allows to adapt this entity. Intuitively, a Jem resource can be seen as a device (*rawdata*) and the driver (*adapter*) that is used to handle it. Basically this adapter enables to : 1) identify the resource: what is its name ? 2) locate the resource: what is its logical location ? 3) communicate with the resource: what are the services (i.e. inputs and outputs) it offers ?

The Jem platform itself is a middleware that enables the usage of Jem resources. It makes it possible to control and manage Jem resources. Furthermore, the platform provides a set of high level services that extend the features of the basic resources. To achieve that goal, the architecture of Jem is based on the concept of a chain of responsibility, a behavioral pattern defined in [7]. The aim of this pattern is to get a simple mechanism to separate functional and non-functional code. Figure 1 illustrates this architecture. Each object of this chain of responsibility implements a specific feature of the platform – remote access, concurrency, structuring, etc. –. The only manner for a user to access a Jem resource is through these different layers. All the objects of the chain are what we call *resource handlers* and they all implement the same interface. This interface is shown Prog. 1.1.

Prog. 1.1 Interface of the Components of the Chain of Responsibility.

```
public interface ResourceHandler extends Serializable{
    public void handleRequest(Request request);
}
```

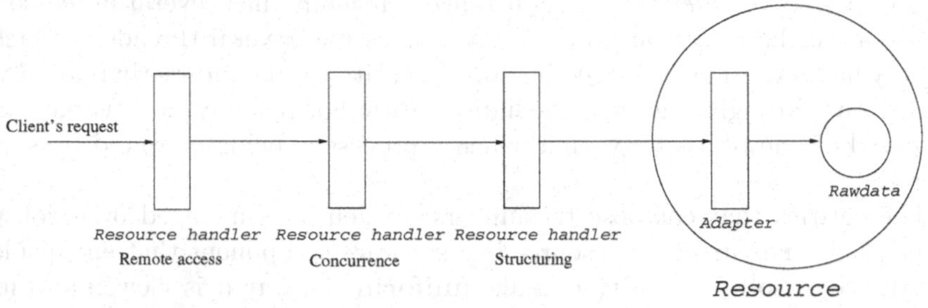

Fig. 1. Chain of Responsibility to Access a Jem Resource.

3.3 Jem for Global Computing

At the highest level, Jem provides a layer of execution that allows the usage of the computational features of the resources connected to the network and that make the platform. This support of execution is built upon a distributed multi-processing multi-user virtual machine that we have built based on a standard Java virtual machine. Our virtual machine will not be described here and the reader is referred to [17] for further details. Nevertheless it is worth giving a flavor of how Jem enables computation to be optimized by distributing control and data: this is basically done by migrating objects, remotely creating threads and using asynchronous calls on services provided by remote resources. In fact, a developer does not need to worry about remote access to resources or other aspects like overlapping of communications by computations. This is managed automatically by the platform.

This support makes it possible to achieve global computing types of applications using Jem.

4 Support for Dynamism and Mobility

The notion of mobility is becoming more and more important for both software and hardware resources. Regarding software, mobility comes from code migration features and the usage of systems of mobile agents. Concerning hardware, wireless devices - e.g. PDAs or PCs with IRDA – are becoming commonplaces, and the fact that they belong to a network is purely virtual and very dynamic. Therefore, Jem provides some features that make it possible to support such software and hardware components.

4.1 Smart Proxies

One uses Jem resources through a piece of software called a *Jem proxy.* To work with a resource, an application must first obtain its proxy. The concept of Jem proxy is halfway between the stub concept used by CORBA or RMI and the proxy concept used by Jini. The Jem proxy holds a remote reference to a resource that remains correct at any location: an object containing the Jem proxy of a resource can migrate – using serialization – towards a new destination and still have a valid reference to the resource. Figure 2 illustrates the migration of an object containing a Jem proxy.

Furthermore, Jem proxies are more than simple descriptors that would refer to remote resources. In fact, these objects can be considered smart in that they integrate high level features. For instance they can collect information about the caller to allow subsequent controls, they can look for a replacement resource if the resource they refer to is unreachable, etc.

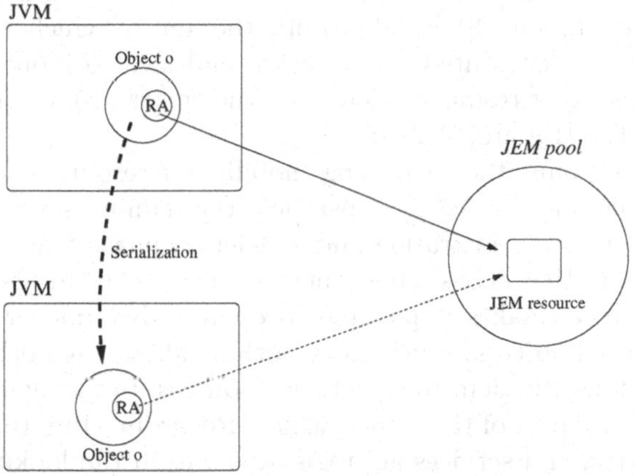

Fig. 2. Migration of an Object Containing the Jem Proxy (RA) of a Resource.

4.2 A Dynamic Naming Service

When designing a system for global computing built upon a network of personal devices, the possible mobility of these devices – mobile phones for instance – and the dynamism of some components – e.g. threads – are important factors that must not be neglected. Several types of resources would be very difficult to integrate into an architecture with a static naming service – users login in and out, threads, PDAs, mobile phones, etc. –. To cope with this problem Jem provides *local just in time structuring* that supports dynamism and then mobility. This is achieved as follows: a Jem resource can have a set of resources that directly depend on it, and, in a manner similar to a UNIX directory, produce on demand the up-to-date list of these sub-resources.

A dynamic naming service has two main advantages over a static naming service. First, unless a dedicated – often costly – mechanism is implemented, a passive naming service can contain entries for services that are no more available. These inconsistencies cannot happen within Jem. Second, the mechanisms used to register a resource and to update the naming service when a service is no more available are no more needed.

4.3 Comparison to Related Work

Regarding the main features of Jem we can notice when looking at the related systems presented section 2 that some already exist.

The main difference is that in Jem the notion of homogeneous interface is extended to all types of resources. In other systems, this approach is often used to adress a limited set of resources. This is true for Legion [9], which defines two types of resources implementing two predefined interfaces: processors and storage devices. The idea of abstracting from physical resources is present in other projects like Millennium [2][2] by Microsoft, the aim of which was to make the distinction between distributed computation and local computation irrelevant. Regarding the usage of resources, Jem uses the smart proxy concept, also used in systems like Jini [15] and WebOS [18].

Taking into account the increasing mobility of resources implies the need for a dynamic naming service. For instance, the naming service of CORBA is mainly passive, with a registration and a deletion phase when an object is no more available. CORBA now supports new services to fill this gap. The relations service, for instance, makes it possible to express dynamic or static relations between distributed objects, which know nothing about each other. This is alike the mechanism found in Jem to structure resources, but in Jem these relationships are an integral part of the system which. Regarding Jini, there is no explicit structuring service, and services are registered within the lookup service. However, the lookup service is a standard Jini service and it can be used to built a hierarchy so as to structure services. This hierarchy of resources is very often presented in a UNIX-like manner by a lot of systems, for instance Legion, Inferno and Jem.

5 Conclusion

In this paper, we have presented Jem, a platform for managing and using distributed resources. This middleware offers features well suited for dynamic and mobile resources.

The concept of resource makes it possible to abstract from the type of the entities integrated in the platform. A non-functional layer is then proposed that does not make any assumption about the way the entities will be used. The notion of resource also makes it easier to adapt entities so as to contribute

[2] This project has been abandoned.

to the system by integrating new components in a plug and participate way. Furthermore, cooperation between resources is also enhanced and made easier by the homogeneous interface that they offer. We believe that Jem is well suited for the management of a dynamic environment, i.e. where resources are dynamic and mobile. The main reasons are that: the relationships between entities are used to build the architecture of Jem as seen by the users and the programmers, based on a just in time structuring of the manipulated resources; mobility is abstracted by means of smart proxies that allow programmers to handle remote resources in a seamless manner.

We are now in the process of enhancing Jem features. We are for instance working on load balancing techniques and automatic distribution of applications. One of our major goals is to provide the programmer with an infrastructure allowing to gain substantial development time, still being able to make an optimal usage of a distributed environment. Therefore we intend to provide fault tolerance services and additional tools to make a better usage of the available computational power.

References

[1] K. Arnold, J. Gosling, and D. Holmes. *The Java Programming Language*. Addison-Wesley, third edition, 2000.

[2] W. J. Bolosky, R. P. Draves, R. P. Fitzgerald, M. B. Jones C. W. Fraser, T. B. Knoblock, and R. Rashid. Operating System Directions for the Next Millennium, 1997.
`http://www.research.microsoft.com/research/sn/Millennium/mgoals.html`.

[3] D. Caromel, W. Klauser, and J. Vayssire. Towards seamless computing and metacomputing in Java. In Geoffrey C.Fox, editor, *Concurrency: practice and experience*, volume 10, pages 1043–1061. Wiley and Sons, Ltd., September-November 1998.

[4] H. Casanova and J. Dongarra. NetSolve : A Network-enabled Server for Solving Computanional Science Problems. *The International Journal of Supercomputer Applications and High Performance Computing*, 11(3):212–223, 2000.

[5] S. Chaumette. Du parallélisme massif aux objets distribués, janvier 2000. Université Bordeaux I. Rapport scientifique pour obtenir l'habilitation à diriger des recherches.

[6] I. Foster and C. Kesselman. The Globus Project: A Status Report. Technical report, Argonne National Laboratory and University of Southern California, 1998.

[7] E. Gamma, R. Helm, R. Johnson, and J. Vlissides. *Design Patterns - Elements of Reusable Object-Oriented Software*. Addison-Wesley, 1994.

[8] A. Geist, A. Beguelin, J. Dongarra, W. Jiang, R. Manchek, and V. Sunderam. *PVM Parallel Virtual Machine : A Users'Guide and Tutorial for Networked Parallel Computing*. Scientific and Engineering Computation Series. The MIT Press, third edition, 1996.

[9] A. Grimshaw, A. Ferrari, F. Knabe, and M. Humphrey. Legion : An Operating System for Wide-Area Computing. Technical report, University of Virginia, 1999.

[10] W. Gropp, E. Lusk, and A. Skjellum. *Using MPI: Portable Parallel Programming with the Message Passing Interface*. The MIT Press, second edition, 1999.

[11] E. Korpela, D. Werthimer, D. Anderson, J. Cobb, and M. Lebofsky. SETI@home: Massively Distributed Computing for SETI. *Computing in Science and Engineering*, 2000.
[12] T. J. Lehman, S. W. McLauhry, and P. Wyckoff. TSpaces: The Next Wave. In *Hawaïi International Conference on System Sciences (HICSS-32)*, january 1999.
[13] K. Nyberg. An Analysis of Inferno and Limbo. Technical report, Helsinki University of Technology, February 1997.
[14] SUN Microsystems. *Java Remote Method Invocation Specification*, 1998.
[15] SUN Microsystems. Jini Architectural Overview. Technical report, Sun Microsystems, January 1999.
[16] Sun Microsystems. JavaSpace - Web server. `http://java.sun.com/products/javaspaces/`, 2000.
[17] A. Ugarte. Mise en œuvre d'un environnement multitâche, multi-utilisateur et distribué en Java. *CPRSR, Réseaux et Systèmes Répartis, Calculateurs Parallèles*, 12(5-6):511–535, décembre 2000.
[18] A. Vahdat, T. Anderson, M. Dahlin, E. Belani, D. Culler, P. Eastham, and C. Yoshikawa. WebOS : Operating System Services for Wide Area Applications. In *The Seventh IEEE Symposium on High Performance Distributed Computing.* July 1998.
[19] S. Vinoski. CORBA : Integrating Diverse Applications Within Distributed Heterogeneous Environments. *IEEE Comunications Magazine*, 35(2), February 1997.
[20] Y. Wang, W. Russell, A. Arora, J. Xu, and R. K. Jagannathan. Towards Dependable Home Networking: An Experience Report. In *IEEE International Conference on Dependable Systems and Networks (FTCS)*. Institute of Electrical and Electronics Engineers, Inc., juin 2000.

Implementing an Efficient Java Interpreter

David Gregg[1], M. Anton Ertl[2], and Andreas Krall[2]

[1] Department of Computer Science
Trinity College, Dublin 2, Ireland
David.Gregg@cs.tcd.ie

[2] Institut für Computersprachen, TU Wien
Argentinierstr. 8, A-1040 Wien
{anton,andi}@complang.tuwien.ac.at

Abstract. The Java virtual machine (JVM) is usually implemented with an interpreter or just-in-time (JIT) compiler. JIT compilers provide the best performance, but must be substantially rewritten for each architecture they are ported to. Interpreters are easier to develop and maintain, and can be ported to new architectures with almost no changes. The weakness of interpreters is that they are much slower than JIT compilers. This paper describes work in progress on a highly efficient Java interpreter. We describe the main features that make our interpreter efficient. Our initial experimental results show that an interpreter-based JVM may be only 1.9 times slower than a compiler-based JVM for some important applications.

1 Introduction

The Java Virtual Machine (JVM) is usually implemented by an interpreter or a just-in-time (JIT) compiler. JIT compilers provide the best performance but much of the compiler must be rewritten for each new architecture it is ported to. Interpreters, on the other hand, have huge software engineering advantages. They are considerably smaller and simpler than JIT compilers, which makes them faster to develop, cheaper to maintain, and potentially more reliable. Most importantly, interpreters are portable, and can be recompiled for any architecture with almost no changes.

The problem with existing interpreters is that they run most code much slower than compilers. The goal of our work is to narrow that gap, by creating a highly efficient Java interpreter. If interpreters can be made much faster, they will become suitable for a wide range of applications that currently need a JIT compiler. It would allow those introducing a new architecture to provide reasonable Java performance from day one, rather than spending several months building a compiler.

This paper describes work in progress on a fast Java interpreter. In section 2 we introduce the main techniques for implementing interpreters. Section 3 describes the structure of our interpreter, and our main optimizations. Section 4 presents some preliminary experimental results, comparing our interpreter with other JVMs. Finally, in section 5 we draw conclusions.

B. Hertzberger et al. (Eds.): HPCN Europe 2001, LNCS 2110, pp. 613–620, 2001.

2 Virtual Machine Interpreters

The interpretation of a virtual machine instruction consists of accessing arguments of the instruction, performing the function of the instruction, and dispatching (fetching, decoding and starting) the next instruction. The most efficient method for dispatching the next VM instruction is direct threading [Bel73]. Instructions are represented by the addresses of the routine that implements them, and instruction dispatch consists of fetching that address and branching to the routine (see fig. 1). Direct threading cannot be implemented in ANSI C and other languages that do not have first-class labels, but GNU C provides the necessary features. Implementors who restrict themselves to ANSI C usually use the giant switch approach (see fig. 2): VM instructions are represented by arbitrary integer tokens, and the switch uses the token to select the right routine.

```
void engine() {
  static Inst program[] = { &&add /* ... */ };
  Inst *ip; int *sp;

  goto *ip++;
 add:
  sp[1]=sp[0]+sp[1];  sp++;  goto *ip++;
}
```

Fig. 1. Direct Threading Using GNU C's "labels as values".

```
void engine() {
  static Inst program[] = { add /* ... */ };
  Inst *ip; int *sp;

  for (;;)
    switch (*ip++) {
    case add:
      sp[1]=sp[0]+sp[1];  sp++;  break;
    /* ... */
    }
}
```

Fig. 2. Instruction Dispatch Using `switch`.

When translated to machine language, direct threading typically needs three to four machine instructions to dispatch each VM instruction, whereas the switch method needs nine to ten [Ert95]. The execution time penalty of the switch method is caused by a range check, by a table lookup, and by the branch to the dispatch routine. In addition, indirect branches are more predictable in threaded code interpreters than those in switch based interpreters which leads to far fewer pipeline stalls and shorter running times on current architectures.

3 Implementing a Java Interpreter

We are currently building a fast threaded code interpreter for Java. Rather than starting from scratch, we are building the interpreter into an existing JVM. We started working with the CACAO [KG97,Kra98] JIT compiler based 64-bit JVM. Therefore, we used the infrastructure available from the JIT compiler and implemented a 64-bit interpreter. But it is our intention that it will be possible to plug our interpreter into any existing JVM. For this reason, we are defining a standard interface which describes a set of services that the wider JVM must provide for our interpreter.

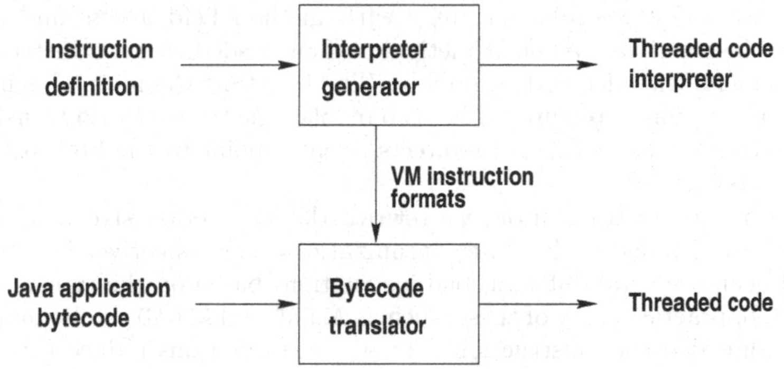

Fig. 3. Structure of the Interpreter System.

Figure 3 shows the structure of our interpreter system. It is important to note that we don't interpret Java byte code directly. Instead, the byte code is translated to threaded code. In the process we also apply optimizations to make the threaded code easier to interpret. Also important is that we do not write all the code in the interpreter ourselves. Instead, we are building an interpreter generator, which constructs an efficient interpreter from a specification of the behavior of each instruction. The following subsections describe the major components in more detail.

3.1 The Byte Code Translator

The main goal of the translator is to remove complex and expensive operations from the interpreter, and instead perform these operations once at translation time[1]. The simplest, and most important example of this is the translation from byte code to threaded code. Like a JIT compiler translates byte code from a

[1] This is the RISC principle of dealing with complex and difficult operations in the compiler, to allow a simpler, faster processor implementation. A processor is a hardware implementation of an interpreter, and many principles of processor design can be applied to interpreters.

complete method to machine code, our translator transforms byte code into threaded code. To interpret a byte code the interpreter must look up the address of the routine that implements the VM instruction in a table. When translating to threaded code, we perform that lookup just once, and replace the byte code with the address. Thereafter, we can interpret the threaded code without any table lookups. Note that the threaded code is larger than the original byte code. An important research question is whether the benefits of these optimizations could be outweighed by an increase in data cache misses, due to threaded code that occupies more memory.

The byte code translator also replaces difficult to interpret instructions with simpler ones. For example, we replace instructions that reference the constant pool, such as `LDC`, with more specific instructions and immediate, in-line arguments. We follow a similar strategy with method field access and method invocation instructions. When a method is first loaded, a stub instruction is placed where its threaded code should be. The first time the method is invoked, this stub instruction is executed. The stub invokes the translator to translate the byte code to threaded code, and redirects itself to point to the first instruction of the threaded code.

In the process of translation, we rewrite the instruction stream to remove some inefficiencies and make other optimizations more effective. For example, the JVM defines several different load instructions based on the type of data to be loaded. In practice many of these, such as `ALOAD` and `LLOAD` can be mapped to the same threaded code instruction. These transformations reduce the number of VM instructions and makes it easier to find common patterns (for instruction combining). Another simple optimization is based on the fact that many instructions in the JVM take several immediate bytes as operands. These are shifted and OR-ed together to form a larger integer operand. We perform this computation once at translation time, and use larger integer immediates in the threaded code.

One implication of translating the original byte code is that the design problems we encounter are closer to those in a just-in-time compiler than a traditional interpreter. Translating also requires a small amount of overhead. Translating allows us to speed up and simplify our interpreter enormously, however. Original Java byte code is not easy to interpret efficiently.

3.2 Optimizations

The CACAO JIT compiler has an analysis to compute the necessary information to remove bound checks. We reused this analysis to implement bound check removal in the interpreter. We defined two sets of array access instructions, one with bound checks and one without bound checks. Depending on the results of the analysis the right instruction is chosen.

The CACAO compiler does null pointer checks using the hardware. This feature is hardware dependent and not suitable for interpreters. To evaluate the cost of software null pointer checks we generated a version of the interpreter without null pointer checks.

3.3 Instruction Definition

The *instruction definition* describes the behavior of each VM instruction. The definition for each instruction consists of a specification of the effect on the stack, followed by C code to implement the instruction. Figure 4 shows the definition of IADD. The instruction takes two operands from the stack (iValue1,iValue2), and places result (iResult) on the stack.

```
IADD ( iValue1 iValue2 -- iResult ) 0x60
{
  iResult = iValue1 + iValue2;
}
```

Fig. 4. Definition of IADD VM Instruction.

We have tried to implement the instruction definitions efficiently. For example, in the JVM operands are passed by pushing them onto the stack. These operands become the first local variables in the invoked method. Rather than copy the operands to a new local variable area, we keep local variables and stack in a single common stack, and simply update the frame pointer to point to the first parameter on the stack. To correctly update the stack and frame pointer on calls and returns using this scheme, one needs to compute several pieces of information about stack heights and numbers of local variables. We compute this information once at translation time, and thereafter the handling of parameters during interpretation is much more efficient.

One complication in our instruction specification is that it was originally designed for Forth. In this language, the number of stack items produced and/or consumed by a given VM instruction is always the same. Java has several VM instructions that consume a variable number of stack items, however. For example, the instruction to create a multidimensional array (MULTIANEWARRAY) takes a number of items from the stack equal to the number of parameters of the array. Similarly, the various method invocation instructions consume a number of stack items equal to the number of parameters. Currently, we have no way to cleanly express this in our instruction definition. One possibility is to create separate VM instructions for each possible number of stack items consumed. This does not scale however, since in theory there can be almost any number of parameters. Our current solution is to manipulate the stack pointer directly in the instruction specification.

3.4 Interpreter Generator

The *interpreter generator* is a program which takes in an instruction definition, and outputs an interpreter in C which implements the definition. The interpreter generator translates the stack specification into pushes and pops of the stack, and adds code to invoke following instructions.

There are a number of advantages of using an interpreter generator rather than writing all code by hand. The error-prone stack manipulation operations can be generated automatically. Optimizations can easily be applied to all instructions. For example, the fetching of the next instruction can be moved up the code (interpreter pipelining [HA00]). It is easy to have both a threaded code and switch-based version of the interpreter. Specifying the stack manipulation at a more abstract level also makes it simpler to change the implementation of the stack. For example, our interpreter keeps one stack item in a register. It is nice to be able to vary the number of cached stack items without changing each instruction specification. The generator also allows us to add tracing and profiling code trivially, and easily disassemble the threaded code.

The main advantage of the generator is that it allows more complicated optimizations such as automatic instruction combining or instruction specialization. Combining replaces a common sequence of VM instructions with a single "super" instruction [Pro95]. For common operands specialization uses different instructions to eliminate the operand decoding overhead. Combining greatly increases the number of VM instructions, making maintenance more difficult if done manually. The generator can create these combinations automatically. Using the stack specifications, it also optimizes stack operations for combined instructions.

4 Preliminary Experimental Results

Our basic thesis is that interpreter based JVMs can be so fast that at least for some applications they are not very much slower than a JIT compiler. To test the performance of a simple, efficient threaded-code interpreter implementation of the JVM we compared it with the CACAO JIT compiler, with an interpreter and two different JIT compilers from COMPAQ, with an interpreter from OSF and with the KAFFE interpreter on different processors. CACAO is one of the fastest available JVM implementations and for computationally intensive programs provides 42% to 82% of the performance of optimized C code [Kra98].

Our two main benchmarks are *javac* and *db* from the SPECjvm98 benchmark suite. The *javac* benchmark is the Java compiler from the JDK 1.0.2. The *db* benchmark performs a sequence of add, delete, find and sort operations on a memory resident database. We also tested two computationally intensive mini-benchmarks *sieve* and *suml*. The first of these is the well-known prime number computation program and *suml* is the single pathological loop `{long i, l1, l2; for (i = l1 = l2 = 0; i > 0; i--) {l1++; l2--;}}` that was designed to maximally stress an interpreter based JVM.

Table 1 shows the run time of our benchmarks relative to the CACAO JIT compiler on Alpha 21064a and Alpha 21164a based workstations. The results for the SPEC benchmarks are surprisingly good, with our interpreter taking not much more than twice the time of the CACAO just-in-time compiler and faster than some other JIT compilers. This result sharply contradicts the widespread belief that interpreter based JVMs are inherently slow and will always perform many times slower than JIT compilers. We examined the proportion of time spent in native functions rather than executing byte codes. We found that both

Table 1. Relative Performance of JIT Compilers and Interpreters.

21064a	javac	db	sieve	suml
CACAO native	1.00	1.00	1.00	1.00
CACAO int	2.12	2.29	10.16	39.91
COMPAQ JVM 1.1.4 native	13.14	23.65	2.04	3.85
COMPAQ JVM 1.1.4 int	15.80	27.50	22.51	104.88
OSF JVM 1.0.1 int			29.06	156.42
21164a	javac	db	sieve	suml
CACAO native	1.00	1.00	1.00	1.00
CACAO int	2.27	2.10	17.54	25.08
COMPAQ JVM 1.3.1 native		5.01	2.52	3.23
KAFFE 1.0.5 int		19.57	63.47	93.21

programs spend about 30% of their time in native functions (synchronization, garbage collection). Both the interpreter and JIT compiler have to spend the same amount of time in native functions. Therefore, even though the JIT compiler executes byte codes about six times as fast as the interpreter for these programs, the overall speedup is not much more than a factor of two.

The results for *sieve* are rather less encouraging, showing the interpreter to be ten to seventeen times slower than the JIT compiler. Not only does the JIT compiler avoid the expensive overhead of dispatching instructions, it is also able to optimize array accesses in the inner loop with pointer arithmetic. This suggests that interpreters are not at all suited to computationally intensive code. Finally, the *suml* benchmark shows the worst case performance of interpreter based JVMs.

Table 2. Effects of Optimizations on Sieve.

	null	bound	null & bound
relative run time	0.995	0.976	0.971

Table 2 shows the effects of different optimizations on the *sieve* benchmark. Whereas the elimination of null pointer checks only gives a small speedup of 0.5%, the elimination of array bound checks gives a speedup of 2.5%. The benefit of these optimizations is quite small compared to speedups of more than 30% achievable by JIT compilers using these optimizations.

Figure 5 shows the potentials of instruction combination. The left graph gives the number of all possible used superinstructions for lengths of up to four instructions. As *javac* is a bigger program more different combinations exist. The right graph shows the speedups for *javac* and *db* on different processors and different length of superinstructions. For all programs *db* has been used as training input for the instruction combiner. The Alpha 21064a has a 16 Kbyte direct mapped instruction cache, the Alpha 21264 has a 64 Kbyte two way set associative instruction cache. For small caches exhaustive instruction combining can lead to cache trashing. For the Alpha 21264 the gap between interpreter and JIT compiler can be reduced to a factor of 2.07 for *javac* and 1.88 for *db*.

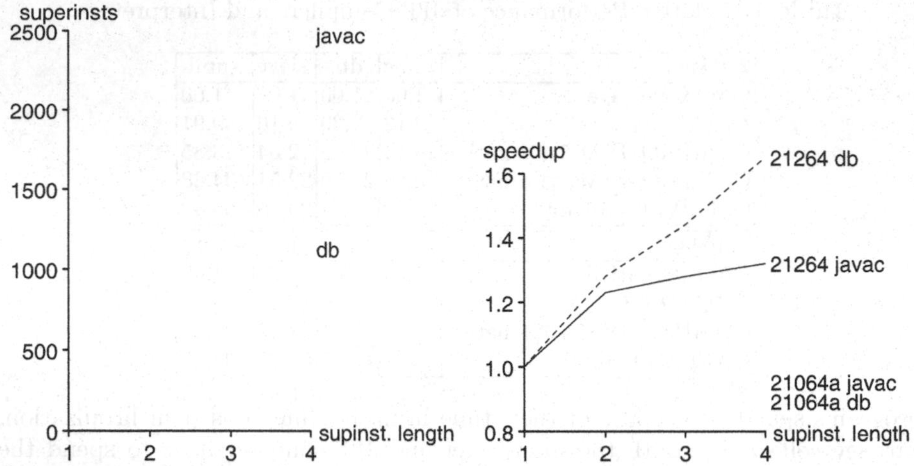

Fig. 5. Left: Number of superinstructions for different maximum sequence lengths. Right: Speedup for different programs and CPUs.

5 Conclusion

We have described an efficient interpreter based implementation of the Java virtual machine. Our interpreter system translates the original Java byte code to threaded code, greatly reducing the interpreter overhead. Our translator also computes constants, targets and offsets at translation time, allowing us to greatly simplify the interpretation of many instructions, such as method invocations. Experimental results show that our interpreter based JVM may be not much more than 1.9 as slow as a good JIT based JVM for some general purpose applications. For scientific code an interpreter cannot replace a compiler.

References

Bel73. J. R. Bell. Threaded code. *Communications of the ACM*, 16(6):370–372, 1973.

Ert95. M. Anton Ertl. Stack caching for interpreters. In *SIGPLAN'95 Conference on Programming Language Design and Implementation*, pages 315–327, 1995.

HA00. Jan Hoogerbrugge and Lex Augusteijn. Pipelined Java virtual machine interpreters. In *Proceedings of the 9th International Conference on Compiler Construction (CC'00)*. Springer LNCS, 2000.

KG97. Andreas Krall and Reinhard Grafl. CACAO - a 64 bit JavaVM just-in-time compiler. *Concurrency: Practice and Experience*, 9(11):1017–1030, 1997.

Kra98. Andreas Krall. Efficient JavaVM just-in-time compilation. In *Proceedings of the 1998 International Conference of Parallel Architectures and Compilation Techniques*, pages 205–212. IEEE Computer Society, October 1998.

Pro95. Todd Proebsting. Optimising an ANSI C interpreter with superoperators. In *Proceedings of Principles of Programming Languages (POPL'95)*, pages 322–342, 1995.

Efficient Dispatch of Java Interface Methods

Bowen Alpern, Anthony Cocchi, David Grove, and Derek Lieber

IBM T. J. Watson Research Center

Abstract. Virtual methods can be dispatched efficiently because the code for corresponding methods reside at the same entries in their respective virtual method tables (VMTs). To achieve efficient interface method dispatch, a fixed-sized *interface method table* (IMT) is associated with each class. Different implementations of the same interface method signature reside at the same entry in their respective IMTs. When a class implements two or more interface methods with the same IMT offset, a *conflict resolution stub* distinguishes between them at run-time. The resulting interface method dispatch is almost as cheap as its virtual counterpart.

1 Introduction

Multiple inheritance is conceptually elegant, however the ambitious C++ realization of this concept has proved inefficient to implement and confusing to use. Interfaces are Java's more limited realization of multiple inheritance. A class can have at most one superclass, but it may *implement* an arbitrary number of interfaces. A class inherits, but can override, the virtual methods of its superclass. However, it must explicitly provide implementations of the method signatures declared by its interfaces. Either way, a class is known to possess a method corresponding to the methods declared by its superclass and by its superinterfaces. (Java does not provide for multiple inheritance of fields.)

Single superclass inheritance makes efficient virtual method dispatch possible. This has led to a widespread misimpression that interface method dispatch is inherently inefficient. This paper presents a scheme for dispatching Java interface methods efficiently. The next section presents some background on the Java virtual machine (JVM) that served as the platform for this research. Section 3 presents a technique for efficiently dispatching interface methods. Section 4 presents performance results. Section 5 reviews related work. Section 6 concludes.

2 Virtual Method Dispatch in Jalapeño

Jalapeño [1] is a virtual machine, primarily designed for Java servers, being developed at IBM's T. J. Watson Research Center. It is written in Java, but it runs directly on PowerPC-based multiprocessors running the AIX operating system rather than running on top of another JVM [2].

Every Jalapeño object contains a two-word header. One of these words is a reference to an array (of objects) called the *type information block* (TIB) for the

B. Hertzberger et al. (Eds.): HPCN Europe 2001, LNCS 2110, pp. 621–628, 2001.

object's type. This TIB contains a pointer to a VM_Type object for the object's class. This class object contains (among other things) a pointer to the class object for the class's superclass and an array of class objects of the interfaces the class implements. Three other slots in the TIB are devoted to data structures used in dynamic type checking [3]; one of these is an array that supports fast checking whether the class implements a particular interface. The TIB also contains slots pointing to the code for each virtual method of the class. This portion of the TIB is referred to as the class's *virtual method table* (VMT).

The VMT is so constructed that the pointer to the code for a method that a class inherits or overrides from its superclass is in the same slot as the corresponding method is in the superclass's TIB. This makes for efficient virtual method dispatch: the TIB is loaded from the object header, the method's code address is loaded at a fixed offset from the TIB, and a branch-and-link is performed to this address.

The difficulty of interface method dispatch is that, while a pointer to the appropriate code exists in the VMT of the object upon which the method is invoked, these pointers may be (and, generally, will be) at different offsets in the VMT's of the different classes that implement the interface.

Jalapeño's implementation methodology [2] requires that the initial implementation of any mechanism be simple, general, and correct. When a mechanism is observed to be a performance bottleneck, a more efficient mechanism can be devised to replace it. The initial implementation of `invokeinterface` in Jalapeño, crawls up the superclass hierarchy looking for a method with a signature that matches the one being dispatched. This mechanism was suspected to impede performance on a number of benchmarks. The replacement for this simple mechanism is described in the next section.

3 Interface Method Tables

Virtual method dispatch is efficient in Jalapeño because the VMT offset of the target method is available as a compile-time constant when the `invokevirtual` bytecode is compiled. Jalapeño devotes a fixed-sized portion of the TIB, called the *interface method table* (IMT), to allow interface methods to be dispatched with similar efficiency.

Every Java method has a *signature* — its name and the types of its parameters and its return type (if any). All methods that implements a particular interface method signature can be reached through the same slot in their class's IMT. Every interface method signature is assigned a unique *id*. Each id is hashed into an *IMT offset*.[1] Whenever a method with this signature is dispatched upon an object using an `invokeinterface` bytecode, the code pointed to by the slot at the IMT offset of the object's IMT is executed.

---

[1] Interface method signature ids are assigned sequentially as new interface method signatures are discovered, either by loading interfaces or by compiling references to interface methods. Currently, ids are mapped directly to IMT slots (modulo the size of the IMT).

```
L     s0, tibOffset(t0); // s0 := TIB of the "this" parameter (in t0)
CMPI  s1, id1;           // compare hidden parameter to id of first method
BNE   4;                 // if not equal, skip
L     r0, offset1, s0    // load VMT entry for first method into reg. 0
MTCTR r0                 // move this address to the count reg.
BCTR                     // branch to it (preserve return address in link reg.)
CMPI  s1, id2;           // compare hidden parameter to id of second method
BNE   4;                 // if not equal, skip
L     r0, offset2, s0    // load VMT entry for second method into reg. 0
MTCTR r0                 // move this address to the count reg.
BCTR                     // branch to it (preserve return address in link reg.)
TI    31, 31, 0xFFFF     // if neither id matches, trap (shouldn't happen)
```

Fig. 1. A Typical Conflict Resolution Stub.

For this scheme to work perfectly, the IMT would have to be big enough that every interface method signature could have a unique IMT offset. Even Jalapeño is not that space insensitive. Consequently, some classes may have two (or more) interface methods with the same IMT offset. Such interface methods are said to *conflict* (or *collide*). Somehow Jalapeño must discriminate between such methods at runtime.

Before Jalapeño branches (and links) to the code at an offset in an IMT, it loads the interface method signature id of the method being invoked into a special scratch register. This *hidden parameter* is used to disambiguate interface method calls upon objects of a class that has conflicts in its IMT slots. Where there is unique method at a slot, the hidden parameter is ignored.

When two or more interface methods of a class share the same IMT offset, the IMT slot points to a custom *conflict resolution stub* (see figure 1). This code successively compares the hidden parameter to the id's of the interface method signatures that share this slot. When a match is found, the VMT offset for the appropriate method is obtained, and control is transfered to the code pointed to at this offset.

What makes conflict resolution stub code tricky is the context in which it must execute. The processor's link register contains the return address in the calling method. The non-volatile registers cannot be used until the callee saves them. The volatile registers cannot be used because they may contain parameters to the method being called. In figure 1, only Jalapeño's three scratch (`s0`, `s1`, and `r0`) are used.[2]

It remains to explain how the IMT's get populated. It would be nice to be able to create the IMT for a class when the class is loaded. Unfortunately, since the class's interfaces cannot loaded with the class, it is not apparent which of the class's public virtual methods *is* an interface method. One could conservatively assume that all these methods are interface methods, but this would lead to many false IMT slot conflicts. A better solution is to load the IMT incrementally. As a class is discovered to implement an interface, its methods are added to the class's

[2] These registers are used by Jalapeño method prologues (and epilogues) to allocate (free) the stack frame for the called method. (They are also used as temporary registers between method calls.) Register 0 (`r0`) is of limited utility since many PowerPC instructions treat what would be a reference to it as a literal 0.

Table 1. The Benchmark Suite. The first seven rows are the SPECjvm98 benchmarks.

Benchmarks	Description	Number of Classes	Size of Class Files (in bytes)
`compress`	Lempel-Ziv compression algorithm	12	17,821
`jess`	Java expert shell system	150	396,536
`db`	Simple memory resident database	3	10,156
`javac`	JDK 1.0.2 Java compiler	175	561,463
`mpegaudio`	Decompression of audio files	54	120,182
`mtrt`	Two-thread ray-tracing algorithm	25	57,859
`jack`	Java parser generator	55	130,889
`opt-compiler`	Jalapeño optimizing compiler	393	1,378,292

IMT. Since there is always a dynamic type check before the interface method is dispatched, the IMT will always have the required methods in it when an interface method is invoked.

4 Performance

Jalapeño can be deployed in a multitude of variations. Key discriminates include: the garbage collector (copying or not, generational or not), the compiler (baseline or optimizing, level of optimization) to be used on methods of classes included in Jalapeño's boot image, the compiler to be used on methods of classes loaded dynamically, whether such methods should be compiled immediately when their class is loaded, and whether, under what circumstances, and how methods should be recompiled.

To simplify presentation, this section uses a variant that currently is consistently among the best performers on the SPECjvm98 benchmarks. It uses a copying, non-generational garbage collector. The optimizing compiler (at optimization level 2) is used to (statically) compile all methods in the boot image. The first time a dynamically loaded method is invoked, it is compiled using the baseline compiler. (The baseline compiler produces inefficient code very quickly.) Methods observed, via online profiling, to be computationally intensive or frequently called are selected for recompilation with the optimizing compiler by Jalapeño's adaptive optimization system [5].

The performance impact of 101 slot IMT tables with conflict resolution stubs using the SPECjvm98 [20] benchmarks (see table 1), and the Jalapeño optimizing compiler [6] is shown in table 2. (Jalapeño performance on these benchmarks roughly matches that of the industry-leading IBM product virtual machine [1].) The best elapsed time from 10 successive runs of the benchmark during a single Jalapeño execution is reported. The results were obtained on an IBM 43P Model 140 with one 333MHz PPC604e processors running AIX v4.3. The system has 512MB of main memory. For the SPECjvm98 benchmarks, the size 100 (large) inputs were used. For the `opt-compiler` benchmark, the time taken for the

Table 2. Execution times in seconds using Jalapeño's old and new implementations of `invokeinterface`. The number in parenthesis is the speedup.

Benchmarks	Prior Invokeinterface	New Invokeinterface
`compress`	42.56	42.89 (0.99)
`jess`	26.26	25.28 (1.04)
`db`	86.40	78.70 (1.10)
`javac`	38.94	36.61 (1.06)
`mpegaudio`	24.56	23.78 (1.03)
`mtrt`	19.84	19.47 (1.02)
`jack`	49.21	46.94 (1.05)
`opt-compiler`	118.11	77.98 (1.51)

Jalapeño optimizing compiler to compile itself (roughly 75,000 lines of Java source code) is reported.

Not surprisingly, the new implementation of interface invocation only measurably improved the execution times of those benchmarks that actually make non-trivial use of interfaces. The results for `compress`, `mtrt`, and, possibly, even `mpegaudio`, are in the measurement noise. There is a small, but probably real, improvement in `jess` and `jack`. The improvements for `javac` and `db` are significant. The effect on the `opt-compiler` itself is dramatic. This reflects the heavy use of interfaces in Jalapeño's optimizing compiler.

5 Related Work

Interface tables, or *itables*, are probably the most commonly used mechanism for interface method dispatch in high performance Java implementations. An itable is a virtual method table for a class restricted to the subset of the class's methods that match those of an interface that the class implements. To dispatch an interface method to an instance of a class, the system must locate the itable that corresponds to the appropriate class/interface pair. Typically, the system stores itables in an array reachable from the class object. Sometimes a JIT compiler can determine statically what itable applies at a particular interface method invocation site. If not, it must search for the relevant itable at dispatch-time [19,12]. In a straightforward implementation, search time increases with the number of interfaces implemented by the class. Thus, basic itable schemes tend to degrade when classes implement more than a handful of interfaces.

The CACAO JVM [16] implements a variant of the basic itable scheme that avoids a dispatch-time search for the right itable. Rather than storing a class's itables in a dense array, it maintains a sparse array of itables for each class indexed by interface id. This sparse array grows down from (the CACAO analog of) the TIB, thus making it easily accessible for dispatching. To dispatch an interface method, CACAO simply loads the TIB from the object, loads the itable for the interface at a constant offset in the TIB, and obtains a pointer to the callee code from a constant offset into the itable. With this mechanism, an interface method dispatch introduces one more dependant load than a virtual method dispatch.

As an alternative to itables, caching techniques originally developed for virtual method dispatch in dynamically typed languages could be adapted for interface dispatch in Java. Dynamic caching [17], inline caching [9], and polymorphic inline caching [14] could all be used to reduce the frequency of full-fledged interface method lookup. In fact, the first edition of The Java Virtual Machine Specification [18] defined a "quick bytecode" that acted as an inline cache by caching history with the invocation site. A feature of *any* caching scheme is that it relies on temporal locality and thus cannot guarantee efficient dispatching for all programs. Polymorphic inline caches are less vulnerable than simple inline caches, but they still can perform poorly at "megamorphic" call sites.

Selector indexed dispatch tables [8] provide a straightforward but space-intensive solution to the interface dispatch problem. Each class maintains a (potentially large) table indexed by interface signature id. Entries that correspond to an interface signature that the class actually implements point to the code for the matching virtual method; all other entries are null. The space impact of selector indexed dispatch tables can be reduced by using sparse data structures [11], modifying the storage allocator to reuse "holes" in the tables [13], or by applying selector coloring [10]. Just as in register allocation [7], the assignment of identifiers to selectors can be viewed as a graph coloring problem. Two selectors can be assigned the same color if they are never implemented by the same class. Using this approach, several algorithms have been proposed that greatly reduce the size of the dispatch tables [10,4,22,21]. Unfortunately, all of these algorithms assume that the set of selectors and the classes that understand them are known *a priori*. Thus, although CACAO has implemented a second interface method dispatching mechanism based on selector coloring [16], traditional selector coloring is a poor match for Java, because it assumes that the JVM has complete *a priori* knowledge of the program to be executed. An optimistic coloring scheme with a recovery mechanism has been considered for CACAO, but has not been implemented [15].

Selector coloring is closely related to the IMT technique presented above. The critical difference is that previous dispatching mechanisms based on selector coloring insisted upon a perfect hash of interface methods in each class while our technique tolerates an occasional collision. It is this ability to tolerate collisions that enables it to be effective in the presence of dynamic class loading.

6 Conclusion

Multiple inheritance is a powerful tool for structuring computer programs. Java interfaces are an adequate, if somewhat pedestrian, embodiment of this technique. Unfortunately, multiple inheritance has a reputation for being inefficient. This paper presents a technique for efficiently dispatching Java interface methods.

A fixed-sized interface method table (IMT) is proposed for each class. All implementations of an interface method signature lie at the same offset in their respective IMTs. This allows interface methods to be dispatched efficiently. A conflict resolution stub disambiguates between interface methods hashed to the same IMT slot.

The extra overhead of a non-conflicted interface method dispatch is a single immediate load (of the interface method signature id). The added cost of an conflicted dispatch is linear in the number of conflicts. This could be reduced to logarithmic by instituting a binary search. The expected number of comparisons might be further reduced using caching or move-to-front techniques, however such schemes would increase the unit cost of each comparison. Trade-offs between the size of the IMT, and the hashing algorithm, and the number of collisions need to be explored. Still, it seems safe to conclude that dispatching an interface method need not be appreciably more expensive than dispatching a virtual method.

Two additional issues must be considered before Java interfaces can be given a certificate of good (performance) health. First, because of Java's dynamic class-loading rules, interface method invocation entails a dynamic type check that is not required for virtual method invocation. Although, the cost of such checks can be quite small [3], it would be nice to be able to eliminate some or all of them. Secondly, an optimizing compiler can often eliminate most of the overhead of a virtual method by inlining it (and checking that the actual method being called is the one inlined). A similar technique for interface methods needs to be worked out.

References

1. B. Alpern, C. R. Attanasio, J. J. Barton, M. G. Burke, P. Cheng, J.-D. Choi, A. Cocchi, S. J. Fink, D. Grove, M. Hind, S. F. Hummel, D. Lieber, V. Litvinov, M. F. Mergen, T. Ngo, J. R. Russell, V. Sarkar, M. J. Serrano, J. C. Shepherd, S. E. Smith, V. C. Sreedhar, H. Srinivasan, and J. Whaley. The Jalapeño virtual machine. *IBM Systems Journal*, 39(1), 2000.
2. Bowen Alpern, Dick Attanasio, John J. Barton, Anthony Cocchi, Derek Lieber, Stephen Smith, and Ton Ngo. Implementing Jalapeño in Java. In *ACM Conference on Object-Oriented Programming Systems, Languages, and Applications*, pages 314–324, 1999.
3. Bowen Alpern, Anthony Cocchi, and David Grove. Dynamic typechecking in Jalapeño. In *Usenix Java Virtual Machine Research and Technology Symposium (JVM'01)*, April 2001.
4. Pascal André and Jean-Claude Royer. Optimizing method search with lookup caches and incremental coloring. In *Proceedings OOPSLA'92*, pages 110–126, October 1992. Published as ACM SIGPLAN Notices, volume 27, number 10.
5. Matthew Arnold, Stephen Fink, David Grove, Michael Hind, and Peter F. Sweeney. Adaptive optimization in the Jalapeño JVM. In *ACM Conference on Object-Oriented Programming Systems, Languages, and Applications*, October 2000.
6. Michael G. Burke, Jong-Deok Choi, Stephen Fink, David Grove, Michael Hind, Vivek Sarkar, Mauricio J. Serrano, V. C. Sreedhar, Harini Srinivasan, and John Whaley. The Jalapeño dynamic optimizing compiler for Java. In *ACM 1999 Java Grande Conference*, pages 129–141, June 1999.
7. G. J. Chaitin, M. Auslander, A. Chandra, J. Cocke, M. Hopkins, and P. Markstein. Register allocation via coloring. *Computer Languages 6*, pages 47–57, 1981.

8. B. J. Cox. *Object Oriented Programming: An Evolutionary Approach.* Addison-Wesley, 1987.
9. L. Peter Deutsch and Allan M. Schiffman. Efficient implementation of the Smalltalk-80 system. In *11th Annual ACM Symposium on the Principles of Programming Languages*, pages 297–302, January 1984.
10. R. Dixon, T. McKee, M. Vaughan, and Paul Schweizer. A fast method dispatcher for compiled languages with multiple inheritance. In *Proceedings OOPSLA '89*, pages 211–214, October 1989. Published as ACM SIGPLAN Notices, volume 24, number 10.
11. Karel Driesen. Selector table indexing & sparse arrays. In *Proceedings OOPSLA '93*, pages 259–270, October 1993. Published as ACM SIGPLAN Notices, volume 28, number 10.
12. Robert Fitzgerald, Todd B. Knoblock, Erik Ruf, Bjarne Steensgaard, and David Tarditi. Marmot: An optimizing compiler for Java. Technical Report MSR-TR-99-33, Microsoft Research, June 1999.
13. Etienne Gagnon and Laurie Hendren. SableVM: A research framework for the efficient execution of Java bytecode. Technical Report Sable Technical Report No. 2000-3, School of Computer Science, McGill University, November 2000.
14. Urs Hölzle, Craig Chambers, and David Ungar. Optimizing dynamically-typed object-oriented languages with polymorphic inline caches. In P. America, editor, *Proceedings ECOOP'91*, LNCS 512, pages 21–38, Geneva, Switzerland, July 15-19 1991. Springer-Verlag.
15. Andreas Krall. Personal Communication, September 1999.
16. Andreas Krall and Reinhard Grafl. CACAO – a 64 bit JavaVM just-in-time compiler. *Concurrency: Practice and Experience*, 9(11):1017–1030, 1997.
17. G. Krasner. *Smalltalk-80: Bits of History, Words of Advice.* Addison-Wesley, 1983.
18. Tim Lindholm and Frank Yellin. *The Java Virtual Machine Specification.* The Java Series. Addison-Wesley, 1996.
19. Ganesan Ramalingam and Harini Srinivasan. Object model for Java. Technical Report 20642, IBM Research Division, December 1996.
20. The Standard Performance Evaluation Corporation. SPEC JVM98 Benchmarks. `http://www.spec.org/osg/jvm98`, 1998.
21. Jan Vitek and Nigel Horspool. Compact dispatch tables for dynamically typed object oriented languages. In *Proceedings of International Conference on Compiler Construction (CC'96)*, pages 281–293, April 1996. Published as LNCS vol 1060.
22. Jan Vitek and R. Nigel Horspool. Taming message passing: Efficient method look-up for dynamically typed languages. In M. Tokoro and R. Pareschi, editors, *Proceedings ECOOP'94*, LNCS 821, pages 432–449, Bologna, Italy, July 1994. Springer-Verlag.

Implementation of a CORBA-Based Metacomputing System*

Y. Cardinale, M. Curiel, C. Figueira, P. García, and E. Hernández

Universidad Simón Bolívar
Departamento de Computación y Tecnología de la Información
Apartado 89000, Caracas 1080-A, Venezuela
{yudith,mcuriel,figueira,emilio}@ldc.usb.ve
http://suma.ldc.usb.ve

Abstract. The access to distributed high performance computing facilities for execution of Java programs has generated considerable interest. A metacomputing system, or metasystem, allows uniform access to heterogeneous resources. Our case study is SUMA, a metasystem defined as a set of CORBA components, offering services for execution of both sequential and parallel applications. This document describes the most important aspects of SUMA design in terms of CORBA services. We present some experimental results related to execution overhead in a campus-wide environment.

1 Introduction

The access to distributed high performance computing facilities for execution of Java programs has generated considerable interest [6,3,17,16]. A metacomputing system, or metasystem, allows uniform access to heterogeneous resources, including high performance computers. This is achieved by presenting a collection of different computer systems as a single computer. A portable execution platform, such as the *Java Virtual Machine* (JVM) makes it potentially easier the implementation of a metasystem. There is increasing interest in using Java as a language for high performance computing [13]. Java provides a portable, secure, clean object oriented environment for application development. Recent results show that Java has also the potential to attain the performance of traditional scientific languages [15,11,2].

SUMA (Scientific Ubiquitous Metacomputing Architecture) [12] is a metacomputing system that executes Java bytecode. It is implemented as a set of CORBA classes, following a three-tier metacomputing system model. SUMA *Execution Agents* (actual platforms where program execution takes place) offer integrated profiling and checkpointing/recovery services, as well as local access to numerical and communication libraries. SUMA architecture additionally provides for self-registration of computing resources, user authentication, efficient job scheduling, and performance analysis [9].

* This work was partially supported by grants from Conicit (project S1-2000000623) and from Universidad Simón Bolívar (direct support for research group GID-25).

B. Hertzberger et al. (Eds.): HPCN Europe 2001, LNCS 2110, pp. 629–636, 2001.

SUMA specification in terms of CORBA IDL has a small number of classes and methods. In this paper we describe the main services provided by SUMA components, accessible from SUMA system software. We also describe the services accessible by an application sent to SUMA, and present some experimental results related to execution overhead in a campus-wide environment.

The rest of the document is organized as follows. Section 2 presents an overall description of SUMA in terms of the services offered by its components. Section 3 describes the services available for application execution in SUMA. Section 4 shows some results on measured overhead using current prototype, implemented in Java and *JacORB* [7], a public domain CORBA implementation. Section 5 summarizes the main differences between SUMA and other approaches related to using Java and CORBA for high performance computing. Section 6 presents some conclusions and future work.

2 CORBA Specification of SUMA

The main goal of SUMA is to extend the run-time JVM model to provide seamless access to distributed high performance resources. SUMA executes sequential Java bytecode and parallel Java bytecode communicating with *mpiJava* [2]. These programs have access to SUMA standard packages, which currently include, apart from Java standard packages, a wrapper for *PLapack* [1].

2.1 General Description

The user executes programs through a client running on her machine that invokes either `execute` or `submit`, which are CORBA services. The `execute` service corresponds to on-line execution mode, while the `submit` service allows off-line execution (batch jobs). Once SUMA receives the request from a client machine, it transparently finds a platform for execution and sends a request message to that platform. An *Execution Agent* at the designated platform starts the execution of the program, communicating directly with the client. Apart from *Execution Agents*, SUMA objects include *Proxies*, that control program execution and represent the clients during off-line executions, a *Scheduler*, for finding appropriate *Execution Agents* and *Proxies* upon client request, and a *User Control* for user registration and authentication. At the client side, an object called *Client Stub* provides services for obtaining classes and files from the client machine. These services are invoked by *Execution Agents* (callbacks).

Before any `execute` or `submit` method is invoked, essential SUMA objects must register themselves with the CORBA name server. Additionally, *Execution Agents* and *Proxies* must register themselves in the *Scheduler*. The sequence of method invocations for `execute` and `submit` are explained below.

`execute` When a client wants to send an execution request to SUMA, it must firstly find a *Proxy*, by invoking the `findProxy` method in *Scheduler*. The *Scheduler* finds an appropriate *Proxy* and returns a CORBA reference to the client. Then the client invokes the `execute` method in its *Proxy* passing the name of the main class as a parameter. This *Proxy* invokes user authentication methods in *User Control* and asks for a suitable *Execution Agent* in *Scheduler*, getting a CORBA reference for the *Execution Agent*. Then the *Proxy* invokes method `execute` in the selected *Execution Agent*, passing all necessary information for the *Execution Agent* to start loading classes and files from the *Client Stub*. This is done by invoking appropriate methods directly in the *Client Stub*. The *Execution Agent* sends the output files to the *Client Stub*. Finally, the *Execution Agent* executes `releaseNode` in the *Scheduler*, indicating it is available again.

`submit` If a client wants to send a `submit` request to SUMA, it invokes the `findProxy` method in *Scheduler*. The *Scheduler* finds an appropriate *Proxy* (note it must have the capacity for acting as a *virtual client*) and returns a CORBA reference to the client. Then the client invokes the `submit` method in its *Proxy*, passing the name of all necessary classes and files as parameters. The selected *Proxy* invokes user authentication methods in *User Control* and asks for a suitable *Execution Agent* in *Scheduler*, getting a CORBA reference for the *Execution Agent*. The *Proxy* gets all classes and files specified in the `submit` command from the *Client Stub*. At this moment, the client can be disconnected from SUMA. The *Proxy* invokes method `execute` in the selected *Execution Agent*, passing necessary information for the *Execution Agent* to start loading classes and files on demand from a *Virtual Client Stub*. This is done by invoking appropriate methods in the virtual client stub associated to it. Finally, the *Execution Agent* executes a `releaseNode`, indicating it is available again. The *Proxy* must have storage capacity for keeping the output files until the client (probably during a later session) invokes method `getResults` in the *Proxy*. Note that getting the results implies that a *Proxy* in the same server has to be contacted; the *Scheduler* keeps a record relating submitted jobs to *Proxy* locations.

Following sections give more detailed explanations about the methods contained in SUMA components. We specify the methods using a *pseudo-IDL*. Most of these methods handle a number of exceptions, whose descriptions are not included for the sake of clarity.

2.2 Execution Agents

The *Execution Agents* provide a single method `execute`.[1] An *Execution Agent* receives an order from a *Proxy* to execute an application, then it loads classes and files dynamically. For parallel platforms, the *Execution Agent* plays the role of the front end. `execute` method has four parameters:

```
RU execute (in EU, in ExecAgentConf, in AbstractClient,
            inout ProfilerConf)
```

Method `execute` returns a data structure (`RU` or *Results Unit*) that contains information about the execution of the program, such as the execution

[1] We are currently working on a checkpointing and recovery facility for SUMA [8], which will include a method `resume`.

node characteristics. Input data structure `EU` (*Execution Unit*) contains information related to the classes and data files needed by the application to be loaded. `ExecAgentConf` contains additional information about the execution request, such as number of nodes, whether an additional facility is required, etc. `ExecAgentConf` is a part of the `Configuration Unit` (CU), a data structure received by the *Proxy* (see Section 2.4). `AbstractClient` is the *Client Stub* reference, used by the *Execution Agent* to get classes and data files. `ProfilerConf` contains information about profiling, when profiling is requested.

2.3 Client Stub

SUMA software includes a package that allows clients to access SUMA services. This package includes a class to serve *Execution Agent* requests (callbacks) to load classes and data dynamically. An important method provided by the *Client Stub* is:

```
sequence<octet> findClass (in string classname);
```

This method returns the class requested by the application. The *Client Stub* contains methods that wrap the Java classes related to file manipulation. These classes are *File, FileDescriptor, FileInputStream, InputStream* (for *System.in*) and *OutputStream* (for both *System.out* and *System.err*). All methods defined in the *Client Stub* are indirectly called by applications running on remote nodes.

2.4 Proxy

A *Proxy* has basically two functions. First, it connects a client with an *Execution Agent* in such a way that these components communicate directly with each other. The second function of a *Proxy* is to implement the `submit` service, explained above (section 2.1). If a program terminates abnormally, the *Proxy* requests a new node and restarts execution. Main *Proxy* methods are:

```
RU execute (in EU, in CU, in AbstractClient, inout ProfilerConf)

Submit_ID submit (in EU, in CU, in AbstractClient)
RU getResults (in Submit_ID, in UsrInfo, out Results)
Status getStatus (in Submit_ID, in UsrInfo)
boolean cancel(in Submit_ID, in UsrInfo)
```

These methods are invoked by clients. The `execute` method provides interactive execution. Parameter structures `EU`, `AbstractClient`, and `ProfilerConf` are the same as in *Execution Agent* `execute` method (section 2.2). The Configuration Unit `CU` contains information about the type of platform and services requested by the user.

Methods `submit`, `getResults`, `getStatus`, and `cancel` are used for off-line execution. A `submit` instantiates a *Virtual Client Stub* at the *Proxy*, then passes its reference to an *Execution Agent*'s `execute`. It returns `Submit_ID`, which identifies the job in SUMA. `getResults` is invoked by SUMA clients and returns the results (in parameter `Results`) of a successfully terminated `submit`. It needs the job id (`Submit_ID`) and information about the user (`UsrInfo`) for authentication. `getStatus` returns the status of a submitted job (*executing, waiting, terminated*). `cancel` allows the user to dequeue a submitted job waiting for execution.

2.5 Scheduler

The *Scheduler* implements methods for *Proxy* and *Execution Agent* selection. The selection is based on application requirements and status information obtained from execution platforms. The *Scheduler* keeps information about SUMA resources, i.e., execution platforms description in terms of their type (clusters, network of workstations, workstations, PCs, etc.), relative power, memory size, available libraries, and average load. *Proxies* and *Execution Agents* register and unregister themselves with the *Scheduler*, supplying the necessary information to take scheduling decisions. Main *Scheduler* methods are:

```
boolean registerProxy (in ProxyInfo);
boolean registerNode (in NodeInfo);
boolean unregisterProxy (in ProxyInfo);
boolean unregisterNode (in NodeInfo);

bytearray findProxy (in EU);
bytearray lookupProxy (in Submit_ID);
bytearray allocateNode (in ExecAgentConf);
boolean releaseNode(in NodeID);
```

The first four methods handle registration and unregistration of *Proxies* and *Execution Agents*. `findProxy` is called by a *Client Stub*, as the first step to access SUMA services. It needs `EU` in order to select a suitable *Proxy* for handling the request (e.g., with storage support for submits). `allocateNode` is called by a *Proxy*, with information about the *Execution Agent* required (`ExecAgentConf`). `releaseNode` is called by the *Execution Agent* to notify that a job has finished and that the *Execution Agent* at node `NodeID` is ready to accept new requests. `lookupProxy` returns a reference to the *Proxy* holding the results for a submit identified by `Submit_ID`.

2.6 User Control

It is in charge of user registration and authentication. It handles capabilities which control access to resources. The principal *User Control* method is:

```
boolean checkAccess(in UserInfo, in ExecAgentConf)
```

`checkAccess` returns *true* if an execution, represented by `ExecAgentConf`, and requested by user described in `UserInfo`, can be granted; otherwise it returns *false*. *User Control* also includes methods for administration tasks, such as adding and removing users, etc.

3 Execution Environment

Previous section described part of the SUMA API, which is intended for developers of SUMA. This section explains the environment for executing applications on SUMA.

Applications that can be executed with the *java* command and only use standard input, standard output, standard error and files for I/O, can be executed on SUMA. The user can invoke the *suma* command passing the same arguments required by the *java* command, plus additional arguments, for instance, to indicate whether a *submit* is requested. A graphical client is also available for program invocation on SUMA.

Class files loaded dynamically should be located in the same directories where the local *java* command can find them. Classes from SUMA standard packages are not loaded from the client, but from local storage at the node where the *Execution Agent* is running. SUMA software includes stubs of SUMA standard packages for compilation on the client's side.

Optionally, an *Execution Agent* may offer checkpointing and profiling services, which have to be invoked explicitly by the user. An *Execution Agent* that provides checkpointing and recovery should be able to execute an extended JVM that implements checkpointing at the thread level [5,8]. The recovery process is controlled transparently by the *Proxy* in charge of the execution. An *Execution Agent* that provides the profiling service executes the application under the control of a profiler, which we call the back-end profiler, and returns a profile data structure to the client machine. This data structure has a single format, regardless of the back-end profiler used. More information on the profiling facility can be found in [10].

4 Experiments

SUMA is currently implemented in Java, using *JacORB*. As Java is an interpreted language, we were concerned about the overhead that SUMA introduces to application execution. We show some results of running applications on SUMA, using the interactive `execute` command.

Currently, SUMA is deployed on a campus-wide network. In this experiment, we use the following platforms: 3 Dual Pentium III, 600 MHz, 256 MB, with Linux 2.2.14-5.0smp and Sun's JDK 1.2.2, all on the same Ethernet; and a Sun UltraSparc 1 at 143 MHz, 64MB, with Solaris 7, and Sun's J2SE (Java 1.3), located in a different building. A single *Execution Agent* runs on one of the Pentium machines, and the rest of SUMA components (*User Control*, *Scheduler*, *Proxy*) are distributed among the two other Pentium machines. A client is run on the UltraSparc.

Applications were chosen from the "Java Grande Benchmark Suite" [18]. Table 1 compares execution times of the application running alone on a Pentium against running through SUMA. Overhead was fairly low, except for application JGFFFTBenchSizeB. The reason is that JGFFFTBenchSizeB uses a larger input file, which is dynamically loaded from the client.

Table 1. SUMA Overhead with Some Java Grande Benchmarks (Time in Seconds).

Application	Execution time (alone)	Execution time (through SUMA)	Overhead
JGFSparseMatmultBenchSizeC	457	484	6%
JGFFFTBenchSizeB	493	563	14%
JGFLUFactBenchSizeC	1242	1268	2%

5 Related Work

There are several projects oriented to remote execution of Java programs, such as Charlotte [3], Ninflet [17] and Javelin [16]. The main difference between these proposals and SUMA is that our metacomputing platform is built as a set of CORBA services. SUMA runs unmodified Java applications and parallel applications communicating through *mpiJava*. Special packages are not necessary, except for stubs of SUMA standard packages, needed for compilation. SUMA is not only a platform for remote execution of bytecode, it also provides for integrated profiling and checkpointing/recovery services.

On the CORBA side, there are several proposals for extending it in such a way that it supports some kind of parallel computing, such as PARDIS [14] and COBRA [4], which extend the standard CORBA IDL. At the moment, SUMA goals do not include support for developing scientific applications based on CORBA. However, SUMA model accepts that processes running on different *Execution Agents* communicate via CORBA.

6 Conclusions and Future Work

This paper describes the most important aspects of SUMA design in terms of CORBA services. A successful implementation of these services has been achieved. Experimental results in a campus-wide environment show that execution overhead is not considerably high, even if long input and output files, loaded from the client on demand, are used. We found that CORBA is powerful enough to implement our design. We are planning to implement SUMA across different institutions. Some issues related to fault tolerance and security must be addressed for a widespread deployment of SUMA.

Acknowledgments

We thank Eduardo Blanco, for his support on the experimental part. Eduardo Baquero, Roberto Bouza, and Luis Berbín, collaborated developing current SUMA prototype.

References

1. Philip Alpatov, Greg Baker, Carter Edwards, John Gunnels, Greg Morrow, James Overfelt, Robert van de Geijn, and Yuan-Jye J. Wu. PLAPACK: Parallel linear algebra package. In *Proceedings of the SIAM Parallel Processing Conference*, 1997.
2. Mark Baker, Bryan Carpenter, Sung Hoon Ko, and Xinying Li. mpiJava: A Java interface to MPI. In *First UK Workshop on Java for High Performance Network Computing, Europar 98*, 1998.
3. A. Baratloo, M. Karaul, Z.M. Kedem, and P. Wyckoff. Charlotte: Metacomputing on the web. *Future Generation Computer Systems*, 15(5–6):559–570, Octuber 1999.
4. P. Beaugendre, T. Priol, and C. Rene. Cobra: A CORBA-compliant Programming Environment for High-Performance Computing. Technical Report Publication interne nro. 1141, IRISA: Institut de Recherche en Informatique et Systemes Aleatoires, December 1997.
5. S. Bouchenak. Making java applications mobile or persistent. In *Proceedings of 6th USENIX Conference on Object-Oriented Technologies and Systems (COOTS'01)*, january 2001.
6. T. Brench, H. Sandhu, M. Shan, and J. Talbot. ParaWeb: towards world-wide supercomputing. In *Proceedings of the 7th ACM SIGOPS European Worshop*, 1996.
7. Gerald Brose. *JacORB - a Java Object Request Broker*, Abril 1997. `http://www.inf.fu-berlin.de/~brose/jacorb/`.
8. Y. Cardinale and E. Hernández. Checkpointing facility in a metasystem. Submitted to Euro-Par 2001, January 2001.
9. M. Curiel, Y. Cardinale, C. Figueira, and E. Hernández. Services for modeling metasystem performance using queuing network models. In *Proc. of the Communication Networks and Distributed Systems Modeling and Simulation Conference (CNDS'01)*, 2000.
10. Carlos Figueira and Emilio Hernández. Profiling facility in a metasystem. In *Proceedings of HPCN 2001*. Springer Verlag, 2001. Accepted for publishing.
11. V. Getov, S. Flynn-Hummel, and S. Mintchev. High-performance parallel programming in Java: Exploiting native libraries. In *ACM 1998 Workshop on Java for High-Performance Network Computing*, 1998.
12. E. Hernández, Y. Cardinale, C. Figueira, and A. Teruel. Suma: A scientific metacomputer. In *Proceedings of the International Conference ParCo99*. Imperial College Press, January 2000.
13. Java Grande Forum. Java Grande Forum Report: Making Java work for high-end computing. Technical Report JGF-TR-1, Java Grande Forum Panel, 1998. Available at `http://www.javagrande.org/sc98/sc98grande.pdf`.
14. K. Keahey. Pardis: Programmer-level abstractions for metacomputing. *"Future Generation Computer Systems"*, 15(5–6):637–647, Octuber 1999.
15. José Moreira. Closing the performance gap between Java and Fortran in technical computing. In *Java for High Performance Computing Workshop, Europar 98*, 1998.
16. Michael O. Neary, Bernd O. Christiansen, Peter Capello, and Klaus E. Schauser. Javelin: Parallel computing on the internet. *Future Generation Computer Systems*, 15(5–6):659–674, Octuber 1999.
17. H. Takagi, S. Matsouka, H. Nakada, S. Sekiguchi, M. Satoh, and U. Nagashima. Ninflet: a migratable parallel object framework using java. In *Proc. of the ACM 1998 Worshop on Java for High-Performance Network Computing*, 1998.
18. The Java Grande Forum Benchmark Suite. `http://www.epcc.ed.ac.uk/javagrande/`. Version 2.0.

JOINT: An Object Oriented Message Passing Interface for Parallel Programming in Java

Eduardo J.H. Yero[1,2*], Marco A. A. Henriques[1],
Javier R. Garcïa[2], and Alina C. Leyva[2]

[1] Department of Computer Engineering and Industrial Automation
School of Electrical and Computer Engineering
State University of Campinas, Sao Paulo, Brasil
[2] Faculty of Mathematics and Computing Sciences, University of Havana, Cuba

Abstract. Message-passing programming interfaces are widely used when programming parallel applications. Systems such as PVM and MPI have been successful at providing the basic capabilities needed to implement parallel applications efficiently. With the advent of Java, efforts have been conducted to define a message-passing interface to be used by applications written in that language. However, current proposals have been designed to stay as close to PVM and MPI as possible, and thus fail to exploit the capabilities offered by the Java platform. This paper introduces JOINT, a message-passing programming interface for parallel applications written in Java. JOINT is designed to be simple, intuitive and to smoothly integrate with the Java environment. The paper also presents a real parallel example implemented using JOINT to illustrate the facilities given by the interface. JOINT is already implemented as part of two different distributed parallel systems: JoiN and mJoiN.

1 Introduction

Parallel applications are normally constructed according to some paradigm. A paradigm specifies, among other things, which are the parallel components of the application, how and by whom they are created and destroyed and how they communicate and synchronize their efforts. Known paradigms for parallel applications include message passing, CSP (Communicating Sequential Processes) and object orientation.

To the present day, message passing has been the most widely accepted and used paradigm, mainly because of its simplicity and efficiency. Systems such as PVM (Parallel Virtual Machine) and implementations of the MPI (Message Passing Interface) standard are all based in this paradigm. PVM and MPI offer a powerful and flexible set of tools to implement parallel applications, either in real parallel machines or in virtual parallel machines formed by common computers connected by a network. However, the programming interfaces of those systems fail to clearly identify the entities that form the message-passing paradigm.

* This work was partially supported by grant 98/305-9 of the Fundação de Amparo à Pesquisa do Estado de São Paulo (FAPESP), Brazil.

B. Hertzberger et al. (Eds.): HPCN Europe 2001, LNCS 2110, pp. 637–644, 2001.

Tasks, for example, are represented as whole programs, and must be compiled to form a standalone application. Messages are implemented as common memory buffers that must be explicitly handled by the application programmer. The interfaces themselves are formed by a single collection of functions, with no modularization whatsoever.

Many researchers realized that these problems with the programming interfaces of PVM and MPI were not inherent to the systems, and could be solved by using better design techniques. Since the popularization of object orientation there were many proposals on how to improve those programming interfaces, such as OOMPI [SML96] and MPI++ [KH95], mainly using C++ as the programming language.

With the advent of Java^{TM}, a fully object oriented and platform independent environment, new efforts have been conducted to allow Java programs to interact with PVM and MPI. These efforts are directed either to provide Java programs with means to interact with them (JavaMPI [MG97] and MPJ [CGJ^{+}00]) or even to fully reimplement them in Java (JPVM [Fer98] and MPIJ [JCS98]).

Although these systems provide a higher level programming interface they are designed to be as close as possible to the original PVM and MPI systems. As a consequence, they still retain some of the original low-level complexities of PVM and MPI, such as the use of memory buffers to define messages and poor task manipulation, and thus fail to take full advantage of the benefits provided by the Java language.

The purpose of this paper is to present JOINT, an object oriented message-passing parallel programming interface for clusters of computers designed to take advantage of the potential of the Java platform. Although some of the guiding ideas in JOINT are inspired from both PVM and MPI, it is not the purpose of this interface to be used as a layer to interact with those systems. Instead, its main goal is to provide a fresh object oriented design guided by the principles listed below.

- Smooth integration with the Java environment.
- High level, simple programming interface.
- Flexible definition and manipulation of tasks and messages.
- Synchronous and asynchronous message handling.
- Facilities for programming data parallel applications.

The remainder of this paper is organized as follows. Section 2 presents the JOINT programming interface. Section 3 introduces an example of a real application implemented using JOINT. Section 4 explains the current implementation status of the interface. Finally, Sect. 5 presents the conclusions and future work.

2 The JOINT Programming Interface

In general terms, an application written using JOINT is not very different from a PVM application. It consists initially of one task that can later create other tasks. Each task is uniquely identified by a Task IDentifier (TID). Any pair of tasks can exchange messages, provided that at least one of them knows the TID of the other.

The JOINT interface is composed by four classes: `AppTask`, `Tasker`, `Message` and `Communicator`, which provide the basic tools for handling tasks and messages.

2.1 AppTask

The class `AppTask` is the one used to define tasks in JOINT. It provides the basic facilities needed to implement a task, such as an initialization funcion (`taskinit`), the body function (`taskrun`) and a finalization function (`taskfinally`). It also provides access to both the `Tasker` and `Communicator` objects, and to internal characteristics of the task such as its own TID.

```
public class AppTask{
    /** Returns a reference to the tasker. */
    protected final Tasker getTasker(){...}
    /** Returns a reference to the communicator. */
    protected final Communicator getCommunicator(){...}
    /** Returns the TID of the task. */
    protected final int getTID(){...}
    /** Returns the TID of the parent of the task. */
    protected final int getParentTID(){...}
    /** Used to pass initial parameters to a task.*/
    protected void taskinit(Serializable [] pars){}
    /** This is where the actual code of the task goes.*/
    protected void taskrun(){}
    /** This function is called whenever a task finishes.*/
    protected void taskfinally(){}
}
```

To define a new task it is only necessary to inherit from AppTask and overwrite the `taskinit, taskrun` and `taskfinally` functions. If any of those functions is not overwritten, the corresponding default empty implementation in AppTask is executed.

2.2 Tasker

The `Tasker` class implements the manipulation of tasks in JOINT. It provides functions to create, monitor and destroy tasks.

```
public interface Tasker{
    /** Start some application tasks. Return an array containing
        the TIDs of the newly created tasks. */
    public int[] startTasks(String className, int copies,
                            Serializable[][]pars);
    /** Kill task.*/
    public void killTask(int tid);
    /** Test if task is alive.*/
```

```
    public boolean isTaskAlive(int tid);
}
```

Although the details are implementation dependent, it is possible to assume that there will be a `Tasker` object on each of the computers forming the cluster. Conceptually, the `Tasker` may be seen as a virtual system-wide object, with representatives on every computer. To create, kill or monitor other tasks, a task must first gain access to a `Tasker` object via the `getTasker()` method in `AppTask` and then make the proper call.

2.3 Message

The `Message` class represents a message in JOINT. A message is identified by a message tag, the destination task and the originating task. The class is shown below.

```
public class Message implements Serializable{
    /** The constructor of the Message. */
    public Message(int tag){...}
    /** Returns the tag of the message.*/
    public final int getTag(){...}
    /** Returns the TID of the task that sent the message. */
    public final int fromTID(){...}
    /** Return the destination TID of the message. */
    public final int destTID(){...}
}
```

In order to define new types of messages, it is sufficient to inherit from `Message` and add the desired fields and functions. Since `Message` instances may be sent through the network, every field added to an inheritor of `Message` must implement the `Serializable` interface provided by Java. All basic types and most of the utility classes provided by Java (such as String) are serializable, but some of them are not. For example, a message may not contain an instance of the class `Thread`.

Since normally a parallel application may have a large number of different messages, JOINT introduces a generic option for the application to send any type of serializable object through the class `ObjectMessage`.

```
public class ObjectMessage extends Message{
    /**Construct a message containing object ''info''. */
    public ObjectMessage(int tag, Serializable info){...}
    /** Return a reference to ''info''. */
    public final Serializable getObject(){...}
}
```

Since arrays are objects in Java, `ObjectMessage` can be used to store an array of objects, and thus it can be used to send and receive any amount of information. The choice between the compactness of using arrays combined with `ObjectMessage` and the clarity and readability of defining new classes for every kind of message must be made by the application programmer.

2.4 Communicator

The Communicator object takes care of sending and receiving messages. Messages are sent synchronously, but can be received synchronously (with or without timeouts) and asynchronously. Methods are also provided to test if a message arrived without removing it from the message queue.

```
public interface Communicator{
    /** Sends a message to a task. */
    public void sendMessage(int tid, Message m);
    /** Replies to a message. */
    public void replyToMessage(Message received,
                               Message newMessage);
    /** Blocks until a message is received. */
    public Message receiveMessage (int sender, int tag);
    /** Receives a message. Blocks until the message is
        received or "millis" milliseconds have passed. */
    public Message receiveMessage (int sender, int tag,
                                   long millis);
    /** Tests whether a message is ready to be received.*/
    public boolean probeMessage(int sender, int tag);
    /** Receives a message if there is one available.*/
    public Message NBreceiveMessage(int sender, int tag)
}
```

Similar to Tasker, the Communicator is a system-wide object with representatives in every computer of the cluster. A task may obtain a reference to this object by making the getCommunicator() call in AppTask.

3 Example: Searching for Prime Numbers

To better illustrate the use of the JOINT interface, this section presents an application that searches for prime numbers in a specified interval. The application consists of an initial task, called PrimeSearch, that creates a number of slave tasks to conduct the search, passes a different subinterval to each one and then waits for their results. The slave tasks are implemented by the class IntervalPrimeSearch, that receives a subinterval as a parameter and performs an exhaustive iteration looking for prime numbers.

```
public class PrimeSearch  extends AppTask{
  public static final int RESULT = Message.MSG_LAST+1;
  private Interval[] split(Interval i, int size)
  {// Split  interval in subintervals of lenght<=size...}
  private void printPrimes(int[] primes)
  { // Print an array of prime numbers...}
  public void taskrun(){
    Interval interval = new Interval(IntervalData.MIN,
                                     IntervalData.MAX);
    Interval []splited = split(interval, IntervalData.SIZE);
    int taskCount = splited.length;
```

```
    // Prepare parameters for task creation
    Interval [][] pars = new Interval[taskCount][1];
    for(int i=0; i<taskCount; i++)
      pars[i][0] = splited[i];
      // Create slave tasks
      Tasker tasker = this.getTasker();
      int [] tids = tasker.startTasks("IntervalPrimeSearch",
                                       taskCount, pars);
      // Wait for response from slaves
      Communicator comm = this.getCommunicator();
      for(int i=0; i<taskCount; i++){
        try{
          Message m = comm.receiveMessage(AppTask.TID_ANY,
                                          Message.MSG_ANY);
          int [] primes = (int[])((ObjectMessage)m).getObject();
          this.printPrimes(primes);
        }catch(Exception e){
           return;
          }
      }
    }
}
public class IntervalPrimeSearch extends AppTask{
  private Interval i;
  private  boolean isPrime(int n)
  { // Determine if n is prime ...}
  public void taskinit(Object [] pars)
  { i = (Interval)(pars[0]);}
  public void taskrun()
  {   // Iterate over interval to search for primes
      int [] answer = new int[i.size()];
      int last =0;
      int parentTID = this.getParentTID();
      for(int j=i.getMin(); j<=i.getMax(); j++){
         if(isPrime(j))
            answer[last++] = j;
      }
      int [] finalAnswer;
      //Compact array and store results in finalAnswer...
      // Send result
      Communicator comm = this.getCommunicator();
      try{
        comm.sendMessage(parentTID,
                         new ObjectMessage(PrimeSearch.RESULT,
                                           finalAnswer));
      }catch(Exception e){
         return;
       }
   }
}
```

This example demonstrates how a parallel application can be implemented with JOINT. The programmer can concentrate on application specific details, since platform specific details can be easily handled through a simple and concise programming interface.

4 Implementation Status

Currently, JOINT is fully implemented in two systems: JoiN [AH99] and mJoiN [GL00]. JoiN is a system intended to perform massively parallel processing using resources scattered in the Internet. It is fully implemented in Java and offers the participants a choice between installing some code on its computer or participating through the Internet using a Java-enabled Web browser. mJoin is also fully based in Java, but aims at smaller, faster clusters of computers, such as those commonly found inside corporations and universities. Both systems use JOINT as their programming interface, although their implementations differ substantially.

JOINT has been used to implement some simple programs, such as the prime searching example shown above and a program to multiply two matrixes. It has also been used to implement more complex problems, such as a DNA Sequencing application [dOLH99] which attempts to reconstruct a large DNA sequence from a set of smaller sequences. Currently it is being used to enhance the DNA sequencing application and to implement parallel versions of known NP Hard problems such as the Traveling Salesman.

5 Conclusions and Future Work

This paper presented JOINT, an object oriented message-passing interface whose main goal is to simplify the work of writing message-passing parallel applications in Java by fully exploiting the potentials of the language. The paper presented the motivations for developing such an interface as well as the main classes forming it. An application implemented using the interface was presented as an example. The JOINT interface has already been implemented in two different systems, and has proven to be a valuable tool for developing parallel applications.

Future work includes studying the possible inclusion of group communication in the interface. Particular attention should be paid to group semantics, since the JOINT interface will be used in a variety of environments where maintaining strict semantics may be difficult. It is also under consideration the inclusion of an asynchronous notification mechanism to complement message passing. This mechanism could be used, for example, to notify a parent task that some child task has failed, and thus help the programmer to write fault tolerant code.

References

[AH99] Marco Aurélio Amaral Henriques. A proposal for java based massively parallel processing on the web. In *Proceedings of The First Annual Workshop on Java for High-Performance Computing, ACM International Conference on Supercomputing*, pages 59–66, Rhodes, Greece, June 1999.

[CGJ+00] Brian Carpenter, Vladimir Getov, Glenn Judd, Anthony Skjellum, and Geoffrey Fox. Mpj: Mpi-like message passing for java. *Concurrency: Practice and Experience*, 12(11):1019–1038, 2000.

[dOLH99] Fabiano de Oliveira Lucchese and Marco Aurelio Amaral Henriques. Aplicação de um computador massivamente paralelo virtual no seqenciamento de cadeias de DNA. Technical Report DCA-RT 02/99, FEEC/UNICAMP, September 1999.

[Fer98] Adam J. Ferrari. JPVM: Network Parallel Computing in Java. In *Proceedings of the ACM 1998 Workshop on Java for High-Performance Network Computing*, Standford University, Palo Alto, California, February 1998.

[GL00] Javier Ramón Garcïa and Alina Castellanos Leyva. mjoin, una màquina paralela virtual distribuida en redes. In *Primer Congreso Internacional de Telematica, CITEL 2000*, Habana, Cuba, 2000.

[JCS98] Glenn Judd, Mark Clement, and Quinn Snell. DOGMA: Distributed Object Group Metacomputing Architecture. *Concurrency Practice and Experience*, 10(11-13):977–983, Sep-Nov 1998.

[KH95] Dennis Kafura and Liya Huang. mpi++: A C++ language binding for MPI. In *Proceedings MPI Developers Conference*, South Bend, Indiana, USA, June 1995. University of Notre Dame.

[MG97] Sava Mintchev and Vladimir Getov. Towards portable message passing in java: Binding mpi. In M. Bubak, J. Dongarra, and J. Wasniewski, editors, *Recent Advances in PVM and MPI. Lecture Notes in Computer Science*, volume 1332. Springer Verlag, 1997.

[SML96] Jeffrey M. Sqyres, Brian C. McCandless, and Andrew Lumsdaine. Object Oriented MPI. A Class Library for the Message Passing Interface. In *Proceedings of the 1996 Parallel Object-Oriented Methods and Applications Conference*, Santa Fe, New Mexico, February 1996.

A Framework for Opportunistic Cluster Computing Using JavaSpaces[1)]

Jyoti Batheja and Manish Parashar

Electrical and Computer Engineering, Rutgers University
94 Brett Road, Piscataway, NJ 08854
{jbatheja,parashar}@caip.rutgers.edu

Abstract. Heterogeneous networked clusters are being increasingly used as platforms for resource-intensive parallel and distributed applications. The fundamental underlying idea is to provide large amounts of processing capacity over extended periods of time by harnessing the idle and available resources on the network in an *opportunistic* manner. In this paper we present the design, implementation and evaluation of a framework that uses JavaSpaces to support this type of opportunistic adaptive parallel/distributed computing over networked clusters in a non-intrusive manner. The framework targets applications exhibiting coarse-grained parallelism and has three key features: (1) portability across heterogeneous platforms, (2) minimal configuration overheads for participating nodes, and (3) automated system state monitoring (using SNMP) to ensure non-intrusive behavior. Experimental results presented in this paper demonstrate that for applications exhibiting coarse grained parallelism, the opportunistic parallel computing framework can provide performance gains. Furthermore, the results indicate that monitoring and reacting to current system state minimizes intrusiveness.

1 Introduction

This paper presents the design, implementation and evaluation of a framework that uses JavaSpaces [1] to aggregate networked computing resources, and non-intrusively exploits idle resources for parallel/distributed computing. Traditional High Performance Computing (HPC) is based on massively parallel processors, supercomputers or high-end workstation clusters connected by high-speed networks. These resources are relatively expensive, and are dedicated to specialized parallel and distributed applications. Exploiting available idle resources in a networked system can provide a more cost effective alternative for certain applications. However, there are a number of challenges that must be addressed before such opportunistic adaptive cluster computing can be a truly viable option. These include: 1) **Heterogeneity:** Cluster environments are typically heterogeneous in the type of resources, the configurations and capabilities of these resources, and the available software, services and tools on the

[1] The research presented in this paper is based upon work supported by the National Science Foundation under Grant Number ACI 9984357 (CAREERS) awarded to Manish Parashar.

B. Hertzberger et al. (Eds.) : HPCN Europe 2001, LNCS 2110, pp. 647-656, 2001.

systems. This heterogeneity must be hidden from the application and addressed in the seamless manner, so that the application can uniformly exploit available parallelism. 2) **Intrusiveness:** Inclusion of the framework must minimize modifications to any existing legacy code or standard practices. Furthermore, a local user should not be able to perceive that local resources are being stolen for foreign computations. 3) **System configuration and management overhead:** Incorporating a new resource into the cycle stealing resource cluster may require system configuration and software installation. These modifications and overheads must be minimized so that the cluster can be expanded on the fly to utilize all available resources. 4) **Adaptability to system and network dynamics:** The availability and state of system and network resources in a cluster can be unpredictable and highly dynamic. These dynamics must be handled to ensure reliable application execution. 5) **Security and privacy:** Secure and safe access to resources in the cluster must be guaranteed so as to provide assurance to the users making their systems available for external computations. Policies must be defined and enforced to ensure that external application tasks adhere to the limits and restrictions set on resource/data access and utilization.

Recent advances in opportunistic cluster computing have followed two approaches, *Job level parallelism* and *Adaptive parallelism*[2]. In the job level parallelism approach, entire application jobs are allocated to available idle resources for computation, and are migrated across resources as resources become unavailable. The Condor [10] system supports cluster-based job level parallelism. In the adaptive parallelism approach, the available processors are treated as part of a dynamic resource pool. Each processor in the pool aggressively competes for application tasks. This approach targets applications that can be decomposed into independent tasks. Adaptive computing techniques can be *cluster based* or *web based*. *Cluster based* systems exploit available resources within a local networked cluster. *Web based* approach extends this model to resources over the Internet. Systems supporting adaptive parallelism include cluster-based systems such Piranha [11][12], Atlas [6], and ObjectSpace/Anaconda [9], and web-based systems such as Charlotte [4][5], Javelin [8], and ParaWeb [7].

This paper presents the design, implementation and evaluation of a framework for adaptive and opportunistic cluster computing based on JavaSpaces that address the issues outlined above. The framework has three key features: (1) portability across heterogeneous platforms, (2) minimal configuration overheads and runtime class loading at participating nodes, and (3) automated system state monitoring (using SNMP [2][3]) to ensure non-intrusive behavior.

The rest of this paper is organized as follows. Section 2 describes the architecture and operation of the proposed framework. Section 3 presents an experimental evaluation of the framework. Section 4 presents our conclusions and outlines current and future work.

---

[2] To best of our knowledge, the term 'adaptive parallelism' was coined by the Piranha project [12].

2 A Framework for Opportunistic Parallel Computing on Clusters

The framework presented in this paper employs JavaSpaces to facilitate master-worker parallel computing on networked clusters. JavaSpaces is a Java implementation of a tuple-space system [13], and is provided as a Jini service [18]. JavaSpaces technology provides a programming model that views applications as a collection of processes cooperating via the flow of objects into and out of one or more spaces. A space is a shared, network accessible repository for objects [14]. In the presented framework, parallel workload is distributed across the worker nodes using the bag of task model with the master producing independent application tasks into the space, and the worker consuming these tasks and computing on them. Results are returned to the space. This model offers two key advantages. (1) The model is naturally load-balanced. Load distribution in this model is worker driver. As longs as there work to be done, and the worker is available to do work, it can keep busy. (2) The model is naturally scalable. Since the tasks are relatively independent, as longs as there are a sufficient number of task, adding workers improves performance.

The framework and underlying parallel computing model supports applications that are sufficiently complex and require parallel computing, that are divisible into relatively coarse-grained subtasks that can be solved independently, and where the subtasks have small input/output sizes.

2.1 Framework Architecture

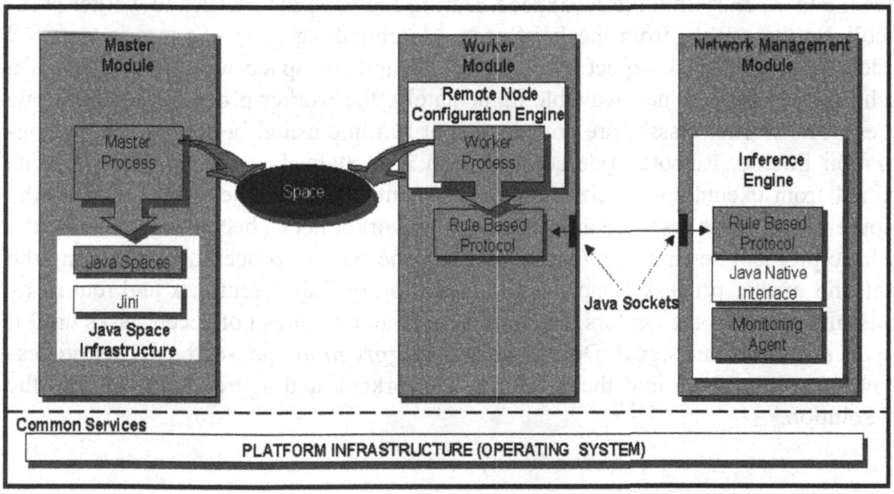

Fig. 1. Framework Architecture.

A schematic overview of the framework architecture is shown in Fig. 1. It consists of 3 key components: the Client-side (Master) components, the Server-side (Worker) components and the Network Management Module.

Master Module: The Master component defines the problem domain for a given application. The application domain is broken down into sub tasks that are JavaSpace enabled.[3] The master also contains the JavaSpace and registers it as a Jini service. It relies on Jini for remote lookup during the discovery phase. The JavaSpace is used to handles all communication issues.

Worker Module: The worker component provides the solution content for the application domain. In an effort to minimize the overheads of deploying worker code, we have implemented a remote node configuration mechanism that facilitates remote loading of the worker implementation classes at runtime.

Network Management Module: In order to exploit idle resources while maintaining non-intrusiveness at the remote nodes, it is critical that the framework monitors the state of the worker nodes, and uses this state information to drive the scheduling of tasks on workers. The *Network Management Module* performs this task. It monitors the state of registered workers and uses defined policies to decide on the workers availability. The policies are maintained by the Inference Engine component and enforced using the Rule Base Protocol.

2.2 Implementation and Operation

The framework implements the master-worker pattern with JavaSpaces as the backbone. The overall operation of the framework consists of three potentially overlapping phases, viz. task-planning, compute, and result-aggregation. During the *task-planning phase*, the master process first decomposes the application problem into sub tasks. It then iterates through the application tasks, creates a task entry for each task, and writes the tasks entry into the JavaSpace. During the *compute phase,* the worker process collects these tasks from the JavaSpace. Matchmaking in JavaSpaces is achieved by identifying each task object by a unique ID and the space where it resides. If a matching task object is not available immediately, the worker process waits until one arrives. The worker classes are downloaded at runtime using the Remote Node Configuration Engine. Remote node configuration is explained in section 2.2.1. Results obtained from executing the computations are put back into the space. During the compute-phase, if the resource utilization on the worker nodes becomes intolerable the rule base protocol sends a stop/pause signal to the worker process. On receiving the signal, the worker process completes the execution of the current task and returns its results into space. It then enters the stop/pause state and does not accept tasks until it receives a start/resume signal. During the *result aggregation* phase, the master process removes results written into the space by the workers, and aggregates them into the final solution.

---

[3] JavaSpace required the Objects being passed across the Space to be in a Serializable format. In order to transfer an entry to or from a remote space, the proxy to the remote space implementation first serializes the fields and then transmits it into the space.

2.2.1 Remote Node Configuration

The required classes for remote configuration of the worker nodes are easily downloadable from the web server residing at the master in the form of executable jar files. The application implementation classes are loaded at runtime from within the configuration classes, and the appropriate method to start the worker application thread is invoked. Our modification of the network launcher [15] provides mechanisms to intercept calls from the inference engine (the network management module) and interpret them as signals to the executing worker code. This interaction is used to enable the worker to react to system state as explained below.

2.2.2 Rule Base Protocol

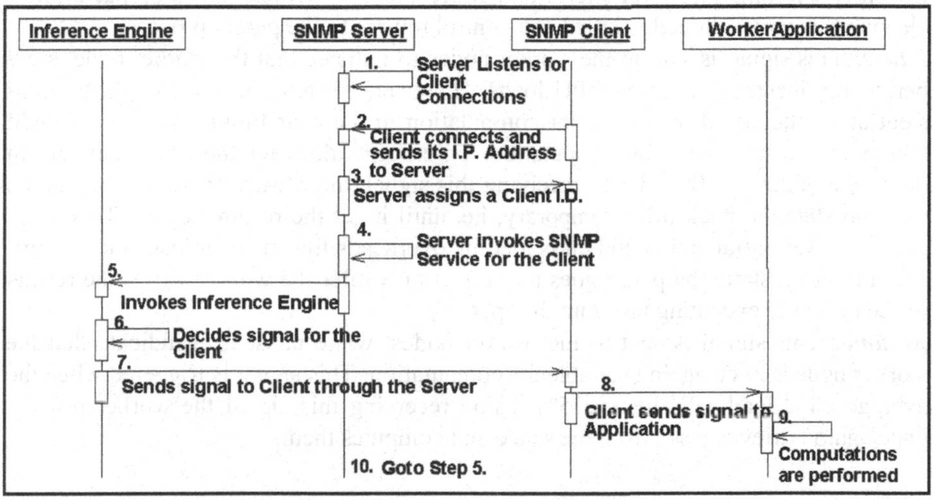

Fig. 2. Sequence Diagram for the Rule Base Protocol.

The rule base protocol defines the interaction between the network management module and the worker module (see Fig. 2) to enable the worker to react to changes in its system state. It operation is as follows:

The SNMP client, which is part of the worker module, initiates the workers participation in the parallel computation by registering with the SNMP server at the network management module. The inference engine, also at the network management module, maintains a list of registered workers. It assigns a unique ID to the new worker and adds its IP address to the list. The SNMP server then continues to monitor the state of workers in its list.

The SNMP parameter monitored is the average worker CPU utilization. As these values are returned they are added to the respective entry in the server list. Based on this return value and programmed threshold ranges, the inference engine makes a decision on the workers current availability status and passes an appropriate signal back to the worker. Threshold values are based on heuristics. The rule base currently defines 4 types of signals in response to the varying load conditions at a worker; viz. *start, stop, pause* and *resume.*

Start: This signal is sent to the worker nodes to signify that the worker node is now idle and can start the parallel processing job. The threshold average CPU load values for this state are in the range of 0% - 25%. On receiving this signal, the worker initiates a new runtime process that loads the application worker classes and starts work execution.

Stop: This signal is sent to the worker nodes to indicate that the worker node can no longer be used for computations. This may be due to a sustained increase in CPU load caused by a higher priority (possibly interactive) job being executed. The cutoff threshold value for the stop state is an average CPU utilization greater than 50%. On receiving the stop signal, the executing worker thread is interrupted and shutdown/cleanup mechanisms are initiated. The shutdown mechanism ensures that the currently executing task completes and its results are written into the space. After cleanup the worker thread is killed and control returns to the parent process.

Pause: This signal is sent to the worker nodes to indicate that the worker node is experiencing increased average CPU loads. However, the load increase might be transient and node could be reused for computation in the near future. Hence it should temporarily not be used for computation. Threshold values for the pause state are in the range of 25% - 50%. Upon receiving this signal, the worker backs off, but unlike the stop state the back off is temporary, i.e. until it get the resume signal. This minimizes worker initialization and class loading overheads for transient load fluctuations. As in the stop state, the pause goes into effect only after the worker writes the results of the currently executing task into the space.

Resume: This signal is sent to the worker nodes, while paused, to indicate that the worker node is once again available for computation. This signal is triggered when the average CPU load falls below 25%. Upon receiving this signal the worker process once again retrieves tasks from the space and computes them.

3 Framework Evaluation

We evaluated the JavaSpaces-based opportunistic cluster-computing framework with a real world financial application that uses Monte Carlo (MC) simulation for Option Pricing. An option is a derivative, that is, its pricing value is derived from something else. Complications such as varying interest rates and complex contingencies can prohibit analytical computation of options and other derivative prices. Monte Carlo (MC) simulation [17] using statistical properties of assumed random sequences is an established tool for pricing derivative securities. An option is defined by the underlying security, the option type (call or put), the strike price, interest rate, volatality and the expiration date. These financial terms are explained in greater depth in [16]. The main MC simulation based on the input parameters is the core parallel computation in our experiments. Input parameters may be defined using a GUI as provided in our implementation. The simulation domain is divided into tasks of size 100 each and MC simulations are performed in parallel on these tasks. High and low pricing estimates are obtained over a wide range of simulations.

3.1 Scalability Analysis

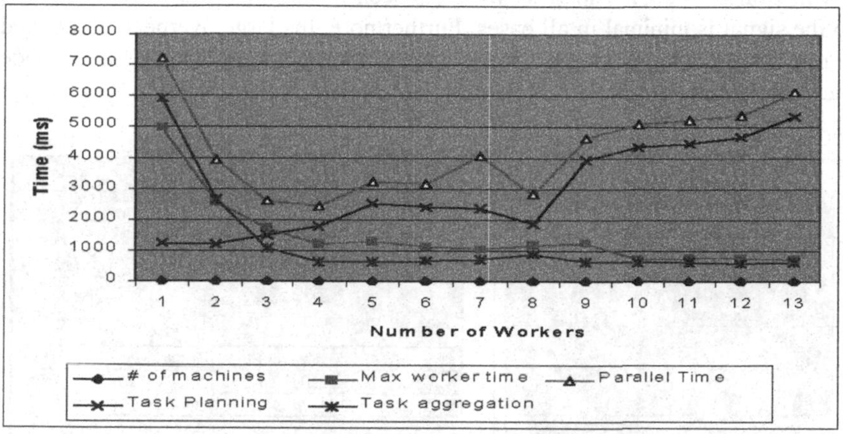

Fig. 3. Application Scalability.

This experiment measures the overall scalability of the application and the framework. Results for this experiment are plotted in Fig. 3. As shown in the figure an initial speedup is obtained as the number of workers is increased. During this part of the curve the total parallel time closely follows the maximum worker time. As the number of workers increases the model spreads the total tasks more evenly across the available workers. Hence the maximum (Max) worker time evens out as the number of workers increases. However, after a point we notice that the total parallel time is dominated by the Task Planning time. That is the workers are able to complete the assigned task and return it to the space much before the master gets a chance to plan a new task and put it into the space. Hence the workers remain starved until the task is made available. As a result the scalability deteriorates. This indicates that the framework favors coarse-grained tasks that are compute intensive. As expected the task aggregation curve closely follows the maximum worker time.

3.2 Adaptation Protocol Analysis

In this experiment, we provide a time analysis to illustrate the overhead involved in signaling worker nodes and adapting to their current CPU load. As a part of the experimental setup, we built two sets of load simulators: load simulator 1 was designed to raise the CPU usage level on the worker to 30% to 50% utilization. The second load simulator (load simulator 2) raised the CPU utilization of the worker machines to 100%. Fig. 4(a) and Fig. 4(b) depict the worker behavior under the simulation conditions. Fig. 4(a) captures the CPU usage history on the worker host throughout the run. We identify the peaks where the worker reacts to the signals sent. The first peak at 80% CPU usage occurs when the worker is started. This sudden load increase is attributed to the remote loading of application classes at the worker. Next, load simulator 2 is started which sends the CPU usage to 100%. This causes a Stop signal to be sent to the worker node. The load simulator 2 is then stopped and load simulator 1 is started which raises the CPU load to 46%. As seen in Fig. 4(b) the worker reaction times to the signal is minimal in all cases. Furthermore, the large overhead associated with remote class loading is avoided in the case of transient load increase at the node using the pause/resume states.

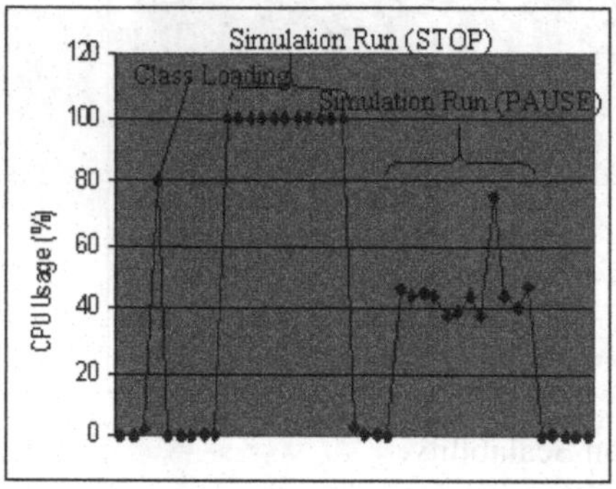

Fig. 4(a). Worker CPU Usage.

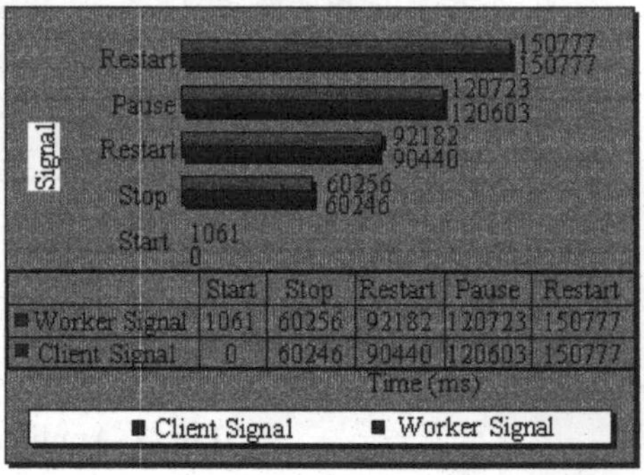

Fig. 4(b). Worker Reaction Times.

Simulation Signal Triggers: Start - Stop - Restart - Pause - Resume

3.3 Dynamic Behavior Patterns under Varying Load Conditions

This experiment consists of three runs: In the first run none of the workers were loaded. In the second and third runs, the load simulator 2 was run to simulate high CPU loads on 3 and 6 workers respectively. As seen in Fig. 5(a), as the number of worker hosts being loaded increases, the total parallel computation time increases. The computational tasks that would have been executed normally at a worker are now off loaded and picked up by other executing workers. The task planning and aggregation times also increase since the master has to wait for the worker with the maximum number of tasks to return its results back into the space. The maximum master overhead and the maximum worker time remains the same across all three runs as expected. Fig. 5(b) illustrates how task scheduling adapts to load current load conditions. It shows that the number of task executed by each worker depends on its current load. Loaded workers execute fewer tasks causing the available workers to execute larger number of tasks.

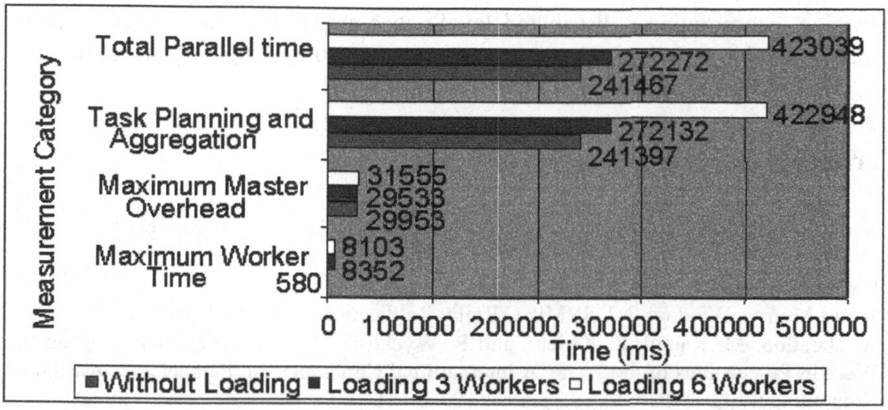

Fig. 5(a). Execution Time Measurements for 12 Workers.

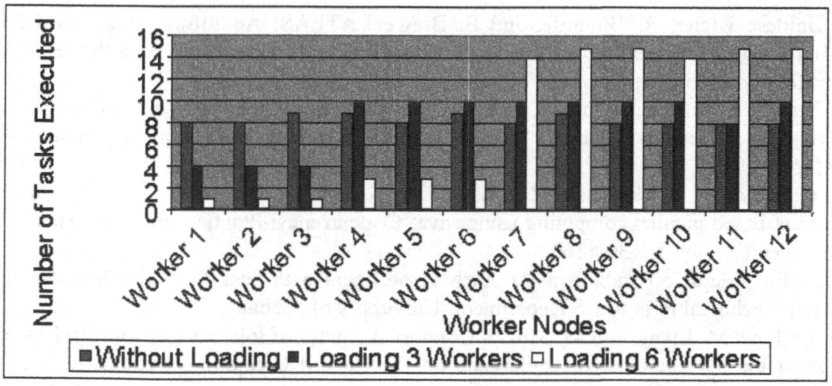

Fig. 5(b). Number of Tasks Executed per Worker for 12 Workers.

4 Conclusions

This paper presented the design, implementation and evaluation of a framework for opportunistic parallel computing on networked clusters using JavaSpace. It provides support for global deployment of application code and remote configuration management of worker nodes, and uses an SNMP system state monitor to ensure non-intrusiveness. The experimental evaluation, using an option pricing application, shows that the framework provides good scalability for coarse-grained tasks. Furthermore, using the system state monitor and triggering heuristics the framework can support adaptive parallelism and minimize intrusiveness. The results also show that the signaling times between the worker and network management modules and the overheads for adaptation to cluster dynamics is insignificant. We are currently investigating ways to reduce the overheads during task planning and allocation phases. As future work, we envision incorporating a distributed JavaSpaces model to avoid a single point of resource contention or failure. The Jini community is also investigating this area.

References

1. Sun Microsystems. Javaspaces, http://www.javasoft.com/products/javaspaces/specs/ (1998).
2. SNMP Documentation, http://www.snmpinfo.com.
3. James Murray, Windows NT SNMP, O'Reilly Publications (January 1998) .
4. A. Baratloo, M. Karaul, Z. Kedem, and P. Wyckoff. Charlotte: Metacomputing on the Web. In Proceedings of the 9th ISCA International Conference on Parallel and Distributed Computing Systems (PDCS), (September 1996) 181-188.
5. A. Baratloo, M. Karaul, H. Karl, Zvi M. Kedem. An Infrastructure for Network Computing with Java Applets. Concurrency: Practice and Experience, Vol. 10, (September 1998) 1029-1041.
6. J. Baldeschwieler, R. Blumofe, and E. Brewer. ATLAS: An Infrastructure for Global Computing. In Proceedings of the 7th ACM SIGOPS European Workshop: Systems support for Worldwide Applications (September 1996) 165-172.
7. T. Brecht, H. Sandhu, J. Talbott, and M. Shan. ParaWeb: Towards world-wide supercomputing. In Proceedings of the 7th ACM SIGOPS European Workshop (September 1996) 181-188.
8. B. Christiansen, P. Cappello, M. Ionescu, M. Neary, K. Schauser, and D. Wu. Javelin: Internet-based parallel computing using Java. Concurrency: Practice and Experience, Vol. 9 (November 1997) 1139-1160.
9. P. Ledru. Adaptive Parallelism: An Early Experiment with Java™ Remote Method Invocation. Technical Report, CS Department, University of Alabama (1997) .
10. M. Lizkow, M. Livny, and M. Mukta. Condor: A hunter of idle workstations. In Proceedings of the 8th International conference on Distributed Computing Systems (June1998) 104-111.
11. N. Carriero, E. Freeman, D. Gelernter, and D. Kaminsky. Adaptive parallelism and Piranha. IEEE Computer, Vol. 28, No. 1 (January 1995) 40-49.
12. D. Gelernter and D. Kaminsky. Supercomputing out of recycled garbage: Preliminary Experience with Piranha. In Proceedings of the 6th ACM International Conference on Supercomputing (July 1992) 417-427.
13. The Linda Group. http://www.cs.yale.edu/HTML/YALE/CS/Linda/linda.html.
14. E. Freeman, S. Hupfer, K. Arnold. JavaSpaces Principles, Patterns, and Practice. Addison Wesley (June 1999) .
15. M. Noble. Tonic: A Java TupleSpaces Benchmark Project. http://hea-www.harvard.edu/~mnoble/tonic/doc/.
16. Glossary of Financial terms http://www.centrex.com/terms.html.
17. Broadie and Glasserman MC algorithm for Option Pricing http://www.puc-rio.br/marco.ind/monte-carlo.html.
18. W. Keith Edwards. Core Jini, Addison Wesley (October 2000).

Scientific Computation with JavaSpaces

Michael S. Noble[1] and Stoyanka Zlateva[2]

[1] Harvard-Smithsonian Center for Astrophysics
[2] Boston University

Abstract. JavaSpaces provides a simple yet expressive mechanism for distributed computing with commodity technology. We discuss the suitability of JavaSpaces for implementing different classes of concurrent computations based on low-level metrics (null messaging and array I/O), and present performance results for several parametric algorithms. We found that although inefficient for communication intensive problems, JavaSpaces yields good speedups for parametric experiments, relative to both sequential Java and C. We also outline a dynamic native compilation technique, which for short, compute-intensive codes further boosts performance without compromising Java portability *or* extensive algorithm recoding. Discussion and empirical results are presented in the context of our public benchmark suite.

1 Introduction

To date much of the study of Java for high performance computation can be grouped into one or more of a few broad categories: lifting language constraints [2], optimizing communication [6,8,9], extending numerical capability [3], and various forms of native compilation [7,8,9]. These and related contributions demonstrate clearly that Java platform can be used for numerical work, but by and large they retain a research orientation in that performance gains are obtained through highly specialized means (e.g. custom JVMs and compilers, pre-compiled native binaries). Notwithstanding the increasing need for computational power, the complexity of parallel and distributed programming tools and the related costs for deployment and administration still hinders their broader adoption in the scientific community.

This paper attempts to bridge the gap between high-performance Java research and its utilization beyond the research community, by investigating the use of commodity tuplespace implementations for high performance computation. Since their introduction by Gelernter in the 1980's, tuplespaces have been thoroughly described in the literature and exist in a wide variety of implementations [5,1,11]. The central idea is that of *generative communication*, whereby processes do not communicate directly, but by adding and removing tuples (ordered collections of data and/or code) to and from a *space*, (process-independent storage with shared-memory-like characteristics, accessed by associative lookup). This communication model is defined by adding only a handful of primitives

B. Hertzberger et al. (Eds.): HPCN Europe 2001, LNCS 2110, pp. 657–666, 2001.

(write, read, take, eval) to a base language, and gives rise to distinctive properties such as time-uncoupled anonymous communication (sender and receiver need not know each other's identity, nor need they exist simultaneously), and networked variable sharing (tuples may be viewed as distributed semaphores). The simplicity and clean semantics of tuplespaces allow natural expressions of problems awkward or difficult to parallelize in other models [5]. We focus on JavaSpaces [1] as a general purpose implementation, but have also investigated TSpaces [11] and will contrast the two briefly in our I/O metrics discussion.

2 Low-Level Metrics

Experiments were conducted on the machines described in Table 1. Communication between the first 11 machines required 1 hop across a 100Mbps ethernet LAN. Packets between this subnet and the 32-node BU SGI Origin2000 traveled 4 hops across the 100Mbps campus network, while packets between BU and the final two Harvard CFA machines traveled 9 hops over the local internet segment. The space was hosted on the 296Mhz server because it had the most available memory to offer the 4 concurrent JVMs used by JavaSpaces, combined with the least user interaction. The order in which the client machines are listed reflects how they were used in the computations: A single 440 Mhz Solaris workstation was used for sequential Java and C tests, while the second 440Mhz host was added for N=2 workers, the Win98 host for N=3, and so forth. We used the intermediate-speed hosts for sequential testing to reflect the typical network containing machines of varying speed, and expressly avoided the use of dedicated hardware. Benchmark numbers are peak times culled from successive invocations, using JavaSpaces 1.1 and JDK1.3 with native threads, and GCC 2.91.6x with -O3 optimization.

Table 1. Summary of Testbed Cluster.

Platform	Processors	Memory	Java	No. of Hosts
Sol 2.6	296Mhz	256 Mb	1.3	1
Sol 2.7	440 Mhz	128 Mb	1.3	2
Windows98	750 Mhz	128 Mb	1.3	1
x86 Linux	666 Mhz	128 Mb	1.3	5
Sol 2.6	333 Mhz	256 Mb	1.3	2
SGI O2k	32 195 Mhz	8192 Mb	1.2.2	1
Sol 2.8	2 450 Mhz	2048 Mb	1.3 beta	1
Sol 2.6	2 400 Mhz	2048 Mb	1.3 beta	1

The canonical NULL message benchmark cannot be directly implemented within JavaSpaces because the Java null value is used as a wildcard for matching fields *within* a tuple, and is thus prohibited from residing in the space *as* a tuple. Instead, the best we can do is create an empty NullEntry subclass of the Jini/JavaSpaces Entry interface. To avoid excess serialization a snapshot template was used for *both* write() and take() operations. This means that tuples were *fully* serialized only once per series of N write()s or take()s, with the client-side space proxy essentially informing the space on the 2nd through Nth calls that the new tuple duplicates an existing one in the space. For retrieval the benchmark tests two approaches: a) result = space.take(null,...,...) and b) result = (NullEntry)space.take(snapshot,...,...). The first method shows that, in contrast with writing, tuples may be retrieved by matching on a null object. The first lines of each entry in Table 2 indicate that this is the fastest technique, arguably because the space manager can avoid costly field matching operations and simply return all available tuples. We suggest the second approach demonstrates the cost of field matching in the space and object casting in the client (necessary because tuples are returned as Entry objects, not subclass instances) and that for small objects it can lengthen retrieval by as much as 35%. It is

Table 2. Peak Null Tuple Rates.

nobj	writes/sec	sec/w	takes/sec	sec/t
Between 296Mhz space server, 440Mhz client				
100	161.0306	0.0062	205.3388	0.0049
100	167.2241	0.0060	163.6661	0.0061
1000	175.9944	0.0057	246.2448	0.0041
1000	176.0253	0.0057	182.0167	0.0055
5000	171.9631	0.0058	248.9792	0.0040
5000	175.4755	0.0057	180.4077	0.0055
Between 296Mhz space server, Win98 client				
5000	194.4769	0.0051	277.4695	0.0036
5000	194.0994	0.0052	211.2379	0.0047
Dual-450Mhz Sparc Ultra80, hosting space and client				
5000	279.9866	0.0036	381.3883	0.0026
5000	280.0022	0.0036	283.0936	0.0035

interesting to note that our best LAN round-trip time exceeds 8 ms, more than triple the 2.6 ms latency reported 15 years ago [4] with Linda. We are aware that this comparison is not completely apt: our testbed is a loosely coupled cluster, while the Linda result was generated on a tightly-coupled bus-based S/Net multicomputer; JavaSpaces uses a centralized space, while Linda was distributed; efficient communication is achieved in Linda with kernel and compiler optimizations, while these are not an option in JavaSpaces; Linda takes advantage of the connectionless UDP protocol, while JavaSpaces, due to current Jini/RMI implementations, is fettered to the heavier, connection-oriented TCP. Despite these differences, however, neither should the comparison be blithely dismissed. The 296Mhz and 750Mhz hosts which yielded our peak ethernet result are 30x-90x faster than the 8Mhz M68000 S/Net nodes, and the peak S/Net bus bandwidth is quoted at only 80% of Fast Ethernet. Even when both space and client were the only live user processes on the dual-450Mhz CPU host (outfitted with a crossbar switch reported to yield 1.8 Gb/sec peak memory bandwidth) — completely avoiding ethernet — null latency still more than doubled that of Linda. The fact that the fastest client on the subnet produced the greatest throughput suggests higher I/O rates may be achieved by hosting the space on a faster server. Using the NetPerf tool we determined that for 1 byte messages the peak TCP send/receive rate on our LAN was 4538 transactions/sec, 40x greater than the null tuple throughput. To estimate the null tuple size a Sizeof class was written, which instantiates an object and initiates a mock serialization to count the number of bytes that would be wire-transferred. We found that NullEntry classes and objects serialize to roughly 37 and 300 bytes, respectively, despite NullEntry being an empty subclass of an empty interface. Only by sending messages 2 orders of magnitude greater in size (32768 bytes) were we able to obtain a TCP request/response rate in the same neighborhood as our peak tuple I/O rate. Similar tests were conducted for array tuples of various sizes, with the re-

Table 3. 100 Double Array Tuples, with Null Template for Retrieval.

#elem	writes/sec	sec/w	Kbytes/sec	takes/sec	sec/t
Between 296Mhz server, Win98 client, Fast Ethernet					
10	151.51	0.0066	82.12	204.08	0.0049
10	151.51	0.0066	76.35	227.27	0.0044
100	140.85	0.0071	175.37	200.0	0.0050
100	151.52	0.0066	129.62	227.27	0.0044
1K	113.64	0.0088	940.49	166.67	0.0060
1K	129.87	0.0077	567.69	181.82	0.0055
5K	44.44	0.0225	1756.73	90.91	0.011
5K	69.93	0.0143	1398.33	121.95	0.0082
localhost communication on multiple-450Mhz CPU Ultra 80					
100	218.8184	0.0046	273.0956	290.6977	0.0034
100	221.2389	0.0045	189.9112	305.8104	0.0033
1K	155.521	0.0064	1287.6045	241.5459	0.0041
1K	183.4862	0.0055	802.5731	268.0965	0.0037
5K	65.4022	0.0153	2585.3039	135.8696	0.0074
5K	100.9082	0.0099	2018.0649	198.8072	0.0050
20K	20.9249	0.0478	3279.2801	50.9424	0.0196
20K	36.8596	0.0271	2896.8954	89.6861	0.0112
50K	8.6222	0.116	3372.0624	20.9512	0.0477
50K	15.2022	0.0658	2976.2887	41.6493	0.024

sults shown in Table 3. The second line of each entry represents the cost of transmitting *strided* arrays, which are employed within algorithms where only portions of an array need be communicated (e.g. Red-Black SOR, where nodes exchange stripes of alternating elements). Our benchmarks indicate that the savings of communicating smaller arrays is worth the extra duplicative work in the sender: while byte throughput is lower, tuple throughput is observably higher. This benchmark also suggests that, as an alternative to simple iteration, it would be useful to overload the Java System.arraycopy() method to support striding.

While most of our effort focused on JavaSpaces, we did perform a cursory review of TSpaces from IBM. TSpaces was easier to configure but added more overhead (a null tuple was ca. 377 bytes), which lead to lower individual I/O rates: 141.5 writes/sec for 1000 null tuples and 76.6 writes/sec for 100 array tuples. The TSpaces *multiWrite()*, and *consumingScan()* batch operators, however, yielded significantly higher throughput: 2347.43 writes/sec for 1000 null tuples and 1818.18 writes/sec for 100 array tuples of 1000 double elements.

3 Parallel Algorithms

Carriero and Gelernter provide in [5] a Linda master/slave prime number generator, and indicate that for a problem size of 216K (all primes below 3e6) a speedup of 5.6 was achieved on 8 IBM RT nodes loosely connected over 10Mbps ethernet. Workers search for primes within GRAIN-sized chunks of consecutive integers. For each chunk a worker writes a new batch of primes to the space, which are retrieved en masse, stored in the master prime list, then *serially rewritten* back to the space (each tuple being a distinct distributed array element) so that each worker may update its local table of primes. It is this series of *out()* calls, equivalent to JavaSpaces write(), which effectively prohibits an identical JavaSpaces implementation. Firstly, a JavaSpaces tuple adds a minimum of ca. 300 bytes to the transmission of each prime, or 75 parts overhead for *each* part data (4-byte int). Further, with peak write() performance of 280 operations/sec, a minimum of 216K/280 = 771 sec would be required to populate a distributed JavaSpaces array of equal size. Lastly, out-of-memory exceptions consistently occurred after writing roughly 190K null tuples. Launching the space JVM with -Xmx192m confirmed that 12+ minutes are needed to write 216K tuples when both space/client are hosted on the dual-CPU Ultra80, with ethernet execution taking quite a bit longer (ca. 20 minutes). In contrast, a sequential C invocation of the given prime-finder needed only 11.8 seconds (440 Mhz Solaris host), and a Java BitSet implementation of the Sieve of Erastosthenes only 0.91 seconds, to find the same 216K primes.

Iterative grid solvers provide another example where the poor communication performance of JavaSpaces prevents efficient solutions. In this class of problems the domain is discretized along each dimension in N intervals, yielding a multidimensional mesh or grid, and an approximation is computed at each mesh point. Efficient implementations of similar algorithms have been reported in Java [8] as well as Linda [4,5]. As an example let us consider the 2-dimensional *Laplace equation* approximated with a finite difference over a four-neighborhood

$$\nabla^2 u \approx \frac{1}{h^2} u(x+h,y) + u(x-h,y) + \quad u(x,y+h) + u(x,y-h) - 4u(x,y) \quad (1)$$

which is iteratively evaluated at each mesh point until a sufficiently small discrepancy is seen between successive calculations. Because neighboring points can straddle processors, adjacent processors must exchange boundary information, the messaging costs of which are incurred *per iteration* and thus create a performance bottleneck. To compare JavaSpaces with sequential C we tested Jacobi relaxation, with times for C (and Java) ranging from 0.01 sec (0.067 sec) and 347 iterations for a 10x10 mesh, to 131.37 sec (174.76 sec) and 28249 iterations for a 120x120 mesh. Taking into account the space access times in Table 3 it becomes clear that a JavaSpaces version will be inefficient: for a 10x10 mesh two workers would need to exchange at least 300 array tuples, or ca. 1.8 seconds in our testbed, a factor of 180 times slower than sequential C. As JavaSpaces lacks broadcast/gather/reduce communication operators, which are heavily used in message-passing mesh implementations, we implemented limited versions and used them to test Jacobi relaxation: the resulting performance was at least an order of magnitude slower than sequential C.

4 Parametric Experiments

Our results so far show that the benefits of JavaSpaces communication are not to be found in improved performance but rather in simplicity and the uncoupling of the communicating agents in space and time. Thus concurrent computations that are loosely coupled and/or have high computation-to-communication ratios stand to gain most from JavaSpaces. An illustrative example is PI digit finding by discretization of [0,1] into I subintervals. Each worker takes from the space a tuple containing the discretized algorithm, I, the total number of tasks T, and the rank R of the current task (where $1 <= R <= T$), and upon completion writes a partial sum tuple back to the space, with the master performing a sum-reduction upon these to generate the final result. Performance figures given in Table 4 were gathered from timestamps in the master, and include the cost of tuple instantiation, task distribution, and results collection. Note that two sets of figures are reported for N workers: The first represents a partitioning into as many tasks as workers ($T = N$); the second represents a finer decomposition into more tasks than workers ($T > N$), which permits faster machines to consume more work and results in superlinear speedups relative to sequential Java (though we ensured that each worker consumed at least one task). The optimum tuple decomposition granularity could not be established apriori or even automatically at runtime, however, but rather only through trial and error. Each of the first 8 workers was run on a single subnet host, while workers 9 through 12 were distinct processes on both of the 2-node CFA machines, and workers 13 through 16 were processes on 4 nodes of the SGI O2K. The approach used to compute PI can be applied to a large class of problems known as *parametric experiments.* In parametric computations a given task is executed repeatedly with different inputs, or *parameters.* In the case of PI these are I and T (R is a function of T, thus not a distinct parameter). Each node executes a *self-contained algorithm* whose runtime behavior is entirely determined by its inputs, avoiding the need

Table 4. PI Estimation Runtimes (Seconds) Using 10e6 Subintervals.

Sequential Times			
gcc	JDK1.3	JDK1.2.2	MIPS C
10.71	24.23	19.688	11.333
JavaSpaces Times			MIPS C/OpenMP #pragma
N	T = N	T > N	N = # Processors
2	12.36	na	5.836
4	6.273	4.726 (6 tasks)	3.205
6	4.263	2.97 (10 tasks)	2.156
8	3.209	2.35 (13 tasks)	1.775
16	2.363	2.03 (20 tasks)	1.119

to communicate with other nodes during computation. We will discuss two parameterized Monte Carlo algorithms: particle shielding and light propagation in multilayer tissue.

The shielding experiment simulates the collision of K particles with a W-unit thick shield constructed of material with density D. The model assumes a subatomic particle, say a neutron, will collide C times with atoms within the shield, each time bouncing off in a random direction and retaining M percentage of its incident momentum. The horizontal distance traveled between collisions, deltaX, is inversely related to the material density by a constant of proportionality P. Coded in C the algorithm body is as follows:

```
deltaX=P/D;
for (i=0; i<K; i++) {
    x=deltaX; momentum = 1.0;
    for (k=1; k < C; k++){
        x += cos(3.1415926*randFloat())* deltaX*momentum;
        if (x > W) { penetrated++;   break;}
        else if (x <= 0.0) break;
        momentum *= M; }
}
```

Though this example is contrived for illustrative purpose it is dissimilar only in *degree*, not *kind*, from simulations that *are* regularly conducted. [1] The host/worker allocation for this experiment followed that of the PI computation, and yielded the favorable results given in table 5. To ensure that the cross-language experiments each traced the same computation we avoided the ANSI C rand() function and the Java Random class (each yields different sequences), in favor of defining our own *randFloat()* function, based upon the linear congruential method given in vol. 2 of Knuth's "The Art of Computer Programming". To simplify evaluation we held the parameters W, C, D, M, and P constant as N and T increased,

[1] Both prior and subsequent to launch in 1999, for example, thousands of ray-trace experiments were conducted at the CFA to simulate the behavior of the Chandra X-Ray telescope optics and instrumentation.

Table 5. Shielding (50e6 Neutrons) and Light Propagation (500K Photons) Runtimes

Sequential Times (sec)					
Particle Shielding			Light Propagation		
gcc	JDK1.3	JDK1.2.2	gcc	JDK1.3	JDK1.2.2
171.25	265.52	392.527	47.76	79.78	247.6
JavaSpaces Times (sec)					
N	T = N	T > N	T > N		
2	128.547	na	41.342		
4	64.256	35.237 (8 tasks)	17.088		
6	42.975	20.315 (14 tasks)	10.558		
8	32.461	14.353 (20 tasks)	8.945		
12	24.815	13.62 (22 tasks)	6.642		

while K varied with T as ceil(50e6/T). Our second parametric simulation, light propagation in multilayer tissue, is a *real* algorithm published by Prahl et al. at the Oregon Medical Laser Center [10]. Here the parameters are the number of photons P, a scattering coefficient S, the thickness of each tissue layer L, and an absorption coefficient A. Our results are given in table 5, and again we held all but P constant for each trial.

5 Dynamic Compilation

We now describe a method which extends the code-delivery and parallelization mechanisms of JavaSpaces to boost worker performance with native compilation. Our idea generalizes the longstanding observation that compute-intensive algorithms often spend much of their time in concise regions of code, and we seek to demonstrate that for *short codes* taking better advantage of the native architecture can outweigh compilation and JNI costs if executed on machines of sufficient power. Computations flow as described earlier, except now our tasks contain C sourcecode. Workers continue to retrieve and execute tasks, oblivious to inner algorithmic details, which in this case include compilation and invocation through JNI. The emphasis upon short codes is important: they decrease the time needed for the compiler to build the shared library; they also enhance portability by increasing the likelihood the code will compile sans error on the (potentially unknown) target system, as short numerical codes tend to avoid exotic system calls, graphics routines, and other sources of platform incompatibilities. By distributing C code in textual form and dynamically generating platform-specific libraries we skirt the use of precompiled binaries and avoid compromising Java portability — the most unpalatable aspect of JNI — and maintain our heterogeneity criterion. Moreover, if the programmer brings existing C code to the table the effort to recode algorithms in Java can in some cases be avoided. With the help of a short utility script, for example, it took only minutes to wrap the light propagation code in a Java class suitable for JNI invocation. We intend to investigate runtime conversion of Java bytecodes to native binary, which has not been pursued yet due to the lack of wide-availability of open tools such as GCJ, and the fact that they tend not to support newer Java features needed by JavaSpaces. The overhead of dynamic compilation was measured by executing the Dynamic task with 0 photons, and typically ranged from 0.68 to 0.71 seconds, of which ca. 0.58 seconds were consumed by compilation, and the rest accounted for by JNI and other Java overheads. The results in table 6 show that these overheads are indeed small enough to yield a sharp performance increase, relative to pure JavaSpaces, for small numbers of workers.

Table 6. Dynamic Compilation: 500K Photon Light Propagation.

# Workers	1	2	3	4	6	8	10
Execution Time	44.177	23.313	15.968	12.061	8.281	7.26	5.194

6 Conclusions

We have shown that JavaSpaces exacerbates well-known problems with the Java platform, most notably communication latency, and that this renders JavaSpaces unsuitable for several algorithms, including one with an efficient Linda implementation. We gave evidence that for parametric experiments they yield good speedups (figure 1), relative to both sequential Java and C. A novel approach for further speedup through native-compilation was also described, which utilizes JNI but avoids compromising the portability of Java by only invoking code compiled on the target platform at runtime. These results have significance beyond

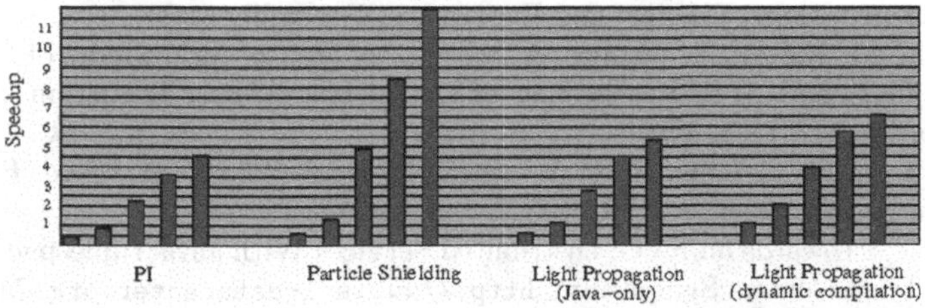

Fig. 1. Speedups of JavaSpaces algorithms versus sequential C implementation compiled with GCC -O3, for N=1, 2, 4, 6, and 8 workers.

the computer science research community, for several reasons. First, we have employed *only* commodity software, and almost exclusively commodity hardware.

Second, while young our benchmark package is readily available [13] and provides a computation framework which complements the simplicity of JavaSpaces and may be used to quickly assess the performance of JavaSpaces or TSpaces on a given network or hardware platform or adapted for further development. Finally, because our benchmarks utilize only vanilla functionality (e.g. avoiding event notification, custom Jini services, or Java-aware C compilers) we believe that laymen in other fields, with historical aversions to the complexity of other parallel programming models, might thus find Java tuplespaces an approachable alternative. Future work will include more detailed comparisons to TSpaces and an exploration of the role of Java tuplespaces within virtual observatories [12].

References

1. E. Freeman, S. Hupfer, K. Arnold, "JavaSpaces: Principles Patterns, and Practice", Addison-Wesley, 1999
2. J. Moreira, S. Midkiff, M. Gupta, "A Standard Java Array Package for Technical Computing", Proc. of the Ninth SIAM Conf. on Parallel Processing for Scientific Computing, March 1999
3. P. Wu, S. Midkiff, J. Moreira, M. Gupta, "Efficient Support for Complex Numbers in Java", Proc. of the ACM Java Grande Conference, June 1999
4. N. Carriero, D. Gelernter, J. Leichter "Distributed Data Structures in Linda", Proc. ACM Symp. Principles of Prog. Languages, 1986
5. N. Carriero, D. Gelernter, "How to Write Parallel Programs: A Guide to the Perplexed", ACM Comput. Surv. 21, 3, Sep. 1989
6. M. Phillipsen, B. Haumacher, C. Nester, "More Efficient Serialization and RMI for Java", Concurrency: Practice and Experience, 12(7):495-518, May 2000
7. Yelick, Semenzato, Pike, Miyamoto, Liblit, Krishnamurthy, Hilfinger, Graham, Gay, Colella, Aiken, "Titanium: A High-Performance Java Dialect", ACM 1998 Workshop on Java for High-Performance Network Computing
8. R. van Nieuwpoort, J. Maassen, H. Bal, T. Kielmann, R. Veldema, "Wide-area parallel computing in Java", Proc. ACM 1999 Java Grande Conference
9. M. Welsh, D. Culler, "Jaguar: Enabling Efficient Communication and I/O from Java", Concurrency: Practice and Experience, December 1999
10. S. Prahl, M. Keijzer, S. Jacques, A. Welch, "A Monte Carlo Model of Light Propagation in Tissue, SPIE Proceedings of Dosimetry of Laser Radiation in Medicine and Biology, Vol. IS 5, 1989
11. T. Lehman, S. McLaughry, P. Wyckoff, TSpaces: The Next Wave, Hawaii Intl. Conf. on System Sciences (HICSS-32), Jan. 1999
12. M.Noble, "Towards an NVO: Distributed Services With Java Tuplespaces", Fourth Science Data Centers Symposium, `http://www.sci-datacenter.org`, March 2001
13. M.Noble, The Tonic Benchmark Package: Scientific Computation with Java TupleSpaces, `http://hea-www.harvard.edu/\%7Emnoble/tonic/doc`.

Computational Communities: A Marketplace for Federated Resources

Steven Newhouse and John Darlington

Imperial College Parallel Computing Centre
Department of Computing, Imperial College of Science
Technology and Medicine, London, SW7 2BZ, UK

Abstract. We define a grid middleware comprising federated resources that facilitates a globally optimal mapping of applications to the available resources while satisfying the goals of both users and resource providers. Applications are annotated with performance and behavioural information to enable the 'best' resources to be found automatically. A computational currency is used by resource providers and consumers to express their goals (e.g. completion time, resource utilisation, etc.) enabling a globally optimal mapping of applications to resources. We describe a prototype implementation of this architecture using Java and Jini.

1 Introduction

The accelerating proliferation of high-performance computing resources and the emergence of high-speed wide area networking has led to much interest in the development of Computational Grids. A Computational Grid is defined as the combination of geographically distributed heterogeneous hardware and software resources to provide a ubiquitous transparent computing environment [1]. Such infrastructures are gaining acceptance outside the traditional high performance computing community as computational and data intensive applications become commonplace in science and commerce. Early experiments in Grid construction have generally involved the explicit connection of supercomputers or scientific instruments, requiring a high degree of expertise and involvement from resource providers and users [2].

The federation of heterogeneous resources under the administrative control of different organisations is achieved through a *grid middleware* that must mask any heterogeneity and provide:

- **Information.** Effective application scheduling requires information on the available hardware, software, storage and networking resources.
- **Security and Control.** Organisations will only federate their resources if they retain control and are able to ensure the needs of their local users.
- **Effective Resource Exploitation.** The *best* resources for an application will depend on the user's and resource provider's goals.

B. Hertzberger et al. (Eds.): HPCN Europe 2001, LNCS 2110, pp. 667–674, 2001.

To achieve a transparent and ubiquitous grid computing environment it is necessary to automate resource selection, but to do so within the constraints and goals of the user and resource provider. We use a computational currency to enumerate these goals allowing us to achieve a balance between the needs of the individual and the community. For example, a user can choose to pay for better resources to reduce their execution time while a resource provider can select a job mix that maximises their revenues and utilisation.

2 A Computational Economy

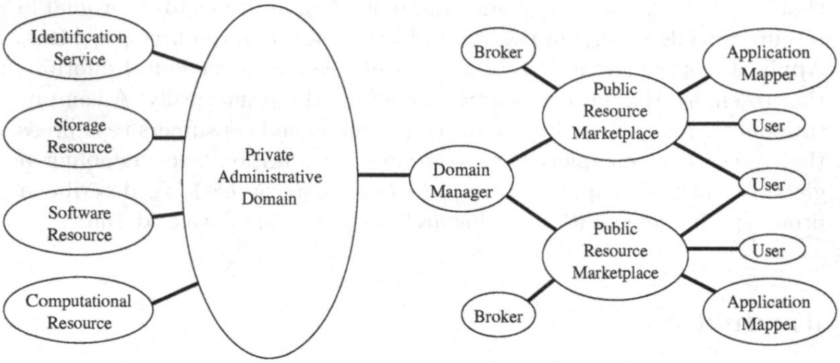

Fig. 1. Building a Resource Marketplace through Federated Resources.

2.1 Overview

Our federated computational economy has four major components that interact through a public resource marketplace (see figure 1):

- **Organisations** contribute resources under a locally defined access control policy to a public resource marketplace. Any organisation can participate by informing their local Domain Manager of the 'well known' URL of the marketplace's Jini Lookup Service and the resources they wish to contribute. Within a local area network the lookup service can be discovered using Jini's automatic protocols.
- **Users** are able to utilise the resources and services within a marketplace by connecting to its Jini Lookup Service through a 'well known' URL or by automatic discovery within a local area network. To access these resources the user's organisation must have an established 'trust' relationship with the resource provider's organisation. This may range from a known public key certification authority to a specific resource sharing agreement.

- **Application Mappers** generate a set of execution plans matching the application's requirements to the resources that are currently available. These represent feasible execution strategies optimised with regard to performance but may not make the *best* economic use of the resources.
- **Brokers** negotiate between organisations and users to find the *best* economic execution plan from those generated by an application mapper.

An organisation's resources are managed through the local Administrative Domain. These resources are made publicly available through the Domain Manager which places this information into one or more public resource marketplaces and enforces the specified local access control policy.

This model allows organisations to federate when they see mutual benefit in doing so by advertising their resources, alongside others in that community, in a common marketplace. It is important to note that the same resources may appear in different marketplaces with different constraints and that users are able to advertise their needs in several marketplaces. This ensures there is no single point of failure in the provision of resources and allows competition between different marketplaces which may have different broker or mapper implementations.

2.2 Building a Computational Community

Local resources will only be federated into a larger computational community if the usage conditions governing remote access are explicit and strictly enforced. For instance, in an academic environment staff may be given a higher priority than students but a student with an upcoming deadline may be given higher priority than most staff. Likewise, remote users may only be allowed to use the resources if they are idle but collaborators may be given priority over other remote users. Being able to express these usage policies is a key motivation for our infrastructure.

Our resources may have both static (e.g. operating system release, architecture, etc.) and dynamic attributes (e.g. current load, available licences, etc.). These attributes are advertised in the computational community and used during resource selection. The resources are registered as Java objects within the Jini lookup service. Persistence of these attributes (between power cycles and unexpected failures) is maintained through an XML syntax that describes the resource and its characteristics.

- **Computational Resources.** We currently access our own local computational hardware through a batch scheduler abstraction with implementations for NQS, PBS [3] and Condor [4]. Each computational resource executes its own segment of an XML defined execution plan passed to it by the Domain Manager.
- **Storage Resources.** The user must be able to automatically and securely access their storage space from any location.
- **Software Resources.** Our current implementation only represents unlimited use software libraries but the execution of a licensed library or application has to be scheduled in the same manner as a computational resource to ensure that a licence is available.

These resource abstractions could encapsulate access to existing infrastructures such as Globus [2] or Legion[5].

The *Domain Manager* allows organisations to contribute their resources into a federated computational community while retaining fine-grained control of how non-local users are permitted to use these resources. As the only route between the private and public areas of the middleware it provides:

- **Authentication.** Authentication of the user, groups and organisation uses a public key infrastructure and is delegated to the Identification Service. Each organisation may act as its own certification authority (CA) and define which organisations (and CA's) it trusts to use its resources.
- **Authorisation.** Access to individual resources is controlled through conventional access control lists that recognise three entities: individuals, groups and organisations. This allows the Domain Manager to implement fine-grained access control policies governing resource usage.
- **Promotion.** Resources and their associated access control polices are published in one or more computational communities by the Domain Manager. Published information can be restricted to hide specific resources and the details of the local access policy.
- **Execution.** The Domain Manager validates all resource requests in a user's execution plan before passing them onto the individual resources.

2.3 Resource Discovery and Selection

Effective automatic resource selection requires information relating to the performance of the application on different architectures (encapsulated within a performance model) and the requirements of the user. This is in addition to any source code or binary annotations such as those regarding execution environment that are currently used in the Condor ClassAds 'matchmaking' system [6]. We are extending our previous work in program composition using skeletons [7] to defining HPC applications as a network of software components which will automatically generate the application's structure and overall performance model from its components [8].

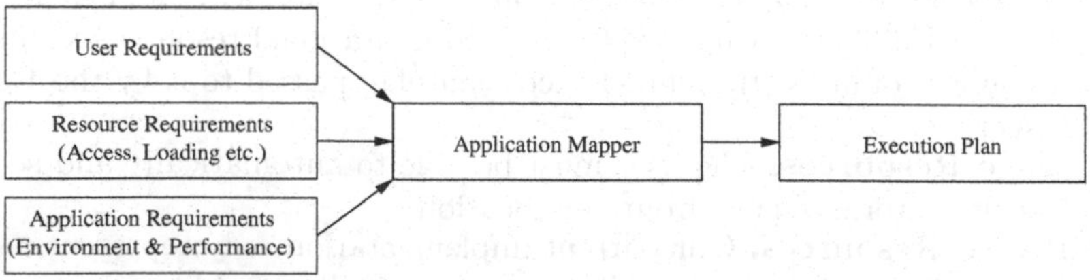

Fig. 2. Application Mapper.

The *Application Mapper* automates the user's selection of the *best* resources from those that are currently available. It uses the requirements specified by the user (e.g. job completion time), the application (e.g. run-time environment and execution time) and the resource provider (e.g. resource access and loading) to generate a set of viable execution plans for final selection by the user (see figure 2). Multiple Application Mappers can provide the user with a variety of feasible execution plans generated using different approaches. One approach to application partitioning is to define the resource and user constraints as a linear programming problem and maximise a user supplied objective function [9].

2.4 Computational Supply and Demand

The Application Mapper ensures effective resource selection and utilisation for an individual user's application but it does not ensure effective resource utitilisation for the resource provider. For example, two applications request the use of a 16 processor PC cluster and that from each application's performance model the optimal number of processors is determined to be nine. If the first job starts with nine processors, the second job would have to wait until the first job has completed and nine processors become available. An alternative approach is for both applications to use eight processors and both to begin immediately. The latter situation is preferred by the resource provider as it increases utilisation but it may not always be feasible (or desirable) to adjust a user's application in this way.

The wishes of the resource provider (to maximise their resource utilisation) are elegantly expressed by charging a user different costs for different resource configurations. It may be acceptable to the user to use eight processors if the cost were considerably less than that of using nine processors (as the resource provider would be charging for the seven 'wasted' processors). To support resource trading we will use a trusted computational currency as an exchange medium. Resource providers will need to recognise and convert between several currenicies which may be backed by a specific resource provider or generic micropayment scheme.

The *Resource Broker* negotiates a cost for the execution plans with all valid resource providers and presents these to the user. The resource provider prices the execution plan according to its own economic priorities (which may be dependent on the individual, group or organisation wishing to use their resource) and attempts to maximise their resource utilisation by consideration of, say, revenue stream or job throughput. Job priority, from the perspective of either a user and resource provider, is elegantly and simply expressed through these market mechanisms. By maximising the pay-off functions we can find the 'best' global allocation of resources to jobs in the computational community by balancing the needs of the users, the applications, and the resource providers over all requests rather than each individual request. Our framework makes no attempt to define what is the 'best' pay-off function other than it will be defined by the user and resource provider in terms of time and money.

3 Implementation

An initial proof of concept prototype of this architecture has been completed at Imperial College using a Java and Jini environment [10]. The architecture described in this paper represents an extension of this work. We exploit Java's portability and its rich API's to simplify many of the development tasks [11]. We use Jini as the primary service infrastructure as it supports dynamic registration, look-up and connection between the Java objects that represent our grid services and resources [12]. As all grid resources are effectively transient this ability to connect and reconnect over time is a highly desirable feature. The Jini leasing mechanism also allows unexpected failures to be handled gracefully. We use the look-up server to represent the public resource marketplaces and private administrative domains. This implementation is currently being extended to meet the needs of our user community in High Throughput Computing (e.g. Particle Physics, Bioinformatics and Medical Image) and Distributed High Performance Computing applications (e.g. solar coronal mass ejection simulations and coupled fluid-structure acoustics).

4 Related Work

Our approach is a combination and logical extension of two leading grid infrastructure projects: Globus and Legion. Globus provides a toolkit of services (information management, security, communication etc.) to integrate heterogeneous computational resources into a single infrastructure [2]. Legion uses a uniform object model for both applications and resources allowing users and administrators to subclass generic interfaces to their specific local needs [5].

Java has been used to provide a homogeneous distributed computing environment across heterogeneous resources (e.g. Javelin [13] and other projects). While Jini has been used to form a meta-computing infrastructure [14]. However, these projects have not yet addressed the policy issues regarding the access of remote users to local resources, which is fundamental in our approach.

The skeleton approach to program composition defines an application as a composition of components which are assembled using pre-defined structural forms of known semantics (e.g. pipe, farm) [7,15]. The mapping of these compositions onto target architectures is guided by analytical performance models, developed with each component, allowing decisions regarding efficient implementation to be made quantitatively and systematically. Our experiences with structured coordination languages is now being used in the context of conventional software components by their annotation with XML encoded meta-data relating to how they can be used, deployed and perform [8]. This compositional approach yields the performance models of the overall application and its substructures allowing the application to systematically and efficiently map the composition to the target architectures.

Application oriented schedulers (or mappers) such as AppLeS select the optimal number of processors from a computational resource for a particular problem size by using static or stochastic computational and networking parameters and standard linear programming techniques [9,16]. Our Application Mapper will extend this work to find an optimal execution plan that considers all the available resources and assumes an application is defined by sequence of inter-dependent tasks, e.g. input/output file staging. Consideration of input and output file staging on a fast heavily used resource means it may be quicker to execute the application on a slower computational resource that has good network connectivity.

The Spawn system has demonstrated how different funding ratios could be used to guide resource allocation and usage [17]. Nimrod/G uses historical execution times and heterogeneous resource costs to implement fixed budget and deadline scheduling of multiple tasks [18]. The resource costs are obtained through standard auctioning techniques (e.g. English, Dutch, Hybrid and Sealed Bid auctions [19]) and incorporated into the linear programming model used by the application mapper when finding an optimal application mapping.

5 Conclusions and Future Work

Computational grids will eventually, like the Internet, change the way we work. However, to effectively exploit the computational potential of the grid we need to articulate the needs of the users, their applications and the resource providers. From this information we can automatically deploy an application to a resource that will satisfy the stated requirements of the user and the resource provider.

Our architecture, through the federation of resources to build computational communities, the use of application mappers to effectively match applications to resources and brokers to make the best economic use of the available resources, address some of the weaknesses in current grid infrastructures. To implement this system we exploit Jini's fault tolerant and decentralised infrastructure and Java's inherent cross-platform portability.

Our prototype implementation is now being re-engineered to fully conform to the model described in this paper and we foresee its deployment over our local test bed during Summer 2001. We also see scope for expanding the software resource model to provide a software service (a software and hardware combination provided by an application service provider) and even deployable single use software libraries. The computational economy could also be extended to include the speculative purchasing of resources (futures) and other market based actions.

References

1. I. Foster and C. Kesselman, editors. *The Grid: Blueprint for a New Computing Infrastructure.* Morgan Kaufmann.
2. I. Foster and C. Kesselman. The Globus Project: A Status Report. In *Proc. IPPS/SPDP '98 Heterogeneous Computing Workshop*, pages 4–18, 1998.
3. `http://www.openpbs.org`.

4. http://www.cs.wisc.edu/condor.
5. A. S. Grimshaw and W. A. Wulf *et.al.* The Legion vision of a worldwide virtual computer. *Communications of the ACM*, 40:39–45, 1997.
6. R. Raman, M. Livny, and M. Solomon. Matchmaking: Distributed Resource Management for High Throughput Computing. In *Proceedings of the Seventh IEEE International Symposium on High Performance Distributed Computing*, pages 28–31, July 1998.
7. J. Darlington *et al.* Parallel Programming using Skeleton Functions. In *Lecture Notes in Computer Science*, volume 694, pages 146–160.
8. S. Newhouse, A. Mayer, and J. Darlington. A Software Architecture for HPC Grid Applications. In *Euro-Par 2000*, pages 686–689, 2000.
9. H. Dail, G. Obertelli, F. Berman, R. Wolski, and A. Grimshaw. Application-Aware Scheduling of a Magnetohydrodynamics Application in the Legion Metasystem. In *Proceedings of the 9th Heterogeneous Computing Workshop*, May 2000.
10. N. Dragios. Java Metacomputer. Master's thesis, Imperial College, Department of Computing, 2000.
11. K. Arnold, J. Gosling, and D. Holmes. *The Java Programming Language, (Third Edition).* Addison-Wesley.
12. http://java.sun.com/jini/.
13. M. O. Neary, B. O. Christiansen, P. Cappello, and K. E. Schauser. Javelin: Parallel computing on the internet. In *Future Generation Computer Systems*, volume 15, pages 659–674. Elsevier Science, Amsterdam, Netherlands, October 1999.
14. Z. Juhasz and L. Kesmarki. JINI-Based Prototype Metacomputing Framework. In *Euro-Par 2000*, pages 1171–1174, 2000.
15. J. Darlington, M. Ghanem, Y. Guo, and H. W. To. Guided Resource Organisation in Heterogeneous Parallel Computing. *Journal of High Performance Computing*, 4(10):13–23, 1997.
16. F. Berman and J. M. Schopf. Stochastic scheduling. In *Supercomputing*, 1999.
17. C. Waldsburger *et al.* Spawn: A Distributed Computational Economy. *IEEE Transactions on Software Engineering*, February 1992.
18. R. Buyya, D. Abramson, and J. Giddy. Nimrod/G: An Architecture for a Resource Management and Scheduling System in a Global Computational Grid. In *The 4th International Conference on High Performance Computing in Asia-Pacific Region (HPC Asia 2000).* IEEE Computer Society Press, USA, 2000.
19. D. Ferguson, C. Nikolaou, J. Sairamesh, and Y. Yemini. Economic models for allocating resources in computer systems. In Scott Clearwater, editor, *Market-Based Control: A Paradigm for Distributed Resource Allocation.* World Scientific, Hong Kong, 1996.

A Compiler Infrastructure for High-Performance Java*

Neil V. Brewster and Tarek S. Abdelrahman

Department of Electrical and Computer Engineering
University of Toronto, Toronto, Ontario, Canada M5S 3G4
{brewste,tsa}@eecg.toronto.edu

Abstract. This paper describes the zJava compiler infrastructure, a high-level framework for the analysis and transformation of Java programs. This framework provides a robust system, guaranteeing under transformations both the consistency of its internal structure and the syntactic correctness of the represented code. We address several challenges unique to Java, which have not been addressed by earlier frameworks. These include automatic maintenance of complex symbol scope information under transformations, insertion of implicit code to accurately model the source program, incorporation of compiled code into the representation, and representation of the complex control flow of exception handling constructs. We include support for the sharing of information between compiler passes, and a framework for interprocedural analysis. We believe that the features we introduce in the zJava compiler infrastructure will result in a means of rapidly prototyping new Java compiler analyses. We give a number of examples illustrating the use and utility of the infrastructure.

1 Introduction

There has been considerable research during the past decade on parallelizing compilers and automatic parallelization of programs. Traditionally, this research focused on scientific applications that consist of loops and array references, typical of Fortran programs [6,9]. Regrettably, this focus has limited the widespread use of automatic parallelization in industry, where the majority of programs are written in C, C++, or more recently in Java. These programs extensively use pointer-based dynamic data structures such as linked-lists and trees, and often use recursion. These features make it difficult to directly utilize parallelizing compiler technology developed for array structures and simple loops.

The goal of the *zJava* (pronounced "zed Java") compiler project, developed at the University of Toronto, is to investigate automatic parallelization technology for programs that use pointer-based dynamic data structures and recursion, written using the Java programming language. The zJava compiler infrastructure was developed in support of this research. The infrastructure has a number of features which facilitate quick and efficient prototyping of new compiler optimizations for Java. These features address challenges in the representation of

* This work is supported by NSERC (Canada) and CITO (Ontario) grants.

B. Hertzberger et al. (Eds.): HPCN Europe 2001, LNCS 2110, pp. 675–684, 2001.

Java programs at the source level, including automatic maintenance of complex symbol scope information under transformations, insertion of implicit code to accurately model the source program, incorporation of compiled code into the representation, and representation of the complex control flow of exception handling constructs. These challenges have not been addressed by similar previous compiler infrastructures.

The remainder of this paper is organized as follows. Section 2 gives a general overview of the infrastructure. Section 3 describes the main components of the architecture. Section 4 describes some implementation details. Section 5 presents examples of the use of the infrastructure. Section 6 reviews and contrasts related work. Finally, Section 7 provides some concluding remarks.

2 Overview

The zJava compiler infrastructure utilizes both a high level and a low level intermediate form. A front-end parses Java source code into the zJava High Level Intermediate Representation (HLIR), which can be converted back to human-readable Java source. Code generation converts HLIR into the zJava ByteCode Intermediate Representation [10] (BCIR), which provides an interface to the `.class` file format. Program analyses and transformations take place primarily on the high-level form. A high-level representation has the advantage of retaining the syntactic structure of high-level constructs, which benefits analyses such as dependence analysis, array structuring, loop parallelization and transformation, and induction variable elimination.

There are a number of interesting features of the zJava HLIR. We use extensive error checking and reporting to speed up the process of prototyping new compiler passes. These checks guarantee that the representation is maintained consistent under transformations. Both the syntactic correctness of the represented Java program and the correctness of internal HLIR structures are enforced. In addition, internal structures are automatically updated under transformations. The core of the representation is a flat list of statements for each method body. High-level symbol information, including a precise representation of scoping, is included. We augment the represented program to include implicit code, which is not necessarily explicit in the source code but is required [8]. A control flow graph is built on the representation, including the modeling of Java exception flow. We provide mechanisms to save the results of an analysis for use by later compiler passes. When source code is not available, skeletal HLIR can be constructed from `.class` files. Finally, a "rich" class file format facilitates interprocedural analysis, even when source code may not be available.

3 Architecture

3.1 Robust Framework

HLIR is designed minimize the time required to prototype new compiler passes, by detecting and reporting errors as early in the development process as possible.

The API is designed such that most errors can be detected at compile-time[1]. Where that is not possible, run-time checks and Java exceptions are used to detect and report errors. We define an error as any transformation which would leave the IR in an inconsistent state, both internally and in terms of the syntactic correctness of the represented Java code.

Syntactic Consistency. The syntactic consistency of the representation is primarily enforced by controlling the ways in which lists of statements are created and modified. The constructors for creating statement lists require that all parts of a construct be specified at once. For example, it is not possible to create a *while* loop with no body. Additionally, the constructors for each Java statement type require that all necessary components be present. It is not possible, for example, to create a *while* loop header without specifying the loop conditional expression.

The statement list implementation also strictly controls the way in which statements are added or removed. It is not possible to incrementally build a program by adding one statement at a time, even if the final result is syntactically correct. Instead, the representation is built by adding statement lists into other statement lists. Removals which would leave a construct in a syntactically invalid state (such as removing the *then-part* of an *if* construct) are not permitted. Removal of constructs is made possible by the fact that removal of a header statement results in the removal of the entire construct. For example, removing a *try-header* statement will result in the removal of the *try block*, and, if present, the *catch clauses* and the *finally clause*. Methods are provided to remove components of constructs, such as individual *case clauses* in a *switch* statement.

Internal Consistency. We have designed HLIR to automatically maintain its internal structures under transformations. For example, when statements are added to or removed from a statement list, the symbol table information is automatically updated, and the types of all symbols and expressions in the new statements are resolved. This relieves users of the need to understand the details of the scope representation in HLIR, and of the need to maintain it correctly themselves. Additionally, HLIR disallows any modifications (such as creating, adding, or removing statements or other objects) which would leave the internal representation in an inconsistent state. For example, removal of an *end* statement, which is used internally to represent the end of a structured construct, is prohibited.

The core functionality of an intermediate representation involves building objects to represent parts of a program (for example, statements), and using these to populate higher-level objects (for example, methods and classes). In this kind of environment, one common programmer error is object *sharing*, where the same object is used as a component of multiple IR objects. For example,

[1] We use the term *compile-time* to refer to the time at which a zJava compiler pass is compiled, and the term *run-time* to refer to the time at which the zJava compiler executes.

several *symbol* objects, representing different integer variables, might each have a reference to a *type* object representing the integer type. Should all of these *symbol* objects reference the same *type* object, a programmer who wished to change the type of one symbol might unwittingly change the type of them all. The concept of object *ownership*, as used in the Collections hierarchy of the Polaris project [6], allows HLIR to to prevent such object sharing. When an object is inserted into a list, the list gains "ownership" of that object. An object can be owned by only zero or one list, and a list may not contain any objects it does not own. For example, attempting to add the same statement object to the body of a method more than once will cause a run-time error.

Method bodies are represented in HLIR as flat lists of non-recursive statement structures. This means that compiler passes can simply iterate over the statements of a method body, without specific knowledge of the structure of each statement. A sequential list of statements is also convenient when dividing a method body into basic blocks for the control flow representation.

In HLIR, each construct begins with a *header* statement and ends with a generic *end* statement. Links are included, specific to the construct; for example, an *if-header* contains links to the *then-part*, the *else-part*, and the *if-end*.

3.2 Symbol Table Representation

Representing all information available in the source code also involves capturing scope information. The zJava HLIR creates a symbol table for each unique scope, and chains it to the enclosing scope. A symbol table tree is constructed in parallel with HLIR, and is automatically updated whenever symbol information changes. Symbol lookups automatically proceed up the chain until the highest level scope is encountered. This design implicitly implements variable hiding; the closest declaration of a variable is seen first, so local variables can shadow instance variables. The type of every symbol and expression is resolved before any user compiler passes are invoked.

Every statement represented in HLIR includes a link to the symbol table for the scope in which it resides. Thus, when examining any statement in the program, HLIR presents what appears to be a single symbol table containing all symbols visible to that statement. Qualified symbols (eg. `Foo.a`) are resolved by querying the symbol table of the qualifying type. Symbol table lookups are designed to also automatically examine available superclasses for accessible symbols, loading them from `.class` files if necessary.

3.3 External Class Resolution

Interprocedural analyses often require information about classes for which the source code is not available. For example, the Java class libraries are usually only available in compiled (`.class` file) form. The zJava infrastructure includes the ability to construct skeletal HLIR from `.class` files generated by any compiler. Using the zJava ByteCode Intermediate Representation [10] (BCIR), zJava automatically locates and loads `.class` files when needed The resulting representation includes all information available in the `.class` file, except for the actual bodies of methods–we make no attempt to decompile bytecodes.

3.4 Modularity of Analyses

The zJava infrastructure is designed to allow compiler passes to build on the results of previous analyses, without requiring that each original analysis be repeated. It is important not to equate the concept of modular analysis to the ability to run several independent compiler passes prior to code generation; the latter is a feature of most program restructuring systems, including zJava.

We include support for passing user-defined directives to the compiler in the form of special comments in the source code, and for including such annotations in the output source code. Each directive is encapsulated inside a an attribute and attached to a list on the appropriate HLIR object. This technique is best suited to storing the results of an analysis (eg. "this loop is parallel"), rather than the actual data structures generated and populated by the analysis (eg., dependence graphs). While the latter would be possible, it could result in unacceptably large blocks of comments in the source code.

The `.class` file specification [11] includes support for user-defined attributes, which are ignored by standard classloaders. We have developed a "rich" class format, designed to store the results of compiler analyses in the form of special attributes. Bytecode generation can store the results of an analysis by including custom attributes while generating BCIR. This technique is very useful for classes which are not typically available in source form. For example, the standard Java library classes could be analysed once, and compiled into augmented class files. Later analysis of user code which calls into these libraries would thus have access to the results of analysing the library classes.

4 Implementation Details

4.1 zJava Compiler Front-End

The zJava compiler infrastructure is implemented entirely in Java. The front-end of the zJava compiler uses the Java Tree Builder (JTB), developed at Purdue University, to automatically generate abstract syntax tree (AST) class definitions based on the input grammar. JTB also generates a Visitor design pattern. The Java Compiler-Compiler (JavaCC) package from Sun Microsystems is used to automatically generate a parser that builds the AST defined by JTB. Finally, visitor classes are defined to generate the high-level representation.

4.2 Collections

At the root of the HLIR class hierarchy is the class `zObject`. Any object of type `zObject` can be cloned, converted to source code, and flagged as either owned or unowned. All HLIR class constructors check the ownership flags of `zObjects` passed into them. If the constructor expects to have ownership of a particular object, but finds that it is already owned, it will throw `zObjectOwnedError`.

Much of the representation involves lists of objects, and this provides another mechanism for ownership enforcement. The classes `zList` and `zRefList` are derived from `java.util.LinkedList` for this purpose. A `zList` may only contain objects of type `zObject`, and any `zObject` may be contained in at most one `zList`. The list manipulation methods are overridden to enforce these constraints. The `zRefList` class exists for cases where objects in the list are expected to be owned by some other list. For example, a *switch* statement references a list of *case* statement objects, each of which is in fact owned by the enclosing statement list. Thus, the *switch* statement utilizes a `zRefList` to store the references to its *case* statements. Any object inserted into a `zRefList` must already be owned. Methods in `zRefList` which provide access to its elements first check that the element is still owned by another list. If an element in a `zRefList` is removed from the `zLinkedList` that owns it, any subsequent attempts to access that element through the `zRefList` will cause a `zObjectUnownedError`.

4.3 HLIR Classes

The `Program` class is the root of HLIR, representing a set of source files compiled on the same command line. `Program` contains a `compilationUnit` to represent the information in each Java source file specified, the core of which is the list of classes declared in the file. Each class is represented by a `classObject`, which contains information specific to the class, in addition to a list of the methods declared. Each method is represented by a `methodObject`, and contains a reference to a `stmtList` object. The `stmtList` contains the list of statements representing the method body. The `zStatement` hierarchy implements classes to represent the various types of Java statements, and the `zExpression` hierarchy defines a class to represent each kind of expression. Certain expressions contain symbols, which are represented in the classes of the `zSymbol` hierarchy. Additional classes are implemented to represent the remaining information available in the source code, such as types and access modifiers.

5 Examples

In this section we detail three examples of HLIR in use. The process of implementing these examples has not only verified the functionality of the infrastructure and the completeness of the API documentation, but has also demonstrated that our goal of rapid prototyping has been achieved.

5.1 Statement Insertion

This simple example shows the insertion of a `while` loop into an existing method body. The source code equivalent of the loop is given below.

```
int i = 0;
while (i < 10) {
  System.out.println("Hello World!"); ++i;
}
```

The Java code given in Figure 1 demonstrates the construction and insertion of this loop. It is assumed that `method_body` references list of statements currently in the method, and that `idx` is the index into that list at which the new loop is to be inserted. The declaration of `i` will be translated into a symbol table entry which will be merged with the existing information, and each use of `i` will be resolved to point to this entry. The expression `System.out.println` will be separated into three variable expressions and then combined as two field accesses, and the type of each will be resolved by examining the appropriate `.class` files.

```
javaType int_t = new javaType(javaType.t_int);
varSymbol loopVar = new varSymbol("i", int_t, new fieldModifier());
loopVar.setInitializer(new varExpression("5"));
LinkedList declsList = new LinkedList(); declsList.addLast(loopVar);

// Loop conditional: i < 10
varExpression leftExpr = new varExpression(new String("i"));
varExpression rightExpr = new varExpression(new String("10"));
binaryExpression conditionalExpr = new binaryExpression(leftExpr,
        binaryExpression.op_lessthan, rightExpr);

// Loop body:
LinkedList argsList = new LinkedList();
argsList.addLast(new StringLiteralExpression("Hello World!"));
naryExpression arguments = new naryExpression(argsList);
varExpression methExpr = new varExpression("System.out.println");
methodInvocation invokeExpr = new methodInvocation(methExpr, arguments);
stmtList bodyList = new stmtList(new exprStatement(invokeExpr));
preUnaryExpression incrExpr = new preUnaryExpression(
        unaryExpression.op_plusplus, new varExpression("i"));
bodyList.addLast(new stmtList(new exprStatement(incrExpr)));

// Loop insertion:
whileStatement whileStmt = new whileStatement(conditionalExpr);
stmtList newList = new stmtList(new declStatement(declsList));
newList.addLast(new stmtList(whileStmt, bodyList));
method_body.add(idx, newList);
```

Fig. 1. Using HLIR to Construct a `while` Loop.

5.2 Data Structure Visualization

The Data Structure Visualizer (DSV) provides a means of graphically representing the objects and pointers in a program, dynamically displaying common data structures (linked lists, binary trees, etc) as a program executes. A zJava compiler pass has been implemented which restructures a Java program, inserting appropriate calls to the DSV library. This pass makes use of zJava source directives to inform the compiler which variables represent nodes to be displayed, which nodes represent various types of pointers, and which classes make use of the node class. Much of the functionality of HLIR is used, including augmenting classes with the addition of static initializer blocks, class fields, and entire methods. As well, each expression is examined, to identify those which assign to a node or allocate a new node. Statements are then automatically added around each of these expressions, calling into the DSV library to update the display.

5.3 Software Architecture Visualization

The Portable Bookshelf (PBS), developed at the University of Toronto, is an implementation of the Software Bookshelf [7], a web-based framework for the presentation and navigation of information representing large software systems. The PBS system is used to generate landscapes of each subsystem, displaying objects within the subsystem and the relationships between them. The zJava compiler infrastructure has been used to implement the first stage of landscape creation–the extraction of file-level facts from source programs. An HLIR pass extracts the set of low-level relationships defined in [1], such as *methodDefBy* (Method **A** is defined by Type **B**), *castsTo* (Method **A** casts an expression to Type **B**), and *arrayType* (Array **A** is an array of Type **B**). Existing tools are then used to infer higher-level relationships, and to generate the landscapes.

6 Related Work

High-level program manipulation and analysis frameworks exist for several languages. Those which represent programs at a high level and support source-to-source transformations are closest to our work. The Polaris Fortran compiler [6] includes automatic enforcement of syntactic consistency, and incremental maintenance of IR structures, including control flow information. The Sage++ toolkit [5], uses a common high-level IR to provide a framework for building Fortran, C, and C++ restructuring systems. The Paraphrase-2 automatic parallelizing compiler [12] compiler uses the combination of a data dependence graph, control flow graph, and call graph to represent C and Fortran programs. The SUIF [9] research framework is targeted to imperative languages such as Fortran and C, providing an intermediate format and a set of common optimization passes. OSUIF [4] is an extension of SUIF2.0, introducing support for object-oriented optimizations. Score [13] and Vortex [3] support both high-level and low-level language constructs. The Illinois Concert system [2] is an optimizing compiler for the concurrent object-oriented programming model ICC++.

The zJava infrastructure is unique in a number of aspects. Automatic consistency maintenance, demonstrated by Polaris for Fortran programs, is more complex in a Java representation. The conversion to a flat representation is complicated by the introduction of more complicated structures. The representation of symbol scope is made more difficult by inheritance relationships and by the possibility of variable declarations anywhere in a block. The semantics of the Java exception handling constructs greatly complicates the control flow representation. In order to accurately model a Java program, HLIR must implicitly insert code which is not present in the source. To facilitate interprocedural analysis, we provide functionality to augment `.class` files with analysis results, and the ability to incorporate compiled code into the high level representation.

Table 1 summarizes the functionality provided by several of the systems discussed above, including zJava. The comparison criteria are: automatic syntactic consistency enforcement under transformation; support for modular analysis; the ability to convert the IR into source code; the representation of both high and low level constructs; and the set of languages supported by the compiler.

Table 1. Comparison of various high-level compiler frameworks. A filled circle implies that the system has the associated functionality.

	Automatic Consistency	Modular Analyses	Source-to-Source	High and Low Level	Languages Supported
zJava	•	•	•	•	Java
Polaris	•	•	•	○	Fortran
SUIF	○	•	•	•	Fortran, C, C++
Score	○	○	○	•	N/A
Sage++	○	•	•	○	Fortran, C, C++
Paraphrase-2	○	○	•	○	Fortran, C
Concert	○	•	○	○	IC++
Vortex	○	○	○	•	Cecil, C++, Modula-3, Java bytecode

7 Concluding Remarks

The zJava infrastructure is a program analysis and source-to-source transformation system for Java. It provides a robust framework for the prototyping of new compiler analyses and optimizations. To facilitate rapid development of new compiler passes, we have designed the high level intermediate representation to detect misuse as early in the process as possible. HLIR prohibits transformations which would leave the representation in an inconsistent state, both in terms of its internal structures and in terms of the syntactic correctness of the represented Java program. We have included support for annotations in both the source code and the bytecode, and the ability to construct the IR from compiled code.

We have successfully verified the functionality of the zJava infrastructure as a program analysis and manipulation tool through the construction of control flow information, the extraction of low-level facts, and the augmentation of user programs to interface with other libraries. Further, a self-compile test and the compilation of the SPECjvm98 benchmarks have verified that our infrastructure produces correct code, and does not alter the semantics of the original program under direct source-to-source transformation.

The zJava infrastructure has been released in the public domain as a tool for researchers to analyse and restructure Java programs. Our current research involves the incorporation of the Omega dependence test in zJava for the parallelization of loops, and the implementation of path expression analysis for the parallelization of Java programs at the method level.

References

1. I. T. Bowman. Architecture recovery for object-oriented systems. Master's thesis, University of Waterloo, 1999.
2. A. Chien, J. Dolby, B. Ganguly, V. Karamcheti, and X. Zhang. Supporting high level programming with high performance: The Illinois concert system. In *Second International Workshop on High-level Parallel Programming Models and Supportive Environments*, 1997.
3. J. Dean, G. DeFouw, D. Grove, V. Litvinov, and C. Chambers. Vortex: An optimizing compiler for object-oriented languages. In *ACM Conference on Object Oriented Programming Styles*, 1996.
4. A. Duncan, B. Cocosel, C. Iancu, H. Kienle, R. Rugina, U. Hölzle, and M. Rinard. OSUIF: SUIF 2.0 with objects. In *2nd SUIF Compiler Workshop*, 1997.
5. F. Bodin et al. Sage++: An object-oriented toolkit and class library for building Fortran and C++ restructuring tools. In *OONSKI*, 1994.
6. K. A. Faigin, J. P. Hoeflinger, D. A. Padua, P. M. Petersen, and S. A. Weatherford. The Polaris internal representation. Technical Report CSRD-1317, University of Illinois at Urbana-Champaign, February 1994.
7. P. Finnigan, R. Holt, I. Kalas, S. Kerr, K. Kontogiannis, H. Muller, J. Mylopoulos, S. Perelgut, M. Stanley, and K. Wong. The software bookshelf. *IBM Systems Journal*, 36(4):564–593, 1997.
8. J. Gosling, B. Joy, and G. Steele. *The Java Language Specification*. Addison-Wesley, 1996.
9. M. Hall, J. Anderson, S. Amarasinghe, B. Murphy, S. Liao, E. Bugnion, and M. Lam. Maximizing multiprocessor performance with the SUIF compiler. *IEEE Computer*, 29(12):84–89, 1996.
10. D. Lew. BCIR: A framework for the representation and manipulation of Java bytecode. Master's thesis, University of Toronto, 2000.
11. T. Lindholm and F. Yellin. *The Java Virtual Machine Specification*. Addison-Wesley, 1999.
12. C. D. Polychronopoulos, M. B. Girkar, M. R. Haghighat, C. L. Lee, B. P. Leung, and D. A. Schouten. The structure of Parafrase-2: An advanced parallelizing compiler for C and Fortran. In *LCPC Workshop*, pages 423–453, 1989.
13. G. E. Weaver, K. S. McKinley, and C. C. Weems. Score: A compiler representation for heterogeneous systems. In *Heterogeneous Computing Workshop*, 1996.

Optimizing Java-Specific Overheads: Java at the Speed of C?

Ronald S. Veldema, Thilo Kielmann, and Henri E. Bal

Division of Mathematics and Computer Science
Faculty of Sciences, Vrije Universiteit
{rveldema,kielmann,bal}@cs.vu.nl
http://www.cs.vu.nl/manta/

Abstract. Manta is a highly optimizing compiler that translates Java source code to binary executables. In this paper, we discuss four Java-specific code optimizations and their impact on application performance. We assess the execution time of three application kernels, comparing Manta with the IBM JIT 1.3.0, and with C-versions of the codes, compiled with GCC. With all three kernels, Manta generates faster code than the IBM JIT. With two kernels, the Manta versions are even faster than their C counterparts.

1 Introduction

Java has become increasingly popular as a general-purpose programming language. Key to Java's success is its intermediate "byte code" representation that can be exchanged and executed by Java Virtual Machines (JVMs) on almost any computing platform. Along with Java's widespread use, the need for a more efficient execution mode has become apparent. A common approach is to compile byte code to executable code. Modern JVMs come with a just-in-time compiler (JIT) that combines Java's platform independence with improved application speed [4]. A more radical approach is to completely avoid byte-code interpretation by statically compiling byte code to executable programs [1,7,12]. Combinations of JIT and static compilation are also possible [11].

A disadvantage of byte-code compilation, however, is the computational model of the byte code which implements a stack machine [14]. Mapping this (virtual) stack machine on existing CPUs (that are register based) is harder than directly generating register-oriented code. Our Manta compiler thus follows a different approach and translates Java *source code* to executable programs. Manta performs code optimizations across the boundaries of Java classes and source files by using a temporary database of the intermediate code of all classes of an application program. With this large base of information, the Manta compiler can generate very efficient code. In addition to generating efficient sequential code, a source-level compiler allows us to add slight language extensions, like new special interfaces, for purposes of parallel computing [8,13].

Until recently, many scientific codes have been implemented in the C language, mainly for reasons of efficiency. From a software-development point-of-view, it is desirable to use Java instead of C, allowing to use features like threads, objects, exceptions, runtime type checks, and array bounds checks in order to ease programming and

B. Hertzberger et al. (Eds.): HPCN Europe 2001, LNCS 2110, pp. 685–692, 2001.

debugging. Unfortunately, several of these features introduce performance overheads, making it difficult to obtain the same sequential speed for Java as for C. In this paper, we discuss to what extent these Java-specific overheads can be optimized away. We implemented a range of existing compiler optimizations for Java, and study their performance impact on application kernels.

2 The Manta Compiler

The Manta compiler is part of the Manta high-performance Java system. Intended for parallel computing, Manta also provides highly-efficient communication mechanisms like remote method invocation (RMI) [9] and object replication [8]. Figure 1 illustrates the compilation process. Manta directly translates Java source code to executables for the Intel x86 platform. The supported operating system is Linux. For extensive program analysis and optimization, the Manta compiler relies on its intermediate code data base. The availability of intermediate code for all classes of an application allows optimizations across the borders of classes and source files. Manta's efficient RMI mechanism comes with a compatibility mode to Sun's RMI, allowing Manta programs to communicate with other JVMs. For this purpose, the exchange of byte code is necessary. Manta uses *javac* to generate byte code which is also stored in the executable program for the sole purpose of sending it to a JVM along with an RMI. For receiving byte code, Manta executable programs contain a just-in-time compiler that compiles and dynamically links new classes into running programs. The treatment of byte code is described in [9]. In this paper, we focus on the generation and optimization of sequential programs by the Manta compiler.

The Manta compiler implements several standard code optimization techniques like common-subexpression elimination and loop unrolling [10]. The intermediate-code data base (see Figure 1) allows extensive, inter-procedural analyses, even across several classes of a Java application. In the following, we discuss four code optimization techniques (object inlining, method inlining, escape analysis, and bounds-check elimination) and two programmer assertions (closed-world assumption and bounds-check deactivation) that are related to Java's language features. Manta allows its optimizations and assertions to be turned on and off individually via command-line options.

2.1 Closed-World Assumption

Many compiler optimizations require knowledge about the complete set of Java classes that are part of an application. Java's polymorphism, in combination with dynamic class loading, however, prevents such optimizations. In this case, the programmer has to explicitly annotate methods as *final* in order to enable a large set of optimizations.

However, the *final* declaration has only limited applicability as it selectively disables polymorphism. Its use for improving application performance furthermore contradicts its original intention as a means for class-hierarchy design. Fortunately, many (scientific) high-performance applications consist of a fixed set of classes and do not use dynamic class loading at all. Such applications can be compiled under a *closed-world assumption*: all classes are available at compile time. The Manta compiler has a

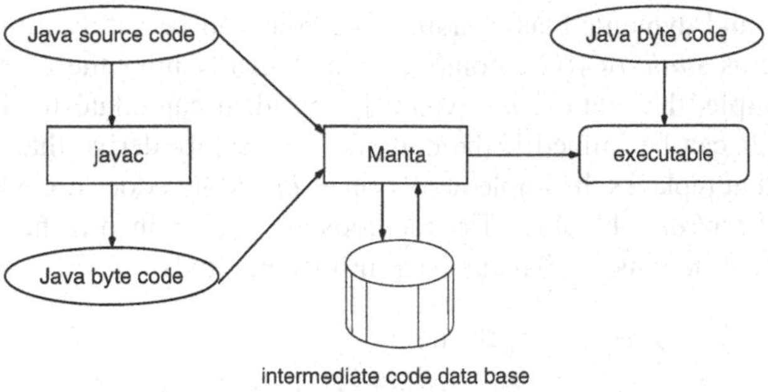

Fig. 1. The Manta Compilation Process.

command line option by which the programmer can assert or deny the validity of the closed-world assumption.

2.2 Object Inlining

Java's object model leads to many small objects with references to other objects. For improving application performance it is desirable to aggregate several small objects to a larger one, similar to C's struct mechanism. Such an aggregation is called *object inlining* [6]. Performance is improved by reducing overheads from object creation, garbage collection, and pointer dereferencing. The following code fragment shows an example of object inlining. When Manta can derive (via proper *final* declarations or the closed-world assumption) that the array a is never reassigned in objects of class A (left), then the array can be statically inlined into objects of class A (right). Note that the shown optimization can not be implemented manually because Java lacks a corresponding syntactical construct for arrays. In order to allow proper use, objects are inlined along with their header information, including, for example, vtables for method dispatch. For feasibility of the analysis, Manta inlines only objects created directly in field initializations.

`class A {` `    int [] a = new int[10]; }`	`class A {` `    int a[10]; }`
original class A with separate array object	A with inlined array

2.3 Method Inlining

In C-like languages, function inlining is a well-known optimization for avoiding the costs of function invocation. In object-oriented programs it is common to have many, small methods, making method inlining desirable for Java. For any object, the methods of its most-specific class have to be invoked, even after casting an object reference to a less specific class. These program semantics, which are at the core of object-oriented programming, prevent efficient method inlining. Only proper *final* declarations or the closed-world assumption enable efficient method inlining. However, in the presence of polymorphism and dynamic class loading, methods can be safely inlined if they are either declared as *static* or if the compiler can statically infer the object type. In the following example, the method *inc* would be an ideal candidate for inlining due to its small size. It can be inlined if the compiler can safely derive that there exists no subclass of *A* that replaces the implementation of *inc*. Manta does not inline methods in the presence of *try/catch* blocks. Also, methods are only inlined if they do not exceed 200 assembly instructions or 20 recursively inlined methods.

```
class A {  int a;
           void inc() { a++; }
           void other() { inc(); }  }
```

2.4 Escape Analysis

Escape analysis considers the objects created by a given method [2,5]. When the compiler can derive that such an object can never *escape* the scope of its creating thread (for example, by assignment to a static variable), then the object becomes unreachable after the method has terminated. In this case, object allocation and garbage collection can be avoided altogether by creating the object on the stack of the running thread rather than via the general-purpose (heap) memory. In the case of creation on the stack, method-local objects can be as efficient as function-local variables in C.

One problem with stack allocation of objects is the limited stack size. To resolve this, Manta maintains separate *escape stacks* that can grow independently of the call stack, while keeping the local objects apart from the general allocation and garbage-collection mechanism. To further minimize the size of the escape stacks, Manta detects repetitive creation of temporary objects in loop iterations and re-uses escape stack entries if possible. In general, Manta implements escape analysis using backward inter-procedural data-flow analysis combined with heap analysis.

2.5 Array Bounds-Check Deactivation and Elimination

The violation of array boundaries is a frequently occuring programming mistake with C-like languages. To avoid these mistakes, Java requires array bounds to be checked at runtime, possibly causing runtime exceptions [3]. This additional safety comes at the price of a performance penalty. A simple-minded, but unsafe optimization is to suppress the code generation for array-bounds checks altogether. The idea is that boundary violations will not occur after some successful, initial program testing with bounds checks activated. Completely deactivating array-bounds checks thus gives the unsafety of C at the speed of C. Manta has a command-line switch by which the programmer can assert that array-bounds checks can be completely deactivated.

A safe alternative to bounds-check deactivation is implemented in the Manta compiler. Inside the code of a method, Manta can safely eliminate those bounds checks that are known to repeat already successfully passed checks. For example, if a method accesses $a[i]$ in more than one statement, then only the first access needs a bounds check. For all other accesses, the checks can safely be omitted as long as Manta can derive from the code that neither the array base a nor the index i have been changed in the meantime. For this purpose, Manta performs a data-flow analysis, keeping track of the array bases and related sets of index identifiers for which bounds checks have already been issued [10]. The current implementation, however, does not yet perform this data-flow analysis across method boundaries. Currently, method invocation between array accesses empties the affected sets of already-checked array identifiers but does propagate this information to the called methods.

3 Application Kernel Performance

We investigated the performance of the code generated by Manta using three application kernels. For each kernel, we used two versions, one written in Java and one written in C. Both versions were made as similar to each other as possible. For each application, we compare the runtimes of the Manta-compiled code (with all assertions and optimizations turned on) with the same code run by the IBM JIT 1.3.0 and with the C version, compiled by GCC 2.95.2 with optimization level "-O3". We thus compare Manta with the (Linux) standard C compiler and the most competitive JIT. Unfortunately, performance numbers for other static Java compilers were not available to us.

Manta's best application speeds have been obtained asserting the closed-world assumption, enabling object inlining, method inlining, and escape analysis, while deactivating array-bounds checks. For evaluating the impact of the individual optimizations, we also provide runtimes with one optimization disabled while keeping all others turned on. Because the closed-world assumption has a strong impact on the other optimizations, we present runtimes for the other optimizations both with and without closed-world assumption. For array-bounds checks, we present runtimes comparing deactivation (included in the best times) with elimination, and general activation. All runtimes presented have been measured on a Pentium III, running at 800 MHz, using the Linux operating system (RedHat 7.0).

3.1 Iterative Deepening A* (IDA*)

Iterative Deepening A* (IDA*) is a combinatorial search algorithm. We use IDA* to solve random instances of the 15-puzzle (the sliding-tile puzzle). The search algorithm maintains a large stack of objects describing possible moves that have to be searched and evaluated. While doing so, many objects are created dynamically. Table 1 lists the runtimes for IDA*. The code generated by Manta, with all assertions and optimizations turned on, needs about 8.3 seconds, compared to 19.5 seconds with the IBM JIT. For IDA*, we actually implemented two C versions. One version keeps all data in static arrays and completes in 4 seconds. A second version emulates the behavior of the Java version by eagerly using *malloc* and *free* to dynamically allocate the search data structures. This version needs 15.6 seconds, almost twice as long as the Manta version. The comparison of the two C versions shows that applications written without object-oriented structure can be much faster than a Java-like version of the same code. However, with similar behavior of the application versions, Java programs can run as fast as C, or even faster.

As can also be seen from Table 1, the most efficient optimization for IDA* is object inlining which drastically reduces the number of dynamically created objects. Under the closed-world assumption, the other optimizations have hardly any impact, except for deactivating array bounds checking. The versions in which array bounds are either active or partially eliminated need 0.5 seconds longer than the best version that completely deactivates all array-bounds checks. Without the closed-world assumption, object inlining is not effective, so the runtimes are in the same order as with completely disabling object inlining.

Table 1. IDA* (15-Puzzle), Application Runtimes in Seconds.

compiler	best time	Manta, optimizations turned off individually	closed world	no closed world
IBM JIT 1.3.0	19.524	no object inlining	26.486	26.516
Manta, best	8.277	no method inlining	8.935	26.661
GCC 2.95.2 -O3	3.959	no escape analysis	8.915	26.561
GCC, malloc/free	15.591	bounds-check elimination	8.717	26.458
		bounds-check activation	8.850	26.631

3.2 Traveling Salesperson Problem (TSP)

TSP computes the shortest path for a salesperson to visit all cities in a given set exactly once, starting in one specific city. We use a branch-and-bound algorithm which prunes a large part of the search space by ignoring partial routes that are already longer than the current best solution. We run TSP with a 17-cities problem.

The compute-intensive part of the application is very small. It consists of two nested loops that iterate over an array containing the visited paths. As can be seen from Table 2, method inlining is the most effective optimization, by inlining the method calls inside the nested loops. The elimination of array bounds checks produces code that is as fast as the unsafe version with deactivated bounds checks. However, keeping all (unnecessary) bounds checks active, adds more than one second to the completion time. With its optimizations, the Manta-generated code needs only 4.5 seconds, compared to 4.7 seconds of the C version and 6.3 seconds with the JIT. The TSP example shows that Java programs can be optimized to run at a speed that is competitive to C.

3.3 Successive Overrelaxation (SOR)

Red/black SOR is an iterative method for solving discretized Laplace equations on a grid. We run SOR with 1000×1000 grid points. The compute-intensive part of the code runs a fixed number of array convolutions with floating-point arithmetic. As Table 3 shows, the most effective optimization is (safe) elimination of array-bounds checks. All other optimizations have hardly any impact on the overall speed. Manta's code completes after 1.5 seconds, slightly faster than the C version with 1.7 seconds, and substantially faster than the JIT version, which needs 4.4 seconds for the same problem.

Table 2. TSP (17 Cities), Application Runtimes in Seconds.

compiler	best time	Manta, optimizations turned off individually	closed world	no closed world
IBM JIT 1.3.0	6.307	no object inlining	4.513	4.530
Manta, best	4.511	no method inlining	7.158	7.181
GCC 2.95.2 -O3	4.696	no escape analysis	4.567	4.530
		bounds-check elimination	4.511	4.525
		bounds-check activation	5.612	5.612

Table 3. SOR (1000×1000 Grid points), Application Runtimes in Seconds.

compiler	best time	Manta, optimizations turned off individually	closed world	no closed world
IBM JIT 1.3.0	4.387	no object inlining	1.583	1.542
Manta, best	1.541	no method inlining	1.547	1.548
GCC 2.95.2 -O3	1.768	no escape analysis	1.579	1.541
		bounds-check elimination	1.569	1.548
		bounds-check activation	2.380	2.379

4 Conclusions

The Java language has several properties that make efficient execution more challenging than for C. Java programs typically use many small, dynamically created objects and short methods, resulting in high overheads. Also, Java is a safe language and thus requires array-bounds checking. In this paper, we investigated to what extent these overheads can be eliminated using compiler optimizations. We implemented four existing optimizations in the same compiler framework (Manta, a native source-to-binary compiler). We studied the impact of the optimizations on three application kernels. The results show that, with all optimizations switched on, two of the three applications run faster than C. Object inlining and method inlining each had a high impact on one application. Array bounds-check elimination saved about 35 % for the third application. This safe elimination leads to code which is almost as efficient as unsafe bounds-check deactivation for all three applications. Escape analysis was less effective. Allowing the compiler to use a closed-world assumption (i.e., disallow dynamic class loading) was shown to be essential for certain optimizations like object inlining. To summarize, the performance of application programs is typically dominated by a "bottleneck" which differs from program to program, and which has to be addressed by a specific optimization. In general, to execute Java programs at the speed similar to C versions, many optimizations, both standard techniques and Java-specific optimizations, have to be provided by a compiler.

Interesting future work will assess larger benchmarks (like the SCIMARK suite). Further tests will evaluate elements of Manta's runtime system like the thread package and the garbage collector.

Acknowledgments

The development of Manta is supported in part by a USF grant from the Vrije Universiteit. We thank Rutger Hofman, Ceriel Jacobs, Jason Maassen, and Rob van Nieuwpoort for their contributions to the Manta compiler and runtime system. We thank John Romein and Kees Verstoep for keeping our computing platform in good shape.

References

1. G. Antoniu, L. Bougé, P. Hatcher, M. MacBeth, K. McGuigan, and R. Namyst. The Hyperion system: Compiling multithreaded Java bytecode for distributed execution. *Parallel Computing*, 2001. To appear.
2. B. Blanchet. Escape Analysis for Object Oriented Languages. Application to Java. In *Proc. OOPSLA'99*, pages 20–34, Denver, CO, Nov. 1999.
3. R. Bodik, R. Gupta, and V. Sarkar. ABCD: Eliminating Array Bounds Checks on Demand. In *Proc. ACM SIGPLAN 2000 Conference on Programming Language Design and Implementation (PLDI 2000)*, pages 321–333, Vancouver, BC, June 2000.
4. M. Burke, J.-D. Choi, S. Fink, D. Grove, M.Hind, V. Sarkar, M. Serrano, V. C. Sreedhar, H. Srinivasan, and J. Whaley. The Jalapeño Dynamic Optimizing Compiler for Java. In *ACM 1999 Java Grande Conference*, pages 129–141, San Francisco, CA, June 1999.
5. J. Choi, M. Gupta, M. Serrano, V. Sreedhar, and S. Midkiff. Escape Analysis for Java. In *Proc. OOPSLA'99*, pages 1–19, Denver, CO, Nov. 1999.
6. J. Dolby. Automatic inline allocation of objects. In *Proc. 1997 ACM SIGPLAN Conf. on Programming Language Design and Implementation*, pages 7–17, June 1997.
7. R. Fitzgerald, T. Knoblock, E. Ruf, B. Steensgard, and D. Tarditi. Marmot: An optimizing compiler for Java. Technical report 33, Microsoft Research, 1999.
8. J. Maassen, T. Kielmann, and H. E. Bal. Parallel Application Experience with Replicated Method Invocation. *Concurrency and Computation: Practice and Experience*, 2001.
9. J. Maassen, R. van Nieuwpoort, R. Veldema, H. E. Bal, and A. Plaat. An Efficient Implementation of Java's Remote Method Invocation. In *Seventh ACM SIGPLAN Symposium on Principles and Practice of Parallel Programming (PPoPP'99)*, pages 173–182, Atlanta, GA, May 1999.
10. S. Muchnick. *Advanced Compiler Design and Implementation*. Morgan Kaufmann Publishers, 1997.
11. M. Serrano, R. Bordawekar, S. Midkiff, and M. Gupta. Quicksilver: A Quasi-Static Compiler for Java. In *Proc. OOPSLA'00*, pages 66–82, Minneapolis, MN, Oct. 2000.
12. V. Seshadri. IBM High Performance compiler for Java. *AIXpert Magazine*, Sept. 1997. `http://www.developer.ibm.com/library/aixpert`.
13. R. V. van Nieuwpoort, T. Kielmann, and H. E. Bal. Satin: Efficient Parallel Divide-and-Conquer in Java. In *Proc. Euro-PAR 2000*, number 1900 in Lecture Notes in Computer Science, pages 690–699, Munich, Germany, Aug. 2000. Springer.
14. B.-S. Yang, S.-M. Moon, S. Park, J. Lee, S. Lee, J. Park, Y. C. Chung, S. Kim, K. Ebcioğlu, and E. Altman. LaTTe: A Java VM Just-in-Time Compiler with Fast and Efficient Register Allocation. In *Proc. Int. Conf. on Parallel Architectures and Compilation Techniques (PACT'99)*, 1999.

Combining Batch and Streaming Paradigms for Metacomputing Applications

Torsten Fink[1] and Stephan Kindermann[2]

[1] Freie Universität von Berlin, Germany
tnfink@computer.org
[2] Friedrich-Alexander-Universität Erlangen-Nürnberg, Germany
snkinder@cs.fau.de

Abstract End users need practical, abstract tools to specify and control multi-component applications using the heterogeneous, distributed resources of metacomputing platforms. We present an abstract component based programming model for the Amica metacomputing infrastructure implemented with Java and CORBA. In this model multi-component applications can be simply built using control and data flow component composition mechanisms. We concentrate on the new streaming extension of Amica. The basic streaming model is introduced along with its concrete implementation. An example illustrates its use.

1 Introduction

Application development in wide area networks is frequently based on the composition of legacy application components. Thus the expressiveness of the composition mechanisms available is essential for meta-application design on computational grids. Two composition mechanisms are commonly used separately:

- control flow-based gluing of components, which allows to specify and control batch processing (e.g. bag of task-like) activities abstractly,
- data flow-wiring of components based on a streaming concept, which provides pipeline parallelism and computation/communication overlapping in an abstract way.

Existing approaches to the abstract design of component based meta-application often concentrate on one of these composition mechanisms (e.g.[6,1]). Others provide no generic abstractions of the mechanisms they use and are aimed at specific application environments (e.g. [12]).

In this paper we describe the integration of streaming composition mechanisms in our abstract meta-application design environment Amica[1]. First, we give a short introduction to the abstract component-based meta-application design front-end and our prototyping metacomputing infrastructure Amica (Section 2). In Section 3, we concentrate on the description of the streaming concept and its integration in Amica. In Section 4, we present an application example of

[1] **A**bstract **M**etacomputing **I**nfrastructure for **C**oarse Grained **A**pplications.

B. Hertzberger et al. (Eds.): HPCN Europe 2001, LNCS 2110, pp. 693–700, 2001.

the new streaming connectors and their integration into control and data flow based meta-application graphs. Finally, we give an overview of related work and provide a brief outlook on future extensions.

2 Combining Batch and Streaming Paradigms

The ability to use collections of powerful computers transparently as metacomputers for, e.g., complex scientific and engineering applications leads to the generation of unprecedented amounts of data. Thus a pure computation-centric view of distributed activities in a metacomputer (reflected, e.g., in the specification of the sequence in which components are activated in control flow graphs) has to be extended to specify data pools and data flow. In such a data-centric view, the specification of data sources, sinks, and data streaming connections dictates the resource allocation in the distributed metacomputing infrastructure.

Both views have to be supported jointly in a general meta-application programming model as many applications are most naturally described as a combination of control and data flow based component interactions.

The abstract programming model of the Amica infrastructure is based on an abstract vocabulary of components and connectors glued together to constitute an application graph. Components specify the usage of computation and data storage services while connectors define the control flow and data access. For more details see [8].

This basic model is now extended by streaming connectors reflecting an abstract streaming model of data exchange. The resulting abstract vocabulary to build meta-applications for Amica is illustrated in Fig. 1.

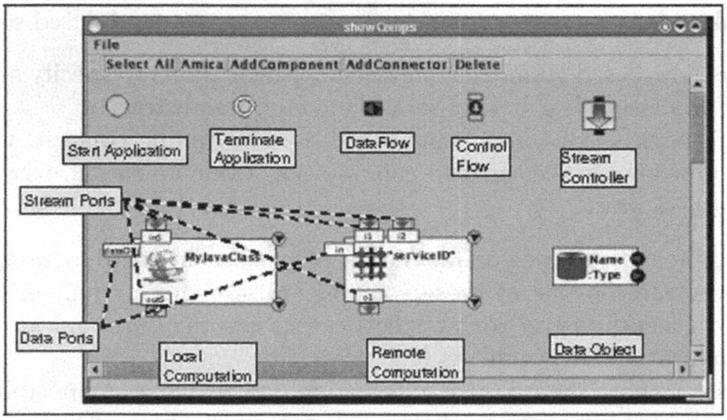

Figure 1. The Elements of the Amica Programming Model.

The composition of local and remote computation/storage components is done using a graphical editor (see the screen-shot in section 4). Control flow connectors specify the forwarding of control tokens, and the activation and deactivation of components. Data flow connectors link computation and data storage components. A stream controller connects data-stream producers to stream consumers. It supports multiple producers/consumers according to different stream forwarding policies.

3 Metacomputing in Amica

Amica provides a CORBA/Java based prototyping infrastructure to transparently access distributed, heterogeneous computing and storage resources. Computing resources are made available through service factories. Data storage resources are accessed through data objects residing in a transparent network of data stores. This transparent data network supports data caching and replica management. For a more detailed description of the Amica infrastructure see [3].

Meta-applications on top of the Amica infrastructure are given as component compositions forming an application graph. This graph is internally represented with the architecture description language Acme [5]. This allows for static analysis techniques, prior to evaluation (see [8]). As illustrated in figure 2, the evaluation is done by an interpreter, which forwards abstract data storage and computing requests specified by components in the application graph to concrete service providers.

Service providers can be local computations performed by Java objects or remote computation and storage made available through a broker mechanism and service factories taking into account the current load in the distributed system. For a more detailed description see [3].

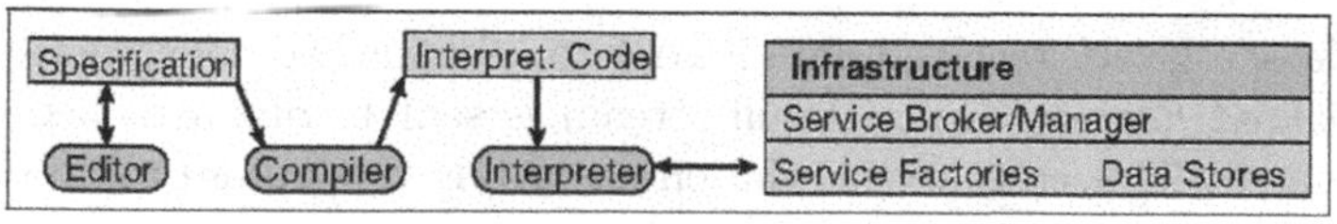

Figure 2. An Overview of Amica.

Two activation mechanisms for components in an application graph are provided:

- either *explicitly* according to their composition via control flow connectors, starting with a specific start connector, or
- *implicitly* by the first arrival of data via stream connectors.

When a component is activated it looks for an appropriate service factory with the help of a broker and instantiates a new remote service with the given parameters. This service accesses its input and output data by CORBA interfaces to the associated data objects or by stream controllers. After deactivation it fires an event to its outgoing control flow port, which activates all components attached to it with a control flow connector.

In the following we concentrate on the integration of the data streaming connectors into the Amica infrastructure. We first describe the abstract streaming model and then its concrete implementation.

3.1 The Streaming Concept in Amica

A stream consists of a series of data of the same type. Stream connectors combine arbitrary many incoming streams from producer components and send the resulting outgoing streams to attached consumer components.

Currently a push data-driven model is supported where producers write data into streaming connectors and consumers are automatically informed of incoming data. Additionally, consumers can access streaming data via a polling interface. Producers are blocked in the case data cannot be forwarded. If the consumer has no buffer space left to store an incoming stream element, its read method blocks until the data can be stored.

In general a streaming connector can be parameterized with policies to define the semantics of multiple stream handling. Two policies manage multiple incoming streams:

- *Mix streams:* Every arriving data from a consumer is directly sent to the output policy. Incoming streams are mixed together and form new streams.
- *FCFS stream:* An incoming stream is completely sent to the output policy. All other streams which want to send their data are blocked. Full streams are consumed in the order of arrival.

Similarly, there are two policies to handle outgoing streams:

- *Broadcast:* Every received data is sent to all attached consumers.
- *Distributed Round Robin:* A full stream is sent to one attached consumer component. This consumer is chosen according to a round robin scheme.

3.2 Implementation of the Streaming Concept

A stream controller connector is implemented as a CORBA object. For every stream data type **T** a pair of IDL interfaces `TStreamConsumer`, `TStream-Producer` as well as a `TStreamController` interface are defined.

Stream controllers are generated at runtime on demand, i.e., the first time a component asks for a reference to the controller. It is created for a specific element type and for an input and an output policy. These parameters cannot be changed after generation. They are generated by and at data stores because data stores are always close to remote services. This allows us to find an efficient location for a controller at runtime in the Amica infrastructure.

Components which use remote services handle data streams automatically. The application programmer has only to specify the needed service and its parameters. Figure 3 depicts the classes used to access data streams from local components written in Java. Application programmers derive their classes from `Base`. If they need access to an incoming stream attached to a port labeled `id` they can get a reference to an associated `StreamReceiver` helper object by invoking `getStreamReceiver(id)`.

For every stream type there is a derived subclass of `StreamReceiver`. In the class diagram only `StringStreamReceiver` is depicted. There are two approaches to receive data. First, a `get` method can be used which blocks until data arrives or the stream ends. In the later case `null` is returned. Second, an object can register itself at the `StreamReceiver` as `StreamEventListener`. Every incoming data is then automatically forwarded to all registered listeners.

The `StreamReceiver` object is configured to one of these modes by `setExplicitUsage()` and `setEventUsage()`. It buffers incoming data. When it is switched from explicit to event mode, all buffered data is sent to the listeners. Outgoing streams are accessed using the `StreamSender` helper class. Its send-method directly sends the data to the attached stream.

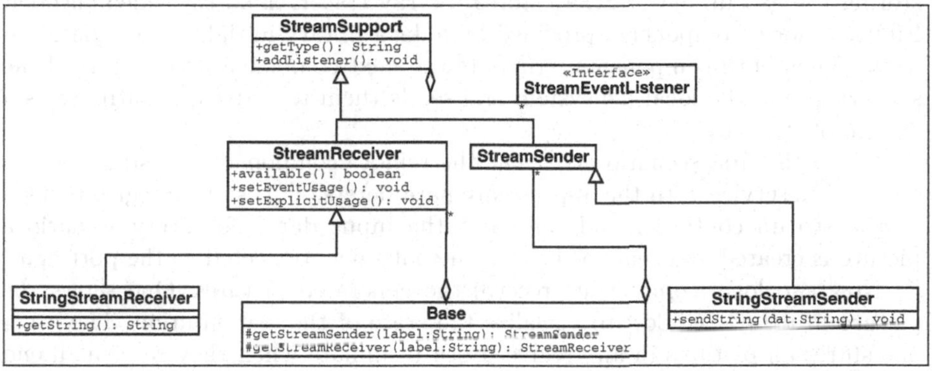

Figure 3. Class Diagram of the Stream Support Classes for Local Access.

4 An Example

The following example is used to illustrate the meta-application development for Amica based on control- and data-flow, and streaming compositions of application components. It consists of a distributed application which creates a movie using off-the-shelves tools to render a picture from a 3D scenario and to generate an MPEG movie. An Amica service was used which allows for placing an executable binary in an Amica data object and executing it on a remote resource. The implementation of this service is described in depth in [4]. We extended the service to handle streaming data.

The service needs a parameter which defines how to start the executable. This parameter is a string comprising parameters for the executable itself and some special commands defining pre- and post-operations. For example the value `"-i $<(#in,inF) -o $>(#out,outF)"` would

- save the contents of the data object attached to port `in` of the component to the remote file system in a file named `inF`,
- start the executable with command line parameters `"-i inF -o outF"`,
- and store after termination of the executable the file `outF` in the data object connected to port `out`.

The adaptation to streams is straightforward. If there is a stream controller attached to a referenced port instead of a data object, every stream data invokes an execution of the service. Instead of the contents of a data object the single stream data is used. If there is more than one stream producer connected to the service an invocation is performed when stream data from every connected producer is available. An end signal from any producer terminates the service. If there is pending stream data it is ignored.

Figure 4 shows the application of this case study. There are two data objects storing the executables. `PovRay` stores the ray tracer, `MPEG` the video encoder. First two local components provided by a Java library initialize these data objects. When both components terminate the next component starts to read the scenarios from the local file system and sends them to a stream controller as a stream of strings.

When the first scenario arrives at the remote component it instantiates an `execution` service with the appropriate parameters. This service registers itself at the stream controller and computes the input data. For every scenario a picture is created and sent to the stream controller connected to the port `tga`.

Now two local components receive the generated pictures. One shows the picture in a Java window to visualize the state of the computation. The other one stores all pictures in an archive. Both terminate when they receive an end signal.

This activates the next remote component. It loads and extracts the archive in a remote file system, generates an MPEG video file, and puts it to a data object. Finally, this video is stored in the local file system by a local component, and the application terminates.

5 Related Work

Component based metacomputing application development is currently very restricted in terms of available component composition mechanisms.

Most often some form of coarse grain data-flow graph model is used, in which components are activated as soon as all available input data is available. Examples include UNICORE [1], GIS [12] and WebFlow [6]. Whereas here some form of task descriptors are used for abstract component description, the CAT [10] provides a general component model for metacomputing application modules. Nevertheless also CAT does not define any module composition framework for end-users.

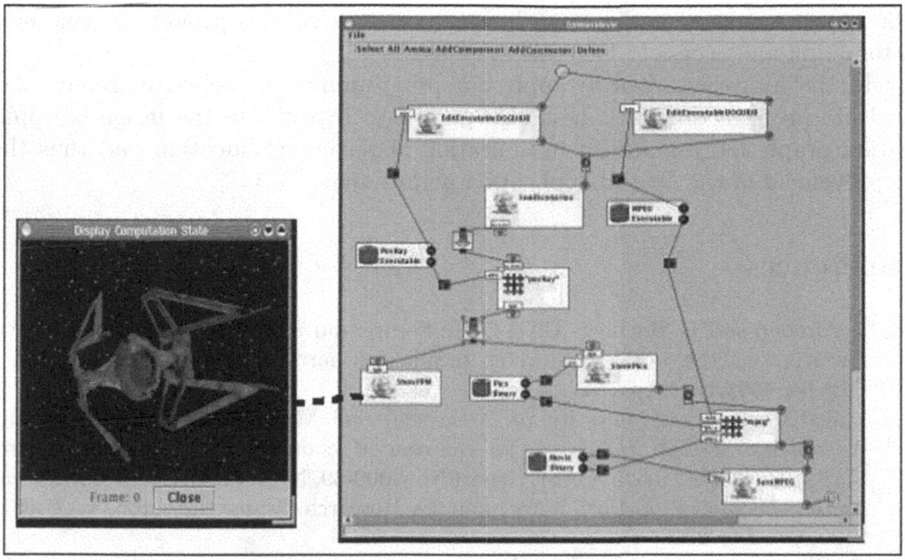

Figure 4. Visual Frontend for Amica.

More general and flexible systems allow the specification of metacomputing applications based on well defined component libraries, like Arcade [2], the problem solution environment presented in [9,11] and the virtual distributed computing environment VDCE [7]. Whereas these frameworks are very flexible with respect to the underlying metacomputing infrastructure, the available component composition tools (visual and script based) are restrictive and unflexible, providing basic control structs such as loops and conditionals.

In contrast to these approaches Amica provides a formal extensible foundation for component composition based on a connector library. In this paper we concentrated on the extension towards streaming-composition missing in current metacomputing application development environments.

6 Conclusions and Future Work

To summarize, we have shown an extension of the component-graph based programming model of Amica to support streaming applications. Thus, means to express and control batch style remote computing are combined with means to control stream based data-flow computing. Both mechanisms are accessible through an intuitive abstract, visual programming interface supporting component based application design.

The Amica infrastructure provides a location-transparent access to the basic computation and storage services, thus freeing the application developer to concentrate on the high-level aspects of application design. Non-remote, local components written in Java can be easily integrated and provide flexible user interaction and access to local file systems.

In the future we plan to apply our programming model to more complex real life problem domains. It is also planed to investigate the usage of application graph analysis for an optimization of resource allocation and thus the improvement of the overall application performance.

References

1. J. Almond and D. Snelling. UNICORE: Secure and uniform access to distributed resources via the world wide web. Technical report, Forschungszentrum Jülich, October 1998.
2. Zhikai Chen, Kurt Maly, Piyush Mehrotra, and Mohammad Zubair. Arcade: A web-Java based framework for distributed computing. Technical Report NASA/CR-2000-210545 ICASE Report No. 2000-39, NASA Langley Research Center, ICASE Mail Stop 132C NASA Langley Research Center Hampton, VA 23681-2199, October 2000.
3. T. Fink and S. Kindermann. First steps in metacomputing with Amica. In *Euromicro-PDP 2000*, pages 197–204. IEEE Computer Society, 2000.
4. Torsten Fink. Integrating MPI components into metacomputing applications. In Jack Dongarra, Peter Kacsuk, and Norbert Podhorszki, editors, *EuroPVM/MPI'2000*, number 1908 in LNCS, pages 208–215. Springer, 2000.
5. D. Garlan, R.T. Monroe, and D. Wile. Acme: An architecture description interchange language. In *Proceedings of CASCON '97*, 1997.
6. T. Haupt, E. Akarsu, and G. Fox. WebFlow: A framework for Web based metacomputing. In *HPCN Europe '99*, number 1593 in LNCS. Springer, 1999.
7. E. Houstis, R. Bramley, and E. Gollopoulos, editors. *Problem Solving Environments*, chapter A Problem Solving Environment for Network Computing. IEEE Computer Society, 1998.
8. Stephan Kindermann and Torsten Fink. An architectural meta-application model for coarse grained metacomputing. In A. Bode, T. Ludwig, W. Karl, and R. Wismüller, editors, *Euro-Par 2000, Parallel Processing*, number 1900 in LNCS. Springer, 2000.
9. Matthew S. Shields, Omer Rana, David W. Walker, Maozhen Li, and David Golby. A Java/CORBA-based visual program composition environment for PSEs. *Concurrency - Practice and Experience*, 12(8):687–704, 2000.
10. J. Villacis, M. Govindaraju, D. Stern, A. Withaker, F. Berg, P. Deuskar, T. Benjamin, D. Gannon, and R. Bramley. CAT: A high performance, distributed component architecture toolkit for the grid. In *Proceedings of the High Performance Distributed Computing Conference*, 1999.
11. David W. Walker, Maozhen Li, Omer Rana, Matthew S. Shields, and Y. Huang. The software architecture of a distributed problem-solving environment. *Concurrency - Practice and Experience*, 12(15):1455–1480, 2000.
12. D. Webb, A. Wendelborn, and K. Maciunas. Process Networks as a High-Level Notation for Metacomputing. In *IPPS'99, workshop on Java for Distributed Computing*, April 1999.

Identification and Quantification of Hotspots in Java Grande Programs

John Waldron and David Gregg

Trinity College Dublin

Abstract. In this paper we present a platform independent analysis of the dynamic profiles of the the Java Grande Forum benchmark suite. A useful platform independent metric for analysing eventual running time is to use a cost center design pattern to "charge" each bytecode to the appropriate method. This necessitates accessing an item in a hash table dictionary each time round the interpreter inner loop and incrementing a counter. The measurements from the experiment are significantly different for either static or dynamic method frequencies and show that for the Grande programs in the Java Grande Forum Benchmark suite 66% of the execution time, as measured by bytecode use, is spent in the top two methods.

1 Introduction

The Java paradigm for executing programs is a two stage process. Firstly the source is converted into a platform independent intermediate representation, consisting of bytecode and other information stored in class files. The second stage of the process involves hardware specific conversions, perhaps by a JIT compiler for the particular hardware in question, followed by the execution of the code. The problem addressed by this research is that while there exist static tools such as class file viewers to look at this intermediate representation, there is currently no easy way of studying the dynamic behaviour at this point in the program. This research therefore sets out to perform dynamic analysis at the platform independent level and investigate whether or not useful results can be gained. In order to test the technique, the Java Grande Forum's Benchmark suite was used.

Specifically this research sets out to establish where most of the execution time for a Grande program is spend in a platform independent way without knowledge of the underlying platform specific phase. It extends existing work [1] by dynamically totaling the bytecodes executed by every method. Both static and dynamic method call frequencies are measured, but it is concluded that there is not necessarily any correlation between the number of times a method is called (i.e. its frequency) and the percentage of execution time that will go to that method. For example, java/lang/Object<init> may be called often, but may only execute one bytecode, whereas another method may only be called once but execute many more bytecodes in total. At the platform independent

B. Hertzberger et al. (Eds.): HPCN Europe 2001, LNCS 2110, pp. 701–710, 2001.

level it is assumed that within a method there will be a good correlation between the number of bytecodes that would be executed if interpreted, and the ultimate time that would be spent in the method when executed in some platform specific way.

In order to study dynamic method usage it was necessary to modify the source code of a Java Virtual Machine. Kaffe [2] is an independent implementation of the Java Virtual Machine which was written from scratch and is free from all third party royalties and license restrictions. It comes with its own standard class libraries, including Beans and Abstract Window Toolkit (AWT), native libraries, and a highly configurable virtual machine with a JIT compiler for enhanced performance. Kaffe is available under the Open Source Initiative and comes with complete source code, distributed under the GNU Public License. Versions 1.0.6 was used for these measurements.

In order to modify the Kaffe virtual machine to accumulate dynamic platform independent statistics, most of the alterations are made in the machine.c file. The simplest measurement, how many of the bytecodes are in the API, can be made in the interpreter loop in the runVirtualMachine() function. Of course since it is in the inner loop, it will impact execution speed when the measurement is being performed. In order to measure dynamic method call frequencies, it is necessary to use a hash table dictionary with method names as keys. This can be called once per method in the virtualMachine() function. The best metric, however, for estimating eventual running time is to use a cost center design pattern to "charge" each bytecode to the appropriate method. This would necessitates accessing an item in the hash table dictionary each time round the interpreter inner loop in the run runVirtualMachine() function and incrementing a counter, which would slow down execution significantly (about 8 times interpreted speed) while the measurement is being performed. To improve performance, when a method is invoked, a local variable in the virtual machines stack frame can be initialised to point to the counter for that method, requiring only one hash table access per method invocation and only doubling the running time relative to interpretation.

2 Grande Programs Measured

A *Grande* application is one which uses large amounts of processing, I/O, network bandwidth or memory. The Java Grande Forum Benchmark Suite [3] is intended to be representative of such applications, and thus to provide a basis for measuring and comparing alternative Java execution environments. It is intended that the suite should include not only applications in science and engineering but also, for example, corporate databases and financial simulations.

- The **moldyn** benchmark is a translation of a Fortran program designed to model the interaction of molecular particles. Its origin as non object-oriented code probably explains its relatively unusual profile, with a few methods which make intensive use of fields within the class, even for temporary and

loop-control variables. This program may still represent a large number of Grande type applications that will initially run on the JVM
- The **search** benchmark solves a game of connect-4 on a 6×7 board using alpha-beta pruning. Intended to be memory and numerically intensive, this is also the only application to demonstrate an inheritance hierarchy of depth greater than 2.
- The **euler** benchmark solves a set of equations using a fourth order Runge-Kutta method. This suite demonstrates a considerable clustering of functionality in the `Tunnel` class, as well as a comparatively high percentage of methods with very large local variable requirements.
- The **raytracer** measures the performance of a 3D ray tracer rendering a scene containing 64 spheres. It is represented using a fairly shallow inheritance tree, with functionality (as measured in methods) fairly well distributed throughout the classes.
- The **montecarlo** benchmark is a financial simulation using Monte Carlo techniques to price products derived from the price of an underlying asset. Its use of classical object-oriented get and set methods accounts for the relatively high proportion of methods with no temporary variables and 1 or 2 parameters (including the `this`-reference).

Version 2.0 of the suite (Size A) was used. The default Kaffe maximum heap size of 64M was sufficient for all programs except *mon* which needed a maximum heap size of 128M. The *ray* application failed its validation test when interpreted, but as the failure was by a small amount, it was included in the measurements.

Table 1. *Measurements of total number of method calls including native calls by Grande applications compiled using SUNs javac compiler, Standard Edition (JDK build 1.3.0 C). Also shown is the percentage of the total which are in the API, and percentage of total which are in API and are native methods.*

Program	Total methods	API %	API native %
eul	3.34e+07	58.0	12.6
mol	5.49e+05	22.7	19.9
mon	8.07e+07	98.7	37.4
ray	4.58e+08	3.1	1.6
sea	7.12e+07	0.0	0.0
ave	1.29e+08	36.5	14.3

3 Dynamic Method Execution Frequencies

In this section we present our dynamic profile of the Grande programs studied. Here we partition the execution profiles based on methods, since these provide both a logical level of modularity at source-code level, as well as a likely unit

Table 2. *Measurements of Java method calls excluding native calls made by Grande applications compiled using SUNs javac compiler, Standard Edition (JDK build 1.3.0-C).*

Program	Java method calls		bytecodes executed	
	number	% in API	number	% in API
eul	2.92e+07	51.9	1.46e+10	0.5
mol	4.40e+05	3.4	7.60e+09	0.0
mon	5.05e+07	97.9	2.63e+09	38.0
ray	4.50e+08	1.5	1.18e+10	0.1
sea	7.12e+07	0.0	7.10e+09	0.0
ave	1.20e+08	30.9	8.75e+09	7.7

Table 3. *Breakdown of Java (non-native) API method dynamic usage percentages by package for Grande applications compiled using SUNs javac compiler, Standard Edition (JDK build 1.3.0-C). None of the applications used methods from the applet, awt, beans, math, security or sql packages.*

	eul	mol	mon	ray	sea	ave
io	2.4	2.9	0.0	0.0	3.0	1.7
lang	97.6	82.3	2.3	100.0	80.2	72.5
net	0.0	0.8	0.0	0.0	1.1	0.4
text	0.0	0.3	0.0	0.0	0.0	0.1
util	0.0	13.7	97.6	0.0	15.7	25.4

Table 4. *Breakdown of Java (non-native) API bytecode percentages by package for Grande applications compiled using SUNs javac compiler, Standard Edition (JDK build 1.3.0-C). None of the applications used methods from the applet, awt, beans, math, security or sql packages.*

	eul	mol	mon	ray	sea	ave
io	7.6	1.2	0.3	0.0	1.2	2.1
lang	92.2	69.5	2.0	99.3	69.6	66.5
net	0.0	1.1	0.0	0.0	1.3	0.5
text	0.0	0.6	0.0	0.0	0.0	0.1
util	0.1	27.6	97.7	0.7	28.0	30.8

of granularity for hotspot analysis. It should be noted that these figures are not the usual *time-based* analysis, but are based on the more platform-independent method frequency and bytecode usage analyses.

Table 1 shows measurements of the total number of method calls including native calls by Grande applications [1]. For the programs studied, on average 14.3% of methods are API methods which are implemented by native code. As the benchmark suite is written in Java it is possible to conclude that any native methods are in the API. This paper is confined to studying how the Java methods execute. Table 1 must be interpreted carefully as it is a method frequency table, without reference to bytecode usage, and so may not correlate with eventual

Table 5. Static *Method Call Frequencies (Including Native Methods).*

Method	static freq
eul	
euler/Statevector/<init>	10.6
java/lang/Math/sqrt	5.6
java/io/PrintStream/println	4.9
java/lang/Math/pow	4.2
java/lang/Math/log	4.2
java/lang/Math/abs	4.2
euler/Vector2/magnitude	4.2
euler/Vector2/dot	4.2
mol	
java/lang/StringBuffer/append	9.5
java/lang/Math/sqrt	7.9
moldyn/random/seed	4.8
java/text/NumberFormat/setMinimumFractionDigits	4.8
java/text/NumberFormat/setMaximumFractionDigits	4.8
java/text/NumberFormat/getInstance	4.8
java/lang/Object/<init>	4.8
moldyn/random/update	3.2
mon	
java/lang/StringBuffer/append	15.3
montecarlo/DemoException/<init>	8.4
java/lang/StringBuffer/<init>	5.0
java/lang/StringBuffer/toString	4.8
montecarlo/Universal/set_prompt	3.2
montecarlo/Universal/set_DEBUG	3.2
montecarlo/PathId/<init>	2.5
montecarlo/PathId/get_dTime	1.8
ray	
raytracer/Vec/<init>	12.6
java/lang/StringBuffer/append	11.7
java/lang/Object/<init>	6.8
raytracer/Vec/dot	4.9
raytracer/Vec/normalize	4.4
java/lang/StringBuffer/toString	2.9
java/lang/StringBuffer/<init>	2.9
raytracer/Scene/addLight	2.4
sea	
java/lang/StringBuffer/append	30.4
java/lang/StringBuffer/toString	7.6
java/lang/StringBuffer/<init>	7.6
search/Game/wins	5.4
search/TransGame/transpose	3.3
search/TransGame/hash	3.3
search/SearchGame/ab	3.3
search/Game/makemove	3.3

Table 6. *Dynamic Method Execution Frequencies (Including Native Methods†).*

Method	Frequency
eul	
java/lang/Math.abs	24.5
java/lang/Object.<init>	19.6
euler/Statevector.<init>	19.6
euler/Statevector.svect	19.2
java/lang/Math.sqrt†	11.5
euler/Vector2.dot	1.8
euler/Vector2.magnitude	1.4
java/lang/Math.pow†	0.3
mol	
java/lang/Math.sqrt†	19.2
moldyn/particle.velavg	18.6
moldyn/particle.mkekin	18.6
moldyn/particle.force	18.6
moldyn/particle.domove	18.6
moldyn/random.update	1.4
java/lang/String.indexOf	1.0
moldyn/random.seed	0.6
mon	
java/util/Random.next	31.5
java/lang/Math.log†	18.6
java/util/Random.nextDouble	15.8
java/util/Random.nextGaussian	12.4
java/lang/Math.exp†	12.4
java/lang/Math.sqrt†	6.2
java/lang/StringBuffer.append	0.3
java/lang/System.arraycopy†	0.2
ray	
raytracer/Vec.dot	47.0
raytracer/Vec.sub2	23.2
raytracer/Sphere.intersect	22.8
java/lang/Math.sqrt†	1.6
java/lang/Object.<init>	1.3
raytracer/Vec.<init>	0.7
raytracer/Vec.normalize	0.6
raytracer/Isect.<init>	0.6
sea	
search/Game.wins	46.5
search/SearchGame.ab	10.3
search/Game.makemove	10.3
search/Game.backmove	10.3
search/TransGame.hash	9.3
search/TransGame.transpose	5.3
search/TransGame.transtore	4.0
search/TransGame.transput	4.0

Table 7. *Dynamic Method Bytecode Percentages.*

Method	Percentage
eul	
euler/Tunnel.calculateR	51.3
euler/Tunnel.calculateDamping	16.0
euler/Tunnel.doIteration	8.7
euler/Tunnel.calculateG	6.6
euler/Tunnel.calculateF	6.6
euler/Tunnel.calculateStateVar	4.1
euler/Tunnel.calculateDeltaT	3.3
euler/Statevector.svect	1.5
mol	
moldyn/particle.force	99.6
moldyn/particle.mkekin	0.1
moldyn/particle.domove	0.1
moldyn/md.runiters	0.1
moldyn/random.update	0.0
moldyn/random.seed	0.0
moldyn/random.<init>	0.0
moldyn/particle.velavg	0.0
mon	
montecarlo/ReturnPath.computeVariance	19.0
java/util/Random.next	17.4
java/util/Random.nextGaussian	12.4
montecarlo/ReturnPath.computeMean	10.6
montecarlo/MonteCarloPath.computeFluctuationsGaussian	10.3
montecarlo/MonteCarloPath.computePathValue	8.0
montecarlo/RatePath.getReturnCompounded	7.6
java/util/Random.nextDouble	7.3
ray	
raytracer/Vec.dot	32.8
raytracer/Sphere.intersect	29.5
raytracer/Vec.sub2	19.8
raytracer/RayTracer.intersect	14.0
raytracer/Vec.normalize	1.0
raytracer/RayTracer.shade	1.0
raytracer/Vec.<init>	0.3
raytracer/Vec.comb	0.3
sea	
search/Game.wins	32.6
search/SearchGame.ab	30.7
search/TransGame.hash	8.2
search/Game.makemove	8.1
search/Game.backmove	7.9
search/TransGame.transpose	7.3
search/TransGame.transput	3.8
search/TransGame.transtore	0.6

running times. For example, there is no guarantee that API methods have the same bytecode frequencies or execution times as non-API methods.

Table 2 shows measurements of the Java method calls excluding native calls. Java method *execution* is mostly (92% on average) in the non-API bytecodes of the programs. This is a significant difference from traditional Java applications such as applets or compiler type tools which spend most of the time in the API [4]. Mixed compiled interpreted systems which precompile the API methods to some native format will therefore not be as effective at speeding up Grande applications like these. The finding that API usage is very low may imply that the benchmark suite may not be fully representative of a broad range of Grande applications (see Table 2). It is interesting to observe that while 98% of Java methods are API for the *mon* benchmark, only 38% of the bytecodes executed. Again, this point highlights the greater information provided by a bytecode level analysis. All measurements in this paper were made with the Kaffe API library, which may differ from other Java API libraries.

Table 3 shows dynamic measurements of the Java API package method percentages and Table 4 shows API bytecode percentages. The figures in these two tables are broadly similar, implying the API methods each execute the same number of bytecodes. As would be expected for the programs considered, the applet and awt packages are not used at all as graphics has been removed from the benchmarks. Of interest is that the math package is not used by the benchmarks which simply use the java.lang.Math class. java.math contains only the two classes BigDecimal and BigInteger, which are not that common in Grande applications. A Grande application should use large amounts of processing, I/O, network bandwidth or memory, yet it is interesting to note how little of the API packages are dynamically used by this benchmark suite.

Table 5 shows Static method call frequencies including native methods, Table 6 dynamic method execution frequencies including native methods and Table 7 dynamic bytecode percentages for the most heavily used methods for the Grande applications. The static data was collected by counting the number of invoke instructions in the program bytecode files, and so will not include API methods which call other API methods. Instances of this will however be recorded dynamically. A gradation can be seen across the three tables – generally the static data Table 5 gives a reasonably high frequency for all methods shown, whereas dynamic execution frequencies Table 6 tend to concentrate in the top 4. This can be most clearly seen with *mol* where the static frequency is relatively evenly distributed. When the measurements for bytecode percentages are taken, the situation is even more compressed. In the most extreme case, 99.6% of bytcodes for *mol* are in one method moldyn/particle.force.

As expected this experiment produces dramatic differences. For example, java/lang/StringBuffer/append appears at very high frequencies statically — its the top method for *mol*, *mon* and *sea* applications. It accounts for 30% statically of *sea* method calls. Yet java/lang/StringBuffer/append only appears once in the dynamic method execution frequencies (Table 6), and then only at 0.3% for *mon*. In terms of execution time on some platform, which should correlate reasonably with Table 7 dynamic bytecode percentages, java/lang/StringBuffer/append does not appear at all in the significant methods.

Table 7 shows that, on average, for the Grande programs studied 66% of the execution time, as measured by bytecode use, is spent in the top two methods.

4 Conclusions

This paper set out to investigate platform independent dynamic Java Virtual Machine analysis using the Java Grande Forum benchmark suite as a test case. This type of analysis, of course, does not look in any way at hardware specific issues, such as JIT compilers, interpreter design, memory effects or garbage collection which may all have significant impacts on the eventual running time of a Java program, and is limited in this respect. It has been shown above however that useful information about a Java program can be extracted at the intermediate representation level, which can be partly used to understand their ultimate behaviour on a specific hardware platform.

Platform independent dynamic analysis has been shown to be a useful tool for studying the Grande benchmark suite. For Grande applications Java method execution time is shown to be predominantly in the non-API bytecodes of the programs (92% average). This has been shown to be because the API methods typically do not feature in the hotspots. This is a significant difference from traditional Java applications such as applets or compiler type tools which spend most of the time in the API. Since a Grande application should use large amounts of processing, I/O, network bandwidth or memory, it is interesting to note how little of the API packages are dynamically used by this benchmark suite. Precompiling the API to some native representation therefore will not yield significant speedup.

A very useful dynamic platform independent metric for analysing eventual running time and identifying hotspots is to use a cost center design pattern to "charge" each bytecode to the appropriate method. This necessitates accessing an item in a hash table dictionary each time round the interpreter inner loop in the runVirtualMachine() function and incrementing a counter. The measurements from the experiment are significantly different from either static or dynamic method frequencies and show that for the Grande programs in the Java Grande Forum Benchmark suite 66% of the execution time, as measured by bytecode use, is spent in the top two methods.

References

1. C. Daly, J Horgan, J Power and J. T. Waldron, *Platform Independent Dynamic Java Virtual Machine Analysis: the Java Grande Forum Benchmark Suite*, Proceedings of Joint ACM Java Grande - ISCOPE (International Symposium on Computing in Object-oriented Parallel Environments) 2001 Conference, Stanford University, June 2-4, 2001
2. T.J. Wilkinson, *KAFFE, A Virtual Machine to run Java Code*, `http://www.kaffe.org`, URL last accessed on 20/10/2000
3. Bull M, Smith L, Westhead M, Henty D and Davey R. *Benchmarking Java Grande Applications*, Second International Conference and Exhibition on the Practical Application of Java, Manchester, UK, April 12-14, 2000.
4. J. Waldron, C. Daly, D. Gray and J. Horgan, *Comparison of Factors Influencing Bytecode Usage in the Java Virtual Machine*, Second International Conference and Exhibition on the Practical Application of Java, Manchester, UK, April 12-14, 2000.

HEPGRID2001: A Model of a Virtual Data Grid Application

Koen Holtman

California Institute of Technology
Mail code 256-48, 1200 E. California Blvd.
Pasadena, CA 91125, USA
koen@hep.caltech.edu

Abstract. Future high energy physics experiments will require huge distributed computational infrastructures, called data grids, to satisfy their data processing and analysis needs. This paper records the current understanding of the demands that will be put on a data grid around 2006, by the hundreds of physicists working with data from the CMS experiment. The current understanding is recorded by defining a model of this CMS physics analysis application running on a 'virtual data grid' as proposed by the GriPhyN project. The complete model consists of a hardware model, a data model, and an application workload model. The main utility of the HEPGRID2001 model is that it encodes high energy physics (HEP) application domain knowledge and makes it available in a form that is understandable for the CS community, so that architectural and performance requirements for data grid middleware components can be derived.

1 Introduction

In several areas of science, the growth in scale of the data generation and processing activities out-paces Moore's law. One example is the CMS high energy physics (HEP) experiment at the CERN laboratory [1] which plans to generate 1000 TB of raw data per year from 2006 on, with an estimated 10,000 CPUs all over the world being used around the clock in data analysis activities [2]. In the past, large physics experiments often developed their data processing facilities using in-house knowledge. To cope with the increase in scale and complexity, several experiments have recently joined up with computer science groups in research projects like GriPhyN [3] and the European DataGrid [4], which are tasked with researching and developing the necessary future software infrastructure. For these projects to be successful, it is now necessary that the experimentalists communicate their requirements in the language of computer science, rather than internally using language of their own scientific domain. The model in this paper was created as part of this communication effort to computer scientists.

The HEPGRID2001 model reflects the current understanding of the demands that will be put on a the CMS data grid system around 2006, by the hundreds

B. Hertzberger et al. (Eds.): HPCN Europe 2001, LNCS 2110, pp. 711–720, 2001.

of physicists working with data from the CMS experiment. The main reference source used is [2], which records the state of knowledge about the quantitative aspects of the requirements in early 2001. The model is composed of three parts. The first part is a model of the peta-scale distributed grid hardware configuration expected to be available to CMS in 2006. The second part is a CMS data model using the concept of virtual data as proposed by the GriPhyN project [3]. The third part is a workload model, which is complex enough to capture the essential challenges and opportunities faced by the virtual data grid catalogs, schedulers and optimizers.

The 'HEP' in the name 'HEPGRID2001' is the common abbreviation for high energy physics, the number 2001 signifies that it encodes the current understanding of a future application, an understanding that will evolve over time.

2 Hardware Model

The CMS experiment is run by a 'virtual organization', the CMS collaboration, in which over 140 institutions world-wide participate. The funding and manpower constraints involved have the result that the CMS hardware will be a world-wide distributed system, not a centralized one. The hardware consists of a central site called the 'tier 0' center, 5 regional sites called 'tier 1' centers, and 25 more local sites called 'tier 2' centers. The tier 0 center is located at the CERN laboratory, the location of the CMS experiment detector. The tier 1 centers each have a 2.5 Gbit/s network link to the tier 0. Each tier 1 center has 5 tier 2 centers connected to it. Tier 1-2 connections are also 2.5 Gbit/s network links. It should be stressed that the actual link capacity available to CMS in 2006 cannot be estimated very accurately, mainly because of uncertainties about long-term developments in the international telecom market. In any case, it is expected that the effective throughput for grid application level data transport will be only about half of the raw link capacity. This is due to protocol overheads and some other traffic on the same link (interactive sessions, videoconferencing), but also because the need for reasonable round trip times (reasonably short router queues) implies that the links cannot be saturated to full capacity.

The individual center characteristics are as follows:

For each single center at	CPU capacity	Nr of CPUs	Storage space
Tier 0	600,000 SI95	3000	2300 TB
Tier 1	150,000 SI95	750	900 TB
Tier 2	30,000 SI95	150	70 TB

The above numbers were taken from [2], they reflect estimates for 2006. The capacity of a single CPU used in 2006 is estimated to be 200 SI95.

3 Data Model

Figure 1 shows the HEPGRID2001 data model, which is defined in terms of a data flow arrangement inside the grid system.

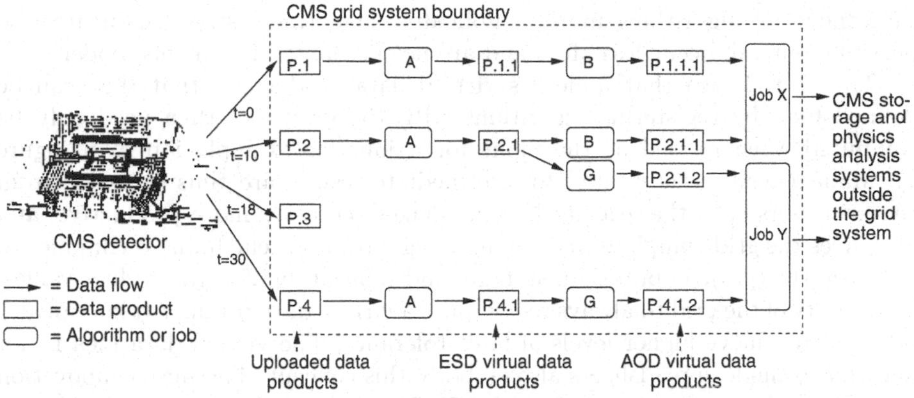

Fig. 1. HEPGRID2001 Data Model, in Terms a Data Flow Inside the CMS Grid.

Any piece of data in the grid is called a 'data product' in this model. A data product is the smallest piece of data that the grid needs to handle. Figure 1 shows how products (the boxes with square corners) are generated and used. There are two types of data product: uploaded data products and virtual data products.

Each *uploaded data product* contains the CMS detector output for a single *event*. An event is defined here as a set of particle collisions occurring simultaneously inside the CMS detector. Events will occur at a rate of 40 Mhz inside the detector: a real-time filter selects only some 100 events per second for storage, for uploading into the grid. All uploads happen at CERN, the location at the tier 0 site. Given the event size and the duty cycle of the detector, this yields 1000 TB of uploaded data products per year.

A *virtual data product* is the output of some data processing algorithm that has been registered with the grid. By having the algorithm registered with the grid, the virtual data product value can be computed on demand. The algorithm will use the value of another (virtual or uploaded) data product as its input. Therefore, each uploaded product $P.1, P.2, \ldots$ is the top node of a tree of virtual data products which can be derived from it by applying algorithms (A, B, G). These tree structures are modeled in more detail further below. Each product has a unique identifier (UID), for example $P.1, P.4.1.2$. From the UID of a virtual data product, the grid can determine which algorithms and inputs are needed to compute (derive) the product value.

The data flow in figure 1 is a somewhat simplified representation of the several processing steps that physicists will perform to extract information from the detector measurements. The timing of these steps, and the feedback loop that occurs while making them, is discussed in section 4. The virtual data products obtained directly from the uploaded (raw) data products are generally called ESD (event summary data) products by CMS physicists, those obtained from ESD products are generally called AOD (analysis object data) products. Arrangements of algorithms more complicated than this 2-stage chain are also possible, but will occur less often and are not accounted for in this model.

The central idea that underlies virtual data products is that they can be defined first, by registering algorithms with the grid, and then need only be computed when needed as the input for a job (for example job X in figure 1). Management tasks related to intermediate results are thus offloaded from the grid users onto the grid itself. This makes the users more productive, and also gives the grid complete freedom in using advanced scheduling techniques to optimize the (pre)computation, storage, and replication of data product values. The ability of the grid to always re-compute a virtual data product value can also be used to achieve higher levels of fault tolerance. The virtual data idea is not new, for example spreadsheets also embody this concept. The major innovation of the virtual data grid work of the GriPhyN project is the scale at which these services will be provided.

The uploaded and virtual data products (figure 1) are modeled more formally as being the nodes in 10^9 trees. Each tree has, as its top node, one uploaded data product with the UID $P.e$, for $1 \leq e \leq 10^9$. The e here is an event identifier, a number which uniquely identifies the corresponding event in the detector. The branching structure of the trees reflects the use of different (improved) versions of derivation algorithms over time. Algorithm development is iterative: the next version is developed based on experience using the current version. The branching structure is as follows. Below the top node in each tree are 5 ESD virtual data products, these get UIDs $P.e.x$ with $1 \leq x \leq 5$. Below each ESD there are 20 AOD virtual data products, these get UIDs $P.e.x.y$ with $1 \leq y \leq 20$. Each virtual data product can only be derived by applying some specific algorithm to the value of the product that is its parent in its tree.

Product characteristics are as follows:

CMS name	Type	Level in tree	Size	CPU power needed to derive	CPU power needed to analyze	CPU time needed to derive	CPU time needed to analyze
RAW	uploaded	1	1 MB	–	3000 SI95s	–	15 s
ESD	virtual	2	500 KB	3000 SI95s	25 SI95s	15 s	0.125 s
AOD	virtual	3	10 KB	25 SI95s	10 SI95s	0.125 s	0.05 s

Here, the 'time to derive' is the runtime of the algorithm that computes the product value, and the 'time to analyze' is the time spent in a job to analyze the product value. The numbers and terminology were taken from [2].

The structure, size, and final destination of the job output in figure 1 are not captured in this model. They are not captured both because the output is currently less well understood, and because optimizations on the output side are considered less crucial to the successful operation of the physics grid system. Additional work on modeling the job output will likely be done in the next few years. In general the job output is smaller, often significantly smaller, than the job input. Some jobs compute one single output value (with a size below a few KB) based on their input, others will, for each event in the request set, compute and output a derived physics data (DPD) structure with a size as large as 10 KB per event. It is therefore more important to run (sub)jobs close to their input datasets, than to run them close to the destination of their output. Moving both the input datasets and the (sub)jobs close to the destination is expected to be an interesting optimization, but not one that is crucial to the successful operation of the grid.

4 Workload Model

Multi-user physics analysis workloads have a complicated structure. They can be modeled at various levels of detail: the level of detail for HEPGRID2001 is very high, in order to capture the essential challenges and opportunities faced by the virtual data grid catalogs, schedulers and optimizers. This work takes a two-step approach to modeling the workload. As the first step, this paper gives a high-level overview of the shape of the workload, and discusses the factors that determine its shape. For reasons of style and space however, the high-level overview does not contain all statistical and morphological details that would be needed by researchers who want to use this workload model in simulations. Therefore, as the second step, a workload generator has been defined that can be used in such simulations. The workload generator encodes some additional domain knowledge needed to generate a properly stochastic workload. The generator, available at [6], computes a one-year grid workload, containing 124695 jobs and 12565 'hints', following the tree structures defined below. The output of the workload generator is an ASCII file with each line describing a single job or hint.

4.1 Job Model

Physicists get work done from the CMS virtual data grid by submitting jobs to it, see figure 1. A HEPGRID2001 job definition consists of two things: the *job request set*, which is a set of data product UIDs, and the *job code*, which is a parallel program that can be executed by the grid. The grid needs to execute the job code, and deliver the values of all data products in the request set to this code for further analysis. The job code will use grid services to deliver its output to the user. An example of a job request set (for job Y in figure 1) is the set with the product UIDs $P.2.1.2$ and $P.4.1.2$, this set can also be written as

$$\bigcup_{e \in \{2,4\}} P.e.1.2$$

In the HEPGRID2001 model, job request sets always have the general form

$$\bigcup_{e \in E} P.e.X$$

where E is a set of event identifiers and X is a (possibly empty) sequence of integers.

The job code is a parallel program, that is run as a set of subjobs. In this set there will be several 'worker' subjobs (the number to be decided by the grid schedulers) and one 'aggregation' subjob. Communication between subjobs is very minimal: at the end of its run every worker subjob uses the grid services to send a single, relatively small, package of information to the aggregation subjob. The aggregation subjob creates the final job output and sends it outside the grid. To execute the job, the grid schedulers may partition the job request set in any way, and feed the different parts to as many worker subjobs that may run in multiple locations. The products in the request set may be delivered to the worker subjobs in any order. This allows for massive parallelism in virtual data product derivation and job execution. The grid also has complete freedom in choosing when to create (derive) and delete virtual data product values, and in replicating and migrating product values over the grid sites.

4.2 Properties of Physics Analysis Workloads

The properties of the grid workloads produced by physics analysis are determined by three major interacting factors: the methodology of high energy physics as an experimental science, the way in which physicists collaborate and divide their work, and the need to maximize the utility of the available computing resources.

The goal in a physics experiment is to observe new physics phenomena, or observe phenomena with new levels of accuracy, in the particle collisions occurring inside the detector. The physics of two colliding particles is highly stochastic. The collision creates a highly localized concentration of energy, in which new particles may be created, with different rates of probability. Most 'interesting' particles will be created with extremely low probabilities: for example in the CMS experiment the (so far only theoretically predicted) creation of a Higgs boson force carrier particle might occur only once in 10^{12} collisions. The most resource and time-consuming task in physics analysis is therefore to recognize and isolate, from all events, only those events with a collision in which a sought-after phenomenon occurred. To decide whether an event e is interesting, whether it fits the sought-after phenomenon, the uploaded (raw) data product of event e is run through a chain of feature extraction algorithms. Examples of features that are extracted are the tracks (trajectories) of any photons and electrons emanating from the collision point). Then, several 'cut predicates' are applied to the extracted features of event e. The cut predicates select for the sought-after phenomenon, only the events which satisfy all cut predicates are left as 'interesting'. An example of a cut predicate is `n_elec==2`, which is an abbreviation for 'the observed number of electrons produced by the collisions in the event, and emanating from the collision point, is 2'. Due to the stochastic nature of collision physics, the probability that an event satisfies a set of cut predicates is uncorrelated with the time at which the event occurred.

Physics analysis, the development of feature extraction algorithms and cut predicates, is an iterative process, in which subsequent versions are refined until their effects are well-understood. The grid jobs run during this process can be compared to the compile-and-run steps in iterative software development. The grid job 'locate the Higgs events and calculate the Higgs mass from them' is highly atypical: it is the final job at the end of a long analysis effort. A much more typical job is 'run this next version of the system I am developing to locate the Higgs events, and create a plot of these parameters that I will use to determine the properties of this version'.

In an analysis effort to isolate a particular phenomenon, the cut predicates are generally developed one after each other. Each cut predicate is developed and refined by using it inside grid jobs and studying the output of these jobs. The request sets of these jobs always consist of one data product for each of the events that satisfy all cut predicates so developed so far. The cut predicates are developed by individual physicists, with the exception of a first 'group' level cut predicate that defines the 'channel' that a group of physicists is interested in.

The feature extraction algorithms that produce the virtual data products of section 3 are not written by individual physicists, but by groups of specialists. In the HEPGRID2001 workload model, a new ESD derivation algorithm (and associated detector calibration constants) is released 5 times per year, a new AOD derivation algorithm 75 times per year.

4.3 Workload Details

The 137260 jobs and hints in the workload, as created by the workload generator, are arranged as the nodes in five trees called the 'workload trees'. Each tree is 5 levels deep. Only the leaf nodes at level 5 represent actual physics analysis jobs. All non-leaf nodes represent 'hints' for the grid scheduler. Physicists can submit these hints to the scheduler to help it anticipate and optimize the future workload. Each hint node encodes a prediction about the job request sets of the jobs at the leaf nodes below. This prediction takes the form of a job request set over a set of events ES, where it is guatanteed that this ES a superset of the sets of events in all job request sets below. In current practice, physicists supply similar hints to their computing system operators and management boards, who use them to allocate resources and to perform 'production efforts', in which large sets of data products are pre-computed and stored for later use.

The tree properties are summarized by the following table, then discussed in more detail further below.

Level	Type	Fan-out	Interpretation of this (sub)tree in CMS
1	hint	5 trees	Each represents the use of a different ESD
2	hint	20 subtrees	Each represents the actions of a physics group
3	hint	25 subtrees	Each represents an analysis effort of a physicist
4	hint	3-5 subtrees	Each represents a phase in an analysis effort
5	job	5-20 leaf nodes	Each represents a job in an analysis effort

The jobs and hints are not submitted to the grid all at once, but over a period of about a year, following a 'job sequence' order. This job sequence roughly sweeps from left to right through the workload trees as illustrated in the leftmost plot of figure 2.

At level 1, the highest level of the workload model, each of the five trees in the workload corresponds to the joint use by all physicists of new ESD derivation

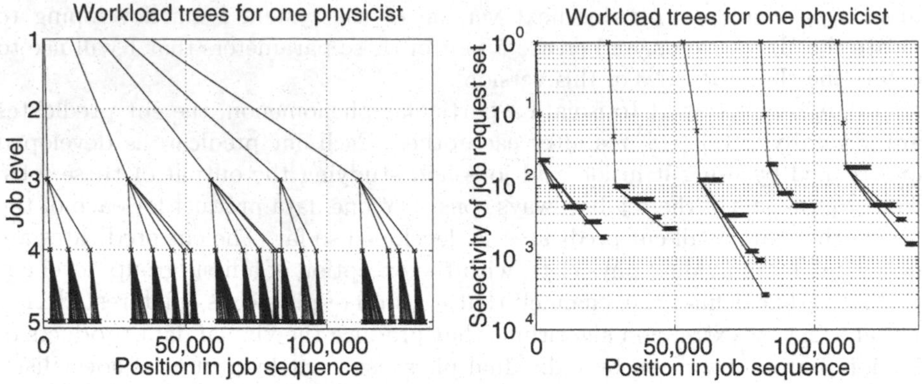

Fig. 2. Partial visualizations of the workload trees, as created by the workload generator [6]. Only the jobs and hints for a single physicist in a single group are shown. Each job or hint is plotted with a small 'X', these are connected by lines to show the tree structure. The left hand plot shows level and position of in the job sequence. The right hand plot shows exactly the same trees, but this time the y axis shows the selectivity (fraction of all events included) of the job and hint request sets.

algorithm. A new ESD derivation algorithm (and associated detector calibration constants) is released 5 times per year. The release of new AOD derivation algorithms is not reflected directly in the workload trees, but modeled in a different way in the workload generator output.

At level 2 of the trees, the workload model reflects that CMS physicists will organize themselves into 20 groups. Each group will decide on a 'first' cut predicate, with a selectivity from 4%–15%, that represents a very rough selection of the events that the group is interested in. In every group subtree the hint and job request sets are always over sets of events that satisfy at least the first group cut predicate. This group coordination, and the associated level 2 hints, provide important resource saving opportunities for the grid.

At level 3 of the trees, the model reflects that each group has 25 physicists in it, each physicists will perform an independent analysis effort on the events selected by the first group cut predicate. At level 3 every physicist develops a second, private cut predicate with a selectivity of 20%–25%, this predicate is combined with the group predicate to select the events considered in the analysis effort. The level 3 hint notifies the grid of this sub-selection.

At level 4, the different iterative phases in the activity of a physicist are modeled: each subtree represents a phase. A single phase models the development of a single new cut predicate. Going from one phase to the next, the physicist adds the newly developed predicate to the set used for the subsequent hints and jobs, increasing the selectivity with 30%–70%.

At level 5, the leaf nodes represent the actual physics analysis jobs run by the physicists. Each set of jobs under a single level 4 parent represents the iterative development of a single predicate. All these jobs will share the same request set. Job request sets generally contain AOD products. In the later phases of a physicist effort however, when the event set left is relatively small, the physicist is more likely to select larger products (ESD or even RAW products) for the job request sets. This reflects both the decreased runtime penalty of using larger products, and the increased need to use larger products because the information present in the smaller products has been exhausted already, as a means for event selection, by previous cut predicates.

The different groups and physicists all work independently and in parallel: this means that the level 3 physicist subtrees in any workload tree will overlap in time. Each individual physicist has a sequential think-submit-wait cycle. Therefore, in the job sequence, the hints and jobs of each physicist subtree appear in strict tree traversal order.

4.4 Workload Statistics

The following table gives some statistical properties of the HEPGRID2001 workload.

Time span covered by workload in the model	1 year
Number of physicists submitting jobs	500
Number of jobs	341/day
Average size of a job request set	10^7 products
Average size of a job request set	1.3 TB
CPU capacity needed to analyze requested products in jobs	960,000 SI95
RAW products requested by all jobs	$4.5*10^9$/year
ESD products requested by all jobs	$3.0*10^{11}$/year
AOD products requested by all jobs	$9.4*10^{11}$/year
Average number of times that a single product is requested	40
CPU capacity needed to derive all requested products once	433,000 SI95
Different virtual data products defined	105/event/year
Different virtual data products derived at least once	31/event/year
ESD products derived if all derived only once	$4.3*10^9$/year
AOD products derived if all derived only once	$2.7*10^{10}$/year
Size of RAW products	1000 TB/year
Size of ESD products derived if all derived only once	2166 TB/year
Size of AOD products derived if all derived only once	269 TB/year

5 Conclusions

This paper records the current understanding of the demands that will be put on a virtual data grid around 2006, by the hundreds of physicists working with data from the CMS experiment. Related work on the modeling and simulation of CMS computing has been done in the MONARC project [5]. In comparison to this work, the MONARC models generally contain more hardware details and less workload details. Also, in stead of hints, the MONARC workloads have explicit 'production jobs' submitted by production managers, jobs that compute and store large sets of virtual data product values for further analysis. The GriPhyN [3] and European DataGrid [4] projects are currently both going through application requirements gathering cycles, and this work is part of that effort. An important contribution of this work is that it encodes resolutions to many detailed modeling issues, based on domain knowledge in CMS, resolutions that are needed to do simulations with realistic workloads. Encoding this domain knowledge is a necessary step towards collaborating more closely with modern computer science.

The CMS data grid has many requirements in common with other grid applications: for example security and sharing policies for creating a virtual organization, fault tolerance, and the handling of differences between hardware platforms. This paper focuses on those requirements that might be unique to high energy physics: the scale of the problem, the structure of the virtual data products, and the nature of the workload. It is not known currently how unique these high energy physics requirements are. From the standpoint of the CMS experiment, it would be preferable if commonalities and new abstractions could be found that show that the requirements are less unique than thought, so that there can be greater sharing with grid related software development occurring in other efforts.

Acknowledgements

Most quantitative elements of the hardware and data models are due to the MONARC project [5] and the recent LHC computing review efforts in the CMS collaboration [2]. Thanks go to Paolo Capiluppi, Ian Foster, Irwin Gaines, Iosif Legrand, Harvey Newman, and Kurt Stockinger for their comments and feedback in creating the HEPGRID2001 model.

References

1. CMS Computing Technical Proposal. CERN/LHCC 96-45, CMS collaboration, 19 December 1996. See also: `http://cmsinfo.cern.ch/Welcome.html`
2. S. Bethke et al. Report of the steering group of the LHC computing review. CERN/LHCC/2001-004, CERN/RRB-D 2001-3, 22 February 2001. Available from `http://lhc-computing-review-public.web.cern.ch/lhc-computing-review-public/`
3. `http://www.griphyn.org/`
4. `http://www.eu-datagrid.org/`
5. `http://monarc.web.cern.ch/MONARC/`
6. `http://kholtman.home.cern.ch/kholtman/hepgrid2001/`

Grid Services for Fast Retrieval on Large Multidimensional Databases

Peter Baumann[1]

Active Knowledge GmbH
Kirchenstr. 34, D-81675 Munich, Germany
Dial-up: voice +49-89-458677-30, fax -39, mobile +49-173-5644078
peter.baumann@tive-knowledge.de

Abstract. Fast, user-centric access to and evaluation of large supercomputing results is a well-recognised problem among compute and data service providers.
In the ESTEDI initiative, research and industry from Europe and beyond co-operate to overcome this obstacle. The basic approach is to augment the high-volume data generators with a database system for the management and extraction of spatio-temporal data. Led by ERCOFTAC (European Research Community on Flow, Turbulence and Combustion), a detailed requirements analysis has been carried out which forms the basis for the specification of a High-Performance Computing (HPC) database platform. Implementation of this platform relies on the multidimensional database system RasDaMan. Evaluation covers all major application fields by HPC centres with in-depth experience in their resp. field.
We introduce the ESTEDI project and give an overview of the data management platform under development.

1 Introduction

Satellites and other sensors, supercomputer simulations, and experiments in science and engineering all generate arrays of some dimensionality, spatial extent, and cell semantics. While such data differ in aspects such as data density (from dense 2-D images to highly sparse high-dimensional accelerator data) and data distribution, they usually share the property of extreme data volumes, both per data item and in quantity of data items. Usually, user access nowadays is accomplished in terms of files containing (part of) the information required, encoded in a sometimes more, sometimes less standardised data exchange format chosen from a rich variety of options. This implies several shortcomings.

First, access is done on an inappropriate semantic level. Applications accessing HPC data have to deal with directories, file names, and data formats instead of accessing spatio-temporal data in terms of, say, simulation space-time and other user-oriented terms.

---

[1] Research supported by the European Commission under grant no. IST-11009.

B. Hertzberger et al. (Eds.) : HPCN Europe 2001, LNCS 2110, pp. 721-730, 2001.

Second, data access is inefficient. Data are stored according to their generation process, for example in time slices, as opposed to a retrieval-driven organisation. All access pertaining to different criteria, such as spatial coordinates, requires data-intensive extraction processes and, hence, suffers from severe performance penalties.

Third, search across a multitude of data sets is hard to support. Evaluation of search criteria usually requires networks transfer of each (large) candidate data item to the client, implying a prohibitively immense amount of data to be shipped. Hence, many interesting and important evaluations currently are impossible.

All the aforementioned access efficiency problems are substantially intensified as the Grid community grows, as in the absence of optimisation methods obviously networks load grows linearly with the number of users.

In summary, a major bottleneck today is fast, user-centric access to and evaluation of supercomputing results.

Recently the ESTEDI project[2] (*European Spatio-Temporal Data Infrastructure for High-Performance Computing*) has set out to comprehensively collect requirements for HPC data management and to develop a standard architecture and best-practice knowledge for the efficient combination of HPC data processing and management, including fast and flexible retrieval, which is accepted as a broad consensus among the European HPC community. Among the various data structures on hand, ESTEDI focuses on n-dimensional raster data, so-called *Multidimensional Discrete Data* (MDD), as they comprise a central information category encountered in supercomputing.

The envisaged common data management platform is currently under implementation and will be thoroughly evaluated through in-practice application in all major HPC application fields. An existing multidimensional database management system (DBMS), RasDaMan[3] [1, 2, 3, 7], is used at the core, as RasDaMan already offers several important features.

In the remainder of this contribution, we will first outline the ESTEDI project (Section 2) and then give a brief overview of the RasDaMan DBMS in the context of HPC data (Section 3).

2 The ESTEDI Project

ESTEDI started in February 2000 and will be funded until January 2003; currently, the specification phase is completed and the first implementation phase is under way. The project addresses the delivery bottleneck of large HPC results to the users by augmenting the high-volume data generators with a flexible data management and extraction tool for spatio-temporal data. The observation underlying this approach is that, whereas transfer of complete data sets to the client(s) is prohibitively time consuming, users actually do not always need the whole data set; in many cases they require either some subset (e.g., cut-outs in space and time), or some kind of summary data (such as thumbnails or statistical evaluations), or a combination

---

[2] See http://www.estedi.org.

[3] See http://www.rasdaman.com.

thereof. Consequently, it is expected that an intelligent spatio-temporal database server can drastically reduce networks traffic and client processing load, leading to increased data availability. For the end user this ultimately means improved quality of service in terms of performance and functionality.

The project is organised as follows. Under guidance of ERCOFTAC[4] (European Research Community on Flow, Turbulence and Combustion), represented by University of Surrey, a critical mass of large European HPC centres plus the CFD package vendor Numeca s.a. perform a thorough requirements elicitation. In close cooperation with these partners and based on the requirements, the database experts of FORWISS (financial/administrative project coordination) and Active Knowledge GmbH (technical/scientific project management) specify the common data management platform.

The implementation platform RasDaMan (see Section 3) has been installed at all sites in the HPC partners' particular environment. In the current first implementation phase, data loaders and accessors are being implemented. Each partner addresses one specific area where the lab has special expertise and users:

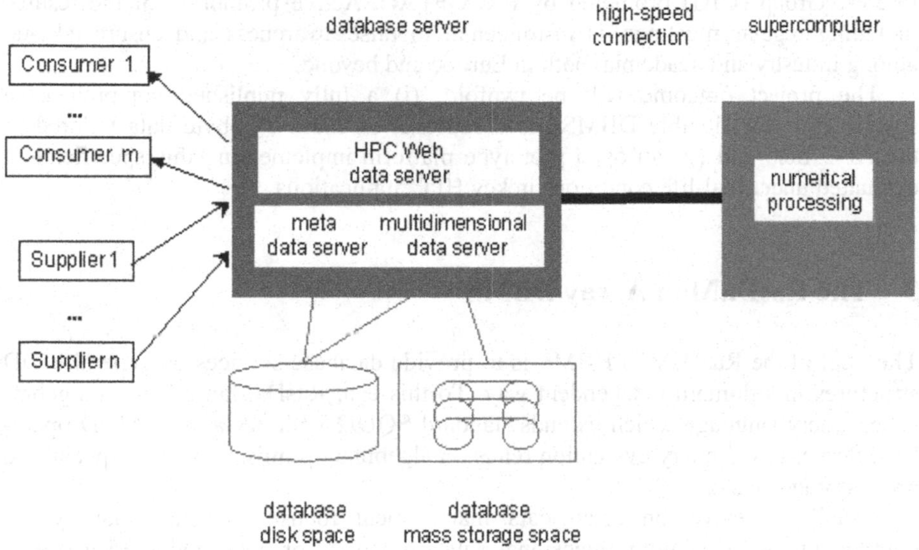

Fig. 1. The Generic ESTEDI Data Management Platform.

[4] See http://imhefwww.epfl.ch/Imf/ERCOFTAC.

- climate modelling by CCLRC (Council of the Central Laboratory of the Research Councils, UK) and MPIM (Max Planck Institute for Meteorology, DE);
- cosmological simulation by CINECA (Interuniversity Consortium of the North Eastern Italy for Automatic Computing, IT);
- flow modelling of chemical reactors by CSCS (Swiss Center for Scientific Computing, CH);
- satellite image retrieval and information extraction by DFD-DLR (German Remote Sensing Data Center, DE);
- simulation of the dynamics of gene expression by IHPC&DB (Institute for High-Performance Computing and Databases, RU);
- computational fluid dynamics (CFD) post-processing by Numeca International s.a. (BE).

In the next step starting in autumn 2001, the resulting application pilots will be operated under real-life conditions for evaluation.

In parallel to the HPC application development going on, the database developers extend RasDaMan with features determined necessary (see Section 3.2).

All development is in response to the user requirements crystalised by the User Interest Group (UIG) promoted by ERCOFTAC. Active promotion of the results, including regular meetings, is instrumental to raise awareness and ensure take-up among industry and academia, both in Europe and beyond.

The project outcome will be twofold: (i) a fully published comprehensive specification for flexible DBMS-based retrieval on multi-Terabyte data tailored to the HPC field and (ii) an open prototype platform implementing this specification, evaluated under real-life conditions in key HPC applications.

3 The RasDaMan Array DBMS

The goal of the RasDaMan DBMS is to provide database services on general MDD structures in a domain-independent way. To this end, RasDaMan offers an algebra-based query language which extends standard SQL92 with declarative MDD operators. Server-based query evaluation relies on algebraic optimisation and a specialised array storage manager.

Usually, research on array data management focuses on particular system components, such as multidimensional data storage [8] or data models [5, 6]. RasDaMan, conversely, is a complete array DBMS. It is generic in that functionality and architecture are not tied to some particular application area, but apply equally well, e.g., to 2-D satellite maps and 4-D climate simulations.

3.1 Modelling and Querying Array Data in RasDaMan

The conceptual model of RasDaMan centers around the notion of an n-D array (in the programming language sense) which can be of any dimension, spatial extent, and array cell type. As cell types, all valid C/C++ types and nested "structs" are admissible; the only exception is pointers, which are replaced by the database concept of persistent OID (object identifier) references. Following the classical relational database paradigm, RasDaMan also supports sets of arrays. Hence, a RasDaMan database can be conceived as a set of tables where each table contains a single column of array-valued attributes, together with a system-provided unique OID. These OIDs can be used to reference particular MDD objects from elsewhere in the database.

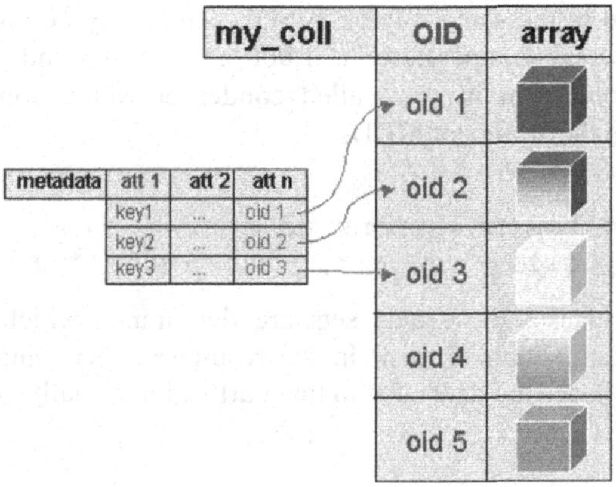

Fig. 2. The RasDaMan conceptual model: tables of n-D data cubes which can be referenced from elsewhere in the database.

Based on a specifically designed array algebra [1], the RasDaMan query language, RasQL, offers MDD primitives embedded in the SQL query paradigm [2]. The expressiveness of RasQL enables a wide range of signal processing, imaging, and statistical operations. To give a flavour of the query language, we present some small examples.

Let us consider a table `ClimateModels` of 4-D climate models with dimensions `x`, `y`, `z`, and `t` containing model variables `T` (temperature) and `vx`, `vy`, `vz` for the wind components.

The following query retrieves, from all climate models stored, the T component:

```
select cs.T
from   ClimateSimulations as cs
```

The result is a set of 4-D cubes containing only the `T` variable. The principle behind is that expressions on cells are applied simultaneously to all cells, in RasQL called *induced operations*. For example, let us retrieve the absolute of the wind speed:

```
select sqrt( cs.vx*cs.vx + cs.vy*cs.vy + cs.vz*vz )
from   ClimateSimulations as cs
```

Spatial subsetting is done by so-called trim expressions where, for each dimension, the lower and upper bound of the desired result is specified; a wildcard "*"denotes the array's current bound. A section (projection) at a specific position along a dimension is specified by indicating not the lower/upper bound pair, but the section point. As an example, let us extract at time `t=42` the layers from ground up to 1,000 m (assuming meters as unit):

```
select cs[ *:*, *:*, *:1000, 42 ]
from   ClimateSimulations as cs
```

To select a subset from a table whose items fulfil a certain predicate, the `where` clause can be used in the same way as in SQL. The query below retrieves all those models where average temperature in 1,000 m over ground exceeds 5° C; the `avg_cells()` operation is a so-called condenser which computes the average value of all cells in the argument MDD:

```
select cs
from   ClimateSimulations as cs
where  avg_cells( cs[ *:*, *:*, 1000, *:* ] ) > 5.0
```

In the where clause, those data sets are determined which fulfil the average temperature criterion. Each element in the result set, then, undergoes a so-called trimming which reduces the data cube to that part which actually is desired.

3.2 Application Development and Client/Server Communication

Application programming interfaces supported are a Java and a C++ binding, both compliant to the object database standard ODMG 3.0[5]. The client libraries perform client/server communication, query preparation, and handling of MDD objects. For example, the following Java code piece creates a query object, prepares it, sends it to the server call, and receives the result set:

```
OQLQuery myQu = myApp.newOQLQuery();
myQu.create("select avg_cells(a) from " + coll + " as
a");
DSet result = (DSet) myQu.execute();
```

MDD objects by default are returned in the main memory format which the client code expects, depending on CPU and compiler. Hence, arrays can immediately be processed with the means of the programming language. If desired, objects can be packaged by the server into particular data formats; for example, the following query would return JPEg images ready for usage in a browser-based Web application:

```
select jpeg( cs[ *:*, *:*, 1000, 42 ] )
from   ClimateSimulations as cs
```

The C++ interface uses RPC as the underlying communication protocol which allows the server a sign of life monitoring. The Java binding relies on http as its communication protocol. While this does not permit a sign of life technique, it has

[5] See http://www.odmg.org.

the advantage that it is compatible with firewalls. Hence, a Java client can safely be admitted in high-security environments.

3.3 Physical Array Storage and Processing

Internally, RasDaMan employs a storage structure based on the partitioning of an MDD object into tiles, i.e., sub-arrays. Subdivision does not have to be a regular grid, but can be defined to consist of arbitrary non-aligned tiles (see Fig. 5). To quickly determine the tiles affected by a query, a spatial index is employed [3]. Optionally tiles are compressed when stored; moreover, result data can be compressed for transfer to the client. Both tiling strategy and compression form tuning parameters invisible at the query level, but under control of the database developer and administrator.

Each tile is stored as a BLOB (Binary Large Object) in a relational database comprising a safe persistent store. As the resulting structure is very simple – an MDD object is mapped to a set of BLOBs –, any relational DBMS can serve as the underlying storage manager. Actually, even an object-oriented DBMS, O2, has been coupled with RasDaMan this way. An immediate advantage of this technique is the reduced administration overhead (only one database has to be maintained for both meta and array data) and the guaranteed consistency as opposed to the conventional mixture of relational meta data and file-based array management.

A series of optimisation rules is applied to a query prior to its execution to achieve an optimal access and processing pattern [7]. Of the 150 heuristic rewriting rules, 110 are actually optimising while the other 40 serve to transform the query into canonical form.

In the course of ESTEDI, a series of enhancements to RasDaMan will be done. To accommodate data volumes beyond disk capacity, tape archives will be coupled to RasDaMan, including staging techniques which take into account multidimensional spatial neighbourhood of tiles.

Tiles form not only the units of storage access, but obviously also form natural units for query parallelisation. Therefore, the RasDaMan server will be modified so as to evaluate queries by passing each tile to a separate processor. Among the problems to be addressed is overlapping tile borders in operations such as filters requiring access to the neighbourhood of each cell. Techniques known in parallel computing, such as halo exchange [4], seem promising.

Optimisation of queries is already done extensively in RasDaMan to minimise disk access. The result is that on the average query processing is CPU-bound as opposed to I/O-bound which is usual with conventional database systems. The next step now is to optimise complex imaging and statistical queries with involved operation sequences per cell causing high processing workloads.

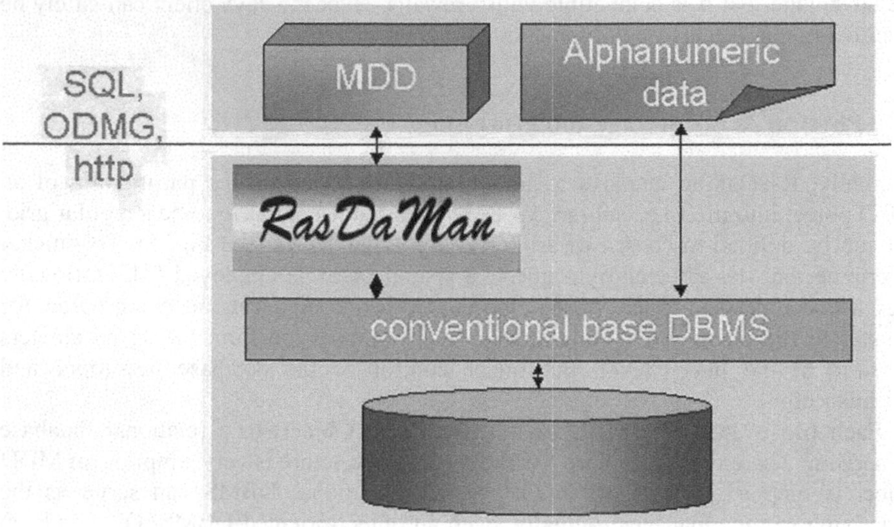

Fig. 3. Embedding of RasDaMan into Overall Data Management.

4 Conclusion

What is unique about the ESTEDI project is the combination of both HPC and database expertise. In contrast to other approaches also striving to overcome the HPC data delivery bottleneck, ESTEDI does so by providing database query support at a high semantic level, internally supported by transparent optimisation techniques. Two main advantages result from this: firstly, it enables the users to more concisely state the data they need, leading to a more focused result data set. Secondly, the query describes the whole task to the database server, and this opens up a wide field for internal optimisation.

Hence, we feel that such an interdisciplinary approach has considerable potential to overcome the current HPC data management bottleneck and, at the same time, provide a substantially new quality of service to the users. In this sense, we see the ESTEDI initiative as an essential step towards the envisaged Grid community.

Acknowledgement

The author wishes to express his joy about the team of outstanding ingenuity and spirit that has met to form the ESTEDI consortium. It is such kind of active involvement which makes wheels rolling.

References

1. P. Baumann: A Database Array Algebra for Spatio-Temporal Data and Beyond. Proc. Next Generation Information Technology and Systems NGITS'99, Zikhron Yaakov, Israel, 1999, pp. 76 - 93.
2. P. Baumann, P. Furtado, R. Ritsch, N. Widmann: Geo/Environmental and Medical Data Management in the RasDaMan System. Proc. VLDB'97, Athens, Greece, 1997, pp. 548-552.
3. P. Furtado, P. Baumann: Storage of Multidimensional Arrays Based on Arbitrary Tiling. Proc. ICDE '99, Sydney, Australia 1999, pp. 480-489.
4. J. Hague: Halo Exchange in Mixed Shared and Distributed Memory Processors. 9th Workshop on the Use of High Performance Computing in Meteorology – Developments in Teracomputing, European Centre for Medium-Range Weather Forecast, Reading, UK, 2000.
5. L. Libkin, R. Machlin, and L. Wong: A Query Language for Multidimensional Arrays: Design, Implementation, and Optimization Techniques. Proc. ACM SIGMOD'96, Montreal, Canada, 1996, pp. 228 - 239.
6. A.P. Marathe, K. Salem: Query Processing Techniques for Arrays. Proc. ACM SIGMOD '99, Philadelphia, USA, 1999, pp. 323-334.
7. R. Ritsch: Optimization and Evaluation of Array Queries in Database Management Systems. PhD Thesis, Technische UniversitäMüchen, 1999.
8. S. Sarawagi, M. Stonebraker: Efficient Organization of Large Multidimensional Arrays. Proc. ICDE'94, Houston, USA, 1994, pp. 328-336.

Appendix

The following images show sample HPC application data retrieved from RasDaMan databases. All visualisations have been done with rView, the RasDaMan visual frontend.

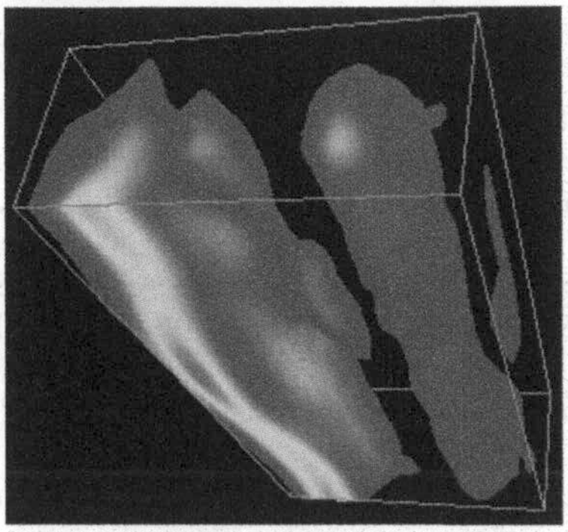

Fig. 4. Visible Human excerpt (3-D); data courtesy of Visible Human Project, see http://www.nlm.nih.gov/research/.

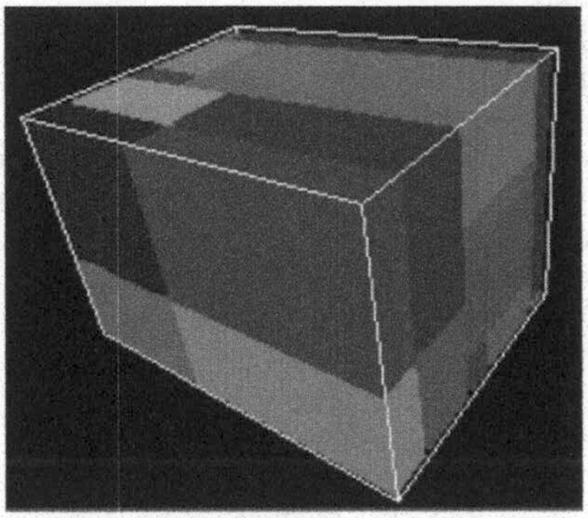

Fig. 5. Human brain activitation map (3-D); data courtesy Karolinska Institutet, Stockholm, Sweden.

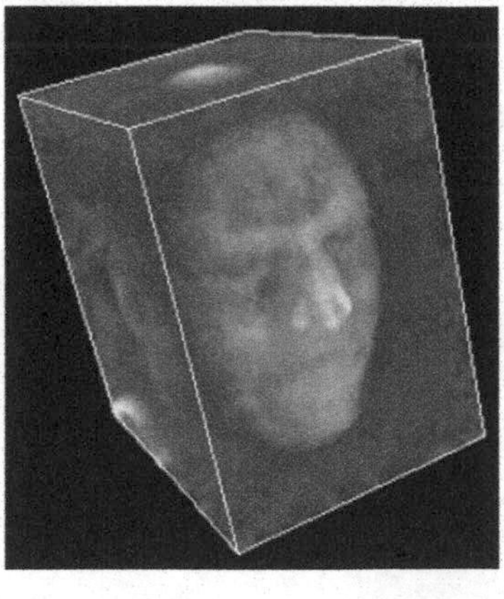

Fig. 6. Climate simulation (3-D retrieval result from 4-D climate model); data courtesy German Climate Research Centre.

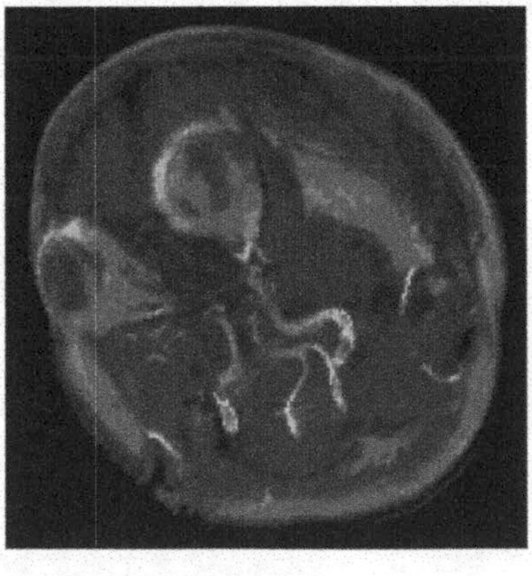

Fig. 7. Visualisation of the internal tiling structure of a 3-D object.

Author Index